AF319265

Nuclear and Particle Physics with Cosmology, Volume 2

Particle physics and cosmology

Online at: https://doi.org/10.1088/978-0-7503-5032-7

Nuclear and Particle Physics with Cosmology, Volume 2

Particle physics and cosmology

Jyotirmoy Guha
Department of Physics, Santipur College, Santipur, Nadia, West Bengal, India

IOP Publishing, Bristol, UK

ISBN   978-0-7503-5032-7 (ebook)
ISBN   978-0-7503-5030-3 (print)
ISBN   978-0-7503-5033-4 (myPrint)
ISBN   978-0-7503-5031-0 (mobi)

DOI   10.1088/978-0-7503-5032-7

Version: 20241101

IOP ebooks

British Library Cataloguing-in-Publication Data: A catalogue record for this book is available from the British Library.

Published by IOP Publishing, wholly owned by The Institute of Physics, London

IOP Publishing, No.2 The Distillery, Glassfields, Avon Street, Bristol, BS2 0GR, UK

US Office: IOP Publishing, Inc., 190 North Independence Mall West, Suite 601, Philadelphia, PA 19106, USA

My PhD Guide

late Dr Satyabrata Biswas

Ex-Professor of Physics

University of Kalyani, West Bengal, India.

An excellent human being in whose company I realised how little I know … he helped me to navigate in the world of physics.

Contents

16 Geometry of space-time   16-1

17 Cosmic microwave background radiation   17-1

Preface

In the earlier volume we studied nuclear physics. In this volume we address particle physics and cosmology.

The aim of this volume is to provide substantial knowledge and information about the two subjects of particle physics and cosmology. We try to present things in the form of a comprehensive, self-contained course with elaborate coverage of essential features. However, we will be extra careful not to render things drab and gloomy by overloading things with mathematics. The primary target is to develop in the mind of the reader what the two subjects stand for and to create interest regarding them in their mind. The material presented in this volume can be treated as a class note on the two subjects making up a full course. We are starting from scratch. We steadily develop the subjects in a student-friendly manner through brief but logical arguments and with plenty of examples and analogies. We present problems that are fully worked out so that readers do not have to search for solutions.

22 July, 2024
Santipur College
Santipur, NadiaWest Bengal, India

Dr Jyotirmoy Guha
jgsantipurcollege@gmail.com
http://youtube.com/@jgphysics

Acknowledgements

I express my sincere gratitude to IOP Publishing Ltd for publishing this second volume.

I wish to thank all my departmental colleagues Dr Atreyi Paul, Dr Dipankar Bhattacharyya, Dr Anita Gangopadhyay, Dr Palash Das, Professor Chintaharan Majumder, Dr Dibyendu Biswas and Professor Subhau Choudhury for their continuous support and words of encouragement.

Author biography

Jyotirmoy Guha

 Dr Jyotirmoy Guha is a faculty of Santipur College, West Bengal, India in the Department of Physics. He is the Head of the department and serves as Associate Professor.

Dr Guha has a brilliant academic record, having secured first class first both in undergraduate and postgraduate examinations. He did his doctoral work on quantum cosmology and has published works in international journals of repute.

Dr Guha has authored several books of which those that need special mention are the following. *Nuclear and Particle Physics with Cosmology, Volume 1: Nuclear Physics* published by IOP publishing, *Solid state Physics, Theory, Problems and Solutions* published by Books and Allied, Kolkata; *Quantum Mechanics: Theory, Problems and Solutions* published by Books and Allied, Kolkata; *Modern Physics Volume I & II* published by Techno World, Kolkata. Dr Guha is a co-author of the book *Quantum Optics and Quantum Computation: An introduction* published by IOP publishing.

Several of his students are well established in various institutes of repute and carrying on research around the globe. Dr Guha regularly contributes to his YouTube channel http://youtube.com/@jgphysics.

Purpose/scope/basic approach of the book

This volume on particle physics and cosmology can be treated as an introductory course on particle physics and cosmology. The book has been divided into a set of seventeen modules or chapters, and each module or chapter has been subdivided into sections. Two or three sections can be looked upon as a single lecture.

The first twelve modules or chapters focus on particle physics and the remaining five modules or chapters focus on cosmology.

The book follows a simple style, lucid approach and step-by-step explanation. All efforts have been made to describe things in a manner that is understandable without any trouble. Care has been taken to give illustrations whenever applicable. The book will impart basic knowledge on the two subjects of particle physics and cosmology and help generate further interest in them.

IOP Publishing

Nuclear and Particle Physics with Cosmology, Volume 2
Particle physics and cosmology
Jyotirmoy Guha

Chapter 1

Introduction to particle physics

In this chapter we describe detection of the electron and determination of its specific charge, detection of the proton and its mass determination, detection of the neutron and its mass determination. We discuss cosmic rays, their detection and origin. Composition of cosmic rays, primary and secondary components, hard and soft components are discussed. We describe the variation of cosmic ray intensity with altitude, latitude and direction. Cosmic ray shower, bremsstrahlung, pair production, hadronic shower and electromagnetic shower are described. We also elaborate on the discovery of the positron, discovery of pair production and pair annihilation, discovery of the muon, discovery of the pion and discovery of the anti-proton. The muon paradox is resolved. Detection of the electron anti-neutrino is dealt with.

1.1 Introduction

Particle physics originated from two sources. One is nuclear physics. The other is cosmic rays.

Cosmic rays contribute considerably to particle physics and many particles have been discovered while studying cosmic rays. As cosmic rays have very high energy they can produce many particles when they collide with atmospheric nuclei. Photographic emulsions were sent in balloons or in spacecraft to record particle tracks. By looking through a microscope and studying the tracks, many particles were discovered. Study of neutrino physics is still based on cosmic rays.

The next phase of advancement in particle physics was when particle accelerators took over.

Particle physics grew out of nuclear physics. In the microscopic world, particles like protons and neutrons cling together to form a nucleus. In volume 1 we discussed the physics of the nucleus. Electrons and nucleus couple to form an atom. Atoms combine and we get a molecule. And molecules assemble to form matter. Clearly, constituents come together to form gradually bigger objects in the microscopic scale. In the first few chapters of particle physics in this volume we discuss physics of particles.

doi:10.1088/978-0-7503-5032-7ch1

On the other hand, in the macroscopic world it has been observed that stars and galaxies are drifting away from each other towards the edge of our visible Universe.

In this volume we will be discussing the essential features of the microscopic world of elementary particles in the branch called particle physics (chapters 1–12) while the essential features of the macroscopic world of stars and galaxies will be dealt with in the branch called cosmology (chapters 13–17) which is the physics of the Universe.

Study of nature in the microscopic and macroscopic scale by physicists is done with the aim of obtaining a reasonable understanding of the structure of the Universe and the forces that govern various phenomena in the Universe. Through this study we obtain vital and reliable information regarding how the Universe evolved as time passed.

Let us briefly mention the subject matter of the following branches of study.

1.1.1 Particle physics

Particle physics is often referred to as elementary particle physics as well as high energy physics, since high energy is needed to probe the basic constituents of matter.

Several hundreds of subatomic particles have been discovered and there is a way to study such a huge number of particles in this branch of physics. In particle physics we study the properties, relationships, interactions, origin, life and decay of elementary particles.

1.1.2 Astronomy

Astronomy refers to observing the sky and studying natural objects in space outside Earth's atmosphere such as the Sun, planets and stars. Astronomers observe positions, luminosities and motions of stars, planets and other celestial bodies in space.

1.1.3 Astrophysics

Astrophysics is concerned with the evolution of stars and other celestial bodies. This branch tries to interpret astronomical observations using the laws and theories of physics and chemistry. In this branch of study we explore topics such as birth, life and death of a star, planet, galaxy, nebula etc.

1.1.4 Cosmology

Cosmology is the study of the history of the Universe or cosmos (kosmos means harmony or order). In cosmology we study the Universe as a system, analyze the large-scale structure and properties with the help of equations and governing broad principles.

We study and put forward theories regarding its origin, its development or expansion and its ultimate fate. Cosmology is the study of the evolution of space and time. In cosmology one tries to address how the Universe began, perhaps through a violent explosion (big bang) that signaled the beginning of space-time.

We discuss the dynamics of expansion of the Universe and also address the ultimate fate of the Universe—whether it would come to an end or not and if it died, how it would die.

1.2 Connection between nuclear physics and particle physics

We now highlight the connection between nuclear physics and particle physics by pointing out the things that are discussed under these terminologies.

Nuclear Physics

Matter is made up of molecules, which are built out of atoms. Again, an atom consists of a massive, dense, compact, central nucleus and electrons. The intermediate space is empty of matter but filled with electric and magnetic force fields.

In nuclear physics we study the properties or characteristics of the atomic nucleus as a whole.

Here focus is upon the nucleus which is one constituent of an atom.

We develop or build a reasonable model of nucleus to study the interactions (or the reactions) that nucleus performs, under various situations.

Nucleus involves neutrons (n) and protons (p) and we study their interactions (nn, pp, np) to investigate the properties of nucleus. So in nuclear physics we place our attention on neutrons, protons and the nucleus.

In nuclear physics the size of objects studied is $\sim 10^{-15}\,m \equiv 1\,fm$

eg. radius R of uranium nucleus $_{92}U^{235}$ is

$$R = R_0 A^{1/3} = (1.2\,fm)(235)^{1/3}$$
$$= 7.4\,fm.$$

(Atom size $\sim 10^{-10}\,m = 1\,\text{Å}$,
Molecular size $\sim 10^{-9}\,m = 1\,nm$)

Nuclear physics investigates the laws of nucleus and the forces operative in a nucleus.

Particle Physics

In particle physics we study the properties or characteristics of various elementary particles such as the following.

✓ Protons and neutrons, which are the constituents of a nucleus, also called nucleons.

✓ Electrons that revolve around nucleus.

✓ Particles produced by smashing or colliding nuclei with proton, neutron, electron etc that are accelerated in particle accelerators to high energies.

✓ Particles produced in cosmic rays.

Here focus is upon particles which are the fundamental constituents or building blocks of the Universe.

We study the interactions performed by various high energy (relativistic) elementary particles.

In particle physics we study various elementary particles.

We study also the structure of proton and neutron as they are built out of smaller particles called quarks.

In particle physics one studies particles which are of very small size.

Size or radius of electron $\sim 2.8\,fm$ (*exercise 1.1*), proton $\sim 0.8\,fm$,
pion $\sim 0.46\,fm$,
quark $\sim 0.43 \times 10^{-18}\,m$.

Particle physics investigates the laws of nature and the forces operative in the Universe.

Various elementary particles were detected and their existence confirmed in various experiments.

We now discuss detection of a few elementary particles like electron e^-, proton p, neutron n, positron e^+, muon μ, pion π, anti-proton $\bar{p}$, electron anti-neutrino $\bar{\nu}_e$.

1.3 Detection of the electron (particle) by J J Thomson (1897)

J.J.Thomson discovered electron in the cathode ray experiment.

J J Thomson used a cathode ray tube (figure 1.1), which is a big glass tube, sealed all over. The air within was pumped out (pressure $\sim 0.01\ mm$ of Hg). Two metal pieces, cathode K and anode A (with a central hole) were connected to a power source creating high voltage of $\sim 10\,000\ V$, cathode K being negative and the perforated anode A being positive.

A ray shot from cathode K, traveled in a straight line across the tube and hit point O of the tube where there was a phosphor (ZnS) coating and so the point O glowed (implying it was a particle beam). Actually, electrons from cathode K were attracted to the positively charged anode A and proceeded fast across the length of the tube.

To verify if the ray, coming out from K, was charged Thomson put two metal plates P, Q on either side of the cathode ray tube, as shown in figure 1.1(a). When plate P was made positive and plate Q negative, applying an electric field $\vec{E} = E\,\hat{PQ} = E\hat{z}$, the force on the cathode ray was the Coulomb force

$$\begin{aligned}
\vec{F_e} &= -|e|\vec{E} \\
&= -|e|E\hat{z} = |e|E(-\hat{z}) = |e|E\hat{QP}
\end{aligned} \tag{1.1}$$

which means upward ($Q \to P$) deflection. So the cathode ray is bent up towards the positively charged plate P following the path $\overset{\frown}{MU}$. This demonstrates that the ray from cathode K was negatively charged.

✓ To verify further, Thomson surrounded the cathode ray tube by a magnet NS, applying magnetic induction $\vec{B} = B\hat{NS} = B\hat{x}$, as shown in figure 1.1(b).

Now the ray from cathode K bends down, following the path $\overset{\frown}{ML}$, just as a negative charge would do in a magnetic field. This is because the force on the cathode ray moving with velocity $\vec{v} = v\hat{KO} = v\hat{y}$ is the magnetic Lorentz force given by

$$\begin{aligned}
\vec{F_m} &= -|e|\,\vec{v} \times \vec{B} \\
&= -|e|v\hat{y} \times B\hat{x} = -|e|vB\,\hat{y} \times \hat{x} = |e|vB\hat{x} \times \hat{y} = |e|vB\,\hat{z}
\end{aligned} \tag{1.2}$$

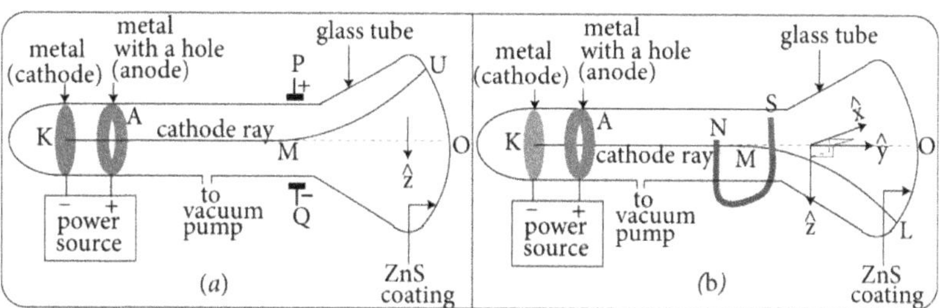

Figure 1.1. Cathode ray tube experiment. (a) Under electric field the cathode ray deflects up along the path $\overset{\frown}{MU}$. (b) Under magnetic induction the cathode ray deflects down along the path $\overset{\frown}{ML}$.

which means downward deflection. So the cathode ray follows the path $\overarc{ML}$. This confirms that the ray from cathode K was negatively charged.

Repeat of the experiment with different metal cathodes (K, Zn, Cu etc) led to the same negatively charged particles (i.e. the same cathode ray). This means the negatively charged particles exist in Zn, Cu etc and hence in all matter.

The mass of the negative charges of the cathode ray was found to be 1835 times smaller than hydrogen mass (*exercise 1.4*) which is the smallest atom. So particles smaller than atoms, were discovered. This means these negative charges were coming from inside the atoms. In other words these negative charges were a constituent of an atom.

1.3.1 Conclusions made by J J Thomson

The following conclusions follow from J J Thomson's experiment.
- Cathode rays are negatively charged material particles called electrons.
- Matter is composed of atoms. Electrons are one of the constituents or building blocks of an atom.
- Due to high voltage between cathode and anode, electrons from atoms of metal cathodes (Zn, Cu etc) shoot off or are released in the cathode ray experiment.
- Electrons (i.e. negative charges) are present in atoms and since atoms are electrically neutral they must also contain an equal amount of positive charge to balance out the negative charges carried by electrons. (Rutherford's α scattering experiment proved that an atom contains a concentrated mass called nucleus, carrying an equal amount of positive charge).

1.4 Determination of specific charge $\dfrac{|e|}{m}$ of the electron by J J Thomson

The charge-to-mass ratio is referred to as specific charge.

Thomson determined the ratio $\dfrac{\text{charge}}{\text{mass}}$ called specific charge of electron by studying deflection of cathode rays in electric and magnetic fields.

In the absence of any electric or magnetic field the electron beam from metal cathode K, proceeds through perforated anode A and strikes the point O of a fluorescent ZnS screen, as shown in figure 1.1.

The electric field $\vec{E} = E\,\hat{PQ} = E\hat{z}$ between P and Q is applied (figure 1.1(a)). Under the effect of the uniform electric field $\vec{E}$ the electron beam experiences electric force given in equation (1.1) viz. $\vec{F}_e = -|e|\vec{E} = -|e|E\hat{z}$ and strikes point U, travelling along the path $\overset{\frown}{MU}$. The energy gained by an electron, upon travelling through the field is

$$|e|V = \frac{1}{2}mv^2$$

$$\frac{|e|}{m} = \frac{v^2}{2V} \tag{1.3}$$

where V is the electric potential between the plates P, Q and $\vec{v} = v\hat{KO} = v\hat{y}$ is the velocity of the cathode ray (electrons).

Applying a suitable magnetic induction $\vec{B} = B\hat{x}$ the electron beam is brought back to point O due to the magnetic Lorentz force given in equation (1.2) viz.

$$\vec{F}_m = -|e|\,\vec{v} \times \vec{B} = -|e|v\hat{y} \times B\hat{x} = |e|vB\hat{z} \text{ (figure 1.1(b))}.$$

The force balance at point O can be expressed as

$$F_e = F_m \tag{1.4}$$

$$|e|E = |e|vB$$

$$v = \frac{E}{B} \tag{1.5}$$

Putting equation (1.5) in equation (1.3) we have

$$\frac{|e|}{m} = \frac{1}{2V}\left(\frac{E}{B}\right)^2 = \frac{E^2}{2VB^2} \tag{1.6}$$

Putting values of E, B, V we can find the specific charge $\frac{|e|}{m}$.

The value of specific charge of electron was obtained to be (*exercise 1.2*)

$$\frac{|e|}{m} = 1.76 \times 10^{11}\frac{C}{kg} \tag{1.7}$$

- In *exercise 1.3* we show how Milikan showed that electric charge is discrete or quantized and obtained its value to be $e = -|e| = -1.6 \times 10^{-19}\ C$.

1.5 Detection of the proton by Goldstein (1886)

Goldstein employed a modified cathode ray tube (figure 1.2) that had a perforated cathode K placed in the middle and anode A at one end. The tube was filled with hydrogen gas at very low pressure. When a high voltage was applied to the electrodes the electrons of the cathode ray flowing from the cathode K to the anode A collided with hydrogen atoms, knocking off the electrons from them. The left-over positive charge of the hydrogen atom was attracted to the negative cathode K. As the positive charges moved from anode A to cathode K they were called anode rays (also called positive rays or canal rays). They passed through the perforation and produced a glow at the fluorescent screen at the other end of the tube and hence got detected.

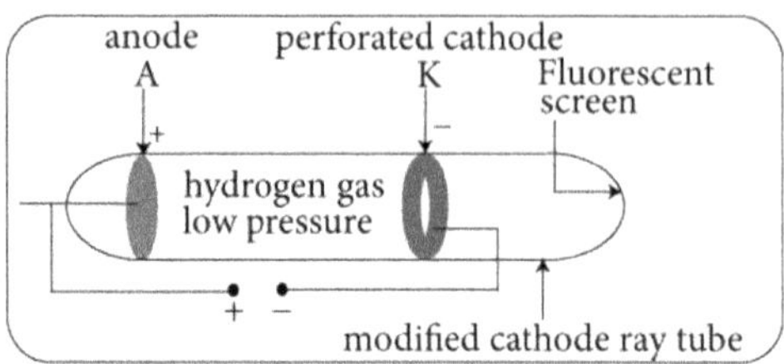

Figure 1.2. Goldstein observed positive rays or anode rays in modified cathode ray tube (1886 experiment).

The following observations were also made.

- ✓ If an object was placed between cathode and fluorescent screen a shadow was formed on the screen, proving that the anode rays travel in straight lines.
- ✓ If a padded wheel is kept in between cathode and fluorescent screen it starts to rotate, implying that the constituents of the positive rays moving from anode A to cathode K have mass.
- ✓ If an electric field was applied from the side of the tube it was observed that these anode rays got bent towards the negative terminal, implying that they are positively charged.
- ✓ Properties of anode rays (or positive rays) were dependent on the type of gas filled within the tube.
- ✓ With a tube filled with hydrogen gas the positive rays were actually hydrogen nuclei that were named protons by Rutherford in 1920.

H atom = H nucleus + 1 electron. So if we remove this sole electron what remains is H nucleus which is a proton:

$$\text{H atom} = \text{proton} - \text{electron} \equiv pe^- \text{ system.}$$

1.6 Naming of the proton by E Rutherford (1920)

In 1917, Rutherford performed the artificially induced nuclear reaction of bombarding nitrogen with $\alpha \equiv {}_2\text{He}^4$ as per the reaction

$$_2\text{He}^4 + {}_7\text{N}^{14} \rightarrow {}_8\text{O}^{17} + {}_1\text{H}^1$$

As a result, a subatomic particle was emitted that was identified as the hydrogen nucleus ${}_1\text{H}^1$.

Hydrogen is the lightest element and its nucleus is the lightest nuclei.

Again, the reaction showed that hydrogen nucleus ${}_1\text{H}^1$ was a part of nitrogen nucleus ${}_7\text{N}^{14}$ and by extension of argument, hydrogen nucleus was part of other nuclei (like ${}_2\text{He}^4$) as well.

Rutherford thus argued that the hydrogen nucleus was a fundamental building block of matter, being present in all elements, and in 1920 Rutherford coined the term proton for hydrogen nucleus. So

$$_1\text{H}^1 \equiv p$$

1.7 Determination of mass of the proton m_p

Goldstein, in his 1886 experiment with anode rays observed that properties of anode rays (or positive rays) were dependent on the type of gas filled within the tube.

So charge-to-mass ratio, i.e. specific charge, was dependent on the type of gas filled within the tube.

With hydrogen gas the specific charge was found to be maximum $=9.54 \times 10^7 \frac{C}{kg}$.

Again, having hydrogen gas in the tube means the anode rays or positive rays are made of protons ($\equiv$ hydrogen nuclei). Also,

$$\left|Q_{\text{proton}}\right| = \left|Q_{\text{electron}}\right| = |e| = 1.6 \times 10^{-19} C$$

Hence

Specific charge of proton is

$$\frac{\text{charge}}{\text{mass}} = \frac{Q_{\text{proton}}}{m_p} = \frac{1.6 \times 10^{-19} C}{m_p} = 9.54 \times 10^7 \frac{C}{kg}$$

$$m_p = \frac{1.6 \times 10^{-19} C}{9.54 \times 10^7 \frac{C}{kg}} = 1.67 \times 10^{-27} kg = \text{mass of proton}$$

1.8 Discovery of the nucleus by E Rutherford (1911)

In 1911, E Rutherford discovered the nucleus in his experiment on the scattering of alpha particles by thin gold foil. It turned out that the positive charge in the atom is not spread out but instead occurs as a concentrated mass at the centre of the atom, called the nucleus. We discussed this elaborately in volume 1.

1.9 Detection of the neutron by Chadwick (1932)

J J Thomson discovered electrons (1897) which were negatively charged, while Milikan measured electronic charge (1909) as detailed in *exercise 1.3*. According to the Bohr model of the atom (1913) electrons revolve in stationary non-radiating orbits around the nucleus. Rutherford's gold foil scattering experiment (1911) established that the positive charges were concentrated at the centre of the atom in a nucleus. The positive charges were the protons.

Protons are positively charged particles and the nucleus contains multiple protons. Since like charges repel each other, the protons in the nucleus mutually repel, leading to violent repulsive interaction within the nucleus. The question is how the protons stay together within the nucleus despite their mutual repulsion. Their staying together must be assisted by the presence of some other particle. In fact, strong interaction binds protons together, but the presence of neutrons helps reduce the mutual repulsion. Presence of neutrons in the nucleus was predicted by Chadwick in 1932.

When masses of helium and hydrogen atoms were compared it turned out that one helium (having two protons) was heavier than two hydrogen (having two protons, there being one proton in each hydrogen); but four hydrogen atoms (having four protons) balance out one helium atom (having two protons).

Mass of proton in atomic mass unit is $m_p = 1$ *amu*. Now the following discrepancies were observed.

Helium has two protons but its mass is 4 *amu*, lithium has three protons but its mass is 7 *amu*, beryllium has four protons but its mass is 9 *amu*, boron has five protons but its mass is 11 *amu* and so on. This means nuclei of elements contain some other constituent in addition to the protons, that make them more massive.

In 1932 Chadwick performed an experiment the schematic diagram of which is shown in figure 1.3.

Polonium $_{84}\mathrm{Po}^{210}$ was used as a source of $\alpha \equiv {}_2\mathrm{He}^4$, the corresponding reaction being

$$_{84}\mathrm{Po}^{210} \rightarrow {}_{82}\mathrm{Pb}^{206} + {}_2\mathrm{He}^4 \tag{1.8}$$

The emitted $_2\mathrm{He}^4 = \alpha$ was made to strike beryllium$_4\mathrm{Be}^9$ and an unknown particle, say x, is emitted according to the following reaction

$$_2\mathrm{He}^4 + {}_4\mathrm{Be}^9 \rightarrow {}_6\mathrm{C}^{12} + x \tag{1.9}$$

The beam x coming out of the $_4\mathrm{Be}^9$ target was found to remain undeflected in electric and magnetic fields, as it produces no track in the ionization chamber. So it is uncharged (like γ radiation).

To test whether this unknown x beam was γ radiation or consisted of a beam of massive particles, it was made to strike paraffin as per the following reaction

$$x + \text{paraffin} \rightarrow p + x \tag{1.10}$$

When this beam x was passed through paraffin wax (which is made of hydrogen nuclei, i.e. protons $_1\mathrm{H}^1 \equiv p$), a proton p was emitted.

Now γ, being massless, cannot dislodge the massive protons from paraffin. It was thus established that the unknown beam was not γ radiation.

Also, it was clear that the x beam was built out of particles that are massive—having mass comparable to the mass of proton, i.e. $m_X \sim m_p$ so as to be able to dislodge protons (mass $= m_p$) from paraffin. Clearly, the unknown beam was proved to be uncharged and massive (like protons).

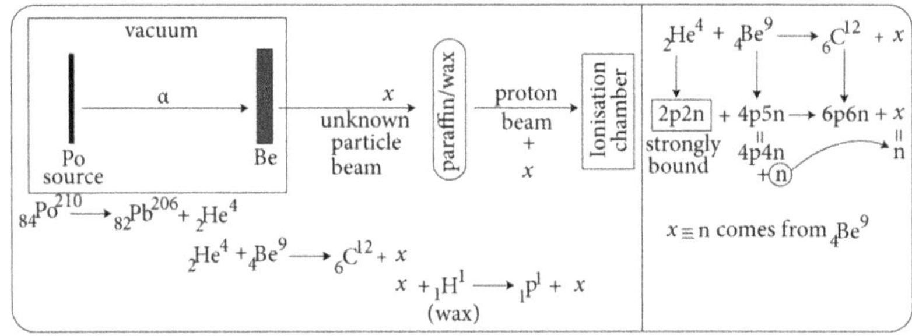

Figure 1.3. Schematic diagram of experimental arrangement of Chadwick for detection of neutrons.

Chadwick studied the interactions and from analysis of energy and momentum conservation proposed the existence of a new particle called the neutron $x \equiv n$ that had the following properties.

(i) It carries no charge $Q_n = 0$.

(ii) It is equally as massive as the proton, i.e. $m_n \cong m_p$.

(iii) Mass number of unity was attributed to n so that we can write $= {}_{Z=0}n^{A=1} \equiv {}_0n^1$.

The interactions represented by equations (1.8), (1.9), and (1.10) can thus be rewritten with $x \equiv {}_0n^1$ as follows

$$_{84}\mathrm{Po}^{210} \rightarrow {}_{82}\mathrm{Pb}^{206} + {}_2\mathrm{He}^4 \tag{1.11}$$

$$_2\mathrm{He}^4 + {}_4\mathrm{Be}^9 \rightarrow {}_6\mathrm{C}^{12} + {}_0n^1 \tag{1.12}$$

$$_0n^1 + {}_1\mathrm{H}^1 \rightarrow {}_1p^1 + {}_0n^1 \tag{1.13}$$

Writing the equation (1.12) viz. ${}_2\mathrm{He}^4 + {}_4\mathrm{Be}^9 \rightarrow {}_6\mathrm{C}^{12} + {}_0n^1$ in terms of proton p and neutron n we have:

$$2p2n + 4p5n \rightarrow 6p6n + n \tag{1.14}$$

A look at equation (1.12) and figure 1.3 shows that this new particle neutron n was coming from ${}_4\mathrm{Be}^9$ nucleus meaning that the nucleus contains a neutron.

It was thus established that the nucleus contains two types of particles:

Proton (charged particle) $Q_p = |e| = 1.6 \times 10^{-19}\,C$, mass $m_p = 1.6 \times 10^{-27}\,kg$ and

Neutron (uncharged particle) $Q_n = 0$, mass $m_n \cong m_p$, i.e. having almost the same mass as proton.

1.10 Mass determination of neutron by Chadwick and Goldhaber (1934)

Deuteron ($d \equiv {}_1\mathrm{H}^2 \equiv np$ system) was bombarded by γ rays of energy 2.62 MeV from $ThC'' \equiv {}_{81}\mathrm{Th}^{208}$ (thallium).

This energy is greater than the binding energy of d and hence can cause disrupture of deuteron to emit proton $p \equiv {}_1\mathrm{H}^1$ and neutron ${}_0n^1$ as

$$\gamma + {}_1\mathrm{H}^2 \rightarrow {}_1\mathrm{H}^1 + {}_0n^1 + T_p + T_n$$

or alternatively

$$\gamma + d \rightarrow p + n + T_p + T_n$$

where T_p and T_n are kinetic energies of p and n, respectively.

Conservation of mass-energy leads to

$$E_\gamma + m_d c^2 = (m_p c^2 + T_p) + (m_n c^2 + T_n)$$

$$m_n c^2 = E_\gamma + m_d c^2 - m_p c^2 - (T_p + T_n) \tag{1.15}$$

where the formula used for total energy of a particle is

$$E = T + mc^2 = \text{kinetic energy} + \text{rest energy} \tag{1.16}$$

$$m_n = \frac{E_\gamma}{c^2} + m_d - m_p - \frac{T_p + T_n}{c^2} \tag{1.17}$$

Using

$$1 \ amu = 932 \ \frac{MeV}{c^2}$$

$$\frac{1}{932} \ amu = \frac{MeV}{c^2}$$

and taking

$$m_d = 2.013\ 553 \ amu, \quad m_p = 1.007\ 276 \ amu$$

$$E_\gamma = 2.62 \ MeV$$

$$\frac{E_\gamma}{c^2} = 2.62\frac{MeV}{c^2} = \frac{2.62}{932} \ amu.$$

Also,

$$T_p + T_n = 0.45 \ MeV$$

$$\frac{T_p + T_n}{c^2} = 0.45\frac{MeV}{c^2} = \frac{0.45}{932} amu.$$

With this we have from equation (1.17)

$$m_n = \frac{2.62}{932} \ amu + 2.013\ 553 \ amu - 1.007\ 276 \ amu - \frac{0.45}{932} \ amu$$

$$m_n = 1.008\ 605 \ amu$$

Clearly $m_n \approx m_p$ (actually slightly greater).

We would take for convenience

$$m_n = 1.675 \times 10^{-27} \ kg = 939 \ \frac{MeV}{c^2} = 1.008\ 665 \ amu \text{ as neutron mass.}$$

1.11 Cosmic rays and elementary particles

Cosmic rays are highly penetrating high energy radiations (mostly protons) entering into Earth's atmosphere from interstellar space being incident in all directions, hitting Earth's atmosphere (at the rate $\sim$ 1000 per square meter per second) and generating showers of particles. The term cosmic is indicative of the fact that these high-altitude rays do not originate in our atmosphere.

Cosmic rays/particles were discovered by Hess in 1912 who sent electroscopes in balloons to measure radiation in the Earth's atmosphere. Hess was awarded the 1936 Nobel Prize in Physics for discovery of cosmic rays.

Milikan coined the term cosmic rays.

In *exercise 1.14* we elaborate the importance of studying cosmic rays.

1.12 Gold leaf electroscope

A gold leaf electroscope is an instrument that is used for detecting the presence of electric charge. A schematic diagram of a gold leaf electroscope and its working is shown in figure 1.4. The extremely thin conducting gold foils are flexible and respond very quickly to small electrostatic forces.

Suppose a gold leaf electroscope is negatively charged through conduction from a negatively charged bar and the spacing between leaves is *AB* as shown in figure 1.4(a).

If a negatively charged rod is brought near it (figure 1.4(b)) then the spacing between leaves increases to *CD* while if a positively charged rod is brought near it (figure 1.4(c)) then spacing between leaves decreases to *PQ* because discharge occurs.

1.13 Discovery of cosmic rays/particles

Suppose a properly insulated and charged gold leaf electroscope is left to itself. If air around the leaves is ionized then the electroscope gets discharged.

Hess arranged high-altitude balloon flights with electroscopes and recorded ionization of air at different altitudes above the Earth (figure 1.5). Ionization

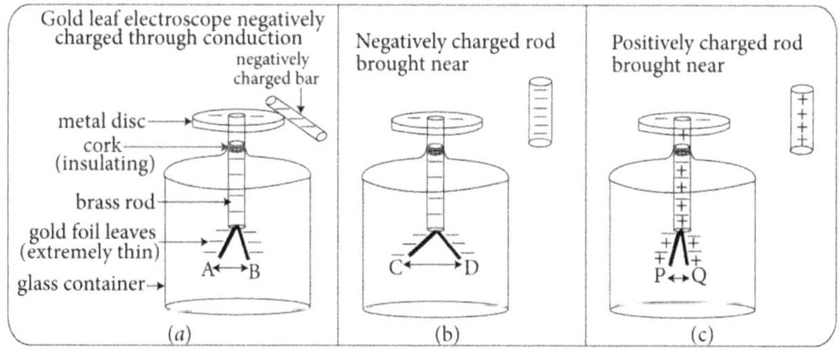

Figure 1.4. Gold leaf electroscope.

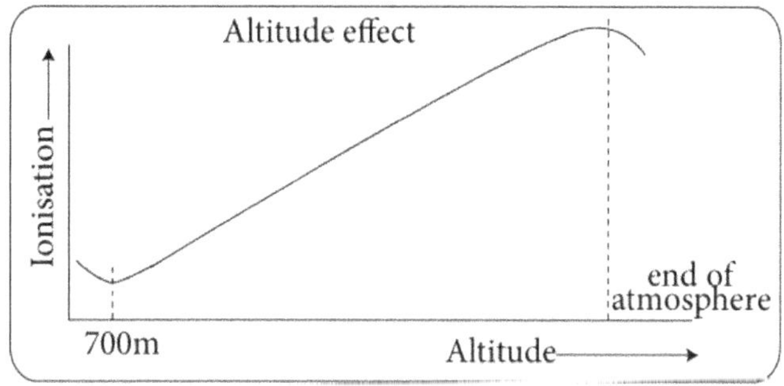

Figure 1.5. Variation of ionization with altitude done by Hess (1912).

showed an initial small decrease (below 700 m) and beyond that steadily increased to large values at higher altitudes (*exercise 1.13*).

These ionizations were not due to radioactive substances on Earth since if that was so then electroscope discharge would have diminished far away from Earth at higher altitudes as well as over sea, where radioactivity is negligible. But this did not occur, thereby proving that the source of atmospheric ionization is cosmic.

1.14 Composition of cosmic rays/particles

Cosmic rays can be classified into two types based on their origin. They are:
1) Primary cosmic rays.
2) Secondary cosmic rays.

1.15 Primary cosmic rays/particles

Primary cosmic rays are high energy cosmic rays or ionized nuclei.

They are beyond Earth's atmosphere and are incident upon the outer boundary of Earth's atmosphere.

1.15.1 Constituents of primary cosmic rays

About 90% of primary cosmic ray are protons, 9% helium nuclei i.e. α particles and 1% consists of positively charged heavy atomic nuclei e.g. Li, C, F, O, Co, Fe, Si etc.

Energy of primary cosmic rays ranges from MeV to energy greater than 10^{20} GeV.

They interact with Earth's atmosphere to produce secondary cosmic rays.

1.16 Secondary cosmic rays/particles

Secondary cosmic rays are produced when primary cosmic rays interact with the atmosphere through two main processes viz:

Cosmic ray shower or burst or cascade shower, and
Disruption of nuclei.
Below an altitude of 20 km all cosmic radiation is secondary.

1.16.1 Constituents of secondary cosmic rays at sea level

70% of secondary cosmic rays are pions, 29% electrons and positrons, 1% heavy particles, α particles, neutrons, γ radiations (photons).

1.17 Hard and soft components of cosmic rays/particles

At sea level, cosmic rays consist of two components:
1) Hard component.
2) Soft component.

Figure 1.6 shows the absorption curve for hard and soft components of cosmic rays.

The intensity of soft component decreases quickly while larger thickness of Pb is required for the intensity of hard component to decrease.

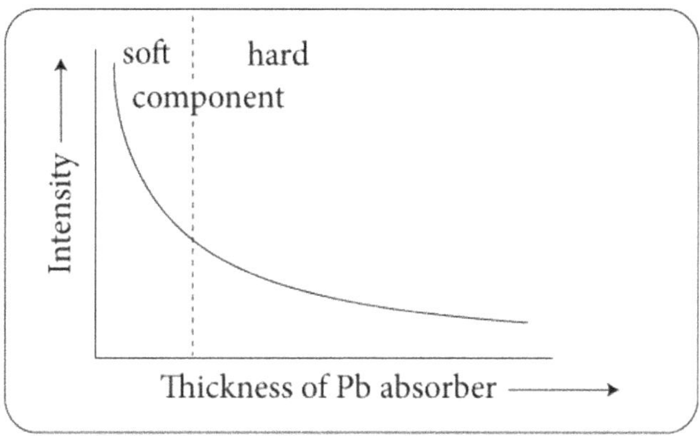

Figure 1.6. The absorption curve for soft and hard components of cosmic rays.

1.17.1 Hard component

This is more penetrating and not easily absorbed in matter, i.e. it needs larger length of matter ($\sim$ several metre thickness of *Pb*) to get absorbed. It is thus slowly absorbed.

It consists of muons.

1.17.2 Soft component

It is less penetrating and easily absorbed in matter ($\sim$ 10 *cm* Pb thickness or less).

It consists of electrons, positrons, photons.

1.18 Characteristics of cosmic rays/particles

Cosmic ray particles are free particles.

Most cosmic ray particles are relativistic. They possess energy $\sim MeV - GeV$ or more. Such high energy particles cannot be produced in the laboratory.

Very few of the cosmic ray particles have ultra-relativistic energies extending up to 10^{20} *eV* (i.e. about 20 *J*).

Radioactivity offers α particles of energy up to 5 *MeV*. But cosmic rays with energy $\sim$1 *GeV* are common and a detector may even record energy value $\sim 10^{11}$ *GeV*. Such huge energy cannot be created in colliders.

The rate at which cosmic ray particles strike a detector or probe is very low (since the screen of the detector/probe is very small in extension). So they are difficult to detect.

Cosmic ray particles are completely uncontrollable.

Cosmic ray constituents are predominantly particles (matter) and about 0.01% are anti-particles (anti-matter) and the latter are generated due to high energy particle collisions. So there is matter—anti-matter asymmetry in cosmic ray components.

1.19 Variation of cosmic ray intensity with altitude, latitude and direction

The incidence of cosmic rays on the surface of Earth is not uniform but depends on various factors that we mention in the following.

1.19.1 Altitude effect

This is shown in figure 1.5. Above the surface of Earth, after an initial small decrease (below 700 m) ionization increases with altitude at any latitude and becomes maximum beyond which there is a small decrease. This decrease occurs at the atmospheric limit (i.e. at the top of atmosphere) because of absorption of secondary cosmic rays produced by collision of primary cosmic rays with atmospheric particles. Only energetic secondary particles are able to reach Earth.

1.19.2 Latitude effect

This was detected by Clay (1927) who showed that ionizing flux is latitude dependent.

It follows from figure 1.7(a) that at sea level cosmic ray intensity increases from the equator ($\lambda = 0°$ latitude) to about $\lambda = 40°$N latitude by nearly 14%. Beyond 40°, latitude intensity is almost steady.

At higher altitude (say 1000 m) cosmic ray intensity increases from the equator ($\lambda = 0°$ latitude) to about $\lambda = 55°$N latitude by a greater amount $\sim$ nearly 22%. Beyond 55°, latitude intensity is almost steady.

In a nutshell, intensity of cosmic ray is maximum and uniform at the poles and small along the equatorial line.

1.19.3 Explanation

Latitude effect occurs due to Earth's magnetic field. A particle having sufficient kinetic energy can overcome the deflecting Lorentz force in Earth's magnetic field and reach Earth (figure 1.7(b)).

At the equator, particles move perpendicularly to the magnetic field and so experience maximum deflecting force, since then

$$\overrightarrow{F} = q\,\overrightarrow{v}\times\overrightarrow{B} \sim qvB \sin 90° = qvB = \text{maximum}$$

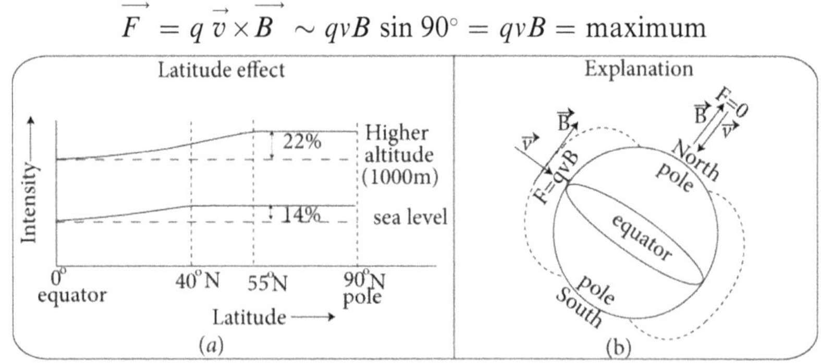

Figure 1.7. Variation of cosmic ray intensity with latitude.

So a small number of them reach the equator.

At the poles, particles move parallel to the magnetic field and so experience small or no deflecting force, since then

$$\vec{F} = q\,\vec{v} \times \vec{B} \sim qvB \sin \pi = 0 = \text{minimum}$$

So a large number of them reach the pole.

1.19.4 Directional effect or east–west asymmetry

At any azimuthal angle ϕ the number of cosmic rays per second coming from the west side is found to be more than that coming from the east side. This is referred to as east–west asymmetry.

This effect is maximum at the equator, where 14% more particles arrive from the west.

The effect is minimum at the poles.

This east–west asymmetry shows that primary cosmic rays are composed predominantly of positively charged particles, as explained in figure 1.8.

Earth's magnetic field is $\vec{B}$ and a particle of positive charge q moving with velocity $\vec{v}$ experiences a deflecting Lorentz force

$$q\vec{v} \times \vec{B} \sim qvB \sin \frac{\pi}{2}\,\hat{\phi}$$

i.e. along the east. So positively charged particles appear to come from the west.

1.20 Origin of cosmic rays

The following facts are taken into account regarding the source of cosmic rays:
- Cosmic rays come from outside Earth's atmosphere.
- Cosmic ray influx is not significantly affected by solar activity.
- Solar activity refers to the variable short-lived disturbances on the Sun like ejection of mass from Sun, solar wind, sun spots etc.
- The bulk of cosmic rays come from outside the Solar System but from within the Milky Way Galaxy.

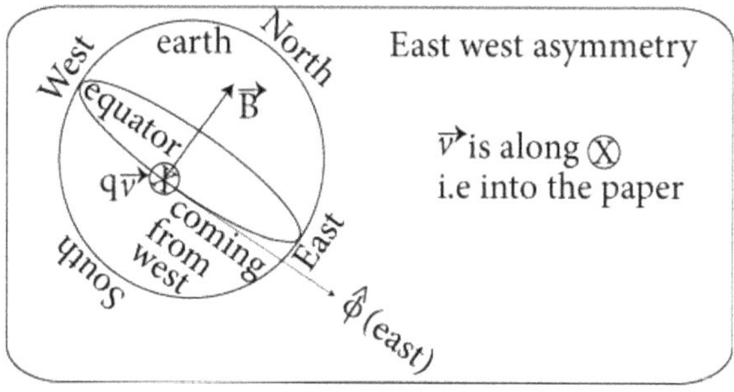

Figure 1.8. Variation of cosmic ray intensity due to east–west asymmetry.

- There is rapid variation of influx of a small portion of cosmic rays during violent events in the Sun. This shows that a small portion of cosmic rays come from within the Solar System.
- A small portion of cosmic rays, having very high energy, have gyro-radii in the galactic magnetic field larger than the size of the Galaxy. This shows that a small portion of cosmic rays come from outside the Milky Way.

1.20.1 Energy of cosmic rays

Kinetic energy of cosmic rays comes from gravitational potential energy due to collapse of astrophysical objects. They might be remaining or residual parts of a supernova.

The energy of cosmic ray particles ranges from MeV to energy greater than $10^{20}\ GeV$. They are absorbed in Earth's atmosphere through generation of showers of particles.

1.21 Cosmic ray shower or burst or cascade or electronic shower (hadronic and electromagnetic)

There are occasional sudden short-duration significant increases in cosmic ray intensity in the upper atmosphere and a consequent sudden burst in ionization (as can be recorded in an ionization chamber). The primary particle (that initiates the avalanche or shower) suffers some collision, creates secondary particles and the latter create more particles. The process is cumulative and there is a burst or significantly huge rise in cosmic ray intensity because of production of a large number of particles in a very very short span of time. This phenomenon is called a cosmic ray shower or burst or cascade shower, as shown in figure 1.9.

The number of particles in a cosmic shower is called the size of shower.

We can explain this phenomenon using Bhabha–Heitler theory and Carlson–Oppenheimer theory (1937).

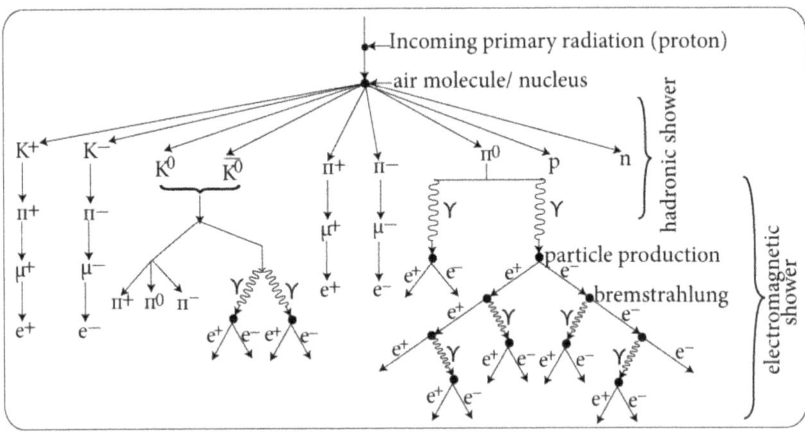

Figure 1.9. Cosmic ray shower or burst or cascade shower or electronic shower.

A cosmic ray (protons) of high energy enters the atmosphere and collides with air molecules or any nucleus. Several secondary particles are produced, like pions ($\pi^{\pm}$, π^0), kaons ($K^{\pm}$, K^0, $\overline{K^0}$), neutrons, protons. These particles either decay and/or suffer further collision with air molecules or any nucleus. A few decay schemes are as follows.

$$K^+ \rightarrow \pi^+ + \pi^0$$

$$K^- \rightarrow \pi^- + \pi^0$$

$$\pi^+ \rightarrow \mu^+ + \nu_\mu$$

$$\pi^- \rightarrow \mu^- + \overline{\nu_\mu}$$

$$\pi^0 \rightarrow \gamma + \gamma$$

The product particles (e.g. $\mu^{\pm}$) further decay, for instance

$$\mu^+ \rightarrow e^+ + \nu_e + \overline{\nu_\mu}$$

$$\mu^- \rightarrow e^- + \overline{\nu_e} + \nu_e$$

Figure 1.9 shows a cosmic ray shower or burst. The following two processes occur continuously while the cosmic shower proceeds.
- Bremsstrahlung,
- Pair production (materialization of energy).

1.21.1 Bremsstrahlung or braking radiation

When a moving charged particle suddenly encounters a nucleus or molecule it loses energy and slows down (decelerates) and there is emission of energy in the form of γ radiation. This phenomenon is called bremsstrahlung or braking radiation or deceleration radiation.

1.21.2 Pair production (materialization of energy)

When a γ photon carrying energy

$$E_\gamma = h\nu \geqslant 1.02 \; MeV \tag{1.18}$$

strikes a nucleus, it gets converted into an electron–positron pair (i.e. energy is transformed to matter), i.e.

$$\gamma + \text{Nucleus} \rightarrow e^+ + e^- + \text{Nucleus} \tag{1.19}$$

The nucleus serves to conserve linear momentum. Here h is Planck's constant and ν is frequency of γ.

Energy conservation requires that the γ should possess energy $\geqslant 1.02 \; MeV$ as shown in the following

$$h\nu \geqslant m_{e^+}c^2 + m_{e^-}c^2 \tag{1.20}$$

As $m_{e+} = m_{e-} = m_e = 0.511\frac{MeV}{c^2}$ we have

$$hv \geqslant 2m_e c^2 = 2 \times 0.511 \; MeV$$

$$hv \geqslant 1.02 \; MeV \tag{1.21}$$

We discuss the interaction in more detail in *exercise 1.20*.

1.21.3 Hadronic shower or cascade

Kaons and pions (which are particles called hadrons) are generated. These particles (hadrons) decay or collide further. The corresponding production of particles or cascade is referred to as a hadronic shower or cascade (figure 1.9).

1.21.4 Electromagnetic shower or cascade

The sudden production of numerous photons and electron–positron pairs through processes like bremsstrahlung, pair production and collision results in a sudden cascade of particles referred to as an electromagnetic shower or cascade (figure 1.9).

Eventually the energies of e^+, e^- and γ are reduced so much that no further creation of energetic photons and other particles is possible. The low energy photons produced now lose their energy by Compton scattering and the low energy electrons and positrons lose energy by ionization.

- In *exercise 1.15* we obtain an estimate of the photon energy of an electromagnetic shower according to the Heitler model of cascade.
- We discuss how pair production was discovered by Blackett and Occhialini in 1933 in section 1.23.

1.22 Discovery of the positron by Anderson (1932)

A macroscopic particle has de Broglie wavelength $\lambda \ll$ size of particle.

But microscopic or small particles like p, n, e^- etc have de Broglie wavelength $\lambda \sim$ size of particle and their behavior is described using quantum mechanics.

The positron was predicted in 1928 by Dirac in his electron theory based on relativistic quantum mechanics which is quantum mechanics incorporating the special theory of relativity.

Dirac's theory (chapter 11) describes the behavior of electrons. But Dirac's equation posed the following problem.

Dirac's equation had two solutions. One was for an electron with positive energy states. The other one was for an electron with negative energy states. However, energy of a particle cannot be negative. Dirac interpreted his equation leading to negative energy states as follows (*exercise 1.27*).

For every particle there is a corresponding anti-particle that has the same mass as the particle but carries charge that is opposite to that of the particle. Accordingly, Dirac mathematically predicted the existence of an anti-particle of the electron called an anti-electron or positron that would have the same mass as the electron but carries a positive charge of the same magnitude as that of the electron. So the positron was supposed to be identical to the electron except for opposite charge.

Further, the two (electron and positron) would annihilate each other upon interaction—a process called pair annihilation.

- Anderson photographed tracks of cosmic rays in a cloud chamber placed between the poles of an electromagnet ($B \sim 1.5\ T$).

  A cloud chamber is a chamber with super saturated fluid (alcohol) and when a tiny charged particle goes through, it leaves a track. We discuss more about cloud chambers in chapter 3.

- Consider a positively charged particle $q > 0$, as shown in figure 1.10(a), travelling with velocity $\vec{v} = v\,\hat{LO} = v\hat{y}$ and entering a region of magnetic induction $\vec{B} = B\hat{z}$, $\hat{z}$ being the direction pointing down into the paper away from the reader. At entry point O the Lorentz force acts and is given by

$$\vec{F} = q\,\vec{v} \times \vec{B} = qv\hat{y} \times B\hat{z} = qvB\hat{x} = qvB\,\hat{OC}.$$

So force is towards point C.

$\overset{\frown}{OM}$ is the trajectory of the particle which is a portion of a circle. The particle, at every point of its trajectory $\overset{\frown}{OM}$ feels such a deflecting force towards point C which is being balanced by the centripetal force as

$$\frac{mv^2}{r} = qvB \tag{1.22}$$

$$r = \frac{mv}{qB} \tag{1.23}$$

So radius r of circular trajectory depends on mass m and charge q of the particle. Clearly a positive charge will travel along the circular arc $\overset{\frown}{OM}$.

By similar arguments a negative charge $q < 0$ would experience a force towards point D and would follow the circular arc $\overset{\frown}{ON}$.

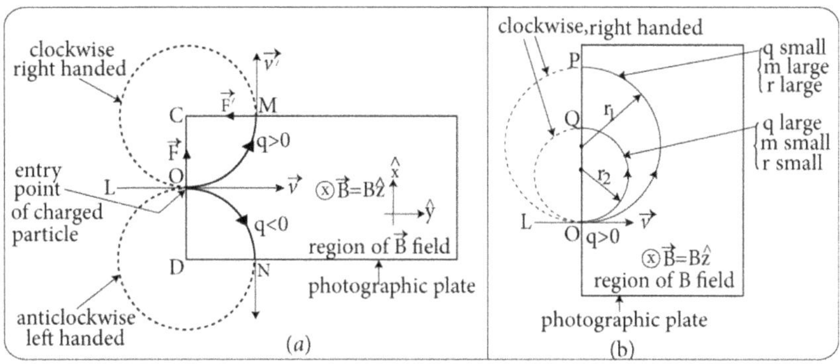

Figure 1.10. Charged particle tracks in a magnetic field. From the direction in which the path curves one can deduce the sign of charge carried by the particle.

Also, as depicted in figure 1.10(b), it follows from the relation $r = \frac{mv}{qB}$ that the more massive (hence more inertia), the less charged the particle is, the less will be the amount of deflection and the less will be the curvature of path followed. In other words, the particle path will be more straight and so the radius will be large (as for the circular track $\overset{\frown}{OP}$ of radius r_1). On the other hand, the less massive (hence less inertia), the more charged the particle is, the more will be the amount of deflection and the more will be the curvature of path followed. In other words, the particle path will be more curved and so the radius will be small (as for the circular track $\overset{\frown}{OQ}$ of radius r_2).

- Anderson (in 1932) observed occassional occurrence of tracks like $\overset{\frown}{AGS}$ in a cloud chamber, as shown in figure 1.11.

  The track had two parts.
  One part is $\overset{\frown}{AG}$ (which is a part of the larger circle AGH with centre at J) lying below the Pb sheet (6 *mm*).
  The other part is $\overset{\frown}{GS}$ (which is a part of the smaller circle GST with centre at K) lying above the Pb sheet.
  The nature of bending of the path of the charged particle indicated that it is positively charged, $q > 0$.
  The path of the particle was identified to be $\overset{\frown}{AGS}$. In *exercise 1.16* we argue why the Pb sheet was used and why the path is identified to be $\overset{\frown}{AGS}$ and not $\overset{\frown}{SGA}$.
  The radius of the track, say $\overset{\frown}{GS}$, was measured and found to be the same as the one which an electron would have created if it had a positive charge. This means the particle that produced the track $\overset{\frown}{AGS}$ was positively charged and had all the properties or characteristics (like mass, charge magnitude etc) identical to that of an electron. In *exercise 1.17* we discuss if this track could be due to a proton.

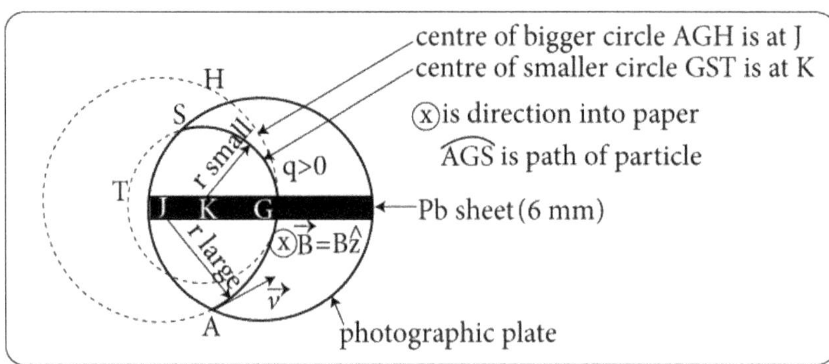

Figure 1.11. Charged particle tracks in a cloud chamber led Anderson to discover the positron in 1932.

Anderson named the particle positron e^+ possessing the following properties.

Charge $Q_{e^+} = |e| = 1.6 \times 10^{-19}C$, mass $m_{e^+} = m_{e^-} = 9.1 \times 10^{-31}kg = 0.511\frac{MeV}{c^2}$.

This positron was actually the anti-electron predicted by Dirac in 1928 and was the anti-particle of the electron.

- We discuss in detail in chapter 2, how a positively charged particle will bend and trace a left-handed circle (clockwise motion) and a negatively charged particle will bend and trace a right-handed circle (anticlockwise motion) w.r.t the magnetic field when projected perpendicular to the field. They will trace helical paths when projected obliquely to the magnetic field.

1.23 Discovery of pair production by Blackett and Occhialini (1933)

There was occasional occurrence of pairs of tracks in a cloud chamber, as shown in figure 1.12, from a point S (say). The tracks had equal and opposite radii of curvature. This can be interpreted as follows.

A high energy γ (which does not produce a track in a cloud chamber) collides with an electron at a point, say S (perhaps the position of an atom) and simply shoots off along SH. The energy that γ carries with it is released and converted to mass, the process is called pair production or pair creation. The circular track $\overset{\frown}{SP}$ is traced by a positron e^+ and the circular track $\overset{\frown}{SQ}$ is traced by an electron e^- in the $\vec{B} = B\hat{z}$ field.

The interaction is represented as (equation 1.19)

$$\gamma + \text{nucleus} \rightarrow e^+ + e^- + \text{nucleus}$$

The minimum or threshold energy that the γ should possess for pair production is

$$E_\gamma = m_{e^+}c^2 + m_{e^+}c^2 = 1.02 \ MeV$$

as shown in equation (1.21).

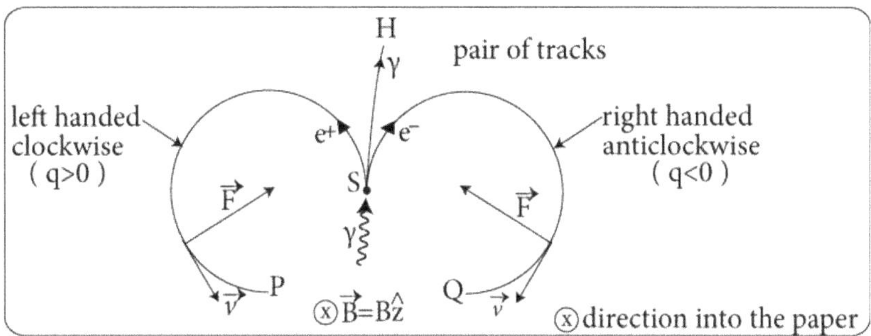

Figure 1.12. Blackett and Occhialini discovered pair production in 1933 by taking photographs of a pair of tracks, beginning at the same point, one of e^+ and the other of e^-.

1.23.1 Pair annihilation

When a particle meets its anti-particle, interaction occurs between them. They destroy each other. In other words, matter (i.e. particle–anti-particle pair) is converted to energy.

For instance, when a positron (mass m_{e^+}) meets its anti-particle electron (mass m_{e^-}) they mutually annihilate, i.e. destroy each other, and in the process radiation, $h\nu$, i.e. γ photons are emitted. Energy conservation gives

$$e^+ + e^- \rightarrow \gamma + \gamma \tag{1.24}$$

Pair production is not strictly the opposite process of pair annihilation because in pair production ($\gamma + \text{nucleus} \rightarrow e^+ + e^- + \text{nucleus}$) one photon γ is transformed into two particles (e^+, e^-) while pair annihilation ($e^+ + e^- \rightarrow \gamma + \gamma$) transforms two particles (e^+, e^-) into at least two photons ($\gamma\gamma$).

- In *exercise 1.20* we show that a freely moving γ cannot disintegrate into $e^\pm$ but presence of a nucleus is necessary in the vicinity of which pair production occurs.
- In *exercise 1.21* we discuss why an electron–positron pair cannot annihilate into a single photon.
- In *exercise 1.22* we discuss whether it is possible for an isolated free electron to absorb or emit a photon.
- In *exercise 1.23* we argue why it is impossible for a photon to transfer all its energy to a free electron.

1.24 Discovery of muon by Anderson and Neddermeyer (1937)

At sea level, cosmic ray components are made up of a soft component and a hard component.

The hard component was found to be highly penetrating (can penetrate several meter thickness of Pb) and hence possesses high energy.

It was noted that they pass alone and there was no cascade electronic shower associated. This means the possiblity of their being electrons, positrons and photons was ruled out.

A cloud chamber photograph of tracks of cosmic ray particles of the hard component was studied.

The highly penetrating radiation was allowed to traverse a lead plate 3.5 cm thick (inserted in a cloud chamber).

Their range was about 4 cm and the radius of curvature of the track was found to be about 7 cm. Had they been electrons ($m_e = 9.1 \times 10^{-31}\,kg$) they would have suffered much more energy loss. Since energy loss in the lead plate was low, these particles were more massive than electrons. So they were not electrons.

From the measured range of 4 cm the momentum P was determined. The magnetic induction applied was $B = 0.79\,T$. Now equating Lorentz force and centripetal force we have

$$|\vec{F}| = |q\,\vec{v} \times \vec{B}| = \frac{mv^2}{r}$$

$$qvB \sin 90° = \frac{mv^2}{r} \quad (\vec{v} \perp \vec{B} \text{ situation})$$

$$r = \frac{mv}{qB} = \frac{p}{qB} \tag{1.25}$$

Had the cosmic ray particle producing the track been a proton ($m = m_p$), then with $q = q_p = |e|$ one would get $r = \frac{p}{qB} = 20\,cm$ and not $7\,cm$ as observed. Hence they were not protons.

Anderson and Neddermeyer concluded that the track of the hard cosmic ray component created in the cloud chamber was that of a new particle whose mass was between m_e and m_p and calculation showed that the mass of this particle was $200m_e$. Also the hard component that was highly penetrating was minimally ionizing and it was shown that they carry one electronic charge. It was named a negative muon μ^- and had the following properties.

$$m_{\mu^-} = 200m_e$$

$$Q_{\mu^-} = -|e|.$$

Muons are not stable particles, but decay with a lifetime of $2\,\mu s$ as follows

$$\mu^- \rightarrow e^- + \overline{\nu_e} + \nu_\mu$$

where $\overline{\nu_e}$ is an electron anti-neutrino and ν_μ is a muon neutrino.

1.25 Muon paradox

A negative muon that is created in the upper atmosphere travels with a velocity

$$v = 0.998c \tag{1.26}$$

$$= 0.998\left(3 \times 10^8 \frac{m}{s}\right) = 2.994 \times 10^8 \frac{m}{s} \tag{1.27}$$

It is unstable and decays as $\mu^- \rightarrow e^- + \overline{\nu_e} + \nu_\mu$ with a lifetime of

$$t_0 = 2\,\mu s = 2 \times 10^{-6}\,s \tag{1.28}$$

Suppose the muon is formed at a height $6.5\,km = 6.5 \times 10^3\,m = 6500\,m$ from Earth's surface. So it can travel a distance of

$$D = vt_0$$

$$= \left(2.994 \times 10^8 \frac{m}{s}\right)(2 \times 10^{-6}\,s) = 599\,m < 6500\,m. \tag{1.29}$$

Clearly muons should not be able to reach Earth's surface. However, observation shows that approximately 1 muon reaches 1 cm^2 of Earth's surface per minute. This apparent puzzle is referred to as the muon paradox.

1.25.1 Resolution to the paradox

Resolution of the paradox is possible using the special theory of relativity.

The lifetime $t_0 = 2\ \mu s$ is w.r.t. the muon frame of reference.

In Earth's frame of reference the lifetime is obtained by the time dilation formula of the special theory of relativity, namely

$$
t = \frac{t_0}{\sqrt{1 - \frac{v^2}{c^2}}}
$$

$$
= \frac{2\ \mu s}{\sqrt{1 - (0.998c)^2/c^2}} = \frac{2\ \mu s}{\sqrt{1 - (0.998c)^2}} = 31.6\ \mu s
$$

(1.30)

This is the dilated time.

The distance that a muon travels in Earth's frame is

$$
D = vt = \left(2.994 \times 10^8 \frac{m}{s}\right)(31.6 \times 10^{-6} s\,) = 9461\ m > 6500\ m.
$$

(1.31)

Clearly, thus muons would certainly reach Earth. So the muon paradox is resolved by the time dilation concept of the special theory of relativity.

Suppose we focus attention upon the muon i.e. travel with the muon.

In the muon frame the lifetime of muon is $t_0 = 2\ \mu s = 2 \times 10^{-6}\ s$. Then the distance 6500 m would appear contracted in the muon frame of reference. The distance that muons would see in their frame is given by the length contraction formula of the special theory of relativity, namely

$$
l = l_0 \sqrt{1 - \frac{v^2}{c^2}}
$$

(1.32)

$$
= 6500\ m\sqrt{1 - (0.998c)^2/c^2} = 6500\ m\sqrt{1 - (0.998c)^2} = 411\ m < 599\ m.
$$

Clearly, muons seem to travel a contracted distance to reach Earth. So the muon paradox is resolved by the length contraction concept of the special theory of relativity.

1.26 Discovery of the pion by Lattes, Occhialini and Powell (1947)

Yukawa in 1935 predicted the existence of the pion as a force carrying particle mediating strong force.

Nuclear photographic emulsion plates were exposed to cosmic radiation at mountain altitude. Several tracks were observed in emulsion plates.

Investigation of the tracks proved the existence of positive and negative pions $\pi^\pm$. We discuss this in the following with reference to figure 1.13.

Consider the track AB of figure 1.13(a).

Detailed analysis of track AB revealed that they were created by a particle of mass 273 m_e and charge $|e|$ that was called a positive pion π^+.

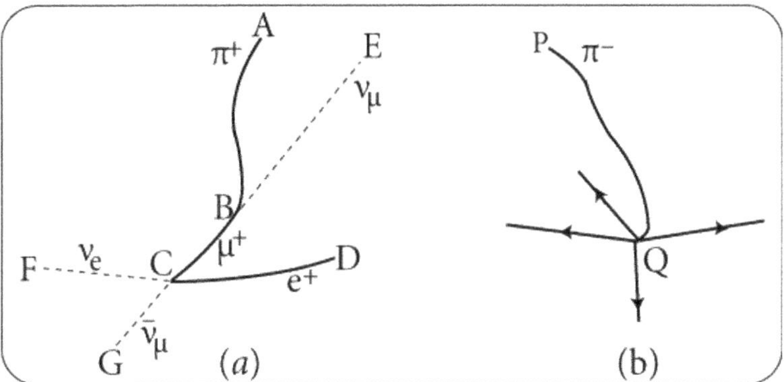

Figure 1.13. Tracks of pion in nuclear emulsion plate.

The particle comes to a halt at point B and the track that emerges BC corresponds to another particle that was confirmed to be a positive muon μ^+ with mass 207 m_e and charge $|e|$. Clearly, π^+ decays to generate μ^+. Along with it, another particle is generated that cannot produce any track in the emulsion. So we have shown it by a dashed line BE. Also, since there are no e^+, e^- pairs along the path it cannot be a γ photon. Actually, the positive pion π^+ decays to a positive muon μ^+ plus a muon neutrino ν_μ as follows.

$$\pi^+ \rightarrow \mu^+ + \nu_\mu \ (\text{lifetime } 10^{-8} \ s)$$

This muon neutrino ν_μ is uncharged and hence cannot produce any track in the emulsion. The kinetic energy available to μ^+, ν_μ is

$$m_\pi c^2 - m_\mu c^2 = 273 \ m_e c^2 - 207 \ m_e c^2$$
$$= 66 \ m_e c^2 = 66 \times 0.511 \ MeV = 33 \ MeV$$

Of this μ^+ takes away $T_\mu = 4 \ MeV$ as its kinetic energy (in all $\pi \rightarrow \mu$ decay) and the rest 29 MeV is taken by ν_μ. As ν_μ takes away a major part of the energy it follows that ν_μ is nearly massless.

The track of μ^+ ends at point C. This means μ^+ is unstable and decays at point C to positron e^+. The track of e^+ is distinctively observed to be CD. Kinetic energy of e^+ was different in different cases of decay and satisfied the condition $T_{e^+} \leqslant 50 \ MeV$.

Analysis of μ^+-e^+ energy spectrum showed that two non-identical, neutral particles are accompanied with $\mu^+ \rightarrow e^+$ decay. They cannot be γ quanta as evident from the absence of e^+-e^- pairs near the decay site. One of the particles is a muon neutrino ν_μ and the other is an electron anti-neutrino $\overline{\nu_e}$.

So a positive muon μ^+ decays to a positron e^+, muon anti-neutrino $\overline{\nu_\mu}$ and an electron neutrino ν_e and we can represent the interaction as follows

$$\mu^+ \rightarrow e^+ + \nu_e + \overline{\nu_\mu} \ (\text{lifetime } 2 \ \mu s)$$

As ν_e, $\overline{\nu_\mu}$ are neutral they do not leave any track and we have indicated them by symbolic lines CF (for ν_e) and CG (for $\overline{\nu_\mu}$).

Similarly, for a negative pion π^- the following decay schemes have been confirmed

$$\pi^- \to \mu^- + \overline{\nu_\mu} \quad \text{(lifetime } 10^{-8}\,s)$$

$$\mu^- \to e^- + \overline{\nu_e} + \nu_\mu \quad \text{(lifetime } 2\,\mu s)$$

We now mention another type of track registered in the nuclear photographic emulsion, as shown in figure 1.13(b).

Consider the star of figure 1.13(b).

Track PQ corresponds to a particle of charge $|e|$ and mass 273 m_e. The track ends or terminates at point Q. From this point Q of termination of the track, several charged particles fly out generating their respective tracks in the emulsion and forming a star consisting of several rays as shown.

This was interpreted as the capture of a negative pion π^- by a nucleus Q, resulting in the splitting of the nucleus which is observed in the emulsion as a star.

The mass of π^- was calculated from here, taking into account the kinetic energy, binding energy of released particles from point Q of the star and was obtained to be 273 m_e.

We note that this splitting of nucleus could be done by negative pion π^- which can come very close to the nucleus and get absorbed by it. On the other hand, the positive pion π^+ cannot come close to the nucleus on account of Coulomb repulsion between π^+ and positively charged nucleus. π^+ decays to μ^+, ν_μ and μ^+ decays to e^+, ν_e, $\overline{\nu_\mu}$.

The decay scheme of pion and muon are shown in figure 1.14.

It is clear that experimental observation of the tracks (figure 1.14) revealed the existence of pions $\pi^\pm$ and established that they interact strongly with matter.

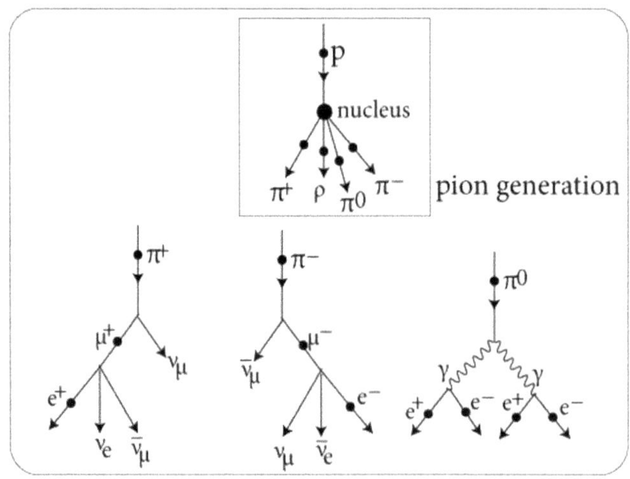

Figure 1.14. Decay scheme of pion and muon.

1.26.1 The neutral pion π^0.

The neutral pion π^0 is more difficult to observe than the charged pions $\pi^\pm$ because a neutral particle will not leave a track in emulsion. Its existence was confirmed from its decay products in the cosmic ray.

The neutral pion π^0 was identified at the Berkeley cyclotron in 1950 through its decay into two photons as per the following scheme:

$$\pi^0 \rightarrow \gamma + \gamma \ \ (\text{lifetime } 10^{-16}\, s)$$

and in the cosmic ray balloon experiment at Bristol university, UK, also in 1950.

$$\pi^0 \text{ has mass of } 264\, m_e = 264 \times 0.51\, MeV = 135\, MeV.$$

After decay of $\pi^0 \rightarrow \gamma\gamma$ each photon carries energy $\sim 70\, MeV$. This means that neutral pion π^0 is its own anti-particle.

1.27 Discovery of anti-proton by Segre and Chamberlin (1955)

The following are important facts characterizing proton.

$$m_p = 1835\, m_e, \ \ Q_p = |e|.$$

A proton is a fermion with $s_p = \frac{1}{2}$. Magnetic moment $\mu_p = 2.792\,75\,\mu_N$

Dirac predicted the existence of an anti-particle for every particle. The anti-particle of proton p is anti-proton $\bar{p}$.

Anti-proton $\bar{p}$ should have charge opposite to the charge carried by a proton but of the same magnitude i.e.

$$Q_{\bar{p}} = -|e| = -Q_p \text{ (negatively charged)}, \ \ m_{\bar{p}} = m_p.$$

In an encounter of a proton–anti-proton pair (i.e. particle–anti-particle pair) they should destroy, i.e. annihilate, each other.

An Anti-proton $\bar{p}$ can be created through collision between two protons, i.e. through p–p collision at high energy according to the scheme

$$p + p \rightarrow p + p + (p + \bar{p})$$

The first proton p can be in the beam and the second proton p may be sitting fixed in a copper target. In *exercise 1.25* we have estimated the minimum energy E_p of an incident proton needed to collide with a stationary or fixed target proton so as to make a search for anti-proton $\bar{p}$ which turns out to be $E_p = 6.6\, GeV$.

Figure 1.15 shows the experimental set-up.

A copper target was bombarded with $6.2\, GeV$ protons from a particle accelerator —the proton synchrotron (called Bevatron) at Berkeley.

Anti-protons were produced due to collision of incident $6.2\, GeV$ proton and fixed target proton.

About 50 000 negative pions π^- per anti-proton $\bar{p}$ are also produced in the process. All these are produced in the Bevatron magnetic field.

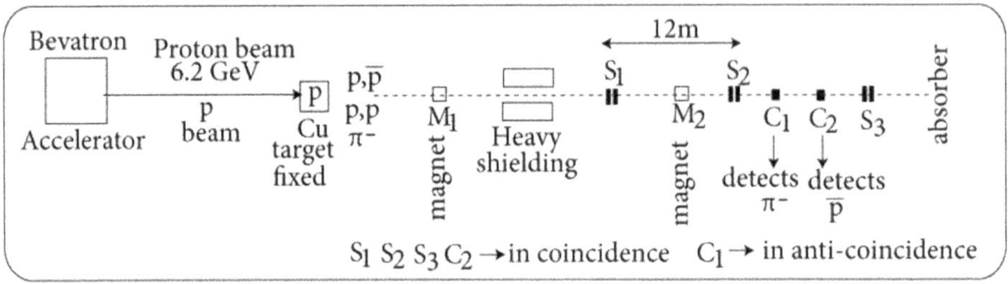

Figure 1.15. Experimental arrangement to detect anti-proton $\bar{p}$.

Anti-proton $\bar{p}$ and π^- are negatively charged and follow a path having curvature opposite to the curvature of proton path. So $\bar{p}$ and π^- emerge out of the copper target.

A series of magnets were employed to deflect away all unwanted positively charged particles so that only negatively charged particles ($\bar{p}$ and π^-) could be focused upon.

The magnets M_1, M_2 select particles of a definite momentum $1.19\frac{GeV}{c}$ to pass through an aperture in the heavy shielding. The speed of anti-proton $\bar{p}$ was $v_{\bar{p}} = 0.79c$ and the speed of negative pion π^- was $v_{\pi^-} = 0.99c$.

Scintillation counters S_1, S_2, S_3 and Cerenkov counters C_1, C_2 are employed.

C_1 detects particles of speed $v > 0.79c$. So C_1 detects the π^- only.

C_2 detects particles of speed between $v = 0.75c$ and $0.79c$. So C_2 detects the $\bar{p}$ only.

Distance between S_1 and S_2 was 12 m. A pion took 40 *ns* to cross this distance but the anti-proton took 51 *ns*, i.e. longer time, to cross the same distance.

S_1, S_2, S_3, C_2 were connected in coincidence while C_1 was in anti-coincidence.

So a pion, detected by C_1, did not produce any coincidence pulse.

Only the coincidence produced by anti-protons could be detected by this arrangement. In other words, the recorded coincidence pulses indicated presence of an anti-proton.

The rest mass of anti-protons was estimated from experiment to be same as that of a proton, i.e. $m_{\bar{p}} = m_p$.

Photographic emulsions exposed to anti-protons were bombarded with high energy protons. Distinctive star bursts resulting from the annihilation of anti-protons by protons in photographic emulsions was observed.

In *exercise 1.26* we discuss the differences between annihilation of a proton–anti-proton pair with annihilation of an electron–positron pair.

1.28 Detection of the electron anti-neutrino $\bar{\nu}_e$ by Reines and Cowan (1956)

Detection of the electron anti-neutrino $\bar{\nu}_e$ was done based on the inverse beta decay, namely the reaction

$$p + \bar{\nu}_e \rightarrow n + e^+$$

The probability of such a reaction is extremely small and so the reaction has to be studied or observed in a region where there is intense anti-neutrino flux. A nuclear reactor can be treated as a source of intense anti-neutrino flux $\bar{\nu}_e$.

A schematic arrangement for detection of an anti-neutrino is shown in figure 1.16.

An intense $\bar{\nu}_e$ beam enters from a nuclear reactor to the target tank B that serves as a source of hydrogen (proton) p which is the target.

Inverse beta interaction between p and $\bar{\nu}_e$ occurs, and a neutron n and a positron e^+ are created.

Just after formation, the positron e^+ gets annihilated by electron e^- through the process of pair annihilation viz.

$$e^+ + e^- \rightarrow 2\gamma$$

and so two photons are produced.

These photons γs produce scintillations in the scintillation counters S_1, S_2 and generate two electrical pulses in the photomultipliers P_1, P_2 that are fed to a coincidence circuit. The pulse obtained due to positron annihilation is a prompt γ pulse.

The neutron n loses energy through repeated elastic collisions with a proton in the target tank and gets slowed down.

An aqueous solution of $CdCl_2$ is taken in the tank B. The slow n gets captured or absorbed by a Cd nucleus (that occurs in the tank) and a γ photon is emitted as per the following reaction

$$^{112}Cd(n, \gamma)Cd^{113}$$

The γ emitted is registered in a detector as a delayed pulse with a delay of 1–25 μs since the first n is created and then it slows down and gets absorbed by Cd.

Occurrence of prompt and delayed pulses due to e^+ annihilation and Cd capture of n is a proof that inverse beta decay had really occurred. In other words an anti-neutrino $\bar{\nu}_e$ took part in the interaction and could be detected.

The entire set-up was shielded against neutrons and gamma radiation by surrounding it with a lead–paraffin cover and placing it deep underground.

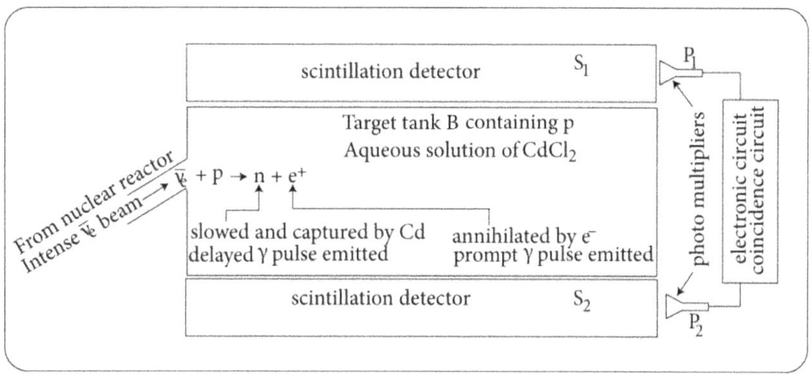

Figure 1.16. Detection of the electron anti-neutrino $\bar{\nu}_e$ by Reines and Cowan (1956).

1.29 List of a few discoveries of various particles and phenomena

Year	Discoverer	Particle or phenomena discovered/detected
1886	Goldstein	Discovered proton
1896	Becquerel	Discovered that uranium salt affects photographic plates (radioactivity discovered)
1897	J J Thomson	Discovered electron (particle)
1898	Schmidt, Marie Curie	Showed that thorium also affects photographic plates
1898	Pierre Curie, Marie Curie	Discovered polonium, radium Marie Curie coined the name radioactivity
1899	Rutherford	Discovered α ray, β ray Invented concept of half-life
1900	Villard	Discovered γ ray
1902	Rutherford, Soddy	Showed that radioactivity involves disintegration of nuclei
1907	Rutherford, Royds	Showed that α are doubly ionized helium atoms
1908–1910	Geiger, Marsden	Performed experiment of α particle scattering from gold foil
1909	Milikan	Determined electron charge though oil drop experiment
1911	Rutherford	Gave nuclear model of atom Showed that hydrogen nucleus is an elementary particle
1913	Soddy, Fajans	Discovered radioactive displacement law
1913	Bohr	Explains atomic line spectra using quantum theory
1913	Moseley	Proved that the charge on the nucleus is equal to the atomic number
1917	Soddy, Hitchins	Discovered existence of isotopes
1920	Rutherford	Coined the name proton
1930	Dirac	Predicted existence of positron
1930	Pauli	Predicted existence of neutrino
1932	Chadwick	Discovered neutron
1932	Anderson	Discovered positron e^+ as predicted by Dirac in 1930
1933	Blackett, Occhialini	Discovered pair production $\gamma \to e^+ + e^-$
1935	Yukawa	Predicted existence of pion as strong force carrying particle
1937	Anderson, Neddermeyer	Discovered muons (μ^+, μ^-) (They were mistaken as strong force mediator that Yukawa had predicted)
1945–46	Conversi, Pancini, Piccioni	Proved that muon is not the strong force mediator that Yukawa had predicted

(*Continued*)

1947	Powell, Occhialini, Lattes	Discovered charged pions $\pi^{\pm}$ and their decay chain. This was the strong force mediator that Yukawa had predicted
1950	Bjorklund *et al*	Discovered neutral pion π^0 and its decay
1955	Segre, Chamberlain	Detected anti-proton $\bar{p}$
1956	Reines, Cowan	Detected anti-neutrino
1956	Cork, Lumberton, Piceioni and Wentzel	Detected anti-neutron
1964	Gell-Mann, Zweig	Proposed quarks
1978	DESY	Discovery of gluon
1983	CERN	Discovery of W and Z intermediate vector boson
1995	Fermi lab	Top quark discovered

1.30 Exercises

Exercise 1.1 *Estimate the radius of an electron.*

[Ans.] An electron carries a charge $-|e| = -1.6 \times 10^{-19}\, C$ and a mass $m = 9.1 \times 10^{-31}\, kg$. The rest energy of the electron is

$$mc^2.$$

Consider an electron to be a spherical charge distribution of radius R.

If we assume charge to be distributed uniformly throughout the volume of a sphere then the electrostatic energy it possesses is

$$\frac{1}{4\pi\varepsilon_0}\frac{3}{5}\frac{|e|^2}{R} \quad \text{(classical formula)}$$

On the other hand, if we assume charge to be distributed uniformly over the surface of a sphere then the electrostatic energy it possesses is

$$\frac{1}{4\pi\varepsilon_0}\frac{|e|^2}{2R} \quad \text{(classical formula)}$$

For purpose of rough calculation, let us ignore the factors $\frac{3}{5}$ or $\frac{1}{2}$ and take electrostatic energy to be

$$\frac{1}{4\pi\varepsilon_0}\frac{|e|^2}{R}$$

Now we equate the rest energy of an electron to its electrostatic energy to get

$$mc^2 = \frac{1}{4\pi\varepsilon_0}\frac{|e|^2}{R}$$

$$R = \frac{1}{4\pi\varepsilon_0}\frac{|e|^2}{mc^2} = \frac{1}{4\pi\left(\frac{10^{-9}}{36\pi}\frac{F}{m}\right)}\frac{(1.6\times10^{-19}\,C)^2}{(9.1\times10^{-31}\,kg)\left(3\times10^{8}\frac{m}{s}\right)^2}$$

$$= 2.8 \times 10^{-15}\,m = 2.8\,fm$$

This $R = \frac{1}{4\pi\varepsilon_0}\frac{|e|^2}{mc^2} = 2.8\,fm$ is called classical electron radius, which gives the size that an electron would need to have for its mass to be completely due to its electrostatic potential energy.

Exercise 1.2 *Calculate the specific charge of an electron.*
[Ans.] Specific charge of an electron is

$$\frac{|e|}{m} = \frac{1.6 \times 10^{-19}\,C}{9.1 \times 10^{-31}\,kg} = 1.76 \times 10^{11}\frac{C}{kg}.$$

Exercise 1.3 *What did Milikan's oil drop experiment demonstrate? How did Milikan measure electronic charge?*
[Ans.] Milikan determined charge of electron in his oil drop experiment (1909).

Milikan's oil drop experiment demonstrated the discrete or quantized nature of electric charge and gave the correct value of charge on an electron.

The experimental arrangement of Milikan's oil drop experiment is shown in figure 1.17.

Two flat metal plates P_1, P_2 are separated by an ebonite or glass plate and a high potential difference $\sim 10\,000\ V$ is applied. The arrangement is enclosed in a metal chamber to avoid air currents.

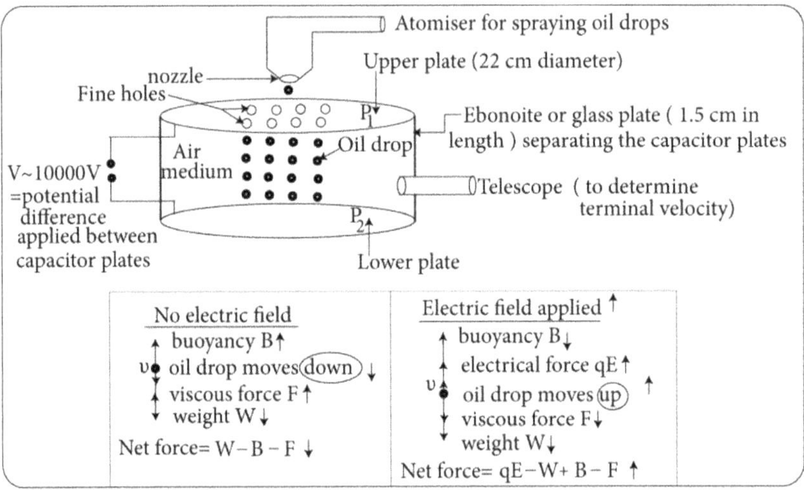

Figure 1.17. Milikan's experimental arrangement to determine the charge of an electron.

A fine spray of non-volatile oil drops is produced by an atomizer above the upper plate P_1. The majority of these drops become charged by friction at the atomizer nozzle during time of spray. Some of the oil drops enter between the plates through fine holes in the upper plate P_1. Some of the oil drops introduced into the space between the plates are positively charged while some are negatively charged and some are uncharged.

When no electric field is applied between capacitor plates, the downward motion of an oil drop is governed by the following forces:

  (i) weight of oil drop acting downwards viz.

$$W = mg = \left(\frac{4}{3}\pi r^3 \rho\right)g$$

where

$m = \frac{4}{3}\pi r^3 \rho$ is mass, ρ is the density and r the radius of oil drop

  (ii) buoyant force on oil drop acting upwards viz.

$$B = m'g = \left(\frac{4}{3}\pi r^3 \sigma\right)g$$

where

$m' = \frac{4}{3}\pi r^3 \sigma$ is mass of displaced medium i.e. air, σ is density of air

  (iii) viscous force acting upwards (against motion of oil drop) given by Stoke's law viz.

$$F = 6\pi\eta r v$$

where v is the velocity of oil drop, η the coefficient of viscosity of air

The net force on an oil drop is given by

$$W - B - F = mg - m'g - 6\pi\eta r v$$

$$= \left(\frac{4}{3}\pi r^3 \rho\right)g - \left(\frac{4}{3}\pi r^3 \sigma\right)g - 6\pi\eta r v = \frac{4}{3}\pi r^3 (\rho - \sigma)g - 6\pi\eta r v \quad (1.33)$$

$$= W_{\text{eff}} - 6\pi\eta r v$$

where

$$W_{\text{eff}} = W - B = \frac{4}{3}\pi r^3 (\rho - \sigma)g = \text{effective weight of oil drop.}$$

Suppose oil drop accelerates downwards. So v increases, $F = 6\pi\eta r v$ increases. So the net force $W - B - F = W_{\text{eff}} - 6\pi\eta r v$ decreases and eventually for some $v = v_t$ (say), net force will reduce to zero, which means rate of change of momentum is zero, i.e.

$$W - B - F \Big|_{v=v_t} = 0$$

$$\frac{dp}{dt}\Big|_{v=v_t} = 0 \text{ (by Newton's law, } p \text{ being momentum)}$$

Then

$$\frac{dv}{dt}\Big|_{v=v_t} = 0$$

This means that the oil drop no longer accelerates but attains a constant terminal velocity v_t downward in this supposed case. Thereafter the oil drop moves downward with this v_t. Hence

$$W - B - F \bigg|_{v=v_t} = w_{\text{eff}} - 6\pi\eta r v_t = 0$$

$$w_{\text{eff}} = 6\pi\eta r v_t \tag{1.34}$$

$$v_t = \frac{w_{\text{eff}}}{6\pi\eta r} = \frac{1}{6\pi\eta r}\frac{4}{3}\pi r^3(\rho - \sigma)g = \frac{2r^2 g}{9\eta}(\rho - \sigma)$$

$$v_t = \frac{2r^2 g}{9\eta}(\rho - \sigma). \tag{1.35}$$

This gives

$$r = \sqrt{\frac{9\eta v_t}{2g(\rho - \sigma)}} \tag{1.36}$$

and hence

$$w_{\text{eff}} = 6\pi\eta r v_t = 6\pi\eta v_t \sqrt{\frac{9\eta v_t}{2g(\rho - \sigma)}} = \frac{2^{1/2}\, 9\pi\eta^{3/2} v_t^{3/2}}{\sqrt{g(\rho - \sigma)}} \tag{1.37}$$

Knowing η, g, ρ, σ and determining v_t we know w_{eff}.

When an electric field $\vec{E}$ is applied between capacitor plates, the force on the oil drop carrying charge q will be $q\vec{E}$. Suppose the electric field is such that

$$qE > mg$$

and so the oil drop now starts to move upwards. Now qE acts upwards, weight downwards, buoyant force is against motion, i.e. downwards, viscous force is against motion, i.e. it acts downwards. Hence net force on oil drop will be

$$qE - W + B - F = qE - (W - B) - F = qE - w_{\text{eff}} - 6\pi\eta r v$$

Let at $v = v_t'$ net force become zero and so the oil drop moves with a constant upward terminal velocity v_t'. Thus we have

$$qE - w_{\text{eff}} - 6\pi\eta r v_t' = 0 \tag{1.38}$$

Using the result of equation (1.34) viz. $w_{\text{eff}} = 6\pi\eta r v_t$ we can write

$$qE - 6\pi\eta r v_t - 6\pi\eta r v_t' = 0$$

$$qE = 6\pi\eta r \left(v_t + v_t'\right).$$

Hence

$$\frac{qE}{w_{\text{eff}}} = \frac{6\pi\eta r \left(v_t + v_t'\right)}{6\pi\eta r v_t} = \frac{v_t + v_t'}{v_t}$$

$$q = \frac{v_t + v_t'}{v_t} \cdot \frac{w_{\text{eff}}}{E} = \frac{v_t + v_t'}{v_t E} \frac{2^{1/2} \, 9\pi\eta^{3/2} v_t^{3/2}}{\sqrt{g(\rho - \sigma)}} \quad \text{(using equation (1.37))}$$

$$q = \frac{2^{1/2} \, 9\pi\eta^{3/2} v_t^{1/2}(v_t + v_t')}{\sqrt{g(\rho - \sigma)} \, E} \tag{1.39}$$

So q can be determined using this equation.

Air between the capacitor plates was exposed to a beam of x-rays due to which some of the air molecules were ionized. Due to thermal collisions between these ionized air molecules and the oil drops, the charge on an oil drop may change say from q to q' and since the electrical force thus changes there is change in terminal velocity also say from v_t' to v_t''. So equation (1.38) viz. $qE - w_{\text{eff}} - 6\pi\eta r v_t' = 0$ changes to

$$q'E - w_{\text{eff}} - 6\pi\eta r v_t'' = 0 \tag{1.40}$$

Subtracting equation (1.38) from equation (1.40) we have

$$(q' - q)E - 6\pi\eta r(v_t'' - v_t') = 0$$

$$q' - q = \frac{6\pi\eta r(v_t'' - v_t')}{E}$$

Using the result $w_{\text{eff}} = 6\pi\eta r v_t$, i.e. $w_{\text{eff}} = 6\pi\eta r = \frac{w_{\text{eff}}}{v_t}$ we can write

$$q' - q = \frac{w_{\text{eff}}}{v_t} \frac{(v_t'' - v_t')}{E}$$

Using equation (1.37) we write

$$q' - q = \frac{v_t'' - v_t'}{v_t E} \frac{2^{1/2} \, 9\pi\eta^{3/2} v_t^{3/2}}{\sqrt{g(\rho - \sigma)}}$$

$$q' - q = \frac{2^{1/2} \, 9\pi\eta^{3/2} v_t^{1/2} \, (v_t'' - v_t')}{\sqrt{g(\rho - \sigma)} \, E} \tag{1.41}$$

Milikan, by observations on a single drop found that its velocity changed repeatedly in the electric field which means repeated changes of its electrical charge. These changes were determined using equation (1.41) and were found to be an integral multiple of a minimum value. This minimum value of the charge was taken to be the charge magnitude of an electron $|e|$.

This shows that an electron carries a discrete (minimum) amount of charge. The charge on an oil drop is thus

$$q = \pm n \, |e|$$

where n is an integer.

Actually, measurements gave oil drop charges to be

$$q_1, q_2, q_3, \ldots.$$

The H.C.F. (highest common factor) of magnitude of these charges is $|e|$. Using

$$\eta_{\text{air}}^{\text{at } 23\,°C} = 0.000\ 183\ 25 \text{ Poise}$$

Milikan obtained

$$|e| = 1.6 \times 10^{-19}\ C$$
$$= \text{electron charge magnitude.}$$

So $e = -|e| = -1.6 \times 10^{-19}\ C =$ electron charge.

Clearly, electric charge is quantized or discrete in nature.

Sources of error in Milikan's experiment

✓ Since $q \propto \eta^{3/2}$ (equation 1.39), accurate value of coefficient of viscosity of air should be used to get accurate value of electronic charge.

✓ Stoke's law $F = 6\pi\eta rv$ which is valid for an infinitely extended medium was employed in the analysis though oil drops traced a medium of finite extent.

✓ Field between the capacitor plates was assumed to be uniform. But there is distortion in the field due to presence of holes in the upper plate of capacitor.

Exercise 1.4 *How was the mass of an electron found?*

Ans. Since specific charge of an electron is by equation (1.7)

$$\frac{|e|}{m} = 1.76 \times 10^{11} \frac{C}{kg}$$

and charge magnitude of the electron is

$$|e| = 1.6 \times 10^{-19}\ C$$

the mass of the electron will be

$$m = \frac{|e|}{|e|/m} = \frac{1.6 \times 10^{-19}\ C}{1.76 \times 10^{11}\ kg^{-1}} = 9.1 \times 10^{-31}\ kg$$

This is the rest mass of the electron.

Again, mass of the H atom is

$$m_{\text{H}} = 1.008u$$

where

$$1u = 1\ amu = 1.66 \times 10^{-27}\ kg.$$

Hence

$$m_{\text{H}} = 1.008 \times (1.66 \times 10^{-27}\ kg) = 1.67 \times 10^{-27}\ kg. \quad \text{Thus}$$

$$\frac{\text{Mass of hydrogen atom}}{\text{Mass of electron}} = \frac{m_{\text{H}}}{m} = \frac{1.67 \times 10^{-27}\,kg}{9.1 \times 10^{-31}\,kg} = 1835$$

Exercise 1.5 *In Milikan's oil drop experiment, an oil drop was used and not a water drop. Why?*

[Ans.] This was done to avoid errors in measurement that would occur due to evaporation of a water drop.

Exercise 1.6 *Could Milikan's oil drop experiment be performed in vacuum?*

[Ans.] In Milikan's oil drop experiment, performed in air medium, one produces balanced oil drops moving with a constant terminal velocity, the balance occurring between the weight of drop, buoyant force and viscous force of air medium.

In vacuum, buoyant force and viscous force would disappear and all oil drops would fall with acceleration g. So one cannot create balanced oil drop in vacuum. Clearly, thus the experiment cannot be performed in vacuum.

Exercise 1.7 *Can electrons be balanced instead of the oil drop in Milikan's oil drop experiment?*

[Ans.] ✓The force of viscosity operating in Milikan's oil drop experiment is

$$F = 6\pi\eta r v$$

where r is the radius of the drop. Instead of the oil drop, if we have an electron then since

$$r_{\text{electron}} \sim 0$$

the force of viscosity will be extremely small ≈ 0. So it will not be able to balance the W–B force. In other words electron will not achieve terminal velocity and will fall with acceleration which cannot be measured with telescope.

✓ Further, an electron cannot be detected in telescope directly.

Exercise 1.8 *Could Milikan's oil drop experiment be influenced by changes in temperature?*

[Ans.] The coefficient of viscosity η is heavily temperature dependent. A slight change in temperature will cause significant change in η.

Since we got in equation (1.39), $q = \dfrac{2^{1/2}\,9\pi\eta^{3/2}v_t^{1/2}(v_t + v_t')}{\sqrt{g(\rho - \sigma)E}} \propto \eta^{3/2}$ such change in η will cause error in the value of charge q. Clearly temperature changes will give wrong results in Milikan's oil drop experiment. So temperature should be kept fixed during the experiment.

Exercise 1.9 *Calculate the electric field in $\frac{volt}{cm}$ required to balance an oil drop of radius $3.6 \times 10^{-5}cm$ carrying an electric charge if density of oil is $0.8\frac{g}{cc}$.*

$\boxed{\text{Ans.}}$ Considering the force balance
Electrostatic force = gravitational force

$$qE = mg$$

$$qE = \left(\frac{4}{3}\pi r^3 \rho\right)g$$

Taking density of oil

$$\rho = 0.8 \ g \ cc^{-1} = 0.8\frac{10^{-3} \ kg}{(10^{-2}m)^3} = 0.8 \times 10^3 \ kg \ m^{-3}$$

we have

$$E = \frac{4\pi r^3 \rho g}{3q} = \frac{4\pi (3.6 \times 10^{-5} \times 10^{-2}m)^3(0.8 \times 10^3 \ kg \ m^{-3})(9.8 \ m \ s^{-2})}{3(1.6 \times 10^{-19} \ C)}$$

$$= 9576 \ N \ C^{-1} = 9576 \ V \ m^{-1}$$

$$= 9576 \ \frac{V}{100 \ cm} = 95.76\frac{V}{cm}$$

Exercise 1.10 *In Milikan's oil drop experiment it was found that the oil drop acquired a downward terminal velocity of $8.58 \times 10^{-4} \ m \ s^{-1}$ when no electric field was present. It acquired a terminal upward velocity of $2.04 \times 10^{-5} \ m \ s^{-1}$ in the presence of an electric field of $3.18 \times 10^5 \ V \ m^{-1}$. What is the electric charge on this drop? The density of oil is $919.9 \ kg \ m^{-3}$ and the viscosity of air is $1.825 \times 10^{-5} \ N. \ s \ m^{-2}$.*

$\boxed{\text{Ans.}}$ We got in equation (1.39)

$$q = \frac{2^{1/2} \ 9\pi\eta^{3/2}v_t^{1/2}(v_t + v_t')}{\sqrt{g(\rho - \sigma)}E}$$

$$q = \frac{2^{1/2} \ 9\pi(1.825 \times 10^{-5} \ N.s \ m^{-2})^{3/2}(8.58 \times 10^{-4} \ m \ s^{-1})^{1/2}(8.58 \times 10^{-4} \ m \ s^{-1} + 2.04 \times 10^{-5} \ m \ s^{-1})}{\sqrt{9.8 \ m \ s^{-2}(919.9 \ kg \ m^{-3} - 0)}(3.18 \times 10^5 \ V \ m^{-1})}$$

$$= 2.6566 \times 10^{-18} \ C$$

$$q = 2.6566 \times 10^{-18} \ C\frac{|e|}{|e|} = \frac{2.6566 \times 10^{-18} \ C}{1.6 \times 10^{-19} \ C} \ |e| \approx 16 \ |e|$$

Exercise 1.11 *Suggest how α particles would have been scattered from gold foil according to the following atomic models*
   (a) *Dalton's atomic model (1808);*
   (b) *Plum pudding model (1904);*
   (c) *Rutherford's nuclear model (1911).*

$\boxed{\text{Ans.}}$ This is shown in figure 1.18.

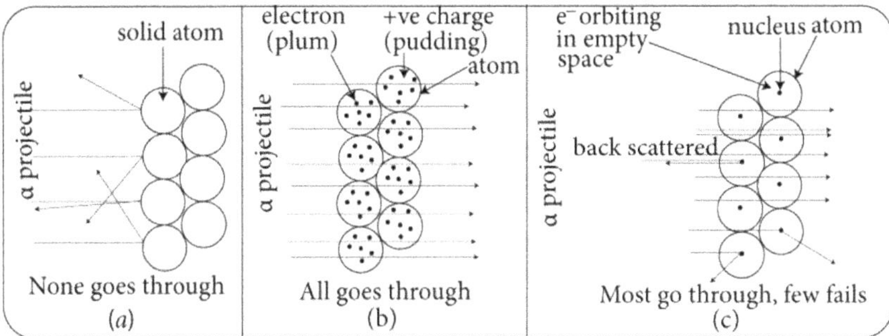

Figure 1.18. α scattering from (a) Dalton's atomic model; (b) plum pudding model, (c) Rutherford's nuclear model.

We briefly mention the atomic models referred to.

(a) Dalton's atomic model (1808)

All matter is made of tiny indivisible movable particles called atoms which are hard, solid and impenetrable. This is not correct because an atom is divisible into electrons and a nucleus. Had this been true the α particles would bounce off the hard atomic balls, none would go through.

(b) Plum pudding model of J J Thomson (1904)

Negatively charged electrons are like 'plums' embedded in a spherical volume of positive charge that can be likened to a 'pudding'. This is not correct as positive charge is not spread out but concentrated at the centre of the atom. Had this been true all the α particles would have passed straight through the positively charged pudding and the tiny plum (electron) cannot influence the speeding heavy αparticles. So α particles would simply fly through the plum pudding atoms.

(c) Rutherford's nuclear model of atom (1911)

An atom has a tiny massive concentrated mass called the nucleus at its centre. Most of the space within the atom is empty. This is correct. It is observed experimentally that most α particles are forward scattered through empty space and very few are back scattered on being ricocheted by the nucleus.

Exercise 1.12 *How was the unknown radiation emitted from $_4\mathrm{Be}^9$, on being hit by α from$_{84}\mathrm{Po}^{210}$ identified? (Why was this emited radiation allowed to fall on paraffin?)*

Hint. Detection of neutron by Chadwick in section 1.9.

Exercise 1.13 *In measurements by Hess, ionization showed an initial small decrease (below 700 m, figure 1.5) and beyond that it steadily increased to large values at higher altitudes. Explain the decrease and the increase.*

Ans. As electroscopes in the balloon ascended it was moving away from the radioactive substances on Earth and hence ionization reduced up to an altitude $\sim700\,m$.

Thereafter, ionization increases because of cosmic radiation.

Exercise 1.14 *What is the importance of studying cosmic rays?*

Ans. Cosmic rays are the major constituent of interstellar space. Through study of cosmic rays one tries to understand how energy is apportioned between thermal gas, magnetic fields and cosmic rays.

Study of origin of cosmic rays can shed light on space or the region they come from.

One studies how cosmic rays interact with thermal gas despite being virtually collisionless.

One studies how cosmic ray spectrum is formed.

The mechanism of how cosmic ray particles are accelerated to high energies is studied.

We get information about particle physics, dark matter and fundamental physics through a study of cosmic rays.

Study of cosmic rays as a particle-discovery tool is an important aspect. Indirect detection of dark matter particles through identification of yields from corresponding annihilation/decay can be attempted to find cosmic ray anti-matter.

Exercise 1.15 *How can one obtain an estimate of the photon energy of an electromagnetic shower according to the Heitler model of cascade.*

Ans. We start with a photon γ having energy E_0. The photon travels through the atmosphere, feels Coulomb interaction of a nucleus of atmosphere and produces a pair $e^{\pm}$. The phenomena of bremsstrahlung and pair production continue and an electromagnetic shower is produced, as shown in figure 1.19. We have shown three steps only of the electromagnetic shower.

The conversion probability is measured by mentioning length of each step, i.e. how much of the medium is traversed before each disintegration, and this is given as

$$\lambda = \text{(density)(length)} = kg\ m^{-3}.\ m = kg\ m^{-2}.$$

Suppose depth or length of the shower in the atmosphere is X.

$$\text{Number of steps } n = \frac{\text{length of shower}}{\text{length of each step}} = \frac{X}{\lambda}$$

At each step each particle produces two particles.

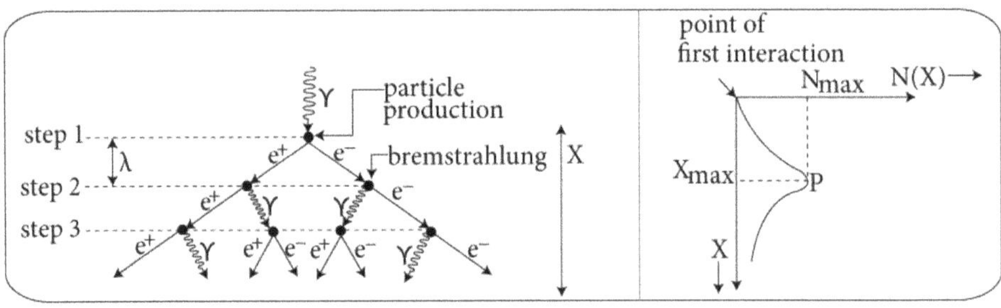

Figure 1.19. Finding number of particles in an electromagnetic shower.

So number of particles at a given X having n steps is

$$N(X) = 2^n = 2^{X/\lambda} \tag{1.42}$$

The mean energy at a given X is (assuming energy to be equally distributed)

$$\langle E(X)\rangle = \frac{\text{Initial energy}}{\text{Number of particles}} = \frac{E_0}{2^n} = \frac{E_0}{2^{X/\lambda}}$$

Large energy is required for occurrence of pair production and bremsstrahlung. And when energy is low collisions will occur only.

Mean energy will decrease fast and get depleted soon if the length of cascade X is large.

Let the characteristic energy or threshold energy be E_c such that

For $\langle E(X)\rangle > E_c$ pair production plus bremsstrahlung occurs.

For $\langle E(X)\rangle < E_c$ collision occurs.

Clearly the maximum number of particles in the shower will be

$$N_{\text{max}} = \frac{E_0}{E_c} \tag{1.43}$$

From equation (1.42) we have

$$N_{\text{max}} = 2^{X_{\text{max}}/\lambda}$$

Hence $N_{\text{max}} = \frac{E_0}{E_c} = 2^{X_{\text{max}}/\lambda}$ (using equation (1.43))

$$X_{\text{max}} = \lambda \log_2 \frac{E_0}{E_c} \tag{1.44}$$

λ and E_c are characteristic of the medium where the cosmic shower is produced. For instance, for air $\lambda = 45\ g\ cm^{-2}$, $E_c \approx 80\ MeV$.

A plot of X versus $N(X)$ is shown in figure 1.19.

If an instrument is able to map point P then X_{max} is predicted and hence using equation (1.44) we can estimate the value of the initial energy E_0.

Exercise 1.16 *Why was a Pb sheet used in Anderson's experiment (1932) on a cloud chamber in the path of a cosmic particle* (figure 1.11)?

$\boxed{\text{Ans.}}$ A Pb sheet was used to determine if the track or trace of the charged particle entering the magnetic field (in a cloud chamber) was travelling along $\overset{\frown}{AGS}$ or $\overset{\frown}{SGA}$ as depicted in figure 1.11. The portion of the track on the lower side of the Pb sheet is $\overset{\frown}{AG}$ that has a small curvature (less bent), larger radius (centre at J), i.e. it has the capacity to travel a longer distance due to its higher energy. On the other hand, the portion of the track on the upper side of the Pb sheet is $\overset{\frown}{GS}$ that has a larger curvature (more bent), smaller radius (centre at K) i.e. it does not have the capacity to travel a large distance as it is low on energy.

In other words, the path portion AG corresponds to higher energy and the path portion GS corresponds to lower energy. Such reduction in energy is due to absorption of energy by the Pb sheet. Clearly a positive charged particle enters the magnetic field at point A, travels the longer arc $\overgroup{AG}$, falls on the Pb sheet, loses energy and then traces the shorter arc $\overgroup{GS}$. So due to the Pb sheet we could identify the path of the charge to be $\overgroup{AGS}$ and not $\overgroup{SGA}$.

Exercise 1.17 *Could the track observed in Anderson's experiment (1932) on a cloud chamber be created by a proton (figure 1.11)?*

    $\boxed{\text{Ans.}}$ The Pb sheet diminished the momentum p of the charged particle from 63 MeV to 23 MeV. A proton that has rest energy 938 MeV, with such small momentum would have created a thicker track of shorter length ($\sim$5 mm). But the observed track length (figure 1.11) in Anderson's experiment (1932) was greater than 50 cm. This proved that the charged particle responsible for creating the track was not a proton.

    In a nutshell, a proton, being a heavy particle, would have created a shorter track. The observed long track was due to a lighter particle identified as a positron.

Exercise 1.18 *Consider Anderson's experiment on the discovery of the positron. Suppose the particle emerging from the lead plate has momentum 23 $MeV\ c^{-1}$. Calculate the kinetic energy of the particle if it is said that the particle is (a) a proton (b) a positron.*

    $\boxed{\text{Ans.}}$ Relation between momentum p, kinetic energy T, total energy E, rest mass m_0 is $E = T + m_0 c^2 = \sqrt{p^2 c^2 + m_0^2 c^4}$

$$T = \sqrt{p^2 c^2 + m_0^2 c^4} - m_0 c^2$$

(a) For a proton

$$m_0 c^2 = 938\ MeV$$

$$p = 23\ MeV\ c^{-1}$$

$$pc = 23\ MeV.$$

$$T = \sqrt{p^2 c^2 + m_0^2 c^4} - m_0 c^2 = \sqrt{(23\ MeV)^2 + (938\ MeV)^2} - 938\ MeV$$

$$T = 0.28\ MeV$$

(b) For a positron

$$m_0 c^2 = 0.511\ MeV$$

$$p = 23\ MeV\ c^{-1}$$

$$pc = 23 \; MeV.$$

$$T = \sqrt{p^2 c^2 + m_0^2 c^4} - m_0 c^2 = \sqrt{(23 \; MeV)^2 + (0.511 \; MeV)^2} - 0.511 \; MeV$$

$$T = 22.5 \; MeV.$$

Exercise 1.19 *If pair annihilation occurs between an electron and a positron having kinetic energy 1 MeV, find the wavelength of each photon produced in the process.*
  Ans. Pair annihilation process is

$$e^- + e^+ \rightarrow \gamma + \gamma.$$

Energy conservation equation is

$$(T + m_0 c^2) \Big|_{e^-} + (T + m_0 c^2) \Big|_{e^+} = h\nu + h\nu$$

$$(1 \; MeV + 0.511 \; MeV) + (1 \; MeV + 0.511 \; MeV) = 2h\nu$$

$$(1 + 0.511 + 1 + 0.511) \; MeV = 2h\nu$$

$$3.022 \; MeV = 2h\frac{c}{\lambda}$$

$$\lambda = \frac{2hc}{3.022 \; MeV}$$

$$= \frac{2(6.626 \times 10^{-34} \; J.s)(3 \times 10^8 \; m \; s^{-1})}{3.022 \; MeV} = \frac{2(6.626 \times 10^{-34} \; J.s)(3 \times 10^8 \; m \; s^{-1})}{3.022 \times 10^6 (1.6 \times 10^{-19} \; J)} = 8.22 \times 10^{-13} \; m$$

$$\lambda = 8.22 \times 10^{-13} \; m.$$

Exercise 1.20 *Show that a γ when moving freely cannot disintegrate into $e^{\pm}$ but can do so in the presence of a nucleus.*
  Ans. Suppose, if a possible photon γ decays in vacuum to $e^{\pm}$ *i. e.*

$$\gamma \rightarrow e^- + e^+$$

Consider four-momentum conservation ($\mu = 0, 1, 2, 3$)

$$(p_\gamma)^\mu = (p_{e^+})^\mu + (p_{e^-})^\mu \tag{1.45}$$

$$(p_\gamma)_\mu = (p_{e^+})_\mu + (p_{e^-})_\mu \tag{1.46}$$

Multiplying equations (1.46) and (1.45) we get

$$(p_\gamma)_\mu (p_\gamma)^\mu = \left[(p_{e^+})_\mu + (p_{e^-})_\mu \right] \left[(p_{e^+})^\mu + (p_{e^-})^\mu \right]$$

$$= (p_{e^+})_\mu (p_{e^+})^\mu + (p_{e^-})_\mu (p_{e^-})^\mu + (p_{e^+})_\mu (p_{e^-})^\mu + (p_{e^-})_\mu (p_{e^+})^\mu \tag{1.47}$$

$$p^\mu = \left(\frac{E}{c}, \vec{p}\right) = (p^0, p^i) \tag{1.48}$$

$$p_\mu = \left(\frac{E}{c}, -\vec{p}\right) = (p_0, p_i) \tag{1.49}$$

Using the relation

$$p_\mu p^\mu = p_0 p^0 + p_i p^i = \frac{E}{c}\frac{E}{c} + (-\vec{p}).(\vec{p}) = \frac{E^2}{c^2} - p^2 = \frac{E^2 - p^2 c^2}{c^2} = \frac{m_0^2 c^4}{c^2}$$

$$p_\mu p^\mu = m_0^2 c^2 \ (m_0 \text{ being the rest mass of corresponding particle}) \tag{1.50}$$

we have

$$(p_\gamma)_\mu (p_\gamma)^\mu = m_{\gamma 0}^2 c^2 = 0 \text{ since } m_{\gamma 0} = 0 \text{ (photon has zero rest mass)}$$

$$(p_{e^+})_\mu (p_{e^+})^\mu = (m_{e^+})_0^2 c^2 = m_0^2 c^2 \ (\text{as } m_{e^+} = m_0)$$

$$(p_{e^-})_\mu (p_{e^-})^\mu = (m_{e^-})_0^2 c^2 = m_0^2 c^2 \ (\text{as } m_{e^-} = m_0)$$

Equation 1.47 gives

$$0 = m_0^2 c^2 + m_0^2 c^2 + (p_{e^+})_\mu (p_{e^-})^\mu + (p_{e^-})_\mu (p_{e^+})^\mu$$

$$0 = 2m_0^2 c^2 + (p_{e^+})_0 (p_{e^-})^0 + (p_{e^+})_i (p_{e^-})^i + (p_{e^-})_0 (p_{e^+})^0 + (p_{e^-})_i (p_{e^+})^i$$

$$0 = 2m_0^2 c^2 + \frac{E_{e^+}}{c}\frac{E_{e^-}}{c} + (-\vec{p}_{e^+}).(\vec{p}_{e^-}) + \frac{E_{e^-}}{c}\frac{E_{e^+}}{c} + (-\vec{p}_{e^-}).(\vec{p}_{e^+})$$

$$0 = 2m_0^2 c^2 + 2\frac{E_{e^+}E_{e^-}}{c^2} - 2\vec{p}_{e^+} \cdot \vec{p}_{e^-}$$

$$0 = m_0^2 c^2 + \frac{E_{e^+}E_{e^-}}{c^2} - p_{e^+}p_{e^-}\cos\theta$$

$$p_{e^+}p_{e^-}\cos\theta = m_0^2 c^2 + \frac{E_{e^+}E_{e^-}}{c^2}$$

$$\cos\theta = \frac{E_{e^+}E_{e^-} + m_0^2 c^4}{c^2 p_{e^+}p_{e^-}}$$

Using the relativistic expression of energy, namely

$$E = \sqrt{p^2 c^2 + m_0^2 c^4}$$

$$\cos\theta = \frac{\sqrt{p_{e^+}^2 c^2 + m_0^2 c^4}\sqrt{p_{e^-}^2 c^2 + m_0^2 c^4} + m_0^2 c^4}{c^2 p_{e^+}p_{e^-}} \geqslant 1$$

This is impossible.

So our assumption $\gamma \to e^- + e^+$ was incorrect. Photon γ cannot decay in vacuum to $e^{\pm}$.

This conclusion directly follows if we apply momentum conservation to $\gamma \to e^- + e^+$ in the centre of mass frame.

Then in C frame we have

$$\vec{p}_\gamma^{\,CM} = \vec{p}_{e^+}^{\,CM} + \vec{p}_{e^-}^{\,CM}$$

Again in C frame centre of mass momentum is zero since centre of mass is always at rest. Hence $\vec{p}_\gamma^{\,CM} = 0$ and so

$$\vec{p}_{e^+}^{\,CM} + \vec{p}_{e^-}^{\,CM} = 0$$

But $\vec{p}_\gamma^{\,CM} = 0$ is not possible as there is no reference frame in which photon γ is at rest since we cannot change the speed of a photon as per Einstein's postulate of the special theory of relativity. So $\gamma \to e^- + e^+$ cannot take place in vacuum.

This interaction can however take place in the vicinity of a nucleus or atom, i.e. in the Coulomb field of a nucleus or atom as follows

$$\gamma + \text{Nucleus} \to e^- + e^+ + \text{Nucleus} \tag{1.51}$$

Now applying conservation of momentum in C frame we have

$$\vec{p}_\gamma^{\,CM} + \vec{p}_{\text{Nucleus}}^{\,CM} = \vec{p}_{e^+}^{\,CM} + \vec{p}_{e^-}^{\,CM} + \vec{p}_{\text{Nucleus}}^{\,\prime\,CM}$$

In C frame, centre of mass momentum is zero and so we have

$$\vec{p}_\gamma^{\,CM} + \vec{p}_{\text{Nucleus}}^{\,CM} = \vec{p}_{e^+}^{\,CM} + \vec{p}_{e^-}^{\,CM} + \vec{p}_{\text{Nucleus}}^{\,\prime\,CM} = 0 \tag{1.52}$$

This is possible.

The nucleus or atom absorbs momentum change and thus momentum is conserved.

Energy conservation applied to equation (1.48) gives

$$E_\gamma + E_{\text{Nucleus}} = \left[T_{e^+} + (m_{e^+})_0 c^2 \right] + \left[T_{e^-} + (m_{e^-})_0 c^2 \right] + E'_{\text{Nucleus}}$$

$$E_\gamma + E_{\text{Nucleus}} = \left[T_{e^+} + m_0 c^2 \right] + \left[T_{e^-} + m_0 c^2 \right] + E'_{\text{Nucleus}} \tag{1.53}$$

Pair production interaction is shown in figure 1.20.

Exercise 1.21 *Discuss why an electron–positron pair cannot annihilate into a single photon.*

Ans. Suppose if possible an electron–positron pair annihilate each other into a single photon as

$$e^+ + e^- \to \gamma \text{ (one photon emission if possible)}$$

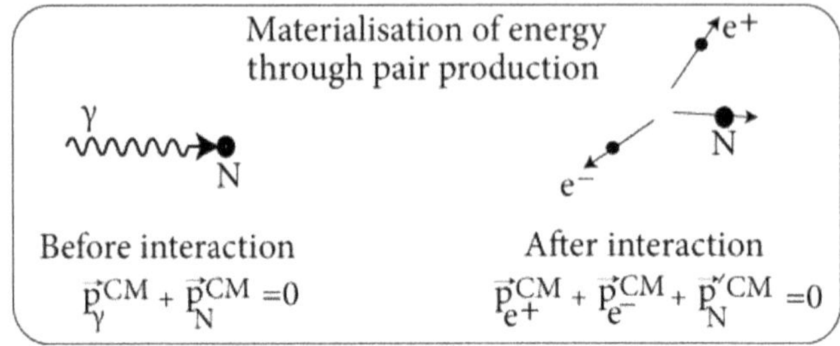

Figure 1.20. Pair production interaction.

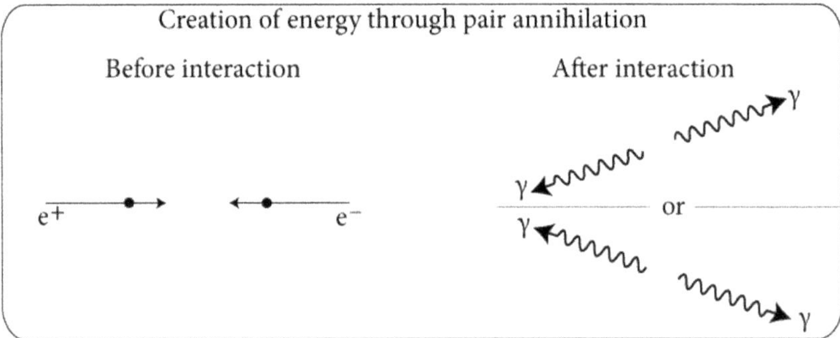

Figure 1.21. Pair annihilation interaction.

Consider momentum conservation in centre of mass frame where centre of mass is at rest always before and after the interaction.

So centre of mass momentum is 0.

$$\vec{p}_{e^+}^{\,CM} + \vec{p}_{e^-}^{\,CM} = \vec{p}_{\gamma}^{\,CM} = 0$$

But $\vec{p}_{\gamma}^{\,CM} = 0$ is not possible as there is no reference frame in which photon γ is at rest.

So the reaction $e^+ + e^- \to \gamma$ cannot take place in vacuum.

Actually, an e^+ and e^- combine, annihilate each other and create two photons as follows

$$e^+ + e^- \to \gamma + \gamma \text{ (two photon emission)}$$

Momentum conservation in C frame becomes

$$\vec{p}_{e^+}^{\,CM} + \vec{p}_{e^-}^{\,CM} = \vec{p}_{\gamma_1}^{\,CM} + \vec{p}_{\gamma_2}^{\,CM} = 0$$

Clearly, the product photons move in opposite directions but with the same magnitude of momentum $p_{\gamma_1} = p_{\gamma_2}$ which means $\frac{E_{\gamma_1}}{c} = \frac{E_{\gamma_2}}{c}$. That is $E_{\gamma_1} = E_{\gamma_2}$. So the product photons have the same energy and same momentum magnitude.

Figure 1.21 shows the pair annihilation process.

Exercise 1.22 *Discuss whether it is possible for an isolated free electron to absorb or emit a photon.*

$\boxed{\text{Ans.}}$ Suppose if possible an isolated free electron e at rest ($v = 0$) emits a photon γ and we now have an electron e' moving with velocity v (figure 1.22).

$$e \rightarrow e' + \gamma$$

$$e - \gamma \rightarrow e' \tag{1.54}$$

Suppose if possible an isolated free electron e at rest absorbs a photon γ and we now have an electron e' moving with velocity v (figure 1.22).

$$e + \gamma \rightarrow e' \tag{1.55}$$

Combining equations (1.54) and (1.55) we write

$$e \pm \gamma = e'$$

where e refers to a free isolated electron at rest ($v = 0$) and e' refers to the product electron after absorption or emission that moves with velocity v.

Let us check the four-momentum conservation relation

$$p_\mu^e \pm p_\mu^\gamma = p_\mu^{e'} \text{ (relation between covariant components)} \tag{1.56}$$

$$p^{e\mu} \pm p^{\gamma\mu} = p^{e'\mu} \text{ (relation between contravariant components)} \tag{1.57}$$

Multiplying equations (1.56) and (1.57) we have

$$(p_\mu^e \pm p_\mu^\gamma)(p^{e\mu} \pm p^{\gamma\mu}) = p_\mu^{e'} p^{e'\mu}$$

$$p_\mu^e p^{e\mu} + p_\mu^\gamma p^{\gamma\mu} \pm (p_\mu^e p^{\gamma\mu} + p_\mu^\gamma p^{e\mu}) = p_\mu^{e'} p^{e'\mu} \tag{1.58}$$

Using equation (1.48) viz. $p^\mu = (p^0, p^i) = (\frac{E}{c}, \vec{p}) = (p^0, p^i)$ and equation (1.49) viz. $p_\mu = (p_0, p_i) = (\frac{E}{c}, -\vec{p})$ and equation (1.50) viz. $p_\mu p^\mu = m_0^2 c^2$ (m_0 being the rest mass of the corresponding particle) we have

$$p_\mu^e p^{e\mu} = m_0^2 c^2 \text{ (denoting rest mass of electron } e \text{ by } m_0)$$

$$p_\mu^\gamma p^{\gamma\mu} = m_{\gamma0}^2 c^2 = 0 \text{ (since } m_{\gamma0} = 0, \text{ photon has zero rest mass)}$$

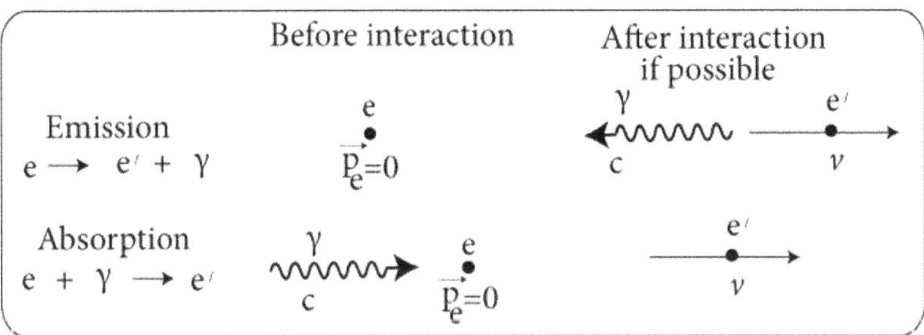

Figure 1.22. Pertaining to *exercise 1.22*. It is not possible for an isolated free electron to either absorb or emit a photon.

$$p_\mu^{e'} p^{e'\mu} = m_0^2 c^2 \text{ (denoting rest mass of electron } e' \text{ by } m_0)$$

From equation (1.58)

$$m_0^2 c^2 + 0 \pm (p_\mu^e p^{\gamma\mu} + p_\mu^\gamma p^{e\mu}) = m_0^2 c^2$$

$$p_\mu^e p^{\gamma\mu} + p_\mu^\gamma p^{e\mu} = 0$$

$$p_0^e p^{\gamma 0} + p_i^e p^{\gamma i} + p_0^\gamma p^{e0} + p_i^\gamma p^{ei} = 0 \tag{1.59}$$

Electron e absorbs or emits a photon from rest and hence its four-component of momentum is

$$p_\mu^e = \left(p_0^e, p_i^e\right) = \left(\frac{E_e}{c}, \ -\vec{p}^{\,e}\right) = \left(\frac{E_e}{c}, \ 0\right)$$

$$p^{e\mu} = (p^{e0}, p^{ei}) = \left(\frac{E_e}{c}, \ \vec{p}^{\,e}\right) = \left(\frac{E_e}{c}, \ 0\right)$$

since for e at rest we have $\vec{p}^{\,e} = 0$. Hence from equation (1.59)

$$\frac{E^e}{c}\frac{E^\gamma}{c} + 0 + \frac{E^\gamma}{c}\frac{E^e}{c} + 0 = 0$$

$$\frac{E^e}{c}\frac{E^\gamma}{c} = 0$$

$$E^e E^\gamma = 0$$

using $E^e = m_e c^2$ where m_e is mass of electron while $E^\gamma = h\nu$ where ν is frequency of emitted or absorbed photon. Hence

$$(m_e c^2)(h\nu) = 0$$

h, c are constants and so

$$m_e \nu = 0.$$

This means either $m_e = 0$, which means an electron is not there or $\nu = 0$ which means a photon is not there. Both of these possibilities are not feasible. This means our assumption was wrong. So we conclude that it is not possible for an isolated free electron to either absorb or emit a photon.

Exercise 1.23 *Argue why it is impossible for a photon to transfer all its energy and momentum to a free electron.*

   Ⓐⓝⓢ Energy of an electron is given by

$$E_e = mc^2 \tag{1.60}$$

Where m is mass of electron, m_0= rest mass of electron, v= velocity of electron and their inter-relation is given by

$$m = \frac{m_0}{\gamma}$$

where $\gamma = \sqrt{1 - \frac{v^2}{c^2}}$

Electron momentum is given by

$$p_e = mv \tag{1.61}$$

Photon energy E_γ and photon momentum p_γ are related as

$$E_\gamma = p_\gamma c \tag{1.62}$$

Suppose if possible a photon transfers all its energy and momentum to a free electron. Then

$$E_\gamma = E_e$$

$$p_\gamma = p_e$$

Taking the ratio we have using equations (1.60), (1.61), and (1.62)

$$\frac{E_\gamma}{p_\gamma} = \frac{E_e}{p_e}$$

$$c = \frac{mc^2}{mv}$$

$$v = c$$

But the electron, being a massive particle cannot travel with speed $v = c$. This means our supposition is wrong. So a photon cannot transfer all its energy and momentum to a free electron.

Exercise 1.24 *Why do you think that anti-proton production in proton–proton collision requires three protons in the final state?*

Ans. Baryon number B has to be conserved in all interactions.

Let us consider anti-proton production through proton–proton collision (considering three protons in the final state)

$$p + p \rightarrow p + p + p + \bar{p} \tag{1.63}$$

Now proton p is a baryon and so $B(p) = +1$ while $\bar{p}$ is an anti-baryon and so $B(\bar{p}) = -1$.

Let us apply baryon number conservation to equation (1.63)

$$B(\text{LHS}) = B(p) + B(p) = +1 + 1 = +2$$

$$B(\text{RHS}) = B(p) + B(p) + B(p) + B(\bar{p}) = +1 + 1 + 1 - 1 = +2$$

Clearly, the presence of three protons in the final state is needed to conserve baryon number and this means that the reaction can proceed.

Exercise 1.25 *Make an estimate of the minimum energy E_p of incident proton needed to collide with a stationary or fixed target proton so as to make a search for anti-proton $\bar{p}$?*

$\boxed{\text{Ans.}}$ The interaction that produces anti-proton $\bar{p}$ is given by equation (1.63) viz.

$$p + p \rightarrow p + p + p + \bar{p}$$

Let us employ the invariance

$$E^2 - p^2 c^2 = m_0^2 c^4 = \text{invariant} \tag{1.64}$$

to equation (1.63) to get

$$(E^2 - p^2 c^2)_{\text{LHS}} = (E^2 - p^2 c^2)_{\text{RHS}} \tag{1.65}$$

Now the incident proton on LHS has energy E_p (to be calculated) and momentum p.

Target proton has energy $m_p c^2$ and momentum zero (as it is stationary i.e. fixed to target).

$$(E^2 - p^2 c^2)_{\text{LHS}} = (E^{\text{incident}} + E^{\text{target}})^2 - (p^{\text{incident}} + p^{\text{target}})^2 c^2$$

$$(E^2 - p^2 c^2)_{\text{LHS}} = (E_p + m_p c^2)^2 - (p + 0)^2 c^2 = E_p^2 + m_p^2 c^4 + 2E_p m_p c^2 - p^2 c^2$$

$$= E_p^2 - p^2 c^2 + m_p^2 c^4 + 2E_p m_p c^2$$

Using the invariance relation of equation (1.64) viz. $E_p^2 - p^2 c^2 = m_p^2 c^4$ we have

$$(E^2 - p^2 c^2)_{\text{LHS}} = m_p^2 c^4 + m_p^2 c^4 + 2E_p m_p c^2$$
$$= 2m_p^2 c^4 + 2E_p m_p c^2 \tag{1.66}$$

Again minimum value of E_p corresponds to when the product particles $(p, p, p, \bar{p})$ are emitted with zero kinetic energy, zero momentum. Hence with $m_{\bar{p}} = m_p$ we can write

$$(E^2 - p^2 c^2)_{\text{RHS}} = (E^2)_{\text{RHS}} = (m_p c^2 + m_p c^2 + m_p c^2 + m_p c^2)^2 = (4m_p c^2)^2$$
$$= 16 m_p^2 c^4 \tag{1.67}$$

Equation (1.65) gives with equations (1.66), (1.67)

$$2m_p^2 c^4 + 2E_p m_p c^2 = 16 m_p^2 c^4$$

$$2E_p m_p c^2 = 14 m_p^2 c^4$$

$$E_p = 7 m_p c^2$$
$$= 7(938 \; MeV) = 6566 \; MeV = 6566 \times 10^6 e\,V = 6.566 \times 10^9 \; eV$$
$$E_p = 6.6 \; GeV$$

Exercise 1.26 *Discuss the differences between annihilation of proton–anti-proton pair with annihilation of electron–positron pair.*

[Ans.] In the annihilation of proton–anti-proton pair, four positive and four negative pions are created so that it looks like a star (figure 1.23). Also protons, deuterons, K mesons may also be produced.

$$p + \bar{p} \rightarrow \pi^{\pm} + \pi^0 + K^{\pm} + K^0 + \eta + \eta' + \ .....$$

In this case very large energy is involved and there is transformation of matter from one form to another.

But in electron–positron annihilation a small amount of energy is involved and gets released in the form of γ radiation. ($e^+ + e^- \rightarrow 2\gamma$). Here transformation of matter to energy or radiation takes place.

Exercise 1.27 *How did Dirac predict the existence of the anti-particle from his hole theory?*

[Ans.] The Dirac equation has a positive energy solution as well as a negative energy solution (figure 1.24)

$$E = \pm\sqrt{p^2 c^2 + m^2 c^4}$$

It is the natural tendency of a physical system to go to the lowest energy state. So even if we start with positive energy state, any perturbation would cause the system

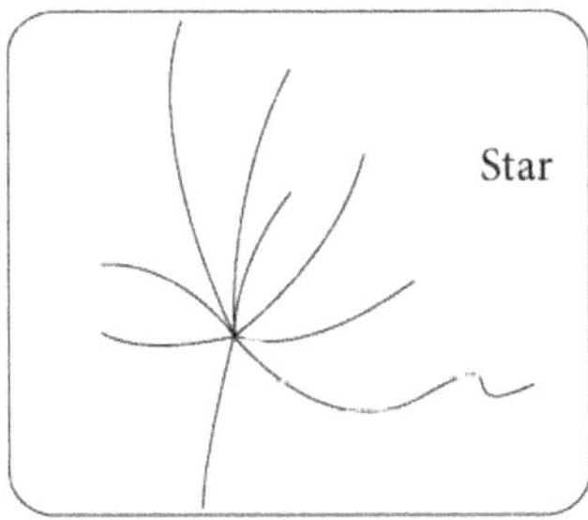

Figure 1.23. Star burst in photographic emulsion or bubble chamber due to annihilation of a proton–anti-proton pair.

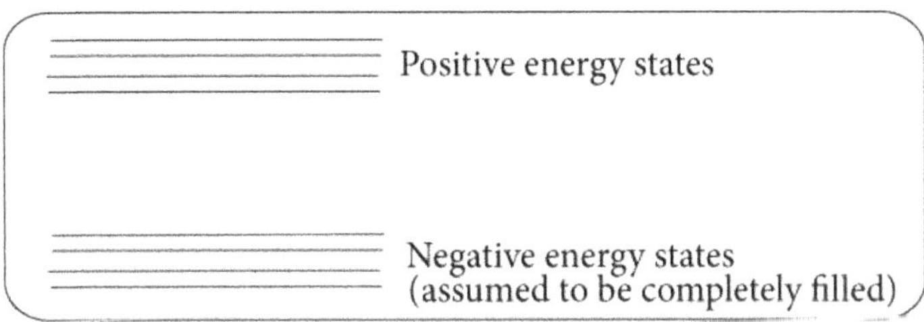

Figure 1.24. Energy spectrum for a free Dirac particle.

to fall to negative energy state as they lie lower—thereby destabilizing the system leading to its collapse. This means prediction of negative energy states leads to instability of the system.

To save the theory Dirac assumed that in the ground state all negative energy states are filled up completely.

The electron, being a fermion, has to obey Pauli exclusion principle. So an electron sitting in a positive energy state cannot go further down and occupy any negative energy state as they are filled. In this way the electron system achieves stability.

However, if enough energy is pumped into a negative energy electron it can go to positive energy state and appear as a positive energy electron, which is a particle. Simultaneously, a hole is created in the negative energy sea of states. So the absence of a negative energy electron (i.e. absence of negative energy and negative charge) can be thought of as a hole having exactly the same mass as the electron with all other quantum numbers opposite. This hole is equivalent to creation of an equal amount of positive energy and positive charge carried by an anti-particle identified as the positron or the anti-electron.

In this way Dirac theory predicts existence of an anti-particle corresponding to every particle.

Exercise 1.28 *Consider the density of interstellar space between two points A and B to be 1 hydrogen atom per cc. Cosmic nuclei pass from A to B through interstellar matter of $2.5\frac{g}{cm^2}$. What is the lifetime of cosmic nuclei?*

$\boxed{\text{Ans.}}$ Amount of interstellar matter traversed $= \frac{\text{Mass}}{\text{volume}}$ length traced

$$= \text{Density(velocity. time)}$$

Hence

$$2.5 \; g \; cm^{-2} = \frac{1 \, \text{H atom}}{cc} \, c. \, t$$

$$2.5 \; g \; cm^{-2} = \frac{1.67 \times 10^{-27} \times 10^3 \, g}{cm^3} \, 3 \times 10^{10} \frac{cm}{s} . \, t$$

$$t = 4.99 \times 10^{13} \, s = \frac{4.99 \times 10^{13}}{365 \times 24 \times 3600} yr = 1.6 \times 10^6 \, yr$$

Exercise 1.29 *Write down the decay scheme of a negative muon. What is the maximum energy available in the decay? What is the average electron energy?*

$\boxed{\text{Ans.}}$ Decay scheme of a negative muon is

$$\mu^- \rightarrow e^- + \bar{\nu}_e + \nu_\mu \; (\text{lifetime 2 } \mu s)$$

Mass of μ^- is 207 m_e. Mass of e^- is m_e. And $\bar{\nu}_\mu$, ν_e have negligible mass. Hence

Maximum available energy is given by

$$m_{\mu^-} - m_e = 207\,m_e - m_e = 206\,m_e = 206 \times 0.51\ MeV$$
$$= 105\ MeV$$

Assume that this energy is divided equally amongst product particles (e^-, $\bar{\nu}_e$, ν_μ). Then average energy of electron will be $\frac{1}{3}105\ MeV = 35\ MeV$.

Exercise 1.30 *Write down the decay scheme of $\pi^\pm$. Product muon $\mu^\pm$ has energy 3.66 MeV. Find kinetic energy of the accompanying neutrino.*
[Ans.] The decay modes of pions are

$$\pi^+ \rightarrow \mu^+ + \nu_\mu \quad (\text{lifetime } 10^{-8}\ s)$$

$$\pi^- \rightarrow \mu^- + \bar{\nu}_\mu \quad (\text{lifetime } 10^{-8}\ s)$$

Mass of $\pi^\pm$ is 273 m_e. Mass of $\mu^\pm$ is 207 m_e. And $\bar{\nu}_\mu$, ν_e have negligible mass. Hence

$$\text{Energy available} = m_{\pi^\pm}c^2 - m_{\mu^\pm}c^2$$
$$= 273\,m_e - 207\,m_e = 66\,m_e = 66 \times 0.51\ MeV = 33.66\ MeV$$

Kinetic energy of muon $\mu^\pm$ is 3.66 MeV.
   Hence kinetic energy of the accompanying neutrino is

$$33.66\ MeV - 3.66\ MeV = 30\ MeV.$$

Exercise 1.31 *Which of the following is the correct value of magnetic moment of neutral pion π^0?*

(a) 0   (b) 0.5 μ_N (c) μ_N (d) $\frac{|e|\hbar}{2m_\pi}$.

Exercise 1.32 *Which of the following is the correct value of masses of π^0, $\pi^\pm$ in MeV?*

(a) 264, 273   (b) 273, 264
(c) 140, 135   (d) 135, 140.

Exercise 1.33 *Which of the following is the correct value of masses of $\pi^\pm$, π^0 compared to electron mass?*

(a) 264, 273   (b) 273, 264
(c) 140, 135   (d) 135, 140.

Exercise 1.34 *Primary cosmic rays are primarily composed of which of the following?*

(a) *meson*   (b) *proton*   (c) *electron*   (d) *neutron.*

Exercise 1.35 *Secondary cosmic rays are primarily composed of which of the following?*

(a) *pion*   (b) *proton*   (c) *electron*   (d) *neutron.*

Exercise 1.36 *The percentage of protons in primary cosmic rays is*

(a) 10   (b) 50   (c) 70   (d) 90.

Exercise 1.37 *The percentage of $\pi^\pm$, π^0 secondary cosmic rays is*

(a) 10   (b) 50   (c) 70   (d) 90.

Exercise 1.38 *Minimum energy of pair production of γ travelling in vacuum is*

(a) 0   (b) 0.51 *MeV*   (c) 1.02 *MeV*   (d) *cannot pair produce.*

Exercise 1.39 *Minimum energy of pair production of γ is*

(a) 0   (b) 0.51 *MeV*   (c) 1.02 *MeV*   (d) 2 *MeV.*

Exercise 1.40 *At sea level which of the following form the hard component?*

(a) *photon*   (b) *electron*   (c) *muon*   (d) *proton*   (e) *pion.*

Exercise 1.41 *At sea level which of the following form the soft component?*

(a) *photon*   (b) *electron*   (c) *muon*   (d) *proton*   (e) *pion.*

Exercise 1.42 *A muon has lifetime of about*

(a) $2\ \mu s$ (b) $10^{16}\ s$ (c) $10^{-16}\ s$ (d) $10^{-6}\ s$.

Exercise 1.43 *Which of the following has the smallest lifetime?*

(a) $\mu^{\pm}$ (b) π^{0} (c) $\pi^{\pm}$ (d) p.

$$\boxed{\text{Answers to multiple choice questions}}$$

1.31 *a*, 1.32 *a*, 1.33 *c*, 1.34 *b*, 1.35 *a*,1.36 *d*, 1.37 *c*, 1.38 *d*, 1.39 *c*, 1.40 *c*, 1.41 *a,b*, 1.42 *a*, 1.43 *b*

1.31 Question bank

Q1.1 Mention areas where nuclear physics finds application.

Q1.2 What is the difference between astronomy, astrophysics, cosmology, nuclear and particle physics?

Q1.3 What is the radius of the following?
Molecule, atom, uranium nucleus, electron, proton, electron, pion, quark.

Q1.4 How did J J Thomson detect the electron in his experiment with a cathode ray tube? Mention the conclusions of his experiment.

Q1.5 What is the specific charge of an electron? How did J J Thomson determine the specific charge of the electron?

Q1.6 Briefly sketch how Goldstein detected the proton in his observation of positive rays in a modified cathode ray tube.

Q1.7 How did Rutherford coin the name proton?

Q1.8 How was mass of the proton estimated?

Q1.9 How did Chadwick detect the neutron in his experiment?

Q1.10 How was mass of the neutron determined by Chadwick and Goldhaber?

Q1.11 What is a cosmic ray?

Q1.12 Who discovered cosmic rays and how?

Q1.13 Mention the working of a gold leaf electroscope.

Q1.14 Mention the composition of cosmic rays.

Q1.15 Mention the composition of primary cosmic rays.

Q1.16 Mention the composition of secondary cosmic rays.

Q1.17 Mention the composition of cosmic rays at sea level.

Q1.18 What are hard and soft components of cosmic rays? Why are they so called?

Q1.19 Mention some characteristics of cosmic rays.

Q1.20 How does cosmic ray intensity vary with altitude, latitude and direction?

Q1.21 A cosmic ray shows east–west asymmetry. Explain.

Q1.22 What is the origin of cosmic rays?

Q1.23 What is a cosmic ray shower or burst?

Q1.24 Explain the following terms.
Bemstrahlung, pair production, hadronic shower, electromagnetic cascade, pair annihilation.

Q1.25 Obtain the energy of γ involved in the process of pair production and pair annihilation.

Q1.26 How did Anderson make the discovery of the positron?

Q1.27 How did Blackett and Occhialini discover the phenomenon of pair production?

Q1.28 How was the muon discovered by Anderson and Neddermeyer?

Q1.29 What is the muon paradox and how was it resolved by the concepts of time dilation and length contraction of the special theory of relativity?

Q1.30 How was the pion discovered by studying tracks in a nuclear emulsion plate?

Q1.31 What is the charge possessed by an anti-proton? How was it discovered?

Q1.32 How was detection of the electron anti-neutrino done by Reines and Cowan?

Further reading

[1] Krane S K 1988 *Introductory Nuclear Physics* (New York: Wiley)

[2] Tayal D C 2009 *Nuclear Physics* (Mumbai: Himalaya Publishing House)

[3] Satya P 2005 *Nuclear Physics & Particle Physics* (New Delhi: Sultan Chand & Sons)

[4] Guha J 2019 *Quantum Mechanics: Theory, Problems & Solutions* 3rd edn (Kolkata: Books and Allied (P) Ltd)

[5] Yung K L 2002 *Problems and Solutions on Atomic, Nuclear and Particle Physics* (Singapore: World Scientific)

IOP Publishing

Nuclear and Particle Physics with Cosmology, Volume 2
Particle physics and cosmology
Jyotirmoy Guha

Chapter 2

Particle accelerators

In this chapter we discuss various types of accelerators such as electrostatic accelerator, oscillating field accelerator. In particular, we discuss the Cockroft–Walton electrostatic accelerator, Van de Graff accelerator, linear accelerator and circular accelerator. The principle, construction, working, condition of operation, advantages and disadvantages of the cyclotron are discussed in detail. We also discuss the synchrocyclotron and betatron. A brief description is given of the Large Electron Positron Collider, the Tevatron and the Large Hadron Collider.

2.1 Introduction

Accelerators are precision instruments that accelerate particles to a well-defined beam of high energy and high speed, i.e. control and constrain the beam motion with high degree of accuracy. They can generate and shoot off sufficient numbers of energetic particles and strike macroscopic targets with a single pulse of the beam.

Accelerators serve as a probing agency for the study of nuclear and particle structure and have helped develop the fields of nuclear and particle physics. The accelerated beam of particles (whether protons, electrons, charged subatomic particles, heavier particles like alpha, gold, uranium etc) are used for a variety of research purposes.

It is clear from Heisenberg's uncertainty principle, namely

$$\Delta x \, \Delta p \sim \hbar \tag{2.1}$$

that large Δp is associated with small Δx, i.e. large momentum transfers correspond to small distances and vice versa. It is clear from equation (2.1) that the more energy a particle has the more deeply it can probe the constituents over a small region such as constituents of a nucleus. Probing short distance behaviour of nuclei and elementary particles thus requires beams with high energy that are able to cause large momentum transfers to target particles.

Accelerators are sources of high energy particles.

Another source of high energy particles is cosmic rays. But the flux of high energy particles in cosmic rays is quite low and so their energies cannot be controlled.

doi:10.1088/978-0-7503-5032-7ch2

In *exercise 2.1* we discuss the need of having particle accelerators.

2.2 Types of accelerators

Particle accelerators are devices that accelerate charged particles. Based on the method of acceleration particle accelerators can be split into two main categories.

Electrostatic accelerators and oscillating field accelerators.

2.2.1 Electrostatic accelerator

Electrostatic accelerators employ electrostatic fields that do not change with time. Hence one needs to employ very large electrostatic fields to accelerate charged particles to useful energies.

Maintenance of such a high electrostatic field is difficult and risky.

Examples of electrostatic accelerators are the Cockcroft–Walton accelerator and the Van de Graff accelerator.

2.2.2 Oscillating field accelerators

Oscillating field accelerators employ electric fields that periodically change with time. This is used to accelerate particles to very high energies.

Examples of oscillating field accelerators are the linear accelerator (LINAC) and circular accelerators like cyclotron, synchrocyclotron, betatron and synchrotron.

2.3 Cockcroft–Walton electrostatic accelerator

Cockcroft and Walton built an electrostatic accelerator in 1932.

2.3.1 Principle of Cockcroft–Walton electrostatic accelerator

Voltage multiplication in cascade is done by suitable arrangement of rectifiers and capacitors.

The basic circuit is shown in figure 2.1(a).

It consists of a high voltage step-up transformer and a voltage multiplier circuit consisting of several voltage doublers connected in cascade. And voltage doublers are built of ideal diodes and capacitors.

(Ideal diode means they have zero resistance when forward biased and infinite resistance when reverse biased)

Consider the first voltage doubler circuit consisting of capacitors C_1, C_2 and rectifiers D_1, D_2.

During negative half cycle (figure 2.1(b)) D_1 is forward biased (i.e. closed and conducts) and D_2 is reverse biased (i.e. open). So current flows through C_1, D_1 charging C_1 to V where V is the secondary voltage.

During positive half cycle (figure 2.1(c)) D_1 is reverse biased (i.e. open) and D_2 is forward biased (i.e. closed and conducts). So current flows through D_2 and C_2 charging C_2 to $2V$ since voltage across C_2 is the sum of the voltages across secondary (which is V) plus voltage across C_1 (which is V). So voltage across C_2 is doubled, i.e. it is double the supply voltage.

Higher voltages of 4, 6, 8, … , $2n$ times the input voltage V can be obtained using, respectively, 2, 3, 4, … , n stages connected in series, i.e. in cascade.

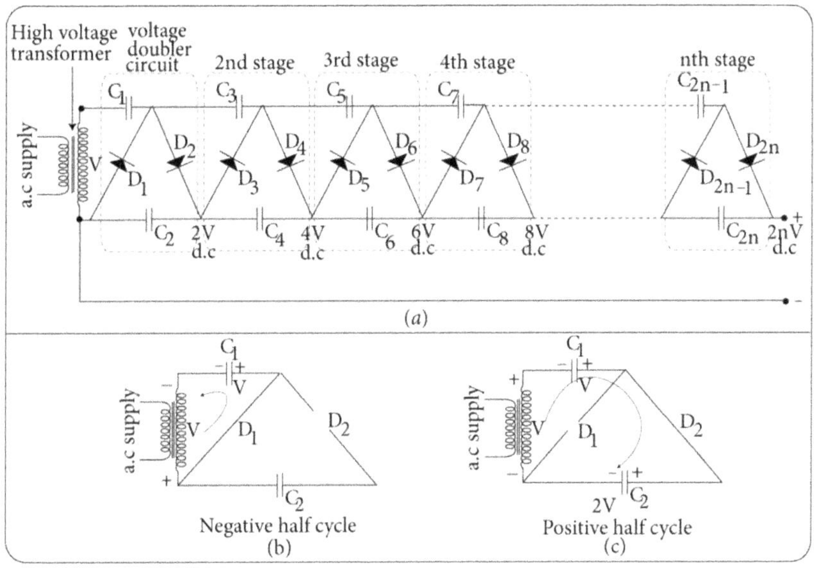

Figure 2.1. Cockcroft–Walton accelerator.

2.3.2 High voltage operation of Cockcroft–Walton electrostatic accelerator

During negative half cycle of supply, odd numbered rectifiers $D_1,D_3,D_5,D_7,...,D_{2n-1}$ are forward biased and conduct. Even numbered rectifiers are reverse biased (i.e. open).

In the subsequent positive half cycle of supply even numbered rectifiers $D_2,D_4,D_6,D_8,...,D_{2n}$ are forward biased and conduct. Odd numbered rectifiers are reverse biased (i.e. open).

If n number of stages are added then peak dc voltage developed will be $2n$ times the peak secondary voltage V of the high voltage transformer.

The individual capacitor in every stage will have potential of $2V$.

The voltages across the even numbered capacitors $C_2,C_4,C_6,C_8,...,C_{2n}$ will, respectively, be $2V,4V,6V,8V,...,2nV$.

As the voltage multiplier gives a constant supply of dc voltage it can be used to accelerate a charged particle to produce a high energy projectile.

The Cockcroft–Walton accelerator is simple having no moving parts.

The maximum energy obtained in a Cockcroft–Walton accelerator is low, about $2\,MeV$.

In principle it is possible to extend the voltage limit indefinitely but leakage across capacitors makes the circuit unsuitable for potentials above $2\,MeV$.

2.4 Van de Graff accelerator (generator)

A Van de Graff generator is depicted in figure 2.2(a).

A Van de Graff accelerator is a device for building up extremely high potential differences of the order of few MeV. It is used to accelerate ions required in experiments and research on nuclear and particle physics.

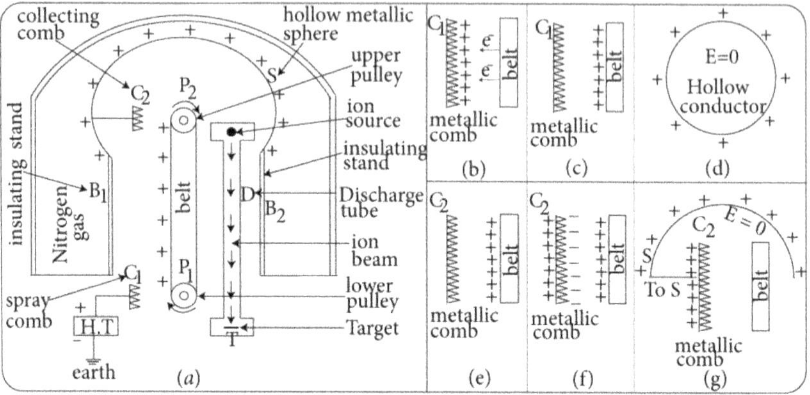

Figure 2.2. Van de Graff accelerator.

2.4.1 Principle of the Van de Graff accelerator

A Van de Graff generator works on the basis of the following principles.
1. Corona discharge:

Charge density σ is proportional inversely to the radius of curvature r, i.e.

$$\sigma \propto \frac{1}{r}$$

So pointed ends will have larger charge density.

C_1 is a metallic comb shaped conductor, as shown in figure 2.2(b), and has several pointed ends. The pointed ends have large charge density as shown.

If an insulating nylon or rubber belt is brought near it then the positive charges would attract electrons from the belt and get neutralized. The belt in the process becomes positively charged, as shown in figure 2.2(c). In other words, positive charge gets transferred from the metallic comb C_1 to the belt. In other words, C_1 sprays charge on the belt. This is referred to as corona discharge.

2. Charge given to a hollow conductor is transferred to the outer surface and is uniformly distributed over it. And the electric field inside the hollow conductor is zero, as indicated in figure 2.2(d).

2.4.2 Construction of a Van de Graff accelerator

As shown in figure 2.2(a), a Van de Graff generator consists of a hollow metallic hemisphere S of few metre radius. It is mounted upon two insulating supporting columns B_1, B_2.

A motor (not shown in the diagram) rotates the pulleys P_1, P_2 and a belt of insulating material like rubber or nylon runs between the two pulleys.

C_1 is a sharp comb called a spray comb. It is fixed near pulley P_1 and connected to high tension battery (10^4 V). C_1 sprays charge on the belt.

C_2 is another sharp comb called a collecting comb. It is fixed near pulley P_2 and connected to the metallic hemisphere S. C_2 collects charge from belt and transfers to S.

D is a discharge tube of insulating material (like glass, porcelain) containing a positive ion source at one end and these ions are accelerated down the tube. At the other end of the discharge tube the target T is placed.

The entire arrangement is enclosed in a steel chamber which is filled with nitrogen gas or methane gas at high pressure.

2.4.3 Working

The spray comb C_1 is given positive potential $\sim 10^4\ V$ w.r.t. ground.

Due to large electrostatic field near the sharp points of the comb C_1 the density of positive charge is large (figure 2.2(b)). Electrons from the belt get attracted and these charges get neutralized. As a result the belt becomes positively charged (figure 2.2 (c)). This phenomenon of transferring or spraying positive charge from comb C_1 to the belt is called corona discharge.

The belt rotates and takes the positive charges up near the collecting comb C_2 (figure 2.2(e)). As the belt is positive, negative charges are induced on C_2 on the belt side and positive charges are induced at the backside (figure 2.2(f)). As C_2 is connected to the hollow metallic sphere S the positive charges get transferred to it. The positive charge on the belt and negative charge on the belt side of C_2 get neutralized (figure 2.2(g)).

The belt rotates, C_1 sprays charge on belt and C_2 collects it and transfers to S.

In this process positive charges accumulate on the metallic hemisphere S. Again the potential V of metallic conductor S is proportional to the accumulated charge Q, i.e.

$$V = \frac{Q}{C} = \frac{Q}{4\pi\varepsilon_0 r} \propto Q \ (r \text{ is radius of } S)$$

As Q increases V also increases and reaches $\sim$ a few MeV s.

Ions from an ion source are subjected to this high potential difference, accelerate and strike the target with high velocity. Proton beams and deuteron beams from such a machine have been used for studying scattering and nuclear reactions.

If we have air (which has breakdown voltage 3×10^6 V m^{-1}) inside the steel chamber then air would ionize and there would be loss of charge from S by leakage. This would prevent building up of the high voltage. So the whole apparatus is enclosed in a gas-tight steel chamber containing a gas having better insulating properties than air like nitrogen gas or methane gas.

2.4.4 Limitations of the Van de Graff accelerator

1. Construction and maintenance of a Van de Graff generator are costly.
2. This device is bulky.
3. It cannot accelerate neutral particles like the neutron etc.

2.5 Linear accelerator or LINAC

A linear accelerator abbreviated as LINAC is used to accelerate charged particles (like proton, electron, alpha etc) to high kinetic energies. The particles move in a straight line and are accelerated by oscillating electric fields. For this reason it is called a linear accelerator.

2.5.1 Principle of the linear accelerator

1. A charged particle accelerates when moving through an electric field.
2. The electric field inside a hollow conductor is zero.

2.5.2 Construction of the linear accelerator

A schematic diagram of a linear accelerator is shown in figure 2.3.

Consider an evacuated cylindrical glass tube as shown. S is ion source.

There are several hollow metallic cylindrical tubes numbered 1,2,3,4,5,... of increasing lengths. These are called drift tubes. The odd numbered tubes 1,3,... are connected to one end of a high frequency (radio frequency) oscillator (time period T) and the even numbered tubes 2,4,... are connected to the other end of this oscillator. So alternate tubes have potentials of opposite sign.

In one half cycle (between time 0 to $\frac{T}{2}$) if tubes 1,3,5,.. are positive then the tubes 2,4,6,... etc will be negative. Polarities are reversed in the next half cycle (between time $\frac{T}{2}$ to T). Polarity of oscillator alters periodically in a time $\frac{T}{2}$. This is shown in the diagram.

When a potential difference is applied between two neighbouring tubes, ions are accelerated in the gap between these tubes. But the electric field inside the hollow conductors being zero the ions travel with constant velocity in the field free space within the tubes. So gaps are regions of acceleration and tubes are regions of constant velocity.

Suppose a positively charged particle (like a proton) of charge q, mass m is in S and drift tube 1 is negative (in one half cycle). It accelerates in the field (in the gap between S and 1) and enters drift tube 1 which is a hollow metallic cylinder and is a region of zero electric field. It travels with constant speed within the drift tube 1 for time $\frac{T}{2}$ and reaches the outlet of drift tube 1. At that instant the polarity changes and drift tube 2 is negatively charged. The positively charged particle accelerates in the field (in the gap between 1 and 2) and enters drift tube 2 to move with constant velocity within drift tube 2. The process repeats. Eventually, the charged particle emerges out of the final drift tube with high kinetic energy and is made to fall upon a target.

Let charged particle q accelerate in the gap where applied potential difference is V. Energy gained is qV.

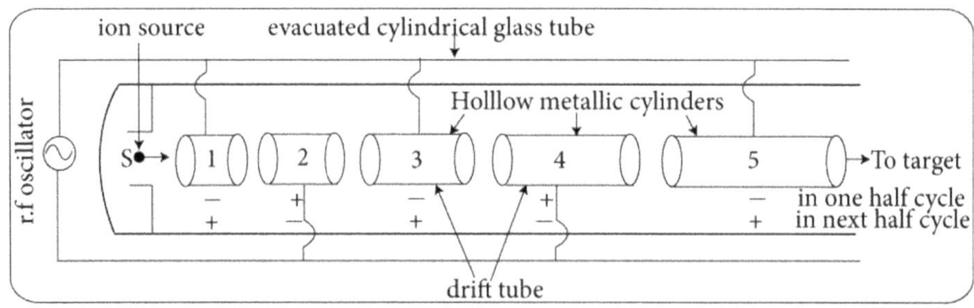

Figure 2.3. Linear accelerator or LINAC.

For n cylinders there are n gaps and energy gained after passing through n gaps is V. Hence final energy of the accelerated particle will be

$$\frac{1}{2}mv_n^2 = nqV$$

where v_n is the velocity after passing through the nth gap.

$$v_n = \sqrt{\frac{2nqV}{m}} \tag{2.2}$$

As $\sqrt{2qV/m}$ is constant it follows that

$$v_n \propto \sqrt{n}$$

So velocity is proportional to square root of natural numbers $n = 1, 2, 3, \ldots$, i.e. if $v_1, v_2, v_3, \ldots$ are the respective velocities of emergence from the drift tubes 1, 2, 3, $\ldots$ then

$$v_1 : v_2 : v_3 : \cdots = \sqrt{1} : \sqrt{2} : \sqrt{3} : \cdots. \tag{2.3}$$

Clearly the ion or charged particle traverses with gradually increasing velocities through successive drift tubes but with constant velocity v_n within the nth drift tube.

2.5.3 Resonance condition of linear accelerator

Since velocities in successive drift tubes are different, times to traverse the drift tubes are also different. Again, for successful operation of the linear accelerator, the time to traverse any drift tube should be equal to half the time period $\frac{T}{2}$ of the r.f accelerating voltage which is the polarity reversal time.

Let L_n be the length of nth tube that is needed to be traversed in time $\frac{T}{2}$ with velocity v_n. Hence

$$v_n \frac{T}{2} = L_n \tag{2.4}$$

Using equation (2.2) we have

$$L_n = \sqrt{\frac{2nqV}{m}} \frac{T}{2} = \sqrt{\frac{nqV}{2m}} T \tag{2.5}$$

As $\sqrt{nqV/2m}$ is constant, it follows that

$$L_n \propto \sqrt{n} \tag{2.6}$$

So drift tube length is proportional to square root of the natural numbers $n = 1, 2, 3, \ldots$, i.e. if $L_1, L_2, L_3, \ldots$ are respective lengths of the drift tubes 1, 2, 3, $\ldots$ then

$$L_1 : L_2 : L_3 : \cdots = \sqrt{1} : \sqrt{2} : \sqrt{3} : \cdots.$$

Using $T = \frac{1}{\nu}$ (ν is frequency of r.f oscillator) we have from equation (2.4)

$$v_n \frac{1}{2\nu} = L_n$$

$$\nu = \frac{v_n}{2L_n} \tag{2.7}$$

This is the resonance condition.

2.5.4 Limitation of the linear accelerator

To accelerate a charged particle to very high energy, the drift tubes have to be of increasing length. The length of the linear accelerator thus becomes large. So the device becomes heavy and bulky and hence difficult to handle.

In *exercise 2.2* we show that the total length of the accelerator is proportional to the wavelength of radio frequency signal.

2.6 Circular accelerator

In a circular accelerator particles are propelled repeatedly through a circular track. With each pass the strength of the electric field increases so as to accelerate the beam of charged particles. When the particles reach the desired energy level a target is placed or implanted into their path. Collision occurs between the beam and the fixed target and a particle detector observes the collision. It records data of the particles produced in the collision between the beam and the fixed target.

Another possibility is not to use a fixed target but to use two beams of particles and make them collide. This is done in a synchrotron such as in the Large Electron–Positron Collider, Tevatron and the Large Hadron Collider.

Particles accelerated in a circular accelerator can run around many times and receive multiple kicks of energy each time and in this process can acquire very high energy even though the accelerator track (which is the circumference) has small length. So circular accelerators achieve higher energy levels compared to linear accelerators.

Circular accelerators do not have accelerators all around the circle. They have a relatively short length of accelerator and the rest of the circle is a series of magnets used to bend the charged particle in a circular path.

In *exercise* 2.3 we compare linear accelerator and circular accelerator.

In *exercise* 2.4 we mention why it is harder to build a high energy circular electron accelerator than a proton accelerator.

2.7 Cyclotron

A cyclotron is a device to accelerate charged particles like protons, deuterons and alpha particles to high energies so that they can be used in atom smashing experiments.

2.7.1 Principle of the cyclotron

The charged particle is bent into a circular path using a magnetic field. Charged particles are accelerated using an electric field, i.e. an accelerating voltage. Multiple revolutions, i.e. passage through such fields, increase the energy of charged particles to a significantly large value.

2.7.2 Construction and working of cyclotron

The cyclotron consists of two D shaped hollow metallic Cu hemispheres called Dees, $D1$ and $D2$. The Dees are connected to an r.f electric oscillator that has a time period say T_0. The two Dees are separated by a gap, as shown in the diagram in figure 2.4.

At the centre an ion source S is placed. The ion source (say deuteron molecules) is bombarded with high energy (100 eV) electrons. Due to collision many positive deuteron ions are formed. They enter the cyclotron through a small hole in the wall of the ion source.

The electrical oscillator establishes a potential difference across the gap between the Dees ($\sim 10^5$ V). The direction of potential difference is made to change sign, i.e. it changes the polarity of the Dees.

The Dees are kept in a magnetic field $\vec{B}$ (~ 1.6 T) whose direction is towards the reader, i.e. away from the plane of the paper/screen (indicated by an encircled dot in the diagram). It is set up by a huge electromagnet. This magnetic field is used to bend the ions around so that they may pass again and again through the same accelerating potential.

The space in which ions move is evacuated to a pressure of 10^{-6} mm of Hg. This is done to prevent ions from colliding with air molecules.

Suppose that a deuteron (charge $q = +ve$, $m = $ mass, $\vec{v} = $ velocity) emerging from ion source finds Dee $D1$ to be $-ve$. It will be attracted and accelerated along a straight path So towards $D1$ and will enter it at point o.

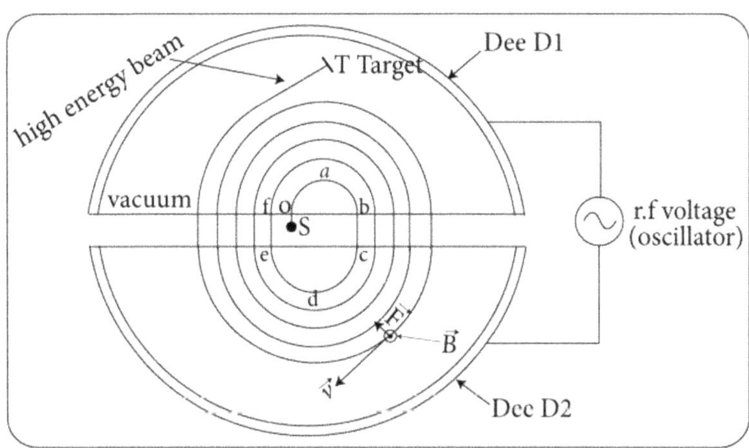

Figure 2.4. Top view of cyclotron.

Once inside it is screened from electrical forces by metal walls of the Dees. The magnetic field $\vec{B}$ is not screened by the Dees and so the ion bends in a circular path oab.

The force equation represents balance of centripetal force $\frac{mv^2}{r}$ and magnetic Lorentz force $q\,\vec{v}\times\vec{B}$ at any point of the circular trajectory of radius r, i.e.

$$\frac{mv^2}{r} = |q\,\vec{v}\times\vec{B}|$$

$$\frac{mv^2}{r} = qvB \tag{2.8}$$

$$r = \frac{mv}{qB} \tag{2.9}$$

Angular velocity w is given by

$$w = \frac{v}{r} = \frac{qB}{m} \tag{2.10}$$

Linear frequency f is given by

$$f = \frac{w}{2\pi} = \frac{qB}{2\pi m} \tag{2.11}$$

Time period T is given by

$$T = \frac{1}{f} = \frac{2\pi m}{qB} \tag{2.12}$$

We note that w, f and T do not depend on speed of particles.

The frequency at which a charged particle or ion circulates in a cyclotron is called cyclotron frequency. So w is angular cyclotron frequency and f is linear cyclotron frequency.

It follows from equation (2.9) that

$$r \propto v$$

which means that fast particles move in large circles and slow particles move in smaller circles—all requiring the same time period T to complete one revolution in the Dee.

After time $\frac{T}{2}$ the charged particle q, upon tracing the semi-circular path oab emerges out of Dee $D1$ at point b. At this instant the accelerating voltage changes sign so that Dee $D2$ is now $-ve$. So q is further accelerated, momentum mv increases, energy $\frac{1}{2}mv^2$ increases and it traces the straight path bc.

Thus q enters Dee $D2$ at point c with greater velocity v and traces a bigger semicircle cde of larger radius r in the same time $\frac{T}{2}$.

At the exit point e from Dee $D2$ accelerating voltage shifts in sign and now Dee $D1$ is $-ve$ helping q to accelerate. Its v increases, energy increases and it enters Dee $D1$ at point f to trace a bigger semicircle in time $\frac{T}{2}$.

The process repeats and after each passage through the accelerating voltage energy of particle increases. To reach say energy of 10 MeV with 10^5 V acelerating voltage the charged particle or ion has to revolve $\frac{10\,MeV}{10^5\,eV} = 100$ times.

When q gains requisite energy it is pulled out by a negatively charged plate and made to strike a target T.

2.7.3 Resonance condition (condition of operation of cyclotron)

The charged particle traces a semicircle (that corresponds to half revolution) in time $\frac{T}{2}$ and arrives at exit point of one Dee when the accelerating voltage should change sign. This means sign alteration of accelerating voltage should also occur in time $\frac{T}{2}$. So the resonance condition is

Time period of charge $T =$ Time period of accelerating voltage T_0

$$T = T_0 = \frac{2\pi m}{qB} \tag{2.13}$$

For a charged particle $\frac{q}{m}$ is fixed. The oscillator is thus designed to work at a single frequency

$$f = \frac{qB}{2\pi m} \tag{2.14}$$

The cyclotron can be tuned by varying B until

$$B = \frac{2\pi m f}{q} \tag{2.15}$$

and an accelerated beam appears. So this is the value of magnetic field needed to accelerate the charged particle.

2.7.4 Calculations at exit of Dee of cyclotron

Number of cycles or revolutions n that the charged particle performs before exit from Dee is given by

$$n = f t_0 \tag{2.16}$$

where t_0 is the circulation time of a charged particle in the cyclotron.

In each revolution a charged particle crosses Dee twice and so the number of crossings N of the charged particle during n revolutions is given by

$$N = 2n \tag{2.17}$$

Work done on charged particle q per crossing where voltage is V is given by

$$W = qV \tag{2.18}$$

Total work done for N crossings will be

$$K = NW \tag{2.19}$$

This is the final kinetic energy gained by a charged particle due to n cycles, N crossings with frequency. Again

$$K = \frac{1}{2}mv_D^2 \tag{2.20}$$

where v_D is the velocity of exit or emergence from Dee. It is given by

$$v_D = \sqrt{\frac{2K}{m}} \tag{2.21}$$

The radius of the path followed by the charged particle is given by

$$r_D = \frac{mv_D}{qB} \tag{2.22}$$

which is similar to equation (2.9).

Also, combining equations (2.17) and (2.18) we have from equation (2.19)

$$K = NW = 2nqV \tag{2.23}$$

From equations (2.20) and (2.23) we can write

$$K = \frac{1}{2}mv_D^2 = 2nqV$$

$$v_D = \sqrt{\frac{4nqV}{m}} \tag{2.24}$$

From equations (2.22) and (2.24) we have

$$r_D = \frac{mv_D}{qB} = \frac{m}{qB}\sqrt{\frac{4nqV}{m}} = \frac{2}{B}\sqrt{\frac{mnV}{q}} \tag{2.25}$$

From equation (2.22) we get the velocity v_D gained in the orbit of radius r_D as

$$v_D = \frac{qBr_D}{m} \tag{2.26}$$

The kinetic energy acquired by the charged particle during exit at Dee radius r_D is given by

$$K = \frac{1}{2}mv_D^2 = \frac{1}{2}m\left(\frac{qBr_D}{m}\right)^2$$

$$K = \frac{q^2B^2r_D^2}{2m} \tag{2.27}$$

Clearly the energy of the particles produced in the cyclotron depends on the Dee radius or the last circular orbit of radius $r = r_D$ before striking the target.

2.7.5 Advantages of a cyclotron

✓ Particles in a circular accelerator circulate multiple times getting multiple kicks of energy.

✓ Though metal Dees are not of great length higher energies can be achieved using the same voltage.

✓ Accelerating particles go around many times. So two oppositely charged particles may be accelerated in opposite directions and made to collide.

✓ Can accelerate alpha particles, deuterons and protons to high energy.

2.7.6 Disadvantages of a cyclotron

✓ For a cyclotron the frequency relation $f = \frac{qB}{2\pi m}$ (equation 2.11) is independent of velocity. But when a charged particle is accelerated to a high speed $\leqslant c$, mass of the particle increases with increase in velocity, i.e. particle gets heavier as per Einstein's relation

$$m(v) = \frac{m_0}{\sqrt{1 - \frac{v^2}{c^2}}} > m_0$$

With this, frequency becomes

$$f = \frac{qB}{2\pi m} = \frac{qB}{2\pi m_0}\sqrt{1 - \frac{v^2}{c^2}} = f(v) < f$$

So cyclotron frequency starts to decrease as velocity v attains relativistic speeds. Thus the resonance condition given by equation (2.13) viz. $T = T_0$ is disturbed and the ions get out of step with the electric oscillator. Cyclotron action fails. Energy of particle does not increase. Function of cyclotron ceases. Clearly in a fixed frequency cyclotron the accelerating voltage gets progressively out of synchronization with the motion of the particles as they reach relativistic speeds.

✓ Magnets used are huge and costly. For a 30 BeV proton ($v \sim 0.99998c$) in a field of $1.5T$ the radius of curvature is 65 m.

✓ Cannot accelerate electrons to high energy since its mass m_0 is very small and very soon it acquires significant speed $\sim c$. At such high speed its mass increases significantly to a large value of $m = \frac{m_0}{\sqrt{1 - \frac{v^2}{c^2}}}$ and it describes a radius $r = \frac{mv}{Bq}$ which is large and falls outside the Dee.

✓ Cannot accelerate neutral particles like a neutron.

2.8 Synchrocyclotron

A synchrocyclotron is a frequency modulated cyclotron. With its help protons and alpha particles can be accelerated up to 300 MeV.

2.8.1 Principle of the synchrocyclotron

The cyclotron frequency is (equation 2.11)

$$f = \frac{qB}{2\pi m}$$

and cyclotron radius is (equation 2.9)

$$r = \frac{mv}{qB}$$

At high speed v mass of charged particle increases as per the relativistic formula

$$m = \frac{m_0}{\sqrt{1 - \frac{v^2}{c^2}}} \quad (m = \text{kinetic mass}, \ m_0 = \text{rest mass}) \tag{2.28}$$

and this affects f, which is inversely proportional to mass ($f \propto \frac{1}{m}$) and r which is directly proportional to mass ($r \propto m$).

With high v kinetic energy T is also high and we have shown in *exercise 2.11* that

$$f = \frac{qBc^2}{2\pi(T + m_0c^2)} \tag{2.29}$$

$$r = \frac{1}{qBc}\sqrt{T(T + 2m_0c^2)} \tag{2.30}$$

where m_0 is rest mass of charged particle.

Clearly cyclotron frequency f decreases and cyclotron radius r increases as kinetic energy of charged particle T increases.

To maintain resonance, frequency of the accelerating voltage of the generator has to be synchronized with cyclotron frequency. This means that with relativistic effect of reduction of cyclotron frequency f the frequency of the accelerating voltage is also simultaneously reduced, keeping magnetic field constant, so that synchronism is maintained.

(Another possibility is to vary the frequency of the magnetic field also so that synchronism is maintained. This is done in the synchrotron.)

2.8.2 Construction of the synchrocyclotron

Construction of a synchrocyclotron is shown in figure 2.5.

Only one Dee is enclosed in a vacuum chamber. The entire unit is placed between pole pieces of a huge electromagnet.

Instead of a second Dee there is an earthed metal plate held opposite to the opening of the Dee. The r.f alternating potential difference is applied between the Dee and the earthed metal plate.

When the charged particle attains relativistic speed its mass varies, cyclotron frequency reduces and the charged particle lags. To maintain sychronization the

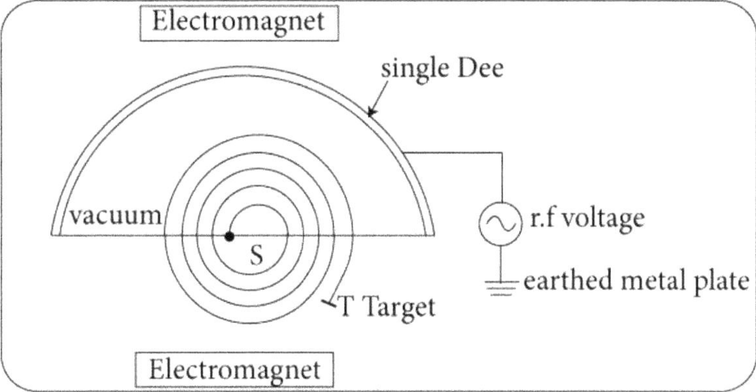

Figure 2.5. Synchrocyclotron.

electric field frequency automatically reduces and gets in step with the particle rotation frequency. This persuades particles to enter the Dee at the right moment to get maximum acceleration.

2.8.3 Limitation of synchrocyclotron

A small fraction of particles or ions coming from source gets synchronized and captured. Intensity of beam output is weak.

In *exercise* 2.5 we have compared synchrocyclotron and cyclotron.

2.9 Betatron

A betatron is a device which is used to accelerate electrons (i.e. β particles) to high kinetic energies $\sim 300\ MeV$. Hence the device is called a betatron.

2.9.1 Principle of the betatron

A betatron uses Faraday's law of electromagnetic induction.

2.9.2 Theory of the betatron

Suppose an electron revolves in a circular orbit of constant radius r in a spatially non-uniform magnetic field (due to an electromagnet powered by ac) perpendicular to the plane of the orbit (figure 2.6). The magnetic field persuades the electron to circulate in a circle of constant radius which is a region of varying magnetic flux.

Let the magnetic flux linked to the orbit be ϕ.

By Faraday's law of electromagnetic induction an electromagnetic field (emf) will be induced in the orbit or circuit given by

$$\varepsilon = -\frac{d\phi}{dt} \tag{2.31}$$

So this alternating magnetic field generates an induced emf which accelerates the charged particle to high energy.

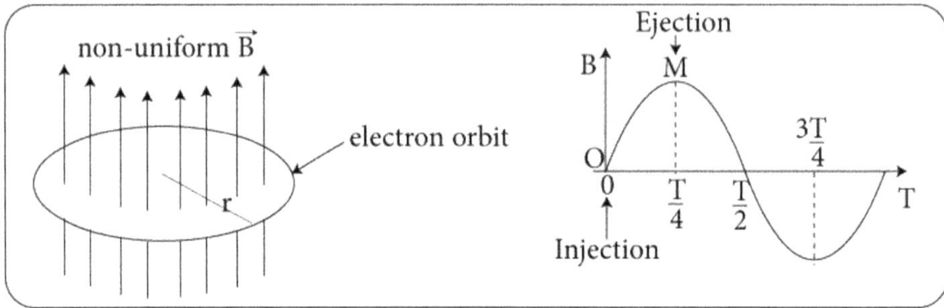

Figure 2.6. Electron rotating in a circle of constant radius in a non-uniform alternating magnetic field.

Work done on an electron in one revolution is

$$W = \text{charge} \times \text{emf}$$

$$W = -|e|\varepsilon = -|e|\left(-\frac{d\phi}{dt}\right) = |e|\frac{d\phi}{dt} \tag{2.32}$$

Again, work done by tangential force F acting on electron over a circular displacement $2\pi r$ is given by

$$W = \text{Force} \times \text{displacement}$$

$$W = F 2\pi r \tag{2.33}$$

Hence from equations (2.32) and (2.33) we have

$$W = F 2\pi r = |e|\frac{d\phi}{dt}$$

$$F = \frac{|e|}{2\pi r}\frac{d\phi}{dt} \tag{2.34}$$

On a circular orbit electron velocity is v and force balance is given by

Centripetal force $=$ Lorentz force

$$\frac{mv^2}{r} = B\,|e|v \tag{2.35}$$

$$mv = B\,|e|r$$

So linear momentum of an electron is given by

$$p = B\,|e|r \tag{2.36}$$

From Newton's second law of motion we can write

$$F = \frac{dp}{dt}$$

$$F = \frac{d}{dt}(B\,|e|r) = |e|r\frac{dB}{dt} \tag{2.37}$$

To maintain constant radius of orbit we equate equations (2.34) and (2.37) to get

$$F = \frac{|e|}{2\pi r}\frac{d\phi}{dt} = |e|r\frac{dB}{dt}$$

$$d\phi = 2\pi r^2 dB \tag{2.38}$$

Integrating

$$\int_0^{\phi} d\phi = \int_0^{\phi} 2\pi r^2 dB$$

$$\phi = 2\pi r^2 B \tag{2.39}$$

This is called betatron condition.

We note that if the field were uniform throughout the orbit then the flux would have been $\phi_0 = $ Field $\times$ area $= B\pi r^2$.

But equation (2.39) says that flux is twice this value. This is needed in order to maintain a stable orbit of constant radius through a proper design of pole pieces.

2.9.3 Construction of the betatron

A betatron consists of a doughnut shaped vacuum chamber (figure 2.7) placed between pole pieces of an electromagnet. The electromagnet is energized by alternating current.

The electrons to be accelerated are produced in an electron gun and injected within the doughnut and allowed to move in a stable circular orbit of constant radius. For this, pole pieces of the electromagnet are properly designed so that a stronger field is created at the central space than at the circumference.

The electrons are injected into the doughnut when the magnetic field is instantaneously zero and just begins to rise at time $t = 0$ (from point O of figure 2.6). The electrons move in an increasing magnetic field which gives rise to induced emf responsible for accelerating and increasing the kinetic energy of electrons. During

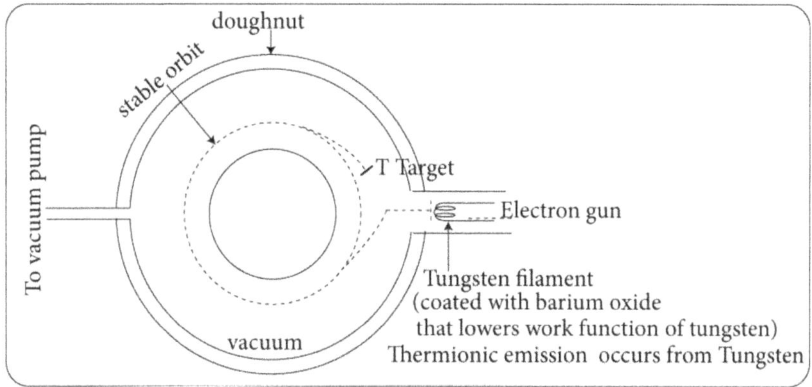

Figure 2.7. Betatron.

the time ($t = 0$ to $\frac{T}{4}$) the magnetic field reaches its peak value, the electrons make several revolutions and get gradually accelerated to high energy. These high energy electrons are ejected or extracted at time $t = \frac{T}{4}$ (at point M of figure 2.6 when magnetic field becomes maximum) and allowed to strike a target T, as shown in figure 2.7.

2.9.4 Advantage of the betatron

The operation of a betatron is independent of the relativistic increase of mass.

2.9.5 Limitations of the betatron

Electrons cannot be accelerated to very high energies (beyond 300 MeV) due to radiation loss.

For higher energies a huge amount of iron is required to make the magnets. This is a costly affair. And so a betatron is not used to accelerate heavy ions. To accelerate a proton up to 30 MeV the radius of the device needs to be eight times that required for accelerating electrons of the same energy.

2.10 Synchrotron

A synchrotron is a cyclic particle accelerator.

It can accelerate charged particles like electrons (as in electron synchrotron) and protons (as in proton synchrotron) to almost light speed.

2.10.1 Principle of synchrotron

Magnetic Lorentz force on a charged particle is $\vec{F} = q\,\vec{v} \times \vec{B}$ and this magnetic force causes the particle to bend. When velocity $\vec{v}$ increases to a high value, larger $\vec{B}$ is needed to bend the particle so as to keep it at the same circular path as before.

When a charged particle of charge q and rest mass m_0 attains relativistic speed, its mass becomes $m = \dfrac{m_0}{\sqrt{1 - \frac{v^2}{c^2}}}$ and kinetic energy T increases hugely, the cyclotron frequency $f = \frac{qB}{2\pi m}$ is reduced to $f = \dfrac{qBc^2}{2\pi(T + m_0 c^2)}$ (*exercise 2.11*) and cyclotron radius $r = \frac{mv}{qB}$ reduces to

$$r = \frac{1}{qBc}\sqrt{T(T + 2m_0 c^2)} \ (exercise\ 2.11)$$

With relativistic effect of reduction of cyclotron frequency, to maintain synchronism, i.e. resonance condition, the frequency of the magnetic field can be varied so that synchronism is maintained.

Also, the frequency of the r.f accelerating voltage may also be varied and made equal to the cyclotron frequency.

2.10.2 Construction and working

A schematic diagram of a synchrotron is shown in figure 2.8.

The charged particle is projected from a gun and is pre-accelerated by an electrostatic accelerator, e.g. by a linear accelerator (LINAC) as shown. A pulsed linear beam comes out of the linear accelerator along the path *ab* and then enters the booster ring at *b*.

At each revolution the energy of the beam increases. So to keep the beam inside the circular tube the strength of the magnets also have to increase to make the beam travel in a circle of constant radius.

The beam is constrained to move in a circular path *cdef* of constant radius in the booster ring through proper adjustment of magnetic field and frequency of accelerating electric field to preserve resonance or synchronous condition, i.e. constant radius requirement. Strength of magnetic field $\vec{B}$ increases with energy of particles to maintain radius. The velocity and hence energy of the charged particle increases after each revolution.

The energetic high velocity beam traverses along the path *pq* and is made to enter into the polygon ring also called storage ring. The charged particle is made to circulate along the path *ABCDEFGH* in this storage ring and is forced to turn around (i.e. change direction) by magnets in a curved path where it radiates energy.

The turning points are *A,B,C,D,E,F,G,H* where magnets are placed.

Acceleration perpendicular to velocity ($\vec{a} \perp \vec{v}$) causes radiation (called synchrotron radiation) to be emitted at each turn. Synchrotron radiation may be produced

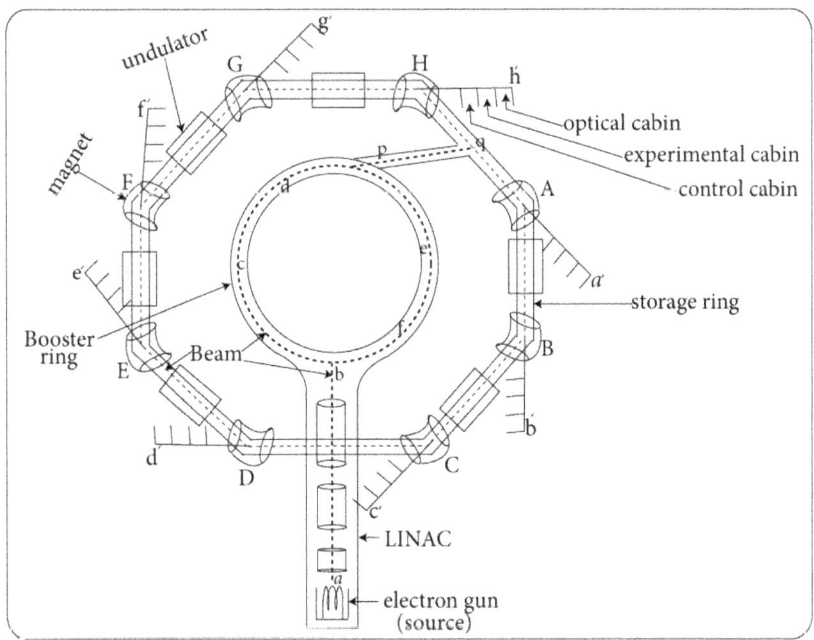

Figure 2.8. Schematic diagram of a synchrotron.

by a fast moving charged particle either artificially (through a device called an undulator as shown, where a charged particle is forced to oscillate and radiate energy) or naturally. The radiation that proceeds along the beam line is received in the three cabins or stations.

Aa′, *Bb′*, *Cc′*, *Dd′*, *etc* are the beam lines.

An optical cabin studies properties of the radiation emitted. In an experimental cabin radiation can be used in experiments. And in a control cabin analysis of data regarding radiation can be done.

2.10.3 Advantages

The synchrotron has variable energy.

Beam extraction efficiency is high. There is narrow angular collimation.

Beam correction is easy due to large spaces between electromagnets.

Spectral range is broad and continuous.

Beam intensity is high.

High degree of polarization.

Synchrotron radiation can be put to use.

There is no need to restrict the number of revolutions executed by a charged particle.

Relativistic effects are managed efficiently and a particle can be accelerated to nearly the speed of light.

2.10.4 Disadvantages

Magnetic field has to be changed in synchronization with the moving particle and so the control mechanism is complex.

The beam from a synchrotron is discontinuous.

The synchrotron requires ample space.

In *exercise* 2.6 we have compared synchrotron and cyclotron.

2.11 Large Electron–Positron (LEP) Collider

Let us mention a few facts about the LEP.

The LEP was built in CERN (European Organisation for Nuclear Research), Geneva Switzerland.

The LEP was a circular lepton collider that collided electrons and positrons.

The LEP was a synchrotron type accelerator.

Tunnel circumference was of length 27 *km*.

LEP was in operation from 1989 to 2000.

Energy generated in the LEP was 200 *GeV* (100 *GeV* per beam) i.e. a 100 *GeV* electron and a 100 *GeV* positron move in circles and collide.

In 1983 *Z* and *W* bosons were discovered and their mass calculated. This enabled a detailed study of electroweak interaction. The LEP established that there are only three generations of particles of matter (three generation of leptons and three generations of quarks). LEP data confirmed the Standard Model and put it on a solid footing.

2.11.1 Generation of particles due to collision of beams in the LEP

Suppose an electron and a positron are accelerated in a field of 470 MeV. Each will have energy of 470 MeV. If they collide head on the total energy of the collision will be 470 $MeV \times 2 = 940$ MeV. This energy will be sufficient to produce a neutron since a neutron has mass of $939\frac{MeV}{c^2}$.

2.12 Tevatron

Let us mention a few facts about the Tevatron.

The Tevatron was a circular particle accelerator located at the Fermilab, USA.

The Tevatron was in operation between 1983 and 2011.

The Tevatron was a synchrotron. It accelerated protons and anti-protons in an underground collider (circular ring) having a circumference of 6.3 km and was used to smash them together at nearly the speed of light.

Energy generated was about 2 TeV (1 TeV proton and 1 TeV anti-proton move in circle and collide).

The resulting subatomic particles would then be studied by scientists and the building blocks of the Universe analyzed.

The Tevatron accelerator was shut down with the arrival of the Large Hadron Collider.

In 1995 top quark was discovered in the Tevatron.

2.13 Large hadron collider (LHC)

Let us mention a few facts about the LHC.

In 2008, CERN laboratories in Switzerland built the LHC.

Tunnel circumference length is 27 km. It is the same tunnel that was used for the LEP.

In *exercise* 2.4 we mention why a circular proton accelerator can generate more energy than a circular electron accelerator using the same track.

The LHC is a synchrotron. It accelerates protons to 99.9999% of the speed of light and makes them collide at huge energies $\sim$14 TeV (7 TeV per beam) and records the results on huge computers. It acts as accelerator and collider of other hadrons too as well as heavy ions.

The LHC has been in operation since 2008.

The LHC discovered the Higgs boson in 2012.

2.13.1 Acceleration of proton beam in a linear accelerator

The energy per beam of the LHC is 7 TeV.

Suppose it is possible to provide 25 $MeV\ m^{-1}$ to a charged particle in a linear accelerator. The track length L would be obtained from the equation

$$\text{Energy} = \frac{\text{Energy}}{\text{length}} \times \text{Track length}$$

$$7\,TeV = 25\frac{MeV}{m} \times L$$

$$L = \frac{7\,TeV}{25\,MeV\,m^{-1}} = \frac{7 \times 10^{12}}{25 \times 10^{6}}m = 280\,000\,m = 280\,km$$

To build such a long track for particle acceleration is not feasible, and it would also be hugely expensive.

So it is better to use a circular track as done in the LHC.

2.13.2 Acceleration of a proton beam in the LHC

In the case of the LHC the accelerator is a circular track of $27\,km$ but the portion of circle that behaves as an accelerator is a length of around $2\,m$ over which the electric field is $8\,MeV\,m^{-1}$. When a proton goes through the acceleration region its energy goes up by

$$8\,MeV\,m^{-1} \times 2m = 16\,MeV$$

With each orbit or cycle the beam gets more and more energy. The proton beam moves with speed virtually at the speed of light. Clearly the beam cannot further speed up but its energy increases.

To gain energy of $7\,TeV$ the number of revolutions the proton makes is

$$\frac{7\,TeV}{16\,MeV} = \frac{7 \times 10^{12}}{16 \times 10^{6}} = 437500$$

Speed of light is $3 \times 10^{8}\,m\,s^{-1}$.

Path length is the circumference of the LHC is $27\,km$.

So number of revolutions the proton makes is

$$\frac{\text{velocity}}{\text{length}} = \frac{3 \times 10^{8}\,m\,s^{-1}}{27\,km} = \frac{3 \times 10^{8}\,m\,s^{-1}}{27 \times 10^{3}m} = 11111\frac{1}{s} \approx 11 \times 10^{3}s^{-1}$$

So to make 437500 revolutions time needed is

$$\frac{437500}{11111\,s^{-1}} = 40\,s$$

So the LHC can accelerate a proton beam to $7\,TeV$ energy in about $40\,s$.

The numbers we arrived at are, however, slightly different due to technical considerations.

As the energy of the beam increases at each revolution the strength of the magnets also has to increase to make the beam travel in a circle of constant radius so as to confine the beam within the circular tube.

In *exercise* 2.10 we have compared the LEP and LHC accelerators.

2.14 List of particle accelerators

Accelerator name	Year	Invented by/place
Cockcroft–Walton electrostatic accelerator	1932	Cockcroft, Walton
Van de Graff electrostatic accelerator	1929	Van de Graff
Linear accelerator (LINAC)	1928	Wideroe
Cyclotron, shape circular	1929	Lawrence and Livingstone
Synchro-cyclotron	1945	E McMillan
Betatron	1941	Kerst
Synchrotron	1944	Veksler, E McMillan
LEP, 100 GeV per beam,	1989–	CERN, Geneva, Switzerland
Study of electroweak physics, Z, W bosons, established standard model	2000	
Tevatron, 1 TeV per beam,	1983–	Fermilab (Fermi National
Discovery of top quark	2000	Accelerator Laboratory), USA
LHC, 7 TeV per beam,	2008	CERN, Geneva, Switzerland
Discovery of Higgs boson		

2.15 Exercises

Exercise 2.1 *What is the justification of having particle accelerators?*
[Ans.] The following are the justifications of having particle accelerators.

- Similar charges repel. But using accelerators we can force similarly charged particles to come in close proximity and study their interaction.

  For instance Rutherford projected high energy α particles at thin pieces of gold foil which actually meant that positively charged α particles were being pushed to the field of positively charged gold nucleus overcoming the electrostatic force of repulsion.

- According to de Broglie wave–particle relation viz.

$$\lambda = \frac{h}{p} \tag{2.40}$$

the more energy/momentum given to a particle the shorter is its de Broglie wavelength λ and therefore greater detail could be investigated or probed by the particle.

For instance, electrons can be accelerated to very high energies and they can be made to smash protons and neutrons revealing charge concentrated at three points which are the quarks.

- When particles collide the energy is re-distributed producing new particles according to the Einstein mass–energy relation

$$E = mc^2 \qquad (2.41)$$

Exercise 2.2 *Show that the total length of a linear accelerator is proportional to the wavelength of radio frequency signal.*

Ans. The length of the nth drift tube of a LINAC (figure 2.3) is given by equation (2.5) viz.

$$L_n = \sqrt{\frac{nqV}{2m}}\, T$$

where charged particle of charge q and mass m is accelerated through inter-tube potential difference V and T is the time period of the r.f oscillator, i.e. time period of the accelerating voltage.

If ν is linear frequency and λ is wavelength of the r.f source then

$$\nu\lambda = c$$

where c is the speed of light in free space. And

$$T = \frac{1}{\nu} = \frac{\lambda}{c}$$

$$L_n = \sqrt{\frac{nqV}{2m}}\, T = \sqrt{\frac{nqV}{2m}}\,\frac{\lambda}{c} = \lambda \ (\text{constant})$$

Also the total length will be

$$L = \lambda \ (\text{constant})$$

Hence the total length of the accelerator is proportional to the wavelength of radio frequency signal.

Exercise 2.3 *Compare linear and circular particle accelerators?*

Ans. We compare linear and circular accelerators in the following.

Linear accelerator

In linear accelerator particles move along a linear or straight track.

Each accelerating segment is used only once as a particle travels along a straight line.

Linear accelerators are used for fixed target experiments.

In a linear accelerator particles reach higher energies without the need for extremely high voltages.

Linear accelerators impart lower energy to particles compared to circular accelerators.

A linear accelerator is simpler and conceptually easier.

In a linear accelerator, after fixed target collision occurs, the remaining uncollided particles are lost.

Particles move along a straight path and hence do not emit electromagnetic radiation.

A linear accelerator with an arrangement of 25 $MeV\ m^{-1}$ would accelerate a charged particle to $\sim$7 TeV (= energy per beam of the LHC) in a track of length 280 km. To build such a long track for particle acceleration would be hugely expensive and is not feasible. Hence circular tracks are preferred.

Linear accelerator takes a longer time and path to reach very high energies.

Circular accelerator

In a circular accelerator particles move through a circular track, i.e. the path is bent.

In a circular accelerator particles are propelled repeatedly through a circular track.

Circular accelerators are used for both a colliding beam experiment as well as a fixed target experiment.

In a circular accelerator, with each pass of the particle the strength of the electric field has to be increased so as to accelerate the beam of a charged particle.

Circular accelerators achieve higher energy levels compared to linear accelerators.

A circular accelerator is complicated.

In a circular accelerator, after collision occurs the remaining uncollided particles keep on circulating and are available for future collisions.

Charged particles move along a bent path and hence emit electromagnetic radiation (synchrotron radiation). So they lose energy. Massive particles like protons lose energy slowly while lighter particles like electrons lose energy fast.

At each revolution the energy of the beam increases. So to keep the beam inside the circular tube the strength of the magnets also has to increase to make the beam travel in a circle of constant radius. So a circular accelerator requires a lot of precision measures for error-free operation.

Exercise 2.4 *Why is it harder to make a high energy circular electron accelerator than a proton accelerator. Comment on the fact that the LEP and LHC make use of the same track.*

|Ans.| An accelerator uses electromagnetic fields to accelerate charged particles to high speeds and keep them concentrated in a well-defined beam.

Charged particles move along a bent or curved path and hence emit electro-magnetic radiation. Acceleration perpendicular to velocity causes synchrotron radiation. So they lose energy.

The amount of synchrotron radiation produced in the process of acceleration is inversely related to the mass of the particle being accelerated. (In fact the total power radiated is inversely proportional to the 4th power of rest mass.)

Electron mass is much less than proton mass.

Less massive particles (electrons) accelerate more quickly than more massive particles (protons). Being accelerated more due to lower mass, electrons lose a large amount of energy as synchrotron radiation and this limits the maximum attainable energy in the accelerator. But being accelerated less due to larger mass, protons lose a smaller amount of energy as synchrotron radiation and this allows them to reach higher energy limits in the accelerator.

In a nutshell, we can say that massive particles like protons lose energy slowly while lighter particles like electrons lose energy faster when accelerated in a synchrotron.

Thus it is much harder to make a high energy circular electron accelerator than a proton accelerator.

- LEP and LHC use the same track to accelerate electrons and protons, respectively. But the energy gained by electrons and protons is widely different.

The LEP could accelerate electrons to an energy of 100 GeV per beam while the LHC can accelerate protons to an energy of 7 TeV per beam. Again

$$\frac{7\ TeV}{100\ GeV} = \frac{7 \times 10^{12}}{100\ \times 10^9} = 70$$

In other words the LHC attains 70 times more energy than the LEP.

Exercise 2.5 *Compare cyclotron and synchrocyclotron.*
 Ans. We compare a cyclotron and a synchrotron in the following.
- Number of Dees used:
  A cyclotron uses two Dees.
  A synchrocyclotron uses one Dee.
- Frequency of accelerating voltage:
  A cyclotron operates at a fixed frequency of accelerating voltage.
  A synchrocyclotron operates at a variable frequency of accelerating voltage.
- Output:
  In a cyclotron output is a continuous beam of high energy charged particles.
  In a synchrocyclotron output is discontinuous bursts (of 10^8 s duration) of high energy charged particle beams.

Exercise 2.6 *Compare a cyclotron and a synchrotron.*
 Ans. We compare a cyclotron and a synchrotron in the following.
- Path traversed:
  As magnetic field is constant a cyclotron accelerates particles in a spiral.
  In a synchrotron particles are kept in a circular orbit by adjusting the magnetic field.
- Cost:
  A cyclotron is expensive.
  A synchrotron is cheaper.
- Field frequency:
  A cyclotron uses ac of constant frequency.
  A synchrotron uses ac that is varying in frequency.
- Field applied:
  A cyclotron uses a constant electric field and a constant magnetic field.
  A synchrotron uses a constant electric field but a varying magnetic field.

- Chamber used:
  A cyclotron is made of a cylindrical or spherical chamber.
  A synchrotron is made of a torus shaped tube.

Exercise 2.7 *Consider a cyclotron having Dee radius of 12.5 m and a magnetic field of 1.3 T and it accelerates protons. Compute the maximum proton energy and the corresponding frequency of the varying voltage.*

$\boxed{\text{Ans.}}$ Maximum energy is given by equation (2.27)

$$K = \frac{q^2 B^2 r_D^2}{2m}$$

$$= \frac{(1.6 \times 10^{-19}\ C)^2 (1.3\ T)^2 (12.5\ m)^2}{2(1.67 \times 10^{-27}\ kg)} = 2.024 \times 10^{-9}\ J = \frac{2.024 \times 10^{-9}}{1.6 \times 10^{-19}}\ eV$$

$$K = 1.265 \times 10^{10}\ eV = 1.265 \times 10^4 \times 10^6\ eV = 12650\ MeV$$

Frequency is given by equation (2.14)

$$f = \frac{qB}{2\pi m} = \frac{(1.6 \times 10^{-19}\ C)(1.3\ T)}{2\pi(1.67 \times 10^{-27}\ kg)} = 19.8 \times 10^6\ Hz$$

Exercise 2.8 *A fixed frequency cyclotron has oscillator frequency of 10 MHz and a Dee radius of 0.5 m. It is used to accelerate deuterons. Calculate the magnetic field needed.*

$\boxed{\text{Ans.}}$ A deuteron is a system of 1 proton 1 neutron, i.e. $d = {}_1H^2 = np$
Deuteron charge is $q = |e|$, deuteron mass is $m = 2\,m_p$
Linear frequency is given by equation (2.14)

$$f = \frac{qB}{2\pi m} = \frac{|e|B}{2\pi\ 2m_p} = \frac{|e|B}{4\pi m_p}$$

$$B = \frac{4\pi m_p f}{|e|}$$

$$B = \frac{4\pi(1.67 \times 10^{-27}\ kg)(10\ MHz)}{1.6 \times 10^{-19}\ C} = \frac{4\pi(1.67 \times 10^{-27}\ kg)(10 \times 10^6\ Hz)}{1.6 \times 10^{-19}\ C}$$

$$B = 1.3\ T$$

Exercise 2.9 *A proton, a deuteron and an alpha move with equal kinetic energy. They enter perpendicularly into a magnetic field. Find the ratio of the radii of the circular paths traced by them.*

$\boxed{\text{Ans.}}$ Kinetic energies are

$$T_p = \frac{1}{2}m_p v_p^2, \quad T_d = \frac{1}{2}m_d v_d^2, \quad T_\alpha = \frac{1}{2}m_\alpha v_\alpha^2$$

$$T_p = T_d = T_\alpha \text{ (given)}$$

$$\frac{1}{2}m_p v_p^2 = \frac{1}{2}m_d v_d^2 = \frac{1}{2}m_\alpha v_\alpha^2$$

$$m_p v_p^2 = m_d v_d^2 = m_\alpha v_\alpha^2 = K \text{ (say)} \tag{2.42}$$

$$v_p^2 = \frac{K}{m_p}$$

$$v_p = \frac{\sqrt{K}}{\sqrt{m_p}}, \quad v_d = \frac{\sqrt{K}}{\sqrt{m_d}}, \quad v_\alpha = \frac{\sqrt{K}}{\sqrt{m_\alpha}} \text{ (similarly)} \tag{2.43}$$

Radii of circular paths are (equation (2.9))

$$r_p = \frac{m_p v_p}{q_p B}, \quad r_d = \frac{m_d v_d}{q_d B}, \quad r_\alpha = \frac{m_\alpha v_\alpha}{q_\alpha B}$$

Hence

$$r_p : r_d : r_\alpha = \frac{m_p v_p}{q_p B} : \frac{m_d v_d}{q_d B} : \frac{m_\alpha v_\alpha}{q_\alpha B}$$

$$= \frac{m_p v_p^2}{q_p v_p} : \frac{m_d v_d^2}{q_d v_d} : \frac{m_\alpha v_\alpha^2}{q_\alpha v_\alpha}$$

$$= \frac{K}{q_p v_p} : \frac{K}{q_d v_d} : \frac{K}{q_\alpha v_\alpha} \quad \text{(using equation (2.42))}$$

$$= \frac{K}{q_p \frac{\sqrt{K}}{\sqrt{m_p}}} : \frac{K}{q_d \frac{\sqrt{K}}{\sqrt{m_d}}} : \frac{K}{q_\alpha \frac{\sqrt{K}}{\sqrt{m_\alpha}}} \quad \text{(using equation (2.43))}$$

$$= \frac{\sqrt{K}\sqrt{m_p}}{q_p} : \frac{\sqrt{K}\sqrt{m_d}}{q_d} : \frac{\sqrt{K}\sqrt{m_\alpha}}{q_\alpha}$$

$$r_p : r_d : r_\alpha = \frac{\sqrt{m_p}}{q_p} : \frac{\sqrt{m_d}}{q_d} : \frac{\sqrt{m_\alpha}}{q_\alpha}$$

Charges are $q_p = |e|$, $q_d = |e|$, $q_\alpha = 2\,|e|$
Masses are m_p, $m_d = 2m_p$, $m_\alpha = 4m_p$ (m_p = proton mass)

$$r_p : r_d : r_\alpha = \frac{\sqrt{m_p}}{|e|} : \frac{\sqrt{2m_p}}{|e|} : \frac{\sqrt{4m_p}}{2\,|e|}$$

$$r_p : r_d : r_\alpha = 1 : \sqrt{2} : 1$$

Exercise 2.10 *Compare LEP and LHC accelerators?*

Ans. We compare LEP and LHC accelerators in the following.

LEP

The LEP was a collider for electrons and positrons.

Tunnel circumference length 27 *km*.

The LEP was in operation from 1989 to 2000.

Energy generated 200 *GeV* (100 *GeV* per beam).

Z, W bosons were discovered and their masses calculated in 1983, detailed study of electroweak interaction was done. LEP established that there are only three generations of particles of matter (three generation of leptons and three generations of quarks). LEP data confirmed standard model and put it on a solid footing.

LHC

The LHC is a collider for protons and other hadrons.

Tunnel circumference length 27 *km* (the same tunnel used for LEP).

The LHC has been in operation from 2008.

Energy generated 14 *TeV* (7 *TeV* per beam).

The LHC discovered the Higgs boson in 2012.

Exercise 2.11 *What is cyclotron frequency? What is cyclotron radius? How do these vary as kinetic energy of the charged particle increases?*

Ans. The frequency of circulation of a charged particle in a magnetic field is the cyclotron frequency given by

$$w = \frac{qB}{m} \text{ (angular frequency)}, \quad f = \frac{w}{2\pi} = \frac{qB}{2\pi m} \text{ (linear frequency)}$$

The cyclotron radius is given by

$$r = \frac{mv}{qB} = \frac{p}{qB}$$

According to Einstein's mass–energy equivalence

$$E = mc^2 = T + m_0 c^2$$

where T is kinetic energy, m_0 is rest mass and m is kinetic mass of the charged particle.

$$m = \frac{T + m_0 c^2}{c^2}$$

The linear cyclotron frequency becomes

$$f = \frac{qB}{2\pi m} = \frac{qBc^2}{2\pi(T + m_0 c^2)} \tag{2.44}$$

Cyclotron frequency decreases as kinetic energy of a charged particle increases.

Relation between energy and momentum under relativistic condition is given by

$$E^2 = p^2 c^2 + m_0^2 c^4 = (T + m_0 c^2)^2$$

$$p^2c^2 + m_0^2c^4 = T^2 + 2Tm_0c^2 + m_0^2c^4$$

$$p^2c^2 = T^2 + 2Tm_0c^2$$

$$p = \frac{1}{c}\sqrt{T^2 + 2Tm_0c^2}$$

$$p = \frac{1}{c}\sqrt{T(T + 2m_0c^2)} \tag{2.45}$$

The cyclotron radius becomes

$$r = \frac{1}{qBc}\sqrt{T(T + 2m_0c^2)} \tag{2.46}$$

Cyclotron radius increases as kinetic energy of the charged particle increases.

Exercise 2.12 *What will be the path of a charged particle executing non-relativistic motion in a constant (time independent) and uniform (space independent) electric field if the particle is projected at angle θ with the field direction?*
What if $\theta = 90°$?
Consider an electron inside a capacitor and find its exit velocity.
What if $\theta = 0°$?
$\boxed{\text{Ans.}}$ Consider motion of a charged particle in an electric field

$$\vec{E} = E\hat{k} \text{ where } E = \text{constant.}$$

Equation of motion of charge q which has displacement $\vec{r}$ at time t under Coulomb force $\vec{F} = q\,\vec{E}$ is given by

$$m\frac{d^2\vec{r}}{dt^2} = q\,\vec{E} \tag{2.47}$$

Suppose a particle is projected at $t = 0$ in the electric field from origin $(0, 0, 0)$ in the YZ plane at angle θ with the Z-axis with velocity (figure 2.9)

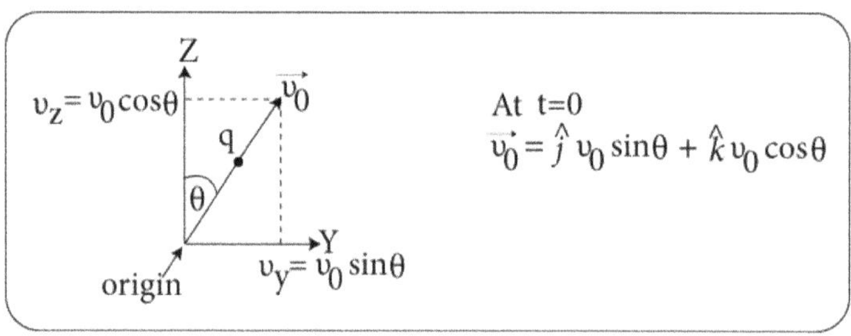

Figure 2.9. A charged particle is projected at $t = 0$ from origin in the YZ plane at angle θ with the Z-axis with velocity $\vec{v_0} = \hat{j}\,v_0 \sin\theta + \hat{k}v_0 \cos\theta$.

$$\vec{v}_0 = \hat{j}\, v_0 \sin\theta + \hat{k} v_0 \cos\theta = \hat{i}\,0 + \hat{j}\, v_y + \hat{k} v_z \text{ (at } t = 0) \tag{2.48}$$

Also

$$\vec{r}_0 = \hat{i}\,0 + \hat{j}\,0 + \hat{k}\,0 = 0 \text{ (at } t = 0) \tag{2.49}$$

With $\vec{r} = x\hat{i} + y\hat{j} + z\hat{k}$ equation (2.47) gives

$$m\frac{d^2}{dt^2}(x\hat{i} + y\hat{j} + z\hat{k}) = q\, E\hat{k} \tag{2.50}$$

This gives

$$m\frac{d^2 x}{dt^2} = 0 \text{ i. e. } \frac{d^2 x}{dt^2} = 0 \tag{2.51}$$

Integrating equation (2.51)

$$\frac{dx}{dt} = v_x = \text{constant} = k \tag{2.52}$$

Let at $t = 0\ v_x = 0$, $k = 0$ (from equation (2.48)). So from equation (2.52) we have

$$v_x = \frac{dx}{dt} = 0 \text{ for all } t \tag{2.53}$$

Integrating equation (2.53)

$$x = \text{constant} = k \tag{2.54}$$

Let at $t = 0$, $x = 0$, $k = 0$ (from equation (2.49)). So from equation (2.54) we have

$$x = 0 \text{ for all } t \tag{2.55}$$

So motion occurs in the YZ plane.

From equation (2.50)

$$m\frac{d^2 y}{dt^2} = 0 \ \Rightarrow\ \frac{d^2 y}{dt^2} = 0 \tag{2.56}$$

Integrating equation (2.56)

$$\frac{dy}{dt} = v_y = \text{constant} = k \tag{2.57}$$

At $t = 0$, $v_y = v_0 \sin\theta = k$ (from equation (2.48)). So from equation (2.57) we have

$$\frac{dy}{dt} = v_y = v_0 \sin\theta \text{ at all } t \tag{2.58}$$

Integrating equation (2.58)

$$y = v_0 \sin\theta t + k \tag{2.59}$$

At $t = 0$, $y = 0$, $k = 0$ (from equation (2.49)). So from equation (2.59) we have

$$y = v_0 \sin \theta t \tag{2.60}$$

$$t = \frac{y}{v_0 \sin \theta} \tag{2.61}$$

From equation (2.50)

$$m\frac{d^2z}{dt^2} = qE \implies \frac{d^2z}{dt^2} = \frac{qE}{m} \tag{2.62}$$

Integrating equation (2.62)

$$\frac{dz}{dt} = v_z = \frac{qEt}{m} + k \ (k = \text{constant}) \tag{2.63}$$

At $t = 0$, $v_z = v_0 \cos \theta = k$ (from equation (2.48)). So from equation (2.63) we have

$$\frac{dz}{dt} = v_z = \frac{qEt}{m} + v_0 \cos \theta \text{ for all } t \tag{2.64}$$

Integrating equation (2.64)

$$z = \frac{qEt^2}{2m} + v_0 \cos \theta t + k \ (k = \text{constant}) \tag{2.65}$$

At $t = 0$, $z = 0$, $k = 0$ (from equation (2.49)). So from equation (2.65) we have

$$z = \frac{qEt^2}{2m} + v_0 \cos \theta t \tag{2.66}$$

Eliminating t from equations (2.61) and (2.66) we get

$$z = \frac{qE}{2m}\left(\frac{y}{v_0 \sin \theta}\right)^2 + v_0 \cos \theta \frac{y}{v_0 \sin \theta}$$

$$z = \frac{qE}{2m(v_0 \sin \theta)^2}y^2 + \cot \theta \, y \tag{2.67}$$

$$z = Ay^2 + By \quad \text{(equation of parabola)} \tag{2.68}$$

Clearly particle traces a parabola in the YZ plane.

From equations (2.53), (2.58), and (2.64) we get

$$v = \sqrt{v_x^2 + v_y^2 + v_z^2} = \sqrt{(0)^2 + (v_0 \sin \theta)^2 + \left(\frac{qEt}{m} + v_0 \cos \theta\right)^2}$$

$$v = \sqrt{v_0^2 + \frac{q^2E^2t^2}{m^2} + 2\frac{qEt}{m}v_0 \cos \theta} \tag{2.69}$$

- Particle thrown at $\theta = 90°$ (particle thrown along the Y-axis)

From equation (2.48) the initial velocity is

$$\vec{v}_0 = \hat{j}\,v_0 \tag{2.70}$$

From equation (2.61) with $\theta = 90°$

$$t = \frac{y}{v_0} \tag{2.71}$$

From equation (2.66) with $\theta = 90°$

$$z = \frac{qEt^2}{2m} \tag{2.72}$$

We eliminate t from equations (2.71) and (2.72) to get

$$z = \frac{qE}{2m}\left(\frac{y}{v_0}\right)^2$$

$$z = \frac{qE}{2mv_0^2}y^2 \tag{2.73}$$

This is a parabola in the YZ plane.

For $q = |e|$ (figure 2.10(a))

$$z = \frac{|e|E}{2mv_0^2}y^2 = Ay^2, \quad A = \frac{|e|E}{2mv_0^2} \tag{2.74}$$

For $q = -|e|$ (figure 2.10(b))

$$z = \frac{-|e|E}{2mv_0^2}y^2 = -Ay^2, \quad A = \frac{|e|E}{2mv_0^2} \tag{2.75}$$

The charged particle describes a parabola.

From equation (2.69) with $\theta = 90°$

$$v = \sqrt{v_0^2 + \frac{q^2E^2t^2}{m^2}} = \sqrt{v_0^2 + \frac{|e|^2E^2t^2}{m^2}} \quad \text{(in both cases } q = \pm|e|\text{)} \tag{2.76}$$

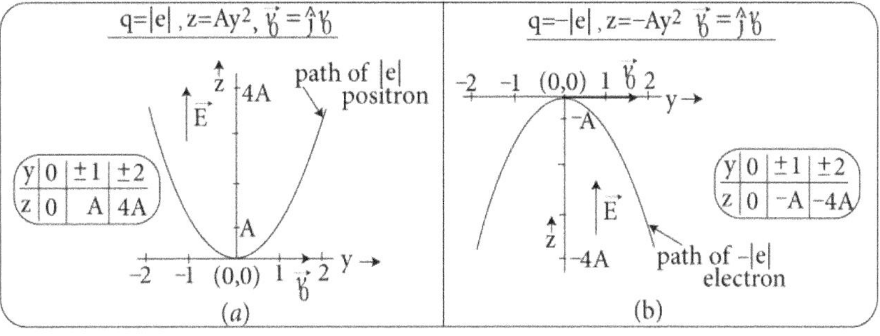

Figure 2.10. Charged particle thrown along $\theta = 90°$.

Application of $\theta = 90°$ case

Consider a parallel plate capacitor with internal field $\vec{E} = E\hat{k}$, length l, inter-plate separation D (figure 2.11).

An electron $q = -|e|$ is projected with initial velocity $\vec{v}_0 = \hat{j}\, v_0$.
Equation of its path is obtained from equation (2.75) to be

$$z = \frac{-|e|E}{2mv_0^2}y^2 \tag{2.77}$$

The path of electron is parabolic inside the capacitor.

Let us find the exit velocity v_{exit}.

At exit point $(0, l, -\frac{D}{2})$ i.e. for $z = -\frac{D}{2}$, $y = l$ we have from equation (2.77)

$$-\frac{D}{2} = \frac{-|e|E}{2mv_0^2}l^2$$

$$v_0^2 = \frac{|e|El^2}{mD} \tag{2.78}$$

$$v_0 = l\sqrt{\frac{|e|E}{mD}} \tag{2.79}$$

This is the velocity of projection for an electron to just make exit at $(0, l, -\frac{D}{2})$

From equation (2.69) we have for $v = v_{\text{exit}}$, $\theta = 90°$, $q = |e|$

$$v_{\text{exit}} = \sqrt{v_0^2 + \frac{|e|^2 E^2 t^2}{m^2}} \tag{2.80}$$

From equation (2.71) we write with $y = l$, using equation (2.79)

$$t = \frac{l}{v_0} = l.\frac{1}{l}\sqrt{\frac{mD}{|e|E}} = \sqrt{\frac{mD}{|e|E}} \tag{2.81}$$

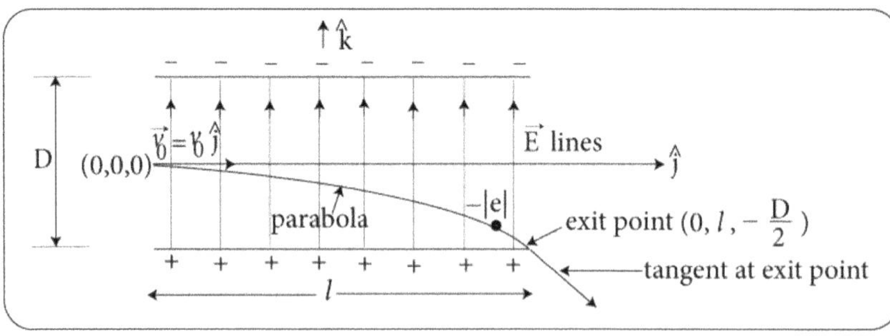

Figure 2.11. Electron path inside capacitor.

Equation (2.80) gives with equations (2.78) and (2.81)

$$v_{\text{exit}} = \sqrt{\frac{|e|El^2}{mD} + \frac{|e|^2 E^2}{m^2}\frac{mD}{|e|E}}$$

$$= \sqrt{\frac{|e|El^2}{mD} + \frac{|e|E}{m}D} \tag{2.82}$$

$$v_{\text{exit}} = \sqrt{\frac{|e|E}{mD}(l^2 + D^2)} \tag{2.83}$$

The electron makes exit from capacitor at point $(0, l, -\frac{D}{2})$ along the tangent to parabola at that point.

- Particle thrown at $\theta = 0°$ (particle thrown along the Z-axis)

From equation (2.48) the initial velocity is $\vec{v}_0 = \hat{k} v_0$
From equation (2.60)

$$y = 0$$

From equation (2.66)

$$z = \frac{qEt^2}{2m} + v_0 t$$

Motion is only along the Z-axis. Particle never deviates from the Z-axis.
If $v_0 = 0$ i.e. a particle is just released then

$$z = \frac{qEt^2}{2m} \tag{2.84}$$

So the particle moves parabolically with time.

Exercise 2.13 *What will be the path of a charged particle $q = \pm|e|$ executing non-relativistic motion in a constant (time independent) and uniform (space independent) magnetic field if the particle is projected at angle θ with the field direction?*
What if $\theta = 90°$?
What is the work done by the charged particle?
Ans. Consider a particle of charge q projected at an angle θ with the magnetic field

$$\vec{B} = B\hat{k} = (0, 0, B)$$

with velocity

$$\vec{v} = \hat{i} v_x + \hat{j} v_y + \hat{k} v_z = v_\perp \hat{\rho} + \hat{k} v_z = (v_x, v_y, v_z) \tag{2.85}$$

where

$$v_\perp = \sqrt{v_x^2 + v_y^2}, \quad v_z = v_{||}. \tag{2.86}$$

Equation of motion of the particle is given by

$$m\frac{d\vec{v}}{dt} = q\,\vec{v} \times \vec{B} = q \begin{vmatrix} \hat{i} & \hat{j} & \hat{k} \\ v_x & v_y & v_z \\ 0 & 0 & B \end{vmatrix} \tag{2.87}$$

$$m\frac{d}{dt}(\hat{i}v_x + \hat{j}v_y + \hat{k}v_z) = q(\hat{i}v_y B - \hat{j}v_x B) \tag{2.88}$$

Comparison of both sides gives the following equations

$$m\frac{dv_x}{dt} = qv_y B \;\Rightarrow\; \frac{dv_x}{dt} = \frac{qB}{m}v_y \tag{2.89}$$

$$m\frac{dv_y}{dt} = -qv_x B \;\Rightarrow\; \frac{dv_y}{dt} = -\frac{qB}{m}v_x \tag{2.90}$$

$$m\frac{dv_z}{dt} = 0 \;\Rightarrow\; \frac{dv_z}{dt} = 0 \;\Rightarrow\; v_z = \text{constant} = A \tag{2.91}$$

Define

$$\frac{qB}{m} = w \tag{2.92}$$

w has dimension of frequency (*exercise 2.14*).

From equations (2.89), (2.90), and (2.91) we have

$$\frac{dv_x}{dt} = wv_y \tag{2.93}$$

$$\frac{dv_y}{dt} = -wv_x \tag{2.94}$$

Assume a trial solution

$$v_x = C\cos(w_0 t + \phi) \tag{2.95}$$

$$v_y = D\sin(w_0 t + \phi) \tag{2.96}$$

Putting the trial solutions in equation (2.93) we get

$$\frac{d}{dt}C\cos(w_0 t + \phi) = wD\sin(w_0 t + \phi)$$

$$-w_0 C\sin(w_0 t + \phi) = wD\sin(w_0 t + \phi)$$

$$-w_0 C = wD \;\Rightarrow\; w_0 = -\frac{wD}{C} \tag{2.97}$$

Putting the trial solutions in equation (2.94) we get

$$\frac{d}{dt} D \sin(w_0 t + \phi) = -wC \cos(w_0 t + \phi)$$

$$w_0 D \cos (w_0 t + \phi) = -wC \sin (w_0 t + \phi)$$

$$w_0 D = -wC \tag{2.98}$$

Putting equation (2.97) in equation (2.98) we get

$$\left(-\frac{wD}{C}\right) D = -wC$$

$$D^2 = C^2 \Rightarrow D = \pm C \tag{2.99}$$

✓ For $q = -|e|$ we choose

$$D = C \tag{2.100}$$

From equation (2.98) we get

$$w_0 C = -wC$$

$w_0 = -w = -\frac{qB}{m}$ (using equation (2.92))

$$w_0 = -w = -\frac{(-|e|)B}{m} = \frac{|e|B}{m} \tag{2.101}$$

✓ For $q = |e|$ we choose

$$D = -C \tag{2.102}$$

From equation (2.98) we get

$$w_0(-C) = -wC$$

$w_0 = w = \frac{qB}{m}$ (using equation (2.92))

$$w_0 = w = \frac{|e|B}{m} \tag{2.103}$$

The frequency $\frac{|e|B}{m}$ is the frequency of rotation of charged particle and is called cyclotron frequency.

To derive the trajectory (circle)
- Let us proceed with $q = -|e|$, $D = C$
    From equations (2.95), (2.96) and (2.91) we have

$$v_x = C \cos(w_0 t + \phi) \tag{2.104}$$

$$v_y = C \sin(w_0 t + \phi) \tag{2.105}$$

$$v_z = A \tag{2.106}$$

$$v^2 = v_x^2 + v_y^2 + v_z^2 = C^2 + A^2 = v_\perp^2 + v_\parallel^2 = \text{constant} \tag{2.107}$$

$$v = \text{constant} \tag{2.108}$$

$$\text{As } v_z = A = \text{constant} = v_\parallel \tag{2.109}$$

$$v_x^2 + v_y^2 = v_\perp^2 = C^2 = \text{constant}, \ \ C = v_\perp \tag{2.110}$$

- For $q = -|e|$

It follows from equations (2.104), (2.105) and (2.110)

$$v_x = \frac{dx}{dt} = v_\perp \cos(w_0 t + \phi) \tag{2.111}$$

$$v_y = \frac{dy}{dt} = v_\perp \sin(w_0 t + \phi) \tag{2.112}$$

Integrating equation (2.111) we have

$$\begin{aligned}
x &= \int v_\perp \cos(w_0 t + \phi) + \text{constant} \\
&= \frac{v_\perp}{w_0} \sin(w_0 t + \phi) + \alpha \ (\alpha = \text{constant})
\end{aligned} \tag{2.113}$$

At $= 0$, $x = x_0$

$$x_0 = \frac{v_\perp}{w_0} \sin \phi + \alpha$$

$$\alpha = x_0 - \frac{v_\perp}{w_0} \sin \phi \tag{2.114}$$

From equation (2.113) we get

$$x - \alpha = \frac{v_\perp}{w_0} \sin(w_0 t + \phi) \tag{2.115}$$

Integrating equation (2.112) we have

$$y = \int v_\perp \sin(w_0 t + \phi) + \text{constant} = -\frac{v_\perp}{w_0} \cos(w_0 t + \phi) + \beta_1 (\beta_1 = \text{constant}) \tag{2.116}$$

At $t = 0$, $y = y_0$

$$y_0 = -\frac{v_\perp}{w_0} \cos \phi + \beta_1$$

$$\beta_1 = y_0 + \frac{v_\perp}{w_0} \cos \phi \tag{2.117}$$

From equation (2.116) we get

$$y - \beta_1 = -\frac{v_\perp}{w_0}\cos(w_0 t + \phi) \qquad (2.118)$$

Combining equations (2.115) and (2.118) we have

$$(x - \alpha)^2 + (y - \beta_1)^2 = \left(\frac{v_\perp}{w_0}\right)^2 = \rho^2 \qquad (2.119)$$

where

$$\rho = \frac{v_\perp}{w_0} \qquad (2.120)$$

So in the XY plane the motion is circular with the centre at (α, β_1).
- For $q = |e|$

It follows from equations (2.104), (2.105) and (2.110)

$$v_x = \frac{dx}{dt} = v_\perp \cos(w_0 t + \phi) \qquad (2.121)$$

$$v_y = \frac{dy}{dt} = -v_\perp \sin(w_0 t + \phi) \qquad (2.122)$$

Integrating equation (2.121) we have

$$x = \int v_\perp \cos(w_0 t + \phi) + \text{constant} = \frac{v_\perp}{w_0}\sin(w_0 t + \phi) + \alpha(\alpha = \text{constant}) \qquad (2.123)$$

At $t = 0$, $x = x_0$

$$x_0 = \frac{v_\perp}{w_0}\sin\phi + \alpha \qquad (2.124)$$

$$\alpha = x_0 - \frac{v_\perp}{w_0}\sin\phi$$

From equation (2.123) we get

$$x - \alpha = \frac{v_\perp}{w_0}\sin(w_0 t + \phi) \qquad (2.125)$$

Integrating equation (2.122) we have

$$y = -\int v_\perp \sin(w_0 t + \phi) + \text{constant} = \frac{v_\perp}{w_0}\cos(w_0 t + \phi) + \beta_2 \ (\beta_2 = \text{constant}) \qquad (2.126)$$

At $t = 0$, $y = y_0$

$$y_0 = \frac{v_\perp}{w_0}\cos\phi + \beta_2$$

$$\beta_2 = y_0 - \frac{v_\perp}{w_0} \cos \phi \tag{2.127}$$

From equation (2.126) we get

$$y - \beta_2 = \frac{v_\perp}{w_0} \cos(w_0 t + \phi) \tag{2.128}$$

Combining equations (2.125) and (2.128) we get

$$(x - X_0)^2 + (y - Y_{02})^2 = \left(\frac{v_\perp}{w_0}\right)^2 = \rho^2 \tag{2.129}$$

where

$$\rho = \frac{v_\perp}{w_0} \tag{2.130}$$

So in the XY plane the motion is circular with the centre at $(\alpha,\ \beta_{1,2})$.

Upon the circle in this XY plane, Lorentz force $qv_\perp B$ and centripetal force $\frac{mv_\perp^2}{\rho}$ are balanced, i.e.

$$\frac{mv_\perp^2}{\rho} = qv_\perp B \tag{2.131}$$

$$mv_\perp = q\rho B$$

$$v_\perp = \frac{qB}{m}\rho = w\rho \tag{2.132}$$

For $q = \pm|e|$ at $t = 0$ we can show from equations (2.111), (2.112) (for $q = -|e|$) and from equations (2.121), (2.122) (for $q = |e|$) that

$v_\perp = \sqrt{v_{x0}^2 + v_{y0}^2}$ = initial velocity in the XY plane along $\hat{\rho}$ (figure 2.12). $v_\perp$ is the component of $\vec{v}$ perpendicular to $\hat{k}$ i.e. along the XY plane, i.e.

$$v_\perp = \vec{v}.\hat{\rho} = v \cos(90° - \theta) = v \sin \theta \tag{2.133}$$

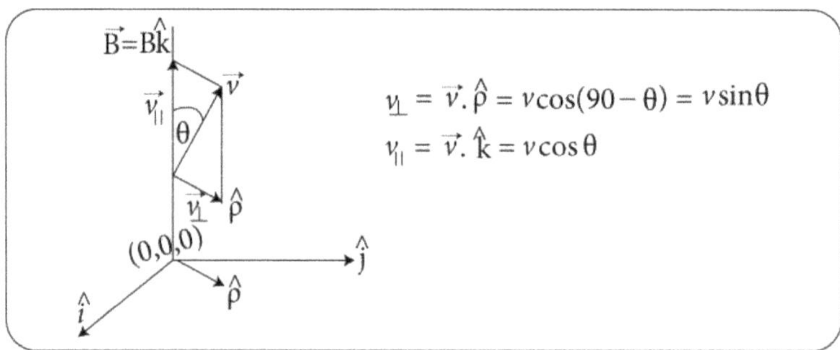

Figure 2.12. Velocity of projection is $\vec{v}$. Its component along XY plane is $\hat{\rho}v_\perp = \vec{v}_\perp$ where $v_\perp = v \sin \theta$ and component along $\hat{k}$ is $\hat{k}v_{\parallel} = \vec{v}_{\parallel}$ where $v_{\parallel} = v \cos \theta$.

From equation (2.91)

$$v_z = \frac{dz}{dt} = A = v_{||} \text{ (from equation (2.108))} \qquad (2.134)$$

So $v_{||}$ is the initial velocity along $\hat{k}$ (figure 2.12). $v_{||}$ is the component of $\vec{v}$ along $\hat{k}$ i.e.

$$v_{||} = \vec{v}.\hat{k} = v \cos\theta \qquad (2.135)$$

Integrating equation (2.134) we have

$$z = v_{||} \int dt + \delta = v_{||}t + \delta$$

Let at $t = 0,\ z = 0, \delta = 0$

$$z = v_{||}t \qquad (2.136)$$

So particle advances along the Z-axis (i.e. along the field direction).

Nature of motion
Motion of the charged particle is the resultant of two motions.
 (1) The particle moves in the XY plane ($\theta = 90°$) in a circular orbit of radius

$$\rho = \frac{v_\perp}{w} = \frac{mv_\perp}{|e|B} = \frac{mv\sin\theta}{|e|B} \text{ (with equation (2.133))} \qquad (2.137)$$

rotating with angular frequency

$$w = \frac{|e|B}{m} \qquad (2.138)$$

linear frequency

$$f = \frac{w}{2\pi} = \frac{|e|B}{2\pi m} \qquad (2.139)$$

time period

$$T = \frac{2\pi}{w} = \frac{2\pi m}{|e|B} \qquad (2.140)$$

as obtained in equations (2.119), (2.129) and shown in figure 2.13(a) (for $q = -|e|$) and in figure 2.13(b) (for $q = |e|$)

 (2) The particle advances along $\vec{B}$ axis with pitch

$$h = z = v_{||}T = v \cos\theta\, T \text{ (with equation (2.135))} \qquad (2.141)$$

This is the distance travelled along z in time period T as obtained in equation (2.136).

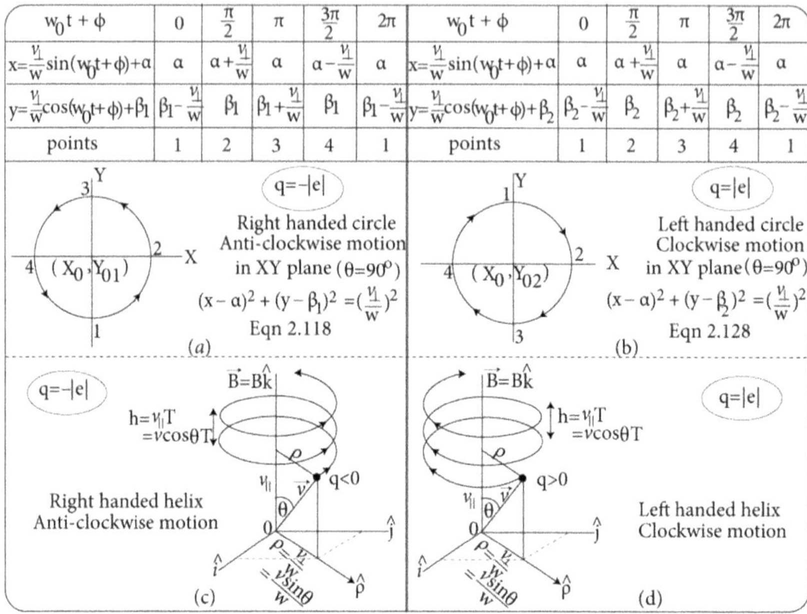

$w_0t+\phi$	0	$\frac{\pi}{2}$	π	$\frac{3\pi}{2}$	2π
$x=\frac{v_\perp}{w}\sin(w_0t+\phi)+\alpha$	α	$\alpha+\frac{v_\perp}{w}$	α	$\alpha-\frac{v_\perp}{w}$	α
$y=\frac{v_\perp}{w}\cos(w_0t+\phi)+\beta_1$	$\beta_1-\frac{v_\perp}{w}$	β_1	$\beta_1+\frac{v_\perp}{w}$	β_1	$\beta_1-\frac{v_\perp}{w}$
points	1	2	3	4	1

$w_0t+\phi$	0	$\frac{\pi}{2}$	π	$\frac{3\pi}{2}$	2π
$x=\frac{v_\perp}{w}\sin(w_0t+\phi)+\alpha$	α	$\alpha+\frac{v_\perp}{w}$	α	$\alpha-\frac{v_\perp}{w}$	α
$y=\frac{v_\perp}{w}\cos(w_0t+\phi)+\beta_2$	$\beta_2-\frac{v_\perp}{w}$	β_2	$\beta_2+\frac{v_\perp}{w}$	β_2	$\beta_2-\frac{v_\perp}{w}$
points	1	2	3	4	1

Figure 2.13. Motion of a charged particle entering obliquely w.r.t the magnetic field.

Obviously the resultant of these two motions in space will be a helix of radius ρ and pitch h with frequency w and time period T, i.e. the circle spreads out into a helix.

For a negatively charged particle ($q=-|e|$) it is a right-handed helix, as explained in figure 2.13(c), and for a positively charged particle ($q=|e|$) it is a left-handed helix, as explained in figure 2.13(d).

✓ Particle projected in the XY plane ($\theta=90°$) with velocity $\vec{v}=\hat{\rho}v$

For $\theta=90°$ we have from equation (2.135)

$$v_\parallel = v\cos\theta = v\cos 90° = 0$$

And from equation (2.136)

$$z = v_\parallel t = 0$$

So there will be no linear motion and only circular motion confined to the XY plane will be there as given by equation (2.119) (for $q=-|e|$, figure 2.13(a)) and by equation (2.129) (for $q=|e|$, figure 2.13(b)). We also note from equation (2.133) that

$$v_\perp = v\sin\theta = v\sin 90° = v$$

So the radius of the circle will be

$$\rho = \frac{v_\perp}{w} = \frac{v}{w} = \frac{mv}{qB} \tag{2.142}$$

Angular frequency will be

$$w = \frac{qB}{m} \tag{2.143}$$

linear frequency will be

$$f = \frac{w}{2\pi} = \frac{qB}{2\pi m} \tag{2.144}$$

and time period will be

$$T = \frac{2\pi}{w} = \frac{2\pi m}{qB} \tag{2.145}$$

- Kinetic energy of charged particle in magnetic field

Kinetic energy $= \frac{1}{2}mv^2 = $ constant (from equation (2.107))

Also, power is

$$\frac{dW}{dt} = \vec{F}.\vec{v} = q\vec{v} \times \vec{B}.\vec{v} \ (W = \text{work})$$
$$= q\vec{v}.\vec{v} \times \vec{B} = q\vec{v} \times \vec{v}.\vec{B} = 0 \tag{2.146}$$

$$W = \text{constant} \tag{2.147}$$

Work involves change in kinetic energy and as work done is constant, kinetic energy does not decrease or increase, i.e. kinetic energy $=$ conserved.

Exercise 2.14 *Show that $\frac{qB}{m}$ has dimension of frequency.*

Ans. q has unit C (Coulomb), B has unit T (tesla), m has unit kg.

$$\text{So } \frac{qB}{m} \text{ has unit } \frac{CT}{kg}.$$

Again from the Lorentz force relation we get

$$\vec{F} = q\,\vec{v} \times \vec{B} \ \Rightarrow \ F \sim qvB$$

$$B = \frac{F}{qv} \text{ has unit}$$

$$T = \frac{N}{C\left(\frac{m}{s}\right)} = \frac{Ns}{Cm}$$

Using $N = kg\ m\ s^{-2}$

$$T = \frac{(kg\ m\ s^{-2})\ s}{Cm} = \frac{kg\ s^{-1}}{C}$$

Clearly

$$\text{So } \frac{qB}{m} \text{ has unit } \frac{CT}{kg} = \frac{C}{kg} \frac{kg \; s^{-1}}{C} = s^{-1}$$

$\frac{qB}{m}$ has unit that of frequency.

Exercise 2.15 *If the maximum value of the r.f is 20 kV find the number of cycles or revolutions of a proton within a cyclotron, before achieving $\frac{1}{5}$ the velocity of light.*

[Ans.] From equation 2.24 we have

$$v_D = \sqrt{\frac{4nqV}{m}}$$

$$n = \frac{mv_D^2}{4qV}$$

$$n = \frac{(1.67 \times 10^{-27}\,kg)(\frac{1}{5}c)^2}{4(1.6 \times 10^{-19}\,C)(20\,kV)} = \frac{(1.67 \times 10^{-27}\,kg)(\frac{1}{5}3 \times 10^8\,m\,s^{-1})^2}{4(1.6 \times 10^{-19}\,C)(20 \times 10^3\,V)} = 469\,s^{-1}$$

Exercise 2.16 *A uniform magnetic field $\vec{B} = 85.3\hat{k}\mu T$ exists in the region $x \geqslant 0$. If an electron enters this field at the origin with a velocity $\vec{v} = 450\hat{i}\ km\ s^{-1}$ find the position where it exits.*

Where would a proton with the same velocity exit?

[Ans.] Entry of a particle is along $\hat{i}$ and magnetic field is along $\hat{k}$. So velocity is perpendicular to the field.

For an electron

$$w = \frac{|e|B}{m} = \frac{(1.6 \times 10^{-19}\,C)(85.3 \times 10^{-6}\,T)}{9.1 \times 10^{-31}\,kg} = 15 \times 10^6\,s^{-1} \quad \text{(equation (2.143))}$$

$$\rho = \frac{v}{w} = \frac{450 \times 10^3\,m\,s^{-1}}{15 \times 10^6\,s^{-1}} = 0.03\,m \quad \text{(equation (2.142))}$$

Electron leaves at point $(0, 2 \times 0.03\,m, 0) = (0, 0.06\,m, 0)$ (figure 2.14)

For a proton

$$w = \frac{|e|B}{m} = \frac{(1.6 \times 10^{-19}\,C)(85.3 \times 10^{-6}\,T)}{1.67 \times 10^{-27}\,kg} = 8 \times 10^3\,s^{-1} \quad \text{(equation (2.143))}$$

$$\rho = \frac{v}{w} = \frac{450 \times 10^3\,m\,s^{-1}}{8 \times 10^3\,s^{-1}} = 56.25\,m \quad \text{(equation (2.142))}$$

The proton leaves at point $(0, 2 \times 56.25\,m, 0) = (0, 112.5\,m, 0)$ (figure 2.14).

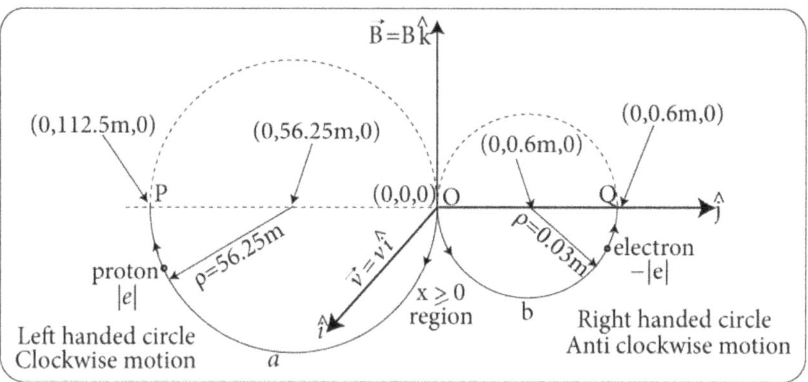

Figure 2.14. Pertaining to *exercise 2.16*.

Exercise 2.17 *A beam of protons with a velocity* $4 \times 10^5\, m\, s^{-1}$ *enters a uniform magnetic field of* 0.3 T. *The velocity makes an angle* 60° *with the field. Find the radius of the helical path taken by the proton beam and the pitch of the helix.*

Ans. Radius of helical path is (equation (2.137)) (left-handed helix for proton, figure 2.13(d))

$$\rho = \frac{v_\perp}{w} = \frac{mv \sin \theta}{|e|B} = \frac{(1.6 \times 10^{-27}kg)(4 \times 10^5 ms^{-1}) \sin 60°}{(1.6 \times 10^{-19}\, C)(0.3\, T)} = 0.0115m = 1.15\ cm$$

Pitch of the helix is (equation (2.141))

$$h = v \cos \theta T$$
$$= v \cos \theta \frac{2\pi}{w} = v \cos \theta \frac{2\pi m}{qB} = v \cos \theta \frac{2\pi m}{|e|B}$$
$$= (4 \times 10^5\, ms^{-1}) \cos 60° \frac{2\pi(1.6 \times 10^{-27}\, kg)}{(1.6 \times 10^{-19}\, C)(0.3\, T)} = 0.0419\ m = 4.19\ cm$$

Exercise 2.18 *A cyclotron has Dees of radius of* 50 cm *which are placed in a magnetic field* 3 $\frac{Wb}{m^2}$. *Calculate the frequency of the oscillator to accelerate particles like alpha particles and protons.*

Also, calculate the maximum obtainable energy in each case.

Ans. Cyclotron frequency is given by equation (2.11) viz. $f = \frac{w}{2\pi} = \frac{qB}{2\pi m}$
Maximum kinetic energy is given by

$$K = \frac{1}{2}mv^2 = \frac{1}{2}m(wr)^2 = \frac{1}{2}m(2\pi fr)^2 = 2\pi^2 mr^2 f^2$$

For an alpha particle

$$f = \frac{qB}{2\pi m} = \frac{(2\,|e|)B}{2\pi(4m_p)} = \frac{|e|B}{4\pi m_p} = \frac{(1.6 \times 10^{-19}\, C)(3\, T)}{4\pi(1.67 \times 10^{-27}\, kg)} = 22.87 \times 10^6\, s^{-1}$$

$$K = 2\pi^2 mr^2 f^2 = 2\pi^2 (4m_p)r^2 f^2 = 8\pi^2 m_p r^2 f^2$$
$$= 8\pi^2 (1.67 \times 10^{-27} \, kg)(50 \times 10^{-2} \, m)^2 (22.87 \times 10^6 \, s^{-1})^2$$
$$= 17.24 \times 10^{-12} \, J = \frac{17.24 \times 10^{-12}}{1.6 \times 10^{-19}} \, eV = 107.75 \times 10^6 \, eV = 107.75 \, MeV$$

For a proton

$$f = \frac{qB}{2\pi m} = \frac{|e|B}{2\pi m_p} = \frac{(1.6 \times 10^{-19} \, C)(3 \, T)}{2\pi(1.67 \times 10^{-27} \, kg)} = 45.75 \times 10^6 \, s^{-1}$$

$$K = 2\pi^2 mr^2 f^2 = 2\pi^2 m_p r^2 f^2$$
$$= 2\pi^2 (1.67 \times 10^{-27} \, kg)(50 \times 10^{-2} \, m)^2 (45.75 \times 10^6 \, s^{-1})^2$$
$$= 17.24 \times 10^{-12} \, J = \frac{17.24 \times 10^{-12}}{1.6 \times 10^{-19}} \, eV = 107.75 \times 10^6 \, eV = 107.75 \, MeV$$

Exercise 2.19 *Deuterons are accelerated in a fixed frequency cyclotron to a maximum D orbit radius of 88 cm. The magnetic field applied is $1.4\frac{Wb}{m^2}$. Calculate the energy of the emerging deuteron.*

    $\boxed{\text{Ans.}}$ Kinetic energy of deuteron is given by equation (2.27)

$$K = \frac{q^2 B^2 r_D^2}{2m} = \frac{|e|^2 B^2 r_D^2}{2(2m_p)} = \frac{|e|^2 B^2 r_D^2}{4m_p}$$

$$K = \frac{(1.6 \times 10^{-19} \, C)^2 (1.4 \, T)^2 (88 \times 10^{-2} \, m)^2}{4 \times 1.67 \times 10^{-27} \, kg} = 5.8 \times 10^{-12} \, J = \frac{5.8 \times 10^{-12}}{1.6 \times 10^{-19}} \, eV = 36.25 \times 10^6 \, eV$$
$$= 36.25 \, MeV$$

Exercise 2.20 *An electron with kinetic energy of 22.5 eV enters a uniform magnetic field B pointing up from the plane of the paper such that the velocity of the electron makes an angle of 60° with the direction of the field. Find the radius of the helical path of the electron and the pitch of the helix.*

    $\boxed{\text{Ans.}}$ Kinetic energy is

$$\frac{1}{2}mv^2 = 22.5 \, eV$$

$$v = \sqrt{\frac{2 \times 22.5 \, eV}{m}} = \sqrt{\frac{2 \times 22.5 \times (1.6 \times 10^{-19} \, J)}{9.1 \times 10^{-31} \, kg}} = 2.8 \times 10^6 \, m \, s^{-1}$$

Radius of the helical path is (equation (2.137)) (right-handed helix for electron, figure 2.13(c))

$$\rho = \frac{v_\perp}{w} = \frac{mv \sin\theta}{|e|B} = \frac{(9.1 \times 10^{-31} \, kg)(2.8 \times 10^6 \, m \, s^{-1})\sin 60°}{(1.6 \times 10^{-19} \, C)B} = \frac{1.4 \times 10^{-5}}{B} \, m \, (B \text{ in Tesla})$$

Pitch of the helix is (equation (2.141))

$$h = v\cos\theta\, T$$

$$= v\cos\theta\frac{2\pi}{w} = v\cos\theta\frac{2\pi m}{qB} = v\cos\theta\frac{2\pi m}{|e|B}$$

$$= (2.8 \times 10^6\ ms^{-1})\cos 60°\frac{2\pi(9.1 \times 10^{-31}kg)}{(1.6 \times 10^{-19}\ C)B} = \frac{5 \times 10^{-5}}{B}\ m\ (B\ \text{in Tesla})$$

Exercise 2.21 *A cyclotron having Dees of radius* 40 *cm accelerates hydrogen nuclei. The polarity of Dee are reversed in* 30×10^6 *times a second. Find the energy of the hydrogen nuclei.*

Ans. Kinetic energy of a proton is given by

$$K = \frac{1}{2}mv^2 = \frac{1}{2}m(wr)^2 = \frac{1}{2}m(2\pi fr)^2$$

$$K = \frac{1}{2}(1.67 \times 10^{-27}\ kg)\,[\,2\pi(30 \times 10^6\ s^{-1})(40\ cm)]^2$$

$$= \frac{1}{2}(1.67 \times 10^{-27}\ kg)\,[\,2\pi(30 \times 10^6\ s^{-1})(40 \times 10^{-2}\ m)]^2 = 4.75 \times 10^{-12}\ J$$

$$K = \frac{4.75 \times 10^{-12}}{1.6 \times 10^{-19}}\ eV = 29.69 \times 10^6\ eV = 29.69\ MeV$$

Exercise 2.22 *Consider a uniform magnetic field* $\vec{B} = (0, 0, B) = B\hat{k}$. *A particle with charge q is projected along the Y-axis with a uniform speed* $v_\perp$.

What will be the trajectory of the particle?

Give an expression of the force acting on the particle at any instant.

What will be the work done by this force?

What happens if the particle starts with some additional velocity $v_\parallel$ *parallel to the field?*

Give a rough sketch of the trajectory.

Ans. Magnetic field is given by $\vec{B} = (0, 0, B) = B\hat{k}$

Initial velocity is given by $\vec{v}_\perp = v_\perp\hat{j}$

O is the point of entrance (figure 2.15).

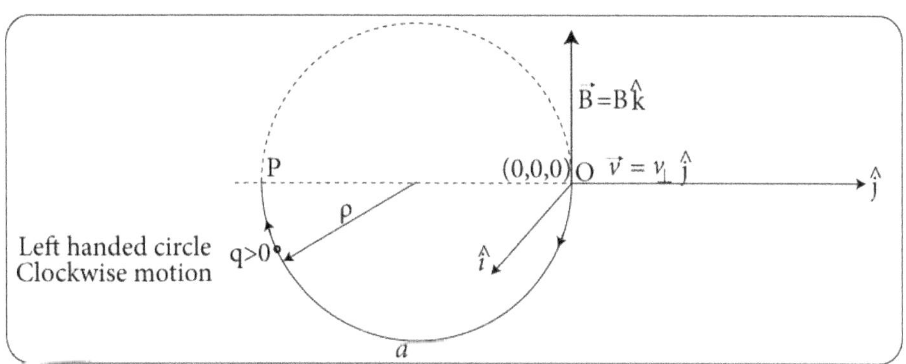

Figure 2.15. Pertaining to *exercise 2.22*.

Motion is confined in the XY plane. Motion of the particle will be circular with radius $\rho = \frac{v_\perp}{w}$ and angular frequency $w = \frac{qB}{m}$ and time period $= \frac{2\pi}{w}$.

Two forces act on the particle viz.

Lorentz force at entry point O is $q\vec{v}_\perp \times \vec{B} = qv_\perp \hat{j} \times B\hat{k} = qv_\perp B\hat{i}$ (bends along X)

Centripetal force $\frac{mv_\perp^2}{\rho}$

On the circle these two forces balance, i.e.

$$\frac{mv_\perp^2}{\rho} = qv_\perp B \implies mv_\perp = q\rho B$$

$$v_\perp = \frac{qB}{m}\rho = w\rho$$

Work done

Power is given by

$$\frac{dW}{dt} = \vec{F} \cdot \vec{v}_\perp = q\vec{v}_\perp \times \vec{B} \cdot \vec{v}_\perp = q\vec{v}_\perp \cdot \vec{v}_\perp \times \vec{B} = q\vec{v}_\perp \times \vec{v}_\perp \cdot \vec{B} = 0$$

$$W = \text{constant.}$$

So kinetic energy is constant.

Work done involves change in kinetic energy. As kinetic energy is constant work done is zero.

If the particle starts with some additional velocity $v_\parallel$ parallel to the field $\vec{B} = B\hat{k}$.

Initial velocity vector is $\vec{v} = v_\perp \hat{j} + v_\parallel \hat{k}$

And velocity $v = \sqrt{v_\perp^2 + v_\parallel^2}$

Then resultant motion is superposition or a combination of two motions.

Motion of the particle in circular orbit of radius $\rho = \frac{v_\perp}{w}$, angular frequency $w = \frac{qB}{m}$, time period $T = \frac{2\pi}{w}$.

The particle advances along the $\vec{B}$ axis with pitch

$$h = z_{at\ t\ =\ T} = v_\parallel T$$

So the actual path traced by an electron will be a left-handed helix of radius $\rho = \frac{v_\perp}{w}$, pitch $h = v_\parallel T$, angular frequency $w = \frac{qB}{m}$, time period $T = \frac{2\pi}{w}$. This clockwise motion is shown in figure 2.13(d).

Exercise 2.23 *Electrons are accelerated to get maximum kinetic energy in a cyclotron in a magnetic field 3.2 T. Calculate the frequency of revolution of an electron.*

Ans. Frequency of revolution

$$f = \frac{qB}{2\pi m} = \frac{(1.6 \times 10^{-19}\,C)(3.2\,T)}{2\pi(9.1 \times 10^{-31}\,kg)} = 8.95 \times 10^{10}\,s^{-1}$$

Exercise 2.24 *Which is true in relation to a charged particle in an accelerator?*
  (a) Magnetic fields are used to accelerate and electric fields are used to bend;
  (b) Magnetic fields are used to bend and electric fields are used to accelerate;
  (c) Magnetic fields can be used to bend and accelerate;
  (d) Electric fields can be used to bend and accelerate.

Exercise 2.25 *In a LINAC the length of drift tubes are:*
  (a) directly proportional to the natural numbers;
  (b) inversely proportional to the natural numbers;
  (c) directly proportional to the square root of natural numbers;
  (d) inversely proportional to the square root of natural numbers.

Exercise 2.26 *A cyclotron is accelerating an alpha particle in a magnetic field of* 1.6
T. The alternating voltage across the Dees is 30 *kV. Circulation time is* 20.37 μs*. Find
the following in connection to acceleration of an alpha particle.*

  *Frequency, number of revolutions, number of cycles or revolutions, number of
crossing of Dees, work done on* α *per crossing, total work done, final velocity, i.e.
velocity of emergence, final radius of the path followed by* α *particle.*

  Ans.

An alpha particle $\alpha \equiv 2n2p$ has charge $2|e|$ and mass $4m_p$
Frequency is given by equation (2.14)

$$f = \frac{qB}{2\pi m} = \frac{2\,|e|B}{2\pi 4m_p} = \frac{|e|B}{4\pi m_p} = \frac{(1.6 \times 10^{-19}\,C)(1.6\,T)}{4\pi(1.67 \times 10^{-27}\,kg)} = 12.199 \times 10^6\,s^{-1}$$

Number of cycles or revolutions is given by equation (2.16)
  $n = ft_0$ (t_0 is the circulation time of α in the cyclotron)

$$n = (12.199 \times 10^6\,s^{-1})(20.37\,\mu s) = 12.199 \times 10^6 \times 20.37 \times 10^{-6} = 248$$

Number of crossing of Dees is given by equation (2.17)

$$N = 2n$$

$$N = 2(248) = 496$$

Work done on the α particle per crossing is given by equation (2.18)

$$W = qV = 2\,|e|V$$

$$W = 2(1.6 \times 10^{-19}\,C)(30\,kV) = 2 \times 1.6 \times 10^{-19} \times 30 \times 10^3\,J = 9.6 \times 10^{-15}\,J$$

Total work done, i.e. the kinetic energy gained, is given by equation (2.19)

$$K = NW$$
$$= 496 \times 9.6 \times 10^{-15}\,J = 4.76 \times 10^{-12}\,J$$

Final velocity, i.e. velocity of emergence, is given by equation (2.21)

$$v = \sqrt{\frac{2K}{m}}$$

$$= \sqrt{\frac{2K}{4m_p}} = \sqrt{\frac{2(4.76 \times 10^{-12}\,J)}{4(1.67 \times 10^{-27}\,kg)}} = 37.75 \times 10^6 \; m\,s^{-1}$$

Final radius of the path followed by the α particle is given by equation (2.21)

$$r_D = \frac{mv}{qB}$$

$$= \frac{4m_p v}{2\,|e|\,B} = \frac{2m_p v}{|e|\,B} = \frac{2(1,67 \times 10^{-27}\,kg)(37.75 \times 10^6\,m\,s^{-1})}{(1.6 \times 10^{-19}\,C)(1.6\,T)} = 0.49\,m = 49\,cm.$$

Exercise 2.27 *Energy generated per beam in the LHC is which of the following?*

(*a*) 7 TeV   (*b*) 14 TeV   (*c*) 7 GeV   (*d*) 14 TeV.

Exercise 2.28 *Higgs boson was discovered by which of the following?*

(*a*) Tevatron   (*b*) LHC   (*c*) LEP   (*d*) Van de Graff accelerator.

Exercise 2.29 *Charged particle traces a stable orbit in which of the following?*

(*a*) Synchrotron   (*b*) Cyclotron   (*c*) Betatron   (*d*) Synchrocyclotron.

Exercise 2.30 *Synchrotron radiation corresponds to which of the following? v is velocity, a is acceleration.*

(*a*) $a \| v$ (*b*) $a > v$ (*c*) relativistic v (*d*) $a \perp v$.

Answers to multiple choice questions

2.24*b*, 2.25*c*, 2.27*a*, 2.28*b*, 2.29*a,c*, 2.30*d*.

2.16 Question bank

Q2.1 High energy particles can probe small regions. Show that this conclusion follows from Heisenberg's uncertainty principle.

Q2.2 What are electrostatic accelerators? Give an example of such accelerators.

Q2.3 What are oscillating field accelerators? Give an example of such accelerators.

Q2.4 What is the principle of the Cockcroft–Walton accelerator? Give a sketch of the device. Explain its working.

Q2.5 What is a corona discharge?

Q2.6 What is the principle of the Van de Graff accelerator? Give a sketch of the device. Explain its working. What are the limitations of the device?

Q2.7 What is the resonance condition in a linear accelerator?

Q2.8 Show that the velocity of emergence from drift tubes in a LINAC is proportional to the square root of natural numbers.

Q2.9 How do you explain that in a LINAC the gaps between drift tubes are regions of acceleration and the region within drift tubes are regions of constant zero acceleration.

Q2.10 Give the principle and working of a linear accelerator.

Q2.11 Justify that circular accelerators do not have accelerators all around the circle but have a series of magnets.

Q2.12 How is a charged particle accelerated in a linear accelerator? How is a charged particle accelerated in a circular accelerator?

Q2.13 What is the principle of a cyclotron? Obtain its condition of operation.

Q2.14 Find the energy that a charged particle attains while leaving the Dee of a cyclotron.

Q2.15 Obtain the relation between Dee radius and the velocity of exit of a charged particle from a cyclotron.

Q2.16 What are the disadvantages of a cyclotron operation?

Q2.17 Can a cyclotron accelerate electron to high energy. Justify your answer.

Q2.18 What is the principle of a synchrocyclotron? Discuss its construction and operation.

Q2.19 Give the principle and theory of a betatron. What is betatron condition?

Q2.20 How is a stable orbit maintained in a betatron?

Q2.21 What is the principle of a synchrotron? Give a schematic diagram of construction of a synchrotron and explain its working.

Q2.22 Mention the successful discoveries made in the LEP, Tevatron and LHC.

Q2.23 Show that the LHC can accelerate a proton beam to 7 MeV energy in about 40 s.

Q2.24 Show that energy generated in the LEP is able to produce a neutron.

Further reading

[1] Krane S K 1988 *Introductory Nuclear Physics* (New York: Wiley)

[2] Tayal D C 2009 *Nuclear Physics* (Mumbai: Himalaya Publishing House)

[3] Satya P 2005 *Nuclear Physics & Particle Physics* (New Delhi: Sultan Chand & Sons)

[4] Guha J 2019 *Quantum Mechanics: Theory, Problems & Solutions* 3rd edn (Kolkata: Books and Allied (P) Ltd)

[5] Yung K L 2002 *Problems and Solutions on Atomic, Nuclear and Particle Physics* (Singapore: World Scientific)

IOP Publishing

Nuclear and Particle Physics with Cosmology, Volume 2
Particle physics and cosmology
Jyotirmoy Guha

Chapter 3

Particle detectors

We discuss various types of detectors. We give the principle, construction and working of gas-filled detectors. Ionization current versus applied voltage of detector is described in detail. We give details of the ionization chamber, proportional counter, Geiger–Müller (GM) counter, semiconductor diode (diffused junction type and surface barrier type). Scintillation detector or counter and photomultiplier tube are described. We also discuss a cloud chamber, bubble chamber, spark chamber, nuclear emulsion detector and Cherenkov detector.

3.1 Introduction

When radiation interacts with matter, positively charged ions and electrons may be created. Nuclear radiation detectors are devices that can make an estimate of the amount of ionization. They can detect or trace tracks formed by particles as a result of nuclear interactions on photographic plates. So nuclear radiation detectors can analyze the nature of radiation, measure its intensity and can study energy spectrum of particles, track the disintegration of unstable particles and investigate interactions between particles and atomic nuclei. The clues and information gathered by a detector regarding the speed and charge of a detected particle are used by scientists to work on the particle, analyze and make further predictions to be verified through experiments.
   We note the following facts.

- ✓ Ionization occurs when a charged particle moves through matter. The greater the ionization the thicker the tracks and the slower the moving particle.
- ✓ If magnetic field is applied perpendicular to the direction of travel, then positive and negative particles curve in opposite directions.
- ✓ The greater the momentum the less curved are the tracks. As particles lose energy they spiral inwards. From a measurement of curvature, momentum can be calculated.
- ✓ When particles collide, ions are produced that can be accelerated by electric fields and detected. So the path of the particle can be detected.

doi:10.1088/978-0-7503-5032-7ch3

3.2 Types of detectors

Nuclear detectors are instruments to detect ionizing radiation due to α, β particles.

Detectors can be classified into two categories, namely electrical detectors and optical detectors (figure 3.1).

Detectors like ionization chamber, proportional counter, GM counter, scintillation counter and semiconductor detector fall into the category of electrical detector.

Detectors like cloud chamber, bubble chamber, spark chamber and photographic emulsion fall into the category of optical detectors.

An ideal detector should have the following characteristics. Efficiency of the detector depends on the nature of the working medium and the design of the detector.

  ✓ It should possess 100% detection efficiency.

  ✓ It should be able to make high speed counting of particles.

  ✓ Response should be linear.

  ✓ It should show very good energy resolution of detected particles.

  ✓ It should be able to detect and register all types of particles and radiation.

  ✓ It should have the capability to detect and register a very high energy particle or radiation.

  ✓ It should possess reasonably large solid angle of acceptance.

A nuclear detector or counter is used

  ✓ To detect charged particles.

  ✓ To count the number of charged particles.

  ✓ To compare ionizing power of charged particles.

  ✓ To compare momenta and kinetic energy of charged particles.

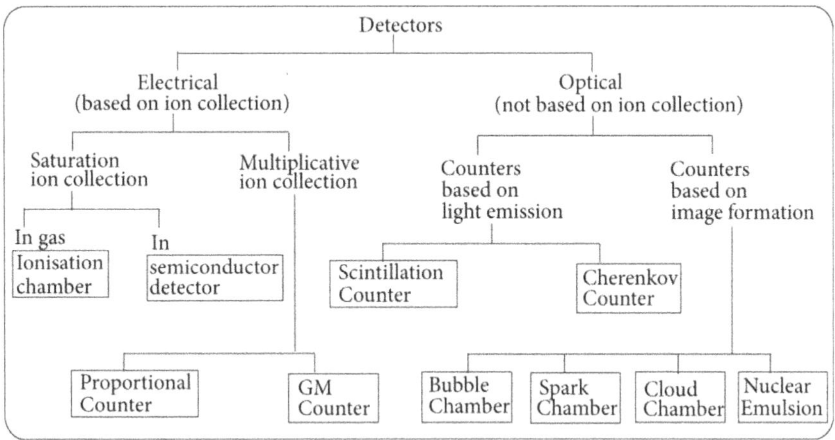

Figure 3.1. Classification of detectors.

3.3 Gas-filled detector

3.3.1 Principle

When a charged particle passes through a gas it collides and transfers energy to the electron of an atom and so electrons get knocked out and the atom becomes positively ionized. In other words gas is ionized.

$$\text{Incident charged particle} + \text{atom} \rightarrow \text{electron(anion)} + \text{ion(cation)}$$

3.3.2 Construction and working

Construction of a gas-filled detector is shown in figure 3.2.

A metal cylinder is connected to the negative of supply denoted by N and acts as a cathode. An axial wire serves as the central anode wire and is connected to P through resistance R. The metallic cylinder or container contains a gas like argon or neon etc, at atmospheric pressure. The gas gets ionized.

In the absence of an electric field the ion pairs thus created will recombine.

As potential difference is applied, negative ions flow towards the anode (central wire) and positive ions flow towards the cathode (metal cylinder) and there is current pulse across resistor R giving rise to current in the external circuit. The current pulse is fed to an amplifier for amplification and then to a counter for counting and detection of particles.

The output pulse height is proportional to the number of ions collected by the anode. It depends upon two factors:

    (i) applied external voltage;

    (ii) ionizing power of the incident charged particle.

Figure 3.3 depicts the plot of pulse height against applied voltage for two different types of events in the detector viz. passage of low ionizing β (event 1) and passage of highly ionizing α (event 2).

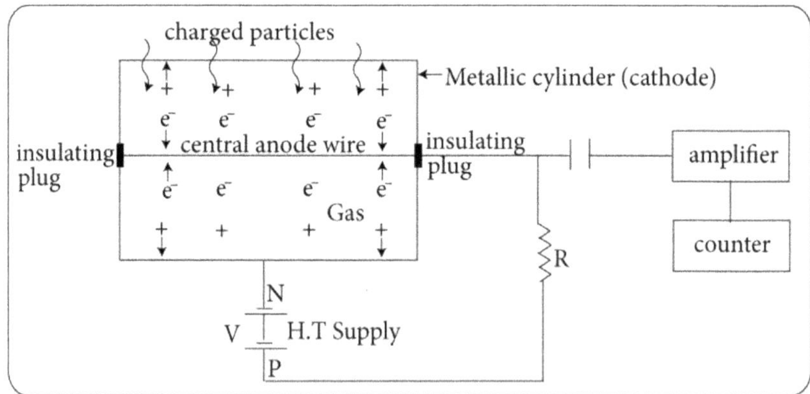

Figure 3.2. Gas-filled detector chamber.

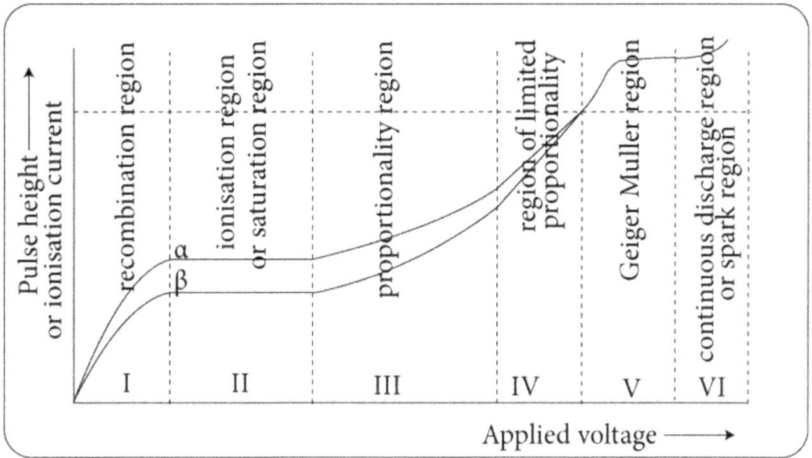

Figure 3.3. Ionization current versus applied voltage of a chamber or detector.

The plot has six regions.
(I) Recombination region:
For very low applied voltage the positive and negative ions have sufficient energy to get attracted to each other and recombine to form neutral atoms. Due to such recombination of positive and negative ions the ion pairs will not be detected.

As the electric field is increased ions move faster and there is less and less time for recombination. So more and more ions reach electrode escaping recombination. And pulse height increases. We get a region called the recombination region.

(II) Ionization region:
As voltage is increased, loss of ions due to recombination is almost negligible. Almost all ion pairs are collected by electrodes. Hence this region is called the ionization region.

Ionization current becomes constant, i.e. saturates. Hence this region is a saturation region. The pulses are of constant height in this region.

In this region the detector chamber has application as an ionization chamber for detection of different particles.

(III) Proportionality region:
Voltage is sufficiently large and the electrons liberated by a primary ionizing event get sufficient energy to cause further ionization due to collision with neutral atoms of gas in the course of their passage to the electrode. This is called secondary ionization.

The total ionization current is thus due to primary ions plus secondary ions.

The secondary ionization results in an increase in the collected charge. With increase in applied voltage the total ionization current linearly increases and is proportional to the initial number of primary ions produced in the detector chamber. Hence this region is called the proportionality region.

The greater the ionizing power of the initial ionizing particle the greater will be the pulse height.

β will produce a smaller number of primary ions compared to α, i.e. pulse heights for α particles will be larger. So we can distinguish the particles on the basis of the pulse height they produce.

Primary ions produce secondary ions and this process is referred to as gas multiplication. The number of secondary ions produced per primary ion is called gas amplification factor M.

In this region the detector chamber has application as a proportional counter for detection of different particles.

(IV) Region of limited proportionality:

With increase in voltage, the production of secondary ions is huge and dominates. The proportional behaviour of the ionization current with initial number of primary ions is lost. In other words the proportionality region breaks down and we have a region of limited proportionality.

In this region pulse height increases, but study of this region is not useful as a link to primary ionization is not evident or detectable.

(V) Geiger–Müller region:

Secondary ions are produced so rapidly that ionization current becomes completely independent of the primary ionization. This region involves a very large amplification of the primary ionization and the pulse height or size is large which can be used for counting of particles. This region is called the Geiger–Müller region.

In this region the detector chamber has application as a Geiger–Müller counter for detection of different particles. As there is no need of further amplification, counting circuits are simplified.

(VI) Continuous discharge region:

For further increase in voltage there is onset of continuous electrical discharge. Large current is obtained even without a primary particle. This region cannot be employed to detect particles.

3.4 Ionization chamber

An ionization chamber is a gas-filled nuclear detector.

An ionization chamber operates in the ionization region of the curve shown in figure 3.3.

It is used to detect charged particles having high ionizing power such as α particles.

It cannot detect charged particles having low ionizing power such as β, γ particles.

3.4.1 Principle of operation

When a charged particle passes through a gas the atoms of gas ionize since energy is given to them and this energy knocks out electrons and the atom becomes a positively charged ion.

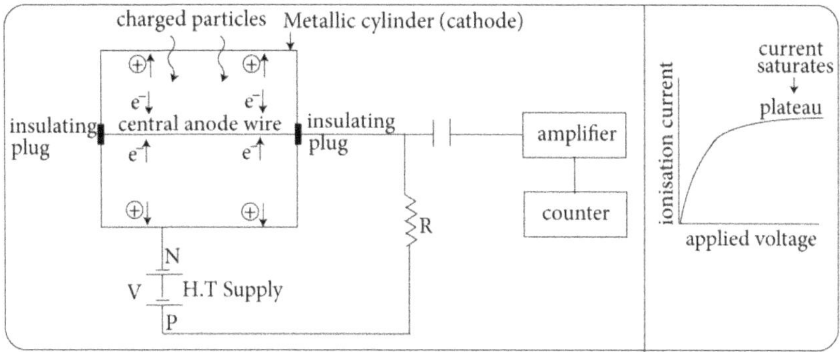

Figure 3.4. Construction of an ionization chamber.

Construction of an ionization chamber is shown in figure 3.4.

A metal cylinder is connected to the negative of supply N. It acts as a cathode. An axial tungsten wire serves as the central anode wire and is connected to the positive end of supply P through resistance R. A suitable gas like argon or neon or air is taken within the chamber at one atmospheric pressure.

An insulating plug avoids direct contact between the central anode wire and the metallic cylinder.

The gas gets ionized.

In the absence of an electric field the ion pairs thus created will recombine and there is no current in the external circuit.

As potential difference is applied, negative ions flow towards the anode and positive ions flow towards the cathode and there is current pulse across resistor R giving rise to current in the external circuit. The current pulse is fed to an amplifier for amplification and then to a counter for counting and detection of particles.

In this region of operation ion-recombination is prevented. Also, gas multiplication is not allowed to start. The ionization current against applied voltage is shown in figure 3.4.

Current pulse depends upon two factors:

   (i) For an incident charged particle having high ionizing power, the ionization current pulse will be large.

(ii) Applied voltage decides the spread of saturation plateau region.

3.4.2 Limitations

✓ In the case of charged particles having low ionizing power such as β, γ particles, the number of ions produced is low and hence the pulse height is so small that they are not detectable.

✓ An ionization chamber cannot detect particles continuously. It has to wait for $\sim ms$ after detection of one charged particle has been done. In other words it has a slow response time.

3.5 Proportional counter

A proportional counter is a gas-filled detector that is used to detect charged particles, especially those which are of low ionizing power like β, γ, meson.

A proportional counter operates in the proportionality region of the curve of figure 3.3.

3.5.1 Principle

In an ionization chamber, interaction between a charged particle and gas atoms produces primary ions.

A proportional counter works on the principle that when voltage between cathode and anode is sufficiently increased the primary ions gain requisite energy to further collide with gas atoms and produce secondary ions. In this situation current pulses with significant height are produced at the output, which are detectable.

3.5.2 Construction and working

A proportional counter consists of a metallic cylinder of length 20 cm and diameter 2 cm. There is an axial central fine wire of tungsten of radius 0.01 cm. The metallic chamber is connected to negative of supply and functions as a cathode. The central wire is connected to the positive of supply and acts as an anode. This anode wire and the metallic cathode are not in direct contact due to the insulating plug. The chamber contains 90% argon and 10% methane. Gas amplification occurs as about 10^3 secondary ions are produced per primary ion (gas amplification factor being $M = 10^3$) (figure 3.5).

The capacitor C blocks any noise that gets generated. The amplifier amplifies pulse that goes to the counter and is detected. The height of current pulse is a measure of the energy of incident particles entering the counter. The proportional counter is thus able to distinguish directly the α, β and γ radiations by virtue of their pulse heights since they differ too much in their ionizing powers.

Let us calculate the gas amplification factor for a proportional counter.

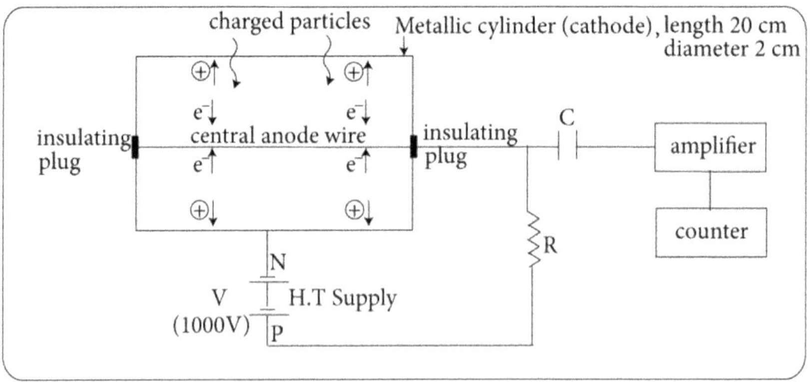

Figure 3.5. Proportional counter.

Let n be the number of secondary electrons produced per primary ion.

The secondary ions excite argon gas atom which subsequently de-excites through emission of photons. These photons strike the cathode wall and photoelectric effect occurs, thereby producing electrons.

P is the probability of emission of a photoelectron by a secondary electron.

The total number of photoelectrons produced by n secondary electrons is nP.

The additional number of secondary electrons produced by nP photoelectrons is

$$n(nP) = n^2 P$$

The total number of photoelectrons produced by $n^2 P$ secondary electrons is

$$n^2 P \, P = n^2 P^2$$

The additional number of secondary electrons produced by $n^2 P^2$ photoelectrons is

$$n(n^2 P^2) = n^3 P^2$$

Gas multiplication is given by the geometric progression

$$M = n + n^2 P + n^3 P^2 + \ldots \; = n(1 + nP + n^2 P^2 + \ldots) = n\frac{1}{1 - nP}$$

If the number of electrons increases arbitrarily then there will be discharge of gas. So gas multiplication factor should not be unlimited. This is limited by use of methane. Actually, methane absorbs the photon and so photoelectric effect is less and the number of secondary electrons gets limited.

3.5.3 Limitation

In a proportional counter gas amplification factor M depends on applied voltage and this applied voltage has to be maintained within narrow limits for proper operation of the proportional counter.

3.6 Geiger–Müller counter

A Geiger–Müller counter is a type of gas-filled detector which is used to detect and measure ionizing radiations or charged particles like α, β etc.

A Geiger–Müller counter operates in the Geiger–Müller region of the curve shown in figure 3.3.

3.6.1 Principle

A Geiger–Müller counter is operated by applying a voltage that exceeds a certain limit so that the detector operates in the Geiger–Müller region (region V) of the ionization current versus applied voltage plot. In this region charge collected is no longer proportional to the initial ionization. The gas multiplication factor M is very large in this region. The total number of ions produced becomes independent of the initial number of ions formed by the entering particle. Since applied voltage is large, a single ion pair produced by radiation entering the chamber can produce an electric discharge inside the chamber.

A charged particle or radiation when passing through argon gas ionizes it. The ion pair produced during ionization gets accelerated towards the electrode. Because of the shape of the electrodes, the electrostatic field is radial and very strong near the anode wire. Under a high potential difference, further ionization is produced. This is secondary ionization. In fact the process is cumulative and an avalanche of electrons or ions is produced called Townsend avalanche. The operating voltage is sufficiently high for the electrons to raise some gas atoms to excited states and when they come down they emit photons which strike the metallic cathode to produce photoelectric effect. So a series of photons are emitted that cause further ionization and further avalanches are generated. The avalanche spreads rapidly in the entire volume of the tube in contrast to the localized action in a proportional counter. Eventually, a large number of ion pairs are produced. This ionization of atoms, excitation of atoms, avalanches and ion-pair production leads to discharge of a GM tube called Geiger discharge.

3.6.2 Construction and working

A Geiger–Müller tube consists of a copper cylinder 20 *cm* in length, diameter 2 *cm*. It is connected to negative of supply and acts as cathode. The cylinder consists of argon inert gas and a quenching gas of ethyl alcohol in the ratio of 10:1 at 0.1 atmospheric pressure.

The axial central anode wire is of radius 0.01 *cm* and is made of tungsten and connected to the positive end of supply. The insulating plug avoids any direct contact between them.

An amplifier is used to amplify current pulse, and to reduce unwanted noise a discriminator can be used. A counter is used to count the pulses (figure 3.6).

3.6.3 Quenching

The aim of the counter is to correctly count pulses. Such counting of current pulses is possible accurately if single entry of a particle leads to a single pulse and not multiple pulses. In fact if one Geiger discharge triggers multiple Geiger discharges then single

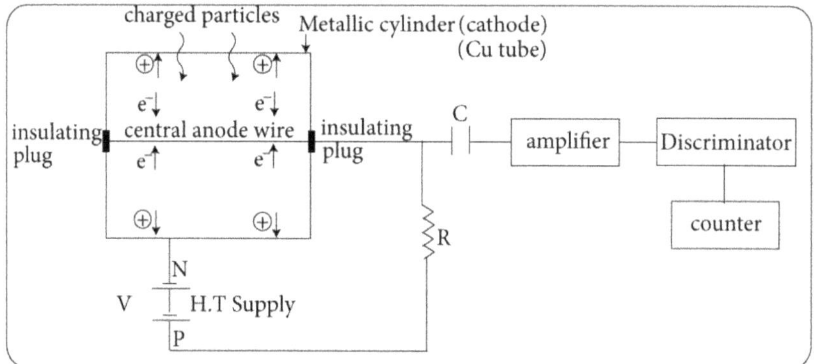

Figure 3.6. GM counter.

particle entry would cause multiple pulses which is not desirable. The tube should recover quickly from this dead state and switch over to an active state so as to record the next entering particle, i.e. record the pulse due to the next entering particle. This is done through quenching.

An electric field is directed from anode to cathode and is quite intense in the region near the anode. Ionization of gas atoms produces electrons and positive ions in the tube. The electrons being light, quickly (in time $10^{-7}\ s$) reach the anode. The massive positive ions move slowly towards the cathode and form a sheath around the anode that slowly drifts towards the cathode. Only after the positive ions reach the cathode does operation of the Geiger–Müller tube resume. In other words, the operation of the Geiger–Müller tube remains virtually suspended (i.e. dead or inoperative) for this period of time which is $\sim$100 μs called dead time. During this time, if any charged particle or radiation makes entry into the tube it will not be detected. So no further pulse is detected in this time. The period during which the ionization remains suspended and the instrument becomes inoperative is called dead time.

The process of removal of the sheath of positive ions around the anode wire is called quenching. This would remove the spurious pulses (i.e. the undesirable secondary pulses) and count the main required pulse. The time interval during which the Geiger–Müller counter becomes ready to produce full size pulse is called recovery time. It is $\sim$100 μs.

There are two methods of quenching called internal quenching and external quenching.

3.6.4 Internal quenching

Inert gas argon along with a quenching gas viz. ethyl alcohol are taken in the ratio of 10:1 at 0.1 atmospheric pressure in the chamber. Argon gas atoms do not interact with atoms of ethyl alcohol. The quenching gas has ionization potential lower than the inert gas.

During their passage to cathode the inert gas ions get neutralized by transferring their charge to alcohol ions. The alcohol ions produced capture electrons from the cathode and are neutralized. Hence there is no multiple pulsing and discharge is quenched soon after initial ionization. The neutralized alcohol ions dissociate and after 10^8 counts the amount of alcohol in the counter is exhausted. This then results in poor characteristics of the Geiger–Müller tube.

3.6.5 External quenching

In this method a large series resistance is used in the circuit. Secondary emission produces large current pulse. A large voltage drop across R is produced. This lowers potential difference between the electrodes of the counter so that further gas ionization is avoided. This technique requires a large recovery time and is not employed.

Dead time plus recovery time is referred to a resolving time $\sim$200 μs.

3.6.6 Advantages

A Geiger–Müller tube efficiently counts α, β particles, γ radiation.
Output pulse height is very large and so further external amplification is not required.
It is very handy and the cheapest kind of nuclear radiation detector.

3.6.7 Disadvantages

The Geiger–Müller counter has resolving time $\sim 200\,\mu s$ and remains dead for $\sim 100\,\mu s$ after each ionizing event. So it has low counting rates. It cannot handle particle counting rates greater than 10^4 particles per second.

It cannot provide information about type of ionizing particle which may have produced a count since it depends upon secondary ionization and not on primary ionization.

Life of the counter is short due to dissociation of quenching gas.

3.7 Semiconductor diode

A semiconductor detector is a solid state detector in which semiconducting material is utilized to detect ionizing radiation.

3.7.1 Principle

When ionizing radiation falls upon a semiconductor detector it breaks the covalent bonds within the semiconductor crystal and so electron–hole pairs are produced. These electron–hole pairs are collected with the help of an external electric field and an electric pulse is produced. The pulse height is proportional to the energy of the ionizing radiation. This is because the higher the energy of ionizing radiation the higher is the number of electron–hole pairs produced.

3.7.2 Construction and working

Semiconductor materials are of two types.
  (1) p-type semiconductor created by doping trivalent impurity like B, In in a pure Si semiconductor crystal leading to majority carriers as holes and minority carriers as electrons and there are immobile negative acceptors.
  (2) n-type semiconductor created by doping pentavalent impurity like P, As in a pure Si semiconductor crystal leading to majority carriers as electrons and minority carriers as holes, and there are immobile positive donors.

When a single crystal is doped properly so that one end is p type and another end is n type then a two terminal device called p–n junction diode is created.

Diffusion of majority carriers occur because of difference in their concentration in the p-side and n-side. After diffusion, recombination of carriers occurs at the junction.

As a result, depletion region is formed at the p–n junction. It is depleted of mobile charge carriers and is the region of an internal electric field or built-in electric field due to the uncovered immobile negative donors in the n side and positive acceptors in the p side. This is depicted in figure 3.7(a).

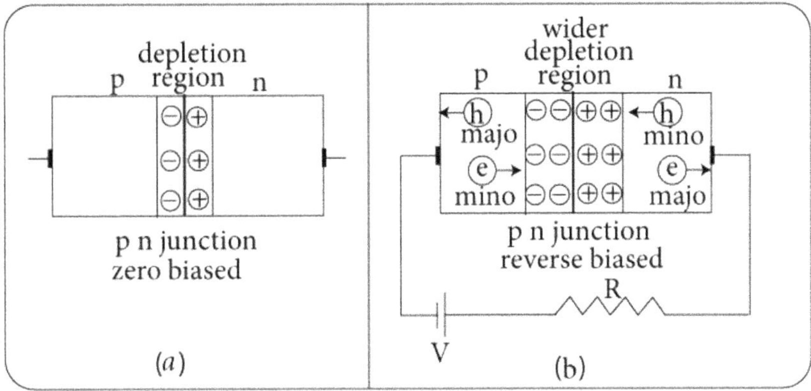

Figure 3.7. (a) *p–n* junction diode (b) reverse biased *p–n* junction diode.

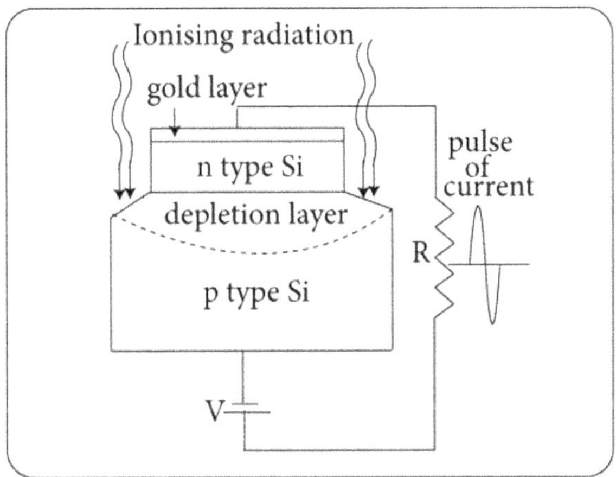

Figure 3.8. Diffused junction semiconductor detector.

When a reverse voltage is applied across the *p–n* junction (figure 3.7), the depletion region gets widened, the majority carriers are pulled by the battery terminals (holes of *p* side move towards negative battery terminal and electrons of *n* side move towards positive battery terminal) and as minority carriers are favourably biased they flow across the *p–n* junction to form a minority reverse leakage current which is very very small. This is shown in figure 3.7(b). Any pulse or increase in current can now be easily detected.

3.8 Diffused junction semiconductor detector

Figure 3.8 shows a diffused junction semiconductor detector that is formed using a slab of *p* type Si and a thin layer of *n* type Si atop it as shown. Upon *n* type Si a thin layer of gold is evaporated for making a good electrical contact. Under reverse bias negligible current (reverse leakage current) flows.

When ionizing radiation is made to be incident on the depletion region, electron–hole pairs are produced. They are collected by electric field caused due to reverse bias voltage. Appreciable current starts flowing in the circuit causing voltage drop across R. It produces an electric pulse whose height is a measure of energy of ionizing radiation.

3.8.1 Advantages

The device is compact and small in size.

Energy resolution is excellent.

Counting rate is very high about 10^9 counts per second.

3.9 Surface barrier semiconductor detector

A surface barrier semiconductor detector is shown in figure 3.9.

A surface barrier semiconductor detector can be produced by etching the n type Si wafer (having resistivity of 1000–80 000 Ω-cm) with a chemical called $CP4A$ which is a mixture of three chemicals viz. HNO_3, HF, CH_3COOH (acetic acid) in the ratio 5: 3: 3. After chemical etching of the surface of the n type Si wafer it is exposed to air. As a result the etched surface gets oxidized, i.e. an oxidized layer is produced. This oxidized layer formed on the etched surface acts like a very thin p type layer. This device thus acts like a p-n junction diode. Thin gold film is evaporated on the p type surface and thin aluminum film is evaporated on the back surface of the n type Si wafer to establish good electrical contact.

The $p - n$ junction is kept reverse biased when negligible reverse leakage current flows. When ionizing radiation falls on the depletion region electron–hole pairs are produced. They are collected by electric field caused by reverse bias voltage.

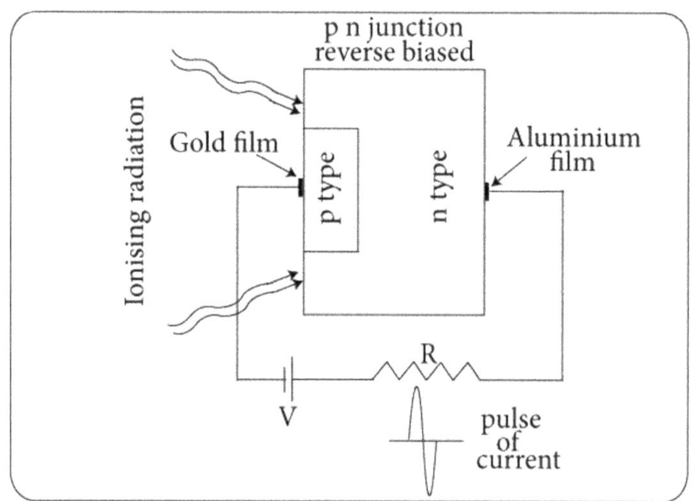

Figure 3.9. Surface barrier semiconductor detector.

Appreciable current starts flowing in the circuit causing voltage drop across R. It produces an electric pulse whose height is a measure of energy of ionizing radiation.

The fabrication of a surface barrier semiconductor detector is easier than a diffused junction semiconductor detector.

3.9.1 Advantages

(1) A semiconductor detector has very high conversion efficiency. The number of electron–hole pairs produced in a semiconductor detector is about 10 times the number of ion pairs produced in an ionization chamber. This is because energy required to produce an ion pair in a gas is about 35 eV while the energy required to produce an ion pair in Si is 3.7 eV and in Ge is 3 eV. Such relatively easy production of free holes and electrons in solids results from the close proximity of atoms which causes many electrons to exist at energy levels just below the conduction band. In contrast, in gases the atoms are virtually isolated and electrons are relatively tightly bound.

(2) The energy resolution is very good.

(3) Semiconductor detector has a high response speed. This is because the carrier mobilities are high about 1500 cm^2 V^{-1} s^{-1} for electron and about 500 cm^2 V^{-1} s^{-1} for holes.

3.9.2 Disadvantages

1) We cannot detect neutrons, i.e. it is insensitive to neutrons.

2) This detector is susceptible to damage from exposure to vapours and hence requires careful handling.

3.10 Scintillation detector or counter

Scillintation detector is a device used to detect and measure radiation like x-ray, γ-ray etc.

3.10.1 Principle

When charged particles like α particles and radiation like x-rays and γ-rays fall on certain materials called scintillators or phosphors they produce flashes of light called scintillations.

A block diagram of the construction of a scintillation detector is depicted in figure 3.10.

When ionizing radiation falls on a scintillator the ground state atoms of scintillator materials get energized by absorbing the ionizing radiation and transit to excited states. Subsequently, the atoms revert back, i.e. de-excite to the ground state by emitting photons, i.e. emits flashes of light.

A photomultiplier tube is coupled to a scintillator. The light flashes (or the scintillations) enter into the photomultiplier tube and get converted into electrical pulse.

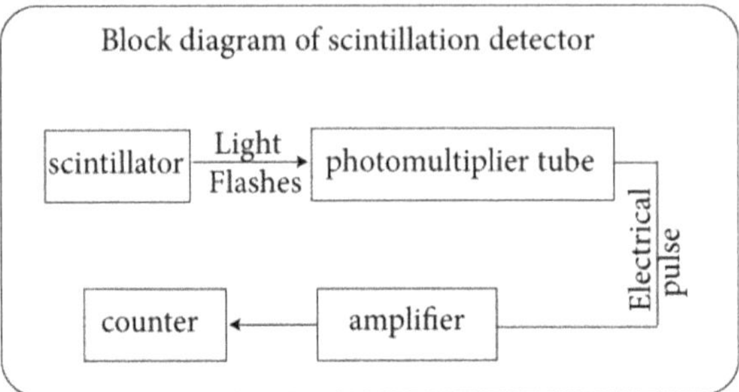

Figure 3.10. Block diagram of scintillation counter and photomultiplier tube.

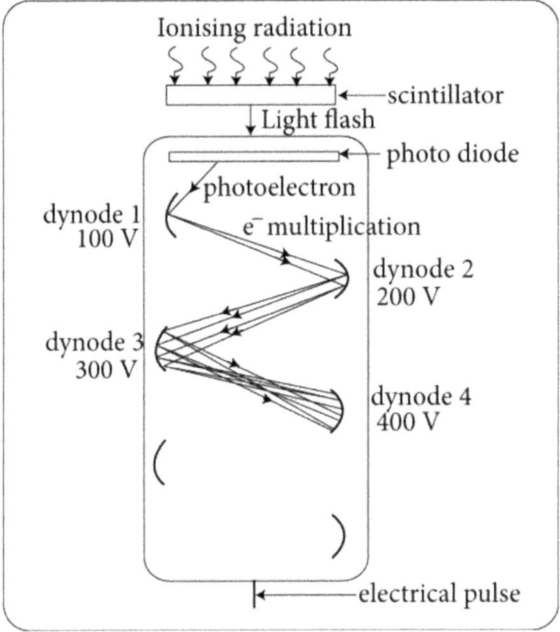

Figure 3.11. Photomultiplier tube.

3.11 Photomultiplier tube

A photomultiplier tube is an essential component of a scintillation counter.

It is a cylindrical shaped tube as shown in figure 3.11.

Ionizing radiation falls upon the scintillator and excites the atoms of the scintillator material to excited states. As the atoms of the scintillator material return to their ground state, light flashes called scintilllations are emitted. In other words the energy of ionizing radiation gets converted to light flashes.

The light flashes fall upon the photodiode or photocathode, which has coating of photosensitive material Cs_3Sb. Photoelectric effect occurs. Photoelectrons are emitted.

A large number of electron multipliers called dynodes are arranged in the photomultiplier tube. Each dynode is held at a more positive voltage than the previous one and multiplies electrons through the process of secondary emission.

An electron from the photocathode accelerates towards the first dynode (at 100 V).

On striking the first dynode, more low energy electrons are produced. They are accelerated by the electric field towards the second dynode (200 V).

The process repeats for an arrangement of around 10 dynodes.

The geometry of the chain of dynodes is such that a cascade occurs with an ever increasing number of electrons being produced at each stage.

Finally, the anode is reached where the accumulation of 10^6 electrons results in a sharp current pulse after a time $\sim 10^{-6}\,s$ of entry of photon at the photocathode.

Obviously the amplification factor of a photomultiplier tube is about 10^6.

This electrical pulse is amplified by an amplifier and counted by a counter indicating arrival of photons at the photocathode (figure 3.10).

3.11.1 Advantages

Scintillation detectors are used to measure energy of x-rays, γ-rays, α particles and β particles.

They have high efficiency.

They are rugged and can withstand mechanical jerks.

Scintillation detectors have very short dead time, $\sim 10\,s$, and have high counting rates.

3.11.2 Limitations

Energy resolution of these detectors is poor.

These detectors are hygroscopic and may get damaged if they absorb moisture.

3.12 Cloud chamber (of Wilson)

A cloud chamber is an instrument that makes possible the visual observation of tracks of charged particles (α, β) in their passage through matter (figure 3.12).

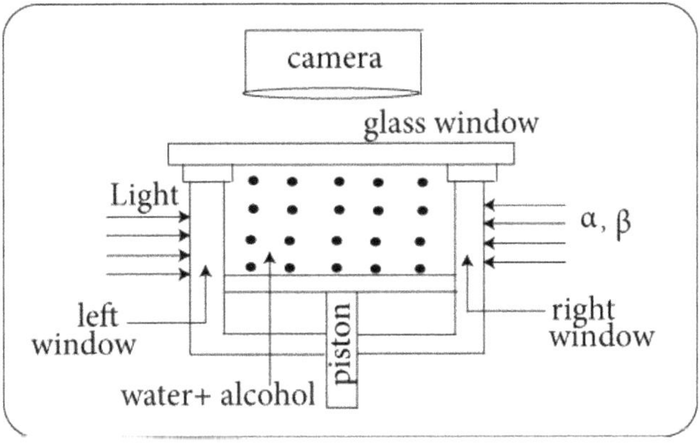

Figure 3.12. Cloud chamber of Wilson.

It is a cylindrical chamber fitted with a piston (on the lower part) and the upper part is covered with a glass window. The chamber is filled with saturated vapour of liquid (water plus a small amount of alcohol).

There are two windows, one on the left and the other on the right end.

The left window is illuminated by light from a mercury vapour lamp.

The charged particle (say α, β, source may be a radioactive sample or cosmic ray) enters into the chamber through the right window.

A photographic camera is set on top to record pictures.

Saturated vapour is a state that contains a large amount of vapour (it is bursting with vapour so to say) at a given temperature and pressure so that the vapour can easily condense and form droplets, when it gets a chance to do so.

The piston is pulled down to increase the volume of the chamber, causing the vapour to expand and do work. This is a (sudden) adiabatic expansion (involving no heat transfer) and so there is decrease in internal energy of the gas. As internal energy is related to temperature of gas this means adiabatic expansion causes cooling. So the saturated vapour tries to condense if it gets proper condensation centres.

If a charged particle (say α, β) passes through the chamber at that point of time then it will interact with the vapour and ionize it along its trajectory. So ions are formed and these ions along the track of the charged particle will act as condensation centres for the surrounding vapour leading to formation of a close array of droplets, i.e. a linear cloud track is thus formed that can be photographed and studied.

Clearly the path of the charged particle is thus detected in the cloud chamber.

Positive and negative charged particles curve in different directions under application of electric and magnetic fields and thus particles can be detected.

3.12.1 Advantages

When subjected to electric and magnetic fields a cloud chamber is used to find the charge on the ionizing particles and their momentum.

The range of high energy particles can be easily determined.

The nature (width) of track indicates whether the track is due to α or β.

The discovery of the positron (e^+), discovery of muons in secondary cosmic rays, discovery of K^0 mesons by observing their decay in flight ($K^0 \rightarrow \pi^+ + \pi^-$), discovery of the hyperon Λ^0 by observation of its decay in flight ($\Lambda^0 \rightarrow p + \pi^-$) and determination of rest mass of the muon were all done by a cloud chamber.

3.12.2 Limitations

Because of low density of gas it is not possible to observe high energy particles.

A highly energetic ionizing particle may fail to stop in the cloud chamber and if so we would not get full information about the particle.

The recovery time of the cloud chamber is relatively very long $\sim$10–60 s after the expansion and hence some ionizing particles may be missed.

3.13 Bubble chamber

A cloud chamber uses super saturated vapour while a bubble chamber uses superheated liquid to display tracks of ionizing particles. The tracks in a cloud chamber consist of tiny droplets of liquid while the tracks in a bubble chamber consist of a series of closely spaced bubbles and so it is called a bubble chamber.

3.13.1 Principle

Working of a bubble chamber is based upon the property of superheated liquid.

At NTP liquid boils at normal boiling point. Under high external pressure it is possible to heat a liquid without making it boil (i.e. without bubble formation) well above its normal boiling point. With sudden release of external pressure the liquid remains in a superheated state for some time ($\sim$a few seconds). When such a superheated liquid is exposed to ionizing particles, ion pairs are produced. As liquid starts to form bubbles (i.e. boiling starts) these ions act as condensation centres for the formation of vapour bubbles along the path of the particle. This track of ionizing particles is photographed and analyzed.

3.13.2 Construction and working

Figure 3.13 shows a schematic diagram of a bubble chamber.

A bubble chamber consists of a thick-walled glass chamber filled with liquid hydrogen at $-245\,°C \equiv 28.12\,K$ at a high pressure in a superheated state (since its normal boiling point is $-252.9\,°C \equiv 20.2\,K$). To maintain the chamber at constant temperature it is surrounded by liquid nitrogen. The box is in communication with a pressure system. The chamber is illuminated with a floodlight (or a mercury vapour lamp).

External pressure is released suddenly and the pressure now is atmospheric pressure and the liquid is now in a superheated state for around 40–$50\,s$ without boiling. High energy particles enter the chamber through a side window in this time. It interacts with liquid molecules and ionizes. Positive and negative ions are produced. Now the liquid starts boiling, turning into vapour that gets condensed in the form of bubbles on the ions formed by an ionizing particle.

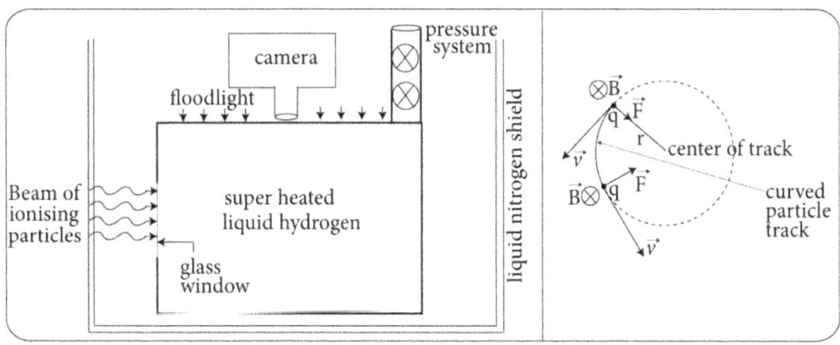

Figure 3.13. Bubble chamber.

A photographic camera is adjusted on its upper end. Tracks of ionizing particles are illuminated by floodlight and photographs of the tracks formed are obtained by a camera.

Generally, a bubble chamber is subjected to a strong magnetic field to determine the sign of charge on the ionizing particle and to measure the momentum from the radius of curvature of the bubble tracks.

The liquid used should have low surface tension to avoid collapse of bubbles and high vapour pressure to speed up enlargement of bubbles.

3.13.3 Advantages

Due to high density of the liquid, even high energy cosmic rays can also be recorded in the bubble chamber.

Energy, momentum and range determination is easy.

The bubble chamber is sensitive to both high and low ionizing particles.

As bubbles grow rapidly the tracks formed in a bubble chamber are sharp and undistorted.

Any experimental liquid can be used.

A number of mesons and baryons were discovered using the bubble chamber.

3.13.4 Limitations

Time during which a bubble chamber is sensitive is only a few microseconds. Photographing the tracks formed must take place during this short time.

Bubble chambers are costly detectors.

3.13.5 Momentum and energy determination of a charged particle in a bubble chamber

A bubble chamber is placed in a magnetic field $\vec{B} = B\otimes$ where $\otimes$ denotes direction downwards away from the reader into the plane of the paper. This $\vec{B}$ acts on charged particle q that enters the bubble chamber. Suppose a charged particle moves with velocity $\vec{v}$ as shown in figure 3.13.

Magnetic Lorentz force on q is

$$\vec{F} = q \;\; \vec{v} \times \vec{B} = qvB \sin 90° \hat{F} = qvB\hat{F}$$

Due to this force $\vec{F}$ at each point of the track the charged particle bends along a circle, i.e. the charged particle moves along a circular curved track as depicted.

Let r be the radius of curvature of the track. Force balance at each point of the track is due to equality of centripetal force $\frac{mv^2}{r}$ and the magnetic Lorentz force qvB where m is mass of particle. So we have

$$\frac{mv^2}{r} = qvB$$

$$mv = qBr$$

Hence momentum is given by

$$p = mv = qBr$$

Energy is given by

$$E = \frac{p^2}{2m} = \frac{(qBr)^2}{2m} = \frac{q^2B^2r^2}{2m}$$

Values of B are known, r is measured from the track. Knowledge of q, m gives momentm p and energy E of the ionizing charged particle.

3.14 Spark chamber

A spark chamber is a particle detector for detecting electrically charged particles studied in high energy particle physics.

A spark chamber consists of a stack of metal plates alternately connected to a source of direct voltage. There are about 25–100 plates, each about 1 mm thick and 1 m square and placed about 6 mm apart. These are placed in a sealed box filled with an inert gas like helium, neon or their mixture at 1 atm pressure (figure 3.14).

When a high energy charged particle enters the spark chamber, a detector triggers the high voltage to the plates and opens the camera shutter.

The charged particle travels through the box and ionizes the gas between the plates. If a high voltage ~ 10–$15\ kV$ is applied between adjacent plates causing a very high electrical field across the gaps, a series of sparks are produced along the trajectory followed by the ray. The charged particle in effect becomes visible as the line of spark. The sensitivity time is $\sim 0.5\ \mu s$. Then the high voltage is cut off suddenly by the discharge which disappears in 0.1 μs. A field is then applied to sweep away the ions from the chamber and now the chamber is ready for the next event.

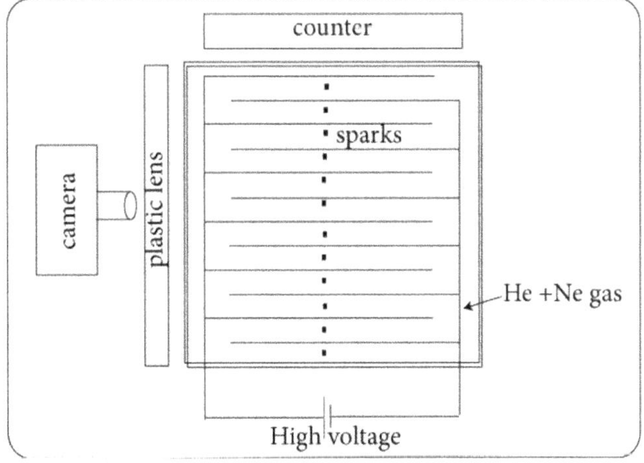

Figure 3.14. Spark chamber.

A large plastic lens makes it possible to photograph light from sparks between the plates.

3.14.1 Drawback

The spark chamber pictures lack the resolution and details of other detectors such as cloud chambers.

3.15 Nuclear emulsion detector

A nuclear emulsion detector is a photographic plate with a particularly thick emulsion layer of silver halide crystals.

When ionizing radiation or a charged particle pass through an emulsion they ionize the atoms of silver halide (say silver bromide AgBr) crystals and leave their tracks at the microscopic level. In other words, the incident radiation or particle interacts with silver halide grains present on the photographic plate. When developed the affected silver halide grains change into black grains of metallic silver.

After exposing the nuclear emulsion to ionizing radiation or charged particles, the emulsion plate is developed and fixed. In other words, these tracks can be developed, viewed under a microscope and analyzed. In this way it is possible to determine the trajectory and type of particles passed through detector.

It has been found experimentally that optical photographic emulsions are not suitable for qualitative work with nuclear radiation. The sensitivity of ordinary photographic plates is low and the tracks formed do not show clear range because the developed crystal grains are large and widely spaced. In nuclear emulsions the grain size of silver halide is kept much smaller, about 0.1–0.6 micron (diameter) compared to that in optical emulsions, which is about 1–3.5 micron (diameter). The thickness of nuclear emulsions is kept large about 50–2000 micron compared to that of optical emulsions which is about 2–4 micron. The photosensitive material AgBr has to be present in much larger concentration (about 3–4 times as large) than in an ordinary photographic plate.

3.15.1 Advantages

This technique has high spatial resolution and provides very precise information about the path of particles.

The emulsion plates are very light in weight.

The method is suitable for study of cosmic rays at high altitudes.

For exposing the emulsion plates to ionizing radiation or charged particles, no electronic circuitry is required.

The emulsions are extensively used in cosmic ray studies. They can be exposed to cosmic rays in the upper atmosphere using balloon flights.

From the observed range of ionizing particles in nuclear emulsion their energy can be easily calculated.

Different ionizing particles form tracks which are markedly different. A study of these tracks reveals the nature of interacting particles.

Because of high stopping power it can record short lived particles also.

Determination of pion rest mass (273 m_e), discovery of K mesons and antilambda hyperon was done by nuclear emulsion technique.

3.15.2 Disadvantages

The process of developing and analyzing the emulsion plate is time consuming.

Measurement of tracks has to be done manually as no automation is possible.

The sensitivity of nuclear emulsions depends on age of nuclear emulsion.

The sensitivity of nuclear emulsions is affected by atmospheric conditions, for instance temperature, humidity etc.

The length of the track is relatively small compared to the tracks recorded in a cloud chamber or bubble chamber. Hence measurement is difficult.

3.16 Cherenkov detector

Cherenkov effect refers to emission of pulse of visible electromagnetic radiation by charged particles when they pass through an optically transparent dielectric medium at speeds greater than the phase velocity of light in that medium, i.e. a particle emitting Cherenkov radiation must move with velocity

$$v > \frac{c}{n}$$

where n is refractive index of the medium and c is speed of light in free space.

The radiation is anisotropic and is emitted in a direction making angle θ with the incident particle in a forward cone, as indicated in figure 3.15. Suppose in one second a particle goes from P to Q and the radation travels from P to R in the medium under consideration. Then

$$\cos \theta = \frac{PR}{PQ} = \frac{c/n}{v} = \frac{1}{n\beta}$$

where $\beta = \frac{v}{c}$

$$\theta = \cos^{-1} \frac{1}{n\beta}$$

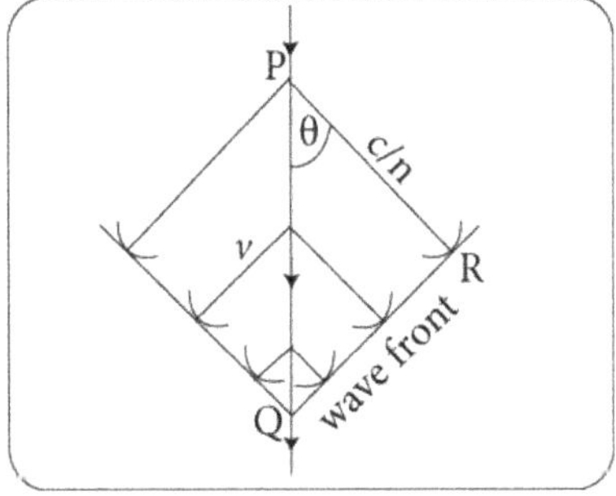

Figure 3.15. Emission of Cherenkov radiation.

The largest value of angle of emission will correspond to

$$v = c, \ \beta = 1$$

$$\theta_{\max} = \cos^{-1}\frac{1}{n}$$

For

$$\theta = 0°, \cos\theta = \cos 0° = 1$$

for which

$$\frac{1}{n\beta_{\text{th}}} = 1 \ \text{ i.e. } \ \beta_{\text{th}} = \frac{1}{n}$$

and the corresponding threshold velocity will be

$$v_{\text{th}} = \frac{c}{n}$$

Clearly θ increases with β such that θ possesses a maximum $\theta_{\max}$ for relativistic particles $\beta = 1$.

Due to passage of a charged particle dielectric medium gets polarized. As the particle moves forward the polarized molecules de-excite and electromagnetic waves are emitted. When the speed of the particle is low the electromagnetic impulses cancel one another by destructive interference. When velocity of the particle exceeds the velocity of light in the medium the optical interference between successive wavelets is constructive in the forward direction. The net electric dipole moment developed for high velocity particles emitting electromagnetic waves in the form of Cherenkov light as the medium returns to its original condition.

If the Cherenkov radiation emitted during passage of an energetic charged particle is collected by a photomultiplier then optical impulses are converted into short-time sharp electrical pulses. This type of device is called a Cherenkov counter. The Cherenkov effect is used for particle identification in particle detectors.

Cerenkov radiation is emitted in a forward cone of angle $\theta = \cos^{-1}\frac{1}{n\beta}$ and hence Cerenkov counters are directional counters. The direction of an incident particle can thus be measured with this counter.

Charged particles above a certain velocity emit Cherenkov radiation. Hence low velocity particles are automatically rejected in the process of counting.

Particles of a given velocity emit radiation along the forward cone of a given definite angle. So by recording Cherenkov light from given angles velocity of an incoming charged particle can be measured employing the relation

$$v = \beta c = \frac{c}{n\cos\theta}$$

3.17 List of particle detectors

Year	Invented by	Name of detector
1911	Wilson	Cloud chamber
1940	Frisch	Ionization chamber
1950	Mckay	Semiconductor detector
1968	Charpak and collaborators	Proportional counter
1908	Geiger	GM counter
1953	Packard	Scintillation counter
1934	Cherenkov	Cherenkov radiation (counter)
1952	Glaser	Bubble chamber
1948	Keuffel	Spark chamber
1910	Kinoshita	Nuclear emulsion detector

3.18 Exercises

Exercise 3.1 *The substance used in a solid state detector is which of the following?*
(a) an alloy (b) an insulator (c) a semiconductor (d) a conductor.

Exercise 3.2 *A Cherenkov counter is useful when velocity v of a particle in dielectric is*
(*a*) equal to $\dfrac{c}{n}$ (*b*) less than $\dfrac{c}{n}$ (*c*) greater than $\dfrac{c}{n}$ (*d*) greater than c.

Exercise 3.3 *Consider a GM counter that collects 10^7 electrons per discharge. What will be the average current if the counting rate is 500 counts/min?*

$\boxed{\text{Ans.}}$ Count rate $= 500 \frac{1}{\min}$

Total number of electrons collected in *1 min* is

$$n = 500 \frac{1}{\min} \times 10^7 = 5 \times 10^9$$

Average current is

$$\frac{\text{charge}}{\text{time}} = \frac{n\,|e|}{t} = \frac{(5 \times 10^9)\,(1.6 \times 10^{-19}\,C)}{60\,s} = 1.33 \times 10^{-11}\,A$$

$\boxed{\text{Answers to multiple choice questions}}$

3.1*c*, 3.2*c*.

3.19 Question bank

Q3.1 Mention the use of nuclear detectors or counters.

Q3.2 What are the various types of detectors used to track and study particles?

Q3.3 What are the characteristics of an ideal detector?

Q3.4 What is the principle of a gas-filled detector? How does it work?

Q3.5 Show in a plot of ionization current versus applied voltage of a chamber the following regions and describe their significance.

Recombination region, ionization region, proportionality region, Geiger–Müller region

Q3.6 An ionization chamber is a gas-filled detector. Mention its principle and its operation. What are its limitations?

Q3.7 Mention why the proportional counter is so called. Describe its principle, construction and working. Calculate gas amplification of the counter. How is the gas multiplication limited by use of methane?

Q3.8 Give the principle, construction and operation of a GM tube. What is quenching?

Q3.9 What is dead time? What is meant by internal and external quenching?

Q3.10 What are the advantages and disadvantages of a GM tube?

Q3.11 What is a semiconductor diode? How does it detect particles? What are the advantages and disadvantages of using a semiconductor diode.

Q3.12 What is the difference between diffused junction semiconductor diode and surface barrier semiconductor diode?

Q3.13 Give a block diagram of a scintillation detector. Describe construction and working of a photomultiplier tube.

Q3.14 How does a cloud chamber work. Describe its success in detecting particles. What are its advantages and limitations?

Q3.15 How does a bubble chamber work. Describe its success in detecting particles. What are its advantages and limitations?

Q3.16 How are momentum and energy of a charged particle determined in a bubble chamber?

Q3.17 What is a spark chamber?

Q3.18 How does a nuclear emulsion detector work? Mention its advantages and disadvantages.

Q3.19 How is the concept of Cherenkov radiation used in a Cherenkov counter?

Further reading

[1] Krane K S 1988 *Introductory Nuclear Physics* (New York: Wiley)

[2] Tayal D C 2009 *Nuclear Physics* (Mumbai: Himalaya Publishing House)

[3] Prakash S 2005 *Nuclear Physics & Particle Physics* (New Delhi: Sultan Chand & Sons)

[4] Guha J 2019 *Quantum Mechanics : Theory, Problems & Solutions* 3rd edn (Kolkata: Books and Allied (P) Ltd)

[5] Lim Y K 2002 *Problems and Solutions on Atomic, Nuclear and Particle Physics* (Singapore: World Scientific)

IOP Publishing

Nuclear and Particle Physics with Cosmology, Volume 2
Particle physics and cosmology
Jyotirmoy Guha

Chapter 4

Units of particle physics

In this chapter we introduce the units used in particle physics. The advantages of natural units are explained. We discuss the Planck system of units as well as reduced Planck units. Planck scale plays an important role in quantum gravity and string theory. Various solved problems based on units used in particle physics are presented.

4.1 Units used in particle physics

The fundamental dimensions studied in physics are those of mass, length and time which have units *g* or *kg*, *cm* or *m* and *s*, respectively. However, these units are arbitrary to some extent—since realistically, a meter is the distance between two marks on a platinum–iridium bar kept in Paris and a kilogram is the mass of a cylinder of platinum–iridium kept in Paris.

The fundamental units of physics are not convenient to use in particle physics. Instead we build fundamental units and dimensions by making the following choice.

In particle physics the system of units preferred is known as natural units.

In natural units we take the speed of light in free space to be unity, i.e.

$$c = 3 \times 10^8 \, \text{m s}^{-1} = 1 \tag{4.1}$$

and the reduced Planck's constant or Dirac's constant to be unity, i.e.

$$\hbar = \frac{h}{2\pi} = \frac{6.626 \times 10^{-34} \, J.\, s}{2\pi} = 1.055 \times 10^{-34} \, J.\, s = 1 \tag{4.2}$$

It is more advantageous to use natural units in particle physics, i.e. take

$$\hbar = c = 1 \tag{4.3}$$

Under such a choice all formulae become simpler as evident from the following discussion.

doi:10.1088/978-0-7503-5032-7ch4

4.2 Advantage of natural units

Consider the relation between energy E and total mass m of a particle, namely

$$E = mc^2 \tag{4.4}$$

With $c = 1$ and writing in terms of dimension we have

$$[E] = [m] \tag{4.5}$$

where [ ] indicates the dimension of the quantity referred to.

✓ So energy and mass have the same dimension.

Again, the relation between energy E, momentum p and rest mass m (let us denote mass also by m) is

$$E = \sqrt{p^2 c^2 + m^2 c^4} \tag{4.6}$$

With $c = 1$

$$E = \sqrt{p^2 + m^2} \tag{4.7}$$

As $[E] = [m]$ (equation 4.5) it follows that p also has the same dimension. So

$$[E] = [m] = [p] \tag{4.8}$$

✓ Hence energy, mass and momentum have the same dimension.

Again

$$E = h\nu = \hbar w \tag{4.9}$$

$$w = \frac{2\pi}{T} \tag{4.10}$$

and so

$$E = \frac{2\pi\hbar}{T} \tag{4.11}$$

With $\hbar = 1$ we have

$$[E] = \left[\frac{2\pi}{T}\right] \sim \frac{1}{[T]} \tag{4.12}$$

This equation (4.12) also follows from the $\hbar$. The unit of $\hbar$ is $J.\,s$ i.e. (energy) (time). So

$$\hbar = [\text{energy}][\text{time}] \tag{4.13}$$

and since $\hbar = 1$ we have

$$[\text{energy}][\text{time}] = 1$$

$$[\text{energy}] = [\text{time}]^{-1}$$

i.e. $[E] = \frac{1}{[T]}$ (equation 4.12)

✓ So energy and time^{-1} have the same dimension.

Again c is velocity, i.e $c = \frac{length}{time}$. Since $c = 1$ we have

$$\frac{[\text{length}]}{[\text{time}]} = 1$$

$$[\text{length}] = [\text{time}] \text{ i. e.}$$

$$[L] = [T] \tag{4.14}$$

✓ So length and time have the same dimension.

Collecting the basic results, i.e. combining equations (4.8), (4.12), (4.14), we have in natural units $\hbar = c = 1$

$$[E] = [m] = [p] = \frac{1}{[T]} = \frac{1}{[L]} \tag{4.15}$$

✓ So energy, mass, momentum, time^{-1}, length^{-1} all have the same dimension.

Clearly there is only one basic dimension and we can take that to be length, i.e. $[L]$. From this all the other dimensions can be derived. This is a great advantage of using natural units since all formulae then get simplified.

- Unit of energy is eV. Unit of mass is $\frac{eV}{c^2} = eV$ (since $c = 1$ in natural units).

  So mass and energy have the same unit eV. Again, $1\,eV = 1.6 \times 10^{-19}J$. Also,

$$MeV = 10^6 eV, \quad GeV = 10^9 eV, \quad TeV = 10^{12} eV \tag{4.16}$$

It follows from equation (4.15) that energy and length have an inverse relationship, $[E] = \frac{1}{[L]}$, and we have shown in *exercise 4.1* that

$$(1\,fm)^{-1} = 197\,MeV \tag{4.17}$$

$$1\,GeV = \frac{1}{0.197}\,fm^{-1} \tag{4.18}$$

In figure 4.1 we have shown some important lengths and their corresponding energies. The details of calculation have been done in *exercise 4.7*.

System	Atom	Nucleon	Length scale reached	Planck scale
Length	10^{-10}m	10^{-15}m	10^{-19}m	10^{-35}m
Energy	2 keV	200 MeV =0.2 GeV	2 TeV =2000GeV	10^{19}GeV

Figure 4.1. Some important length scales and their corresponding energy. Calculation done in *exercise 4.7*.

4.3 Planck system of units

We now discuss what the Planck system of units is. The most basic or fundamental constants of Nature are

$$c = 3 \times 10^8 \, m \, s^{-1} = \text{speed of light in free space}$$

$$\hbar = \frac{h}{2\pi} = \frac{6.626 \times 10^{-34} J. \, s}{2\pi} = 1.055 \times 10^{-34} J. \, s = \text{Dirac's constant or the quantum of action}$$

$$G = 6.67 \times 10^{-11} \, m^3 \, kg^{-1} \, s^{-2} = \text{Newtonian gravitational constant}$$

These constants c, $\hbar$, G can be suitably combined, as shown in *exercise 4.8*, and one can construct the following quantities having dimension of mass, length and time.

4.3.1 Planck mass

$$m_{Pl} = \sqrt{\frac{\hbar c}{G}} = \sqrt{\frac{\left(1.055 \times 10^{-34} J \, s\right)\left(3 \times 10^8 \, m \, s^{-1}\right)}{6.67 \times 10^{-11} \, m^3 \, kg^{-1} \, s^{-2}}} = 2.18 \times 10^{-8} \, kg \qquad (4.19)$$

4.3.2 Planck length

$$l_{Pl} = \sqrt{\frac{G\hbar}{c^3}} = \sqrt{\frac{\left(6.67 \times 10^{-11} \, m^3 \, kg^{-1} \, s^{-2}\right)\left(1.055 \times 10^{-34} \, J \, s\right)}{\left(3 \times 10^8 \, m \, s^{-1}\right)^3}} = 1.61 \times 10^{-35} \, m \qquad (4.20)$$

4.3.3 Planck time

$$t_{Pl} = \sqrt{\frac{G\hbar}{c^5}} = \sqrt{\frac{(6.67 \times 10^{-11} \, m^3 \, kg^{-1} \, s^{-2})(1.055 \times 10^{-34} \, J \, s)}{(3 \times 10^8 \, m \, s^{-1})^5}} = 5.38 \times 10^{-44} \, s \qquad (4.21)$$

We note that

$$\frac{\text{Planck length}}{c} = \frac{l_{Pl}}{c} = \frac{1.61 \times 10^{-35} \, m}{3 \times 10^8 \, m \, s^{-1}} = 5.37 \times 10^{-44} \, s \cong t_{Pl} = \text{Planck time}$$

The set

$$m_{Pl} = \sqrt{\frac{\hbar c}{G}}, \quad l_{Pl} = \sqrt{\frac{G\hbar}{c^3}}, \quad t_{Pl} = \sqrt{\frac{G\hbar}{c^5}} \qquad (4.22)$$

constitute Planck units.

To achieve simplification in writing equation we set

$$\hbar = 1, \quad c = 1, \quad G = 1 \qquad (4.23)$$

in Planck units. Then from equation (4.22) we have

$$m_{Pl} = 1, \ l_{Pl} = 1, \ t_{Pl} = 1 \tag{4.24}$$

Planck scale is the scale of quantum gravity or string theory.

We have shown in *exercises 4.7 and 4.10* that Planck energy is

$$E_{Pl} = \sqrt{\frac{\hbar c^3}{G}} = 10^{19} \, GeV \tag{4.25}$$

4.4 Reduced Planck units

Replacing G by $8\pi G$ in equation (4.22) we have reduced Planck units as discussed in *exercise 4.9*

$$m_{Pl}^{\text{reduced}} = \sqrt{\frac{\hbar c}{8\pi G}}, \ l_{Pl}^{\text{reduced}} = \sqrt{\frac{8\pi G \hbar}{c^3}}, \ t_{Pl}^{\text{reduced}} = \sqrt{\frac{8\pi G \hbar}{c^5}} \tag{4.26}$$

To achieve simplification we set

$$\hbar = 1, \ c = 1, \ 8\pi G = 1 \tag{4.27}$$

in reduced Planck units, i.e. in equation (4.26). Then

$$m_{Pl}^{\text{reduced}} = 1, \ l_{Pl}^{\text{reduced}} = 1, \ t_{Pl}^{\text{reduced}} = 1 \tag{4.28}$$

In fact, setting $8\pi G = 1$ in Einstein's field equation eliminates G.
- We note that $\hbar$ represents quantum mechanics and G represents gravity.
  Actually, c, $\hbar$, G correspond to the natural length scale and time scale where gravity as well as quantum mechanics are expected to play a role.

4.5 Notion of space-time in Planck region

It follows from equations (4.19), (4.20), (4.21) that $m_{Pl} \sim 10^{-8} kg$, $l_{Pl} \sim 10^{-35} m$, $t_{Pl} \sim 10^{-44} s$. Actually, we do not know what occurs below these limits because concepts of mass, length, time may not even be definable continuously below these limits (up to zero). So the notion of space-time breaks down beyond this limit.

The physical laws we work with may not operate below these limits. So there is problem of quantizing gravity.

The shortest distance that can be experimentally probed is $\sim 10^{-19} \, m$, the shortest time scale that has been dealt with is $\sim 10^{-23} \, s$ which is much above the l_{Pl}, t_{Pl} values, respectively (equations 4.20, 4.21).

Planck scale plays a very important role in quantum gravity and string theory.

4.6 Exercises

Exercise 4.1 *Prove that* $(1\,fm)^{-1} = 197\,MeV$, $1\,GeV = \frac{1}{0.197}\,fm^{-1}$

$\boxed{\text{Ans.}}$ Let us consider the quantity $\hbar c$ and evaluate its value

$$\hbar c = \frac{6.626 \times 10^{-34}J.\,s}{2\pi}(3 \times 10^8\,m\,s^{-1}) = 3.16 \times 10^{-26}J.\,m$$

$$\hbar c = \frac{3.16 \times 10^{-26}eV.\,10^{15}\,fm}{1.6 \times 10^{-19}} = 197.5 \times 10^6\,eV.\,fm$$

$$\hbar c = 197\,MeV.\,fm \tag{4.29}$$

In natural units $\hbar = c = 1$

$$197\,MeV.\,fm = 1 \tag{4.30}$$

$(1\,fm)^{-1} = 197\,MeV$ (equation 4.17)

Again from equation (4.30)

$$1\,eV = \frac{1}{197\,M.\,fm} = \frac{1}{197 \times 10^6.\,fm}$$

$$1\,GeV = 10^9\,eV = 10^9\,\frac{1}{197 \times 10^6\,fm} = \frac{1}{0.197fm}$$

$1\,GeV = \frac{1}{0.197}\,fm^{-1}$ (equation 4.18)

Exercise 4.2 *Show that* $(1\,GeV)^{-2} = 0.388\,mb$ $(b = barn)$, $1\,mb = 2.58\,GeV^{-2}$.

$\boxed{\text{Ans.}}$ From *exercise 4.1* we get

$$1\,GeV = \frac{1}{0.197}\,fm^{-1}$$

$$(1\,GeV)^{-2} = \left(\frac{1}{0.197}\,fm^{-1}\right)^{-2} = \left(\frac{1}{0.197}\right)^{-2}fm^2$$

$$= 0.0388\,fm^2 = 0.0388 \times (10^{-15}m)^{\,2} = 0.0388 \times 10^{-30}\,m^2$$

$$(1\,GeV)^{-2} = 0.388 \times 10^{-3} \times 10^{-28}\,m^2$$

As $1b = 1\,barn = 10^{-28}\,m^2$, $10^{-3} = milli \equiv m$ we can write

$$(1\,GeV)^{-2} = 0.388\,mb \tag{4.31}$$

$$1\,mb = \frac{(1\,GeV)^{-2}}{0.388} = 2.58\,GeV^{-2} \tag{4.32}$$

Exercise 4.3 *Show that* $1\,GeV = 5\,fm^{-1}$, $1\,fm = 5\,GeV^{-1}$.

$\boxed{\text{Ans.}}$ From equation (4.17) viz. $(1\,fm)^{-1} = 197\,MeV$ we have

$$1\,fm = \frac{1}{197\,MeV}$$

$$1\,MeV = \frac{1}{197\,fm} \tag{4.33}$$

$$1\,GeV = 10^9\,eV = 10^3\,MeV = 10^3\frac{1}{197\,fm}$$

$$1\,GeV = \frac{5}{fm} = 5\,fm^{-1} \tag{4.34}$$

$$1\,fm = \frac{5}{GeV} = 5\,GeV^{-1} \tag{4.35}$$

Exercise 4.4 *Show that* $1\,fm^2 = 9.7\,mb$.
 Ans. From *exercise 4.3* we have

$$1\,fm = 5\,GeV^{-1}$$

Hence

$$1\,fm^2 = (5\,GeV^{-1})^2 = 25\,GeV^{-2}$$

Using equation (4.31) i.e. $(1\,GeV)^{-2} = 0.388\,mb$ we have

$$1\,fm^2 = 25(0.388\,mb\,) = 9.7\,mb$$

Exercise 4.5 *What is the Compton wavelength of a proton and a pion in natural units?*
 Ans. Compton wavelength of a particle of mass m is

$$\lambda = \frac{h}{mc}$$

and so using $\lambdabar = \frac{\lambda}{2\pi}$, $\hbar = \frac{h}{2\pi}$ we have

$$\lambdabar = \frac{\hbar}{mc}$$

In natural units $\hbar = c = 1$

$$\lambdabar = \frac{1}{m}$$

✓ For a proton ($m_p = 938\,MeV$) the Compton wavelength is

$$\lambdabar_p = \frac{1}{m_p} = \frac{1}{938\,MeV}$$

Using equation (4.33) i.e. $1\,MeV = \frac{1}{197\,fm}$ we have

$$\lambda_p = \frac{1}{938\,MeV} = \frac{197\,fm}{938} = 0.21\,fm$$

✓ For a pion ($m_\pi = 140\,MeV$) the Compton wavelength is

$$\lambda_\pi = \frac{1}{m_\pi} = \frac{1}{140\,MeV}$$

Using equation (4.33), i.e. $1\,MeV = \frac{1}{197\,fm}$, we have

$$\lambda_\pi = \frac{1}{140\,MeV} = \frac{197\,fm}{140} = 1.41\,fm.$$

Exercise 4.6 *What is the Compton wavelength of a particle of mass–energy* $197\,MeV$*?*
[Ans.] A particle having mass–energy $mc^2 = 197\,MeV$ has a Compton wavelength of $\lambda = \frac{\hbar}{mc} = \frac{\hbar c}{mc^2} = \frac{197\,MeV.fm}{197\,MeV} = 1fm$ (using equation (4.30))

Exercise 4.7 *Refer to* figure 4.1. *Verify the data given in the figure.*
[Ans.] We use the result of *exercise 4.1* viz. $(1\,fm)^{-1} = 197\,MeV$ i.e. $\frac{1}{1\,fm} = 197\,MeV$ or

$$1\,fm = \frac{1}{197\,MeV} \sim \frac{1}{200\,MeV}\ \text{(take)}$$

✓ For an atom the characteristic length is

$$1\,\text{Å} = 10^{-10}\,m = 10^5 10^{-15}\,m = 10^5\,fm$$
$$= 10^5 \frac{1}{200\,MeV} = \frac{1}{10^{-5}\times 200\times 10^6\,eV} = \frac{1}{2000\,eV} = \frac{1}{2\,keV}$$

So energy corresponding to length 1 Å is $2\,keV$.
✓ For a nucleon the characteristic length is

$$1\,fm = \frac{1}{200\,MeV} = \frac{1}{0.2\times 10^3\,MeV} = \frac{1}{0.2\,GeV}$$

So energy corresponding to length $1\,fm$ is $200\,MeV = 0.2\,GeV$
✓ Length scale reached is

$$10^{-19}\,m = 10^{-4} 10^{-15}\,m = 10^{-4}\,fm$$
$$= 10^{-4}\frac{1}{200\,MeV} = \frac{1}{10^4\times 200\times 10^6\,eV} = \frac{1}{2\times 10^{12}\,eV} = \frac{1}{2\,TeV} = \frac{1}{2000\times 10^9\,eV} = \frac{1}{2000\,GeV}$$

So energy corresponding to length $10^{-19}m$ is $2\,TeV = 2000\,GeV$.

✓ Characteristic length of Planck scale is

$$10^{-35}\,m = 10^{-20}10^{-15}\,m = 10^{-20}\,fm$$

$$= 10^{-20}\frac{1}{200\,MeV} = \frac{1}{10^{20}\times 200\times 10^6\,eV} = \frac{1}{2\times 10^{19}\times 10^9\,eV}$$

$$\sim \frac{1}{10^{19}\,GeV}$$

So energy corresponding to length $10^{-35}\,m$ is $10^{19}\,GeV$ (equation 4.25). This explains figure 4.1.

Exercise 4.8 *The fundamental constants of nature are $\hbar$, c and G. Construct from these, quantities having dimension of mass, length and time.*

Ans. We can construct quantities having dimensions of mass, length and time from the fundamental constants of Nature viz. $\hbar$, c and G as follows.

- Let us construct a quantity m_{Pl} having dimension of mass as

$$m_{Pl} = \hbar^{\alpha}c^{\beta}G^{\gamma} \tag{4.36}$$

We are to find the values of α, β, γ through dimensional analysis.

We note that

$$\hbar \sim (\text{energy})(\text{time})$$

$$c \sim (\text{length})(\text{time})^{-1} \text{ and}$$

$$\text{force} \sim G\frac{(\text{mass})^2}{(\text{length})^2} \text{ i.e. } G \sim [\,MLT^{-2}.\,L^2.\,M^{-2}] = [L^3M^{-1}T^{-2}]$$

Let us put the dimensions in equation (4.36) to get

$$M = (MLT^{-2}L.\,T)^{\alpha}(LT^{-1})^{\beta}(L^3M^{-1}T^{-2})^{\gamma}$$

$$M = L^{2\alpha+\beta+3\gamma}M^{\alpha-\gamma}T^{-\alpha-\beta-2\gamma}$$

$$2\alpha + \beta + 3\gamma = 0$$

$$\alpha - \gamma = 1$$

$$-\alpha - \beta - 2\gamma = 0$$

Solving, we have

$$\alpha = \tfrac{1}{2}, \beta = \tfrac{1}{2}, \gamma = -\tfrac{1}{2}$$

Hence equation (4.36) gives

$$m_{Pl} = \sqrt{\tfrac{\hbar c}{G}} = 2.18 \times 10^{-8}\,kg \text{ (by equation 4.19)}$$

- Let us construct a quantity l_{Pl} having dimension of length as

$$l_{Pl} = \hbar^\alpha c^\beta G^\gamma \tag{4.37}$$

We are to find the values of α, β, γ through dimensional analysis. Putting dimensions in equation (4.37) we have

$$L = (MLT^{-2}L.\ T)^\alpha (LT^{-1})^\beta (L^3 M^{-1} T^{-2})^\gamma$$

$$L = L^{2\alpha+\beta+3\gamma} M^{\alpha-\gamma} T^{-\alpha-\beta-2\gamma}$$

$$2\alpha + \beta + 3\gamma = 1$$

$$\alpha - \gamma = 0$$

$$-\alpha - \beta - 2\gamma = 0$$

Solving, we have

$$\alpha = \tfrac{1}{2},\ \beta = -\tfrac{3}{2},\ \gamma = \tfrac{1}{2}$$

Hence from equation (4.37) we arrive at

$$l_{Pl} = \sqrt{\frac{G\hbar}{c^3}} = 1.61 \times 10^{-35}\, m \text{ (by equation 4.20)}$$

- Let us construct a quantity t_{Pl} having dimension of time as

$$t_{Pl} = \hbar^\alpha c^\beta G^\gamma \tag{4.38}$$

We are to find the values of α, β, γ through dimensional analysis.

$$T = (MLT^{-2}L.\ T)^\alpha (LT^{-1})^\beta (L^3 M^{-1} T^{-2})^\gamma$$

$$T = L^{2\alpha+\beta+3\gamma} M^{\alpha-\gamma} T^{-\alpha-\beta-2\gamma}$$

$$2\alpha + \beta + 3\gamma = 0$$

$$\alpha - \gamma = 0$$

$$-\alpha - \beta - 2\gamma = 1$$

Solving, we have

$$\alpha = \tfrac{1}{2},\ \beta = -\tfrac{5}{2},\ \gamma = \tfrac{1}{2}$$

Hence from equation (4.38) we arrive at

$$t_{Pl} = \sqrt{\frac{G\hbar}{c^5}} = 5.38 \times 10^{-44} s \text{ (by equation 4.21)}$$

Exercise 4.9 *Find the reduced Planck units. (Hint: Replace G by $8\pi G$ in Planck units.)*

$\boxed{\text{Ans.}}$ Planck units are given in equations (4.19), (4.20), (4.21) as

$$m_{Pl} = \sqrt{\frac{\hbar c}{G}}, \quad l_{Pl} = \sqrt{\frac{G\hbar}{c^3}}, \quad t_{Pl} = \sqrt{\frac{G\hbar}{c^5}}$$

Replacing G by $8\pi G$ we have reduced the Planck units as follows.

$$m_{Pl}^{\text{reduced}} = \sqrt{\frac{\hbar c}{8\pi G}} = \sqrt{\frac{1}{8\pi}}\sqrt{\frac{\hbar c}{G}} = \sqrt{\frac{1}{8\pi}}\, m_{Pl} = \sqrt{\frac{1}{8\pi}}(2.18 \times 10^{-8}\, kg) = .44 \times 10^{-8}\, kg$$

$$l_{Pl}^{\text{reduced}} = \sqrt{\frac{8\pi G\hbar}{c^3}} = \sqrt{8\pi}\sqrt{\frac{G\hbar}{c^3}} = \sqrt{8\pi}\, l_{Pl} = \sqrt{8\pi}(1.61 \times 10^{-35}\, m) = 8.07 \times 10^{-35}\, m$$

$$t_{Pl}^{\text{reduced}} = \sqrt{\frac{8\pi G\hbar}{c^5}} = \sqrt{8\pi}\sqrt{\frac{G\hbar}{c^5}} = \sqrt{8\pi}\, t_{Pl} = \sqrt{8\pi}(5.38 \times 10^{-44}\, s) = 27 \times 10^{-44}\, s$$

Exercise 4.10 *The fundamental constants of Nature are $\hbar$, c and G. Express Planck energy E_{Pl} in their terms and find its value.*

$\boxed{\text{Ans.}}$ Let us express energy E_{Pl} in terms of the fundamental constants of nature viz. $\hbar$, c, G as

$$E_{Pl} = \hbar^\alpha c^\beta G^\gamma \tag{4.39}$$

We are to find the values of α, β, γ through dimensional analysis. Putting dimensions in equation (4.39) we have

$$MLT^{-2}.\, L = (MLT^{-2}L.\, T)^\alpha (LT^{-1})^\beta (L^3 M^{-1}T^{-2})^\gamma$$

$$L^2 MT^{-2} = L^{2\alpha+\beta+3\gamma} M^{\alpha-\gamma} T^{-\alpha-\beta-2\gamma}$$

$$2\alpha + \beta + 3\gamma = 2$$

$$\alpha - \gamma = 1$$

$$-\alpha - \beta - 2\gamma = -2$$

Solving, we have

$$\alpha = \frac{1}{2}, \ \beta = \frac{5}{2}, \ \gamma = -\frac{1}{2}$$

Hence from equation (4.39) we get

$$E_{Pl} = \sqrt{\frac{\hbar c^5}{G}} = \sqrt{\frac{\frac{1}{2\pi}(6.626 \times 10^{-34} J.s)(3 \times 10^8 ms^{-1})^5}{6.67 \times 10^{-11} N.m^2\, kg^{-2}}} = 1.96 \times 10^9 \sqrt{\frac{N.m.s.m^5.s^{-5}}{N.m^2.kg^{-2}}}$$

$$= 1.96 \times 10^9 kg\, m^2 s^{-2} = 1.96 \times 10^9 J = \frac{1.96 \times 10^9 eV}{1.6 \times 10^{-19}} = 1 \times 10^{28} eV = 10^{19} 10^9 eV$$

$$E_{Pl} = 10^{19} GeV \ (\text{equation } 4.25)$$

Exercise 4.11 *The choice $\hbar = c = 8\pi G = 1$ is called which of the following?*

(a) S.I unit   *(b) natural unit*   *(c) reduced Planck unit*   *(d) Planck unit*

Exercise 4.12 *Which of the following is correct in natural units?*

(a) $[E] = [m] = [p] = \frac{1}{[T]} = [L]$ *(b)* $[E] = [m] = \frac{1}{[p]} = \frac{1}{[T]} = \frac{1}{[L]}$

(c) $[E] = [m] = [p] = [T] = \frac{1}{[L]}$ *(d)* $[E] = [m] = [p] = \frac{1}{[T]} = \frac{1}{[L]}$

Exercise 4.13 *Which of the following is correct in natural units?*

(a) $(1\,fm)^{-1} = 197\,MeV$ *(b)* $1\,GeV = \frac{1}{0.197}\,fm^{-1}$

(c) $(1\,fm)^{-1} = \frac{1}{0.197}\,MeV$ *(d)* $1\,GeV = 197\,fm^{-1}$

Exercise 4.14 *Which of the following is correct in natural units?*

(a) $1\,GeV = 5\,fm^{-1}$ *(b)* $1\,fm^2 = 9.7\,mb$

(c) $1\,fm = 5\,GeV^{-1}$ *(d)* $1\,fm^2 = 25\,GeV^{-2}$

Exercise 4.15 *Planck scale corresponds to which value of energy?*

(a) $10^{19}\,MeV$ *(b)* $10^{15}\,GeV$

(c) $10^{19}\,GeV$ *(d)* $10^{15}\,MeV$

Exercise 4.16 *Which of the following is/are true regarding Planck length?*

(a) $1.61 \times 10^{-35}\,m$ *(b) c* (Planck time)

(c) $\sqrt{\frac{G\hbar}{c^3}}$ *(d)* $\sqrt{\frac{G\hbar}{c^5}}$

Answers to multiple choice questions
4.11*c*, 4.12*d*, 4.13*a,b*, 4.14*a,b,c,d*, 4.15*c*, 4.16*a,b,c*.

4.7 Question bank

Q 4.1 What are natural units? What is the advantage of using natural units?
Q 4.2 Show that in natural units energy and time^{-1} are of the same dimension.

Q 4.3 Show that in natural units energy, mass and momentum are of the same dimension.

Q 4.4 Show that in natural units length and time are of the same dimension.

Q 4.5 What is the justification of introducing natural units in particle physics?

Q 4.6 Mention the length scale reached and its corresponding energy.

Q 4.7 What is the characteristic length and energy associated with an atom and a nucleus?

Q 4.8 Define Planck scale and where it is used.

Q 4.9 What is the justification of setting $8\pi G = 1$ in reduced Planck units?

Q 4.10 Where does the notion of space-time break down?

Q 4.11 Mention the scale or region where quantum gravity or string theory is discussed.

Further reading

[1] Perkins H 2000 *Introduction to High Energy Physics* (Cambridge: Cambridge University Press)

[2] Griffiths D 2008 *Introduction to Elementary Particles* (New York: Wiley-VCH)

[3] Tayal D C 2009 *Nuclear Physics* (Mumbai: Himalaya Publishing House)

[4] Satya P 2005 *Nuclear Physics & Particle Physics* (New Delhi: Sultan Chand & Sons)

[5] Guha J 2019 *Quantum Mechanics: Theory, Problems & Solutions* 3rd edn (Kolkata: Books and Allied (P) Ltd)

[6] Yung K L 2002 *Problems and Solutions on Atomic, Nuclear and Particle Physics* (Singapore: World Scientific)

IOP Publishing

Nuclear and Particle Physics with Cosmology, Volume 2
Particle physics and cosmology
Jyotirmoy Guha

Chapter 5

A brief introduction to the Standard Model

We introduce various aspects of the Standard Model. We discuss the fundamental forces of nature viz. the strong, electromagnetic and weak interaction. The aspect of unification of forces is mentioned. Elementary fermions like leptons and quarks are introduced. We describe the leptons like the electron, electron neutrino, muon, muon neutrino, tauon, tauon neutrino. We introduce the eightfold way. We define colour charge. We describe the quarks like up, down, strange, charm, bottom and top quark. Gauge bosons as force carriers are described like photons, intermediate vector bosons, gluons. The Higgs boson and Higgs mechanism and the concept of field in the Standard Model are elaborated.

5.1 The Standard Model

The Standard Model has emerged as the standard theory describing nature though it describes three fundamental interactions, i.e. it describes three forces of nature viz. strong interaction, electromagnetic interaction, weak interaction. The Standard Model does not include gravitational interaction.

Various research works and improvement of accelerators led to the discovery of numerous different particles and we needed a structure, pattern or a mechanism to arrange the particles in an ordered way. And the Standard Model classifies all the fundamental elementary particles.

The model explains how material particles interact through force exchange.

The particles of the Standard Model as well as the force carriers are listed in figure 5.1.

In fact, most of the known facts of particle physics have been tied up into a package called the Standard Model.

Let us present a very brief introduction to the various interactions and the various particles of the Standard Model.

The Standard Model of particle physics was developed in stages through the work of several scientists around the globe and was finalized around 1970 after

　　　

	Fermions Half-integer spin Obey F-D statistics 3 generations of matter			Bosons Integer spin Obey B-E statistics Force carriers	
	1st generation	2nd generation	3rd generation	Vector bosons Gauge bosons 3 kinds viz. g, γ and $W^{\pm}, Z^0$	Scalar bosons (one kind)
Quarks (Have color charge) (electro weak int.)	u 1968 Stanford linear Accelerator Center $m=2.2$ MeV/c^2 $Q=\frac{2}{3}, s=\frac{1}{2}$	c 1974 Stanford linear Accelerator Center, Brookhaven National Laboratory $m=1.28$ GeV/c^2 $Q=\frac{2}{3}, s=\frac{1}{2}$	t 1995 Fermilab $m=173$ GeV/c^2 $Q=\frac{2}{3}, s=\frac{1}{2}$	g 1979 PETRA collider $m=0$ $Q=0, s=1$ 8 types of gluons strong int	H^0 2012 CERN, LHC $m=125$ GeV/c^2 $Q=0, s=0$ Gives mass to particles
	d 1968 Stanford linear Accelerator Center $m=4.7$ MeV/c^2 $Q=-\frac{1}{3}, s=\frac{1}{2}$	s 1968 Stanford linear Accelerator Center $m=96$ MeV/c^2 $Q=-\frac{1}{3}, s=\frac{1}{2}$	b 1977 Fermilab $m=4.18$ GeV/c^2 $Q=-\frac{1}{3}, s=\frac{1}{2}$	γ 1905 Work of Einstein $m=0$ $Q=0, s=1$ em int	
Leptons (no color charge) (strong and electroweak int.)	e 1897 J J Thomson $m=0.51$ MeV/c^2 $Q=-1, s=\frac{1}{2}$	μ 1937 Anderson, Neddermeyer $m=106$ MeV/c^2 $Q=-1, s=\frac{1}{2}$	τ 1978 Stanford linear Accelerator Center $m=1.78$ GeV/c^2 $Q=-1, s=\frac{1}{2}$	W 1983 CERN $m=80$ GeV/c^2 $Q=\pm1, s=1$ weak int	
	ν_e 1930 Pauli 1956 Cowan, Reines $m< 1$ eV/c^2 $Q=0, s=\frac{1}{2}$	ν_μ 1962 Brookhaven National Lab Lederman, Schwartz and Steinberger $m<0.17$ eV/c^2 $Q=0, s=\frac{1}{2}$	ν_τ 2000 DONUT expt Fermilab $m< 18$ MeV/c^2 $Q=0, s=\frac{1}{2}$	Z^0 1983 CERN $m=91$ GeV/c^2 $Q=0, s=1$ weak int	

Figure 5.1. The fundamental particles of the Standard Model showing symbol, year of discovery, discoverer or the place of discovery and the particles' mass, charge and spin.

experimental confirmation of the existence of quarks. Later experimental confirmation of top quark (1995), tau neutrino (2000) and Higgs boson (2012) further strengthened our faith on the standard model.

In this chapter, we introduce the various statements of the Standard Model very briefly. In later chapters we will describe them more elaborately.

5.2 Fundamental forces of nature

There are four fundamental forces of nature. They govern the entire Universe. We briefly introduce them here and will describe them in greater detail in later chapters.

5.3 Strong interaction

It is the strongest of all forces known to us. It is a force that binds the quarks to form the proton and neutron. Again, proton and neutron form the nucleus. Beyond a distance of 10^{-15} m strong force is zero. So it has a short range.

5.4 Electromagnetic interaction

This is $\frac{1}{137}$ times the strong force, i.e. 137 times weaker than the strong force. The factor $\frac{1}{137}$ is called fine structure constant. Electromagnetism binds nuclei to the electron and forms the atom. Atoms are bound to form molecules. Molecules bind together to form solid, liquid and gas which are the various forms of matter. Forces involved in elasticity, viscosity and surface tension are all manifestations of electromagnetism. It always acts and is never zero. In other words it is a long-range interaction. It has infinite range as evident from $\frac{1}{r^2}$ dependence of Coloumb force.

5.5 Weak interaction

Weak force is a disruptive force as it causes decay of all elementary particles and nuclei.

Weak interaction plays another major role in the Universe. The thermonuclear fusion reaction which powers the Sun is due to weak interaction.

It is a short-range interaction having range $10^{-17}\,m$.

5.6 Gravitational interaction (not part of the Standard Model)

Gravity governs the Universe at large. Gravity binds the planets to the Sun. It binds the stars to form Galaxies. Gravity binds the galaxies to form galactic clusters and is responsible for large-scale phenomena of the Universe. It has infinite range because of Newton's law of gravitational force, which is $\sim\frac{1}{r^2}$.

We note that electromagnetism is stronger than gravity. But electromagnetic force can be made zero by accumulating enough number of positive and negative charges in equal numbers. This is because electromagnetism has both attraction and repulsion. But gravity is always attractive. So the higher the mass the greater the attraction.

5.7 Unification of forces

According to the Standard Model, electromagnetism and weak force do not stand alone—they get unified into the electroweak force, as sketched in figure 5.2.

The electroweak force and strong force are understood through the Standard Model.

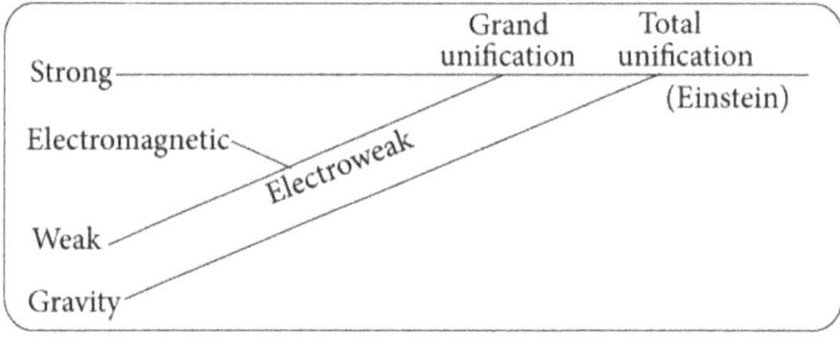

Figure 5.2. Unification of forces.

The electroweak force and the strong force are alike. Many theories attempt at unification of the electroweak force and strong force, which is called grand unification.

In this theory the strong force operating within the nuclei and within the nucleons as well as the weak forces (which is the cause of the phenomenon called radioactivity) can be looked upon as a generalization of the electrodynamics of Faraday and Maxwell.

However, it requires clinching experimental evidence to be confirmed.

The dream of Einstein was to unify all the forces—achieving total unification. One thus aspires to unify gravity also.

The Standard Model is consistent with quantum mechanics, i.e. understood quantum mechanically. But gravity has not been understood quantum mechanically. In other words, there is no quantum gravity.

5.8 Elementary fermions as constituents of matter

The constituents of matter are called fermions. These particles have half-integral spins $(\frac{1}{2}, \frac{3}{2}, ...)$. They obey Fermi–Dirac statistics. They are of two types: leptons and quarks. We present a brief discussion on them now.

5.9 Leptons

We cite the various leptons, referring to their discovery and properties.

Leptons are characterized by a quantum number called lepton quantum number L which is of three types.

    a) L_e is electron lepton quantum number;
    b) L_μ is muon lepton quantum number;
    c) L_τ is tauon lepton quantum number.

For non-leptons, lepton quantum numbers L_e, L_μ, L_τ are all zero.
In all interactions, lepton numbers L_e, L_μ, L_τ are all conserved.
Leptons have zero colour charge.
Leptons take part in weak interactions and also in electromagnetic interaction if they are charged.
Leptons do not take part in strong interaction.
We discuss more about leptons in chapter 7.

5.10 Electron

The search for the carrier of electrical properties in matter and experiments with cathode rays culminated in the discovery of the electron in 1897 by J J Thomson.

The electron was the first fundamental particle to be discovered. And so it is referred to as the first generation lepton.

We discuss what we mean by generation of particles in *exercise 5.5*.

The electron e^- has electrical charge $Q = -1\ |e|$. We denote $\frac{Q}{|e|} \equiv Q_{e^-} = -1$.

The electron e^- has mass $m_{e^-} = 0.51 \frac{MeV}{c^2}$.

The electron e^- is a fermion with spin $s = \frac{1}{2}$.

The electron e^- has no colour charge (we will discuss colour quantum number later).

The anti-particle of the electron e^- is the positron e^+ $(m_{e^+} = 0.51\frac{MeV}{c^2}, \; Q_{e^+} = +1)$.

The electron e^- is characterized by the following lepton quantum numbers

$$L_e(e^-) = +1, \quad L_\mu(e^-) = 0, \quad L_\tau(e^-) = 0.$$

Lepton quantum numbers of the positron e^+ are

$$L_e(e^+) = -1, \quad L_\mu(e^+) = 0, \quad L_\tau(e^+) = 0$$

Clearly electron e^-, positron e^+ are characterized by electron lepton number L_e.

Stability of the electron follows from charge conservation principle as has been proved in chapter 8 section 8.17.

5.11 Electron neutrino

The electron neutrino was theorized by Pauli in 1930 to account for the missing momentum and energy in beta decay.

In 1956 Reines and Cowan discovered the electron neutrino in their experiment.

The electron neutrino ν_e is an elementary particle in the first generation of leptons.

The electron neutrino ν_e has $Q_{\nu_e} = 0$.

The electron neutrino ν_e has a small but non-zero mass $m_{\nu_e} < 1\frac{eV}{c^2}$.

The electron neutrino ν_e is a fermion with spin $s = \frac{1}{2}$.

The electron neutrino ν_e has zero colour charge.

The anti-particle of the electron neutrino ν_e is the electron anti-neutrino $\bar{\nu}_e$.

The electron neutrino ν_e is characterized by the lepton quantum numbers

$$L_e(\nu_e) = +1, \quad L_\mu(\nu_e) = 0, \quad L_\tau(\nu_e) = 0.$$

Lepton quantum numbers of the electron anti-neutrino $\bar{\nu}_e$ are

$$L_e(\bar{\nu}_e) = -1, \quad L_\mu(\bar{\nu}_e) = 0, \quad L_\tau(\bar{\nu}_e) = 0.$$

✓ Whenever a positron e^+ is emitted it is associated with emission of an electron neutrino ν_e to conserve electron lepton number L_e since then

$$L_e(\text{interaction}) = L_e(e^+) + L_e(\nu_e) = -1 + 1 = 0.$$

✓ Whenever an electron e^- is emitted it is associated with the emission of an electron anti-neutrino $\bar{\nu}_e$ to conserve electron lepton number L_e since then

$$L_e(\text{interaction}) = L_e(e^-) + L_e(\bar{\nu}_e) = +1 - 1 = 0.$$

Since ν_e, $\bar{\nu}_e$ are associated with the emission of e^+, e^-, respectively, ν_e, is named an electron neutrino and $\bar{\nu}_e$ is named an electron anti-neutrino.

Neutrinos travel with nearly the speed of light and are stable.

The electron neutrino ν_e is left-handed. The electron anti-neutrino $\bar{\nu}_e$ is right-handed.

We will discuss helicity or handedness in chapter 8 section 8.44 and in chapter 11 section 11.29. Chirality is defined in chapter 11 section 11.30.

5.12 Muon

The muon was discovered by Anderson and Neddermeyer in 1937 while studying cosmic radiation.

Muon properties are similar to that of the electron but it is heavier. So the muon is referred to as a second generation lepton.

The negative muon μ^- has electrical charge $Q = -|e|$ i.e. $\frac{Q}{|e|} \equiv Q_{\mu^-} = -1$.

The negative muon μ^- has mass $m_{\mu^-} = 106\frac{MeV}{c^2}$.

The muon is a fermion with spin $s = \frac{1}{2}$.

The muon has zero colour charge.

The anti-particle of the negative muon μ^- is the positive muon μ^+ $(m_{\mu^+} = 106\frac{MeV}{c^2}, Q_{\mu^+} = +1)$.

The negative muon μ^- is characterized by lepton quantum numbers

$$L_e(\mu^-) = 0, \ L_\mu(\mu^-) = +1, \ L_\tau(\mu^-) = 0.$$

The lepton numbers of the positive muon μ^+ are

$$L_e(\mu^+) = 0, \ L_\mu(\mu^+) = -1, \ L_\tau(\mu^+) = 0.$$

Muon lifetime is $2 \times 10^{-6}s \equiv 2\mu s$.

The decay scheme of the muon is discussed in chapter 6 (*exercise 6.3*).

5.13 Muon neutrino

The muon neutrino was discovered in 1962 by Lederman, Schwartz and Steinberger in an experiment at Brookhaven National Laboratory.

The muon neutrino ν_μ is an elementary particle in the second generation of leptons.

The muon neutrino ν_μ has electrical charge $Q_{\nu_\mu} = 0$.

The muon neutrino ν_μ has a small but non-zero mass $m_{\nu_\mu} < 0.17\frac{eV}{c^2}$.

The muon neutrino ν_μ is a fermion with spin $s = \frac{1}{2}$.

The muon neutrino ν_μ has zero colour charge.

The anti-particle of the muon neutrino ν_μ is the muon anti-neutrino $\bar{\nu}_\mu$.

The muon neutrino ν_μ is characterized by the lepton quantum numbers

$$L_e(\nu_\mu) = 0, \ L_\mu(\nu_\mu) = +1, \ L_\tau(\nu_\mu) = 0.$$

Lepton quantum numbers of the muon anti-neutrino $\bar{\nu}_\mu$ are

$$L_e(\bar{\nu}_\mu) = 0, \ \ L_\mu(\bar{\nu}_\mu) = -1, \ \ L_\tau(\bar{\nu}_\mu) = 0.$$

Whenever a positive muon μ^+ is emitted it is associated with the emission of a muon neutrino ν_μ to conserve muon lepton number L_μ since then

$$L_\mu(\text{interaction}) = L_\mu(\mu^+) + L_\mu(\nu_\mu) = -1 + 1 = 0.$$

Whenever a negative muon μ^- is emitted it is associated with the emission of an electron anti-neutrino $\bar{\nu}_\mu$ to conserve muon lepton number L_μ since then

$$L_\mu(\text{interaction}) = L_\mu(\mu^-) + L_\mu(\bar{\nu}_\mu) = +1 - 1 = 0.$$

Since $\nu_\mu, \bar{\nu}_\mu$ are associated with the emission of μ^+, μ^-, respectively, ν_μ is named the muon neutrino and $\bar{\nu}_\mu$ is named the muon anti-neutrino.

Neutrinos travel with nearly the speed of light and are stable.

The muon neutrino ν_μ is left-handed. The muon anti-neutrino $\bar{\nu}_\mu$ is right-handed.

5.14 Tauon

The tauon was anticipated by Tsai in 1971 and was discovered by Martin Lewis Perl *et al* in 1978 at Stanford Linear Accelerator Center (SLAC).

The tauon is hugely massive compared to masses of electron and muon. So tauon is referred to as third generation lepton.

The negative tauon τ^- has electrical charge $Q = -|e|$ i.e. $\frac{Q}{|e|} \equiv Q_{\tau^-} = -1$.

The negative tauon τ^- is the heaviest lepton having mass $m_{\tau^-} = 1.78\frac{GeV}{c^2}$.

The negative tauon τ^- is a fermion with spin $s = \frac{1}{2}$.

The negative tauon τ^- has no colour charge.

The anti-particle of the negative tauon τ^- is the positive tauon τ^+ $(m_{\tau^+} = 1.78\frac{GeV}{c^2}, \ Q_{\tau^+} = +1)$.

The negative tauon τ^- is characterized by lepton quantum numbers

$$L_e(\tau^-) = 0, \ \ L_\mu(\tau^-) = 0, \ L_\tau(\tau^-) = +1.$$

The lepton numbers of positive tauon τ^+ are

$$L_e(\tau^+) = 0, \ \ L_\mu(\tau^+) = 0, \ L_\tau(\tau^+) = -1.$$

Tauon lifetime is $2.9 \times 10^{-13} \, s$.

5.15 Tauon neutrino

The tauon neutrino was discovered in 2000 by the DONUT (Direct Observation of the Nu Tau) collaboration experiment conducted at Fermi National Accelerator laboratory (Fermilab).

The tauon neutrino ν_τ is an elementary particle in the third generation of leptons.

The tauon neutrino ν_τ has $Q_{\tau^-} = 0$.

The tauon neutrino ν_τ has a mass ($m_{\nu_{\tau^-}} < 18\frac{MeV}{c^2}$).

The tauon neutrino ν_τ is a fermion with spin $s = \frac{1}{2}$.
The tauon neutrino ν_τ has zero colour charge.
The anti-particle of the tauon neutrino ν_τ is the tauon anti-neutrino $\bar{\nu}_\tau$.
The tauon neutrino ν_τ is characterized by lepton quantum numbers

$$L_e(\nu_\tau) = 0, \quad L_\mu(\nu_\tau) = 0, \quad L_\tau(\nu_\tau) = +1.$$

Lepton quantum numbers of the tauon anti-neutrino $\bar{\nu}_\tau$ are

$$L_e(\bar{\nu}_\tau) = 0, \quad L_\mu(\bar{\nu}_\tau) = 0, \quad L_\tau(\bar{\nu}_\tau) = -1.$$

Whenever a positive tauon τ^+ is emitted it is associated with the emission of a tauon neutrino ν_τ to conserve tauon lepton number since then

$$L_\tau(\text{interaction}) = L_\tau(\tau^+) + L_\tau(\nu_\tau) = -1 + 1 = 0.$$

Whenever a negative tauon τ^- is emitted it is associated with the emission of a tauon anti-neutrino $\bar{\nu}_\tau$ to conserve tauon lepton number since then

$$L_\tau(\text{interaction}) = L_\tau(\tau^-) + L_\tau(\bar{\nu}_\tau) = +1 - 1 = 0.$$

Since ν_τ, $\bar{\nu}_\tau$ are associated with the emission of τ^+, τ^-, respectively, ν_τ is named the tauon neutrino and $\bar{\nu}_\tau$ is named the tauon anti-neutrino.
The tauon neutrino ν_τ is left-handed. The tauon anti-neutrino $\bar{\nu}_\tau$ is right-handed.

5.16 Quarks

An atom is not the smallest indivisible enitity but is made up of protons (discovered by Goldstein in 1886) and neutrons (discovered by Chadwick in 1932). Protons and neutrons together form a nucleus. Electrons orbit the nucleus.

In 1953 Gell-Mann and Nishijima introduced the strangeness quantum number to describe processes involving elementary strange particles (like Σ, Λ etc). We will discuss the strangeness quantum number in chapter 8 section 8.28.

Gell-Mann in 1961 spotted a pattern in the quantum numbers characterizing the elementary particles and tried to explain this on the basis of a structure attributed to them. The pattern observed by Gell-Mann was as follows.

Elementary particles that take part in strong interaction, i.e. the hadrons, were grouped into two categories:

(a) mesons, which are particles that are bosons,
(b) baryons, which are particles that are fermions.

Gell-Mann was able to arrange mesons and anti-mesons by their electrical charge and strangeness quantum number and an octet, i.e. a group of eight, was obtained, as shown in figure 10.8.

    ✓ Gell-Mann was able to arrange the light baryons by their electrical charge and strangeness quantum number and an octet for baryons was obtained, as shown in figure 10.11.

✓ A separate octet for anti-baryons was similarly obtained.

✓ Heavier baryons like delta baryons were found. Gell-Mann was able to arrange them into a decuplet, i.e. into a group of ten. The arrangement was a triangle with a vacant vertex. In other words, he was able to fit nine baryons in the decuplet structure and predicted the properties of the particle missing in the vertex of the decuplet, as shown in figure 10.16. The missing particle was named omega Ω^- and was predicted to have strangeness quantum number $S = -3$. In the decuplet structure masses increased in fixed steps and it was also possible to make a prediction of mass of Ω^-, as discussed in section 10.6.

This Ω^- baryon was discovered in 1964. The mass of the discovered particle Ω^- was $m_\Omega = 1670\frac{MeV}{c^2}$ and this agreed remarkably well with the predicted mass of Ω^-.

5.17 The eightfold way

Clearly studying the symmetrical properties of elementary particles, Gell-Mann was able to classify them into groups, the number of members of each group being 1, 8, 10 etc. This organizational scheme is referred to as the eightfold way. This is discussed in section 10.15.

To explain such a pattern and structure of the group of baryons and mesons, i.e. to describe the eightfold way, Gell-Mann and Zweig independently proposed the quark model in 1964.

In this model there were three quarks named u, d, s (u stands for up quark, d stands for down quark and s stands for strange quark).

All hadrons are made of quarks, i.e. quarks are the building blocks of hadrons—which are mesons and baryons.

Mesons were composed of quark–anti-quark pair $q\bar{q}$. And this is consistent with the fact that the meson octet contains mesons (particles) and anti-mesons (anti-particles).

Baryons have a quark structure of qqq which is a bound system of three quarks. Anti-baryons have a quark structure of $\bar{q}\,\bar{q}\,\bar{q}$ which is a bound system of three anti-quarks. And this is consistent with the fact that the baryon octets are separate for baryons and anti-baryons.

Successful explanation required quarks to possess fractional electric charge.

$$Q(u) = \frac{2}{3}\,|e|,\ \ Q(d) = -\frac{1}{3}\,|e|,\ \ Q(s) = \frac{2}{3}\,|e|$$

A particle should have integral electric charge $\pm n\,|e|$, n being any integer for free existence. So it follows that quarks cannot be observed, i.e. a sole quark cannot be knocked out of a composite system of quarks.

Though proton and neutron are seen experimentally, quarks are not seen experimentally. This is because the quarks are confined within hadrons—i.e. within a region of $10^{-15}\,m$. This is quark confinement hypothesis.

According to the quark model, if we impart energy to a hadron which is a bound state of quarks and/or anti-quarks, then another bound state of quarks and/or anti-quarks is produced and we cannot pull out a quark into free existence.

The quark model explained the structure of the eightfold way.

The quark takes part in weak, electromagnetic and strong interactions.

5.18 Colour

Consider the delta double plus baryon Δ^{++} whose electric charge is $Q(\Delta^{++}) = +2\,|e|$. It has a quark structure $\Delta^{++} \equiv uuu$ (figure 10.17). Now its electrical charge is

$$Q(\Delta^{++}) = Q(u) + Q(u) + Q(u) = \frac{2}{3}\,|e| + \frac{2}{3}\,|e|\frac{2}{3}\,|e| = 2\,|e|$$

Again, delta double plus baryon Δ^{++} has spin $s = \frac{3}{2}$ and so we can write

$$s(\Delta^{++}) = s(u) + s(u) + s(u) = \frac{1}{2} + \frac{1}{2} + \frac{1}{2} = \frac{3}{2}$$

This means the delta double plus baryon $\Delta^{++} \equiv uuu$ is in state $\uparrow\,\uparrow\,\uparrow$. But if all the u quarks point up (i.e. have same spin $\frac{1}{2}$ up state) then Pauli exclusion principle will be violated which is not acceptable. Quarks being fermions have to obey Pauli exclusion principle.

This problem of violation of Pauli exclusion principle was overcome by proposing another quantum number possessed by quarks called colour quantum number or colour charge or colour.

The name colour is just a label and does not reflect different wavelengths of visible light. It is a convenient way of representing quantum charges. Clearly quarks do not have any physical colour.

Any quark assumes colour charge, either red R or blue B or green G, while anti-quarks possess anti-colour charges, anti-red $\overline{R}$ or anti-blue $\overline{B}$ or anti-green $\overline{G}$.

So the quarks u, u, u sitting inside Δ^{++} possess different colour charges, for instance

$$\Delta^{++} = \text{Red } u \text{ Blue } u \text{ Green } u$$

And so Pauli exclusion principle is respected.

According to the quark model, hadrons are overall colourless.

So mesons $q\bar{q}$ have a colour structure $R\overline{R}$ or $G\overline{G}$ or $B\overline{B}$.

And baryons qqq have colour structure RGB.

5.19 Up quark

Existence of the up quark was postulated in 1964 by Gell-Mann and Zweig in order to explain hadron structure in the eightfold way scheme.

The up quark was first observed in 1968 in experiments at SLAC.

The up quark u is an elementary particle which is a first generation quark. It is a constituent of a hadron.

The up quark u has electrical charge $Q = +\frac{2}{3}|e|$ i.e. $\frac{Q}{|e|} \equiv Q_u = +\frac{2}{3}$.

The up quark u is the lightest of all quarks having mass $m_u = 2.2\frac{MeV}{c^2}$.

The up quark u is a fermion with spin $s = \frac{1}{2}$.

The up quark u is the carrier of isospin quantum number $I = \frac{1}{2}$, $I_3 = \frac{1}{2}$ i.e.

$$|u> \ =|I\ I_3> \ =|\frac{1}{2}\frac{1}{2}> \ .$$

The up quark u has colour charge R or B or G.
The anti-particle of up quark u is the anti-up quark $\bar{u}$.

5.20 Down quark

Existence of the down quark was postulated in 1964 by Gell-Mann and Zweig to explain hadron structure in the eightfold way scheme.

The down quark d was first observed in 1968 in experiments at SLAC.

The down quark is an elementary particle which is a first generation quark. It is a constituent of a hadron.

The down quark d has charge $Q = -\frac{1}{3}|e|$ i.e. $\frac{Q}{|e|} \equiv Q_d = -\frac{1}{3}$.

The down quark d has mass $m_d = 4.7\frac{MeV}{c^2}$.

The down quark d is a fermion with spin $s = \frac{1}{2}$.

The down quark d is the carrier of isospin quantum number $I = \frac{1}{2}$, $I_3 = -\frac{1}{2}$ i.e.

$$|d> \ =|I\ I_3>=|\frac{1}{2}\ -\frac{1}{2}\rangle$$

The down quark d has colour charge R or B or G.
The anti-particle of the down quark d is the anti-down quark $\bar{d}$.

5.20.1 Nucleon (proton, neutron) structure

Hey used an electron accelerator and performed an experiment on e^-p scattering and cross-section was measured. High energy electrons were fired at protons.

If a proton was a structureless uniform blob of positive charge, the electron would be deflected by a small amount (just like scattering of alpha particles from protons in the Rutherford scattering experiment)—it would be an elastic scattering. On the other hand, if the proton had a structure inside it, then the scattering cross-section would be very different.

In the experiment a far larger momentum transfer was seen and a huge deviation from elastic scattering. This was clear evidence that the proton has a structure consistent with the quark model. So this experiment confirmed that the proton has an internal structure.

A proton p is built out of two up quarks and one down quark, i.e. $p \equiv uud$.

A neutron is built out of two down quarks and one up quark, i.e. $n \equiv udd$.

5.21 Strange quark

Existence of the strange quark was postulated in 1964 by Gell-Mann and Zweig to explain hadron structure in the eightfold way scheme.

The strange quark was first observed in 1968 in experiments at SLAC.

The strange quark s is an elementary particle which is a second generation quark. It is a constituent of a hadron.

The strange quark s has charge $Q = -\frac{1}{3}|e|$ i.e. $\frac{Q}{|e|} \equiv Q_s = -\frac{1}{3}$.

The strange quark s has mass $m_s = 96\frac{MeV}{c^2}$.

The strange quark s is a fermion with spin $s = \frac{1}{2}$.

The strange quark s is the carrier of strangeness quantum number $S = -1$.

The strange quark s has colour charge R or B or G.

The anti-particle of the strange quark s is the anti-strange quark $\bar{s}$.

5.22 Charm quark

Existence of the charm quark c was predicted in 1970 by Glashow, Iliopoulos and Maiani.

In 1974 the charm quark c was discovered through detection of a meson named J/ψ meson at Brookhaven National Laboratory (by Ting *et al*) and SLAC (by Richter *et al*).

The charm quark c is an elementary particle which is a second generation quark. It is a constituent of a hadron.

J/ψ particle had a mass $m_{J/\psi} = 3.1\frac{GeV}{c^2}$ and a mean lifetime of $7.2 \times 10^{-21}\,s$. This lifetime was about 1000 times longer than what was predicted by Gell-Mann's eightfold way. It was explained by assuming that there was a new type of quark—the charm quark denoted as c quark. The quark structure of J/ψ meson was $\frac{J}{\psi} \equiv c\bar{c}$. So the c quark was discovered.

The charm quark c has electric charge $Q = \frac{2}{3}|e|$ i.e. $\frac{Q}{|e|} \equiv Q_c = \frac{2}{3}$.

It is the third most massive quark with mass $m_c = 1.28\frac{GeV}{c^2}$.

The charm quark c is a fermion having spin $=\frac{1}{2}$.

The charm quark c carries charmness quantum number $C = +1$.

The charm quark c has colour charge R or B or G.

Lifetime of the charm quark c is $10^{-20}\,s$.

The anti-particle of the charm quark c is the anti-charm quark $\bar{c}$.

5.23 Bottom quark

The bottom quark was predicted to exist by Kobayashi and Maskawa in 1973.

The bottom quark was discovered in 1977 by Lederman *et al* in Fermilab.

The bottom quark b is also known as beauty quark. It is a third generation quark. It is a constituent of a hadron.

The bottom quark b has electric charge $Q = -\frac{1}{3}|e|$ i.e. $\frac{Q}{|e|} \equiv Q_b = -\frac{1}{3}$.

The mass of bottom quark b quark is $m_b = 4.18\frac{GeV}{c^2}$.

The bottom quark b is a fermion with spin $=\frac{1}{2}$.

The bottom quark b is the carrier of bottomness quantum number $B' = -1$.

The bottom quark b has colour charge R or B or G.

The anti-particle of the bottom quark b is the anti-bottom quark $\bar{b}$.

The bottom quark b has lifetime $\sim 10^{-12}\,s$.

5.24 Top quark

The top quark t was predicted to exist by Kobayashi and Maskawa in 1973.

It was discovered in 1995 by the CDF (Collider Detector at Fermilab) and a collaborative experiment named $D0$ experiment at Tevatron collider at Fermilab.

The top quark t is also known as the truth quark. It is a third generation quark.

The top quark t has electric charge $Q = +\frac{2}{3}\,|e|$ i.e. $\frac{Q}{|e|} \equiv Q_t = +\frac{2}{3}$.

The top quark t is the most massive of all elementary particles and has mass $m_t = 173\frac{MeV}{c^2}$.

The top quark t is a fermion with spin $s = \frac{1}{2}$.

The top quark t is the carrier of topness quantum number of value $T = +1$.

Top quark t has colour charge R or B or G.

The anti-particle of the top quark t is the anti-top quark $\bar{t}$.

Because of its enormous mass, the top quark is extremely shortlived having a lifetime of $5 \times 10^{-25}\,s$, which is faster than the lifetime for strong interaction and so it does not get time to form hadrons as other quarks do. In other words, the decay time of the top quark is shorter than hadronization time. It decays into b, s, d quarks.

5.25 Gauge bosons as force carriers

The constituents of matter are held together due to exchange of certain elementary bosons between them. In other words, there are elementary bosons which act as force carrying agencies and inter-connect or bind the two building blocks (elementary fermions) of a system.

Bosons are particles having integral spins (0, 1, 2). They obey Bose–Einstein statistics.

The exchanged particles responsible for the interactions are called gauge particles or gauge bosons. In other words, the bosons that establish and mediate interaction between elementary fermions are called gauge bosons. They have spin $s = 1$ and are hence vector bosons.

The gauge bosons are of three types: photon γ, intermediate vector bosons or intermediate vector bosons (IVBs) $W^{\pm}$, Z^0, gluon g.

The gauge bosons have virtual existence. They occur over a time in which uncertainty principle allows violation of energy and momentum. They cannot be observed or detected experimentally because their real existence would violate conservation of energy and momentum.

5.26 Photon

The concept of the photon γ was given by Planck in 1900 through his quantum hypothesis in his black body radiation spectrum and by Einstein in 1905 in the quantum explanation of photoelectric effect through electromagnetic energy quanta $h\nu$ called photons. Einstein explained photoelectric effect with his quantum hypothesis of the photon.

The photon is the quantum of light. It is the minimum discrete energy packet of electromagnetic radiation.

The photon is the force carrier of electromagnetic interaction, i.e. quantum of electromagnetic field or field quanta. The photon is a gauge boson in the electroweak interaction.

The photon γ has zero electric charge $Q_\gamma = 0$.

The photon γ has zero rest mass $m_\gamma = 0$.

The photon γ is a vector boson having spin $s = 1$ (i.e. $1\hbar$).

The J^π value of the photon is 1^-. Here π is parity which is -1.

We will discuss parity in chapter 8.

The C parity of the photon γ is -1.

We will discuss C parity in chapter 8.

The photon γ has zero colour charge.

The photon γ is its own anti-particle.

The photon moves with the speed of light in free space and cannot decay.

The photon is a boson obeying B–E statistics.

The photon is abelian.

We will discuss abelian property in chapter 9.

A photon cannot emit or absorb another photon, i.e. it does not have self-interaction.

Interactions of light with matter and interaction between charged particles are described through relativistic the quantum field thery called QED i.e. quantum electrodynamics.

5.27 Intermediate vector boson (IVB)

$W^\pm, Z^0$ are the IVBs. They were discovered at the Super Proton Synchrotron (SPS) at CERN in 1983.

$W^\pm$ bosons, which are charged bosons were named after weak interaction. Z^0 boson, which is a neutral boson, was named by Weinberg.

IVBs are called weak bosons. They mediate weak interactions and hence are gauge bosons.

$W^\pm$ bosons are charged bosons $Q = \pm|e|$ i.e. $Q_{W^\pm} = \pm 1$ and have magnetic moment.

Z^0 boson is a neutral boson $Q_{Z^0} = 0$ and does not have magnetic moment.

The mass of charged IVBs $W^\pm$ is $m_{W^\pm} = 80\frac{GeV}{c^2}$ and the mass of neutral IVBs Z^0 is $m_{Z^0} = 91\frac{GeV}{c^2}$.

IVBs have spin $s = 1$, hence they are vector bosons.

We explain the terms scalar, pseudoscalar, vector, pseudovector in chapter 8, section 8.41.

Half-life of $W^\pm$, Z^0 is about $3 \times 10^{-25}s$, i.e. they are extremely shortlived.

W^+ boson is the anti-particle of W^- boson.

Z^0 boson is its own anti-particle.

5.28 Gluon

The gluon was discovered at the electron–positron collider PETRA (Positron Electron Tandem Ring Accelerator) of DESY (German National Laboratory), Germany in 1979.

The gluon mediates strong interaction between quarks and binds them within hadrons according to Quantum Chromo Dynamics (QCD), i.e. they are exchange particles for strong interaction.

There are eight types of gluons.

Gluons are denoted by g. They do not carry any electrical charge, i.e. $Q_g = 0$.

Gluons are massless $m_g = 0$.

Gluons are vector bosons having spin $s = 1$.

Gluons carry colour charge.

Gluons are non-abelian gauge particles for strong interactions.

A gluon can emit and absorb another gluon, i.e. the gluon has self-interaction.

The theory of strong interactions between quarks and gluons with the gauge group $SU(3)$ is referred to as QCD.

We will discuss group theory and $SU(3)$ in chapter 9.

5.29 Higgs boson

When the Universe began, no particle had mass and being massless these particles travelled with the speed of light.

The present form of the Universe consisting of stars and planets could be possible as particles gained mass from a fundamental field called the Higgs field. The quantum of excitation of the Higgs field is the Higgs boson. The mass-giving interaction between particles and the Higgs field is called the Higgs mechanism, which we outline in section 5.38.

The existence of this mass-giving Higgs field, the Higgs particle H^0 along with the Higgs mechanism of how massive particles acquire mass due to interaction with the Higgs field via the Higgs boson H^0 was predicted in 1964 by Peter Higgs.

It was confirmed in 2012 when the particle called the Higgs boson H^0 was discovered by the ATLAS (A Toroidal LHC Apparatus) and CMS (Compact Muon Solenoid) experiments at the LHC (Large Hadron Collider) at CERN (European Council for Nuclear Reseach).

Here two proton beams were made to travel with $v \sim c$ and they collided. Such smashing of high energy beams led to the production of various particles and the relic of the original Higgs field, i.e. the Higgs boson H^0 was discovered in LHC.

The Higgs particle H^0 has zero electric charge $Q_{H^0} = 0$.

The Higgs particle H^0 has mass $m_{H^0} = 125\frac{MeV}{c^2}$.

The Higgs particle H^0 is a scalar boson having zero spin $s = 0$ and even parity. The Higgs particle H^0 has zero colour charge.

The Higgs boson is unstable and can decay into particles predicted by the Standard Model.

5.30 Classifications in the Standard Model

We show the masses of the particles of the Standard Model in figure 5.3.

We show the charges of the particles of the the Standard Model in figure 5.4.

Figure 5.5 shows the classifications done in the Standard Model regarding particles, i.e. the fundamental constituents of the Universe. Figure 5.5 also depicts the forces that regulate various interactions.

We show the spins of the particles of the Standard Model in figure 5.6.

u $2.2\ \mathrm{MeV}/c^2$	c $1.28\ \mathrm{GeV}/c^2$	t $173\ \mathrm{GeV}/c^2$	g 0	H^0 $125\ \mathrm{GeV}/c^2$
d $4.7\ \mathrm{MeV}/c^2$	s $96\ \mathrm{MeV}/c^2$	b $4.18\ \mathrm{GeV}/c^2$	γ 0	
e $0.51\ \mathrm{MeV}/c^2$	μ $106\ \mathrm{MeV}/c^2$	τ $1.78\ \mathrm{GeV}/c^2$	$W^\pm$ $80\ \mathrm{GeV}/c^2$	
ν_e $< 1\ \mathrm{eV}/c^2$	ν_μ $<0.17\ \mathrm{eV}/c^2$	ν_τ $<18\ \mathrm{MeV}/c^2$	Z^0 $91\ \mathrm{GeV}/c^2$	

Figure 5.3. Masses of fundamental particles of the Standard Model.

Particles					Antiparticles				
u $\frac{2}{3}\lvert e\rvert$	c $\frac{2}{3}\lvert e\rvert$	t $\frac{2}{3}\lvert e\rvert$	g 0	H^0 0	$\bar{u}$ $-\frac{2}{3}\lvert e\rvert$	$\bar{c}$ $-\frac{2}{3}\lvert e\rvert$	$\bar{t}$ $-\frac{2}{3}\lvert e\rvert$	g 0	H^0 0
d $-\frac{1}{3}\lvert e\rvert$	s $-\frac{1}{3}\lvert e\rvert$	b $-\frac{1}{3}\lvert e\rvert$	γ 0		$\bar{d}$ $\frac{1}{3}\lvert e\rvert$	$\bar{s}$ $\frac{1}{3}\lvert e\rvert$	$\bar{b}$ $\frac{1}{3}\lvert e\rvert$	γ 0	
e^- $-\lvert e\rvert$	μ^- $-\lvert e\rvert$	τ^- $-\lvert e\rvert$	W^- $-\lvert e\rvert$		e^+ $\lvert e\rvert$	μ^+ $\lvert e\rvert$	τ^+ $\lvert e\rvert$	W^+ $+\lvert e\rvert$	
ν_e 0	ν_μ 0	ν_τ 0	Z^0 0		$\bar{\nu}_e$ 0	$\bar{\nu}_\mu$ 0	$\bar{\nu}_\tau$ 0	Z^0 0	

Figure 5.4. Charges of the particles of the the Standard Model.

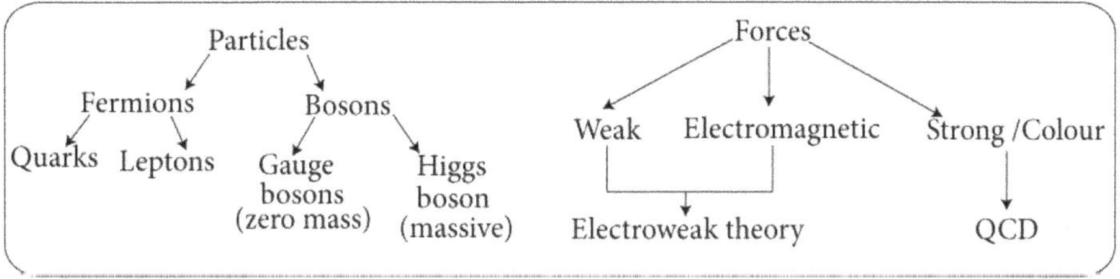

Figure 5.5. Classification of forces and particles in the the Standard Model.

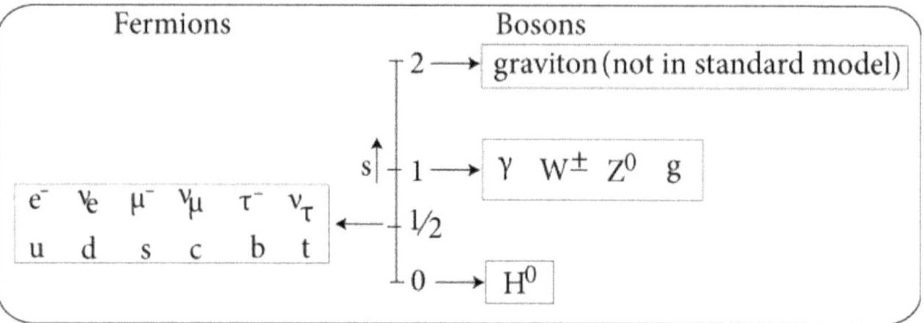

Figure 5.6. Spin s of particles of the Standard Model (except the graviton).

5.31 Concept of field in the Standard Model

The Standard Model of high energy physics is a composite of two different theories —the electroweak theory and QCD.

Electroweak theory is the unification of the dynamics of electromagnetic interaction and weak interaction denoted as $SU(2) \otimes U(1)$.

We discuss $U(1)$ and $SU(2)$ in chapter 9.

QCD is a theory for strong interaction. It is based on $SU(3)$, which is the new name for strong interaction. It is based on Lie group. Strong force is referred to as colour force that acts between the quarks.

Quantum mechanics and the special theory of relativity have been combined leading to quantum field theory which has been used in constructing the Standard Model.

The field concept is the essence of the Standard Model.

5.32 Electromagnetic field theory

We start with electrodynamics as we wish to generalize the laws of electrodynamics. Maxwell's equations are in Gaussian units

$$\checkmark \;\; \vec{\nabla} \cdot \vec{E} = 4\pi\rho$$

This is Gauss's law of electrostatics.

$$\checkmark \;\; \vec{\nabla} \cdot \vec{B} = 0$$

This is Gauss's law of magnetostatics and it says that there is no free magnetic charge.

$$\checkmark \;\; \vec{\nabla} \times \vec{E} + \frac{1}{c}\frac{\partial \vec{B}}{\partial t} = 0$$

This is Faradays's law of electromagnetic induction. According to it a time varying magnetic field generates electric field and if there is a piece of metal a current will flow.

$$\checkmark \;\; \vec{\nabla} \times \vec{B} - \frac{1}{c}\frac{\partial \vec{E}}{\partial t} = \frac{4\pi}{c}\vec{J}$$

This is Ampere's law. It tells us how magnetic field is generated by current distribution i.e. by an electric field.

These four laws of electrodynamics are complete and consistent and govern the behaviour of the electromagnetic field. Here $\vec{E}$ is electric field, $\vec{B}$ is magnetic field, ρ is charge density, $\vec{J}$ is current density, $\frac{1}{c}\frac{\partial \vec{E}}{\partial t}$ is displacement current.

Electromagnetic field concept was first introduced by Faraday.

Electromagnetic waves exist as a consequence of these four equations. Maxwell calculated the speed of the electromagnetic wave to be $3 \times 10^8\, m\, s^{-1}$.

Light also travels at that speed. Hence Maxwell concluded that light is an electromagnetic wave. This established the electromagnetic nature of light.

Hertz generated electromagnetic wave and Maxwell theory was confirmed.

The conflict between Newton's particle dynamics and Maxwell's field dynamics was settled by Einstein upholding the Maxwell's equations and modifying Newton's equations of motion for relavistic velocities giving rise to the special theory of relativity. Theory of relativity and quantum mechanics kept Maxwell's equations of electrodynamics unchanged.

Quantum mechanics made a profound change in the interpretation of the fields $\vec{E}$ and $\vec{B}$ and the energy of the electromagnetic field.

In classical Faraday–Maxwell theory the electromagnetic field energy is continuous—it can be of any value. But in the quantum version the energy is discrete or quantized in terms of a certain basic value known as electromagnetic quanta or photon energy $h\nu$. This gave birth to the quantum field theory.

Quantum field theory is the basic language of the Standard Model and hence the basic language of the elementary particles.

5.32.1 Concept of interaction of two charged particles according to classical theory

Coulomb interaction between two charged particles, say an electron e^- and a proton p, can be explained in the classical theory as follows.

We say that a proton p is surrounded by an electric field and if we place an electron e^- in this electric field then the e^- feels the electric field, i.e. experiences a force due to proton p. If proton p is in motion there will be a magnetic field also and that also will be felt by the electron e^-. This is how the two charges e^- and proton p execute electromagnetic interaction according to the classical theory.

5.32.2 Concept of interaction of two charged particles according to quantum theory

According to quantum field theory, interaction occurs in a different manner. The charged particle proton p emits the quanta of electromagnetic field called photon γ denoted by the wavy line in figure 5.7. This photon is captured by the other charged particle electron e^- situated in the neighbourhood. So the two charged particles p and e^- interact electromagnetically through constant emission and absorption of photons.

In fact, all interactions between various particles occur through exchange of particles or quanta.

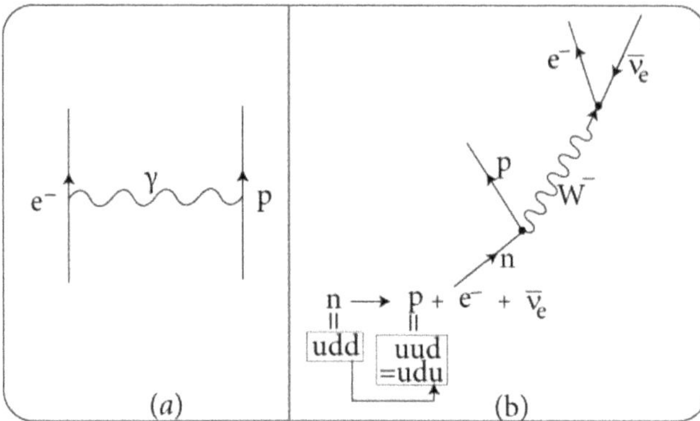

Figure 5.7. (a) Interaction between electron e^- and a proton p in Feynman diagram. (b) Transformation of neutron n to proton p in Feynman diagram.

This has been represented in the Feynman diagram of figure 5.7(a) that helps visualize the process in which elementary particles interact.

So the concept of interaction is different in classical field theory and quantum field theory.

We will discuss Feynman diagrams in chapter 6.

Newton introduced the concept of force but in Lagrangian and Hamiltonian dynamics it is the potential that is used for study. Force is the space derivative of potential.

In quantum field theory it is the emission and absorption of quanta that are studied.

5.33 Relation between range and mass of quanta of force

Range is inversely proportional to the mass of the quantum (elementary boson) exchanged between two elementary constituents (elementary fermions) of matter.

For electromagnetic interaction, range is ∞, mass of the quantum called photon is $m_\gamma = 0$.

(In quantum theory of gravity, since the range of gravitational interaction is ∞ we should have the mass of the quantum called the graviton to be $m_G = 0$.)

Strong interaction has finite range. From the range–mass relationship Yukawa calculated the mass of quanta executing strong interaction to be about $280m_e$ which is about the mass of a pion $\pi^{\pm,\,0}$. Yukawa's theory is not the complete theory of strong interaction. The Standard Model gives the complete theory of strong interaction. The quantum of strong interaction or the mediating quanta are the gluons.

Weak interaction also has a finite range. It is mediated by particles of finite mass. Weak interaction theory is given by the Standard Model. The quantum of weak interaction or the mediating quanta are the intermediate vector bosons $W^\pm, Z^0$.

The Standard Model gives the idea of the Higgs boson.

5.34 Laws of electroweak dynamics

Electrodynamics is based on electric field $\vec{E}$ and magnetic field $\vec{B}$. There are two vectors. Each has three components.

Electroweak dynamics is a four-times generalization of electrodynamics or electromagnetic field.

The generalized electromagnetic field $\vec{E}_i$, $\vec{B}_i$ with $i = 1,2,3,4$ corresponds to four kinds of fields representing γ, W^+, W^-, Z^0.

The index 1 refers to γ which is the photon, i.e. the quanta of the electromagnetic field.

The indices 2,3,4 refer to W^+, W^-, Z^0, which are quanta of weak field.

We can describe the transformation of neutron n into proton p through weak decay, as shown in figure 5.7(b), in which the quanta W^- of weak field play a major role.

In the decay of n to p, electron e^- and electron anti-neutrino $\bar{\nu}_e$ are generated. The neutron and proton are not elementary particles as they are made of three quarks viz. $n = udd$ and $p = uud$. In the process of weak interaction d changes to u (as indicated in the figure) and in the process a W^- is emitted virtually and this W^- generates an e^- and $\bar{\nu}_e$. Clearly particle number also changes since from one particle (n) we end up with three particles (p, e^-, $\bar{\nu}_e$).

The dynamics of the four electroweak fields, namely

$$(\vec{E}_i, \vec{B}_i): i = 1, 2, 3, 4 = \gamma, W^+, W^-, Z^0$$

is given by the following equations that are very similar to the Maxwell's equations (apart from the series of dots).

$$\vec{\nabla} \cdot \vec{E}_i + \ldots = 4\pi\rho_i$$

$$\vec{\nabla} \cdot \vec{B}_i + \ldots = 0$$

$$\vec{\nabla} \times \vec{E}_i + \frac{1}{c}\frac{\partial \vec{B}_i}{\partial t} + \ldots = 0$$

$$\vec{\nabla} \times \vec{B}_i - \frac{1}{c}\frac{\partial \vec{E}_i}{\partial t} + \ldots = \frac{4\pi}{c}\vec{J}_i$$

These are in fact generalized Maxwell's equations except for the index.

The role of the quanta in electroweak interactions is indicated in figure 5.8.

✓ In figure 5.8(a) Coulomb interaction is shown through exchange of γ.

✓ In figure 5.8(b) quark d ($Q_d = -\frac{1}{3}$) changes to quark u ($Q_u = \frac{2}{3}$) by emitting a W^- ($Q_{W^-} = -1$) and this W^- gets absorbed by electron neutrino ν_e ($Q_{\nu_e} = 0$) and the neutrino gets converted to electron e^- ($Q_{e^-} = -1$). In fact,

$$Q_d - Q_u = Q_{W^-}$$

$$-\frac{1}{3} - \left(-\frac{2}{3}\right) = -1.$$

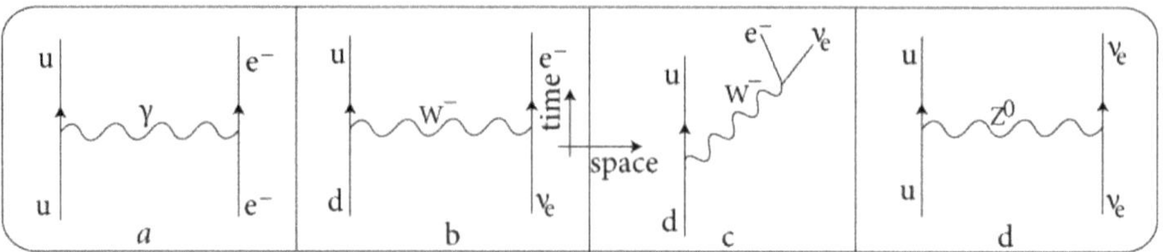

Figure 5.8. Role of quanta in electroweak interactions.

✓ In figure 5.8(c) quark d ($Q_d = -\frac{1}{3}$) changes to quark u ($Q_u = \frac{2}{3}$) by emitting a W^- ($Q_{W^-} = -1$) and this W^- gets converted to electron e^- ($Q_{e^-} = -1$) and electron neutrino ν_e ($Q_{\nu_e} = 0$). Here too $Q_d - Q_u = Q_{W^-}$, i.e. $-\frac{1}{3} - (-\frac{2}{3}) = -1$.

✓ In figure 5.8(d) quark u interacts with electron neutrino ν_e. The u emits a Z^0 which is absorbed by ν_e.

Electrodynamics can be summarized by the following statement.

Every particle that has a charge is capable of emitting or absorbing a photon.

In electrodynamics we speak of a single charge corresponding to photon quanta. Electric charge is a number—actually an attribute described by a number given to every particle, e.g. for a proton it is positive, for an electron it is negative and for neutrino ν it is zero. And these numbers are characterized by the fact that they commute, i.e. obey the relation

$$ab = ba$$

for two numbers a and b.

The algebra of commuting objects is called abelian algebra. Maxwellian field is abelian field.

We discuss abelian algebra in chapter 9.

In electroweak dynamics there are four charges corresponding to the four electroweak fields

$$\gamma, \ W^+, \ W^-, \ Z^0.$$

Quantum field theory generalizes the concept of potential and it explains how particles are created, i.e. emitted and here the particle nature changes.

5.35 Laws of strong interaction or quantum chromodynamics

Quantum chromodynamics is an eight-times generalization of electrodynamics or the electromagnetic field.

The generalized electromagnetic field

$$(\vec{E}_\alpha, \ \vec{B}_\alpha): \alpha = 1 \text{ to } 8$$

corresponds to eight kinds of gluons g_α, i.e. eight quantum chromodynamics fields.

Gluons are interchanged between quarks and due to these interchanges quarks are tightly bound within protons and neutrons, i.e. hadrons. In other words, gluons glue the quarks together.

The dynamics of the eight quantum chromodynamic fields is given by the following equations that are very similar to Maxwell's equations (apart from the series of dots).

$$(\vec{E}_\alpha,\ \vec{B}_\alpha):\ \alpha = 1 \text{ to } 8$$

$$\vec{\nabla}\cdot\vec{E}_\alpha + \ldots = 4\pi\rho_\alpha$$

$$\vec{\nabla}\cdot\vec{B}_\alpha + \ldots = 0$$

$$\vec{\nabla}\times\vec{E}_\alpha + \frac{1}{c}\frac{\partial\vec{B}_\alpha}{\partial t} + \ldots = 0$$

$$\vec{\nabla}\times\vec{B}_\alpha - \frac{1}{c}\frac{\partial\vec{E}_\alpha}{\partial t} = \frac{4\pi}{c}\vec{J}_\alpha$$

According to quantum chromodynamics and the Standard Model there are 8 charges corresponding to 8 gluons which we express as g_α with $\alpha = 1$ to 8.

We discuss gluon colour charge more in chapter 10, section 10.17.

5.35.1 Non-abelian generalization

So there are 12 charges (γ, W^+, W^-, Z^0, g_α with $\alpha = 1$ to 8) in this generalized dynamics that are very different from the electric charge. The 12 charges in generalized dynamics are not numbers and do not commute, i.e.

$$ab \neq ba$$

for two charges a and b.

The algebra of non-commuting objects is called non-abelian algebra.

In other words the generalization is a non-abelian generalization of the abelian Maxwelian field, i.e. the 4 electroweak fields and the 8 quantum chromodynamic fields are non-abelian field.

The idea of gauge invariance has been generalized in the Standard Model. Actually, electroweak and QCD are non-abelian gauge fields. We have 4 non-abelian gauge fields in electroweak theory and 8 non-abelian gauge fields in QCD.

The theory of non-abelian gauge fields was fully developed by Yang and Mills in 1954 and was confirmed later.

Every object which is charged interacts with a photon. But the photon itself is not charged. So a photon cannot interact with itself. So electrodynamics is a linear theory.

The 12 charges (4 weak quanta and the 8 gluons) are carried by the fields themselves. And hence they have interaction—they emit or absorb, i.e. interact (figure 5.9).

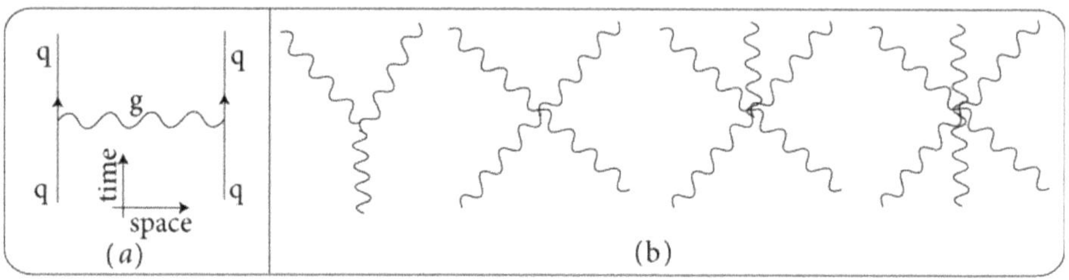

Figure 5.9. (a) Quark-quark interaction through a gluon. (b) Non-abelian gauge field (Yang–Mills field).

5.35.2 Gauge fields

The gauge bosons are

one photon γ, two W^+, W^- bosons, one Z^0 boson and eight gluons g.

Corresponding to these gauge bosons there are 12 gauge fields. They are

γ, W^+, W^-, Z^0, g_α with $\alpha = 1$ to 8.

The gauge bosons are all spin 1 (vector) particles and obey Bose–Einstein statistics.

There is another particle called the Higgs boson H^0 of spin 0 (scalar) which has the role of giving mass to various massive particles.

If we succeed in quantizing gravity the quanta of gravitational field would be graviton of spin 2.

5.36 Elementary building blocks of matter

Matter is made of molecules and a molecule is made of atoms. Atoms consist of nuclei and electrons. Again, the nucleus is made of protons and neutrons. Protons and neutrons are made of up quarks and down quarks, i.e. $p = uud$ and $n = udd$. Clearly u and d are sufficient to form a nucleus. Also, the three particles u, d, e^- are sufficient to build all matter.

If we consider corresponding neutrino ν_e then thermonuclear fusion reactions can be described. In other words, four particles, namely $\begin{pmatrix} \nu_e \\ e \end{pmatrix}$, $\begin{pmatrix} u \\ d \end{pmatrix}$, nearly make up the Universe. This constitutes the first generation of leptons and quarks.

However, to build some more particles, nature repeats this quartet twice and we have the heavier versions, namely $\begin{pmatrix} \nu_\mu \\ \mu \end{pmatrix}$, $\begin{pmatrix} c \\ s \end{pmatrix}$ (second generation) and $\begin{pmatrix} \nu_\tau \\ \tau \end{pmatrix}$, $\begin{pmatrix} t \\ b \end{pmatrix}$ (third generation).

All these particles are fermions having half-integral spin and obey Fermi–Dirac statistics.

5.36.1 Heavier and lighter particles

The Big bang was supposed to have produced the Universe and all these particles were produced in equal numbers. But the heavier particles decayed into lighter particles, which the Universe is now observed to be made of. The heavier particles can be recreated in the laboratory and their existence has been confirmed.

5.37 Particles and anti-particles

Corresponding to the particles there are anti-particles.

When the Universe was formed, particles (or matter) and anti-particles (or anti-matter) were produced in equal numbers. However, the Universe as we see it now, has a pre-dominence of particles over anti-particles, i.e. matter over anti-matter. This is matter–anti-matter asymmetry.

Such matter–anti-matter asymmetry can be built into the Standard Model if the three quartets

$$\begin{pmatrix} \nu_e \\ e \end{pmatrix}; \begin{pmatrix} u \\ d \end{pmatrix}; \begin{pmatrix} \nu_\mu \\ \mu \end{pmatrix}; \begin{pmatrix} c \\ s \end{pmatrix}; \begin{pmatrix} \nu_\tau \\ \tau \end{pmatrix}; \begin{pmatrix} t \\ b \end{pmatrix}$$

are considered, because that will allow the anti-matter to disappear. The repetition of three sets of quarks and leptons is thus very important and each such set is called a generation. So we have three generations of leptons and quarks.

We note that the field consists of quanta which are particles of the field.

Again, in quantum field theory the quarks are quanta of the quark field, electrons are quanta of the electron field, neutrinos are the quanta of the neutrino field and so on. So the particles or quanta involve fields.

In quantum field theory the field concept and the particle concept get unified and it is better to speak in terms of bosons and fermions instead of fields and particles.

5.38 Symmetry breaking interaction

The range of electromagnetism is ∞ while the range of the weak force is $10^{-17}\,m$. Clearly there is vast disparity between electromagnetism and weak force regarding their range and this disparity or divergence in range values is referred to as breakdown of symmetry. In other words if such discrepancy or inconsistency exists the two forces cannot be unified because unification occurs between symmetric objects. So with this value of ranges electroweak symmetry is not possible, i.e. there is breakdown of electroweak symmetry because electromagnetism and weak forces are widcly asymmetric.

This asymmetry in electromagnetism and weak force was achieved in the Standard Model by what is called spontaneous breakdown of symmetry engineered by a mechanism called Higgs mechanism that keeps photon massless but raises the masses of $W^\pm$, Z^0 to finite value viz $80\frac{GeV}{c^2}$, $91\frac{GeV}{c^2}$, respectively. If we approximate the proton mass to be $1\frac{GeV}{c^2}$ then these $W^\pm$, Z^0 are, respectively, 80, 90 times heavier than the proton. The range–mass relation then gives the range to be $10^{-17}m$.

The Higgs mechanism postulates existence of a universal all pervading field called the Higgs field and this field gives masses to $W^\pm$ and Z^0 bosons and also gives masses to all the massive fermions. Thus, masses of quarks and electrons come from the Higgs field.

5.39 Higgs mechanism

We note from figure 5.3 that most elementary particles have mass. Some particles do not have mass (γ, g) while some particles have negligible mass (ν_e, ν_μ, ν_τ). Again, some elementary particles are massive (t, W, Z^0, b, τ, c), some particles have intermediate mass (μ, s) while some have small mass (d, u, e^-).

Higgs studied what causes elementary particles to have mass.

The Higgs field is a scalar field that exists at all points in space throughout the Universe.

Particles that have mass interact with that field. Particles without mass do not interact with the Higgs field.

Figure 5.10(a) shows an approximate representation of the Higgs field ϕ in a 2D plot of V versus ϕ where V is potential energy associated with the Higgs field.

Everything tries to go down to its lowest energy. The state of lowest energy corresponds to the bottom of the curve viz. points c_1, c_2. Actually it is a 3D curve and the lowest energy or minimum potential points are on the circle shown in figure 5.10(b).

According to the theory the massless particles occupy this lowest energy state. And interaction of the particle with the field may cause oscillation of a particle along the paths like ac_1b, $a'c_2b'$ (normal to the circle) which is the Higgs boson and this gives rise to mass of particle.

The Higgs field is mediated by the Higgs boson of mass $125\frac{GeV}{c^2}$ that gives mass to particles. In other words, the quantum of excitation of the Higgs field is the Higgs boson. Particles do not have mass of their own but get their mass by interacting with the Higgs field.

A particle ends up being heavier, i.e. gains a large amount of mass if it interacts with the Higgs field strongly and a paricle ends up being lighter, i.e. gains a small amount of mass if it does not interact with the Higgs field strongly.

A photon is massless because it does not interact with the Higgs field. Elementary particles like electrons, quarks, bosons do interact with the Higgs field and acquire a variety of masses.

The mass-giving interaction between particles and the Higgs field is called the Higgs mechanism.

The Higgs boson is a neutral scalar particle and couples with fermions and gauge bosons with a coupling strength proportional to the mass of the particles.

Figure 5.10(b) shows a simple mechanical example of a ball placed on top of a hill of circular cross-section surrounded by a circular valley. This system has a circular symmetry but the ball is in unstable equilibrium and will roll down into the bottom

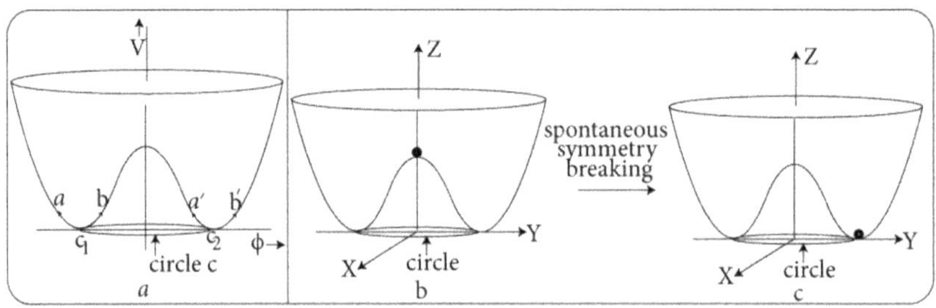

Figure 5.10. (a) Higgs field. (b,c) Spontaneous symmetry breaking.

of the valley where it will reach some point of stable equilibrium. And once it does that, the circular symmetry is broken. This example shows spontaneous symmetry breaking.

5.40 Nobel prizes for the Standard Model

Year	Receipient	Work for which awarded
1979	Glashow, Salam, Weinberg	Theory of unified weak and electromagnetic interaction between elementary particles, including prediction of weak neutral current.
1984	Rubbia, Van der Meer	Discovery of field particles $W^\pm$, Z^0 which are the communicators of weak interaction
1990	Friedman, Kendall, Taylor	Deep inelastic scattering of electrons on protons and bound neutrons proving the existence of quarks
1999	't Hooft, Veltman	Proof of renormalisability of the electroweak theory
2004	Gross, Politzer, Wilczek	Discovery of asymptotic freedom in the theory of strong interaction (Yang–Mills theory of QCD)
2008	Nambu, Kobayashi, Maskawa	Discovery of mechanism of spontaneous symmetry breaking and matter–anti-matter asymmetry and prediting three generations of quarks
2013	Englert, Higgs	Understanding the origin of mass of particles through the Higgs mechanism

5.41 Drawbacks of the Standard Model

The Standard Model is theoretically a self-consistent theory and has made predictions that were confirmed experimentally, which is a great success. However, we cannot call it a complete theory of fundamental interactions due to the following:

- The Standard Model does not explain theory of gravitation, which is the weakest of the four fundamental interactions of nature. The most prevalent force in our daily life is the force of gravity which is not part of the standard model.

  The quantum theory used to describe the microworld and the general theory of relativity used to describe the macroworld do not fit into a single framework. Making the two theories mathematically compatible in the context of the Standard Model has not been possible.

  However, if we are concerned with particle physics in which we are interested with the minuscule scale of particles, gravity can be completely overlooked as it produces a negligible effect.

- The Standard Model does not explain matter–anti-matter asymmetry in the observable Universe though the Big Bang should have produced equal amounts of matter and anti-matter. This problem is called the baryon asymmetry problem. Some physical laws must have acted differently for matter and anti-matter resulting in the imbalance of matter and anti-matter since we live in a Universe that is matter dominated.

- The Standard Model does not explain the accelerating expansion of the Universe, meaning that the velocity at which a distant Galaxy recedes from an observer is continuously increasing with time. In fact, objects in the Universe are moving away from one another at an accelerating rate. This might be due to dark energy.

  Most matter of the Universe does not emit or absorb or scatter light and is referred to as dark matter. However the Standard Model does not contain any component of dark matter.

- Neutrinos are massless in the Standard Model. The Higgs mechanism does not give mass to the neutrinos.

  However, discovery of neutrino oscillation and neutrino mass was made by Super-Kamiokande Observatory and the Sudbury Neutrino Observatories.

  Since neutrinos do have small, i.e. non-zero rest mass, the Standard Model needs modification. This indicates that we have to go beyond the Standard Model.

- According to grand unified theory the proton decays. But there is no experimental evidence of this. We take proton lifetime to be greater than 10^{34} years (chapter 8, section 8.20).

5.42 Beyond the Standard Model

Some theories that go beyond the Standard Model are as follows.

5.42.1 Grand unification theory

In grand unification theory the baryon number is violated. So according to grand unification theory the proton can decay. However, this needs experimental verification.

We shall take proton lifetime $\tau_p > 10^{-34}\,s$.

We shall discuss stability of the proton in section 8.20.

5.42.2 Supersymmetry

Supersymmetry postulates the existence of the boson corresponding to every known fermion and vice versa. This is a very elegant symmetry that leads to a better quantum field theory than the one on which the Standard Model has been built. If it is right, we have to discover a new world of particles equaling our known world.

5.43 Quantum gravity

The biggest loophole in the Standard Model is that gravity has been left out. If we succeed in constructing a theory of quantum gravity then the analog of electric charge will be mass obtained through a comparison of Newton's law and

Coulomb's law though Einstein's general theory of relativity is what we have to study. Here we have to replace mass m by total energy

$$E = \sqrt{p^2c^2 + m^2c^4} \approx mc^2 \text{ if } m^2c^4 \gg p^2c^2 \text{ i. e. } mc^2 \gg pc.$$

Gravitational field or its quanta graviton carries field energy. So the quantum (graviton) itself carries energy and is self-interacting. The graviton can itself emit or absorb a graviton. It is non-abelian.

Though quantum mechanics and special theory of relativity combine leading to quantum field theory, quantum mechanics and the general theory of relativity have basic incompatibility between them.

Gravity is incorporated into the very fabric of space and time, namely the curvature of space and time in Einstein's general theory of relativity, and has resisted all attempts at being combined with the quantum world.

In Yang–Mills theory, addition of a cubic and quadratic vertex makes the theory complete and consistent. But in gravity this does not happen. Here we have to add 4th, 5th, 6th orders..., and there is an infinite number of vertices of all orders. This is the complication faced in quantum gravity and prevents construction of a complete theory of quantum gravity.

5.43.1 String theory

The most successful attempt to construct quantum gravity is the string theory which is being developed.

In string theory, a point particle is replaced by a one-dimensional object called string as the fundamental entity. Its length is about $10^{-35}\,m$.

The various vibrational modes or states of the string correspond to the elementary particles of the Standard Model.

String theory automatically contains quantum gravity.

String theory works only if the number of space dimensions is 9, and including time it is 10. So strings live in a 10-dimensional world. Apart from the $3 + 1$ dimensions there are 6 dimensions which are curled up to form space bubbles at distance scales of $10^{-35}\,m$. Both the string and the extra curled up dimensions will be revealed only when we can access such length scales.

String theory offers much more than the quantum theory of gravity. String theory offers a quantum theory of all forces including gravity (called quantum gravity). It can incorporate the Standard Model of high energy physics or particle physics also within a unifying framework that includes gravity.

Experimental support of string theory is awaited.

In relativistic quantum mechanis there is an inverse relationship between length scale L and the energy E required to probe it viz. $E \propto \frac{1}{L}$. We need TeV energy of LHC to probe a length scale of $10^{-19}\,m$. To probe the length scale of quantum gravity i.e. $10^{-35}\,m$ we require energy of $10^{16}\,TeV$. This is the energy required to experimentally test string theory or any theory of gravity and obviously this is not currently possible. To go to such high energies we may require new principles of particle acceleration, and newer technologies have to be invented.

In other words, through string theory we can incorporate gravity into quantum mechanics but this demands experimental verification for which we have to do experiments in the Planck scale, which is currently not possible.

5.44 Exercises

Exercise 5.1 *Which quark does not hadronise, i.e. which quark does not combine with other quarks to form a hadron?*

(a) charm quark   *(b) top quark*   *(c) bottom quark*   *(d) all quarks form hadrons.*

Exercise 5.2 *There is an opportunity to study behaviour of a bare quark. Which quark?*

(a) charm quark   *(b) top quark*   *(c) bottom quark*   *(d) none.*

Exercise 5.3 *A proton has mass* $938\frac{MeV}{c^2}$. *It is made up of two u quarks of mass* $2.2\frac{MeV}{c^2}$ *and one d of mass* $4.7\frac{MeV}{c^2}$. *How do you explain this ?*

$\boxed{\text{Ans.}}$ Quark structure of a proton is $p = uud$. Hence the sum of masses of constituent quarks is $m_u + m_u + m_d = (2.2 + 2.2 + 4.7)\frac{MeV}{c^2} = 9.1\frac{MeV}{c^2}$ while $m_p = 938\frac{MeV}{c^2}$.

Actually, the quarks are constrained to stay or remain confined within a very small region of space forming a proton. A huge amount of energy E is associated with such confinement of quarks. So by $E = mc^2$ formula the mass of a proton m_p will be much greater than $9.1\frac{MeV}{c^2}$. Hence $m_p = 938\frac{MeV}{c^2}$.

Exercise 5.4 *What are gauge bosons?*

$\boxed{\text{Ans.}}$ A gauge boson is an elementary boson that acts as a force carrier or mediator of interaction between elementary fermions.

Example:

Photon γ is a mediator of electromagnetic interaction.

$W^\pm$, Z^0 are mediators of weak interaction.

Gluons g are mediators of strong interaction.

$\gamma, W^\pm, Z^0, g$ are the gauge bosons of the Standard Model.

Exercise 5.5 *What do we mean by generation of leptons and quarks?*

$\boxed{\text{Ans.}}$ There are three generations of quarks in the Standard Model.

Each generation is heavier than the previous generation, i.e 2nd and 3rd generations are heavier replicas of the 1st generation. They are organized by their increasing masses. And heavier particles decay into lighter particles.

Also, three generations are needed to explain the matter–anti-matter asymmetry and disappearance of anti-matter.

The three generations of leptons and quarks are as follows.

First generation

$$\text{Leptons} \begin{pmatrix} \nu_e \\ e^- \end{pmatrix} \xrightarrow{\text{charge}} \begin{pmatrix} 0 \\ -1 \end{pmatrix} \xrightarrow{\text{mass}} \begin{pmatrix} 0 \\ 0.51\frac{MeV}{c^2} \end{pmatrix} \text{ and quarks} \begin{pmatrix} u \\ d \end{pmatrix} \xrightarrow{\text{charge}} \begin{pmatrix} \frac{2}{3} \\ -\frac{1}{3} \end{pmatrix} \xrightarrow{\text{mass}} \begin{pmatrix} 2.2\frac{MeV}{c^2} \\ 4.7\frac{MeV}{c^2} \end{pmatrix}$$

Second generation

$$\text{Leptons} \begin{pmatrix} \nu_\mu \\ \mu^- \end{pmatrix} \xrightarrow{\text{charge}} \begin{pmatrix} 0 \\ -1 \end{pmatrix} \xrightarrow{\text{mass}} \begin{pmatrix} 0 \\ 106\frac{MeV}{c^2} \end{pmatrix} \text{ and quarks} \begin{pmatrix} c \\ s \end{pmatrix} \xrightarrow{\text{charge}} \begin{pmatrix} \frac{2}{3} \\ -\frac{1}{3} \end{pmatrix} \xrightarrow{\text{mass}} \begin{pmatrix} 1.28\frac{GeV}{c^2} \\ 96\frac{MeV}{c^2} \end{pmatrix}$$

Third generation

$$\text{Leptons} \begin{pmatrix} \nu_\tau \\ \tau^- \end{pmatrix} \xrightarrow{\text{charge}} \begin{pmatrix} 0 \\ -1 \end{pmatrix} \xrightarrow{\text{mass}} \begin{pmatrix} 0 \\ 1.78\frac{GeV}{c^2} \end{pmatrix} \text{ and quarks} \begin{pmatrix} b \\ t \end{pmatrix} \xrightarrow{\text{charge}} \begin{pmatrix} \frac{2}{3} \\ -\frac{1}{3} \end{pmatrix} \xrightarrow{\text{mass}} \begin{pmatrix} 4.18\frac{GeV}{c^2} \\ 173\frac{GeV}{c^2} \end{pmatrix}$$

Exercise 5.6 *Can a 4th generation of matter be accommodated in the basic structure of matter?*

Ans. There are three generations of matter. There are two quarks and two leptons in each of the first, second and third generations, i.e. altogether there are six quarks and six leptons.

We do not expect more than three generations.

We note that each generation is heavier than the previous generation. It can be shown that over and above the three neutrino flavours viz. ν_e, ν_μ, ν_τ if there were a 4th neutrino flavour then it would have to be more than half the mass of the Z^0 boson, i.e. $\sim 45\frac{GeV}{c^2}$ or higher, which is absurd.

So with the discovery of the fundamental particles viz. quarks and leptons it is highly suggestive that we have got the basic structure of matter. And a fourth generation of matter cannot be accommodated in the basic structure of matter.

Exercise 5.7 *What is the difference between quarks and leptons?*

Ans. We compare quarks and leptons in the following.

Quark	Lepton
Fundamental particle.	Fundamental particle.
Structureless, cannot be broken down to further constituent particles.	Structureless, cannot be broken down to further constituent particles.
Fermions	Fermions

There are six quarks (i.e. six flavours)	There are six leptons. (i.e. six flavours)
There are three generations	There are three generations
$\begin{pmatrix}u\\d\end{pmatrix}, \begin{pmatrix}c\\s\end{pmatrix}, \begin{pmatrix}b\\t\end{pmatrix}$	$\begin{pmatrix}\nu_e\\e^-\end{pmatrix}, \begin{pmatrix}\nu_\mu\\\mu^-\end{pmatrix}, \begin{pmatrix}\nu_\tau\\\tau^-\end{pmatrix}$
They have anti-particles called anti-quarks $\bar{u}, \bar{d}, \bar{s}, \bar{c}, \bar{b}, \bar{t}$	They have anti-particles called anti-leptons $e^+, \bar{\nu}_e, \mu^+, \bar{\nu}_\mu, \tau^-, \bar{\nu}_\tau$
Have fractional electrical charge.	Have integral electric charge.
Carry colour charge. But hadrons are colourless.	Carry no colour charge.
Quarks cannot exist freely. Quarks are confined within hadrons by strong force mediated by gluons. Quarks are bound by gluon strings.	Leptons can exist freely.
Quarks are subject to all fundamental forces.	Leptons are subject to all fundamental forces except strong force.

Exercise 5.8 *What are the essential differences between QED and QCD ?*
   Ans. The differences between QED and QCD are as follows.

QED	QCD
QED stands for quantum electrodynamics.	QCD stands for quantum chromodynamics.
QED describes interactions of charged particles with electromagnetic fields.	QCD describes strong interactions between quarks and gluons.
Photon is a mediator in QED.	Gluon is a mediator in QCD.

Exercise 5.9 *How many possible combinations of colour charge can gluons carry?*

(a) 16   *(b)* 12   *(c)* 9   *(d)* 3.

   Hint. Gluons carry both colour charge and anti-colour charge. There are nine possible combinations viz. $R\bar{R}$, $R\bar{B}$, $R\bar{G}$, $B\bar{R}$, $B\bar{B}$, $B\bar{G}$, $G\bar{R}$, $G\bar{B}$, $G\bar{G}$.

Exercise 5.10 *What are the essential differences between quarks and leptons with gauge bosons?*
   Ans. There exist two basic sets of fundamental constituents of matter: the fundamental or elementary particles (quark, lepton) and the elementary bosons also called gauge boson.

Quark and Lepton	Gauge boson
Quarks and leptons are the fundamental particles being constituents of matter.	Quarks and leptons interact by transmitting or exchanging particles called gauge bosons. So gauge bosons cement them together.

(Continued)

They are fermions.	They are bosons.
There are six quarks and six leptons and also six anti- quarks and six anti-leptons.	There are four gauge bosons.
Example of quark:	Example of gauge boson:
up (u), down (d), charm (c), strange (s), top (t), bottom (b) quark.	Photon γ (which is the quantum of light and carries electromagnetic interaction)
Example of lepton:	$W^\pm$, Z^0 (which mediates weak interaction)
electron (e^-), electron neutrino (ν_e) muon (μ^-), muon neutrino (ν_μ) tauon (τ^-), tauon neutrino (ν_τ).	Gluon g (which mediates strong interaction).

Exercise 5.11 *Which of the following interactions of nature, described in the Standard Model have range that has $\frac{1}{r^2}$ dependence ?*

(a) Strong interaction

(b) Weak interaction

(c) Gravitational interaction

(d) Electromagnetic interaction.

Exercise 5.12 *Grand unification theory is unification of*

(a) electroweak force and the strong force

(b) electroweak force and the gravitational force

(c) electromagnetic force and weak force

(d) all the forces of nature.

Exercise 5.13 *According to the Standard Model, electromagnetism and weak force get unified into*

(a) grand unified force

(b) electroweak force

(c) GUT

(d) electroweak magnetic force.

Exercise 5.14 *Which of the following statements is/are true ?*
   (a) Quantum gravity is a part of the Standard Model.
   (b) Quantum gravity is not part of the Standard Model.
   (c) There is no quantum gravity.
   (d) Classical gravity is part of the Standard Model.

Exercise 5.15 *The constituents of matter are called*
   (a) bosons
   (b) fermions

(c) both bosons and fermions

(d) neither bosons nor fermions but nucleons and electrons.

Answers to multiple choice questions
5.1b, 5.2b, 5.9c, 5.10d, 5.11d, 5.12a, 5.13b, 5.14c, 5.15b.

5.45 Question bank

Q 5.1 What is the Standard Model?

Q 5.2 What are the various interactions of nature?

Q 5.3 What is the eightfold way?

Q 5.4 How did Gell-Mann predit quark structure?

Q 5.5 What is the quark structure of mesons and baryons?

Q 5.6 Why do quarks not have free existence?

Q 5.7 What is colour charge? How was it introduced?

Q 5.8 Mention a particle having charm quark as its constituent.

Q 5.9 Which is the heaviest quark? (Answer top quark)

Q5.10 What is the spin of gauge bosons?

Q 5.11 What is the role of the Higgs boson? What is the Higgs mechanism?

Q 5.12 How are range and mass of quanta of force related?

Q 5.13 What do we mean by abelian and non-abelian fields? Give examples.

Q 5.14 What do we mean by string theory?

Q 5.15 Classify the fundamental particles of the Standard Model showing their mass, charge and spin.

Q 5.16 Which forces of nature are unified in the Standard Model?

Q 5.17 Is the Standard Model consistent with quantum mechanics, i.e. understood quantum mechanically ?

Q 5.18 Why is gravity not incorporated into the Standard Model?

Further reading

[1] Perkins H 2000 *Introduction to High energy Physics* (Cambridge: Cambridge University Press)

[2] Griffiths D 2008 *Introduction to Elementary Particles* (New York: Wiley-VCH)

[3] Tayal D C 2009 *Nuclear Physics* (Mumbai: Himalaya Publishing House)

[4] Prakash S 2005 *Nuclear Physics & Particle Physics* (New Delhi: Sultan Chand & Sons)

[5] Guha J 2019 *Quantum Mechanics: Theory, Problems & Solutions* 3rd edn (Kolkata: Books and Allied (P) Ltd)

[6] Lim Y K 2002 *Problems and Solutions on Atomic, Nuclear and Particle Physics* (Singapore: World Scientific)

IOP Publishing

Nuclear and Particle Physics with Cosmology, Volume 2
Particle physics and cosmology
Jyotirmoy Guha

Chapter 6

Study of interactions through Feynman diagram

We discuss the rules of Feynman diagram, defining internal lines, external lines and the vertex. We give several examples and describe in Feynman diagrams phenomena such as electron–electron scattering, positron–positron scattering, pair production, pair annihilation, Compton scattering, negative and positive beta decays, muon decay etc.

6.1 Introduction

Feynman diagram is a pictorial representation of interactions between particles. It shows the exchange particles and their passage in time. In other words, Feynman diagrams are graphical ways to represent exchange forces. It is a nice and easy way to capture complex nuclear interactions. It is a tool to analyse interactions.

Feynman introduced these diagrams in 1948.

A Feynman diagram depicts the possible way for a process to occur and does not represent actual paths of particles and anti-particles in space-time.

Processes that we draw in a Feynman diagram can happen only if energy and momentum conservation holds.

From a Feynman diagram one can calculate cross-section, branching ratio etc.

6.2 Rules of a Feynman diagram

We discuss the rules of a Feynman diagram in the following.

A Feynman diagram consists of lines and vertices.

6.2.1 Line

There are two lines: a straight line and a wiggly or wavy line.

The straight line ab of figure 6.1 represents fermions like e^-, μ^-, u, d, s.

The wiggly or wavy line cd represents bosons, i.e. exchange particles or virtual mediators like γ, $W^\pm$, g.

doi:10.1088/978-0-7503-5032-7ch6

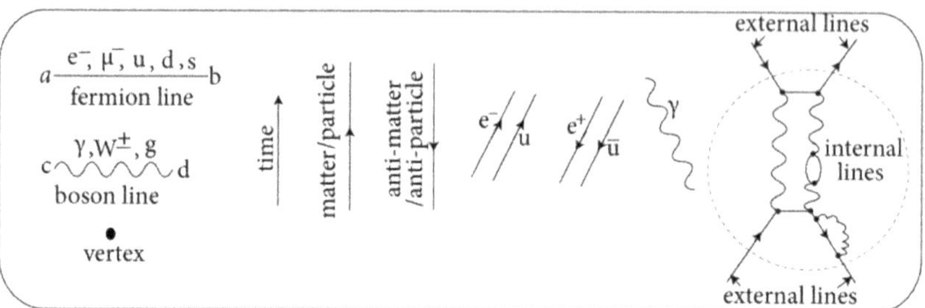

Figure 6.1. Rules of Feynman diagram.

Virtual particles do not have to follow the standard energy–momentum relation and do not have a classical motion in space-time.

Virtual particles are allowed in the representation of a Feynman diagram as a mathematical tool to calculate energy and momentum transfer in interactions.

Feynmann diagrams are purely symbolic. They do not represent particle trajectories.

Let us take the vertical direction to represent time. Time axis determines the order in which events occur.

Horizontal spacing does not correspond to physical separations.

If a fermion moves in the direction of the time axis then it is a particle or matter.

If a fermion moves opposite to the direction of time axis then it is an anti-particle or anti-matter.

Photon is its own anti-particle. So it does not require to be labelled by an arrow. It can be represented as travelling both forward as well as backward in time in the Feynman diagram.

The point where an exchange particle interacts with a particle is referred to as the vertex.

6.2.2 Vertex

Lines meet to form a vertex, which is shown in figure 6.1. We cannot have both arrows of two lines pointing into or away from the vertex. When there is a vertex there should be a charged fermion comimg in and another going out of the vertex. In other words, fermion current has to go through the vertex.

At a vertex charge, lepton number and baryon number are conserved.

We will discuss baryon number in chapter 7.

There are two types of lines in Feynman diagram, external and internal lines.

6.2.3 External lines

Lines which enter or leave the diagram are called external lines. This is shown in figure 6.1. These represent real observable particles. So external lines give information regarding the physical process involved.

6.2.4 Internal lines

Lines which begin and end within the diagram are called internal lines. This is shown in figure 6.1. These represent particles that are not observed. They are virtual particles. So internal lines give information regarding the mechanism involved. In fact, the internal lines are irrelevant as far as the observed process is concerned.

We give various examples in the following, study of which will make clear the interpretation of Feynman diagram.

We also note the following.

✓ We show in *exercise 1.20* that a γ cannot pair produce $e^{\pm}$ in vacuum, i.e.

$$\gamma \neq e^+ + e^-$$

✓ In *exercise 1.21* we show that $e^{\pm}$ cannot annihilate into a single photon γ, i.e.

$$e^+ + e^- \neq \gamma$$

✓ We show in *exercise 1.22* that an isolated $e^{\pm}$ cannot emit or absorb a photon γ, i.e.

$$e^{\pm} \neq e^{\pm} + \gamma$$

$$e^{\pm} - \gamma \neq e^{\pm}$$

The photon γ referred to here is a real photon. And these cannot take place because of violation of energy–momentum.

When constructing a Feynman diagram these processes can be drawn internally within the diagram with a virtual photon as shown in figure 6.2. The internal diagram has no physical relevance as far as the observed process is concerned.

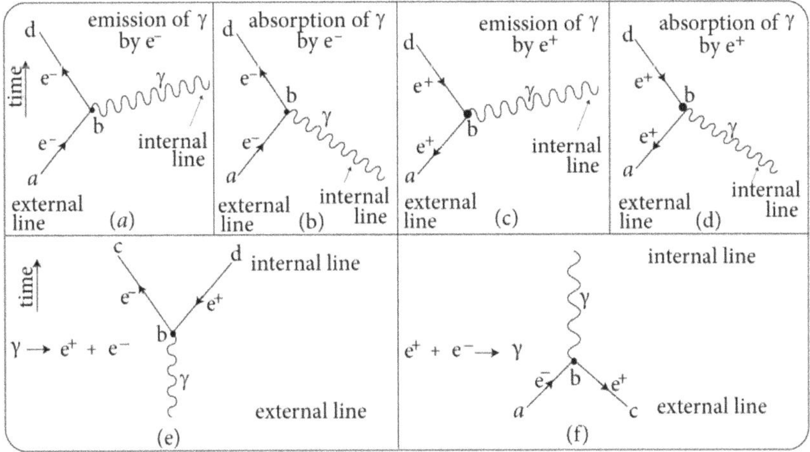

Figure 6.2. (a) e^- emitting a virtual photon, (b) e^- absorbing a virtual photon, (c) e^+ emitting a virtual photon, (d) e^+ absorbing a virtual photon, (e) virtual $\gamma \to e^+ + e^-$, (f) $e^+ + e^- \to$ virtual γ.

Example 6.1. e^- emitting a virtual photon as depicted in figure 6.2(a).

An electron e^- moves along ab forward in time and emits photon γ at b. The e^- then moves along bd forward in time.

Example 6.2. e^- absorbing a virtual photon as depicted in figure 6.2(b).

An electron e^- moves along ab forward in time, encounters a photon γ at b and absorbs it. The e^- then moves along bd forward in time.

Example 6.3 e^+ emitting a virtual photon as depicted in figure 6.2(c).

A positron e^+ moves along ba backward in time and emits photon γ at b. The e^+ then moves along db backward in time.

Example 6.4 e^+ absorbing a virtual photon as depicted in figure 6.2(d).

A positron e^+ moves along ba backward in time, encounters a photon γ at b and absorbs it. The e^+ then moves along db backward in time.

Example 6.5 virtual $\gamma \to e^+ + e^-$ as depicted in figure 6.2(e).

A photon γ (of sufficient energy) disappears at point b and matter, i.e. particle e^-, and anti-matter, i.e. anti-particle e^+, appears thereafter. The e^- travels along bc (forward in time representing particle) while e^+ travels along db (backward in time representing anti-particle).

Example 6.6 $e^+ + e^- \to$ virtual γ as depicted in figure 6.2(f).

We consider electron e^- (particle) moving along ab, forward in time and positron e^+ (anti-particle of electron) moving along bc backward in time. They collide at the vertex b and produce an electromagnetic photon.

6.2.5 Role of the vertex

Each vertex within a Feynmann diagram introduces a factor $\frac{1}{137}$ (the fine structure constant). As this is a small number, diagrams with large numbers of vertices (4, 6, ...) contribute less and less.

6.3 Electron–electron scattering

We have depicted the electron–electron scattering in Feynman diagram of figue 6.3(a).

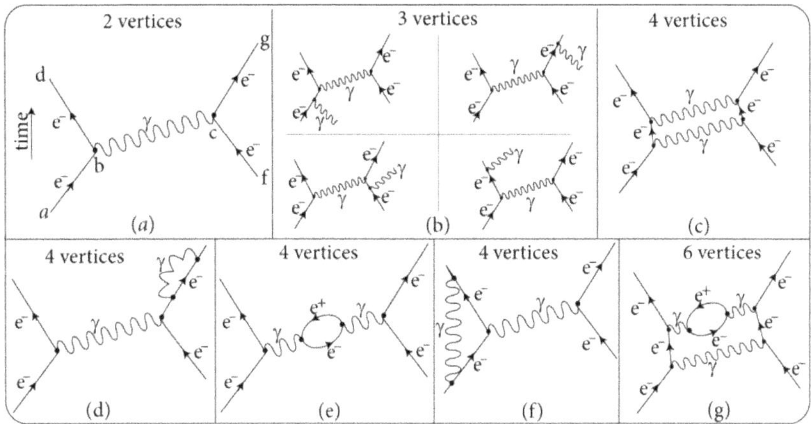

Figure 6.3. Electron–electron scattering in a Feynman diagram.

Suppose two electrons (e^-, e^-) move towards each other over space and time (one along ab, the other along fc, both forward in time). They exchange photon γ represented by the wavy line: one e^- emits γ from b at some time that strikes the other e^- at c and gets absorbed. Then the electrons recoil and move away (one along bd, the other along cg both forward in time). Clearly the electrons e^-, e^- do not bounce off as they are not in direct contact but scatter by exchanging a photon.

We note that after emission or absorption of a photon, e^- follows a different direction to conserve momentum. This is how two electrons e^-, e^- exchange photon.

Other possibilities are shown in figures 6.3(b)–(g). These are possible internal processes that can occur.

Figure 6.3(b) shows that the interacting electrons can emit or absorb additional photons. There are three vertices in this case.

Figure 6.3(c) shows that one of the electrons can emit a second photon which is absorbed by the other electron.

Figure 6.3(d) shows that an electron emits a photon and then absorbs it.

Figure 6.3(e) shows that an emitted photon might cause pair production and then coalesce back into a photon.

Figure 6.3(f) shows one electron emits two photons and one photon is aborbed by itself while the other photon is absorbed by the second electron.

Figure 6.3(g) shows multiple photons are exchanged with some of them creating other particles through pair production and so on.

6.4 Positron–positron scattering

We depict electron–electron scattering in the Feynman diagram of figure 6.4.

Suppose two electrons e^+, e^+ move towards each other over space and time (one along ba, the other along cf both backward in time). They exchange photon γ represented by the wavy line. One e^+ emits γ from b at some time that is absorbed by the other e^+ at c. Then the positrons move away (one along db, the other along gc

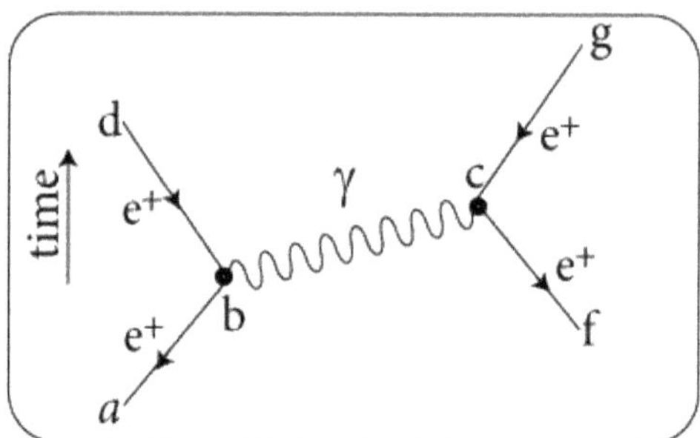

Figure 6.4. Positron–positron scattering through exchange of a photon.

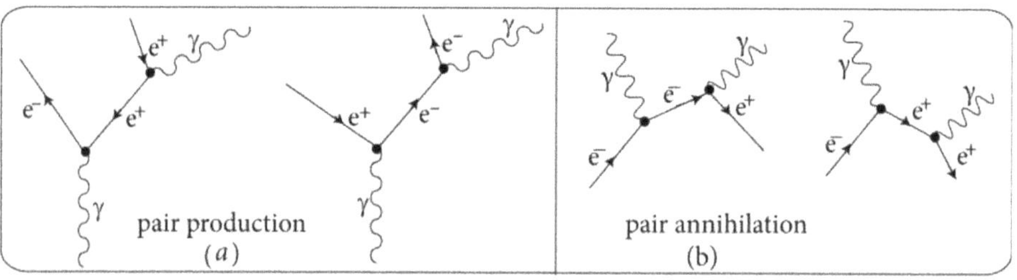

Figure 6.5. Feynman diagram of (a) pair production $\gamma + \gamma \to e^- + e^+$, (b) electron–positron annihilation $e^- + e^+ \to \gamma + \gamma$.

both backward in time). Clearly the electrons e^+, e^+ do not bounce off as they are not in direct contact but scatter by exchanging a photon.

We note that after emission or absorption of a photon, e^+ follows a different direction to conserve momentum.

6.5 Pair production and pair annihilation

We depict pair production

$$\gamma + \gamma \to e^- + e^+$$

and pair annihilation

$$e^- + e^+ \to \gamma + \gamma$$

in the Feynman diagram of figure 6.5(a,b).

6.6 Compton scattering

We depict Compton scattering

$$e^- + \gamma \to e^- + \gamma$$

in the Feynman diagram of figure 6.6.

6.7 Negative beta decay

We depict negative beta, i.e. β^- decay, in figure 6.7(a) as a Feynman diagram.

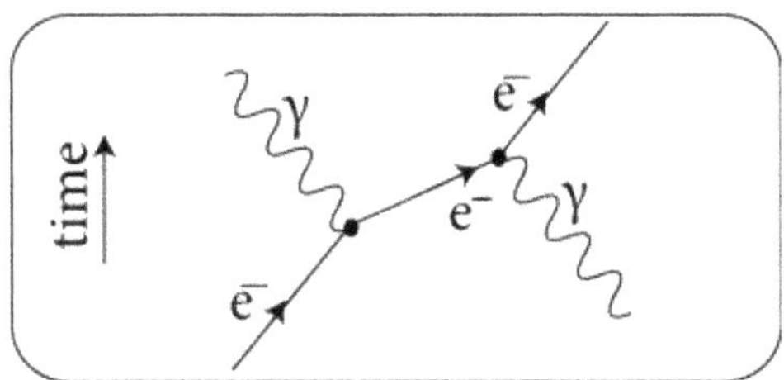

Figure 6.6. Compton scattering in a Feynman diagram.

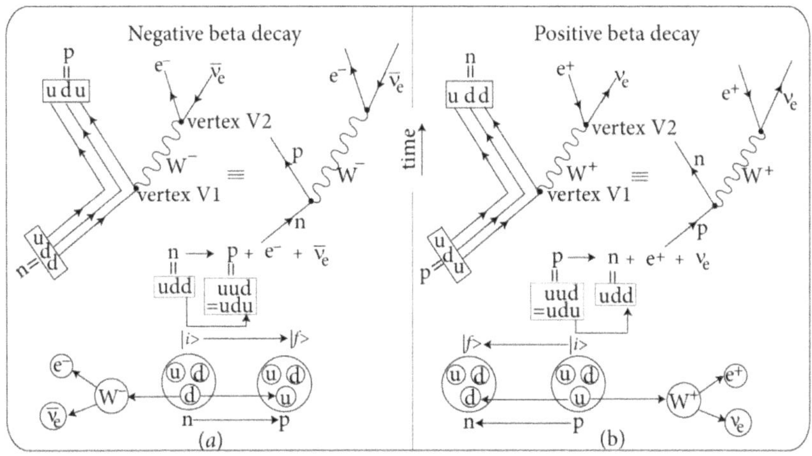

Figure 6.7. (a) Negative beta decay. (b) Positive beta decay.

In negative beta decay $n = udd$ changes to $p = uud$, i.e. one d quark changes to one u quark and an exchange particle W^- is emitted. W^- then decays to e^- (matter) and $\bar{\nu}_e$ (anti-matter). We can represent the scheme $n \rightarrow p + e^- + \bar{\nu}_e$ as follows

$$n \rightarrow p + W^- \text{ or } d \rightarrow u + W^-$$

$$W^- \rightarrow e^- + \bar{\nu}_e$$

Clearly according to the scheme one d quark converts to u quark and W^-. Again, W^- is much more massive than d quark ($m_{W^-} = 80\frac{GeV}{c^2} \gg 4.7\frac{MeV}{c^2}$). This is possible since the W^- generated is virtual and need not have the mass of a real W^- boson. The interaction occurs for an extremely short duration that allows violation of energy conservation as permitted by Heisenberg's uncertainty principle.

6.7.1 Charge conservation

At the vertex $V1$

$$Q(d) = Q(u) + Q(\text{exchange particle})$$

$$Q(\text{exchange particle}) = Q(d) - Q(u) = -\frac{1}{3}|e| - \left(\frac{2}{3}|e|\right) = -|e|$$

This is the charge possessed by the exchange boson W^- only. So negative beta decay is mediated by weak force.

At the vertex $V2$

$$Q_i = Q(W^-) = -|e|$$

$$Q_f = Q(e^-) + Q(\bar{\nu}_e) = -|e| + 0 = -|e|.$$

$Q_i = Q_f$ so charge is conserved.

6.7.2 Baryon number conservation

We will discuss baryon number in chapter 7.

At the vertex $V1$

One quark d enters vertex $V1$ and one quark u leaves it. So quark count or baryon number is conserved at the vertex. Also

$$B_i = B(d) = \frac{1}{3}$$

$$B_f = B(u) = \frac{1}{3}.$$

$B_i = B_f$. So baryon number is conserved.

At the vertex $V2$

No baryons are involved at vertex $V2$.

6.7.3 Electron lepton number conservation

At vertex $V1$.

No leptons are involved at $V1$.

At vertex $V2$.

No leptons enter since W^- is not a lepton. So initial lepton number is

$$L_{ei} = L_{ei}(W^-) = 0 \quad (i \rightarrow \text{initial}).$$

The final lepton number is

$$L_{ef} = L_e(e^-) + L_e(\bar{\nu}_e) = +1 - 1 = 0 \quad (f \rightarrow \text{final})$$

$L_{ei} = L_{ef}$. So electron lepton number is conserved.

6.8 Positive beta decay

We depict positive beta, i.e. β^+, decay in figure 6.7(b) in a Feynman diagram.

In positive beta decay $p = uud$ changes to $n = udd$, i.e. one u quark changes to one d quark and an exchange particle W^+ is emitted. W^+ then decays to e^+ (antimatter) and ν_e (matter). We can represent the scheme $p \rightarrow n + e^+ + \nu_e$ as follows

$$p \rightarrow n + W^+ \text{ or } u \rightarrow d + W^+$$

$$W^+ \rightarrow e^+ + \nu_e$$

Clearly according to the scheme one u quark converts to d quark and W^+. The W^+ is much more massive than the u quark ($m_{W^+} = 80\frac{GeV}{c^2} \gg 2.2\frac{MeV}{c^2}$). We note that the W^+ generated is virtual and need not have the mass of a real W^+ boson. The interaction occurs for an extremely short duration that allows violation of energy conservation as permitted by Heisenberg's uncertainty principle.

6.8.1 Charge conservation

At vertex $V1$:

$$Q(u) = Q(d) + Q \text{ (exchange particle)}.$$

$$Q(\text{exchange particle}) = Q(u) - Q(d) = -\frac{1}{3} |e| - \left(\frac{2}{3} |e|\right) = -|e|$$

This is the charge possessed by the exchange particle W^+. So positive beta decay is mediated by weak force.

At vertex $V2$:

$$Q_i = Q(W^+) = |e|$$

$$Q_f = Q(e^+) + Q(\nu_e) = |e| + 0 = |e|.$$

$Q_i = Q_f$. So charge is conserved.

6.8.2 Baryon number conservation

At vertex $V1$.

One quark u enters vertex $V1$ and one quark d leaves it. So quark count or baryon number is conserved at vertex $V1$ as

$$B_i = B(u) = \frac{1}{3}$$

$$B_f = B(d) = \frac{1}{3}.$$

$B_i = B_f$. So baryon number is conserved.
At vertex $V2$
No baryons are involved at $V2$.

6.8.3 Electron lepton number conservation

At vertex $V1$.
No leptons are involved at $V1$.
At vertex $V2$.

No leptons enter since W^+ is not a lepton. So initial lepton number is

$$L_{ei} = L_{ei}\,(W^+) = 0.$$

The final lepton number is

$$L_{ef} = L_e(e^+) + L_e(\nu_e) = -1 + 1 = 0$$

$L_{ei} = L_{ef}$. So lepton number is conserved.

Free protons are extremely stable and free protons cannot undergo β^+ decay ($p \to n$). It happens only in the nucleus (bound states). We discuss the reason in chapter 8, section 8.20.

6.9 Exercises

Exercise 6.1. *Depict the following interactions in a Feynman diagram and study the conservation laws.*
(a) *Pion decay:* $\pi^+ \to \mu^+ + \nu_\mu$
(b) $p + e^- \to n + \nu_e$
(c) $n + \nu_e \to p + e^-$
(d) $p + \overline{\nu}_e \to n + e^-$.
Ans. We depict the interactions in a Feynman diagram in figure 6.8.
(a) Refer to pion decay as shown in figure 6.8(a) viz.
$\pi^+ \to \mu^+ + \nu_\mu$
π^+ has quark structure $\pi^+ \equiv u\overline{d}$. It gives off W^+ (vertex $V1$) which decays to μ^+ and ν_μ (vertex $V2$).
We study the conservation laws now.
✓ Charge conservation
At vertex $V1$
$Q_i = Q(u) + Q(\overline{d}) = \frac{2}{3}\,|e| + \frac{1}{3}\,|e| = |e|, \quad Q_f = Q(W^+) = |e|.$
$Q_f = Q_i$. So electric charge is conserved at vertex $V1$.
At vertex $V2$: $Q_i = Q(W^+) = |e|, \quad Q_f = Q(\nu_\mu) + Q(\mu^+) = 0 + |e| = |e|$
$Q_f = Q_i$. So electric charge is conserved at vertex $V2$.
✓ Baryon number or quark count conservation
At vertex $V1$: : $B_i = B(u) + B(\overline{d}) = \frac{1}{3} + \left(-\frac{1}{3}\right) = 0, \quad B_f = B(W^+) = 0$

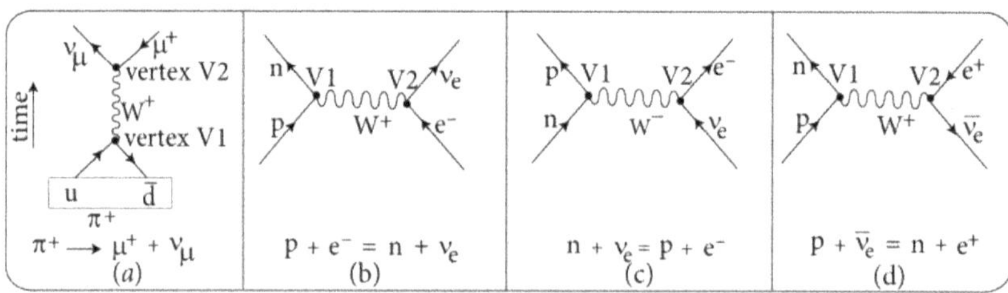

Figure 6.8. (a) A positive pion decays to μ^+ and ν_μ, (b) capture of e^- by p, (c) neutron–neutrino interaction, (d) proton–anti-neutrino interaction.

$B_f = B_i$. So baryon number is conserved at vertex $V1$.
At vertex $V2$: No baryons are involved at vertex $V2$.
✓ Muon lepton number conservation
At vertex $V1$: No leptons are involved at vertex $V1$.
At vertex $V2$: $L_{\mu i} = L_\mu(W^+) = 0$, $L_{\mu f} = L_\mu(\nu_\mu) + L_\mu(\mu^+) = +1 - 1 = 0$.
$L_{\mu f} = L_{\mu i}$. So lepton number is conserved at vertex $V2$.
(b) Refer to the capture of e^- by p shown in figure 6.8(b) viz.
$$p + e^- \to n + \nu_e$$
At vertex $V1$ proton p emits a W^+ and converts to a neutron n. This W^+ is absorbed by electron e^- and a ν_e is given off at vertex.
We study the conservation laws now.
✓ Charge conservation
At vertex $V1$
$Q_i = Q(p) = |e|$, $Q_f = Q(n) + Q(W^+) = 0 + |e| = |e|$.
$Q_i = Q_f$. So electric charge is conserved at vertex $V1$.
At vertex $V2$
$Q_i = Q(W^+) + Q(e^-) = |e| - |e| = 0$, $Q_f = Q(\nu_e) = 0$.
$Q_i = Q_f$. So electric charge is conserved at vertex $V2$.
✓ Baryon number conservation
At vertex $V1$
$B_i = B(p) = 1$, $B_f = B(n) + B(W^+) = 1 + 0 = 1$.
$B_i = B_f$. So baryon number is conserved at vertex $V1$.
At vertex $V2$
No baryons are involved at $V2$.
✓ Electron lepton number conservation
At vertex $V1$.
No leptons are involved at $V1$.
At vertex $V2$
$L_{ei} = L_e(W^+) + L_e(e^-) = 0 + 1 = 1$, $L_{ef} = L_e(\nu_e) = +1$
$L_{ei} = L_{ef}$. So electron lepton number is conserved.
(c) Refer to the neutron–neutrino interaction shown in figure 6.8(c) viz.
$$n + \nu_e \to p + e^-$$
At vertex $V1$ neutron n emits a W^- and converts to proton p. This W^- and ν_e gives off e^-.
We study the conservation laws now.
✓ Charge conservation
At vertex $V1$
$Q_i = Q(n) = 0$, $Q_f = Q(p) + Q(W^-) = |e| - |e| = 0$.
$Q_i = Q_f$. So charge is conserved at vertex $V1$.
At vertex $V2$
$Q_i = Q(W^-) + Q(\nu_e) = -|e| + 0 = -|e|$, $Q_f = Q(e^-) = -|e|$.
$Q_i = Q_f$. So electric charge is conserved at vertex $V2$.
✓ Baryon number conservation

At vertex $V1$

$B_i = B(n) = 1$, $B_f = B(p) + B(W^-) = 1 + 0 = 1$.

$B_i = B_f = 1$. So baryon number is conserved at vertex $V1$.

At vertex $V2$.

No baryons are involved at $V2$.

✓ Electron lepton number conservation

At vertex $V1$.

No leptons are involved at vertex $V1$.

At vertex $V2$

$L_{ei} = L_e(W^-) + L_e(\nu_e) = 0 + 1 = 1$, $L_{ef} = L_e(e^-) = 1$

$L_{ei} = L_{ef}$. So lepton number is conserved at vertex $V2$.

(d) Refer to the proton–anti-neutrino interaction shown in figure 6.8(d) viz.

$$p + \bar{\nu}_e \rightarrow n + e^+$$

At vertex $V1$ proton p emits a W^+ and converts to neutron n. This W^+ and $\bar{\nu}_e$ gives off e^-.

We study the conservation laws now.

✓ Charge conservation

At vertex $V1$

$Q_i = Q(p) = |e|$, $Q_f = Q(n) + Q(W^+) = 0 + |e| = |e|$.

$Q_i = Q_f$. So charge is conserved at vertex $V1$.

At vertex $V2$

$Q_i = Q(W^+) + Q(\bar{\nu}_e) = |e| + 0 = |e|$, $Q_f = Q(e^+) = |e|$

$Q_i = Q_f$. So electric charge is conserved at vertex $V2$.

✓ Baryon number conservation

At vertex $V1$

$B_i = B(p) = 1$, $B_f = B(n) + B(W^+) = 1 + 0 = 1$.

$B_i = B_f = 1$. So baryon number is conserved at vertex $V1$.

At vertex $V2$

No baryons are involved at $V2$.

✓ Electron lepton number conservation

At vertex $V1$

No leptons are involved at vertex $V1$.

At vertex $V2$

$L_{ei} = L_e(W^+) + L_e(\bar{\nu}_e) = 0 - 1 = -1$, $L_{ef} = L_e(e^+) = -1$.

$L_{ei} = L_{ef}$. So electron lepton number is conserved at vertex $V2$.

Exercise 6.2. *Depict the following interactions through Feynman diagram and study the conservation laws.*

    (a) $s \rightarrow u + \bar{u} + d$

    (b) $e^- + e^- \rightarrow e^- + e^-$

    (c) $e^+ + e^+ \rightarrow e^+ + e^+$

    (d) $\nu_e + n \rightarrow \nu_e + n$.

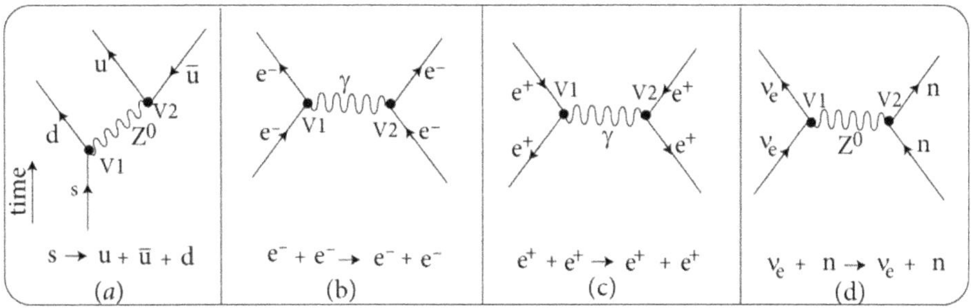

Figure 6.9. (a) s quark gives u, $\bar{u}$, d quarks. (b) Interaction between two electrons through exchange of photon. (c) Interaction between two positrons through exchange of photon. (d) Neutron–neutrino collision mediated by Z^0 boson.

Ans. We depict the interactions in the Feynman diagram in figure 6.9.

(a) Refer to the interaction shown in figure 6.9(a) viz.

$s \rightarrow u + \bar{u} + d$

We study the conservation laws now.

✓ Charge conservation

At vertex $V1$

$Q_i = Q(s) = -\frac{1}{3}|e|, \quad Q_f = Q(d) + Q(Z^0) = -\frac{1}{3}|e| + 0 = -\frac{1}{3}|e|.$

$Q_i = Q_f$. So charge is conserved at vertex $V1$.

At vertex $V2$

$Q_i = Q(Z^0) = 0, \quad Q_f = Q(u) + Q(\bar{u}) = \frac{2}{3}|e| - \frac{2}{3}|e| = 0.$

$Q_i = Q_f$. So charge is conserved at vertex $V2$.

✓ Baryon number conservation

At vertex $V1$

$B_i = B(s) = \frac{1}{3}, \quad B_f = B(d) + B(Z^0) = \frac{1}{3} + 0 = \frac{1}{3}.$

$B_i = B_f = 1$. So baryon number is conserved at vertex $V1$.

At vertex $V2$

$B_i = B(Z^0) = 0, \quad B_f = B(u) + B(\bar{u}) = \frac{1}{3} - \frac{1}{3} = 0$

$B_i = B_f$. So baryon number is conserved at vertex $V2$.

(b) Refer to the interaction shown in figure 6.9(b) viz.

$e^- + e^- \rightarrow e^- + e^-$

We study the conservation laws now.

✓ Charge conservation

At vertex $V1$

$Q_i = Q(e^-) = -|e|, \quad Q_f = Q(e^-) + Q(\gamma) = -|e| + 0 = -|e|.$

$Q_i = Q_f$. So charge is conserved at vertex $V1$.

At vertex $V2$

$Q_i = Q(\gamma) + Q(e^-) = 0 - |e| = -|e|, \quad Q_f = Q(e^-) = -|e|$

$Q_i = Q_f$. So charge is conserved at vertex $V2$.

✓ Electron lepton number conservation

At vertex $V1$

$L_{ei} = L_e(e^-) = 1,\ L_{ef} = L_e(e^-) + L_e(\gamma) = 1 + 0 = 1.$

$L_{ei} = L_{ef} = 1.$ So lepton number is conserved at vertex $V1$.

At vertex $V2$

$L_i = L(\gamma) + L(e^-) = 0 + 1 = 1,\ L_f = L(e^-) = 1.$

$L_i = L_f = 1.$ So lepton number is conserved at vertex $V2$.

(c) Refer to the interaction shown in figure 6.9(c) viz.

$e^+ + e^+ \rightarrow e^+ + e^+.$

We study the conservation laws now.

✓ Charge conservation

At vertex $V1$

$Q_i = Q(e^+) = |e|,\ Q_f = Q(e^+) + Q(\gamma) = |e| + 0 = |e|.$

$Q_i = Q_f.$ So charge is conserved at vertex $V1$.

At vertex $V2$

$Q_i = Q(\gamma) + Q(e^+) = 0 + |e| = |e|,\ Q_f = Q(e^+) = |e|.$

$Q_i = Q_f.$ So charge is conserved at vertex $V2$.

✓ Electron lepton number conservation

At vertex $V1$

$L_i = L(e^+) = 1,\ L_f = L(e^+) + L(\gamma) = 1 + 0 = 1.$

$L_i = L_f = 1.$ So lepton number is conserved at vertex $V1$.

At vertex $V2$

$L_i = L(\gamma) + L(e^+) = 0 + 1 = 1,\ L_f = L(e^+) = 1.$

$L_i = L_f = 1.$ So lepton number is conserved at vertex $V2$.

(d) Refer to the interaction shown in figure 6.9(d) viz.

$\nu_e + n \rightarrow \nu_e + n$

We study the conservation laws now.

✓ Charge conservation

At vertex $V1$

$Q_i = Q(\nu_e) = 0,\ Q_f = Q(\nu_e) + Q(Z^0) = 0 + 0 = 0.$

$Q_i = Q_f.$ So charge is conserved.

At vertex $V2$

$Q_i = Q(Z^0) + Q(n) = 0 + 0 = 0,\ Q_f = Q(n) = 0$

$Q_i = Q_f.$ So charge is conserved.

✓ Electron lepton number conservation

At vertex $V1$

$L_{ei} = L_e(\nu_e) = 1,\ L_{ef} = L_e(\nu_e) + L_e(Z^0) = 1 + 0 = 1.$

$L_{ei} = L_{ef} = 1.$ So lepton number is conserved.

At vertex $V2$

No leptons are involved at V2.

✓ Baryon number conservation

At vertex $V1$

No baryons are involved at $V1$.

At vertex $V2$

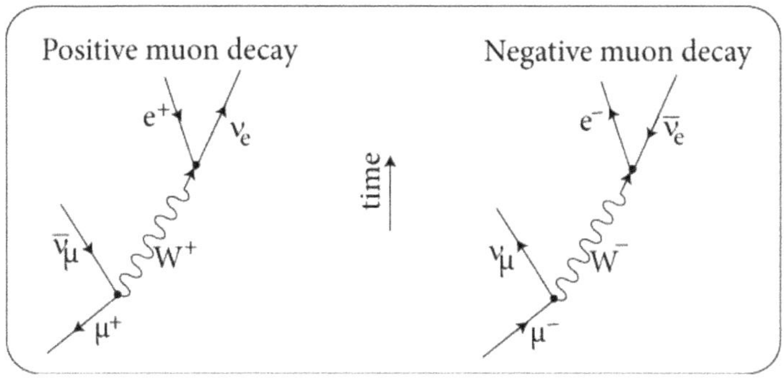

Figure 6.10. Muon decays in a Feynman diagram.

$$B_i = B(Z^0) + B(n) = 0 + 1 = 1, \quad B_f = B(n) = 1$$

$B_i = B_f$. So baryon number is conserved.

Exercise 6.3. *Depict muon decays in a Feynman diagram.*
 Ans. Muon decays are indicated in the Feynman diagram of figure 6.10.

$$\mu^+ \rightarrow e^+ + \overline{\nu}_\mu + \nu_e$$
$$\mu^- \rightarrow e^- + \nu_\mu + \overline{\nu}_e$$

Exercise 6.4. *Show the following interactons in a Feynman diagram.*
  (a) $K^0 \rightarrow \pi^+ + \pi^-$
  (b) $e^- + e^+ \rightarrow \mu^- + \mu^+$
  (c) electron–positron scattering
  (d) $\nu_e + \overline{\nu}_e \rightarrow e^- + e^+$
  (e) $e^- + e^+ \rightarrow q + \overline{q}$
  (f) $e^- + \nu_e \rightarrow e^- + \nu_e$

 Ans. Feynman diagram is shown in figure 6.11.

Exercise 6.5. *Depict the interaction between nucleons through pion exchanges in a Feynman diagram.*
 Ans. Interaction between nucleons through pion exchanges is indicated in the Feynman diagram figure 6.12.

Exercise 6.6. *Interpret the four diagrams of figure* 6.13 *by identifying the interacting particles A, B as e^-, e^+ and C, D as μ^-, μ^+.*

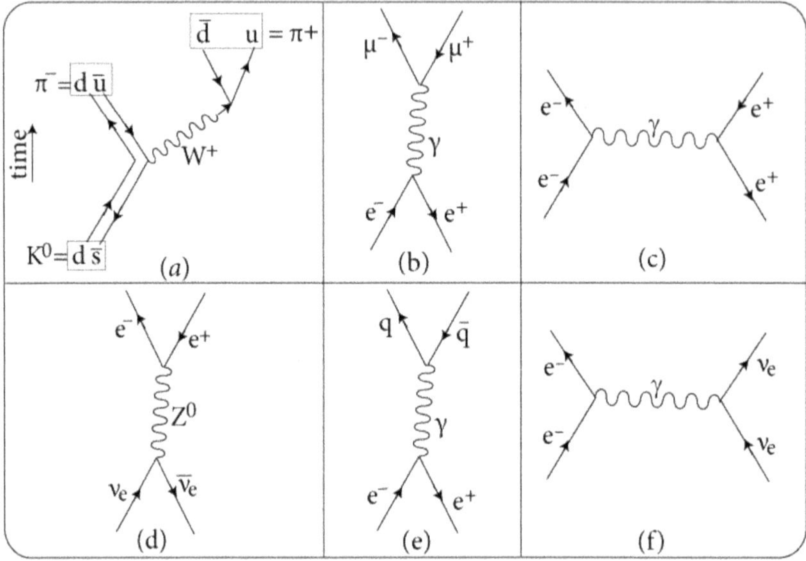

Figure 6.11. Pertaining to *exercise 6.4*.

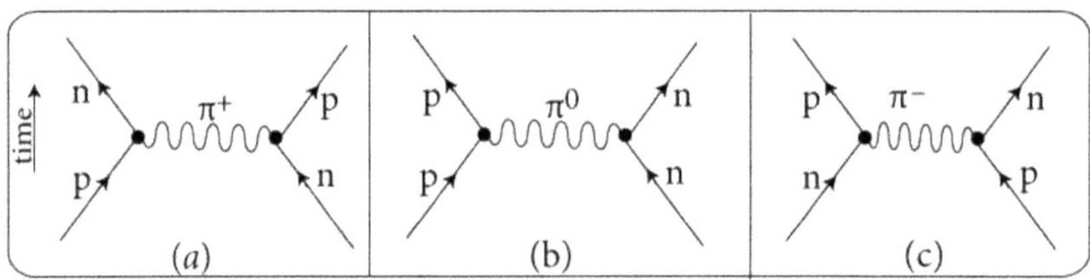

Figure 6.12. (a) Exchange of π^+. (b) Exchange of π^0. (c) Exchange of π^-.

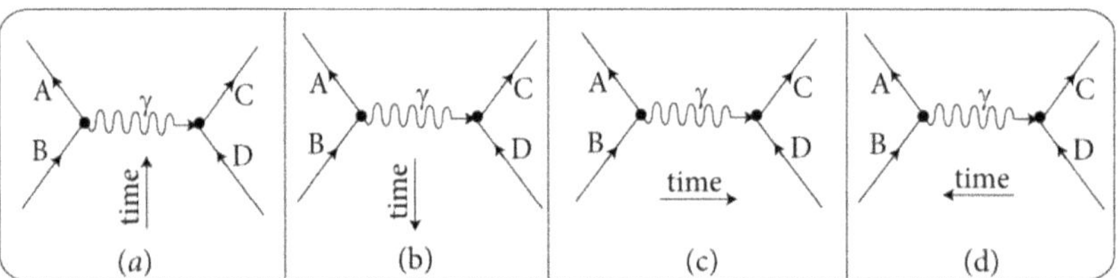

Figure 6.13. Pertaining to *exercise 6.6*.

Ans. (a) $A = e^-$, $B = e^-$, $C = \mu^-$, $D = \mu^-$ ($B + D \to A + C$ i.e. $e^- + \mu^- \to e^- + \mu^-$)

(b) $A = e^+$, $B = e^+$, $C = \mu^+$, $D = \mu^+$ ($A + C \to B + D$ i.e. $e^+ + \mu^+ \to e^+ + \mu^+$)

(c) $A = e^+$, $B = e^-$, $C = \mu^-$, $D = \mu^+$ ($A + B \to C + D$ i.e. $e^+ + e^- \to \mu^- + \mu^+$)

(d) $A = e^-$, $B = e^+$, $C = \mu^+$, $D = \mu^-$ ($C + D \to A + B$ i.e. $\mu^+ + \mu^- \to e^- + e^+$).

It is clear that the same diagram represents four different processes based on the choice of time arrow.

6.10 Question bank

Q 6.1 What is the purpose of a Feynman diagram? What is its utility?

Q 6.2 Are the trajectories in a Feynman diagram real particle trajectories?

Q 6.3 Represent in a Feynman diagram the following interactions or processes.
  (a) electron–electron scattering
  (b) positron–positron scattering
  (c) pair production
  (d) pair annihilation
  (e) Compton scattering
  (f) β^- decay
  (g) β^+ decay
  (h) μ^- decay
  (i) μ^+ decay
  (j) pion exchanges in nucleon–nucleon interaction
  (k) pion decay.

Further reading

[1] Perkins H 2000 *Introduction to High Energy Physics* (Cambridge: Cambridge University Press)

[2] Griffiths D 2008 *Introduction to Elementary Particles* (New York: Wiley-VCH)

[3] Tayal D C 2009 *Nuclear Physics* (Mumbai: Himalaya Publishing House)

[4] Prakash S 2005 *Nuclear Physics & Particle Physics* (New Delhi: Sultan Chand & Sons)

[5] Guha J 2019 *Quantum Mechanics: Theory, Problems & Solutions* 3rd edn (Kolkata: Books and Allied (P) Ltd)

[6] Lim Y K 2002 *Problems and Solutions on Atomic, Nuclear and Particle Physics* (Singapore: World Scientific)

IOP Publishing

Nuclear and Particle Physics with Cosmology, Volume 2
Particle physics and cosmology
Jyotirmoy Guha

Chapter 7

Classification of particles and interactions of nature

In this chapter we deal with the classification of particles and interactions of nature according to the Standard Model. We classify particles according to their charge, mass, spin and other properties. We show the classification of bosons and fermions. We compare the various fundamental interactions. Several examples of such interactions are presented.

7.1 Introduction

Electrons and quarks are structureless and indivisible particles—that cannot be broken or reduced into smaller constituents—they are not composed of other particles. They do not have an internal structure. Obviously with quarks, protons and neutrons are built, combining of which we get a nucleus. And with electrons and a nucleus we have an atom. So electrons and quarks are the basic building blocks of matter.

Fundamental particles as per the Standard Model are shown in figure 7.1.

Figure 7.1 also shows the fundamental particles of nature and their corresponding anti-particles namely quarks and anti-quarks, leptons and anti-leptons, all the gauge bosons and their anti-particles.

Figure 7.2 shows mass, charge, spin of the particles and the corresponding anti-particles as predicted in the Standard Model of particle physics.

Mass and spin of particle and anti-particle are the same.

Electric charge of particle and anti-particle are reversed.

All additive quantum numbers are reversed.

7.2 Classification of elementary particles

Elementary particles are classified on the basis of some parameters.

doi:10.1088/978-0-7503-5032-7ch7   7-1

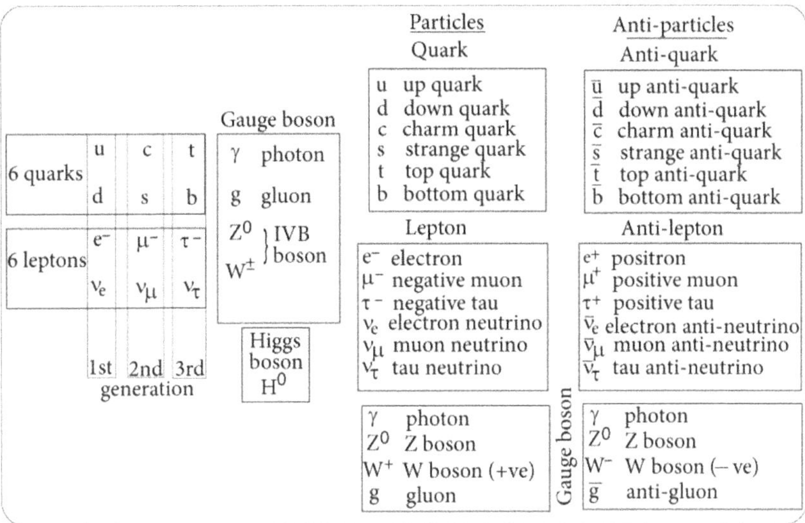

Figure 7.1. Quarks of six flavours, six leptons, four gauge bosons and Higgs boson according to the Standard Model.

Category		Symbol of particle	charge of particle	Symbol of anti-particle	charge of anti-particle	particle / antiparticle mass	spin
Fermions	Quarks	u	$\frac{2}{3}|e|$	$\bar{u}$	$-\frac{2}{3}|e|$	2.2 MeV/c^2	
		d	$-\frac{1}{3}|e|$	$\bar{d}$	$\frac{1}{3}|e|$	4.7 MeV/c^2	
		c	$\frac{2}{3}|e|$	$\bar{c}$	$-\frac{2}{3}|e|$	1.28 GeV/c^2	
		s	$-\frac{1}{3}|e|$	$\bar{s}$	$\frac{1}{3}|e|$	96 MeV/c^2	
		t	$\frac{2}{3}|e|$	$\bar{t}$	$-\frac{2}{3}|e|$	173 GeV/c^2	
		b	$-\frac{1}{3}|e|$	$\bar{b}$	$\frac{1}{3}|e|$	4.18 GeV/c^2	$\frac{1}{2}\hbar$
	Leptons	e^-	$-|e|$	e^+	$|e|$	0.51 MeV/c^2	
		μ^-	$-|e|$	μ^+	$|e|$	106 MeV/c^2	
		τ^-	$-|e|$	τ^+	$|e|$	1.78 GeV/c^2	
		ν_e	0	$\bar{\nu}_e$	0	< 1 eV/c^2	
		ν_μ	0	$\bar{\nu}_\mu$	0	< 0.17 eV/c^2	
		ν_τ	0	$\bar{\nu}_\tau$	0	<18 MeV/c^2	
Bosons	Gauge bosons	γ	0	γ	0	0	
		g	0	g	0	0	$1\hbar$
		Z^0	0	Z^0	0	91 GeV/c^2	
		$W^\pm$	$\pm|e|$	$W^\mp$	$\mp|e|$	80 GeV/c^2	
	Higgs boson	H^0	0	H^0	0	125 GeV/c^2	0

Figure 7.2. Values of charge, mass, spin of particles and anti-particles.

One of the intrinsic quantum characteristics of particles is their intrinsic spin s. Spin also relates to the statistics with which particles are associated. So based on spin and statistics followed by elementary particles, they are classified into two groups.

(1) Boson,
(2) Fermion.

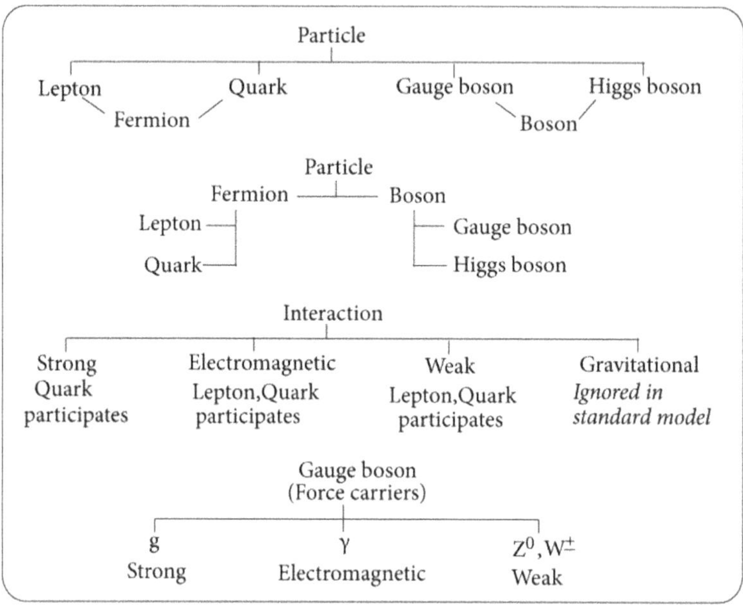

Figure 7.3. Various classifications in particle physics.

All particles fall under any of these two categories: bosons and fermions. We have shown the various classifications done in particle physics in figure 7.3.

7.3 Boson

We mention a few characteristics of the boson.

✓ Bosons follow Bose–Einstein statistics.

✓ Boson has integral spin in units of $\hbar$ (say 0, 1, 2, ... i.e. 0, $\hbar$, $2\hbar$, ...).

✓ Bosons do not obey the Pauli exclusion principle.

Two identical bosons can occupy the same quantum state. There is no exclusion principle for bosons, identical bosons can have the same quantum numbers.

✓ Total number of bosons is not conserved. There is no conservation law controlling total number of bosons.

✓ Bosons favour staying in groups, i.e. they tend to cluster and so in one quantum state there might be a large population of bosons. For instance a LASER is a system in which many photons sit in the same quantum state.

✓ Bosons may be created and may be destroyed.

✓ The force-carrying particles are all bosons, referred to as gauge bosons.

Let $\psi(\vec{r}_1, \vec{r}_2)$ be the probability amplitude of wave function describing two identical particles called '1' and '2' such that '1' is at position $\vec{r}_1$ and '2' is at position $\vec{r}_2$. So the probability of finding two identical particles '1' at position $\vec{r}_1$ and '2' at position $\vec{r}_2$ is $|\psi(\vec{r}_1, \vec{r}_2)|^2$.

Also, $\psi(\vec{r_2},\ \vec{r_1})$ represents probability amplitude of a wave function describing two identical particles such that '1' is at position $\vec{r_2}$ and '2' is at position $\vec{r_1}$, the corresponding probability of which is $|\psi(\vec{r_2},\ \vec{r_1})|^2$.

✓ For bosons, wave function remains unchanged if we swap or interchange, i.e. exchange particle positions. Mathematically we express this as

$$\psi(\vec{r_2},\ \vec{r_1}) = \psi(\vec{r_1},\ \vec{r_2}) \tag{7.1}$$

Clearly bosons are described by a symmetric wave function.

The probability amplitude or wave function of two identical bosons to be found at the same position $\vec{r}$ is obtained by putting $\vec{r_1} = \vec{r_2} = \vec{r}$ in equation (7.1) and we find that

$$\psi(\vec{r},\ \vec{r}) \neq 0 \tag{7.2}$$

✓ In any process number of bosons is not conserved.

7.4 Classification of bosons

Figure 7.4 shows all the elementary particles which are bosons. We have mentioned the spins and masses of the particles also.

7.5 Massless boson

Photon γ, gluon g and graviton are massless bosons.

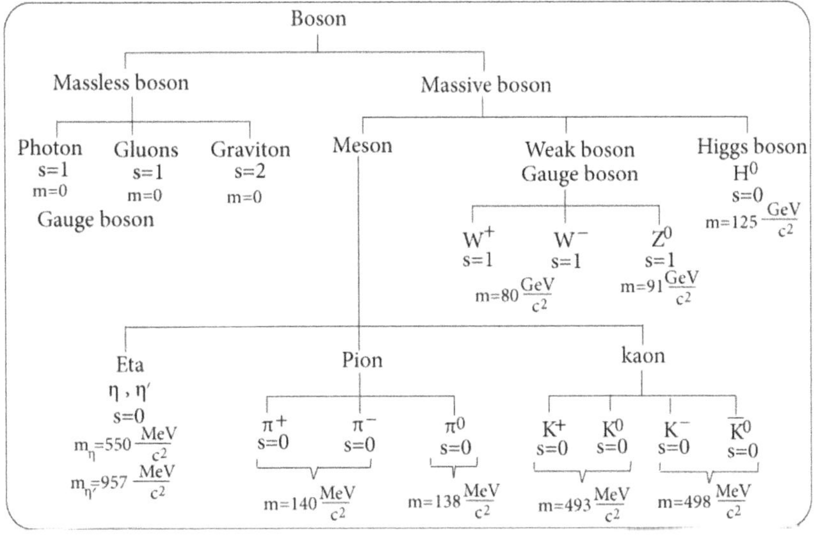

Figure 7.4. All the elementary bosons and their spin and mass values.

7.6 Photon

A photon is the electromagnetic quanta that mediates any electromagnetic interaction. We described properties of a photon in chapter 5, section 5.26.

The term massless refers to the fact that the rest mass of photon is zero, i.e.

$$m_{0\gamma} = 0 \tag{7.3}$$

This follows from the fact that a photon cannot be brought to rest in any frame of reference, i.e. by changing frame of reference as per the postulate of the special theory of relativity. However, a photon that travels with a constant speed of $3 \times 10^8\, m\, s^{-1}$ has energy. By the Einstein mass–energy relation

$$E = mc^2 \tag{7.4}$$

Photon possesses kinetic mass

$$m_\gamma = \frac{E}{c^2} = \frac{h\nu}{c^2} \tag{7.5}$$

Photon has spin $s = 1$.

Photon has momentum

$$p = \frac{E}{c} = \frac{h\nu}{c} = \frac{h}{\lambda}. \tag{7.6}$$

7.7 Gluons

We discussed the gluon, g, in chapter 5, section 5.28. They are massless and mediate interaction between quarks.

Photon and gluon are gauge bosons, i.e. quanta of the gauge field as described in section 5.25.

7.8 Graviton

Like photons of the electromagnetic field, graviton is postulated to be the quantum of the gravitational field. It is thought that gravitational attraction, i.e. interaction between masses is mediated by or carried by graviton.

Graviton rest mass is zero (massless particle).

Graviton is electrically uncharged.

Graviton travels at the speed of light.

Graviton spin is 2 or $2\hbar$.

Graviton has not been experimentally detected yet.

7.9 Massive boson

We now discuss the bosons having non-zero rest mass.

Mesons, gauge bosons and Higgs boson fall into this category.

7.10 Meson

The meson is subject to all the three types of interactions—strong interaction, weak interaction and electromagnetic interaction.

All mesons have spin $s = 0$.

Mass of mesons lies between the mass of leptons and the mass of nucleons, i.e. greater than lepton mass but less than nucleon mass.

Mesons are formed in strong interaction.

Mesons are of three types.

Eta meson, pions, kaons. We discuss them in the following.

7.11 Eta meson

The eta meson is a meson of relatively large mass $m_\eta = 550\ MeV$, $m_{\eta'} = 957\ MeV$ having charge $Q_\eta = 0$, $Q_{\eta'} = 0$. They are bosons of spin $s = 0$. They are members of the meson octet, as shown in chapter 10, figure 10.8.

7.12 Pi meson or pion

There are three pions. They are:

positive pion π^+, negative pion π^- and neutral pion π^0.

These are mesons of relatively less mass

$m_{\pi^\pm} = 140\ MeV$, $m_{\pi^0} = 138\ MeV$.

They have charge $Q_{\pi^\pm} = \pm 1\ |e| Q_{\pi^0} = 0$.

Pions are bosons of spin $s = 0$.

The first particle to be called a meson was μ meson but it is not of the category of mesons at all. It is a fermion (*exercise 7.4*).

Pions are mediators of strong interaction between nucleons in a nucleus according to Yukawa theory.

7.13 K meson or kaon

There are four kaons, namely

$$K^+,\ K^-,\ K^0,\ \overline{K^0}$$

These are mesons of relatively large mass

$$m_{K^+} = 493\ MeV,\ m_{K^-} = 498\ MeV,\ m_{K^0} = 493\ MeV,\ m_{\overline{K^0}} = 498\ MeV$$

They have charge

$$Q_{K^\pm} = \pm 1\ |e|,\ Q_{K^0} = 0,\ Q_{\overline{K^0}} = 0.$$

Kaons are bosons of spin $s = 0$.

7.14 Weak boson

Weak bosons have spin $s = 1$ and are also called Intermediate vector bosons, abbreviated as IVB. We discussed this in chapter 5, section 5.27.

These are massive bosons since they have large mass

$$m_{W^\pm} = 80\frac{GeV}{c^2} \text{ and } m_{Z^0} = 91\frac{GeV}{c^2}.$$

These are gauge bosons, i.e. they are quanta of the gauge field, as described in section 5.25.

7.15 Higgs boson

Higgs boson is a mass-giving boson and we discussed it in chapter 5, sections 5.29 and 5.39.

7.16 Fermion

We mention a few characteristics of the fermion.

✓ Fermion follows Fermi–Dirac statistics.

✓ Fermion has odd half integral spin, i.e. spin is in units of $\frac{\hbar}{2}$ (say $\frac{1}{2}, \frac{3}{2}, \frac{5}{2}, \ldots$ i.e. $\frac{1}{2}\hbar, \frac{3}{2}\hbar, \frac{5}{2}\hbar, \ldots$).

✓ No two identical particles can occupy the same quantum state, i.e. the possibility of two identical fermions (say two electrons) having the same quantum numbers is excluded. This rule is called the Pauli exclusion principle. Fermions are matter particles obeying the Pauli exclusion principle.

✓ Total number of fermions is conserved. In one quantum state there can be either zero fermions or at most one fermion.

Consider two identical particles denoted by 1 and 2. They can occupy positions $\vec{r_1}, \vec{r_2}$ and let ψ be the wave function. Then

$\psi(\vec{r_1}, \vec{r_2})$ is the probability amplitude that '1' is at position $\vec{r_1}$ and '2' is at position $\vec{r_2}$. The corresponding probability is $|\psi(\vec{r_1}, \vec{r_2})|^2$.

$\psi(\vec{r_2}, \vec{r_1})$ is the probability amplitude that '1' is at position $\vec{r_2}$ and '2' is at position $\vec{r_1}$. The corresponding probability is $|\psi(\vec{r_2}, \vec{r_1})|^2$.

✓ For fermions, wave function picks up a minus sign if we swap or interchange or exchange particle positions, i.e.

$$\psi(\vec{r_2}, \vec{r_1}) = -\psi(\vec{r_1}, \vec{r_2}). \tag{7.7}$$

✓ Clearly fermions are described by anti-symmetric wave function.

✓ The probability amplitude or wave function of two identical fermions to be found at the same position $\vec{r}$ is obtained by putting $\vec{r_1} = \vec{r_2} = \vec{r}$ in equation (7.7) and we find that

$\psi(\vec{r}, \vec{r}) = -\psi(\vec{r}, \vec{r})$

$2\psi(\vec{r}, \vec{r}) = 0$

$\psi(\vec{r}, \vec{r}) = 0.$

Vanishing of probability amplitude $\psi(\vec{r}, \vec{r})$ means that putting two identical fermions at the same position is an impossible event. It is clear that possessing anti-symmetric wave function by a fermion is consistent with the Pauli exclusion principle.

✓ Leptons and quarks are fermions.

7.17 Classification of the fermion

Figure 7.5 shows all the elementary particles which are fermions. Their spins and masses are also indicated.

Fermions are of two types:

Leptons and baryons.

Again, quarks which are the constituents of hadrons (mesons and baryons) are fermions.

7.18 Lepton

We briefly discussed leptons in section 5.9.

✓ Leptons are light particles (having small mass).

✓ Leptons are fermions having spin $= \frac{1}{2}$.

✓ Leptons are subject to electromagnetic and weak interaction.

✓ Leptons do not have internal structure.

✓ Leptons carry a conserved internal quantum number called lepton quantum number denoted by L_e, L_μ, L_τ, as given in figure 7.6.

✓ Lepton number is additive. To get the total lepton number we have to add all the lepton numbers of the individual particles.

We discussed in sections 5.10–5.15 the various leptons viz. $e^-, \nu_e, \mu^-, \nu_\mu, \tau^-, \nu_\tau$.

7.19 Baryon

✓ Baryon has mass $\geqslant$ nucleon mass.

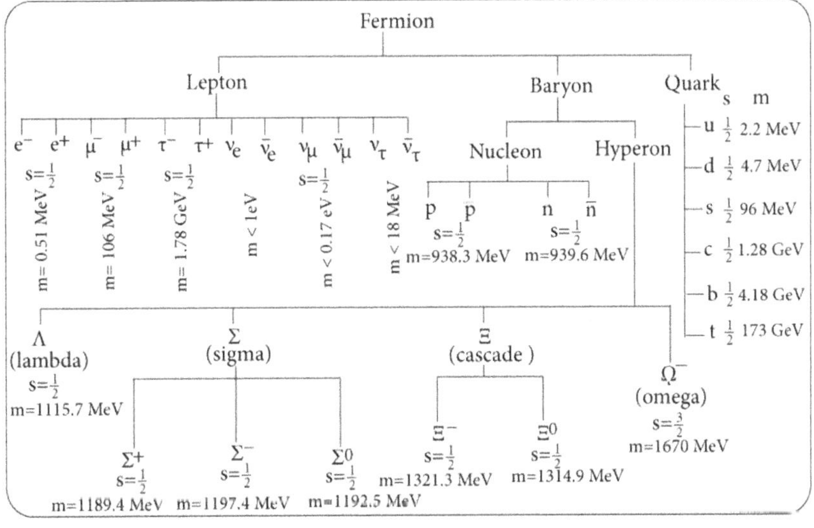

Figure 7.5. All the elementary fermions.

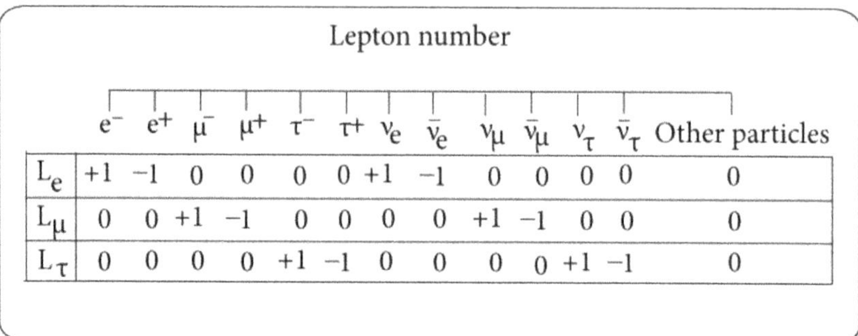

	e^-	e^+	μ^-	μ^+	τ^-	τ^+	ν_e	$\bar{\nu}_e$	ν_μ	$\bar{\nu}_\mu$	ν_τ	$\bar{\nu}_\tau$	Other particles
L_e	+1	−1	0	0	0	0	+1	−1	0	0	0	0	0
L_μ	0	0	+1	−1	0	0	0	0	+1	−1	0	0	0
L_τ	0	0	0	0	+1	−1	0	0	0	0	+1	−1	0

Figure 7.6. Lepton number of leptons.

✓ Baryon is a strongly interacting fermion.
✓ Baryon is subject to all the three types of interaction—strong interaction, weak interaction and electromagnetic interaction.
✓ Baryons have spin $\frac{1}{2}$ or $\frac{3}{2}$, e.g. proton has spin $s = \frac{1}{2}$, Ω^- has spin $s = \frac{3}{2}$.
✓ Baryons have an internal structure—they are made of quarks.
✓ Baryons are characterized by a quantum number called baryon quantum number. In other words, baryons carry a conserved internal quantum number called baryon quantum number denoted by B.
$B = +1$ for baryons
$B = -1$ for anti-baryons
$B = 0$ for non-baryons
✓ Quarks possess a fractional baryon number.

$$B(\text{quarks: } u, d, s, c, b, t) = \frac{1}{3}$$

$$B(\text{anti-quarks: } \bar{u}, \bar{d}, \bar{s}, \bar{c}, \bar{b}, \bar{t}) = -\frac{1}{3}$$

✓ Baryon number is additive. To get the total baryon number we have to add all the baryon numbers of the individual particles.
✓ Figure 7.7 shows baryon number of various baryons, anti-baryons and non-baryons.
✓ Baryons are of two types.

Nucleons and hyperons.

7.20 Nucleon

Protons and neutrons are called nucleons. They make up the nucleus. We have discussed the details of the nucleus in volume 1.

Nucleons include p, n which are proton and neutron while $\bar{p}$, $\bar{n}$ are anti-proton and anti-neutron, respectively.

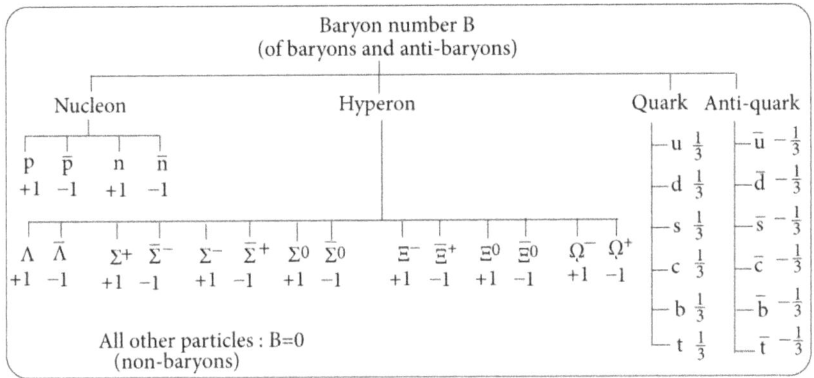

Figure 7.7. Baryon number of various baryons, anti-baryons and non-baryons.

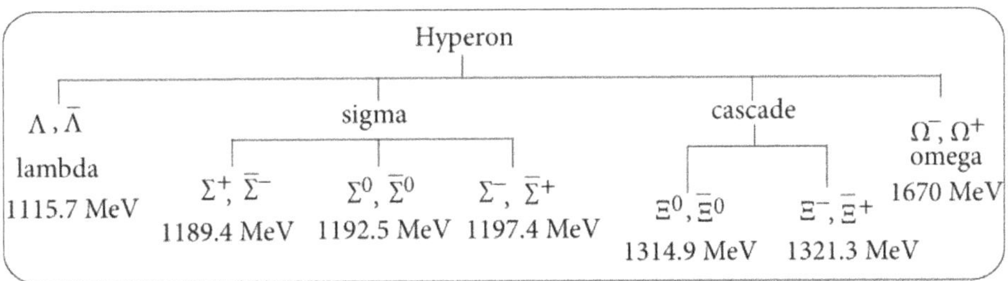

Figure 7.8. Hyperons (strange baryons).

7.21 Hyperon

✓ Baryons heavier than nucleons are collectively called hyperons, i.e. hyperons are baryons having mass greater than nucleons.

✓ Hyperons are strange baryons.

We will discuss strange particles in chapter 8.

✓ Figure 7.8 shows the hyperons along with their masses.

7.22 Fundamental interaction

In order to study the properties and decay mode of various elementary particles it is essential to have knowledge of various fundamental interactions to which they are subjected and governed. All phenomena in high energy physics—all the physical processes and structures in the Universe on all scales (atoms to galaxies)—can be explained in terms of four fundamental interactions. All phenomena of the Universe are governed by four fundamental forces, namely:

(1) gravitational interaction;
(2) electromagnetic interaction;
(3) strong interaction;
(4) weak interaction.

We have mentioned them in chapter 5, section 5.2. We now describe them in greater detail in the following.

The relative strengths of the various interactions vary between 1 and 10^{-39}.

The interactions depend on the distance through which they are effective and this distance is different for different interactions.

Strong interaction is effective when separation between nucleons is $\sim 10^{-15}\,m$, weak interaction is effective over a separation $\sim 10^{-18}\,m$. So strong and weak interactions are short-range forces.

Again, gravitational and electromagnetic interactions vanish when separation between interacting particles is ∞. So these are long-range forces.

As long as we are discussing nuclear physics and interaction between elementary particles the gravitational force is not significant as gravitational force is a very very weak force (10^{-39} times the strong force) and can be overlooked completely.

But if separation between interacting particles is increased beyond the nuclear and atomic dimension, strong force decreases rapidly to zero and electromagnetic and gravitational force operate.

When masses of interacting entities are large, gravitational force dominates, and we can neglect other forces of interaction.

The structure of planets, stars, galaxies are determined by the properties of gravitational interaction.

The structure of nuclei is determined by the properties of strong interaction.

The structure of atoms is determined by the properties of electromagnetic interaction.

The role of weak force in the structure of matter is apparently that of a minor perturbation that occurs whenever there is an inappropriate $\frac{n}{p}$ ratio in the nucleus, and that causes or executes β decay to stabilize the nucleus.

In figure 7.9 we give a comparison of various types of interactions specifying the particles affected, range of interaction, relative strengths and cross-section of

Interaction	Particles affected	Range	Relative strength / cross section	Particle exchanged (mediator)	Lifetime	Role in universe
Gravitational	All particles	∞	10^{-39} —	Graviton	–	Assembles matter into planets, stars, galaxies
Electromagnetic	Charged particles	∞	$\frac{1}{137}$ or 10^{-2} 10^{-4} mb	Photon	10^{-16} s	Determines structure of atoms, molecules, solids, liquids
Weak	Quarks Leptons	$<10^{-18}$ m	10^{-5} 10^{-12} mb	Intermediate boson W^+, W^-, Z^0	10^{-13} s	Mediates transformation of quarks and leptons. Helps to determine composition of nuclei
Strong	Quarks Hadrons	$<10^{-15}$ m	1 10 mb	Gluons (acts like glue) Mesons	10^{-24} s	Holds quarks together to form nucleons. Holds nucleons together to form atomic nuclei.

Figure 7.9. Comparison of various types of interactions.

interaction, mediator or exchange quantum of interaction, lifetime and role of the interaction in the Universe.

7.23 Gravitational interaction

Gravity is always there and manifests in the microscopic plus macroscopic world.

Applications of gravitational interaction are: a stone thrown up falls down; planets revolve around stars; stars form galaxies; the Universe expands.

Strength of gravitational interaction is 10^{39} times smaller than the strength of strong interaction. So in the microscopic scale its contribution is negligibly small. Gravitational interaction has no role or importance in the study of interaction between elementary particles and hence is neglected.

Gravitation is dominant on a large scale.

Gravitational interaction is a long-range attractive interaction. It is always attractive and never repulsive.

It does not depend on velocity, spin, angular orientation, charge, size, colour charge of interacting particles.

Three theories that have been put forward over time to explain gravitational interaction are as follows.

7.23.1 Newton's theory of gravitation or classical theory

Gravitational interaction depends on masses of interacting particles and on the inverse square of the intervening distance.

Consider two nucleons within the nucleus. The magnitude of gravitational interaction between them is given by Newton's law of gravitation

$$F_G = G\frac{m_1 m_2}{r^2}$$

where $G = 6.67 \times 10^{-11}\ Nm^2\ kg^{-2}$ is universal constant of gravitation, m_1, m_2 are nucleon masses $= 1.67 \times 10^{-27} kg$, r is nuclear dimension $= 10^{-15} m$.

$$F_G = (6.67 \times 10^{-11}\ Nm^2\ kg^{-2})\frac{(1.67 \times 10^{-27} kg)^2}{(10^{-15} m)^2} \tag{7.8}$$

$$= 2 \times 10^{-34} N \cong 0 \text{ (very feeble force)}$$

The gravitational potential energy of the system is

$$U = -\frac{Gm_1 m_2}{r}$$

$$= -(6.67 \times 10^{-11}\ Nm^2\ kg^{-2})\frac{(1.67 \times 10^{-27} kg)^2}{10^{-15} m} = -2 \times 10^{-49} J \cong 0 \text{ (negligibly small)}$$

These nearly zero values $F_G \cong 0, U \cong 0$ indicate that the gravitational force plays no role in particle reactions.

7.23.2 Einstein's general theory of relativity

According to Einstein's general theory of relativity gravitational force or interaction propagates by fields.

Space-time curvature is responsible for the gravitational attraction.

7.23.3 Modern theory on the basis of field quanta

Gravitational interaction is mediated by a field particle or quantum called the graviton.

The graviton has zero rest mass and travels with speed $v = c$, the speed of light in free space.

As the gravitational field is extremely weak it is very difficult to detect gravitons. And they are still to be detected.

7.23.4 Coupling constant for gravitational interaction in nuclear phenomena

Strength of any interaction is characterized by a dimensionless coupling constant. Gravitational interaction for nuclear phenomena is characterized by a dimensionless coupling constant given by *exercise 7.6* viz. $\frac{Gm^2}{\hbar c}$ where m is mass of a nucleon, G is universal gravitational constant $\hbar = \frac{h}{2\pi} =$ reduced Planck's constant, $c =$ speed of light in free space. Its value is

$$\frac{Gm^2}{\hbar c} = (6.67 \times 10^{-11} \, Nm^2 \, kg^{-2}) \frac{(1.67 \times 10^{-27} kg)^2}{\frac{1}{2\pi}(6.626 \times 10^{-34} J.s)(3 \times 10^8 m \, s^{-1})}$$

$$= 5 \times 10^{-39} \cong 0 \text{ (negligibly small)}$$

Such a small value of gravitational coupling constant for nuclear interaction indicates that gravitational interaction can be neglected in the case of nuclear effects.

7.24 Electromagnetic interaction

Particles with charge take part in electromagnetic interaction.

Electromagnetic interaction manifests in the microscopic plus macroscopic world. Nucleus and electrons are held together by electromagnetic forces that gives rise to the stable atom.

According to Maxwell's electromagnetic theory, if an electric charge is in motion it produces a time-dependent electric field, then a magnetic field is also produced that varies periodically with space and time. Again, a magnetic field which is time dependent produces an electric field that varies periodically with space and time.

So electric and magnetic phenomena are basically the same and depend on the state of motion of electric charge. Maxwell unified the concepts of electricity and magnetism because these are parts of the same phenomena called electromagnetism.

Electromagnetic interaction depends on the sign of charge, magnitude of charge, and state of motion of the charges carried by the interacting particles.

The interaction reduces to its simplest form for point charges at rest and this is called Coulomb electrostatic interaction, namely

$$F_C = \frac{1}{4\pi\epsilon_0}\frac{q_1 q_2}{r^2} \text{ (force magnitude considered only)}$$

As $r \to \infty$, $F_C \to 0$ which means that it is a long-range force.

Let us compare the Coulomb interaction F_C with gravitational interaction F_G (equation (7.8)) for two protons (charges $q_1 = q_2 = |e|$) at nuclear separation ($r \sim 10^{-15}\,m$)

$$F_C = \frac{1}{4\pi\epsilon_0}\frac{q_1 q_2}{r^2} = \frac{1}{4\pi\left(\frac{10^{-9}}{36\pi}\frac{F}{m}\right)}\frac{(1.6 \times 10^{-19}C)^2}{(10^{-15}m)^2} = 230\,N \tag{7.9}$$

Hence the ratio of Coulomb (or electromagnetic interaction) to gravitational interaction between two protons at nuclear separation is given by

$$\frac{F_C}{F_G} = \frac{230\,N}{2 \times 10^{-34}\,N} = 10^{36} \to \infty \tag{7.10}$$

So electromagnetic force is 10^{36} times larger than gravitational force.

According to field concept, interaction is mediated by a particle. The field quantum of electromagnetic interaction is the massless boson known as the photon, having spin $s = 1$.

The strength of electromagnetic interaction is determined in terms of the dimensionless coupling constant known as fine structure constant that can be expressed as

$$\alpha = \frac{|e|^2}{4\pi\epsilon_0\hbar c} = \frac{1}{4\pi\left(\frac{10^{-9}}{36\pi}\frac{F}{m}\right)} \cdot \frac{(1.6 \times 10^{-19}C)^2}{\frac{1}{2\pi}\left(6.626 \times 10^{-34}\,J.s\right)\left(3 \times 10^8\,m\,s^{-1}\right)} = \frac{1}{137} \tag{7.11}$$

The electromagnetic interaction is charge dependent and it also depends on the z component or 3rd component of isobaric spin, i.e. isospin I_z.

We will discuss the details of isospin in chapter 8.

In electromagnetic interaction isospin is not conserved (figure 8.28).

In electromagnetic interaction quantities such as charge (q), baryon number (B), lepton number (L), hypercharge (Y), parity (π), strangeness (S) remain conserved.

We will discuss strangeness quantum number in chapter 8.

Examples of electromagnetic interactions are

$\gamma + p \to \pi^0 + p$ (a proton captures a photon and produces a neutral pion and proton)

$\gamma + p \to \Lambda + K^+$ (a proton captures a photon and produces a meson Λ and hyperon K^+

$$n + \gamma \to \pi^- + p \text{ (radiactive capture reaction)}$$

Neutral particles π^0, Σ^0 and η decay electromagnetically as follows

$$\pi^0 \to \gamma + \gamma$$

$$\Sigma^0 \to \Lambda + \gamma$$

$$\eta \rightarrow 2\gamma$$

• We discuss in *exercise 7.7* whether the decay of η meson to 3π mesons is permitted electromagnetically.

The process of mutual annihilation of a particle–anti-particle pair, i.e. e^+, e^- pair is an electromagnetic interaction viz.

$$e^+ + e^- \rightarrow 2\gamma$$

with a characteristic lifetime of $10^{-16}\,s$ and characteristic cross-section $10^{-4}\,mb$.

A neutral pion π^0 decays electromagnetically with lifetime $10^{-16}s$ as

$$\pi^0 \rightarrow \gamma + \gamma.$$

Electromagnetic interaction is described by laws like

$$\text{Coulomb's law } \overrightarrow{F} = \frac{1}{4\pi\epsilon_0} \frac{qq'}{r^2} \hat{r}$$

$$\text{Biot-Savart's law } \overrightarrow{B} = \int_C \frac{\mu_0}{4\pi} \frac{i\,\overrightarrow{dr} \times \hat{R}}{R^2},$$

$$\text{Lorentz force } \vec{F} = q\,\vec{v} \times \vec{B} \text{ etc.}$$

7.25 Weak interaction

We studied β decay in volume 1, which is a commonly cited example of weak interaction. In the process of β decay a weak force acts inside nucleus between the nucleons and this weak interaction is responsible for β decay.

The weak interaction is mediated by intermediate vector bosons $W^\pm$, Z^0 of masses

$$m_{W^\pm} = 80\frac{GeV}{c^2}, \quad m_{Z^0} = 91\frac{GeV}{c^2} \text{ respectively.}$$

These quanta are vector particles since they have spin 1 or $\hbar$. Weak interaction manifests in the microscopic scale.

The numerical constant characterizing weak interaction (that can be obtained from Fermi's theory of β decay) is

$$g_F = 1.435 \times 10^{-62} J.\, m$$

The dimensionless coupling constant for weak interaction is given by

$$\left[\frac{g_F^2}{(\hbar c)^2}\right]\left[\frac{m_\pi c}{\hbar}\right]^4 = \left[\frac{(1.435 \times 10^{-62}\,J.m)^2}{\left(\frac{1}{2\pi}(6.626 \times 10^{-34}\,J.s)(3 \times 10^8\,m\,s^{-1})\right)^2}\right]$$

$$\left[\frac{273 \times 10^6 \times (1.6 \times 10^{-19})(3 \times 10^8\,m\,s^{-1})}{\frac{1}{2\pi}(6.626 \times 10^{-34}\,J.s)}\right]^4$$

$$= 5 \times 10^{-4}$$

This coupling constant is negligibly small, indicating that this interaction is weak in strength compared to strong and electromagnetic interactions.

Characteristic lifetime of weak interaction is $10^{-13}\,s$. However, a charged pion decays through weak interaction with lifetime $10^{-8}\,s$ as

$$\pi^- \rightarrow \mu^- + \bar{\nu}_\mu$$

$$\pi^+ \rightarrow \mu^+ + \nu_\mu$$

while a free neutron decays with lifetime of 15 min as

$$n \rightarrow p + e^- + \bar{\nu}_e$$

A characteristic cross-section of weak interaction is $10^{-12}\,mb$.

Range of weak interaction is $<10^{-18}\,m$, i.e. it is a very short-range force and exists between particles inside the nucleus when their separation is less than nuclear diameter $(\sim 10^{-15}\,m)$.

Weak interaction is never attractive—it is always repulsive.

We now give examples of various types of weak interactions.

✓ Leptonic decay

Weak decays in which the final state consists of leptons only are:

$$K^- \rightarrow \mu^- + \bar{\nu}_\mu \;(\mu^-,\; \bar{\nu}_\mu \text{ are leptons})$$

$$\mu^- \rightarrow e^- + \bar{\nu}_e + \nu_\mu \;(e^-,\; \bar{\nu}_e,\; \nu_\mu \text{ are leptons})$$

$$\nu_\mu + e^- \rightarrow \mu^- + \nu_e \;(\mu^-,\; \nu_e \text{ are leptons})$$

$$\mu^+ \rightarrow e^+ + \nu_e + \bar{\nu}_\mu \;(e^+,\; \nu_e,\; \bar{\nu}_\mu \text{ are leptons})$$

$$\nu_e + e^- \rightarrow e^- + \nu_e \;(e^-,\; \nu_e \text{ are leptons}).$$

✓ Semi-leptonic decay

Weak decays in which final state consists of leptons and hadrons are:

$\Lambda \rightarrow p + e^- + \bar{\nu}_e$	(p is a hadron $e^-, \bar{\nu}_e$ are leptons)
$n \rightarrow p + e^- + \bar{\nu}_e$	(p is a hadron $e^-, \bar{\nu}_e$ are leptons). It is nuclear β^- decay.
$K^- \rightarrow \pi^0 + \mu^- + \bar{\nu}_\mu$	(π^0 is a hadron $\mu^-, \bar{\nu}_\mu$ are leptons)
$K^0 \rightarrow \pi^+ + e^- + \bar{\nu}_e$	(π^+ is a hadron $e^-, \bar{\nu}_e$ are leptons)
$K^0 \rightarrow \pi^- + e^+ + \nu_e$	(π^- is a hadron e^+, ν_e are leptons)
$K^+ \rightarrow \pi^0 + e^+ + \nu_e$	(π^0 is a hadron e^+, ν_e are leptons)
$\Sigma^+ \rightarrow \Lambda + e^+ + \nu_e$	(Λ is a hadron e^+, ν_e are leptons)
$\Sigma^- \rightarrow \Lambda + e^- + \bar{\nu}_e$	(Λ is a hadron $e^-, \bar{\nu}_e$ are leptons)
$e^- + p \rightarrow n + \nu_e$	(n is a hadron ν_e is lepton). It is electron capture.
$\mu^- + p \rightarrow n + \nu_\mu$	(n is a hadron ν_μ is a lepton. It is μ^- capture.

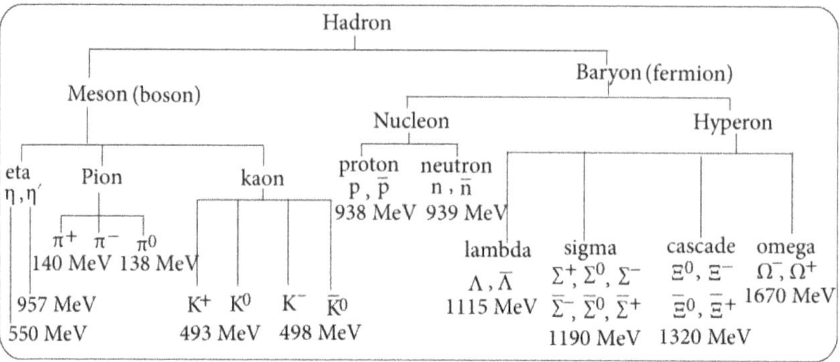

Figure 7.10. List of hadrons.

We give a list of hadrons in figure 7.10.
✓ Non-leptonic decay (final state consists of hadrons only).
Weak decays in which final state consists of hadrons only are

$$\Lambda \rightarrow p + \pi^- \qquad\qquad (p,\ \pi^- \text{ are hadrons})$$
$$K^- \rightarrow \pi^0 + \pi^- \qquad\qquad (\pi^0,\ \pi^- \text{ are hadrons})$$

We note that for strange baryons (K^-, K^0, K^+, Σ^+, Σ^-, Λ) only semi-leptonic and non-leptonic decays are possible because baryon number is strictly conserved. So there must be a baryon in the final state.

Weak interaction can alternately be broadly classified as follows:

(a) Charged current weak interaction

Charged current weak interactions are mediated by W^+ and W^- bosons, i.e. weak force is communicated by $W^\pm$ exchange particles in this interaction. Since this interaction involves transfer of electric charge it is called charged current interaction. As $W^\pm$ have ± 1 electric charge they mediate neutrino absorption and emission inducing electron or positron emission or absorption as well as changing the flavour of a quark along with its electrical charge. For instance

$$\mu^- + W^+ \rightarrow \nu_\mu \qquad\qquad (\mu^- \text{ can absorb a } W^+)$$
$$d \rightarrow u + W^- \qquad\qquad (d \text{ quark converts to } u \text{ quark by emitting } W^-)$$
$$d + W^+ \rightarrow u \qquad\qquad (d \text{ quark absorbs } W^+ \text{ and converts to } u \text{ quark})$$
$$c \rightarrow s + W^+ \qquad\qquad (c \text{ quark converts to } s \text{ quark by emitting } W^+)$$
$$c + W^- \rightarrow s \qquad\qquad (c \text{ quark absorbs } W^- \text{ and converts to } s \text{ quark})$$
$$W^- \rightarrow e^- + \bar{\nu}_e \qquad\qquad (W^- \text{ decays to electron and electron anti-neutrino})$$
$$W^+ \rightarrow e^+ + \nu_e \qquad\qquad (W^+ \text{ decays to positron and electron neutrino})$$

(b) Neutral current weak interaction

Neutral current weak interaction is mediated by a Z^0 boson, i.e. weak force is communicated by a Z^0 exchange particle in this interaction. Since this interaction involves no transfer of electric charge it is called a neutral current interaction. The term current does not refer to electricity but refers to the movement of Z^0 between particles. As Z^0 is electrically neutral, exchange of Z^0 leaves the quantum numbers of interacting particles unaffected. Only linear and angular momenta and energy are transferred. For instance

$$e^- \rightarrow e^- + Z^0 \qquad \text{(electron emits a } Z^0\text{)}$$
$$Z^0 \rightarrow b + \bar{b} \qquad \text{(} Z^0 \text{ boson decays)}$$

- We mention some purely weak interactions.

Neutrino–lepton scattering	$: \nu_\mu + e \rightarrow \nu_\mu + e$
Neutrino–nucleon scattering	$: \nu_\mu + N \rightarrow \nu_\mu + N$

- We mention some interactions that are not purely weak interactions but are accompanied by electromagnetic interactions.

Lepton–lepton scattering	$: e^+ + e^- \rightarrow \mu^+ + \mu^-$
Electron–nucleon scattering:	$e + N \rightarrow e + N$ ($N =$ nucleon)

- We mention some interactions that are not purely weak interactions but are accompanied by electromagnetic and strong interactions
Hadron–hadron interaction: $N + N \rightarrow N + N$.

7.25.1 Weak isospin

Weak isospin is a construct parallel to the idea of isospin under strong interaction.

We can associate the weak interaction with a particular property of a particle called weak isospin.

If weak isospin of a particle is zero then it will not take part in weak interaction. In fact weak isospin is a quantum number relating to the electrically charged part of weak interaction. Weak isospin is a quantum number that describes the symmetry between particles with different weak charge.

We will discuss weak isospin briefly in chapter 10, section 10.19.

7.26 Strong interaction

Strong force manifests in the microscopic scale. It is responsible for holding or binding quarks together inside protons and neutrons. Strong force is responsible for holding or binding protons and neutrons together as nuclei.

Strong interaction acts between a pair of hadrons. Figure 7.10 shows the hadrons along with their masses. Hadrons are strongly interacting particles. In fact mesons (bosons) and baryons (fermions) are called hadrons.

The strong nuclear interaction is always attractive.

Characteristic range of strong interaction is 10^{-15} $m = 1\,fm$, i.e. short range.

Protons and neutrons in a nucleus exchange pions (which is a boson of spin 0) and this is how *pp*, *nn* and *np* interactions occur and we have a stable bound nuclear system. Again, proton and neutron are made of three quarks ($p \equiv uud$, $n \equiv udd$). These quarks are held together, i.e. glued by gluons (spin 0).

Clearly on the basis of the quark model, strong interaction is due to an exchange of colour quanta called gluons. They have zero rest mass and are bosons of spin 1.

The dimensionless coupling constant for strong interaction is the dimensionless number

$$\frac{g^2}{4\pi\hbar c} \cong 14$$

where g plays the same role as charge plays in electromagnetic interaction.

This value is large compared to the value of coupling constant of other interactions. Hence this interaction is called strong interaction as its strength is largest.

In strong interaction baryon number (B), charge (Q), hypercharge (Y), parity (π), isospin (I) and z-component of isospin (I_3) remain conserved.

Characteristic lifetime is 10^{-24} s.

Characteristic cross-section is 10 *mb*

Strong interaction that operates within the nucleus is independent of charge of the nucleon since the force of interaction within a nucleus between *pp*, *nn* and *np* are the same.

• But if a nucleus contains one or more strange particles (say Λ particle) along with nucleons, then such a nucleus is called a hyper nucleus.

Such a nucleus contains n, p, Λ particles. So there will be interactions like $\Lambda - n$, $\Lambda - p$, $n - p$. It is observed that the strength of the interactions $\Lambda - n$, $\Lambda - p$ are unequal. This is against the concept of charge symmetry. Actually, such inequality is because Λ is interacting with n and p which are nucleons of different charge states. Thus the concept that strong interactions are purely charge independent is not absolute.

7.27 Electroweak interaction

We cannot at some stage at high energies separate electromagnetic interaction and weak interaction. Instead we are forced to talk about them as a combined force which is called electroweak interaction. Such mixing or combination of these two is similar to the electromagnetic interaction.

Electromagnetic interaction can be looked upon as a combination of electric and magnetic properties of matter. As a matter of fact the electromagnetic interaction cannot be separated into electric and magnetic forces or interactions. For instance when a charged particle is in motion we cannot separate this electromagnetic interaction into electric and magnetic forces unless we go to very small non-relativistic cases.

In this case also we have to think about a combined electroweak interaction rather than electromagnetic and weak interactions separately.

7.28 Comparison with gravitational interaction

As indicated in figure 7.9, if we consider the strong interaction to be of strength say 1, then electromagnetic interaction is $\frac{1}{100}$ times weaker than this. Weak nuclear interaction is 10^{-5} times weaker than the strong interaction. And gravitational interaction is very very weak $\sim 10^{-39}$ times weaker than strong interaction. So in most of the study of elementary particle physics we neglect the effect of gravitational interaction.

When we consider proton–proton collision at the LHC at 7 or 8 TeV centre of mass–energy we have the effect of strong interaction, weak interaction, electromagnetic interaction. But there will be no effect of gravitational interaction compared to the other interactions. So we do not consider gravitational interaction. Another reason is that we do not have a good model of quantum gravity to fit in. In other words we do not know what gravity is like at very small scales. What we know so far is that even at one fermi scale ($10^{-15}\ m$) we do not have the effect of gravity.

Newtonian gravity is tested up to micron level and at distance just less than that (millimeter or sub millimeter level) gravitational interaction is like Newtonian gravity. So it is very weak at that stage. Beyond that we do not really know. If we take an extension, i.e. extend the idea of gravitational interaction that we have at larger scales then it is going to be very very weak until we reach very very high energies.

Strength of interactions is scale dependent and not strictly a constant. At high energies, say at Planck scale $10^{19}\ GeV$ energy, we expect gravitational interaction may be at a comparable strength to other interactions.

7.29 Force carriers mediating interactions

Figure 7.11 shows the force carriers that mediate various interactions.

✓ Consider an electron interacting with another electron. The interaction occurs through exchange of some particles, i.e. exchange of some information in terms of quanta of the electromagnetic field called photons (massless). In other words a photon is exchanged between the electrons and that is how they interact. Photons carry electromagnetic interaction between any two charged particles. This is diagrammatically depicted in figure 7.11(a).

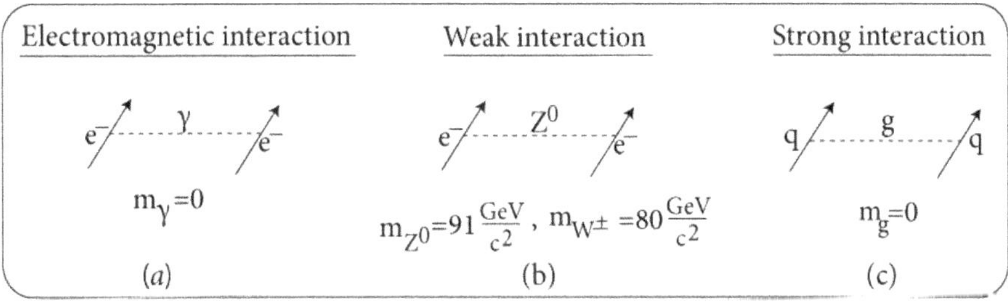

Figure 7.11. Force carriers for various interactions.

✓ For weak interaction between say two electrons, the force exchange occurs similarly but since the interaction is different, the force field is different and hence the carrier is also different. So the force exchange will occur not through photon mediation but through different carrier particles. The quanta of weak field are bosons called Z^0 boson (mass $91\frac{GeV}{c^2}$)and $W^\pm$ bosons (mass $80\frac{GeV}{c^2}$). Figure 7.11(b) shows exchange of Z^0 boson between two weakly interacting electrons.

✓ For strong interaction we show two quarks in figure 7.11(c) (as electrons do not participate in strong interaction). Quarks interact strongly through exchange of quantum of the strong field called gluons. There are eight massless gluons to account for the strong interaction.

• The Standard Model of particle physics (based on gauge quantum field theory) requires a scalar (spin 0) particle called the Higgs boson H to generate masses for the matter particles as well as the masses of the massive force carriers (the gauge bosons). To explain the presence of mass we need to have some mechanism by which masses are generated for the particles. And this needs the presence of the Higgs field. The particle Higgs boson itself has mass of $125\frac{GeV}{c^2}$ that is not given by the Standard Model and is fixed by experiment.

We have discussed the Higgs boson in sections 5.29 and 5.39.

7.30 Exercises

Exercise 7.1 *Which is an elementary particle?*

(a) quark   *(b) electron*   *(c) nucleus*   *(d) atom*   *(e) molecule.*

Exercise 7.2 *Which one is not an elementary particle?*

(a) quark   *(b) electron*   *(c) nucleus*   *(d) proton*   *(e) neutron.*

Exercise 7.3 *Which of the following do not have free existence?*

(a) quark   *(b) electron*   *(c) nucleus*   *(d) proton*   *(e) neutron.*

Exercise 7.4 *Consider the following three statements:*
   (i) Mesons are bosons.
   (ii) The first particle to be called a meson was the μ meson.
   (iii) μ meson is a fermion.

Which of the following is true?
(a) All the statements are incorrect.
(b) Statements (i) and (ii) are correct but (iii) is incorrect.
(c) Statements (i), (iii) are correct but (ii) is incorrect.
(d) Statements (ii) and (iii) are correct but (i) is incorrect.
(e) All the statements are correct.

Exercise 7.5 *Consider the following statements.*
(i) Strong interaction is always larger in magnitude than gravitational interaction.
(ii) Gravitational interaction is always larger in magnitude than strong interaction.

Now identify the correct choice.
(a) Statement (i) is correct.
(b) Statement (ii) is correct.
(c) Statement (i) is correct for interaction within the atom but incorrect for cosmic interaction.
(d) Statement (ii) is correct for interaction within the atom but incorrect for cosmic interaction.

Exercise 7.6 *What is the dimension of $\frac{Gm^2}{\hbar c}$.*
$\boxed{\text{Ans.}}$ $\frac{Gm^2}{\hbar c} \rightarrow \frac{Nm^2 \; kg^2}{kg^2 \; (J.s)ms^{-1}} = \frac{Nm}{J} = \frac{J}{J} = M^0 L^0 T^0$ (dimensionless)

Exercise 7.7 *Discuss whether the decay of η meson to 3π mesons is permitted electromagnetically.*
$\boxed{\text{Ans.}}$ Consider the decay of η meson to 3π mesons
$$\eta \rightarrow \pi^+ + \pi^- + \pi^0$$
Let us check conservation of isospin
$I_{\text{LHS}} = I(\eta) = 0$
$I_{\text{RHS}} = I(\pi^+ + \pi^- + \pi^0) = I(\pi^+) + I(\pi^-) + I(\pi^0) = 1 + 1 + 1 = 3$
$I_{\text{LHS}} \neq I_{\text{RHS}}$
But
$I_z^{\text{LHS}} = I(\eta) = 0$
$I_z^{\text{RHS}} = I(\pi^+ + \pi^- + \pi^0) = I(\pi^+) + I(\pi^-) + I(\pi^0) = 1 - 1 + 0 = 0$
$I_z^{\text{LHS}} = I_z^{\text{RHS}}$
Though I_z is conserved I is not.
Since electromagnetic interaction need not conserve isospin, the decay of η meson to 3π mesons is permitted.

Exercise 7.8 *Consider the following statements.*
   (i) Gravitational force > strong force outside nuclear size.
   (ii) Strong force > gravitational force inside nuclear size.

Which statement is true?
   (a) Statement (i) is correct.
   (b) Statement (ii) is correct.
   (c) Both statements are incorrect.
   (d) Both statements are correct.

Exercise 7.9 *Hadrons are*

(a) fermions   *(b) bosons*   *(c) any one of a and b*   *(d) neither a nor b.*

Exercise 7.10 *Consider the reactions* $\pi^+ \rightarrow \mu^+ + \nu_\mu$, $\pi^- \rightarrow \mu^- + \bar{\nu}_\mu$ *where the* μ *produced has kinetic energy* ~4.2 MeV. *Obtain mass of a charged pion from here (take rest energy of* μ *to be* 105.66 MeV).
   $\boxed{\text{Ans.}}$ Consider the reactions

$$\pi^+ \rightarrow \mu^+ + \nu_\mu$$

$$\pi^- \rightarrow \mu^- + \bar{\nu}_\mu$$

Energy conservation gives

$$E_\pi = E_\mu + E_\nu$$

With

$$E_\pi = m_\pi c^2 + T_\pi = m_\pi c^2 (\text{as } T_\pi = 0)$$

$$E_\nu = m_\nu c^2 + T_\nu = T_\nu (\text{as } m_\nu = 0)$$

$$m_\pi c^2 = E_\mu + T_\nu$$

$$m_\pi c^2 - T_\nu = E_\mu \tag{7.12}$$

Momentum conservation gives (for a pion decaying from rest)

$$p_\mu = p_\nu$$

$$T_\nu = p_\nu c = p_\mu c \tag{7.13}$$

Putting equation (7.12) in equation (7.11)

$$m_\pi c^2 - p_\mu c = E_\mu = \sqrt{p_\mu^2 c^2 + m_\mu^2 c^4}$$

Squaring

$$(m_\pi c^2 - p_\mu c)^2 = p_\mu^2 c^2 + m_\mu^2 c^4$$

$$m_\pi^2 c^4 + p_\mu^2 c^2 - 2m_\pi c^2 p_\mu c = p_\mu^2 c^2 + m_\mu^2 c^4$$

$$m_\pi^2 c^4 - 2m_\pi c^2 p_\mu c = m_\mu^2 c^4$$

$$(m_\pi c^2)^2 - 2(m_\pi c^2)(p_\mu c) - (m_\mu c^2)^2 = 0$$

$$m_\pi c^2 = \frac{-(-2(p_\mu c)) \pm \sqrt{(-2(p_\mu c))^2 - 4(1)(-(m_\mu c^2)^2)}}{2} = p_\mu c \pm \sqrt{p_\mu^2 c^2 + (m_\mu c^2)^2}$$

$m_\pi c^2 = p_\mu c + \sqrt{p_\mu^2 c^2 + m_\mu^2 c^4}$ (we neglect negative sign as it leads to negative value)

$$m_\pi c^2 = p_\mu c + E_\mu \tag{7.14}$$

Now

$$E_\mu = T_\mu + m_\mu c^2 = 4.2 \; MeV + 105.66 \; MeV = 109.86 \; MeV. \tag{7.15}$$

Again

$$E_\mu = \sqrt{p_\mu^2 c^2 + m_\mu^2 c^4}$$

$$p_\mu^2 c^2 = E_\mu^2 - (m_\mu c^2)^2$$

$$p_\mu c = \sqrt{E_\mu^2 - (m_\mu c^2)^2} = \sqrt{(109.86)^2 - (105.66)^2} \; MeV = 30.1 \; MeV \tag{7.16}$$

Putting equations (7.14) and (7.15) in equation (7.13) we get

$$m_\pi c^2 = p_\mu c + E_\mu = 30.1 \; MeV + 109.86 \; MeV = 140 \; MeV$$

This is the mass of a charged pion.

Exercise 7.11 *Suppose the lifetime of a strongly interacting particle is of the order of $10^{-20} s$. How do you estimate its mass?*

Ans. We can use Heisenberg's uncertainty principle to find mass.

$$\Delta E \Delta t \sim h$$

$$\Delta E = \frac{h}{\Delta t} = \frac{6.626 \times 10^{-34}\ J.\ s}{10^{-20}\ s} = 6.626 \times 10^{-14}\ J$$

$$mc^2 = 6.626 \times 10^{-14}\ J = \frac{6.626 \times 10^{-14}\ J}{1.6 \times 10^{-19}\ J\ eV^{-1}} = 0.414 \times 10^6\ eV = 0.414\ MeV$$

$$m = \frac{6.626 \times 10^{-14}\ J}{(3 \times 10^8\ m\ s^{-1})^2} = 7.36 \times 10^{-31}\ kg$$

Exercise 7.12 *Boson and fermion are described by wave function which is*

(a) symmetric, asymmetric

(b) asymmetric, symmetric

(c) odd, even

(d) even, odd.

Answers to multiple choice questions
7.1a, b, 7.2c, 7.3a, 7.4e,7.5d, 7.8d, 7.9c, 7.12a, d.

7.31 Question bank

Q 7.1 Name the quarks, leptons and the gauge boson particles as per the Standard Model.

Q 7.2 What is the Higgs boson? What is its significance?

Q 7.3 Name the interactions as per the Standard Model. What are their range, cross-section, lifetime. Mention the quanta associated with each interaction.

Q 7.4 What are the characteristics of bosons?

Q 7.5 What are the characteristics of fermions?

Q 7.6 Make a list of all bosons according to their spin.

Q 7.7 Make a list of all bosons according to their mass

Q 7.8 Make a list of all fermions according to their spin.

Q 7.9 Make a list of all fermions according to their mass.

Q 7.10 What are leptons? What is lepton number?

Q 7.11 What are baryons? What is baryon number?

Q 7.12 Make a list of all leptons and anti-leptons according to their lepton number.

Q 7.13 Make a list of all baryons and anti-baryons according to their baryon number.

Q 7.14 Show that fermion wave function satisfies the Pauli exclusion principle.

Q 7.15 What are hyperons?

Q 7.16 Make a list of all hyperons according to their mass.

Q 7.17 What are strange baryons called?

Q 7.18 What is electroweak theory?

Q 7.19 What are the coupling constants of:

(a) strong interaction
(b) weak interaction
(c) electromagnetic interaction
(d) gravitational interaction.

Q 7.20 How large is the electromagnetic force compared to gravitational force between two nucleons (i.e. two protons, two neutrons and one proton one neutron) held at nuclear separation?

Q 7.21 Explain with examples the following weak interactions:
(a) leptonic decay
(b) semi-leptonic decay
(c) non-leptonic decay.

Q 7.22 Explain what we mean by
(a) charged current weak interaction
(b) neutral current weak interaction.

Q 7.23 What is isospin for strong interaction? What is weak isospin?

Q 7.24 Make a list of all the hadrons indicating their masses.

Further reading

[1] Perkins H 2000 *Introduction to High Energy Physics* (Cambridge: Cambridge University Press)

[2] Griffiths D 2008 *Introduction to Elementary Particles* (New York: Wiley-VCH))

[3] Tayal D C 2009 *Nuclear Physics* (Mumbai: Himalaya Publishing House)

[4] Prakash S 2005 *Nuclear Physics & Particle Physics* (New Delhi: Sultan Chand & Sons)

[5] Guha J 2019 *Quantum Mechanics: Theory, Problems & Solutions* 3rd edn (Kolkata: Books and Allied (P) Ltd)

[6] Lim Y K 2002 *Problems and Solutions on Atomic, Nuclear and Particle Physics* (Singapore: World Scientific)

IOP Publishing

Nuclear and Particle Physics with Cosmology, Volume 2
Particle physics and cosmology
Jyotirmoy Guha

Chapter 8

Conservation laws and symmetry

In this chapter we describe the meaning of symmetry and deal with various invariance principles relating to translation, rotation in space, time translation, space reflection, exchange symmetry, rotation in iso-space etc. We classify symmetries and describe continuous and discrete symmetries, internal and external symmetries, permutation symmetries, unitary symmetries etc. We discuss conservation of charge, lepton number, baryon number, isospin, strangeness, hypercharge, C, P, T, G parity and discuss stability of particles. Classification of hadrons is done. We discuss the Gell-Mann–Okubo formula, Gell-Mann–Nishijima relation and Gell-Mann–Okubo mass formula. Quark states of pion family, mesons and baryons are obtained. Meson octet, baryon octet and baryon decuplets are plotted. CP violation, CPT symmetry and helicity are discussed.

8.1 Invariance principle

Symmetry and invariance play a very significant role in physics. Let us discuss what is invariance principle.

If a system remains unchanged with the application of some transformation then it is said to be invariant under that transformation. In quantum mechanics such transformation is described by unitary transformation.

Laws of physics are invariant under symmetry operation.

After symmetry operation on a state, the new state is equivalent to the previous state. So the physical quantity related to the symmetry operation does not change and is conserved.

Any conservation principle is due to existence of some symmetry. Symmetry or invariance leads to a conservation principle. Every conservation principle or law is associated with some symmetry or invariance.

Consider the state of a system to be $| \psi >$ and after symmetry operation it changes to $| \psi' >$. Symmetry in the system means that the two states $| \psi >$ and $| \psi' >$ are equivalent to each other. Some quantity related to these states will not change and will be conserved. Clearly thus we state that if physical laws possess symmetry,

doi:10.1088/978-0-7503-5032-7ch8   8-1

existence of an arbitrary state $|\psi\rangle$ would imply that an equivalent state $|\psi'\rangle$ can be formed by symmetry operation.

Example:

An electron with spin component $s_z = +\frac{\hbar}{2}$ can be placed in a magnetic field along $\hat{z}$ as well as in a rotated field along $\hat{z}'$. In principle, both states are possible as characteristic of the electron does not depend on the direction.

8.2 Meaning of symmetry

Symmetry refers to an operation that cannot be observed (figure 9.1).

For instance we cannot say whether rotation by $90°$ has been performed about the centre of a square or not. The positional configuration of the square remains identical irrespective of the $90°$ rotational transformation—performed or not performed.

Measurable quantities are called observables.

There are certain basic quantities that cannot be measured, i.e. observed. These are called non-observables.

The cause of symmetry principle lies in the fact that it is impossible to observe certain basic quantities. So with every non-observable there is a symmetry.

Absolute position cannot be measured and so it is a non-observable. The symmetry associated is invariance under translation of space and this leads to conservation of linear momentum.

Absolute spatial direction is a non-observable and the symmetry associated is invariance under rotation. This leads to conservation of angular momentum.

So far as time is concerned, what we measure is time interval and we cannot measure the absolute time. Absolute time is a non-observable and the symmetry associated is invariance under time translation and this leads to conservation of energy.

We always measure relative velocity and not the absolute velocity. So we take the absolute velocity (uniform) to be a non-observable. Then the symmetry associated is related to invariance under Lorentz transformation performed by 4×4 matrices on the 4D Minkowski space $(x_0,\ x_1,\ x_2,\ x_3)$. The transformation leaves invariant the quantity

$$c^2 t^2 - z^2 - x^2 - y^2 \tag{8.1}$$

8.3 Translation in space

If the Hamiltonian of a system is invariant under displacement then this space translational invariance is due to homogeneity of space and leads to conservation of linear momentum.

The effect of infinitesimal translation δr in space on a wave function ψ is given as

$$\psi' = \psi(r + \delta r) = \psi(r) + \delta r \frac{\partial}{\partial r}\psi(r)$$
$$= \left(1 + \delta r \frac{\partial}{\partial r}\right)\psi(r) = D\psi(r) \tag{8.2}$$

where

$$D = 1 + \delta r \frac{\partial}{\partial r} \tag{8.3}$$

is an infinitesimal space translation operator. The momentum operator is

$$p = -i\hbar \frac{\partial}{\partial r} \tag{8.4}$$

$$\frac{\partial}{\partial r} = \frac{p}{-i\hbar} = \frac{i}{\hbar}p \tag{8.5}$$

So we can express space translation operator in terms of momentum operator as

$$D = 1 + \frac{ip}{\hbar}\delta r \tag{8.6}$$

A finite translation Δr can be obtained by taking n steps in succession so that

$$n\delta r = \Delta r \tag{8.7}$$

$$\delta r = \frac{\Delta r}{n} \tag{8.8}$$

and so

$$D = \left(1 + \frac{ip}{\hbar}\frac{\Delta r}{n}\right)^n \tag{8.9}$$

For $n \to \infty$ we write

$$D = \underset{n \to \infty}{Lt} \left(1 + \frac{ip}{\hbar}\frac{\Delta r}{n}\right)^n \tag{8.10}$$

Using the standard result that

$$\underset{n \to \infty}{Lt} \left(1 + \frac{x}{n}\right)^n = e^x \tag{8.11}$$

we write

$$D = e^{ip\Delta r/\hbar} \tag{8.12}$$

In natural units $\hbar = 1$ and so

$$D = e^{ip\Delta r} \tag{8.13}$$

Clearly

$$D^\dagger D = e^{-ip\Delta r}e^{ip\Delta r} = I \tag{8.14}$$

D is a unitary operator.

So the unitary operator for space translation is $e^{ip\Delta r}$ and the momentum operator p is called the generator of this transformation that leads to conservation of linear momentum.

8.4 Rotation in space

If the Hamiltonian of a system is invariant under rotation then this rotational invariance is due to isotropy of space and leads to conservation of angular momentum.

The effect of infinitesimal rotation $\delta\phi$ in space on a wave function ψ is given as

$$\psi' = \psi(\phi + \delta\phi) = \psi(\phi) + \delta\phi\frac{\partial}{\partial\phi}\psi(\phi)$$

$$= \left(1 + \delta\phi\frac{\partial}{\partial\phi}\right)\psi(\phi) = R\psi(\phi) \tag{8.15}$$

where

$$R = 1 + \delta\phi\frac{\partial}{\partial\phi} \tag{8.16}$$

is an infinitesimal rotation operator. The z component of angular momentum operator is

$$J_z = -i\hbar\left(x\frac{\partial}{\partial y} - y\frac{\partial}{\partial x}\right) = -i\hbar\frac{\partial}{\partial\phi} \tag{8.17}$$

$$\frac{\partial}{\partial\phi} = \frac{J_z}{-i\hbar} = \frac{i}{\hbar}J_z \tag{8.18}$$

So we can express rotation operator in terms of the z component of angular momentum operator as

$$R = 1 + \frac{i}{\hbar}J_z\delta\phi \tag{8.19}$$

A finite rotation $\Delta\phi$ can be obtained by taking n steps in succession so that

$$n\delta\phi = \Delta\phi \tag{8.20}$$

$$\delta\phi = \frac{\Delta\phi}{n} \tag{8.21}$$

and so

$$R = \left(1 + \frac{i}{\hbar}J_z\frac{\Delta\phi}{n}\right)^n \tag{8.22}$$

For $n \to \infty$ we write

$$R = \underset{n\to\infty}{Lt}\left(1 + \frac{i}{\hbar}J_z\frac{\Delta\phi}{n}\right)^n \tag{8.23}$$

Using the standard result of equation (8.11) viz. $\underset{n\to\infty}{Lt}(1 + \frac{x}{n})^n = e^x$ we write

$$R = e^{iJ_z\Delta\phi/\hbar} \tag{8.24}$$

In natural units $\hbar = 1$ and so

$$R = e^{iJ_z \Delta\phi} \tag{8.25}$$

Clearly

$$R^\dagger R = e^{-iJ_z \Delta\phi} e^{iJ_z \Delta\phi} = I \tag{8.26}$$

This R is a unitary operator.

So the unitary operator for rotation is $e^{iJ_z \Delta\phi}$ and the z component of angular momentum operator J_z is called the generator of this transformation that leads to conservation of angular momentum.

8.5 Translation in time

If the Hamiltonian of a system is invariant under time translation then this time translational invariance leads to conservation of energy.

8.6 Reflection in space or spatial inversion

Symmetry under reflection about origin leads to the conservation of parity of the system.

8.7 Exchange of particles

Symmetry under exchange of particle and anti-particle leads to the principle of charge conjugation.

This leads to C parity.

8.8 Rotation in isospace

So far as strong interaction is concerned, if there is an approximate symmetry of u quark and d quark then this leads to the approximate conservation principle of isospin.

8.9 Classifications of symmetries

Symmetries can be classified into various broad categories as follows.

8.10 Space-time symmetry or external symmetry

Space- time symmetry refers to the symmetry of geometrical operations that can be performed in external space like displacement or space translation, space rotation, time translation, spatial inversion or reflection about origin. It involves movement in external space. It is due to external degrees of freedom.

8.11 Continuous space-time symmetry

Examples of continuous space-time symmetries are translation and rotation.

This symmetry is of universal validity. It holds good for all the four known interactions.

This symmetry is never broken.

8.12 Internal symmetry

Symmetry other than space-time symmetry is referred to as internal symmetry. This symmetry is due to internal degrees of freedom and is not related to movement in external space.

Examples are rotation in isospace, spatial inversion, particle–anti-particle conjugation, gauge transformation.

8.13 Permutation symmetry

Examples of permutation symmetry are Bose–Einstein statistics and Fermi–Dirac statistics.

This symmetry is of universal validity. This symmetry is never broken. It holds in all types of interactions.

8.14 Discrete symmetry

Examples of discrete symmetry are space inversion, time reversal, particle–anti-particle conjugation etc.

This symmetry is not of universal validity. In some interactions it is broken, i.e. does not hold in some interactions.

Space inversion in the case of weak interaction is not valid since parity violation takes place in the case of β decay. In other words, space inversion symmetry breaks in the case of weak interaction.

Time reversal symmetry is violated, i.e. it breaks in certain reactions such as in the case of decay of neutral kaons viz.

$$K^0 \leftrightarrows \pi^0 + \pi^0 \leftrightarrows \overline{K^0}$$

$$K^0 \leftrightarrows \pi^+ + \pi^- \leftrightarrows \overline{K^0}$$

(A kaon K^0 cannot be directly transferred to $\overline{K^0}$. It first transforms to π^0, π^0 or π^+, π^- and then to $\overline{K^0}$.)

So symmetry may or may not be of universal validity.

8.15 Unitary symmetries (internal symmetry)

We now mention some symmetries that refer not to external physical space but to internal space and hence are called internal symmetries.

    (i) U(1) symmetry

        $U(1)$ symmetry is related to conservation of electrical charge (Q), baryon number (B), lepton number (L) etc.

        $U(1)$ symmetry is of universal validity as it holds in all types of interactions. We will discuss $U(1)$ symmetry in chapter 9.

    (ii) SU(2) symmetry

        $SU(2)$ symmetry is isospin symmetry and is not of universal validity. We will discuss $SU(2)$ symmetry in chapter 9.

    (iii) SU(3) symmetry.

Invariance / symmetry	Conservation law
Space time invariance	Linear momentum and energy
Rotation invariance	Angular momentum
Reflection in space	Parity
Exchange of particle and anti-particle	Charge conjugation
Symmetry between u and d quarks	Isospin

Figure 8.1. Symmetry and conservation laws.

$SU(3)$ symmetry is colour symmetry (colour is a quantum number). This symmetry is also not of universal validity.

We will discuss $SU(3)$ symmetry in chapter 9.

We show a summary of the conservation laws in figure 8.1.

8.16 Noether's theorem (symmetry theorem)

With every symmetry or invariance a conservation principle is associated.

If there is a symmetry in a system there is an associated conservation law.

The opposite is also true.

If there is a conservation law a system has a symmetry.

Continuous symmetry leads to some conservation laws, e.g. electric charge is a conserved quantum number that comes from current conservation.

Discrete symmetries (global symmetry) lead to some conservation laws, e.g. conservation of baryon number, lepton number etc.

We now discuss some conservation laws associated with internal symmetries.

8.17 Conservation of electric charge

Electric charge is conserved in all interactions.

Stability of electron e^- follows from conservation of electric charge.

8.17.1 Stability of electron

Electron e^- has smallest non-zero rest mass. It follows from mass conservation that its decay can lead to only zero rest mass particles. But all zero rest mass particles are electrically neutral. So such decay would violate electric charge conservation principle and is forbidden. So an electron is a stable particle. Stability of the electron is crucial for the existence of atoms.

8.18 Conservation of lepton number

Figure 7.6 shows the three types of lepton numbers, namely L_e (electron lepton number), L_μ (muon lepton number), L_τ (tauon lepton number) of various particles.

The net lepton number in any process always remains constant.

If only electron e^- is emitted in an interaction then lepton number will not be conserved since an electron carries electron lepton quantum number $L_e = +1$. To neutralize it an electron anti-neutrino $\bar{\nu}_e$ is emitted that carries electron lepton quantum number $L_e = -1$. Hence electron lepton quantum number L_e is conserved.

The following are emitted in pairs.

$$e^-, \ \bar{\nu}_e, \ \text{so that } L_e(e^-) + L_e(\bar{\nu}_e) = 1 - 1 = 0.$$

$$e^+, \ \nu_e, \ \text{so that } L_e(e^+) + L_e(\nu_e) = -1 + 1 = 0.$$

$$\mu^-, \ \bar{\nu}_\mu, \ \text{so that } L_\mu(\mu^-) + L_\mu(\bar{\nu}_\mu) = 1 - 1 = 0.$$

$$\mu^+, \ \nu_\mu, \ \text{so that } L_\mu(\mu^+) + L_\mu(\nu_\mu) = -1 + 1 = 0.$$

$$\tau^-, \ \bar{\nu}_\tau, \ \text{so that } L_\tau(\tau^-) + L_\tau(\bar{\nu}_\tau) = 1 - 1 = 0.$$

$$\tau^+, \ \nu_\tau, \ \text{so that } L_\tau(\tau^+) + L_\tau(\nu_\tau) = -1 + 1 = 0.$$

If L_e, L_μ, L_τ are not conserved then the reaction is forbidden to occur. *Exercises 8.4 and 8.6 give a few illustrations.*

8.19 Conservation of baryon number

Baryons are characterized by a quantum number called baryon quantum number B as described in figure 7.7.

Baryon number is conserved.

Figure 8.2 gives the list of baryons and anti-baryons along with their baryon quantum number and masses.

$B = +1$ for baryons (p, n, Λ, Σ^+, Σ^0, Σ^-, Ξ^-, Ξ^0. Ω^-)

$B = -1$ for anti-baryons ($\bar{p}$, $\bar{n}$, $\bar{\Lambda}$, $\bar{\Sigma}^-$, $\bar{\Sigma}^+$, $\bar{\Sigma}^0$, $\bar{\Xi}^0$, $\bar{\Xi}^+$, Ω^+)

$B = 0$ for all other particles (e.g. $B(\text{meson}) = 0$)

The net baryon number in any process always remains constant.

$$B(\text{quark } q) = \frac{1}{3}, \quad B(\text{antiquark } \bar{q}) = -\frac{1}{3}.$$

Exercise 8.3 shows a few illustrations.

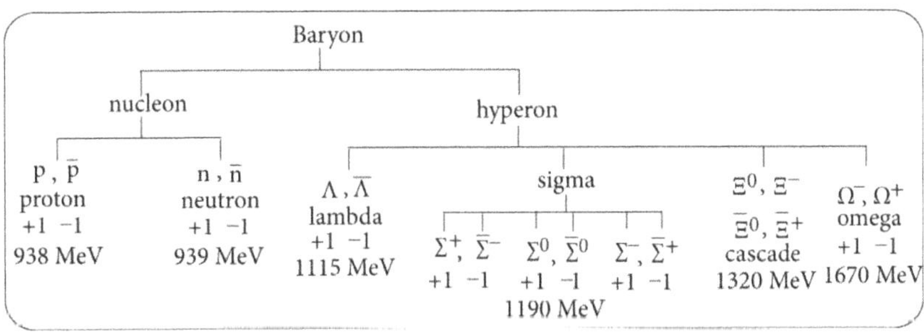

Figure 8.2. List of baryons along with their baryon quantum number and masses.

8.20 Stability of proton

Proton is the least massive baryon having $B = 1$. So it follows from mass energy conservation that the proton decay products cannot contain baryons. But then baryon number is violated. So proton cannot decay. Proton is stable due to baryon number conservation. Stability of protons is crucial for the existence of atoms.

In the grand unification theory (GUT) baryon number conservation is violated and so proton decay is allowed.

8.21 Classification of hadrons

In figure 7.10 we have shown that there are two types of hadrons. One is mesons which are bosons and the other is baryons which are fermions.

A multitude of hadrons has been discovered in high energy particle collisions as well as in cosmic rays. One way of classifying the particles and to understand particle properties is through use of symmetry principles. We try to classify them with the help of symmetry groups.

8.22 Conservation of isospin and Isospin symmetry

We shall assign to the hadrons 2 quantum numbers I (isospin) and I_3 (isospin projection) similar to the ordinary spin angular momentum s and its projection s_z along the direction of quantization direction (i.e. along the z-axis), so that these are conserved in reactions involving hadrons. And it was found that one can consistently and successfully assign isospin to the hadrons that were observed.

Conservation of isospin applies to strongly interacting fundamental particles, i.e. to hadrons (figure 7.10). Let us first consider nucleons.

We note that mass of neutron n and proton p are almost the same ($m_p \cong m_n$) since

$$m_p = 1.672\,621\,92 \times 10^{-27}\ kg,\ m_n = 1.674\,9747 \times 10^{-27}\ kg \text{ (slightly different)}$$

and we can neglect their tiny difference.

We have, however, considered for convenience their approximate values in various calculations viz.

$$m_p = 938\ \frac{MeV}{c^2} = 1.67 \times 10^{-27}\ kg,\ m_n = 939\ \frac{MeV}{c^2} = 1.675 \times 10^{-27}\ kg$$

Except for charge ($Q_p = |e|$, $Q_n = 0$) all other properties are identical like intrinsic spin $s = \frac{1}{2}$ etc.

As electromagnetic force is charge dependent, proton p (m_p, $Q_p = |e|$) and neutron n (m_n, $Q_n = 0$) manifest as different particles.

But nuclear force that is responsible for stability of nucleus is charge independent.

According to Heisenberg's suggestion, if we switch off electric charge of a proton then we cannot distinguish between a proton and a neutron. Also, if we impart electric charge $|e|$ to a neutron then we cannot also distinguish between a proton and a neutron.

So let us consider them as the same particle called the nucleon. Neutron and proton are two states of this entity called nucleon.

The nn, pp, np strong interactions are similar and charge has no role to play. Their electromagnetic interactions are different due to their charge.

We can compare the above case with the case of the electron, which is a spin-half $s = \frac{1}{2}$ fermion. An electron can exist in two states $m_s = \pm\frac{1}{2}$ namely (representation used is $|s\ m_s>$)

$$|\text{spin up electron} > = |s = \frac{1}{2}\ m_s = \frac{1}{2} > = |\alpha > = |\uparrow > = \begin{pmatrix} 1 \\ 0 \end{pmatrix} \tag{8.27}$$

$$|\text{spin down electron} > = |s = \frac{1}{2}\ m_s = -\frac{1}{2} > = |\beta > = |\downarrow > = \begin{pmatrix} 0 \\ 1 \end{pmatrix} \tag{8.28}$$

As there is nothing to distinguish between proton and neutron when their nuclear interaction is discussed, we can state that the 3 interactions nn, pp, np are of equal strength, i.e.

$$V_{nn} = V_{pp} = V_{np} \tag{8.29}$$

This leads to the concept of isospin I.

These particle states p and n are said to form isotopic doublet and this isotopic grouping is described by a new quantum number called isobaric spin or isotopic spin quantum number

$$I = \frac{1}{2} \tag{8.30}$$

And the z projections are $I_3 = m_I \hbar = \pm\frac{1}{2}\hbar$ i.e. $m_I = \pm\frac{1}{2}$. This quantum number facilitates consideration of the consequences of the charge independence of strong nuclear force.

The n and p are the same particle in different charged states, i.e. nucleons can exist in two states, namely

$$|\text{isospin up nucleon} > = |I = \frac{1}{2} m_I = \frac{1}{2} > = |\alpha > = |\uparrow > = \begin{pmatrix} 1 \\ 0 \end{pmatrix} = |\text{ proton}> \tag{8.31}$$

$$|\text{isospin down nucleon} > = |I = \frac{1}{2}\ m_I = -\frac{1}{2} > = |\beta > = |\downarrow > = \begin{pmatrix} 0 \\ 1 \end{pmatrix} = |\text{ neutron}> \tag{8.32}$$

Equations (8.31) are (8.32) are analogous to equations (8.27) are (8.28), respectively.

In other words neutrons and protons have different charge manifestations of the same entity called the nucleon, i.e. a nucleon in charged state is called a proton $|p> = \begin{pmatrix} 1 \\ 0 \end{pmatrix}$ and a nucleon in chargeless state is called a neutron $|n> = \begin{pmatrix} 0 \\ 1 \end{pmatrix}$. They may be considered as two charge states or two isospin states of a doublet called the nucleon $|N>$.

Although isospin has mathematical properties similar to spin, isospin has no direct physical relationship with spin. Both $\vec{I}$ and $\vec{s}$ are angular momentum and the algebra of $\vec{I}$ and $\vec{s}$ are identical.

The z component or 3rd component of isospin viz I_z or I_3 takes $2I + 1$ values $-I$ to $+I$ in unit steps. This $2I + 1$ is referred to as multiplicity of I_3 or m_I.

Since a nucleon can exist in two states $|\,p>$ and $|\,n>$ it follows that

$$2I + 1 = 2$$

leading to

$$I = \frac{1}{2}$$

for a nucleon as we assumed in equation (8.30).

As per convention, we have taken

$$|\,p> \ =|\,I \ I_3 > \ =|\,\frac{1}{2} \ \frac{1}{2}> \tag{8.33}$$

$$|\,n> \ =|\,I \ I_3 > \ =|\,\frac{1}{2} - \frac{1}{2}> \tag{8.34}$$

For a nucleon we define

$$\vec{I} = \frac{\vec{\tau}}{2} = (\frac{\tau_x}{2}, \ \frac{\tau_y}{2}, \ \frac{\tau_z}{2})$$

Isospin quantum number I is associated with hadrons (which suffer strong interaction) but not with leptons. So isospin remains conserved only in strong interactions.

On the other hand, the isospin component I_3 is conserved both in strong and electromagnetic interaction but not in weak interaction.

Let us discuss flavour symmetry of the standard model. It is related to the flavour of the light quarks or the flavour of the nucleons inside the nucleus.

The quark structure of a proton is $p = uud$ and shifting $u \leftrightarrow d$ we get the quark structure of a neutron $n = ddu$. Since

$$\begin{pmatrix} p \\ n \end{pmatrix} = \begin{pmatrix} uud \\ ddu \end{pmatrix} = \begin{pmatrix} u \\ d \end{pmatrix}$$

there is symmetry between p and n, i.e. symmetry between u and d. The name of the symmetry is isospin symmetry.

Therefore, at a fundamental level we think of a single light quark to exist in two states. An $|\,up>$ state which is the u quark or up quark and another $|\,down>$ state which is the d quark or down quark.

And u and d quarks have nearly the same mass

$$m_u = 2.2\frac{MeV}{c^2}, \ m_d = 4.7\frac{MeV}{c^2}.$$

For u quark: $I = \frac{1}{2}$ (spin-half fermion) $I_3 = \frac{1}{2}$

For d quark: $I = \frac{1}{2}$ (spin-half fermion) $I_3 = -\frac{1}{2}$

We note that

$$I_3(p) = I_3(uud) = I_3(u) + I_3(u) + I_3(d) = \frac{1}{2} + \frac{1}{2} - \frac{1}{2} = \frac{1}{2}$$

$$I_3(n) = I_3(ddu) = I_3(d) + I_3(d) + I_3(u) = -\frac{1}{2} - \frac{1}{2} + \frac{1}{2} = -\frac{1}{2}$$

Let us rewrite in the $\mid I\ I_3 >$ representation

$$\mid u > \; = \mid \frac{1}{2}\,\frac{1}{2} > \; = \mid \text{up} > \; = \begin{pmatrix} 1 \\ 0 \end{pmatrix} \Leftarrow u \text{ quark (resembles equation (8.31))}$$

$$\mid d > \; = \mid \frac{1}{2} - \frac{1}{2} > \; = \mid \text{down} > \; = \begin{pmatrix} 0 \\ 1 \end{pmatrix} \Leftarrow d \text{ quark (resembles equation (8.32))}$$

If we make transformation from one state say $\mid p>$ to another state $\mid n>$ which differs in isospin projection, it is not going to make any difference so far as strong interaction is concerned. This led to isospin symmetry. Similar is the situation between $\mid u>$ and $\mid d>$, i.e. between u quark and d quark, i.e. $\begin{pmatrix} p \\ n \end{pmatrix} = \begin{pmatrix} u \\ d \end{pmatrix}$.

Flavour symmetry of the Standard Model is based on the fact that proton and neutron have almost identical masses. Also, masses of u quark and d quark are not the same. Since masses are not exactly identical the symmetry is not perfect. It is thus not a fundamental symmetry of nature and is an artificial construct to describe flavour. Only because their masses are close enough can we think of a symmetry that works if we assume that the up flavour state and down flavour state are two states of the same particle.

This artificial construct is useful because strong force conserves isospin. Looking at isospin symmetry, we can identify whether certain bound states exist and whether some interactions occur or not.

8.23 Charge operator

We can define a charge operator as

$$Q = \mid e \mid \left(I_3 + \frac{1}{2} \right) \tag{8.35}$$

where $\mid e \mid$ is charge on a proton.

For a proton

$$Q \mid p > \; = \mid e \mid \left(I_3 + \frac{1}{2} \right) \mid p > \; = \mid e \mid \left(\frac{1}{2} + \frac{1}{2} \right) \mid p > \; = \mid e \mid \mid p > \tag{8.36}$$

For a neutron

$$Q \mid n > \; = |e|\left(I_3 + \frac{1}{2}\right) \mid n > \; = |e|\left(-\frac{1}{2} + \frac{1}{2}\right) \mid p > \; = 0 \qquad (8.37)$$

So the concept of isospin explains charge of a nucleon.

8.24 Gell-Mann–Okubo formula

The quantum numbers Q, I_3,B attached to p, n are as follows

$$Q \; = 1 \; |e| \rightarrow 1, \; I_3 = \frac{1}{2}, \; B = 1 \text{ (for } p\text{)}$$

$$Q \; = 0 \; |e| \rightarrow 0, \; I_3 = -\frac{1}{2}, \; B = 1 \text{ (for } n\text{)}$$

We wish to relate the quantum number I_z to electric charge Q of a particle.

$$\text{For proton } Q = 1 = \frac{1}{2} + \frac{1}{2} = I_3 + \frac{B}{2}$$

$$\text{For neutron } Q = 0 = -\frac{1}{2} + \frac{1}{2} = I_3 + \frac{B}{2}$$

Clearly we have arrived at the formula

$$Q = I_3 + \frac{B}{2} \qquad (8.38)$$

This is called Gell-Mann–Okubo formula which is the inter-relation between charge Q and the 3rd component of isospin I_3.

We can calculate value of charge Q of a particle from this formula knowing I_3.

8.24.1 Pion family

There are 3 members π^+,π^-, π^0 in the pion family and they have similar masses

$$m_{\pi^\pm} = 140\frac{MeV}{c^2}, \; m_{\pi^0} = 138\frac{MeV}{c^2}.$$

So we can think of putting them in a group.

The 3 members π^+,π^-, π^0 of the pion family are considered as members of a triplet. So I_3 or m_I has multiplicity

$$2I + 1 = 3$$

and so

$$I = 1.$$

So $I = 1$ for π^+,π^-, π^0. The I_3 or m_I values are -1 to $+1$ in unit steps. And so the values are $m_I = -1, 0, +1$. So a pion is an isotriplet. We assign in the representation $\mid I \; I_3>$

$$| I\ I_3 > \ =| 1\ 1 > \ =| \pi^+> \tag{8.39}$$

$$| I\ I_3 > \ =| 1\ 0 > \ =| \pi^0> \tag{8.40}$$

$$| I\ I_3 > \ =| 1 - 1 > \ =| \pi^-> \tag{8.41}$$

Also, pions are mesons, i.e. non-baryons, and so their baryon number is $B = 0$. So the quantum numbers Q, I_3, B attached to π^+, π^-, π^0 are as follows

$$Q = 1\ |e| \to 1, \ \ I_3 = 1, \ \ B = 0 \ (\text{for } \pi^+)$$

$$Q = 0\ |e| \to 0, \ \ I_3 = 0, \ \ B = 0 \ (\text{for } \pi^0)$$

$$Q = -1\ |e| \to -1, \ \ I_3 = -1, \ \ B = 0 \ (\text{for } \pi^-)$$

Now

For π^+: $Q = 1 = 1 + \frac{0}{2} = I_3 + \frac{B}{2}$

For π^0: $Q = 0 = 0 + \frac{0}{2} = I_3 + \frac{B}{2}$

For π^-: $Q = -1 = -1 + \frac{0}{2} = I_3 + \frac{B}{2}$

Obvious that Gell-Mann–Okubo formula given by equation (8.38) holds for nucleonic isodoublet and pionic isotriplet.

8.24.2 Kaon family

We note that we cannot group K^+ with π^+ since their masses are widely different as

$$m_{K^+} = 493\frac{MeV}{c^2}, \ \ m_{\pi^+} = 140\frac{MeV}{c^2}$$

The two members K^+ and K^0 in the kaon family have the same mass

$$m_{K^+} = m_{K^0} = 493\frac{MeV}{c^2}$$

So we can think of putting K^+ and K^0 in a group which is an isodoublet. So I_3 or m_I has multiplicity

$$2I + 1 = 2$$

and so

$$I = \frac{1}{2}.$$

So $I = \frac{1}{2}$ for K^+, K^0. The I_3 or m_I values are $-\frac{1}{2}$ to $+\frac{1}{2}$ in unit steps, i.e. two values viz. $m_I = -\frac{1}{2}, +\frac{1}{2}$. So kaon sub-family K^+, K^0 is an isodoublet. We assign in the representation $| I\ I_3>$

$$| I\ I_3 > \ =| \frac{1}{2}\frac{1}{2} > \ =| K^+> \tag{8.42}$$

$$| I \; I_3 > \; =| \frac{1}{2} - \frac{1}{2} > \; =| K^0> \tag{8.43}$$

The two members $\overline{K^0}$ and K^- in the kaon family have the same masses

$$m_{\overline{K^0}} = m_{K^-} = 498\frac{MeV}{c^2}$$

So we can think of putting them in a group which is an isodoublet. So I_3 or m_I has multiplicity

$$2I + 1 = 2 \text{ and so } I = \frac{1}{2}.$$

So $I = \frac{1}{2}$ for $\overline{K^0}$, K^-. The I_3 or m_I values are $-\frac{1}{2}$ to $+\frac{1}{2}$ in unit steps, i.e. two values viz. $m_I = -\frac{1}{2}, +\frac{1}{2}$. So kaon sub-family $\overline{K^0}, K^-$ is an isodoublet. We assign in the representation $| I \; I_3>$

$$| I \; I_3 > \; =| \frac{1}{2} \frac{1}{2} > \; =| \overline{K^0}> \tag{8.44}$$

$$| I \; I_3 > \; =| \frac{1}{2} - \frac{1}{2} > \; =| K^-> \tag{8.45}$$

We can in a similar fashion ascribe isospin quantum numbers to other families as follows.

8.24.3 Sigma family

The sigma family is an isotriplet $\Sigma^{\pm}$, Σ^0

$$| I \; I_3 > \; =| 1 \; 1 > \; =| \Sigma^+ > \; \left(m_{\Sigma^+} = 1189.4\frac{MeV}{c^2}\right) \tag{8.46}$$

$$| I \; I_3 > \; =| 1 \; 0 > \; =| \Sigma^0 > \; \left(m_{\Sigma^0} = 1192.5\frac{MeV}{c^2}\right) \tag{8.47}$$

$$| I \; I_3 > \; =| 1 - 1 > \; =| \Sigma^- > \; \left(m_{\Sigma^-} = 1197.4\frac{MeV}{c^2}\right) \tag{8.48}$$

8.24.4 Cascade family

The cascade family is an Isodoublet Ξ^-, Ξ^0

$$| I \; I_3 > \; =| \frac{1}{2} \frac{1}{2} > \; =| \Xi^- > \; \left(m_{\Xi^-} = 1321.32\frac{MeV}{c^2}\right) \tag{8.49}$$

$$| I \; I_3 > \; = | \frac{1}{2} - \frac{1}{2} > \; = | \, \Xi^0 > \; \left(m_{\Xi^0} = 1314.9 \frac{MeV}{c^2} \right) \qquad (8.50)$$

8.24.5 Isosinglet (lambda, omega, eta)

Isosinglet refers to a single member group, i.e. a group having one particle as member because their mass does not match with others.

- Lambda Λ is an isosinglet

$$| I \; I_3 > \; = | \, 00 > \; = | \, \Lambda > \; \left(m_\Lambda = 1115.7 \frac{MeV}{c^2} \right) \qquad (8.51)$$

- Omega Ω^- is an isosinglet

$$| I \; I_3 > \; = | \, 00 > \; = | \, \Omega^- > \; \left(m_{\Omega^-} = 1670 \frac{MeV}{c^2} \right) \qquad (8.52)$$

- Eta η is an isosinglet

$$| I \; I_3 > \; = | \, 00 > \; = | \, \eta > \; \left(m_\eta = 550 \frac{MeV}{c^2} \right) \qquad (8.53)$$

We can thus classify the hadrons in terms of I, I_3, i.e. isospin symmetry.

✓ Let us check the Gell-Mann–Okubo formula for strange particle Λ.

Λ is an isosinglet and so its multiplicity m_I is

$$2I + 1 = 1.$$

This gives $I = 0$, $I_3 = 0$.

Λ is a baryon, i.e. $B = 0$. So the quantum numbers Q, I_3, B attached to Λ are $Q = 0 \, |e| \rightarrow 0$, $I_3 = 0$, $B = 1$. Hence for Λ we have

$Q = 0$ and $I_3 + \frac{B}{2} = 0 + \frac{1}{2} = \frac{1}{2}$. Clearly

$$Q \neq I_3 + \frac{B}{2} \text{ for } \Lambda. \qquad (8.54)$$

The Gell-Mann–Okubo formula does not hold for Λ.

The Gell-Mann–Okubo formula does not hold for strange particles and needs modification.

Charge independence of strong nuclear force is expressed in terms of invariance under unitary transformation in the isospin space. When we have 2 states up and down then the symmetry we get is called $SU(2)$. In other words isospin satisfies $SU(2)$ algebra.

We will discuss $SU(2)$ symmetry in chapter 9.

Figure 8.3 shows the various I, I_3 values of hadrons.

8.25 Algebra of isospin

Let us introduce a quantity called isospin $\vec{I} = (I_1, I_2, I_3)$ in isospin space. It is analogous to angular momentum $\vec{J}$.

The operator $\vec{I}$ satisfies the same commutation relations that are satisfied by angular momentum $\vec{J}$.

Let us define raising operator I_+ and lowering operators I_- as

$$I_\pm = I_1 + I_2$$

Eigenvalue equations of I^2 and I_3 are, respectively

$$I^2 |I\ I_3 > \ = I(I+1)\hbar^2|I\ I_3 > > \rightarrow I(I+1)| I\ I_3>$$
(8.55)

$$I_3 |I\ I_3 > \ = m_I \hbar|I\ I_3 > \ \rightarrow m_I\ | I\ I_3>$$
(8.56)

Possible eigenvalues of I are 0, $\frac{1}{2}$, 1, $\frac{3}{2}$, 2...

I_3 has multiplicity $2I + 1$, i.e. assumes values $-I$, $...+I$ in unit steps.

Matrix elements for isospin operators I^2, I_3, I_+, I_- with respect eigenstate $| I\ I_3>$ as basis are given by (with $\hbar = 1$)

$$\langle I\ I_3' |I^2|I\ I_3\rangle \ = I(I+1)\delta_{I_3,\ I_3'}$$
(8.57)

$$\langle I\ I_3' |I_3|I\ I_3\rangle \ = I_3\delta_{I_3,\ I_3'}$$
(8.58)

$$\langle I\ I_3' |I_\pm|I\ I_3 \pm 1\rangle \ = \sqrt{I(I+1) - I_3(I_3 \pm 1)}\,\delta_{I_3,\ I_3 \pm 1}$$
(8.59)

Isosinglet	I	I_3	Q	Isodoublet	I	I_3	Q	Isotriplet	I	I_3	Q
Λ	0	0	0	$N \nearrow$ p	$\frac{1}{2}$	$\frac{1}{2}$	+1	$\pi \nearrow \pi^+$	1	+1	+1
η	0	0	0	$N \searrow$ n	$\frac{1}{2}$	$-\frac{1}{2}$	0	$\pi \rightarrow \pi^0$	1	0	0
Ω^-	0	0	0	$K \nearrow K^+$	$\frac{1}{2}$	$\frac{1}{2}$	+1	$\pi \searrow \pi^-$	1	−1	−1
				$K \searrow K^0$	$\frac{1}{2}$	$-\frac{1}{2}$	0	$\Sigma \nearrow \Sigma^+$	1	+1	+1
				$\overline{K} \nearrow \overline{K}^0$	$\frac{1}{2}$	$\frac{1}{2}$	0	$\Sigma \rightarrow \Sigma^0$	1	0	0
				$\overline{K} \searrow K^-$	$\frac{1}{2}$	$-\frac{1}{2}$	−1	$\Sigma \searrow \Sigma^-$	1	−1	−1
				$\Xi \nearrow \Xi^0$	$\frac{1}{2}$	$\frac{1}{2}$	0				
				$\Xi \searrow \Xi^-$	$\frac{1}{2}$	$-\frac{1}{2}$	−1				

Figure 8.3. Isospin values of hadrons.

Also

$$I_{\pm} \mid I\, I_3 > \; = \sqrt{I(I+1) - I_3(I_3 \pm 1)} \mid I\, I_3 \pm 1> \tag{8.60}$$

Let us define

$$\vec{I} = \frac{1}{2}\,\vec{\tau} \text{ or } I_i = \frac{1}{2}\tau_i \tag{8.61}$$

The fundamental two dimensional representation of isospin rotation group is based on three matrices.

$$\tau_1 = \begin{pmatrix} 0 & 1 \\ 1 & 0 \end{pmatrix}, \quad \tau_2 = \begin{pmatrix} 0 & -i \\ i & 0 \end{pmatrix}, \quad \tau_3 = \begin{pmatrix} 1 & 0 \\ 0 & -1 \end{pmatrix} \tag{8.62}$$

which are identical to Pauli spin matrices σ_1, σ_2, σ_3.

$$[I_i,\ I_j] = i\varepsilon_{ijk}I_k \tag{8.63}$$

$$\left[\frac{\tau_i}{2},\ \frac{\tau_j}{2}\right] = i\varepsilon_{ijk}\frac{\tau_k}{2} \tag{8.64}$$

An explanation for the isospin grouping of hadrons follows from the quark structure. We assign the isospin values to quarks u, d, s and anti-quarks $\bar{u}$, $\bar{d}$, $\bar{s}$ as shown in figure 8.4.

The relative sign between (u, d) and $(\bar{u}, \bar{d})$ is discussed in chapter 10, section 10.11. In figure 8.4 I_q, I_{3q} refer to isospin and isospin projection for a quark while $I_{\bar{q}}$, $I_{3\bar{q}}$ refer to isospin and isospin projection for an anti-quark.

Similar to the ordinary spin angular momentum we can combine different isospin states.

8.26 Quark states of the pion family

Let us try to derive quark states of the pion family using the assigned values of I, I_3 to the quark system u, d and the anti-quark system $\bar{u}$, $\bar{d}$ (figure 8.4).

The quark system has $I = \frac{1}{2} \equiv I_q$ and $I_{3q} = \pm\frac{1}{2}$.

The anti-quark system has $I = \frac{1}{2} \equiv I_{\bar{q}}$ and $I_{3\bar{q}} = \pm\frac{1}{2}$.

quark	q.no. for quark		antiquark	q.no. for antiquark	
q	I_q	I_{3q}	$\bar{q}$	$I_{\bar{q}}$	$I_{3\bar{q}}$
u	$\frac{1}{2}$	$\frac{1}{2}$	$\bar{d}$	$\frac{1}{2}$	$\frac{1}{2}$
d	$\frac{1}{2}$	$-\frac{1}{2}$	$-\bar{u}$	$\frac{1}{2}$	$-\frac{1}{2}$
s	0	0	$\bar{s}$	0	0

Figure 8.4. Isospin values of quarks and anti-quarks.

Let us combine these states (I_q, I_{3q}) and $(I_{\bar{q}}, I_{3\bar{q}})$ using the angular momentum addition rule. The resultant or combined state is (I, I_3). Now

$$(I_q = \frac{1}{2}) \ \oplus \ (I_{\bar{q}} = \frac{1}{2}) = I = 1, 0. \tag{8.65}$$

I_3 values corresponding to I are
$I_3 = 1, 0, -1$ for $I = 1$ and
$I_3 = 0$ for $I = 0$.
✓ Consider $I = 1$ case that gives the state $| I = 1 \ I_3 = 1>$.
    As $I_3 = I_{3q} + I_{3\bar{q}}$ we note that $I_3 = 1$ can be generated from $I_{3q} = +\frac{1}{2}$ and $I_{3\bar{q}} = +\frac{1}{2}$. So

$$| I = 1, \ I_3 = 1 > \ = | I_q = \frac{1}{2} \ I_{3q} = +\frac{1}{2} ; \ I_{\bar{q}} = \frac{1}{2} \ I_{3\bar{q}} = +\frac{1}{2}> \tag{8.66}$$

    Since $| u > \ = | I_q = \frac{1}{2} \ I_{3q} = +\frac{1}{2}>$ and $| \bar{d} > = | I_{\bar{q}} = \frac{1}{2} \ I_{3\bar{q}} = +\frac{1}{2}>$ (from figure 8.4)
    From equation (8.66) we have

$$| RHS > \ = | I_q = \frac{1}{2} \ I_{3q} = +\frac{1}{2} ; \ I_{\bar{q}} = \frac{1}{2} \ I_{3\bar{q}} = +\frac{1}{2} > \ = | u\bar{d}>$$

Since $Q(u) + Q(\bar{d}) = \frac{2}{3} + \frac{1}{3} = 1 = Q(\pi^+)$ we identify

$$| \pi^+ > \ = | u\bar{d}> \tag{8.67}$$

i.e quark structure of π^+ is $u\bar{d}$.
✓ Consider $I = 1$ case that gives the state $| I = 1 \ I_3 = 0>$.
    Invoke the general formula (I_- being a step down operator) using equation (8.60)

$$I_- | I \ I_3> = \sqrt{I(I+1) - I_3(I_3 - 1)} |I \ I_3 - 1>$$
$$= = \sqrt{I^2 + I - I_3^2 + I_3} \ \ | I \ I_3 - 1>$$
$$= \sqrt{(I + I_3)(I - I_3) + (I + I_3)} \ \ | I \ I_3 - 1>$$
$$I_- | I \ I_3> = \sqrt{(I + I_3)(I - I_3 + 1)} \ | I \ I_3 - 1> \tag{8.68}$$

    Applying I_- to both sides of equation (8.66) we get

$$I_- | I = 1, \ I_3 = 1 > \ = I_- \ | I_q = \frac{1}{2} \ I_{3q} = +\frac{1}{2} ; \ I_{\bar{q}} = \frac{1}{2} \ I_{3\bar{q}} = +\frac{1}{2}> \tag{8.69}$$

    Consider the LHS of equation (8.69)

$$I_- | LHS> = I_- | I = 1 \ I_3 = 1 > \ = \sqrt{(1 + 1)(1 - 1 + 1)} \ | 1 \ 1 - 1>$$
$$= \sqrt{2} \ | I = 1 \ I_3 = 0> \tag{8.70}$$

    Consider the RHS of equation (8.69)

$$I_- \,|\mathrm{RHS}\rangle \;=\; I_-|I_q = \tfrac{1}{2}\; I_{3q} = +\tfrac{1}{2}\,;\; I_{\bar q} = \tfrac{1}{2}\; I_{3\bar q} = +\tfrac{1}{2}\rangle$$

$$00 = \sqrt{(I_q + I_{3q})(I_q - I_{3q} + 1)}\;|\,I_q\; I_{3q} - 1\,;\; I_{\bar q}\; I_{3\bar q}\rangle$$

$$+\; \sqrt{(I_{\bar q} + I_{3\bar q})(I_{\bar q} - I_{3\bar q} + 1)}\;|\,I_q\; I_{3q}\,;\; I_{\bar q}\; I_{3\bar q} - 1\rangle$$

$$=\; \sqrt{\left(\tfrac{1}{2} + \tfrac{1}{2}\right)\left(\tfrac{1}{2} - \tfrac{1}{2} + 1\right)}\;|\,I_q = \tfrac{1}{2}\; I_{3q} - 1 = \tfrac{1}{2} - 1\,;\; I_{\bar q} = \tfrac{1}{2}\; I_{3\bar q} = \tfrac{1}{2}\rangle$$

$$+\; \sqrt{\left(\tfrac{1}{2} + \tfrac{1}{2}\right)\left(\tfrac{1}{2} - \tfrac{1}{2} + 1\right)}\;|\,I_q = \tfrac{1}{2}\; I_{3q} = \tfrac{1}{2}\,;\; I_{\bar q} = \tfrac{1}{2}\; I_{3\bar q} = \tfrac{1}{2} - 1\rangle$$

$$I_- \,|\,\mathrm{RHS}\rangle$$

$$= |\,I_q = \frac{1}{2}\; I_{3q} = -\frac{1}{2}\,;$$

$$I_{\bar q} = \frac{1}{2}\; I_{3\bar q} = +\frac{1}{2}\rangle + |\,I_q = \frac{1}{2}\; I_{3q} = +\frac{1}{2}\,;\; I_{\bar q} = \frac{1}{2}\; I_{3\bar q} = -\frac{1}{2}\rangle \tag{8.71}$$

From figure 8.4 we have

$$|\,u\rangle = |\,I_q = \frac{1}{2}\; I_{3q} = +\frac{1}{2}\rangle\,,\quad -|\,\bar u\rangle = |\,I_{\bar q} = \frac{1}{2}\; I_{3\bar q} = -\frac{1}{2}\rangle$$

$$|\,d\rangle = |\,I_q = \frac{1}{2}\; I_{3q} = -\frac{1}{2}\rangle\,,\quad |\,\bar d\rangle = |\,I_{\bar q} = \frac{1}{2}\; I_{3\bar q} = +\frac{1}{2}\rangle$$

With this we have from equation (8.71)

$$I_- \,|\,\mathrm{RHS}\rangle \;=\; |\,d\bar d\rangle - |\,u\bar u\rangle \tag{8.72}$$

Putting equations (8.70) and (8.72) to equation (8.69) we get

$$\sqrt{2}\;|\,I = 1\; I_3 = 0\rangle = |\,d\bar d\rangle - |\,u\bar u\rangle \tag{8.73}$$

$$|\,I = 1\; I_3 = 0\rangle = \frac{1}{\sqrt{2}}(|\,d\bar d\rangle - |\,u\bar u\rangle) \tag{8.74}$$

Since $I = 1$, $I_3 = 0$ corresponds to π^0 we write

$$|\,\pi^0\rangle = \frac{1}{\sqrt{2}}(|\,d\bar d\rangle - |\,u\bar u\rangle) \tag{8.75}$$

This is the quark structure of π^0.

✓ Consider $I = 1$ case that gives the state $|\,I = 1\; I_3 = -1\rangle$. As $I_3 = I_{3q} + I_{3\bar q}$ we note that $I_3 = -1$ can be generated from $I_{3q} = -\tfrac{1}{2}$ and $I_{3\bar q} = -\tfrac{1}{2}$. So

$$|\,I = 1,\; I_3 = -1\rangle = |\,I_q = \frac{1}{2}\; I_{3q} = -\frac{1}{2}\,;\; I_{\bar q} = \frac{1}{2}\; I_{3\bar q} = -\frac{1}{2}\rangle \tag{8.76}$$

Since $\quad |\,d\rangle = |\,I_q = \tfrac{1}{2}\; I_{3q} = -\tfrac{1}{2}\rangle\,,\quad -|\bar u\rangle = |\,I_{\bar q} = \tfrac{1}{2}\; I_{3\bar q} = -\tfrac{1}{2}\rangle \quad$ (from figure 8.4)

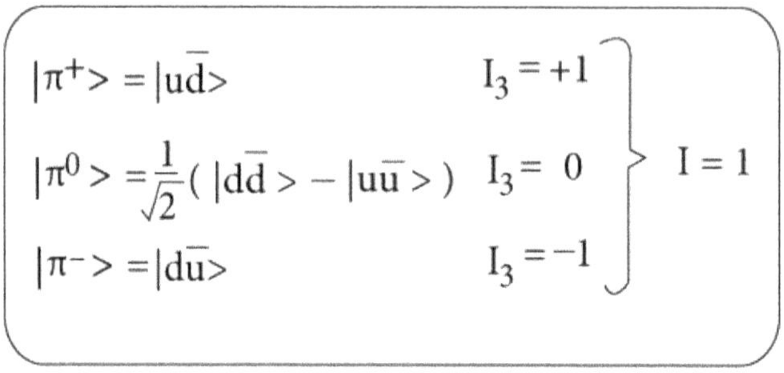

$$|\pi^+> = |u\bar{d}> \qquad\qquad I_3 = +1$$

$$|\pi^0> = \frac{1}{\sqrt{2}}(|d\bar{d}> - |u\bar{u}>) \quad I_3 = 0 \qquad \Big\} \quad I = 1$$

$$|\pi^-> = |d\bar{u}> \qquad\qquad I_3 = -1$$

Figure 8.5. Quark structure for pion family from isospin symmetry.

From equation (8.76) we have

$| \text{RHS} > = | I_q = \frac{1}{2} \, I_{3q} = -\frac{1}{2} \, ; \, I_{\bar{q}} = \frac{1}{2} \, I_{3\bar{q}} = -\frac{1}{2} > = -| \, d\bar{u} > \sim | \, d\bar{u}>$ (overlooking the negative sign)

Since $Q(d) + Q(\bar{u}) = -\frac{1}{3} - \frac{2}{3} = -1 = Q(\pi^-)$ we identify

$$| \pi^- > = | \, d\bar{u}> \tag{8.77}$$

i.e quark structure of π^- is $d\bar{u}$.

We show the derived quark structure for the pion family in figure 8.5.

8.27 Quark states of mesons and baryons

Similarly from figure 8.4 we arrive at the quark structure of other particles (mesons and baryons) such as K^+, K^-, K^0, $\overline{K}^0$ as shown in figure 8.6.

- Whenever there is symmetry we try to associate that symmetry with some group because it is an easy mathematical way of understanding symmetry. For instance rotation can be understood through a special orthogonal group of dimension 3, the $SO(3)$ group.

  And the symmetry that is associated with the isospin symmetry is the symmetry group of $SU(2)$. The doublet combination $\binom{p}{n} = \binom{u}{d}$ are the doublets of $SU(2)$ isospin.

- Strong interaction does not distinguish between the members of a multiplet (i.e. between p and n or between u and d). This behaviour is analogous to the fact that if we have two charged particles A and B both having charge $+1$ and they are placed in an electromagnetic field then their electromagnetic interaction or response will be exactly the same. Possessing the property of identical electric charge prevents distinguishing the particles A and B. So it is the electric charge that distinguishes particles so far as their electromagnetic interaction is concerned.

- Similarly the property that distinguishes particles so far as their strong interaction is concerned is called the colour charge.

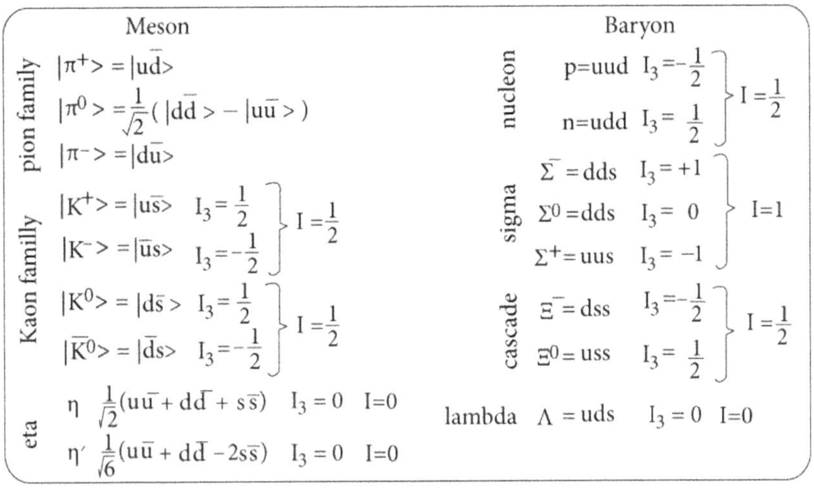

Figure 8.6. Quark structure from isospin symmetry.

8.28 Conservation of strangeness

It was seen that some elementary particles (Λ, K^0 etc) behave in a rather strange way.

We can think of the following two reactions with K^0:

$$\pi^- + p \rightarrow \Lambda + K^0 \text{ (occurs with high probability)} \tag{8.78}$$

$$\pi^- + p \rightarrow n + K^0 \text{ (does not occur)} \tag{8.79}$$

1) Both interactions conserve all the known laws like energy, momentum, electric charge, baryon number, lepton number. But the reaction shown in equation (8.78) occurs while that shown in equation (8.79) does not occur. This means there should be some conservation law that forbids equation (8.79) but allows equation (8.78), i.e. controls behaviour of interactions with particles like K^0.
2) Some particles like Λ, K^0 or Σ^-, K^+ were produced in pairs and not singly which means some quantum number is conserved during their production reaction.
3) These particles were always produced via strong nuclear interaction (having short lifetime) but they did not decay via strong interaction.
4) These particles decayed through weak interaction (exhibiting long lifetime)

So the existing quantum numbers, say baryon number and lepton number, were not sufficient to deal with the world of elementary particles. Even after introduction of baryon number and lepton number there were still some loose ends in the world of elementary particles.

Due to such unexpected or strange behaviour these particles were called strange particles.

So after the discovery of strange particles it was necessary to assign a quantum number. New conservation laws were needed to explain them.

To explain the observations Gell-Mann and Nishijima introduced the strangeness quantum number S defined for all strongly interacting particles, i.e. hadrons.

Strangeness quantum number S is not defined for leptons.

Strangeness quantum number is additive.

Strangeness quantum number is conserved in all processes mediated by strong and electromagnetic interactions but not conserved in weak interactions.

In a reaction, if the strangeness is conserved the reaction has a large cross-section.

Particle and anti-particle have opposite strangeness, i.e.

$$S(\text{particle}) = -S(\text{antiparticle})$$

Figure 8.7 shows the strangeness quantum number of various hadrons—particles and anti-particles. We derive the strangeness quantum number S for hadrons in section 8.31 using the Gell-Mann–Nishijima relation.

We furnish a few examples to describe the strangeness quantum number S.

✓ Consider the reaction in which Σ^- and K^+ are produced

$$\pi^- + p \rightarrow \Sigma^- + K^+$$

$$S(\text{LHS}) = S(\pi^-) + S(p) = 0 + 0 = 0, \quad S(\text{RHS}) = S(\Sigma^-) + S(K^+) = -1 + 1 = 0$$

i.e. $S(\text{LHS}) = S(\text{RHS})$.

✓ Consider the reaction in which Λ and K^0 are produced

Strangeness quantum number of hadrons					
Class	Particle	Strangeness S	Class	Particle	Strangeness S
strange meson	K^+	$+1$	strange baryon	Λ	-1
	K^0	$+1$		$\overline{\Lambda}$	$+1$
	K^-	-1		$\Sigma^+, \Sigma^0, \Sigma^-$	-1
	$\overline{K^0}$	-1		$\overline{\Sigma}^-, \overline{\Sigma}^0, \overline{\Sigma}^+$	$+1$
other meson (π^+ π^0 π^- η^0 η')		0		Ξ^0, Ξ^-	-2
				$\overline{\Xi}^0, \overline{\Xi}^+$	$+2$
				Ω^-	-3
				Ω^+	$+3$
			other baryon ($p, \overline{p}$ $n, \overline{n}$)		0

Figure 8.7. Strangeness quantum number of various particles and anti-particles.

$$\pi^- + p \rightarrow \Lambda + K^0$$

$$S(\text{LHS}) = S(\pi^-) + S(p) = 0 + 0 = 0, \quad S(\text{RHS}) = S(\Lambda) + S(K^0) = -1 + 1 = 0$$

i.e. $S(\text{LHS}) = S(\text{RHS})$.

Strangeness quantum number is conserved in the production reaction of strange particles. These are typical of strong interaction (very small lifetime).

- Decay of Σ^- occurs as follows.

$$\Sigma^- \rightarrow n + \pi^-$$

$$S(\text{LHS}) = S(\Sigma^-) = -1, \quad S(\text{RHS}) = S(n) + S(\pi^-) = 0 + 0 = 0$$

$$S(\text{LHS}) \neq S(\text{RHS})$$

- Decay of K^- occurs as follows.

$$K^- \rightarrow \pi^- + \pi^+ + \pi^- \ (3\pi \text{ decay}), \quad K^- \rightarrow \pi^- + \pi^0 \ (2\pi \text{ decay})$$

$$S(\text{LHS}) = S(K^-) = -1, \quad S(\text{RHS}) = S(2\pi \text{ or } 3\pi) = 0$$

$$S(\text{LHS}) \neq S(\text{RHS})$$

- Decay of Λ occurs as follows.

$$\Lambda \rightarrow \pi^- + p$$

$$S(\text{LHS}) = S(\Lambda) = -1, \quad S(\text{RHS}) = S(\pi^-) + S(p) = 0 + 0 = 0$$

$$S(\text{LHS}) \neq S(\text{RHS}).$$

Strangeness quantum number is not conserved in decay of a strange particle. So the decays are weak decays (long lifetime).

Strange particles are produced by strong interaction but decay by weak interaction. This is because strangeness quantum number S is conserved in strong interaction but violated in weak interaction.

8.29 Flavour symmetry

There is another way of grouping quarks called flavour symmetry. It says that we can put the u quark, d quark and s quark, i.e. the triplet formed with quarks denoted as

$$3 = \begin{pmatrix} u \\ d \\ s \end{pmatrix}$$

in a particular group.

Any transformation within this group is not going to affect strong interaction. With this triplet we can associate the symmetry group called $SU(3)$.

Similarly the triplet formed with anti-quarks $\bar{u}$, $\bar{d}$, $\bar{s}$ denoted as

$$\bar{3} = \begin{pmatrix} \bar{u} \\ \bar{d} \\ \bar{s} \end{pmatrix}$$

is associated with $SU(3)$ also.

Under $SU(3)$, combination of 3 and $\bar{3}$ denoted by $3 \otimes \bar{3}$ is actually represented by a group of 8 and 1. We write this as $8 \oplus 1$, i.e.

$$3 \otimes \bar{3} = 8 \oplus 1 \tag{8.80}$$

This is like combining a spin $\frac{1}{2}$ with another spin $\frac{1}{2}$ and this operation is denoted by

$$\frac{1}{2} \otimes \frac{1}{2}$$

The same can also be written in terms of their multiplicities $2\frac{1}{2} + 1 = 2$ as they are doublets, i.e. combination of a doublet with another doublet is written as

$$2 \otimes 2$$

What we get is a spin 1 particle and a spin 0 particle and this result is written as

$$1 \oplus 0$$

or we can write in terms of their multiplicities.

As spin 1 particle has $2.1 + 1 = 3$ projections 1, 0, -1, i.e. multiplicity 3 and spin 0 particle has 1 projection 0, i.e. multiplicity $2.0 + 1 = 1$ we can write the same as

$$3 \oplus 1.$$

In other words we can write the complete relation as

$$\frac{1}{2} \otimes \frac{1}{2} = 1 \oplus 0 \quad \text{(in terms of spin)}$$

$$2 \otimes 2 = 3 \oplus 1 \quad \text{(in terms of multiplicity)}$$

They add up since $2 \times 2 = 4$ and $3 + 1 = 4$.

In equation (8.80) also we note that $3 \times 3 = 9$ on LHS and $8 + 1 = 9$ on RHS.

$$\text{Consider } \begin{pmatrix} u \\ d \\ s \end{pmatrix} \text{ and } \begin{pmatrix} \bar{u} \\ \bar{d} \\ \bar{s} \end{pmatrix}.$$

Let us find out the different bound states called mesons that can be generated under flavour symmetry.

$$\begin{pmatrix} u \\ d \\ s \end{pmatrix} \otimes \begin{pmatrix} \bar{u} \\ \bar{d} \\ \bar{s} \end{pmatrix} = \begin{pmatrix} u\bar{u} & d\bar{u} & s\bar{u} \\ u\bar{d} & d\bar{d} & s\bar{d} \\ u\bar{s} & d\bar{s} & s\bar{s} \end{pmatrix}$$

Clearly there are 9 combinations. And we can split this into $8 + 1$ combinations. We derived (equations 8.67, 8.77)

$$\pi^+ = \mid u\bar{d} > , \quad \pi^- = \mid \bar{u}d>$$

And also (figure 8.6)

$$K^+ = \mid u\bar{s} > , \quad K^- = \mid s\bar{u} > , \quad K^0 = \mid d\bar{s} > , \quad \overline{K^0} = \mid s\bar{d}>$$

We also derived (equation (8.75))

$$\mid \pi^0 > = \frac{1}{\sqrt{2}} \mid d\bar{d} - u\bar{u}>$$

which is a particle with zero electric charge and is a combination of $u\bar{u}$ and $d\bar{d}$.

Let us build a singlet through a combination of $u\bar{u}$, $d\bar{d}$ and $s\bar{s}$. Construct

$$\mid \eta' > = a \mid u\bar{u} > + b \mid d\bar{d} > + c \mid s\bar{s}> \qquad (8.81)$$

where a, b, c are constants.

Demand that $\mid \eta'>$ should be orthogonal to $\mid \pi^0>$, i.e.

$$\langle \pi^0 \mid \eta' \rangle = 0 \qquad (8.82)$$

$$\frac{1}{\sqrt{2}}(<d\bar{d} \mid - <u\bar{u} \mid)(a \mid u\bar{u} > + b \mid d\bar{d} > + c \mid s\bar{s} >) = 0$$

$$b < d\bar{d} \mid d\bar{d} > - a < u\bar{u} \mid u\bar{u} > = 0$$

$$b - a = 0$$

$$a = b$$

As c is arbitrary we take

$$a = b = c$$

Hence

$$\mid \eta' > = a(\mid u\bar{u} > + \mid d\bar{d} > + \mid s\bar{s} >)$$

Normalising $\mid \eta'>$ we get

$$<\eta' \mid \eta' > = 1$$

$$a^*(<u\bar{u} \mid + <d\bar{d} \mid + <s\bar{s} \mid)(a(\mid u\bar{u} > + \mid d\bar{d} > + \mid s\bar{s} > = 1$$

$$a^*a \, (<u\bar{u} \mid u\bar{u} > + <d\bar{d} \mid d\bar{d} > + <s\bar{s} \mid s\bar{s}>) = 1$$

$$a^*a\,(1 + 1 + 1) = 1$$

$$|a|^2 = \frac{1}{3}$$

$$a = \frac{1}{\sqrt{3}} = \text{normalization constant}$$

$$|\,\eta'\,> \; = \frac{1}{\sqrt{3}}(|u\bar{u}> +|d\bar{d}> +|\,s\bar{s}>\,) \tag{8.83}$$

We find another ket $|\,\eta>$ that is orthogonal to $|\,\eta'>$ and $|\,\pi^0>$. Construct

$$|\,\eta> \; = p\,|u\bar{u}> +q|d\bar{d}> +r\,|\,s\bar{s}>$$

where p, q, r are to be determined.

$$\langle \pi^0 \,|\, \eta \rangle \; = 0 \,(|\,\eta> \ \text{and} \ |\pi^0> \ \text{are orthogonal})$$

$$\frac{1}{\sqrt{2}}(<d\bar{d}\,|-<u\bar{u}|)(p\,|u\bar{u}> +q|d\bar{d}> +r\,|\,s\bar{s}>\,) \; = 0$$

$$q\langle d\bar{d} \,|\, d\bar{d}\rangle - p\langle u\bar{u} \,|\, u\bar{u}\rangle \; = 0$$

$$p = q$$

and so

$$|\,\eta> \; = p\,|u\bar{u}> +p|d\bar{d}> +r\,|\,s\bar{s}>$$

Again $\langle \eta'| \,\eta \rangle \; = 0 \,(|\,\eta> \ \text{and} \ |\eta'> \ \text{are orthogonal})$

$$\frac{1}{\sqrt{3}}(\ <u\bar{u}|+<d\bar{d}\,|+<s\bar{s}|\,)(p\,|u\bar{u}> +p|d\bar{d}> +r\,|\,s\bar{s}>\,) \; = 0$$

$$p <u|u\,|u|u> +p <d|d\,|d|d> +r <s|s\,|\,s|s> \; =0$$

$$p + q + r \; = 0$$

$$r = -2p.$$

Hence

$$|\,\eta> \; = p\,|u\bar{u}> +p\,|d\bar{d}> -2p\,|\,s\bar{s}>$$

Normalise to get

$$<\eta \,|\, \eta> \; =1$$

$$(p^*<u\bar{u}| + p^*<d\bar{d}\,| - 2p^*|\,s\bar{s}>\,)(p\,|u\bar{u}> +p\,|d\bar{d}> -2p\,|\,s\bar{s}>\,) = 1$$

$$|p|^2 < u\bar{u} \mid u\bar{u} > + |p|^2 < d\bar{d} \mid d\bar{d} > + 4\,|p|^2 < s\bar{s} \mid s\bar{s} > = 1$$

$$6\,|p|^2 = 1$$

$$|p|^2 = \frac{1}{6}$$

$$p = \frac{1}{\sqrt{6}}$$

Hence

$$|\,\eta > = \frac{1}{\sqrt{6}}(|u\bar{u} > + |d\bar{d} > -2\,|\,s\bar{s} >)\ \text{i. e.}\quad |\,\eta > = \frac{1}{\sqrt{6}}\,|\,u\bar{u} + d\bar{d} - 2s\bar{s}> \qquad (8.84)$$

Consider a plane of $I_3\,vs\ S$ as shown in figure 8.8 that shows how mesons can be grouped into a group of 8 (octet) and 1 (singlet) considering the flavour symmetry which is associated with the $SU(3)$ group.

We can also think of other mesons and in a similar fashion put them in different groups or multiplets under $SU(3)$ grouping (chapter 9).

Baryons are made of 3 quarks. Under $SU(3)$ we take

$$3 \otimes 3 \otimes 3 = 10 \oplus 8 \oplus 8 \oplus 1$$

So a group of triplets of quarks can be combined with two other groups of triplets of quarks to generate multiplets with 10 members called a decuplet, two multiplets with 8 members called octets and 1 member called a singlet. In other words, 3 triplets generate 1 decuplet, 2 octets and 1 singlet.

The quark content of some baryons is shown in figure 8.9 and when plotted in the I_3 versus S plane we get a multiplet which is a baryon octet.

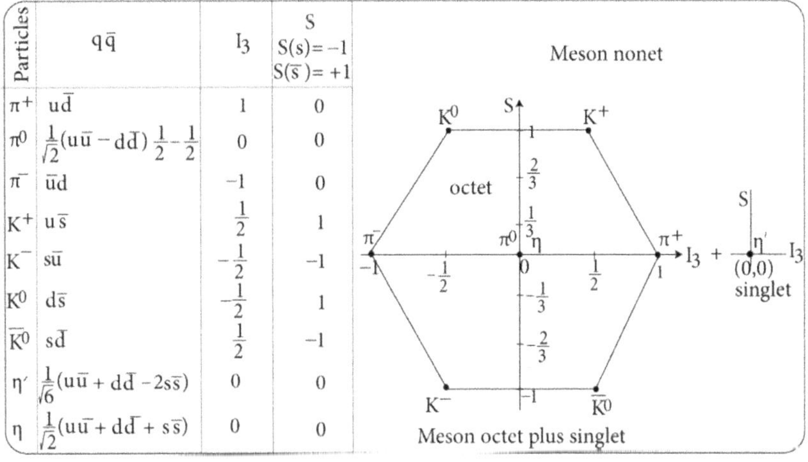

Particles	$q\bar{q}$	I_3	S $S(s)=-1$ $S(\bar{s})=+1$
π^+	$u\bar{d}$	1	0
π^0	$\frac{1}{\sqrt{2}}(u\bar{u}-d\bar{d})\ \frac{1}{2}-\frac{1}{2}$	0	0
π^-	$\bar{u}d$	-1	0
K^+	$u\bar{s}$	$\frac{1}{2}$	1
K^-	$s\bar{u}$	$-\frac{1}{2}$	-1
K^0	$d\bar{s}$	$-\frac{1}{2}$	1
$\bar{K}^0$	$s\bar{d}$	$\frac{1}{2}$	-1
η'	$\frac{1}{\sqrt{6}}(u\bar{u}+d\bar{d}-2s\bar{s})$	0	0
η	$\frac{1}{\sqrt{2}}(u\bar{u}+d\bar{d}+s\bar{s})$	0	0

Figure 8.8. Meson octet plus singlet in I_3 versus S plane.

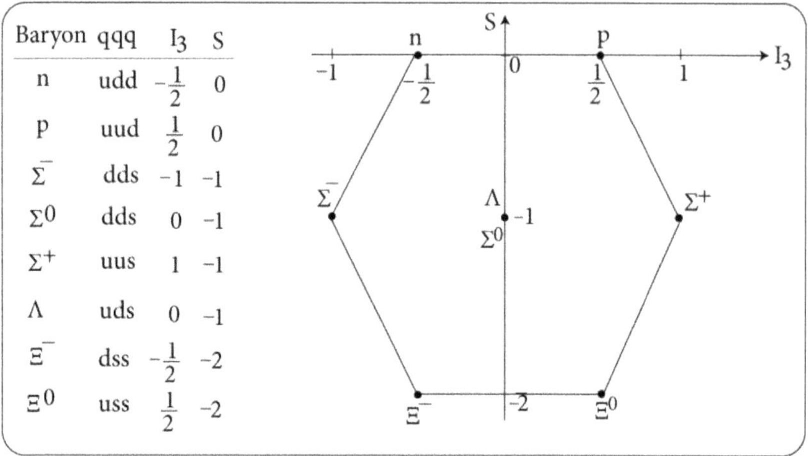

Figure 8.9. Baryon octet in I_3 versus S plane.

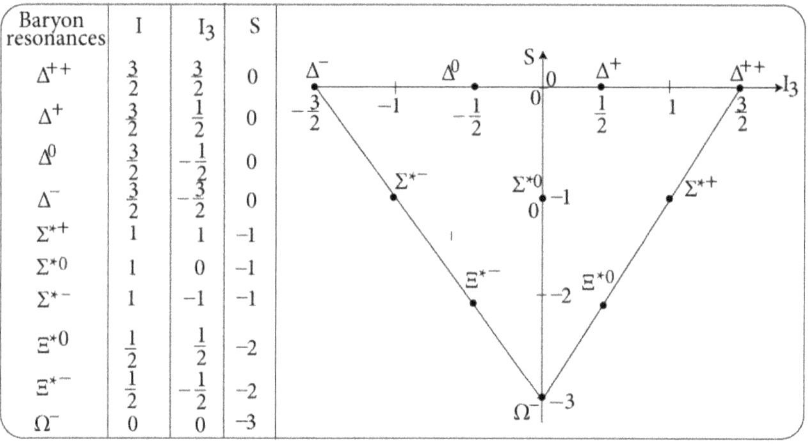

Figure 8.10. Baryon decuplet in I_3 versus S plane.

Figure 8.10 shows a baryon decuplet in the I_3 versus S plane (the quark structure of the relevant baryon resonance particles is shown in figure 10.25).

The set of particles Σ^{*+}, Σ^{*-}, Σ^{*0} (mass 1385 MeV) of the baryon decuplet in figure 8.10 is more massive than the corresponding set of particles Σ^{+}(mass 1189.4 MeV), Σ^{-}(mass 1197.4 MeV), Σ^{0} (mass 1192.5 MeV) of the baryon octet in figure 8.9.

Also, the set of particles Ξ^{*-}, Ξ^{*0} (mass 1520 MeV) of the baryon decuplet in figure 8.10 is more massive than the corresponding set of particles Ξ^{-}(mass 1321.3 MeV), Ξ^{0}(mass 1314.9 MeV) of the baryon octet in figure 8.9.

So the baryon decuplet can be thought as an excited state.

- The octets, decuplets are bigger multiplets or super multiplets compared to the iso-multiplets. The iso-multiplets (based on isospin grouping) are members of the bigger or super multiplets.

For instance the Δ iso-multiplet ($I = \frac{3}{2}$) comprising members Δ^{++}, Δ^{+}, Δ^{0}, Δ^{-}; the Σ^{*} iso-multiplet ($I = 1$) comprising members Σ^{*+}, Σ^{*0}, Σ^{*-} are part of the decuplet. In other words various iso-multiplets suitably combine to generate bigger groups called super multiplets (octet, decuplet etc).

The vertex of the baryon decuplet in the $SU(3)$ grouping in figure 8.10 had the quark structure sss and later the Ω^{-} was discovered in 1964 that filled this vertex. It had $S = -3$, $I = 0$, $I_3 = 0$. This gave an immense boost to the quark model. In fact, discovery of $\Omega^{-} = sss$ after its prediction as a member of the baryon decuplet established $SU(3)$.

8.30 Conservation of hypercharge

Baryon number and strangeness quantum number are conserved in strong interaction. To organize the zoo of particles another quantum number was introduced called hypercharge Y.

Hypercharge Y is defined as the sum of baryon number B and strangeness S, i.e.

$$Y = B + S \tag{8.85}$$

B and S are not defined for the lepton. So $Y = B + S$ is not defined for lepton. For baryon $B = 1$ and so

$$Y = 1 + S \text{ (for baryon)} \tag{8.86}$$

For meson

$$B = 0 \text{ and so}$$

$$Y = S \text{ (for meson)} \tag{8.87}$$

Hypercharge Y is additive.

As B is conserved in all the three interactions viz. strong, electromagnetic, weak interactions while S is conserved in strong and electromagnetic interactions and violated in weak interaction, it follows that Y is conserved in strong interaction and electromagnetic interaction but not in weak interaction.

Figure 8.11 shows hypercharge of hadrons.

8.31 Gell-Mann–Nishijima relation

Gell-Mann–Okubo formula given by (equation (8.38))

$$Q = I_3 + \frac{B}{2}$$

does not hold for strange particles and needs modification.

Hypercharge of hadron							
Meson	B	S	Y=B+S =S	Baryon	B	S	Y=B+S =1+S
π^+	0	0	0	p	1	0	1
π^-	0	0	0	n	1	0	1
π^0	0	0	0	Λ	1	−1	0
K^+	0	+1	+1	Σ^+	1	−1	0
K^-	0	−1	−1	Σ^0	1	−1	0
K^0	0	+1	+1	Σ^-	1	−1	0
$\bar{K}^0$	0	−1	−1	Ξ^0	1	−2	−1
η^0	0	0	0	Ξ^-	1	−2	−1
				Ω^-	1	−3	−2

Figure 8.11. Hypercharge of hadron.

- For Λ (lambda Λ meson is an isosinglet)

$$Q = 0, \quad I = 0, \quad I_3 = 0, \quad B = 1.$$

Consider the formula $Q = I_3 + \dfrac{B}{2}$.

$$\text{LHS} = Q = 0, \quad \text{RHS} = I_3 + \frac{B}{2} = 0 + \frac{1}{2}(1) = \frac{1}{2}.$$

$$\text{LHS} \neq \text{RHS}.$$

The formula fails.

To restore the formula we have to add $-\frac{1}{2}$ to the RHS. So

$$Q = I_3 + \frac{B}{2} - \frac{1}{2} = I_3 + \frac{B-1}{2}$$

We define strangeness $S = -1$ for Λ. So the modified formula is $Q = I_3 + \frac{B+S}{2}$

- For $\Sigma = \begin{pmatrix} \Sigma^+ \\ \Sigma^0 \\ \Sigma^- \end{pmatrix}$ (Sigma Σ family is an isotriplet)

$$Q = \begin{pmatrix} +1 \\ 0 \\ -1 \end{pmatrix}, \quad I = 1, \quad I_3 = \begin{pmatrix} +1 \\ 0 \\ -1 \end{pmatrix}, \quad B = 1.$$

Consider the formula $Q = I_3 + \dfrac{B}{2}$.

$$\text{LHS} = Q = \begin{pmatrix} +1 \\ 0 \\ -1 \end{pmatrix}, \quad \text{RHS} = \begin{pmatrix} +1 \\ 0 \\ -1 \end{pmatrix} + \frac{1}{2}(1) = \begin{pmatrix} +3/2 \\ +1/2 \\ -1/2 \end{pmatrix}.$$

$$\text{LHS} \neq \text{RHS}.$$

The formula fails.

To restore the formula we have to add $-\frac{1}{2}$ to the RHS. So

$$Q = I_3 + \frac{B}{2} - \frac{1}{2} = I_3 + \frac{B-1}{2}$$

So the formula $Q = I_3 + \frac{B+S}{2}$ holds for Σ with $S = -1$ which is defined as the strangeness quantum number for Σ triplet.

- For $\Xi = \begin{pmatrix} \Xi^0 \\ \Xi^- \end{pmatrix}$ (Cascade Ξ family is an isodoublet)

$$Q = \begin{pmatrix} 0 \\ -1 \end{pmatrix}, \quad I = \frac{1}{2}, \quad I_3 = \begin{pmatrix} 1/2 \\ -1/2 \end{pmatrix}, \quad B = 1.$$

Consider the formula $Q = I_3 + \frac{B}{2}$.

$$\text{LHS} = Q = \begin{pmatrix} 0 \\ -1 \end{pmatrix}, \quad \text{RHS} = \begin{pmatrix} 1/2 \\ -1/2 \end{pmatrix} + \frac{1}{2}(1) = \begin{pmatrix} 1 \\ 0 \end{pmatrix}.$$

$$\text{LHS} \neq \text{RHS}.$$

The formula fails.

To restore the formula we have to add -1 to the RHS. So

$$Q = I_3 + \frac{B}{2} - 1 = I_3 + \frac{B-2}{2}$$

So the formula $Q = I_3 + \frac{B+S}{2}$ holds for Ξ with $S = -2$ which is defined as the strangeness quantum number for Ξ doublet.

- For Ω^- (Omega Ω^- is an isosinglet)

$$Q = -1, \quad I = 0, \quad I_3 = 0, \quad B = 1.$$

Consider the formula $Q = I_3 + \frac{B}{2}$.

$$\text{LHS} = Q = -1, \quad \text{RHS} = 0 + \frac{1}{2}(1) = \frac{1}{2}$$

$$\text{LHS} \neq \text{RHS}.$$

The formula fails.

To restore the formula we have to add $-\frac{3}{2}$ to the RHS. So

$$Q = I_3 + \frac{B}{2} - \frac{3}{2} = I_3 + \frac{B-3}{2}$$

So the formula $Q = I_3 + \frac{B+S}{2}$ holds for Ω^- wih $S = -3$ which is defined as the strangeness quantum number for Ω^-.

- For η (Eta η is an isosinglet)

$$Q = 0, \quad I = 0, \quad I_3 = 0, \quad B = 0.$$

Consider the formula $Q = I_3 + \frac{B}{2}$.

$$\text{LHS} = Q = 0, \quad \text{RHS} = 0 + \frac{1}{2}(0) = \frac{1}{2}.$$

So the formula $Q = I_3 + \frac{B+S}{2}$ shows that
$S = 0$ which is defined as the strangeness quantum number for η.

- For $K = \begin{pmatrix} K^+ \\ K^0 \end{pmatrix}$ (isodoublet)

$$Q = \begin{pmatrix} +1 \\ 0 \end{pmatrix}, \quad I = \tfrac{1}{2}, \quad I_3 = \begin{pmatrix} 1/2 \\ -1/2 \end{pmatrix}, \quad B = 0.$$

Consider the formula $Q = I_3 + \frac{B}{2}$.

$$\text{LHS} = Q = \begin{pmatrix} +1 \\ 0 \end{pmatrix}, \quad \text{RHS} = \begin{pmatrix} 1/2 \\ -1/2 \end{pmatrix} + \frac{1}{2}(0) = \begin{pmatrix} 1/2 \\ -1/2 \end{pmatrix}.$$

$$\text{LHS} \neq \text{RHS}.$$

The formula fails.
To restore the formula we have to add $+\frac{1}{2}$ to the RHS. So

$$Q = I_3 + \frac{B}{2} + \frac{1}{2} = I_3 + \frac{B+1}{2}$$

So the formula $Q = I_3 + \frac{B+S}{2}$ holds for $K = \begin{pmatrix} K^+ \\ K^0 \end{pmatrix}$ with $S = +1$ which is defined as the strangeness quantum number for $K = \begin{pmatrix} K^+ \\ K^0 \end{pmatrix}$ doublet.

For $\overline{K} = \begin{pmatrix} \overline{K^0} \\ K^- \end{pmatrix}$ (isodoublet)

$$Q = \begin{pmatrix} 0 \\ -1 \end{pmatrix}, \quad I = \frac{1}{2}, \quad I_3 = \begin{pmatrix} 1/2 \\ -1/2 \end{pmatrix}, \quad B = 0.$$

Consider the formula $Q = I_3 + \frac{B}{2}$.

$$\text{LHS} = Q = \begin{pmatrix} 0 \\ -1 \end{pmatrix}, \quad \text{RHS} = \begin{pmatrix} 1/2 \\ -1/2 \end{pmatrix} + \frac{1}{2}(0) = \begin{pmatrix} 1/2 \\ -1/2 \end{pmatrix}.$$

$$\text{LHS} \neq \text{RHS}.$$

The formula fails.

To restore the formula we have to add $-\frac{1}{2}$ to the RHS. So

$$Q = I_3 + \frac{B}{2} - \frac{1}{2} = I_3 + \frac{B-1}{2}$$

So the formula $Q = I_3 + \frac{B+S}{2}$ holds for $\overline{K} = \begin{pmatrix} \overline{K^0} \\ K^- \end{pmatrix}$ with $S = -1$ which is

defined as the strangeness quantum number for $\overline{K} = \begin{pmatrix} \overline{K^0} \\ K^- \end{pmatrix}$ doublet.

The strangeness quantum number S for hadrons is shown in figure 8.7.

The Gell-Mann–Nishijima relation or formula relates baryon number B, strangeness S, third component of isospin I_3 and electric charge of a particle.

The Gell-Mann–Nishijima relation can be written in terms of hypercharge $Y = B + S$ (equation (8.85)) also as follows

$$Q = I_3 + \frac{Y}{2} = I_3 + \frac{B+S}{2} \tag{8.88}$$

8.32 General definition of hypercharge and the Gell-Mann–Nishijima relation

Incorporating charm quantum number C, bottomness quantum number B' and topness quantum number T, we redefine hypercharge as

$$Y = B + S + c + B' + T \tag{8.89}$$

✓ For strange meson $B = 0$ we have from equation (8.88)

$$Q = I_3 + \frac{B+S}{2} = I_3 + \frac{S}{2} \tag{8.90}$$

✓ For non-strange particles $S = 0$ we have from equation (8.88)

$$Q = I_3 + \frac{B+S}{2} = I_3 + \frac{B}{2} \tag{8.91}$$

With the formula of equation (8.88) we can calculate charge of particle following the Gell-Mann–Nishijima scheme.

Charge distribution of various particles is given in figure 8.12.

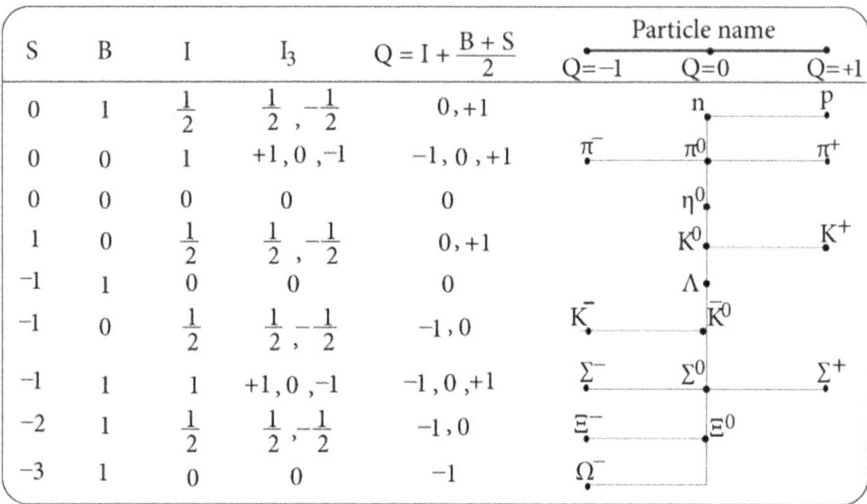

S	B	I	I_3	$Q = I + \dfrac{B+S}{2}$	Particle name		
					$Q=-1$	$Q=0$	$Q=+1$
0	1	$\frac{1}{2}$	$\frac{1}{2}, -\frac{1}{2}$	$0, +1$		n	p
0	0	1	$+1, 0, -1$	$-1, 0, +1$	π^-	π^0	π^+
0	0	0	0	0		η^0	
1	0	$\frac{1}{2}$	$\frac{1}{2}, -\frac{1}{2}$	$0, +1$		K^0	K^+
-1	1	0	0	0		Λ	
-1	0	$\frac{1}{2}$	$\frac{1}{2}, -\frac{1}{2}$	$-1, 0$	K^-	$\bar{K}^0$	
-1	1	1	$+1, 0, -1$	$-1, 0, +1$	Σ^-	Σ^0	Σ^+
-2	1	$\frac{1}{2}$	$\frac{1}{2}, -\frac{1}{2}$	$-1, 0$	Ξ^-	Ξ^0	
-3	1	0	0	-1	Ω^-		

Figure 8.12. Charge distribution of various particles obtained from Gell-Mann–Nishijima scheme.

8.33 Evaluation of charge from Gell-Mann–Nishijima relation

Let us evaluate charge from the Gell-Mann–Nishijima relation given by equation (8.88) for various choices of strangeness S

- For $S = 0$ (non-strange particles)

$$Q = I_3 + \frac{B+S}{2} = I_3 + \frac{B}{2} \tag{8.92}$$

✓ If B = 1 equation (8.92) gives

$$Q = I_3 + \frac{B}{2} = I_3 + \frac{1}{2} \tag{8.93}$$

$$\text{If } I = \frac{1}{2}, \quad I_3 = +\frac{1}{2}, \ -\frac{1}{2}.$$

□ For $I_3 = \frac{1}{2}$ equation (8.93) gives $Q = I_3 + \frac{1}{2} = \frac{1}{2} + \frac{1}{2} = 1$.
  This corresponds to proton p.
□ For $I_3 = -\frac{1}{2}$ equation (8.93) gives $Q = I_3 + \frac{1}{2} = -\frac{1}{2} + \frac{1}{2} = 0$.
  This corresponds to neutron n.

✓ If $B = 0$ equation (8.92) gives

$$Q = I_3 + \frac{B}{2} = I_3 + \frac{0}{2} = I_3 \tag{8.94}$$

□ If $I = 0$, $I_3 = 0$ equation (8.94) gives

$$Q = I_3 = 0$$

This corresponds to neutral meson η^0.

□ If $I = 1$, $I_3 = +1,\ 0,\ -1$ equation (8.94) gives

$$Q = I_3 = +1,\ 0,\ -1$$

This corresponds to pions $\pi^+,\ \pi^0,\ \pi^-$

- For $S = 1$

$$Q = I_3 + \frac{B + S}{2} = I_3 + \frac{B + 1}{2} \tag{8.95}$$

✓ For $B = 0$ equation (8.95) gives

$$Q = I_3 + \frac{B + 1}{2} = I_3 + \frac{0 + 1}{2} = I_3 + \frac{1}{2} \tag{8.96}$$

$$\text{If } I = \frac{1}{2},\ I_3 = +\frac{1}{2},\ -\frac{1}{2}$$

□ For $I_3 = \frac{1}{2}$ equation (8.96) gives

$$Q = I_3 + \frac{1}{2} = \frac{1}{2} + \frac{1}{2} = 1$$

This corresponds to K^+.

For $I_3 = -\frac{1}{2}$ equation (8.96) gives

$$Q = I_3 + \frac{1}{2} = -\frac{1}{2} + \frac{1}{2} = 0.$$

This corresponds to K^0.

- For $S = -1$

$$Q = I_3 + \frac{B + S}{2} = I_3 + \frac{B - 1}{2} \tag{8.97}$$

✓ For $B = 1$ equation (8.97) gives

$$Q = I_3 + \frac{B - 1}{2} = I_3 + \frac{1 - 1}{2} = I_3 \tag{8.98}$$

If $I = 0$, $I_3 = 0$ equation (8.98) gives

$$Q = I_3 = 0.$$

This corresponds to Λ.

✓ For $B = 0$,

$$Q = I_3 + \frac{0 - 1}{2} = I_3 - \frac{1}{2} \tag{8.99}$$

If $I = \frac{1}{2}$, $I_3 = \frac{1}{2}$, $-\frac{1}{2}$

▫ For $I_3 = \frac{1}{2}$ equation (8.99) gives

$$Q = I_3 - \frac{1}{2} = \frac{1}{2} - \frac{1}{2} = 0.$$

This corresponds to $\overline{K}^0$.

✓ For $I_3 = -\frac{1}{2}$ equation (8.99) gives

$$Q = I_3 - \frac{1}{2} = -\frac{1}{2} - \frac{1}{2} = -1.$$

This corresponds to K^-.

If $I = 1$, $I_3 = 1, 0, -1$ equation (8.98) gives

$$Q = I_3 = 1, 0, -1.$$

This corresponds to Σ^+, Σ^0, Σ^-.

• For $S = -2$

$$Q = I_3 + \frac{B + S}{2} = I_3 + \frac{B - 2}{2} \tag{8.100}$$

✓ If $B = 1$ equation (8.100) gives

$$Q = I_3 + \frac{B - 2}{2} = I_3 + \frac{1 - 2}{2} = I_3 - \frac{1}{2} \tag{8.101}$$

$$\text{If } I = \frac{1}{2}, \; I_3 = \frac{1}{2}, \; -\frac{1}{2}$$

▫ For $I_3 = \frac{1}{2}$ equation (8.101) gives $Q = I_3 - \frac{1}{2} = \frac{1}{2} - \frac{1}{2} = 0$.
This corresponds to Ξ^0 hyperon.

☐ For $I_3 = -\frac{1}{2}$ equation (8.101) gives $Q = I_3 - \frac{1}{2} = -\frac{1}{2} - \frac{1}{2} = -1$.
This corresponds to Ξ^- hyperon.

- For $S = -3$

$$Q = I_3 + \frac{B + S}{2} = I_3 + \frac{B - 3}{2} \tag{8.102}$$

✓ If $B = 1, I = 0, I_3 = 0$ then equation (8.102) gives

$$Q = I_3 + \frac{B - 3}{2} = 0 + \frac{1 - 3}{2} = -1.$$

This corresponds to Ω^- hyperon.

The Gell-Mann–Nishijima formula properly explains the shift $Q - I_3$ (i.e. shift between Q and I_3) for various particles. From equation (8.88) it follows that for strange particles

$$Q - I_3 = \frac{1}{2}(B + S) = \frac{Y}{2} \tag{8.103}$$

and for non-strange particles ($S = 0$)

$$Q - I_3 = \frac{B}{2} \tag{8.104}$$

8.34 Extension of isospin symmetry

The carrier of strangeness property is the strange quark or s quark. The masses of u, d, s quarks are (figure 5.3)

$$m_u = 2.2\frac{MeV}{c^2}, \quad m_d = 4.7\frac{MeV}{c^2}, \quad m_s = 96\frac{MeV}{c^2}.$$

Considering that strange quark is a bit more massive than u and d quark we can extend isospin symmetry by considering a quark that can exist in 3 states up or down or strange $|u>, |d>, |s>$.

When we have three states up, down and strange then the symmetry we get is called $SU(3)$.

However, the problem is that the strange quark is noticeably much more massive than u or d quarks. For instance the baryon $\Lambda = uds$ is more massive than $p = uud$ or neutron $n = ddu$.

The symmetry aspect gets worse as we extend further.

If we try to include the charm c quark and extend symmetry to $SU(4)$ the situation is really bad.

Symmetry is even more broken if we extend to $SU(5)$ by including the bottom b quark.

Symmetry is utterly broken if we try to go to $SU(6)$ and include the top quark, because the top quark never actually forms any bound state.

8.35 *CPT* symmetry or invariance

The *CPT* theorem was given by Luders, Pauli and Villars.

There are three very fundamental symmetries in the Standard Model. These are:
- Charge conjugation C (replacing particles by anti-particles),
- Parity P (mirror imaging i.e. replacing $\vec{r}$ by $-\vec{r}$), and
- Time reversal T (reversal of the sense of time).

If we put all of them together it is denoted by *CPT*. Here three operations act simultaneously.

CPT symmetry means simultaneous operation of charge conjugation C (particles replaced by anti-particles), parity P (mirror imaging) and time reversal T (sense of time reversed) in any order leaves all physical laws invariant.

CPT symmetry or invariance holds for all physical phenomena.

8.35.1 *CPT* statement

All interactions are invariant under the succession of operations of C, P and T taken in any order.

The Standard Model is built on relativity and relativity itself actually has symmetries.

CPT is a fundamental symmetry of special relativity.

Together these three symmetries have to be conserved and if not we would have a broken relativity.

8.35.2 Consequences of *CPT*

✓ For a particle we have an anti-particle.
✓ Masses of a particle and an anti-particle should be exactly the same.
✓ Lifetimes of a particle and an anti-particle should be exactly the same.
✓ Magnetic moment of a particle is negative of the magnetic moment of an anti-particle.
✓ *CPT* transformation turns our Universe into its mirror image and vice versa.
✓ *CPT* violation is expected by some string theory models.
✓ There is no fundamental reason why forces in nature should be invariant under *CPT* operation.
✓ Christenson *et al* observed small *CP* violation in kaon weak decay. *T* violation also occurs so that *CPT* is conserved.

We will consider *CP* violation in section 8.54.

Let us next consider the operations C, P, T, CP one at a time.

8.36 Charge conjugation operation, symmetry and conservation

Charge conjugation operation is a discrete symmetry operation.

It is generally represented as C or C parity.

Charge conjugation is a symmetry operation which changes the sign of charge of a particle without affecting any property unrelated to charge. So charge conjugation changes particles to anti-particles, anti-particles to particles, matter to anti-matter and anti-matter to matter.

Dynamical variables like momentum $\vec{p}$, spin s or σ, mass m remain unchanged upon operation by C.

In charge conjugation there is a change of electric charge or magnetic moment, the sign of other quantum numbers like hypercharge Y, baryon number B, strangeness S, lepton numbers L_e, L_μ are also reversed.

$$C \mid Q, p, \sigma > \; = \mid - Q, p, \sigma > \quad (Q \text{ changes sign})$$

$$C \mid Q, I_3, B, Y, L > \; = \mid - Q, -I_3, -B, -Y, -L>$$

Charge conjugation operation carries out transformation between a particle and its anti-particle. If a particle has positive charge, after charge conjugation operation it becomes negative and vice versa. So if C is the charge conjugation operator then

$$C \mid \text{particle} > \; = \mid \text{anti-particle}>$$

$$C \mid \text{anti-particle} > \; = \mid \text{particle}>$$

For instance

$$C \mid p > \; = \mid \bar{p} > , \quad C \mid \bar{p} > \; = \mid p>$$

(proton p will have the same strong interaction as an anti-proton)

$$C \mid e^- > \; = \mid e^+ > , \quad C \mid e^- > \; = \mid e^+>$$

(e^- will have the same electromagnetic interaction as e^+)

$$C \mid \pi^+ > \; = \mid \pi^- > , \quad C \mid \pi^+ > \; = \mid \pi^->$$

(π^- will have the same electromagnetic interaction as π^+)

The resulting states in these examples are not eigenstates of the charge conjugation operator.

Charge conjugation operation remains invariant, i.e. conserved in strong and electromagnetic interactions i.e.

$$[C, H_{\text{strong}}] = 0, \quad [C, H_{\text{em}}] = 0 \tag{8.105}$$

where H_{strong} is the strong interaction Hamiltonian and H_{em} is the electromagnetic interaction Hamiltonian.

Charge conjugation operation is not conserved in weak interaction, i.e. weak interaction distinguishes between a system and its charge conjugate, i.e.

$$[C, \ H_{\text{weak}}] \neq 0 \tag{8.106}$$

where H_{weak} is the weak interaction Hamiltonian. Weak interaction is not invariant under C operation.

Consider a left-handed neutrino ν_L operated by C

$$C \mid \nu_L > \ = \mid \bar{\nu}_L >$$

Left-handed neutrinos ν_L will feel weak interaction because they are left-handed particles. C operation does not flip the helicity state. So C converts left-handed neutrino ν_L to left-handed anti-neutrino $\bar{\nu}_L$ that does not exist. (Anti-neutrinos are always right-handed $\bar{\nu}_R$). So C is violated in weak interaction. Similarly

$$C \mid \bar{\nu}_R > \ = \mid \nu_R >$$

Right-handed anti-neutrinos will feel weak interaction because they are right-handed particles. C operation does not flip the helicity state. So C converts right-handed anti-neutrino $\bar{\nu}_R$ to right-handed neutrino ν_R that does not exist. (Neutrinos are always left-handed ν_L.) So C is violated in weak interaction.

8.36.1 C is a unitary operator

C transforms particle wave function $\mid \psi >$ to anti-particle wave function $\mid \bar{\psi} >$, i.e.

$$C \mid \psi > \ = \mid \bar{\psi} > \tag{8.107}$$

Taking the Hermitian conjugate we get

$$< \psi \mid C^{\dagger} = \ < \bar{\psi} \mid \tag{8.108}$$

Both states $\mid \psi >$ and $\mid \bar{\psi} >$ are normalizable so that

$$\langle \psi \mid \psi \rangle \ = \ \langle \bar{\psi} \mid \bar{\psi} \rangle \ = 1 \tag{8.109}$$

From equations (8.107), (8.108), (8.109) we get

$$\langle \bar{\psi} \mid \bar{\psi} \rangle \ = \ \langle \psi \mid C^{\dagger} C \mid \psi \rangle \ = 1 = \ \langle \psi \mid \psi \rangle$$

Hence we arrive at the relation

$$C^{\dagger} C = 1 \tag{8.110}$$

This means C is a unitary operator.

8.36.2 C is a Hermitian operator

Acting on equation (8.107) by C we have

$$CC \mid \psi > \ = C \mid \bar{\psi} > \ = \mid \psi >$$

$$C^2 \mid \psi > \ = \mid \psi > \tag{8.111}$$

$$C^2 = 1 \tag{8.112}$$

$$C = C^{-1} \tag{8.113}$$

From equations (8.110), (8.113) we have

$$C^{\dagger}C^{-1} = 1$$

$$C^{\dagger} = C \tag{8.114}$$

This means C is a Hermitian operator.

8.36.3 Eigenvalues of C

The eigenvalue equation of the charge conjugation operator C is

$$C \,|\, \psi > \, = \lambda \,|\, \psi > \quad (\lambda \text{ is eigenvlue}) \tag{8.115}$$

Operate by C again to get

$$C^2 \,|\psi> \, = \lambda C |\psi> \, = \lambda^2 \,|\, \psi > \quad (\text{using equation (8.115)})$$

Using equation (8.111) we write

$$C^2 \,|\psi> \, = \lambda^2 |\psi> \, = |\, \psi>$$

This gives the eigenvalues as

$$\lambda^2 = 1$$

$$\lambda = \pm 1 \tag{8.116}$$

C satisfies the following relations

$$CQC^{-1} = -Q \;\Rightarrow\; C \text{ changes sign of } Q$$

$$CYC^{-1} = -Y \;\Rightarrow\; C \text{ changes sign of } Y$$

$$CBC^{-1} = -B \;\Rightarrow\; C \text{ changes sign of } B$$

$$CL_e C^{-1} = -L_e \;\Rightarrow\; C \text{ changes sign of } L_e$$

$$CL_\mu C^{-1} = -L_\mu \;\Rightarrow\; C \text{ changes sign of } L_\mu$$

8.37 Truly neutral particle or self-conjugate particle

If a particle is its own anti-particle then it is called truly neutral or self-conjugate.

There is no change in charge or any property of particle after operation by charge conjugation operator.

Particle is transformed into itself by charge conjugation.

Example of truly neutral particle or self-conjugate particles are

$$\gamma, \;\; \pi^0, \;\; \eta$$

- When C operates on a photon state $|\gamma\rangle$ we get an eigenstate

$$C|\gamma\rangle = -|\gamma\rangle \quad \text{(eigenvalue of } -1 \text{ or } C \text{ parity} -1) \qquad (8.117)$$

(a photon is its own anti-particle)

The reason for negative eigenvalue is because under charge conjugation electric charge switches sign and so do the electric and magnetic fields.

- Again, π^0 can decay via electromagnetic interaction into 2 photons as

$$\pi^0 \to \gamma + \gamma$$

This means the C parity or eigenvalue of $|\pi^0\rangle$ state is $+1$ since C parity is a multiplicative quantum number.

The eigenvalue equation of $|\pi^0\rangle$ state is

$$C|\pi^0\rangle = |\pi^0\rangle \quad \text{(eigenvalue of } +1) \qquad (8.118)$$

(π^0 is its own anti-particle)

- Let us consider operation of C on the multi-particle state $|\pi^-\pi^+\rangle$

$$C|\pi^+\pi^-\rangle = (-1)^l|\pi^-\pi^+\rangle$$

where l is the angular momentum

For ground state $l = 0$ and hence we have

$$C|\pi^+\pi^-\rangle = (-1)^0|\pi^-\pi^+\rangle = |\pi^-\pi^+\rangle \qquad (8.119)$$

- We note that being electrically neutral does not ensure that a particle is truly neutral or self-conjugate.

For instance, neutrino ν is a neutral particle but it is not truly neutral or self-conjugate. When a charge conjugation operator operates upon a free ν it is not transformed into itself since

$$C|\nu\rangle = |\bar{\nu}\rangle.$$

The wave function of a truly neutral particle must be an eigenstate of the charge conjugation operator C. So if ψ is the wave function of a truly neutral particle then

$$C|\psi\rangle = \lambda|\psi\rangle$$

$$\lambda = \pm 1$$

$$\hat{C}|\psi\rangle = \pm|\psi\rangle \qquad (8.120)$$

C is known as the charge parity of the particle.

A particle has positive charge parity, i.e. C is even if

$$C = +1 \qquad (8.121)$$

A particle has negative charge parity, i.e. C is odd if

$$C = -1 \tag{8.122}$$

- C has two eigenvalues ± 1 and so it is a discrete transformation.
- For a single-photon state $| \psi > = | \gamma>$

$$C \,| \gamma > \; = -| \gamma > \quad \text{(equation (8.117))}$$

For a system of n photons $| \psi > = |n\gamma>$

$$C \,| \, n\gamma > \; = (-1)^n | \, n\gamma> \tag{8.123}$$

Thus the n photon state is also an eigenstate of C with eigenvalue $(-1)^n$.
If $n = $ even then $(-1)^n = +1$
This means that for a system of even n photons charge parity is positive.
If $n= $ odd then $(-1)^n = -1$
This means that for a system of odd n photons charge parity is negative.
Clearly for a single-photon state ($n=$ odd) charge parity is negative and for a 2-photon state ($n=$ even) charge parity is positive.

- We discuss in *exercise 8.13* that $\pi^0 \rightarrow 2\gamma$ is possible but $\pi^0 \rightarrow 3\gamma$ is not possible.
Similar examples have been cited in *exercise 7.7*.
- We mention that $| \, \pi^+>$ and $| \, \pi^->$ are not eigenstate of the charge conjugation operator C since

$$C \,| \, \pi^+ > \; = -| \, \pi^-> \tag{8.124}$$

$$C \,| \, \pi^- > \; = -| \, \pi^+> \tag{8.125}$$

Clearly $\pi^\pm$ are not self-conjugate particles.
- Neutrino ν is left-handed polarized ($\nu \equiv \nu_L$) and anti-neutrino $\bar{\nu}$ is right-handed polarized ($\bar{\nu} \equiv \bar{\nu}_R$).

So charge conjugation applied to a free moving neutrino ν results in a process $\nu_L \rightarrow \bar{\nu}_L$ which does not exist in nature. Also, charge conjugation applied to a free moving anti-neutrino $\bar{\nu}$ results in a process $\bar{\nu}_R \rightarrow \nu_R$ which does not exist in nature.

8.38 Space parity

Parity is reversing the direction of all 3 coordinate axes x, y, z, i.e. spatial inversion of coordinates. It is reflection of all coordinates through the origin.

$$(x, y, z) \rightarrow (-x, -y, -z) \tag{8.126}$$

Parity transformation can be compared to a mirror shown in figure 8.13. If we place a vector (along X) in front of a mirror (YZ plane) then the reflected image in the mirror is exactly in the opposite direction (along $-X$). However, vectors along Y,

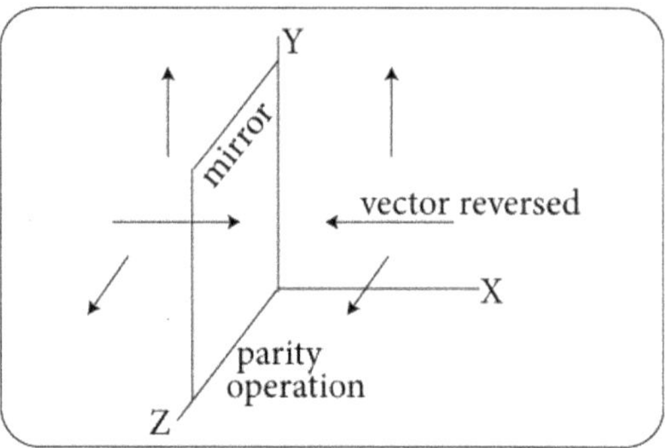

Figure 8.13. Parity operation.

Z axes are not reversed but point in the same direction even after reflection. So mirror is a sort of partial parity transformation.

Consider the displacement vector $\vec{r} = (x, y, z)$. It gets flipped, i.e. reversed under parity transformation since

$$\vec{r} = (x, y, z) \xrightarrow{\text{parity}} (-x, -y, -z) = -\vec{r} \tag{8.127}$$

Velocity $\vec{v}$ is given by

$$\vec{v} = \frac{d\vec{r}}{dt} \xrightarrow{\text{parity}} \frac{d(-\vec{r})}{dt} = -\frac{d\vec{r}}{dt} = -\vec{v} \tag{8.128}$$

Momentum $\vec{p}$ gets flipped, i.e. reversed under parity transformation since

$$\vec{p} = m\frac{d\vec{r}}{dt} \xrightarrow{\text{parity}} m\frac{d(-\vec{r})}{dt} = -m\frac{d\vec{r}}{dt} = -\vec{p} \tag{8.129}$$

Force is given by

$$\vec{F} = m\frac{d^2\vec{r}}{dt^2} \xrightarrow{\text{parity}} m\frac{d^2(-\vec{r})}{dt^2} = -m\frac{d^2\vec{r}}{dt^2} = -\vec{F} \tag{8.130}$$

Electric field

$$\vec{E} = \left(-\frac{\partial V}{\partial x}, -\frac{\partial V}{\partial y}, -\frac{\partial V}{\partial z}\right) \rightarrow \left(-\frac{\partial V}{\partial(-x)}, -\frac{\partial V}{\partial(-y)}, -\frac{\partial V}{\partial(-z)}\right)$$

$$= -\left(-\frac{\partial V}{\partial x}, -\frac{\partial V}{\partial y}, -\frac{\partial V}{\partial z}\right) = -\vec{E}$$

$\vec{r}, \vec{v}, \vec{p}, \vec{F}, \vec{E}$ are polar or true vectors.

There are a certain class of vectors for which this is not true, e.g. orbital angular momentum $\vec{l} = \vec{r} \times \vec{p}$, spin $\vec{s}$ that follows same algebra as $\vec{l}$ since

$$\vec{l} = \vec{r} \times \vec{p} \xrightarrow{\text{parity}} (-\vec{r}) \times (-\vec{p}) = \vec{r} \times \vec{p} = \vec{l} \tag{8.131}$$

It follows from the Lorentz force equation

$$\vec{F} = q\,\vec{v} \times \vec{B} \xrightarrow{\text{parity}} (-\vec{F}) = q\,(-\vec{v}) \times \vec{B}$$

that

$$\vec{B} \xrightarrow{\text{parity}} \vec{B}.$$

$\vec{l},\ \vec{s},\ \vec{B}$ are pseudo or axial vectors.
- Parity operation is an example of discrete transformation such that

$$P\psi(\vec{r}) = \psi(-\vec{r}) \tag{8.132}$$

8.38.1 Eigenvalue of parity operator P

The eigenvalue equation of parity operator P is

$$P\psi(\vec{r}) = \lambda\psi(\vec{r}) \tag{8.133}$$

Operate by parity operator P to get

$$\begin{aligned} PP\psi(\vec{r}) &= \lambda P\psi(\vec{r}) \\ &= \lambda\lambda\psi(\vec{r}) \ \ (\text{using equation (8.131)}) \end{aligned} \tag{8.134}$$

$$P^2\psi(\vec{r}) = \lambda^2\psi(\vec{r}) \tag{8.135}$$

The LHS of equation (8.133) is

$$P^2\psi(\vec{r}) = PP\psi(\vec{r}) = P\psi(-\vec{r}) = \psi(\vec{r}) \tag{8.136}$$

Equating the RHS of equations (8.133) and (8.134) we have

$$\lambda^2\psi(\vec{r}) = \psi(\vec{r})$$

As $\psi(\vec{r})$ is arbitrary we have

$$\lambda^2 = 1 \tag{8.137}$$

So the eigenvalue of parity $\lambda = +1$ corresponds to even parity.
And the eigenvalue of parity $\lambda = -1$ corresponds to odd parity.
Example

$$\checkmark\ \psi(x) = \cos x$$

$$P\psi(x) = P\cos x = \cos(-x) = \cos x = \psi(x)$$

Parity is even.

$$\checkmark\ \psi(x) = \sin x$$

$$P\psi(x) = P\sin x = \sin(-x) = -\sin x = -\psi(x)$$

Parity is odd.

8.39 Effect of parity operation on vectors

There are two types of vectors on the basis of parity operator

8.39.1 Ordinary vector or polar vector or true vector

Consider 2 vectors $\vec{a}, \vec{b}$.

And P is parity operator. Let

$$P(\vec{a}) = -\vec{a}, \quad P(\vec{b}) = -\vec{b} \tag{8.138}$$

Parity operation or transformation changes sign of a vector. We call it ordinary or polar or true vector.

It has odd parity.

8.39.2 Axial vector or pseudovector

Consider the cross product of two ordinary vectors $\vec{a}, \vec{b}$

$$\vec{c} = \vec{a} \times \vec{b}$$

This is a vector. Now parity of $\vec{c}$ is

$$P(\vec{c}) = P(\vec{a} \times \vec{b}) = (-\vec{a}) \times (-\vec{b}) = \vec{a} \times \vec{b} = \vec{c} \tag{8.139}$$

Parity operation or transformation does not change sign of vector. We call it pseudovector or axial vector.

It has odd parity.

8.40 Effect of parity operation on scalars

8.40.1 Ordinary scalar or normal scalar

Consider two polar vectors $\vec{a}, \vec{b}$, i.e.

$$P(\vec{a}) = -\vec{a}, \quad P(\vec{b}) = -\vec{b}$$

We now consider the dot product of these two polar vectors $\vec{a}, \vec{b}$

$$\vec{a} \cdot \vec{b} = m$$

This is a scalar. Now parity of m is

$$P(m) = P(\vec{a} \cdot \vec{b}) = (-\vec{a}) \cdot (-\vec{b}) = \vec{a} \cdot \vec{b} = m \tag{8.140}$$

Parity operation or transformation does not change sign of the scalar. We call it an ordinary scalar or normal scalar.

It has even parity.

8.40.2 Pseudoscalar

Consider three polar vectors $\vec{a}, \vec{b}, \vec{c}$, i.e.

$$P(\vec{a}) = -\vec{a}, \quad P(\vec{b}) = -\vec{b}, \quad P(\vec{c}) = -\vec{c}$$

Consider the scalar triple product of these three polar vectors $\vec{a}, \vec{b}, \vec{c}$ viz.

$$\vec{a} \cdot \vec{b} \times \vec{c} = n$$

This is a scalar. Now parity of n is

$$P(n) = P(\vec{a} \cdot \vec{b} \times \vec{c}) = (-\vec{a}) \cdot (-\vec{b}) \times (-\vec{c}) = -\vec{a} \cdot \vec{b} \times \vec{c} = -n \qquad (8.141)$$

Parity operation or transformation changes sign of the scalar. We call it a pseudoscalar.

It has odd parity.

8.40.3 Conclusion

$$P(S) = S \text{ where } S \text{ is an ordinary or normal scalar (even parity).} \qquad (8.142)$$

$$P(P) = -P \text{ where } P \text{ is a pseudoscalar (odd parity).} \qquad (8.143)$$

$$P(\vec{V}) = -\vec{V} \text{ where } \vec{V} \text{ is an ordinary or polar or true vector (odd parity).} \qquad (8.144)$$

$$P(\vec{A}) = \vec{A} \text{ where } \vec{A} \text{ is a axial or pseudovector (even parity).} \qquad (8.145)$$

8.41 Scalar, pseudoscalar, vector, pseudovector, tensor, pseudotensor particles

8.41.1 Parity eigenvalues

$\lambda = +1$ corresponds to scalar, pseudovector (even parity)
$\lambda = -1$ corresponds to pseudoscalar, vector (odd parity).

8.41.2 Total angular momentum or spin values

$J = 0$ corresponds to scalar, pseudoscalar
$J = 1$ corresponds to vector, pseudovector
$J = 2$ corresponds to tensor, pseudotensor.

8.41.3 Spin parity values

8.41.3.1 Pseudoscalar particles
$J^\pi = 0^-$ particles are pseudoscalar particles. Their wave function transforms as a pseudoscalar. Examples of such particles are:

$$\text{pion, kaon, } \eta, \text{ charmed mesons } D^\pm, D^0, \bar{D}^0 \text{ etc.}$$

8.41.3.2 Scalar particles

$J^\pi = 0^+$ particles are scalar particles. Their wave function transforms as a scalar. Examples of such particles are:

$$\delta(980 \; MeV), \quad S^*(980 \; MeV) \text{ etc.}$$

8.41.3.3 Vector particles

$J^\pi = 1^-$ particles are vector particles. Their wave function transforms as a vector. Examples of such particles are:

photon, $\quad \rho(770 \; MeV), \quad \omega(783 \; MeV), \quad \rho(1020 \; MeV), \quad \psi(3685 \; MeV)$ $J/\psi(3100 \; MeV)$ etc.

8.41.3.4 Axial vector particles

$J^\pi = 1^+$ particles are axial vector particles. Their wave function transforms as an axial vector. Examples of such particles are

$$B(1235 \; MeV), \quad D(1280 \; MeV), \quad E(1420 \; MeV) \text{ etc.}$$

8.41.3.5 Tensor particles

$J^\pi = 2^+$ particles are tensor particles. Their wave function transforms as tensor. Examples of such particles are:

graviton etc.

8.41.3.6 Pseudotensor particles

$J^\pi = 2^-$ particles are pseudotensor particles. Their wave function transforms as tensor, e.g. $A3$ etc.

8.42 Parity of spherically symmetric potential

Consider a system described by spherically symmetric potential wave function defined as

$$\psi(r, \theta, \phi) = R(r) Y_{lm_l}(\theta, \phi) \tag{8.146}$$

$$= R(r) \sqrt{\frac{2l + 1}{4\pi} \frac{(l - m_l)!}{(l + m_l)!}} \; P_l^{m_l}(\cos \theta) e^{im_l \phi} \tag{8.147}$$

Under space inversion we have

$$r \to r, \; \theta \to \pi - \theta, \; \phi \to \pi + \phi \tag{8.148}$$

Hence

$$P\psi(r, \theta, \phi)$$

$$= P[R(r) \; Y_{lm_l}(\theta, \phi)] = R(r)P \; Y_{lm_l}(\theta, \phi)$$

$$= R(r) Y_{lm_l}(\pi - \theta, \pi + \phi)$$

$$= R(r) \sqrt{\frac{2l + 1}{4\pi} \frac{(l - m_l)!}{(l + m_l)!}} \; P_l^{m_l}(\cos(\pi - \theta)) e^{im_l(\pi + \phi)}$$

Using

$$P_l^{m_l}(\cos(\pi-\theta)) = P_l^{m_l}(-\cos\theta) = (-1)^{l+m_l}P_l^{m_l}(\cos\theta)$$

$$e^{im_l(\pi+\phi)} = e^{im_l\pi}e^{im_l\phi} = (e^{i\pi})^{m_l}e^{im_l\phi} = (-1)^{m_l}e^{im_l\phi}$$

Hence

$$P\,Y_{lm_l}(\theta,\ \phi) = \sqrt{\frac{2l+1}{4\pi}\frac{(l-m_l)!}{(l+m_l)!}}\,(-1)^{l+m_l}P_l^{m_l}(\cos\theta)(-1)^{m_l}e^{im_l\phi}$$

$$= (-1)^l\sqrt{\frac{2l+1}{4\pi}\frac{(l-m_l)!}{(l+m_l)!}}\,P_l^{m_l}(\cos\theta)e^{im_l\phi}$$

$$P\,Y_{lm_l}(\theta,\ \phi) = (-1)^l\,Y_{lm_l}(\theta,\ \phi) \tag{8.149}$$

Thus

$$P\psi(r,\theta,\phi) = R(r)\,P\,Y_{lm_l}(\theta,\ \phi) = R(r)(-1)^l\,Y_{lm_l}(\theta,\ \phi)$$

$$P\psi(r,\theta,\phi) = (-1)^l\,\psi(r,\theta,\phi) \tag{8.150}$$

Equation (8.146) suggests that for $l = 0, 2, 4, ...$(even) we have states $s, d, g,...$ of even parity and for $l = 1, 3, 5, ...$ (odd) we have states $p, f, h, ...$ of odd parity.

Thus orbital parity of a particle in an angular momentum state l is $(-1)^l$. So space parity depends on l.

Parity eigenvalue is $(-)^l$.

$$\text{For even,}\ \ (-)^{\text{even}} = +1,\ \psi(\vec{r}) = \psi(-\vec{r})\ \text{and}$$

$$\text{for odd } l,\ (-)^{\text{odd}} = -1,\ \psi(\vec{r}) = -\psi(-\vec{r}).$$

8.43 Intrinsic parity

Parity or orbital parity is independent of species of particles and depends only on the orbital angular momentum states of a system of particles.

A physical state needs to have a definite parity, i.e. it needs to be an eigenstate of parity operator P if the system Hamiltonian H is invariant under parity operation, i.e. if

$$[P\,H] = 0. \tag{8.151}$$

If intrinsic spin $\vec{s}$ of a particle commutes with the Hamiltonian H, i.e. if

$$[s\,H] = 0 \tag{8.152}$$

then the intrinsic spin part of the wave function of a particle has definite parity called intrinsic parity.

Intrinsic parity of a particle is not given by a simple formula involving s but has to be determined relative to another particle whose intrinsic parity has been assigned or fixed by convention.

Let ψ be particle wave function. If

$$\psi \xrightarrow{\ P\ } +\psi \tag{8.153}$$

then ψ is said to have a fixed or definite parity called even parity. If

$$\psi \xrightarrow{\ P\ } -\psi \tag{8.154}$$

then ψ is said to have a fixed or definite parity called odd parity.

If none of the relations of equations (8.151) and (8.152) are obeyed by wave function ψ, then ψ is said to have no fixed or definite parity.

If wave function ψ is such that the relations of equations (8.153) and (8.154) are obeyed in parts, then ψ will be a mixed parity state.

• The convention for setting intrinsic parity is as follows:

Spin $\frac{1}{2}$ particles p, n. e^-, μ^-, ... are assigned positive parity or even parity. (So $J^\pi = \frac{1}{2}^+$ for nucleons).

Spin $\frac{1}{2}$ anti-particles $\bar{p}$, $\bar{n}$, e^+, μ^+, ...are assigned negative parity or odd parity.

Intrinsic parity of other elementary particles is fixed relative to these values.

8.43.1 Pion parity

✓ In *exercise 8.11* we have determined parity of a charged pion and shown that it is negative.

✓ In *exercise 8.12* we have determined parity of a neutral pion and shown that it is negative. So intrinsic parities of π^+, π^0, π^- are determined to be odd for consistency with those fixed as per convention.

8.43.2 Photon parity

A photon is a vector particle as it is represented by vector potential A_μ. Its spin is 1 and its intrinsic parity is -1.

According to quantum field theory, parity of a fermion (half integral spin) must be opposite to that of its anti-particle. Parity of a boson (integral spin) is the same as its anti-particle.

✓ Intrinsic parity is specific to each particle and describes how its state changes under a parity transformation.

✓ Extrinsic parity is determined by the orbital angular momentum.

• Parity is multiplicative quantum number.

8.43.3 Kaon parity

Observation of K^0 decay

$$K^0 \to \pi^0\pi^0, \ \pi^+\pi^- \ \text{(mostly)}$$

$$K^0 \to \pi^+\pi^-\pi^0 \ (10^{-7} \ \text{fraction, i.e. 1 in } 10^7 \ \text{decays to 3 pions})$$

Now pions have odd parity.

Final state is

$$(-1)(-1) = (+1) \text{ for } K^0 \to 2\pi \text{ decay but}$$

$$(-1)(-1)(-1) = (-1) \text{ for } K^0 \to 3\pi \text{ decay.}$$

So K^0 decays to a final state mostly with positive parity but also there is a small decay probability for it to decay into negative parity final state. It appears that K^0 is a particle that cannot be assigned a fixed parity.

Another possibility is that the LHS represents 2 different particles, say τ and θ. The other way to interpret this is that parity is violated in weak interaction since $K^0 \to 2\pi$, $K^0 \to 3\pi$ decays are weak decays. The latter possibility was suggested by Lee and Yang and experimentally verified also by Wu.

Other such decays are

$$K^+ \to \pi^+\pi^0 \quad \text{(even parity)}$$

$$K^+ \to \pi^+\pi^-\pi^+ \quad \text{(odd parity)}$$

8.44 Helicity or handedness

If the spin $\vec{s}$ is opposite to the direction of motion of a particle we say the particle is in the left-handed helicity state (figure 8.14).

Under parity transformation $\vec{p}$ will invert. So motion under parity is flipped but spin is not flipped. So under parity transformation the motion of the particle and its spin both point in the same direction. We then say the particle is in the right-handed helicity state (figure 8.14).

If mass of the particle is zero, it travels at the speed of light and there is no centre-of-mass frame for the particle. And the particle will be in an eigenstate of helicity.

In other words, massless particles are always in helicity eigenstates.

If mass of the particle is not zero then the left- and right-handed helicity states are mixed by the mass of the particle. Massive particles are in a mixture. They can be predominantly left-handed or predominantly right-handed, but there is always a little bit of the other state mixed.

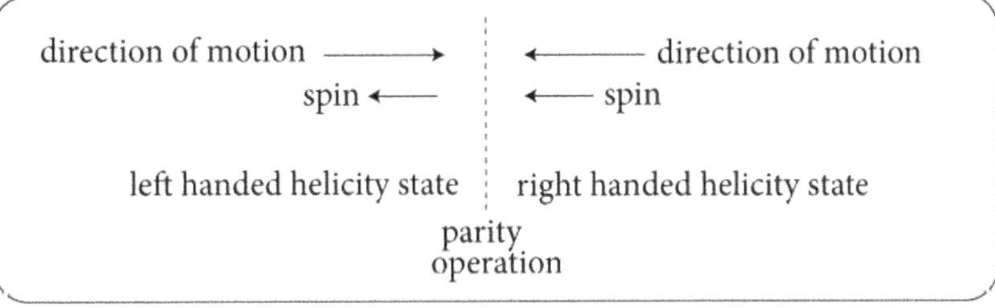

Figure 8.14. Helicity under parity transformation.

Mass of neutrino is nearly zero. Every neutrino ν in the Universe is in left-handed helicity eigenstate $\nu \equiv \nu_L$ and every anti-neutrino $\bar{\nu}$ is in the right-handed eigenstate of helicity $\bar{\nu} \equiv \bar{\nu}_R$.

8.44.1 Condition of parity conservation

If H is a Hamiltonian describing an interaction and if $[\,P,\ H\,] = 0$ then parity is conserved in the said interaction (equation (8.151)). Then we can say that physical laws are invariant under inversion of space coordinates (i.e. replacement of x, y, z by $-x$, $-y$, $-z$), i.e. physical laws do not depend upon left- or right-handedness.

Electromagnetic force and strong force both conserve parity, i.e.

$$[P,\ H_{\text{strong}}] = 0, \quad [P,\ H_{\text{em}}] = 0 \tag{8.155}$$

In strong and electromagnetic interaction parity is a good quantum number. So energy eigenstates are also eigenstates of parity operator.

Parity conservation means particle distribution and its mirror image cannot be distinguished and this is because some quantum number is conserved which is parity.

We assign intrinsic parity to different particles so that parity is conserved in strong and electromagnetic interaction. This was done in section 8.42.

8.45 Parity equation

Consider the reaction

$$a + b \rightarrow c + d \tag{8.156}$$

We define initial state as

$$|\,i> \,=\,|\,a> \,|\,b> \,|\,l> \tag{8.157}$$

where $|a>$ and $|b>$ are the internal states of a, b and $|l>$ describes their relative motion, i.e. relative motion of incident channel with orbital angular momentum l. The parity of the initial state $|i>$ is

$$P(\text{initial}) = P(a)P(b)(-1)^l \tag{8.158}$$

where $P(a), P(b)$ are intrinsic parities of $|a>$ and $|b>$ and $(-1)^l$ is the space parity or orbital parity.

Similarly, we define the final state as

$$|f> \,=\,|\,c> \,|\,d> \,|\,l'> \tag{8.159}$$

where $|c>$ and $|d>$ are the internal states of c, d and $|l'>$ describes their relative motion, i.e. relative motion of exit channel with orbital angular momentum l'. The parity of the final state $|f>$ is

$$P(\text{final}) = P(c)P(d)(-1)^{l'} \tag{8.160}$$

Assuming conservation of parity, we can write from equations (8.158) and (8.160)

$$P(\text{initial}) = P(\text{final})$$

$$P(a)P(b)(-1)^{l} = P(c)P(d)(-1)^{l'} \tag{8.161}$$

8.46 Parity non-conservation in weak interaction

Parity P does not commute with Hamiltonian H describing weak interaction, i.e.

$$[P,\ H_{\text{weak}}] \neq 0 \tag{8.162}$$

So we say that parity is not conserved in weak interaction. Parity violation shows up when we compare particle distribution with its mirror image.

Let us consider the decay of μ^+ polarized along Z as depicted in figure 8.15.

$$\mu^+ \rightarrow e^+ + \nu_e + \bar{\nu}_\mu$$

The e^+ tends to come off preferentially along μ^+ spin direction. So angular distribution of e^+ is not symmetrical about the XY plane. In other words, as depicted in figure 8.15 we are able to distinguish particle e^+ polar plot distribution (which is observed) and the mirrored distribution (which is not observed).

Since distinction between a particle's distribution and its mirrored distribution is possible it follows that some quantum number (called parity) is not conserved.

Angular distribution contains the term

$$\cos\theta = \vec{s}_{\mu^+}.\ \vec{p}_{e^+} \xrightarrow{\text{parity}} \vec{s}_{\mu^+}.\ (-\vec{p}_{e^+}) = -\vec{s}_{\mu^+}.\ \vec{p}_{e^+} \tag{8.163}$$

which is a pseudoscalar and its presence in the Hamiltonian ensures parity violation.

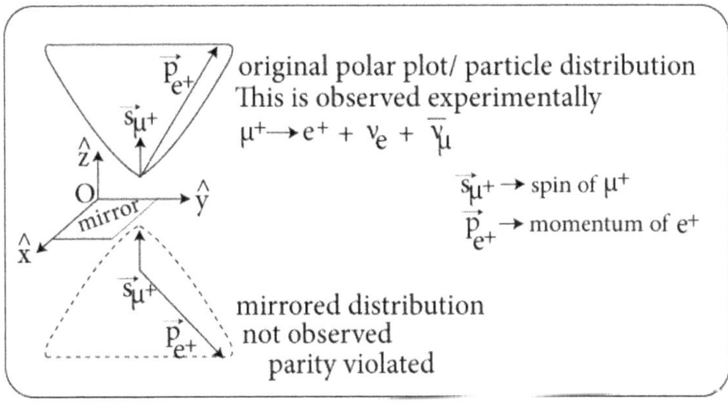

Figure 8.15. Parity violation in weak interaction.

8.47 $\tau - \theta$ puzzle

The following two dominant decay modes for two particles called τ and θ were observed

$$\tau^\pm \rightarrow \pi^\pm + \pi^+ + \pi^-$$

$$\theta^\pm \rightarrow \pi^\pm + \pi^0$$

It was observed that τ, θ had the same mass, same lifetime and same angular momentum, i.e. all properties were found to be the same.

Since τ, θ have all quantum numbers the same, they are the same particle—there is nothing to distinguish between them.

But from the standpoint of parity we have

$$P(\tau) = P(3\pi) = (-)(-)(-) = -ve$$

$$P(\theta) = P(2\pi) = (-)(-) = +ve$$

Clearly $P(\tau) \neq P(\theta)$

Despite being the same particles they differ in parity, which seems absurd. This is called the $\tau - \theta$ paradox. This paradox was explained by Lee and Yang.

8.48 Lee and Yang's explanation

Lee and Yang proposed the following.

In weak interaction parity is not conserved, i.e. parity is not a good quantum number.

τ, θ are identical. They are identified as the particle K^+.

So the decays

$$K^+ \rightarrow 2\pi \text{ (where parity is } + ve) \text{ and}$$

$$K^+ \rightarrow 3\pi \text{ (where parity is } - ve)$$

can occur.

Wu's experiment confirmed Lee and Yang's explanation.

Parity violation in weak interaction can be taken care of by inserting a pseudoscalar term in the Hamiltonian describing weak interaction.

8.49 Experimental evidence of parity non-conservation in weak interaction—Wu's experiment

In 1956 C S Wu cooled a sample of $_{27}\text{Co}^{60}$ to temperature less than 0.03 K to reduce all the joggling of its molecules. It undergoes weak interaction suffering beta decay (figure 8.16)

$$_{27}\text{Co}^{60} \rightarrow (_{28}\text{Ni}^{60}) * + e^- + \bar{\nu}_e \text{ (a pure Gammow-Teller transition)} \qquad (8.164)$$

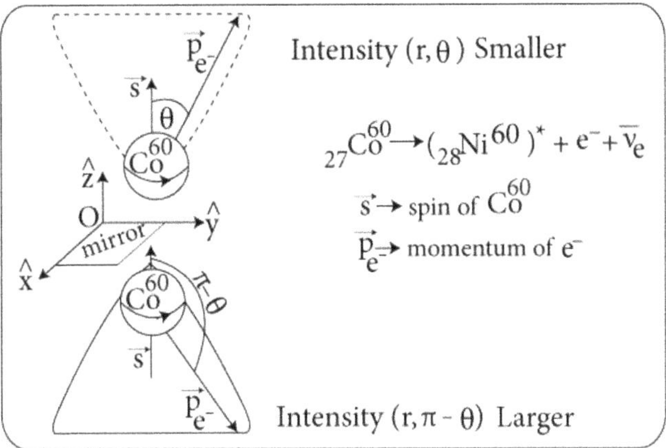

Figure 8.16. Parity violation in weak decay of $_{27}Co^{60}$.

The sample was then placed in a strong magnetic field $\vec{B} = B\hat{z}$. This caused a large fraction of the spins of the $_{27}Co^{60}$ nuclei to line up, i.e. to get polarized with the magnetic field along $\hat{z}$.

Due to weak decay electrons are produced and they fly off from the location of $_{27}Co^{60}$ towards the observation point (r, θ) where it is detected. The problem has got azimuthal symmetry and the azimuthal coordinate ϕ does not come into consideration.

Wu obtained a polar plot of electrons from $_{27}Co^{60}$. Also refer to figure 8.21.

Parity conservation means the probability to detect electron at (r, θ) and $(r, \pi - \theta)$ are the same, i.e.

$$|\psi(r, \theta)|^2 = |\psi(r, \pi - \theta)|^2 \tag{8.165}$$

In other words, if there is mirror symmetry then the electron angular distribution should be the same, i.e. symmetric or indistinguishable at points (r, θ) and $(r, \pi - \theta)$ if parity is conserved.

It was observed that intensity at $(r, \pi - \theta)$ was far greater than intensity at (r, θ).

Electron escapes predominantly against nuclear spin, i.e. the number of electrons emitted along field is far greater than the number of electrons emitted opposite to the field.

This asymmetry means maximum parity violation by the weak force. Parity is not a good quantum number in weak decay.

Parity violation in weak decay is because neutrinos participate in weak interaction in a very special way.

Only left-handed neutrinos ν_L and right-handed anti-neutrinos $\overline{\nu}_R$ take part in weak interaction (as right-handed neutrinos $\overline{\nu}_L$ and left-handed anti-neutrinos ν_R do not exist).

We say that parity is not invariant in weak interaction, i.e. $[P, H_{weak}] \neq 0$ (equation (8.162)).

8.50 Time reversal symmetry

Let us denote the time reversal operator by T.

Time reversal operation is replacing time t by $-t$, i.e. $t \xrightarrow{T} -t$.

Consider a reaction shown in figure 8.17(a) where particles A_i and B_i are at infinity and then start to approach each other initially and after interaction particles A_f and B_f emerge finally. Suppose the process is reversed in time and we get figure 8.17(b) by retracing the steps that build figure 8.17(a). So in figure 8.17(b) we would see that A_f and B_f approach each other and then A_i and B_i would move away from each other to infinity. If both the processes of figure 8.17(a,b) have the same probability then we have time reversal invariance or microscopic reversibility.

Under time reversal, the following occurs

$$\vec{r} \xrightarrow{T} \vec{r} \ \text{(no change in position)}$$

$$\vec{v} = \frac{d\vec{r}}{dt} \xrightarrow{T} \frac{d\vec{r}}{d(-t)} = -\frac{d\vec{r}}{dt} = -\vec{v} \ \text{(velocity changes sign)}$$

$$\vec{p} = m\vec{v} = m\frac{d\vec{r}}{dt} \xrightarrow{T} m\frac{d\vec{r}}{d(-t)} = -m\frac{d\vec{r}}{dt} = -m\vec{v} = -\vec{p} \ \text{(momentum changes sign)}$$

$$\vec{l} = \vec{r} \times \vec{p} \xrightarrow{T} \vec{r} \times (-\vec{p}) = -\vec{r} \times \vec{p} = -\vec{l}, \ \ \vec{s} \xrightarrow{T} -\vec{s} \ \text{(angular momentum changes sign)}$$

$$\vec{E} \rightarrow \vec{E} \ \text{(electric field does not change sign as it is produced by static charges)}$$

$$i = \frac{dq}{dt} \xrightarrow{T} \frac{dq}{d(-t)} = -\frac{dq}{dt} = -i \ \text{(current changes sign)}$$

$$\vec{B} \rightarrow -\vec{B} \ \text{(magnetic field changes sign as source of magnetic field is current)}$$

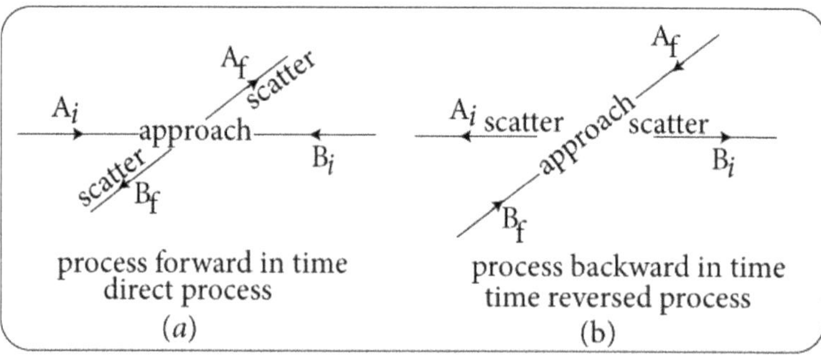

Figure 8.17. Explaining time reversibility.

8.51 Newton's equation under time reversal

Newton's equation is second order in time and is invariant under time reversal operation.

$$\vec{F} = m\frac{d^2\vec{r}}{dt^2} \xrightarrow{T} m\frac{d^2\vec{r}}{d(-t)^2} = m\frac{d^2\vec{r}}{dt^2} = \vec{F} \qquad (8.166)$$

Because Newton's equation is second order in time t we can associate with every solution $\vec{r}(t)$ of Newton's equation another solution

$$\vec{r}\,'(t) = \vec{r}(-t). \qquad (8.167)$$

The correspondence between the two solutions is shown in figure 8.18. Position of particle at time $t = t_0$ in the case of figure 8.18(a) is the same as the position at time $t = -t_0$ in the case of figure 8.18(b) while the velocities and momenta are reversed since

$$\vec{v}\,'(t_0) = \left[\frac{d}{dt}\vec{r}\,'(t)\right]_{t=t_0} = -\left[\frac{d}{d(-t)}\vec{r}(-t)\right]_{t=t_0} \quad \text{(using equation (8.167))}$$

$$= -\left[\frac{d}{dt'}\vec{r}(t')\right]_{t'=-t_0} = -[\vec{v}(t')]_{t'=-t_0} = -\vec{v}(-t_0)$$

8.52 Time reversal in quantum mechanics

Consider a particle of mass m moving in a time-independent potential $V(\vec{r})$. The corresponding time-dependent Schrödinger equation satisfied by the wave function $\psi(\vec{r}, t)$ is

$$i\hbar\frac{\partial}{\partial t}\psi(\vec{r}, t) = H\psi(\vec{r}, t) = \left[-\frac{\hbar^2}{2m}\nabla^2 + V(\vec{r})\right]\psi(\vec{r}, t) \qquad (8.168)$$

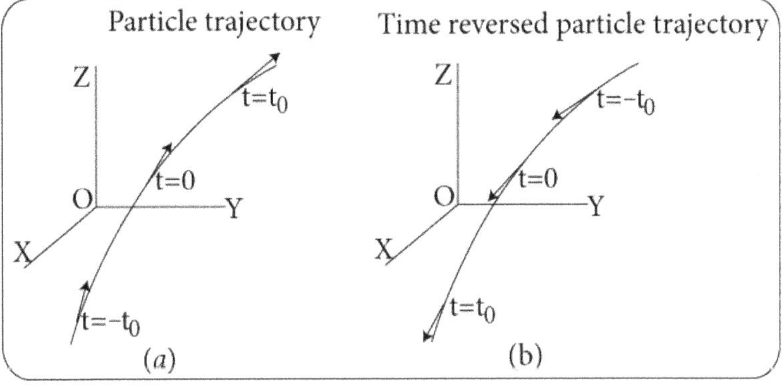

Figure 8.18. Particle trajectory and its time reversed trajectory.

Change t to $-t$

$$-i\hbar\frac{\partial}{\partial t}\psi(\vec{r}, -t) = H\psi(\vec{r}, -t) = \left[-\frac{\hbar^2}{2m}\nabla^2 + V(\vec{r})\right]\psi(\vec{r}, -t) \qquad (8.169)$$

Taking the complex conjugate we get

$$i\hbar\frac{\partial}{\partial t}\psi^*(\vec{r}, -t) = H^*\psi^*(\vec{r}, t) = \left[-\frac{\hbar^2}{2m}\nabla^2 + V^*(\vec{r})\right]\psi^*(\vec{r}, -t) \qquad (8.170)$$

For a time-independent real potential the Hamiltonian is also real and time-independent

$$V = V^*, \quad H = H^*$$

$$i\hbar\frac{\partial}{\partial t}\psi^*(\vec{r}, -t) = H\psi^*(\vec{r}, -t) = \left[-\frac{\hbar^2}{2m}\nabla^2 + V(\vec{r})\right]\psi^*(\vec{r}, -t) \qquad (8.171)$$

This is the Schrödinger wave equation satisfied by the wave function $\psi^*(\vec{r}, -t)$.
We note that $\psi^*(\vec{r}, -t)$ and $\psi(\vec{r}, t)$ are connected by two operations.
One operation is replacing t by $-t$ (let us denote this operation by a unitary operator U) and the other operation is complex conjugation (let us denote this operation by an operator K).
Let us construct the time reversal operator as

$$T = UK$$

where U is replacing t by $-t$ and K is the complex conjugation operator.
We call the wave function $\psi^*(\vec{r}, -t)$ as the time reversed wave function.
So the Schrödinger equation is invariant under time reversal.

$$T\psi(\vec{r}, t) = UK\psi(\vec{r}, t) = U\psi^*(\vec{r}, t) = \psi^*(\vec{r}, -t) \qquad (8.172)$$

Now

$$T^\dagger T = (UK)^\dagger UK = K^\dagger U^\dagger UK = K^\dagger K = I$$

since $U^\dagger U = I$ (unitary). Also since $K^\dagger K = 1$ we have

$$T^\dagger T = I \qquad (8.173)$$

8.52.1 Condition of time reversal invariance of the Schrödinger equation

The time-dependent Schrödinger equation is

$$i\hbar\frac{\partial}{\partial t}\psi(\vec{r}, t) = H\psi(\vec{r}, t) \qquad (8.174)$$

Using equation (8.173) we get

$$i\hbar \frac{\partial}{\partial t} \psi(\vec{r}, t) = HT^{\dagger}T\psi(\vec{r}, t) \qquad (8.175)$$

Operate by $T = UK$

$$UK\left[i\hbar \frac{\partial}{\partial t} \psi(\vec{r}, t)\right] = THT^{\dagger}T\psi(\vec{r}, t)$$

$$U\left[-i\hbar \frac{\partial}{\partial t} K\psi(\vec{r}, t)\right] = THT^{\dagger}T\psi(\vec{r}, t)$$

$$i\hbar \frac{\partial}{\partial t} UK\psi(\vec{r}, t) = THT^{\dagger}T\psi(\vec{r}, t)$$

$$i\hbar \frac{\partial}{\partial t} [T\psi(\vec{r}, t)] = THT^{\dagger}[T\psi(\vec{r}, t)] \qquad (8.176)$$

Equations (8.174) and (8.176) are alike and so we can write upon comparison

$$H = THT^{\dagger}$$

Operating by T from the right

$$HT = THT^{\dagger}T$$

$$HT = TH \text{ (as } T^{\dagger}T = I)$$

$$[T\ H\ ] = 0 \qquad (8.177)$$

This is the condition for time reversal invariance.

8.52.2 The time reversal operator is anti-linear and anti-unitary

$$\text{Consider } T(\alpha\psi + \beta\phi) = UK(\alpha\psi + \beta\phi) = U(\alpha^*K\psi + \beta^*K\phi)$$
$$= \alpha^*UK\psi + \beta^*UK\phi$$

$$T(\alpha\psi + \beta\phi) = \alpha^*T\psi + \beta^*T\phi \qquad (8.178)$$

As T obeys this relation it is called an anti-linear operator.
Consider

$$|T\psi_1> = T|\psi_1> = UK|\psi_1> = U|\psi_1^*>$$

$$|T\psi_2> = T|\psi_2> = UK|\psi_2> = U|\psi_2^*>$$

$$\langle T\psi_1 | T\psi_2 \rangle = \langle \psi_1^* | U^{\dagger}U | \psi_2^* \rangle = \langle \psi_1^* | \psi_2^* \rangle \text{ (as } U^{\dagger}U = 1)$$

$$\langle T\psi_1 | T\psi_2 \rangle = \langle \psi_2 | \psi_1 \rangle^{\star} \qquad (8.179)$$

Since the anti-linear operator T satisfies this condition it is called an anti-unitary operator.

We mention that

$$[T\ H_{\text{strong}}] = 0, \quad [T\ H_{\text{weak}}] = 0 \tag{8.180}$$

$$[T\ H_{\text{weak}}] \neq 0 \tag{8.181}$$

Strong and electromagnetic interactions are time reversal invariants but weak interaction is not a time reversal invariant.

8.53 CP symmetry or invariance

This was proposed by Landau (1957).

CP refers to a combination of C symmetry (charge conjugation) and P symmetry (parity).

If we apply a parity operator on some state of a particle having momentum $\vec{p}$ it changes momentum to $-\vec{p}$. If then we apply a charge conjugation operator it changes a particle to its anti-particle. Collectively thus CP operation acting on a particle state with momentum $\vec{p}$ will give an anti-particle state with momentum $-\vec{p}$.

$$CP\ |\vec{P}_{\text{particle}}> \ = C|-\vec{P}_{\text{particle}}> \ = |-\vec{P}_{\text{anti-particle}}>$$

that is

$$\vec{P}_{\text{particle}} \xrightarrow{\text{CP}} -\vec{P}_{\text{anti-particle}} \tag{8.182}$$

Consider operation of CP on a neutrino. Again, the neutrino is left-handed.

C operates on a left-handed neutrino and turns it into a left-handed anti-neutrino. P then turns the left-handed anti-neutrino to a right-handed anti-neutrino. We show this schematically.

$$PC\ |\nu_L> \ = P|\bar{\nu}_L> \ = |\bar{\nu}_R>$$

$$CP\ |\nu_L> \ = C\ |\nu_R> \ = |\bar{\nu}_R>$$

Consider the observed weak decay, as shown in figure 8.19(a).

$$\pi^+ \rightarrow \mu^+ + \nu_\mu \tag{8.183}$$

In pion rest frame μ and ν shoot off in opposite directions.

To conserve angular momentum both particles must have the same handedness (left say), i.e.

$$\pi^+ \rightarrow \mu_L^+ + \nu_{\mu L} \tag{8.184}$$

Operate equation (8.184) by C to get

$$C\left(\pi^+ \rightarrow \mu_L^+ + \nu_{\mu L}\right) \Rightarrow \pi^- \rightarrow \mu_L^- + \bar{\nu}_{\mu L} \tag{8.185}$$

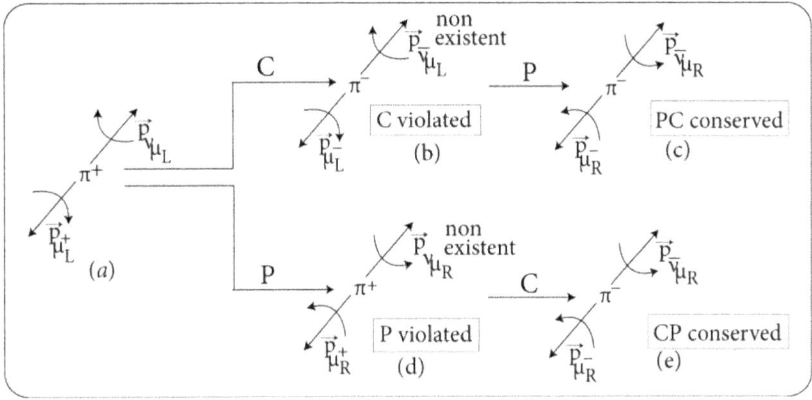

Figure 8.19. C, P, CP operations on weak decay of $\pi^+ \to \mu^+ + \nu_\mu$.

But this reaction does not occur since $\bar{\nu}_{\mu L}$ does not exist. So C is violated in weak decay. (figure 8.19(b))

Operate equation (8.185) by P to get

$$P\left(\pi^- \to \mu_L^- + \bar{\nu}_{\mu L}\right) \Rightarrow \pi^- \to \mu_R^- + \bar{\nu}_{\mu R} \tag{8.186}$$

This is observed. Again equation (8.185) can be rewritten as

$$PC\left(\pi^+ \to \mu_L^+ + \nu_{\mu L}\right) \Rightarrow \pi^- \to \mu_R^- + \bar{\nu}_{\mu R} \tag{8.187}$$

This implies that PC is conserved (figure 8.19(c))

Again operate P on equation (8.186) to get

$$P\left(\pi^+ \to \mu_L^+ + \nu_{\mu L}\right) \Rightarrow \pi^+ \to \mu_R^+ + \nu_{\mu R} \tag{8.188}$$

which does not occur since $\nu_{\mu R}$ does not occur. So P is violated in weak decay (figure 8.19(d)).

Operate equation (8.188) by C to get

$$C\left(\pi^+ \to \mu_R^+ + \nu_{\mu R}\right) \Rightarrow \pi^- \to \mu_R^- + \bar{\nu}_{\mu R} \tag{8.189}$$

This is observed. Again equation (8.189) can be rewritten as

$$CP\left(\pi^+ \to \mu_L^+ + \nu_{\mu L}\right) \Rightarrow \pi^- \to \mu_R^- + \bar{\nu}_{\mu R} \text{ occurs.}$$

This implies that CP is conserved (figure 8.19(e))

Consider the weak interaction

$$\nu_\mu + p \to \mu^- + p + \pi^+ \tag{8.190}$$

Since a muon neutrino ν_μ is left-handed we can write

$$\nu_{\mu L} + p \to \mu^- + p + \pi^+ \tag{8.191}$$

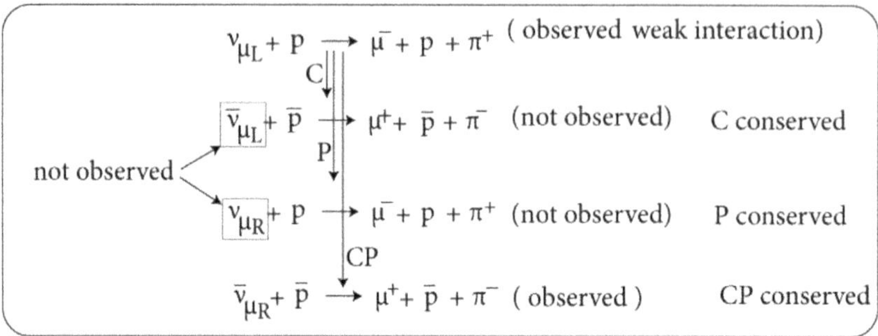

Figure 8.20. Effect of C, P, CP on the weak interaction $\nu_\mu + p \to \mu^- + p + \pi^+$.

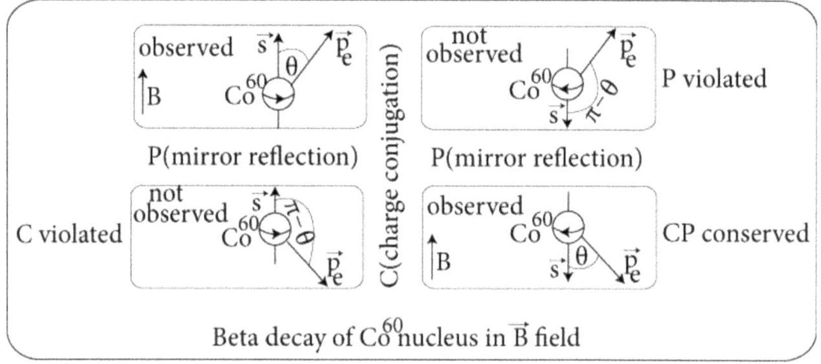

Figure 8.21. Explanation of CP symmetry.

The effect of C, P, CP on equation (8.191) is shown in figure 8.20.

In figure 8.21 application of C, P, CP on the $_{27}Co^{60}$ nucleus (Wu's experiment) is shown.

8.54 CP violation or T violation in neutral kaon decay

As far as strong interactions are concerned, the K^0 and $\overline{K^0}$ particles are distinct. They are distinguished by different values of hypercharge Y viz.

$$Y = 1 \text{ for } K^0 \tag{8.192}$$

$$Y = -1 \text{ for } \overline{K^0} \tag{8.193}$$

Kaons are produced through strong interaction, namely

$$\pi^- + p \to \Lambda + K^0 \tag{8.194}$$

$$\pi^+ + \overline{p} \to \overline{\Lambda} + \overline{K^0} \tag{8.195}$$

If only strong interaction existed, the K^0 and $\overline{K^0}$ mesons would have exactly the same mass and would be stable. However, because the weak interaction does not conserve hypercharge the $| K^0 >$ and $| \overline{K^0} >$ wave functions are coupled and the K^0 and $\overline{K^0}$ particles are not energy eigenstates of the total Hamiltonian (describing strong and weak interactions). In fact, the wave functions of the K^0 and $\overline{K^0}$ particles are linear superposition of the wave functions of the observed particles, namely the short-lived kaon K_S^0 and the long-lived kaon K_L^0 which are identified and distinguished by their different weak decay modes, namely (figure 8.22)

$$K_S^0 \rightarrow 2\pi \ (\tau_S \sim 10^{-10} \, s) \tag{8.196}$$

$$K_L^0 \rightarrow 3\pi \ (\tau_L \sim 10^{-8} \, s) \tag{8.197}$$

Let $t = 0$ be the instant of kaon production.

In time $0 < t < 10^{-10} \, s$ both K_S^0, K_L^0 will be present in kaon beam and so both $K \rightarrow 2\pi$ as well as $K \rightarrow 3\pi$ are observed.

For $t > 10^{-10} s$ only K_L^0 is expected to be found, i.e. $K_L^0 \rightarrow 3\pi$ decay should only be observed.

Neutral kaons have quark structure as follows.

$$| k^0 > \ = | d\bar{s} > \tag{8.198}$$

$$| \overline{K^0} > \ = | \bar{d}s > \tag{8.199}$$

Kaon K^0 can oscillate into anti-kaon $\overline{K^0}$ by either of the box diagrams shown in figure 8.23. It is a doubly weak process, i.e. if we start with a beam of K^0 after some time it will become a mixture of K^0 and some $\overline{K^0}$. A strong interaction can produce a strong eigenstate of K^0 or $\overline{K^0}$ and then as the kaon propagates, weak interaction takes over and we end up with a mixture of K^0 and $\overline{K^0}$ states.

Let S be strangeness operator. It counts strange quarks. Then

$$S \, | K^0 > \ = +1 \, | K^0 > \ \ (\text{strangeness of } K^0 \text{ is } + 1) \tag{8.200}$$

$$S \, | \overline{K^0} > \ = - \, | \overline{K^0} > \ \ (\text{strangeness of } |\overline{K^0} > \text{ is } - 1) \tag{8.201}$$

$| K^0 >$ and $| \overline{K^0} >$ are eigenstates of S.

Construct S in K^0- $\overline{K^0}$ basis. Let us denote

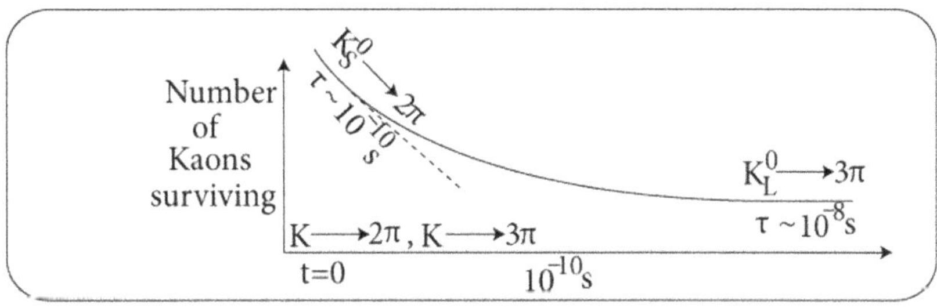

Figure 8.22. Short-lived kaon K_S^0 and long-lived kaon K_L^0.

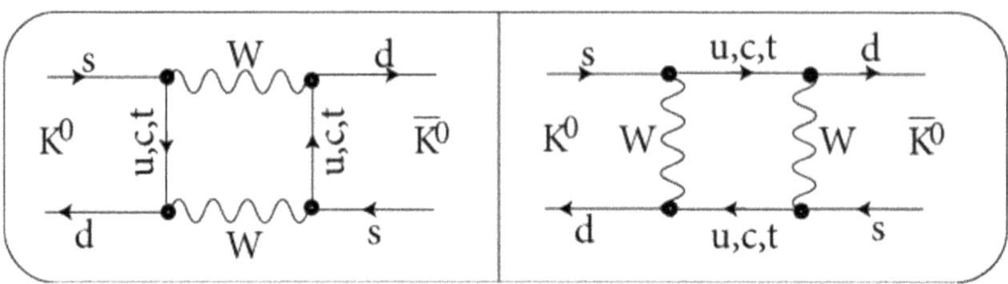

Figure 8.23. Kaon K^0 can oscillate into anti-kaon $\overline{K^0}$.

$$| K^0 > \; = | \, 1> \tag{8.202}$$

$$| \overline{K^0} > \; = | \, 2> \tag{8.203}$$

Hence from equations (8.200) and (8.201) we write

$$S \, | \, 1 > \; = | \, 1> \tag{8.204}$$

$$S \, | \, 2 > \; = - | \, 2> \tag{8.205}$$

Thus from equations (8.204) and (8.205) we have

$$S_{11} = \langle 1 \, | S | 1 \rangle = \langle 1 \mid 1 \rangle = 1$$

$$S_{12} = \langle 1 \, | S | 2 \rangle = -\langle 1 \mid 2 \rangle = 0$$

$$S_{21} = \langle 2 \, | S | 1 \rangle = \langle 2 \mid 1 \rangle = 0$$

$$S_{22} = \langle 2 \, | S | 2 > \; = -\langle 2 \mid 2 \rangle = -1$$

$$S = \begin{pmatrix} 1 & 0 \\ 0 & -1 \end{pmatrix} \text{ in } K^0 - \overline{K^0} \text{ basis.} \tag{8.206}$$

K^0-$\overline{K^0}$ have $J^P = 0^-$, i.e. kaons are pseudoscalar mesons. Hence

$$P \, | \, K^0 > \; = - | \, K^0> \tag{8.207}$$

$$P \, | \, \overline{K^0} > \; = - | \, \overline{K^0}> \tag{8.208}$$

$$C \, | \, K^0 > \; = | \, \bar{K}^0 > \quad (C \text{ replaces particles by anti-particles}) \tag{8.209}$$

$$C \, | \, \overline{K^0} > \; = | \, K^0> \tag{8.210}$$

$$CP \, | K^0 > \; = -C | K^0 > \; = - | \, \overline{K^0}> \tag{8.211}$$

So $| \, K^0>$ is not an eigenstate of CP.

$$CP \, | \, \overline{K^0} > \; = -C \, | \, \overline{K^0} > \; = - | \, K^0> \tag{8.212}$$

So $|\overline{K^0} >$ is not an eigenstate of CP.

Construct CP in K^0-$\overline{K^0}$ basis.

With $| K^0 > = | 1>$ (equation (8.202)) $CP | K^0 > = -| \overline{K^0} >$ (equation (8.211)) is written as

$$CP \, | 1 > = -| 2> \tag{8.213}$$

With $| \overline{K^0} > = | 2>$ (equation (8.203)) $CP | \overline{K^0} > = -| K^0>$ (equation (8.212)) is written as

$$CP \, | 2 > = -| 1> \tag{8.214}$$

Thus from equations (8.213) and (8.214) we have

$$(CP)_{11} = \langle 1 \, |CP|1 \rangle = -<1 \, | 2\rangle = 0$$

$$(CP)_{12} = \langle 1 \, |CP|2 \rangle = -<1 \, | 1\rangle = -1$$

$$(CP)_{21} = \langle 2 \, |CP|1 \rangle = -\langle 2 \, | 2\rangle = -1$$

$$(CP)_{22} = \langle 2 \, |CP|2 \rangle = -\langle 2 \, | 1\rangle = 0$$

$$CP = \begin{pmatrix} 0 & -1 \\ -1 & 0 \end{pmatrix} = -\begin{pmatrix} 0 & 1 \\ 1 & 0 \end{pmatrix} \text{ in } K^0{-}\overline{K^0} \text{ basis.} \tag{8.215}$$

Let us construct normalized eigenstates of CP by constructing a mixture of $| K^0>$ and $| \overline{K^0} >$.

We construct linear combinations of K^0 and $\overline{K^0}$ as

$$| K_S^0 > = \frac{1}{\sqrt{2}} (| K^0 > - | \overline{K^0} >) \tag{8.216}$$

$$| K_L^0 > = \frac{1}{\sqrt{2}} (| K^0 > + | \overline{K^0} >) \tag{8.217}$$

$\frac{1}{\sqrt{2}}$ is normalization constant.

$$CP | K_S^0> = \frac{1}{\sqrt{2}} (CP | K^0 > - CP | \overline{K^0} >) \quad \text{(using equation (8.216))}$$

$$= \frac{1}{\sqrt{2}} (-| \overline{K^0} > + | K^0 >) \quad \text{(using equations (8.211), (8.212))}$$

$$= \frac{1}{\sqrt{2}} (| \overline{K^0} > - \left| \overline{K^0} >\right) = | K_S^0 > \quad \text{(using equation (8.216))} \tag{8.218}$$

$CP | K_S^0 > = | K_S^0>$ means that $| K_S^0>$ is eigenstate of CP with eigenvalue $+1$.

$$CP \mid K_L^0> = \frac{1}{\sqrt{2}}(CP \mid K^0 > + CP \mid \overline{K^0} >) \qquad \text{(using equation (8.217))}$$

$$= \frac{1}{\sqrt{2}}(- \mid \overline{K^0} > - \mid K^0 >) \qquad \text{(using equations (8.211) and (8.212))}$$

$$= -\frac{1}{\sqrt{2}}(\mid \overline{K^0} > + \mid \overline{K^0} > = -\mid K_L^0 > \quad \text{(using equation (8.217))} \qquad (8.219)$$

$CP \mid K_L^0 > = -\mid K_L^0>$ means that $\mid K_L^0>$ is eigenstate of CP with eigenvalue -1. So $\mid K_S^0>$ and $\mid K_L^0>$ are the normalized eigenstates of CP with eigenvalues ± 1. The inverse relations to equations (8.216) and (8.217) are

$$\mid K^0 > \; = \frac{1}{\sqrt{2}}(\mid K_L^0 > + \mid K_S^0 >) \qquad (8.220)$$

$$\mid \overline{K^0} > \; = \frac{1}{\sqrt{2}}(\mid K_L^0 > - \mid K_S^0 >) \qquad (8.221)$$

We now consider the possibility of decay of kaons into pions—either into two pions (two charged pions or two neutral pions) or into three pions (two charged and one neutral pion or three neutral pions), as shown in figure 8.24. So

$$K^0 \to 2\pi$$

$$K^0 \to 3\pi$$

Now pion parity is $P=$ (intrinsic parity)(space or orbital parity)$=(-1)(-1)^l$
For $l = 0$, $P = (-1)(-1)^0 = -1 \equiv -ve$
So for $\mid 2\pi>$: $P = (-1)^2 = +1$ i.e. $P \mid 2\pi > = \mid 2\pi>$ and for $\mid 3\pi>$: $P = (-1)^3 = -1$ i.e. $P \mid 3\pi > = -\mid 3\pi>$

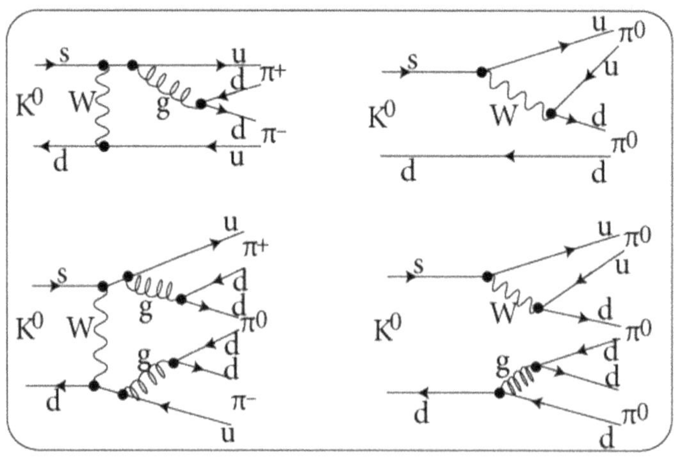

Figure 8.24. Pion production from a kaon.

Also

$$C \mid \pi > \; = \mid \pi>$$

$$C \mid 2\pi > \; = \mid 2\pi>$$

$$C \mid 3\pi > \; = \mid 3\pi>$$

Thus

$$CP \mid 2\pi > \; = C \mid 2\pi > \; = \mid 2\pi> \qquad (8.222)$$

$\mid 2\pi>$ is eigenstate of CP with eigenvalue $+1$

$$CP \mid 3\pi > \; = -C \mid 3\pi > \; = -\mid 3\pi> \qquad (8.223)$$

$\mid 3\pi>$ is eigenstate of CP with eigenvalue -1

Let us collect the main results, i.e. let us make a gist and compare

$CP \mid K_S^0 > \; = \mid K_S^0>$ (equation (8.218)), $CP \mid 2\pi > \; = \mid 2\pi>$ (equation (8.222))
(eigenstate of CP with eigenvalue $+1$)

$CP \mid K_L^0 > \; = -\mid K_L^0>$ (equation (8.219)), $CP \mid 3\pi > \; = -\mid 3\pi>$ (equation (8.223))
(eigenstate of CP with eigenvalue -1)

Clearly $K_S^0 \rightarrow 2\pi$ ($CP = +1$) is possible if CP is conserved.

And $K_L^0 \rightarrow 3\pi$ ($CP = -1$) is possible if CP is conserved.

but $K_L^0 \rightarrow 2\pi$ is impossible if CP is conserved because then $CP(K_L^0) = -1$ but $CP(2\pi) = +1$.

We can produce either K^0 or $\overline{K^0}$ but the weak interaction mixes them and so what propagates is some combination of K_S^0 and K_L^0 states. They combine together to give the right mixture of K^0 and $\overline{K^0}$.

• Mass of kaon $\sim 498 \frac{MeV}{c^2}$, mass of pion $\sim 140 \frac{MeV}{c^2}$.

The mass difference in $\mid K > \; \rightarrow \mid 2\pi>$ decay is $(498 - 2 \times 140)\frac{MeV}{c^2} = 218 \frac{MevV}{c^2}$

The mass difference in $\mid K > \; \rightarrow \mid 3\pi>$ decay is $(498 - 3 \times 140)\frac{MeV}{c^2} = 78 \frac{MevV}{c^2}$.

Larger mass difference between states is unstable and prone to faster decay.

So $\mid K > \; \rightarrow \mid 2\pi>$ (i.e. $K_S^0 \rightarrow 2\pi$) decay occurs faster, has shorter lifetime of $90 \, ps = 9 \times 10^{-11} \, s$ and corresponds to larger phase space while $\mid K > \; \rightarrow \mid 3\pi>$ (i.e. $K_L^0 \rightarrow 3\pi$) decay occurs slower, has longer lifetime of $50 \, ns = 5 \times 10^{-8} \, s$ and corresponds to smaller phase space.

Clearly, if CP is conserved then long-lived kaon state K_L^0 should only decay to 3π maintaining $CP = -1$. $K_L^0 \rightarrow 2\pi$ should not be observed.

• Cronin and Fitch in 1964 smashed protons into a target and generated a kaon beam.

$$\pi^- + p \rightarrow \Lambda + K^0$$

$$p + \overline{p} \rightarrow K^+ + \overline{K^0} + \pi^-$$

All charged particles were removed from the beam and a beam of neutral kaons was prepared. This beam of neutral kaons was made to travel a long distance from the region of their production. On the way the short-lived kaons K_S^0 that would be decaying into 2π disappear from the beam because of their short lifetime. Only long-lived kaons K_L^0 would survive and decay to 3π conserving $CP = -1$. However, in 45 out of 22 700 decay events kaons were found to decay to 2π ($CP = +1$) violating CP. In other words $K_L^0 \rightarrow 2\pi$ was observed though number of such decays was very small. So occurrence of $K_L^0 \rightarrow 2\pi$ hints at slight violation of CP in weak interaction. Hence CP symmetry is slightly broken in weak interaction. This is referred to as CP violation.

We conclude that K_L^0 is not purely CP eigenstate, but a mixture of $CP = -1$ and $CP = +1$ states and involves slight decay of $CP = -1$ state to $CP = +1$ state exhibiting violation of CP in weak interaction.

Detection of two-pion decays by long-lived kaons was an experimental demonstration of the violation of symmetry between matter and anti-matter which is a fundamental principle of physics. Violation of CP symmetry enables us to make a distinction between matter and anti-matter. The laws of physics for matter and anti-matter are different. This has profound implications for cosmology.

The Universe as we observe today has a pre-dominance of matter over anti-matter. This observed imbalance or asymmetry in the matter–anti-matter ratio may have been produced by the occurrence of CP violation in the first seconds after the Big Bang.

- CP violation implies non-conservation of T provided the CPT theorem is valid.
- The CPT theorem is one of the fundamental principles of physics. It states that all interactions are invariant under the combined operation or application of charge conjugation, parity and time reversal in any order.

8.55 G parity

In strong interaction both isospin I and C parity are conserved.

For hadrons we define a new operator called G or G parity as follows.

$G=$ charge conjugation C plus $180°$ rotation about the 2nd axis of isospin space, i.e. about I_2.

Mathematically, we can write this as

$$G = CR_2 = Ce^{i\pi I_2} \tag{8.224}$$

G is a multiplicative quantum number.

For strong interaction G is invariant, i.e. commutes with H_{strong}, i.e.

$$[G, \ H_{\text{strong}}] = 0 \tag{8.225}$$

But for weak and electromagnetic interaction G is not invariant, i.e.

$$[G, \ H_{\text{em}}] \neq 0, \ . \ [G, \ H_{\text{weak}}] \neq 0 \tag{8.226}$$

We have compared the effect of charge conjugation and G parity on pions in exercise 8.7.

8.55.1 G parity for pions

From equation (8.224) we can write for pionic state $|\pi^+>$

$$G\,|\,\pi^+ > \; =CR_2\,|\,\pi^+> \tag{8.227}$$

Rotation about I_2 by 180^0 means I_3 becomes $-I_3$ and hence $I_3 = +1$ becomes $I_3 = -1$ i.e. $|\pi^+>$ transforms to $|\pi^->$. Hence from equation (8.227) we have

$$G\,|\pi^+ > \; =C\,|\pi^- > \; =k\,|\,\pi^+> \tag{8.228}$$

Clearly G leaves $|\pi^+>$ unchanged. So we get an eigen equation for G. Similarly

$$G\,|\,\pi^- > \; =k\,|\,\pi^-> \tag{8.229}$$

Equations (8.228) and (8.229) can be written using quantum numbers $I\ I_3$ noting that

$$|\,\pi^\pm > \; =|\,I\ I_3 > \; =|\,1\ \pm 1 > , \quad |\,\pi^0 > \; =|\,I\ I_3 > \; =|\,1\ 0>$$

$$G\,|\,\pi^\pm > \; =G\,|\,1\ \pm 1 > \; =k\,|\,1\ \pm 1> \tag{8.230}$$

$$G\,|\,\pi^0 > \; =G\,|1\ 0 > \; =k\,|1\ 0> \tag{8.231}$$

In real space $R_2 = e^{i\pi L_2}$, $R_2\,Y_{l0} = (-1)^l\,Y_{l0}$

Isospin function $|I\ I_3>$ has the same property under rotation in isospin space as the spherical harmonics $Y_{lm_l}(\theta,\ \phi)$ in real space. Hence we have

$$R_2\,|I\ 0 > \; =(-1)^I|I\ 0>$$

$$R_2\,|\pi^0 > \; =R_2\,|\,1\ 0 > \; =(-1)^1|1\ 0 > \; =-|\,1\ 0 > \; =-|\,\pi^0 > \;\; \text{(eigenvalue is} - 1)\ (8.232)$$

From equations (8.231) and (8.232) we get

$$G\,|\pi^0 > \; =CR_2|\pi^0 > \; = -C\,|\,\pi^0 > \; = -|\,\pi^0 > \;\; \text{(eigenvalue is} - 1) \tag{8.233}$$

So G parity of π^0 is odd.
G parity of $\pi^\pm$ is arbitrary. As per convention, we take G parity of $\pi^\pm$ to be odd. Hence G parity of pion family is odd. Hence

$$G\,|\,\pi > \; = -|\,\pi> \tag{8.234}$$

For n pion system

$$G\,|\,n\pi > \; =(-1)^n|\,n\pi> \tag{8.235}$$

8.56 Masses of particles in $SU(3)$

We discuss masses of particles in $SU(3)$ flavour symmetry.

We grouped particles into iso-multiplets. Particles constituting iso-multiplets have nearly the same mass.

For instance pion iso-triplet consists of π^+, π^0, π^- having masses

$$m_{\pi^+} = m_{\pi^-} = 140\frac{MeV}{c^2}, \quad m_{\pi^0} = 138\frac{MeV}{c^2}.$$

Kaon iso-doublet consists of members K^+, K^0 having masses

$$m_{K^+} = m_{K^0} = 493\frac{MeV}{c^2}$$

Kaon iso-doublet consists of members $\overline{K^0}$, K^- having masses

$$m_{\overline{K^0}} = m_{K^-} = 498\frac{MeV}{c^2}$$

Nucleon iso-doublet consists of members p, n having masses

$$m_p = 938\frac{MeV}{c^2}, \quad m_n = 939\frac{MeV}{c^2}$$

But when we consider the octet of $SU(3)$ we want to group the pions π^+, π^0, π^- along with the kaons K^+, K^0, $\overline{K^0}$, K^- and the nucleons p, n.

8.56.1 Symmetry breaking

Clearly member particles in the octet do not have the same mass $(m_\pi \sim 140\frac{MeV}{c^2}$, $m_K \sim \frac{500\ MeV}{c^2}$, $m_N \sim \frac{940\ MeV}{c^2})$. This difference of mass is a violation of symmetry and is called symmetry breaking.

In other words, the $SU(3)$ flavour symmetry is not an exact symmetry. The $SU(3)$ flavour symmetry is broken down to $SU(2)$ isospin symmetry and $U(1)$ hypercharge symmetry.

$$SU(3)|_{\text{flavour}} = SU(2)|_I \times U(1)|_Y \tag{8.236}$$

This means the groups with similar masses are grouped under $SU(2)$, i.e. symmetric under $SU(2)$ grouping. Also each iso-multiplet has the same Y value.

$$Y(\pi = \pi^+, \pi^0, \pi^-) = 0$$

$$Y(K^+, K^0) = 1$$

$$Y(\overline{K^0}, K^-) = -1$$

$$Y(N = p, n) = 1$$

8.57 Gell-Mann–Okubo mass formula

Gell-Mann and Okubo derived a general formula called the Gell-Mann–Okubo mass formula that is used to estimate mass of elementary particles. This formula is a sum rule for masses of hadrons within a specific multiplet determined by isospin and hypercharge.

• Gell-Mann and Okubo gave a mathematical representation of this formula for a baryon octet

$$M(I,\ Y) = k_0 + k_1 Y + k_2\left[I(I+1) - \frac{1}{4}Y^2 \right] \tag{8.237}$$

where k_0, k_1, k_2 are constants. If we can determine the constants k_0, k_1, k_2 then unknown masses of the members can be found out.

Let us consider members of a baryon octet, the quantum numbers I (isospin), Y (hypercharge) of which are given in figure 8.25.

To solve for k_0, k_1, k_2 we require three equations that we can set up from figure 8.25 and equation (8.237) as follows.

For p, n from figure 8.25 and equation (8.237) we have

$$M_N = M\left(\frac{1}{2},\ 1\right) = k_0 + k_1.\,1 + k_2\left[\frac{1}{2}\left(\frac{1}{2}+1\right) - \frac{1}{4}1^2 \right] = k_0 + k_1 + \frac{1}{2}k_2 \tag{8.238}$$

For Σ from figure 8.25 and equation (8.237) we have

$$M_\Sigma = M(1,\ 0) = k_0 + k_1.\,0 + k_2\left[1(1+1) - \frac{1}{4}0^2 \right] = \frac{1}{2}k_2 + 2k_2 \tag{8.239}$$

Baryon octet Members		I	Y	Meson octet Members	I	Y
Ξ^-	Ξ	$\frac{1}{2}$	-1	K^+	$\frac{1}{2}$	1
Ξ^0				K^0		
Σ^-				π^+		
Σ^0	Σ	1	0	π^0	1	0
Σ^+				π^-		
Λ	Λ	0	0	$\bar{K}^0$	$\frac{1}{2}$	-1
n				K^-		
	N	$\frac{1}{2}$	1			
p				η	0	0

Figure 8.25. Isospin and hypercharge values of members of a baryon octet.

For Λ from figure 8.25 and equation (8.237) we have

$$M_\Lambda = M(0, 0) = k_0 + k_1. \, 0 + k_2 0 = k_0 \tag{8.240}$$

If we experimentally know the mass M_Λ of Λ then M_0 is determined using equation (8.240) as

$$k_0 = M_\Lambda = M(0, 0) \tag{8.241}$$

If mass M_Σ of Σ is experimentally determined then from equation (8.239) we get the value of M_2

$$k_2 = \frac{1}{2}(M_\Sigma - k_0) = \frac{1}{2}(M_\Sigma - M_\Lambda) \tag{8.242}$$

And from equation (8.238), experimental values of mass M_N of p, n will give the value of M_1.

$$k_1 = M_N - k_0 - \frac{1}{2}k_2 = M_N - M_\Lambda - \frac{1}{2}\left[\frac{1}{2}(M_\Sigma - M_\Lambda)\right]$$

$$k_1 = M_N - \frac{3}{2}M_\Lambda - \frac{1}{4}M_\Sigma \tag{8.243}$$

Equations (8.241),(8.242), (8.243) give values of the constants k_0, k_1, k_2. Now for Ξ from figure 8.25 and equation (8.237) we get

$$M_\Xi = M\left(\frac{1}{2}, -1\right) = k_0 + k_1. \, (-1) + k_2\left[\frac{1}{2}\left(\frac{1}{2} + 1\right) - \frac{1}{4}(-1)^2\right]$$

$$M_\Xi = k_0 - k_1 + \frac{1}{2}k_2 \tag{8.244}$$

Combination of equations (8.234) and (8.240) gives us

$$\frac{M_N + M_\Xi}{2} = \frac{\left(k_0 + k_1 + \frac{1}{2}k_2\right) + (k_0 - k_1 + \frac{1}{2}k_2)}{2} = \frac{2k_0 + k_2}{2} = \frac{2M_\Lambda + \frac{1}{2}(M_\Sigma - M_\Lambda)}{2}$$

$$\frac{M_N + M_\Xi}{2} = \frac{3M_\Lambda + M_\Sigma}{4} \tag{8.245}$$

This formula for a baryon octet was predicted by Gell-Mann. We have verified this relation with experimentally determined mass values for baryons in *exercise 8.8*.

• For a meson octet the formula is

$$M^2(I, Y) = c_0 + c_1 Y + c_2\left[I(I + 1) - \frac{1}{4}Y^2\right] \tag{8.246}$$

where c_0, c_1, c_2 are constants. If we can determine the constants c_0, c_1, c_2 then unknown masses of members can be found out.

Let us consider members of a meson octet the quantum numbers I (isospin), Y (hypercharge) of which are given in figure 8.25. To solve for c_0, c_1, c_2 we require 3 equations that we can set up from figure 8.25 and equation (8.246) as follows.

For K^0 from figure 8.25 and equation (8.246) we have

$$M_{K^0}^2 = M^2\left(\frac{1}{2}, 1\right) = c_0 + c_1. \, 1 + c_2\left[\frac{1}{2}\left(\frac{1}{2} + 1\right) - \frac{1}{4}1^2\right] = c_0 + c_1 + \frac{1}{2}c_2 \quad (8.247)$$

For π^0 from figure 8.25 and equation (8.247) we have

$$M_{\pi^0}^2 = M^2(1, 0) = c_0 + c_1. \, 0 + c_2\left[1(1 + 1) - \frac{1}{4}0^2\right] = c_0 + 2c_2 \quad (8.248)$$

For η from figure 8.25 and equation (8.247) we have

$$M_\eta^2 = M^2(0, 0) = c_0 + c_1. \, 0 + c_2 0 = c_0 \quad (8.249)$$

If we experimentally know the mass M_η of η then M_0 is determined using equation (8.249) as

$$c_0 = M_\eta^2 = M(0, 0) \quad (8.250)$$

If mass M_{π^0} of π^0 is experimentally determined then from equation (8.248) we get the value of M_2

$$c_2 = \frac{1}{2}\left(M_{\pi^0}^2 - c_0\right) = \frac{1}{2}\left(M_{\pi^0}^2 - M_\eta^2\right) \quad (8.251)$$

And from equation (8.247) experimental values of mass M_{K^0} of K^0 will give the value of M_1.

$$c_1 = M_{K^0}^2 - c_0 - \frac{1}{2}c_2 = M_{K^0}^2 - M_\eta^2 - \frac{1}{2}\left[\frac{1}{2}\left(M_{\pi^0}^2 - M_\eta^2\right)\right]$$

$$c_1 = M_{K^0}^2 - \frac{3}{2}M_\eta^2 - \frac{1}{4}M_{\pi^0}^2 \quad (8.252)$$

Equations (8.250), (8.251), and (8.252) give values of the constants c_0, c_1, c_2. Now for $\overline{K^0}$ from figure 8.25 and equation (8.246) we have

$$M_{\overline{K^0}}^2 = M^2\left(\frac{1}{2}, -1\right) = c_0 + c_1. \, (-1) + c_2\left[\frac{1}{2}\left(\frac{1}{2} + 1\right) - \frac{1}{4}(-1)^2\right] \quad (8.253)$$

$$= c_0 - c_1 + \frac{1}{2}c_2$$

Combination of equations (8.243) and (8.249) gives us

$$\frac{M_{K^0}^2 + M_{\overline{K^0}}^2}{2} = \frac{\left(c_0 + c_1 + \frac{1}{2}c_2\right) + (c_0 - c_1 + \frac{1}{2}c_2)}{2} = \frac{2c_0 + c_2}{2} = \frac{2M_\eta^2 + \frac{1}{2}\left(M_{\pi^0}^2 - M_\eta^2\right)}{2}$$

$$\frac{M_{K^0}^2 + M_{\overline{K^0}}^2}{2} = \frac{3M_\eta^2 + M_{\pi^0}^2}{4} \tag{8.254}$$

This formula for a meson octet was predicted by Gell-Mann. We have verified this relation with experimentally determined mass values of mesons in exercise 8.9 and have shown that it holds only approximately.

8.58 Current–current interaction of weak decay

• Consider the weak nuclear β^- decay

$$_{27}\text{Co}^{60} \rightarrow {_{28}\text{Ni}^{60}} + e^- + \bar{\nu}_e$$

At the nuclear level it is conversion of $n \rightarrow p$ and so we can write

$$n \rightarrow p + e^- + \bar{\nu}_e$$

At the quark level since $n \equiv udd$ and $p \equiv uud$ we can write

$$d \rightarrow u + e^- + \bar{\nu}_e$$

We can think of this interaction as current–current interaction as evident from pictorial representation of the interaction in figure 8.26. There are two currents associated with the interaction.

(1) d to u current:

Conversion of $d \rightarrow u$ and production of W^- which is weak gauge boson

$$d \rightarrow u + W^-$$

Charge conservation holds since

$$Q(\text{LHS}) = Q(d) = -\frac{1}{3}, \quad Q(\text{RHS}) = Q(u) + Q(W^-) = \frac{2}{3} - 1 = -\frac{1}{3}. \quad \text{Hence}$$

$$Q(\text{LHS}) = Q(\text{RHS})$$

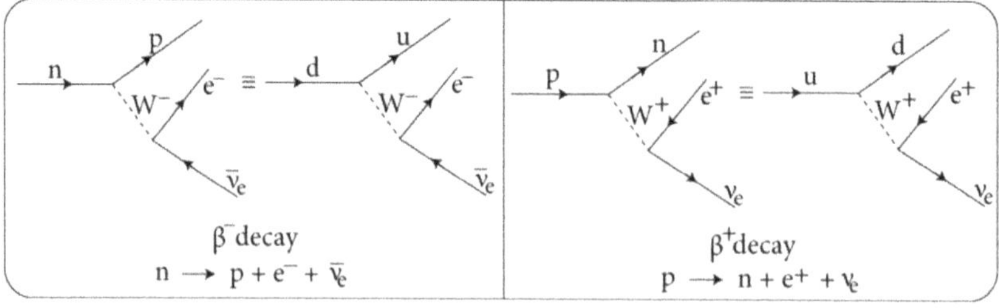

Figure 8.26. Current–current interaction in weak beta decay β^-, β^+.

(2) W^- decays to e^- plus $\bar{\nu}_e$

$$W^- \to e^- + \bar{\nu}_e$$

• Consider the weak nuclear β^+ decay

$$_6\mathrm{C}^{10} \to {}_5\mathrm{B}^{10} + e^+ + \nu_e$$

At the nuclear level it is conversion of $p \to n$ and so we can write

$$p \to n + e^+ + \nu_e$$

At the quark level since $n \equiv udd$ and $p \equiv uud$ we can write

$$u \to d + e^+ + \nu_e$$

We can think of this interaction as current–current interaction as evident from the pictorial representation of the interaction in figure 8.26. There are two currents associated with the interaction.

(1) u to d current:

Conversion of $u \to d$ and production of W^+ which is weak gauge boson

$$u \to d + W^+$$

Charge conservation holds since

$$Q(\mathrm{LHS}) = Q(u) = \frac{2}{3}, \; Q(\mathrm{RHS}) = Q(d) + Q(W^+) = -\frac{1}{3} + 1 = \frac{2}{3}. \text{ Hence}$$

$$Q(\mathrm{LHS}) = Q(\mathrm{RHS})$$

(2) W^+ decays to e^+ plus ν_e

$$W^+ \to e^+ + \nu_e$$

• Other examples of weak decay

$$\checkmark \mu^- \to e^- + \bar{\nu}_e + \nu_\mu$$

We can think of this interaction as current–current interaction as evident from the pictorial representation of the interaction in figure 8.27. There are two currents associated with the interaction.

(1) μ^- to ν_μ current

Conversion of $\mu^- \to \nu_\mu$ and production of W^- which is weak gauge boson

$$\mu^- \to \nu_\mu + W^-$$

(2) W^- decays to e^- plus $\bar{\nu}_e$

$$W^- \to e^- + \bar{\nu}_e$$

$$\checkmark \mu^+ \to e^+ + \nu_e + \bar{\nu}_\mu$$

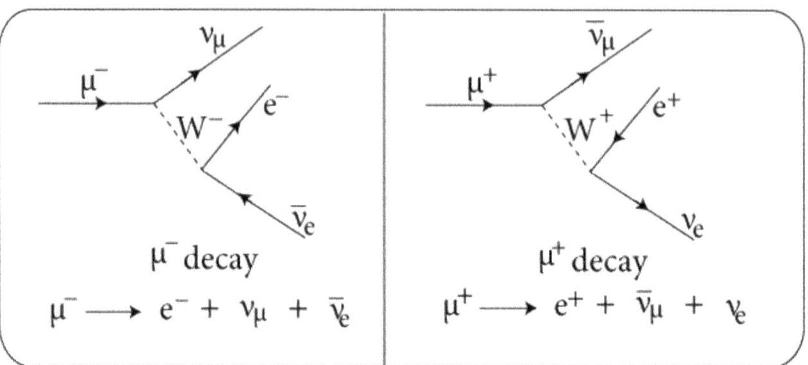

Figure 8.27. Current–current interaction in weak beta decay μ^-, μ^+.

We can think of this interaction as current–current interaction as evident from the pictorial representation of the interaction in figure 8.27. There are two currents associated with the interaction.

(1) μ^+ to $\bar{\nu}_\mu$ current

Conversion of $\mu^+ \to \bar{\nu}_\mu$ and production of W^+ which is weak gauge boson

$$\mu^+ \to \bar{\nu}_\mu + W^+$$

(2) W^+ decays to e^+ plus ν_e

$$W^+ \to e^+ + \nu_e$$

• W^+ and W^- are particle–anti-particle pair.

8.59 Conservation laws

Some conservation laws are valid in all interactions. They are conservation of energy, linear momentum, angular momentum, charge, electron lepton number, muon lepton number, tauon lepton number, baryon number, *CPT*.

The conservation laws in various interactions are shown in figure 8.28.

In *exercise 8.17* we give examples of weak interactions in which:
    (i) only hadrons are present;
    (ii) both leptons and hadrons are present;
    (iii) only leptons are present in the final state.

In figure 8.29 we indicate which quantum numbers are additive, multiplicative. We also indicate the quantum numbers associated with continuous and discrete symmetries.

Figure 8.30 shows the symmetries associated with various conserved quantities.

	Conserved quantities	Strong interaction	Electromagnetic interaction	Weak interaction
4-momentum {	Energy	$\checkmark$	$\checkmark$	$\checkmark$
	Linear momentum	$\checkmark$	$\checkmark$	$\checkmark$
	Angular momentum	$\checkmark$	$\checkmark$	$\checkmark$
	Lepton number	$\checkmark$	$\checkmark$	$\checkmark$
	Baryon number	$\checkmark$	$\checkmark$	$\checkmark$
	Isospin	$\checkmark$	$\times$	$\times$
	Isospin projection	$\checkmark$	$\checkmark$	$\times$
	Strangeness	$\checkmark$	$\checkmark$	$\times$
	Electric charge	$\checkmark$	$\checkmark$	$\checkmark$
	C-parity	$\checkmark$	$\checkmark$	$\times$
	Parity	$\checkmark$	$\checkmark$	$\times$
	Time reversal	$\checkmark$	$\checkmark$	$\checkmark$ } (small violation)
	CP	$\checkmark$	$\checkmark$	$\checkmark$
	CPT	$\checkmark$	$\checkmark$	$\checkmark$
	G-parity	$\checkmark$	$\times$	$\times$

Conservation laws

Figure 8.28. Conservation laws in different interactions.

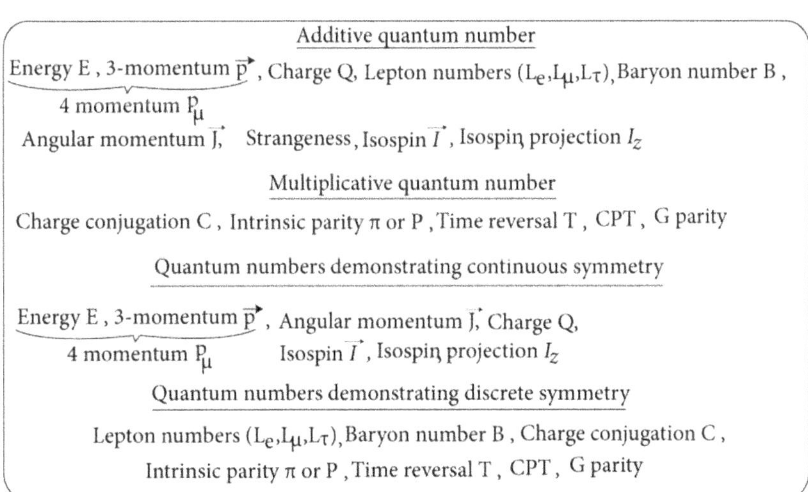

Figure 8.29. Additive, multiplicative, continuous and discrete quantum numbers.

Conserved quantities	Associated symmetries
Energy E	Time translational symmetry
3-momentum $\vec{p}$	Space translational symmetry
4-momentum P_μ	Space time translational symmetry
Angular momentum $\vec{J}$	Rotational invariance in external spin space
Isospin $\vec{I}$, Isospin projection I_z	Rotational invariance in internal spin space (isospace)
Charge conjugation C	Replacing particles by antiparticles
Intrinsic parity π or P	Mirror inversion
Time reversal T	Reversing the sense of time

Figure 8.30. Symmetries associated with various conserved quantities.

8.60 Exercises

Exercise 8.1 *Investigate if hypercharge is conserved in the following reactions.*
(i) $\pi^- + p \rightarrow \Lambda + K^0$
(ii) $\pi^- + p \rightarrow \pi^0 + n$

$\boxed{\text{Ans}}$ (i) In $\pi^- + p \rightarrow \Lambda + K^0$

$$Y(\text{LHS}) = Y(\pi^-) + Y(p) = 0 + 1 = 1$$

$$Y(\text{RHS}) = Y(\Lambda) + Y(K^0) = 0 + 1 = 1$$

Hence $Y(\text{LHS}) = Y(\text{RHS})$.
So hypercharge Y is conserved.
(ii) In $\pi^- + p \rightarrow \pi^0 + n$

$$Y(\text{LHS}) = Y(\pi^-) + Y(p) = 0 + 1 = 1$$

$$Y(\text{RHS}) = Y(\pi^0) + Y(n) = 0 + 1 = 1$$

Hence $Y(\text{LHS}) = Y(\text{RHS})$.
So hypercharge Y is conserved.

Exercise 8.2 *Investigate if strangeness is conserved in the following reactions.*
(i) $\pi^- + p \rightarrow \Sigma^- + K^+$
(ii) $\pi^- + p \rightarrow \Sigma^+ + K^-$

$\boxed{\text{Ans}}$ (i) In $\pi^- + p \rightarrow \Sigma^- + K^+$

$$S(\text{LHS}) = S(\pi^-) + S(p) = 0 + 0 = 0$$

$$S(\text{RHS}) = S(\Sigma^-) + S(K^+) = -1 + 1 = 0$$

Hence $S(\text{LHS}) = S(\text{RHS})$.
So strangeness S is conserved.

(ii) In $\pi^- + p \to \Sigma^+ + K^-$

$$S(\text{LHS}) = S(\pi^-) + S(p) = 0 + 0 = 0$$

$$S(\text{RHS}) = S(\Sigma^+) + S(K^-) = -1 - 1 = -2$$

Hence $S(\text{LHS}) \neq S(\text{RHS})$.
So strangeness S is not conserved.
Reaction is not allowed via strong interaction. But it may proceed via weak interaction.

Exercise 8.3 *Investigate if baryon number is conserved in the following reactions.*
   (i) $n \to p + e^- + \bar{\nu}_e$
   (ii) $\pi^0 + n \to \overline{K^0} + \Sigma^0$
   (iii) $\Lambda \to p + \pi^-$
   (iv) $\overline{\Lambda} \to \bar{p} + \pi^+$
   (v) $\bar{p} \to e^+ + \gamma$
   (vi) $p \to e^+ + \gamma$

$\boxed{\text{Ans}}$ (i) In $n \to p + e^- + \bar{\nu}_e$

$$B(\text{LHS}) = B(n) = 1$$

$$B(\text{RHS}) = B(p) + B(e^-) + B(\bar{\nu}_e) = 1 + 0 + 0 = 1$$

Hence $B(\text{LHS}) = B(\text{RHS})$.
So baryon number B is conserved.
(ii) In $\pi^0 + n \to \overline{K^0} + \Sigma^0$

$$B(\text{LHS}) = B(\pi^0) + B(n) = 0 + 1 = 1$$

$$B(\text{RHS}) = B(\overline{K^0}) + B(\Sigma^0) = 0 + 1 = 1$$

Hence $B(\text{LHS}) = B(\text{RHS})$.
So baryon number B is conserved.
(iii) In $\Lambda \to p + \pi^-$
$B(\text{LHS}) = B(\Lambda) = 1$
$B(\text{RHS}) = B(p) + B(\pi^-) = 1 + 0 = 1$
Hence $B(\text{LHS}) = B(\text{RHS})$.
So baryon number B is conserved.
(iv) In $\overline{\Lambda} \to \bar{p} + \pi^+$

$$B(\text{LHS}) = B(\overline{\Lambda}) = 1$$

$$B(\text{RHS}) = B(\bar{p}) + B(\pi^+) = -1 + 0 = -1$$

Hence $B(\text{LHS}) \neq B(\text{RHS})$.

So baryon number B is not conserved. The reaction cannot proceed.
(v) In $\bar{p} \rightarrow e^+ + \gamma$

$$B(\text{LHS}) = B(\bar{p}) = -1$$

$$B(\text{RHS}) = B(e^+) + B(\gamma) = 0 + 0 = 0$$

Hence $B(\text{LHS}) \neq B(\text{RHS})$.
So baryon number B is not conserved. The reaction cannot proceed.
(vi) $p \rightarrow e^+ + \gamma$

$$B(\text{LHS}) = B(p) = 1$$

$$B(\text{RHS}) = B(e^+) + B(\gamma) = 0 + 0 = 0$$

Hence $B(\text{LHS}) \neq B(\text{RHS})$.
So baryon number B is not conserved. The reaction cannot proceed.

Exercise 8.4 *Investigate if lepton number is conserved in the following reactions.*
(i) $n \rightarrow p + e^- + \bar{\nu}_e$
(ii) $\pi^- \rightarrow \mu^- + \bar{\nu}_\mu$
(iii) $\mu^- \rightarrow e^- + \nu_\mu + \bar{\nu}_e$
(iv) $\gamma \rightarrow e^+ + e^-$
(v) $K^0 \rightarrow \pi^+ + e^- + \bar{\nu}_e$
(vi) $p \rightarrow n + e^+ + \nu_\mu$
(vii) $\pi^+ \rightarrow \mu^+ + \nu_\mu$
$\boxed{\text{Ans}}$ (i) In $n \rightarrow p + e^- + \bar{\nu}_e$

$$L_e(\text{LHS}) = L_e(n) = 0$$

$$L_e(\text{RHS}) = L_e(p) + L_e(e^-) + L_e(\bar{\nu}_e) = 0 + 1 - 1 = 0$$

Hence $L_e(\text{LHS}) = L_e(\text{RHS})$.
So lepton number L_e is conserved.
(ii) In $\pi^- \rightarrow \mu^- + \bar{\nu}_\mu$

$$L_\mu(\text{LHS}) = L_\mu(\pi^-) = 0$$

$$L_\mu(\text{RHS}) = L_\mu(\mu^-) + L_\mu(\bar{\nu}_\mu) = 1 - 1 = 0$$

Hence $L_\mu(\text{LHS}) = L_\mu(\text{RHS})$.
So lepton number L_μ is conserved.
(iii) In $\mu^- \rightarrow e^- + \nu_\mu + \bar{\nu}_e$

$$L_e(\text{LHS}) = L_e(\mu^-) = 0$$

$$L_e(\text{RHS}) = L_e(e^-) + L_e(\nu_\mu) + L_e(\bar{\nu}_e) = 1 + 0 - 1 = 0$$

Hence $L_e(\text{LHS}) = L_e(\text{RHS})$.
So lepton number L_e is conserved.
$L_\mu(\text{LHS}) = L_\mu(\mu^-) = 1$
$L_\mu(\text{RHS}) = L_\mu(e^-) + L_\mu(\nu_\mu) + L_\mu(\bar{\nu}_e) = 0 + 1 + 0 = 1$
Hence $L_\mu(\text{LHS}) = L_\mu(\text{RHS})$.
So lepton number L_μ is conserved.
(iv) In $\gamma \rightarrow e^+ + e^-$

$$L_e(\text{LHS}) = L_e(\gamma) = 0$$

$$L_e(\text{RHS}) = L_e(e^+) + L_e(e^-) = -1 + 1 = 0$$

Hence $L_e(\text{LHS}) = L_e(\text{RHS})$.
So lepton number L_e is conserved.
(v) In $K^0 \rightarrow \pi^+ + e^- + \bar{\nu}_e$

$$L_e(\text{LHS}) = L_e(K^0) = 0$$

$$L_e(\text{RHS}) = L_e(\pi^+) + L_e(e^-) + L_e(\bar{\nu}_e) = 0 + 1 - 1 = 0$$

Hence $L_e(\text{LHS}) = L_e(\text{RHS})$.
So lepton number L_e is conserved.
(vi) In $p \rightarrow n + e^+ + \nu_\mu$

$$L_e(\text{LHS}) = L_e(p) = 0$$

$$L_e(\text{RHS}) = L_e(n) + L_e(e^+) + L_e(\nu_\mu) = 0 - 1 + 1 = 0$$

Hence $L_e(\text{LHS}) = L_e(\text{RHS})$.
So lepton number L_e is conserved.
(vii) In $\pi^+ \rightarrow \mu^+ + \nu_\mu$

$$L_e(\text{LHS}) = L_e(\pi^+) = 0$$

$$L_e(\text{RHS}) = L_e(\mu^+) + L_e(\nu_\mu) = -1 + 1 = 0$$

Hence $L_e(\text{LHS}) = L_e(\text{RHS})$.
So lepton number L_e is conserved.

Exercise 8.5 *Discuss if the following reactions can occur or not for a ρ meson.*

(i) $\rho \rightarrow \gamma\gamma\gamma$ (ii) $\rho \rightarrow \gamma\gamma$ (iii) $\rho \rightarrow \pi^0 + \pi^0$ (iv) $\rho \rightarrow \pi^+ + \pi^-$

$\boxed{\text{Ans}}$ A ρ meson has odd C parity, γ photon has odd C parity, π^0 meson has even C parity.
   (i) $\rho \rightarrow \gamma\gamma\gamma$ is allowed since it preserves C parity, i.e. $C = -1$ on both sides.

(ii) $\rho \to \gamma\gamma$ is not allowed since it does not preserve C parity. The LHS has $C = -1$ but the RHS has C parity $C = +1$.

(iii) $\rho \to \pi^0 + \pi^0$ is not allowed since it does not preserve C parity. The LHS has $C = -1$ but the RHS has C parity $C = +1$.

(iv) $\rho \to \pi^+ + \pi^-$ is allowed since it preserves C parity, i.e. $C = -1$ on both sides.

Exercise 8.6 *Verify if the following reactions occur.*

(i) $e^- + e^- \to \pi^- + \pi^-$
(ii) $\mu^- \to e^- + \gamma$
(iii) $\mu^- \to e^- + e^+ + e^-$
(iv) $p \to \pi^0 + e^+$
(v) $n \to e^- + e^+$
(vi) $p + p \to p + p + \bar{n}$
(vii) $p + n \to p + p + n + \bar{p}$
(viii) $p + n \to p + p + \bar{p}$

Ans (i) In $e^- + e^- \to \pi^- + \pi^-$

$$L_e(\text{LHS}) = L_e(e^-) + L_e(e^-) = +1 + 1 = +2$$

$$L_e(\text{RHS}) = L_e(\pi^-) + L_e(\pi^-)$$

Hence $L_e(\text{LHS}) \ne L_e(\text{RHS})$.
So electron lepton number L_e is not conserved. Reaction forbidden.
(ii) In $\mu^- \to e^- + \gamma$

$$L_\mu(\text{LHS}) = L_\mu(\mu^-) = +1$$

$$L_\mu(\text{RHS}) = L_\mu(e^-) + L_\mu(\gamma) = 0 + 0 = 0$$

Hence $L_\mu(\text{LHS}) \ne L_\mu(\text{RHS})$.
So muon lepton number L_μ is not conserved. Reaction forbidden.
(iii) In $\mu^- \to e^- + e^+ + e^-$

$$L_\mu(\text{LHS}) = L_\mu(\mu^-) = +1$$

$$L_\mu(\text{RHS}) = L_\mu(e^-) + L_\mu(e^+) + L_\mu(e^-) = 0 + 0 + 0 = 0$$

Hence $L_\mu(\text{LHS}) \ne L_\mu(\text{RHS})$.
So muon lepton number L_μ is not conserved. Reaction forbidden.
(iv) In $p \to \pi^0 + e^+$

$$B(\text{LHS}) = B(p) = +1$$

$$B(\text{RHS}) = B(\pi^0) + B(e^+) = 0 + 0 = 0$$

Hence $B(\text{LHS}) \ne B(\text{RHS})$.

So baryon number B is not conserved. Reaction forbidden.
(v) In $n \to e^- + e^+$

$$B(\text{LHS}) = B(n) = +1$$

$$B(\text{RHS}) = B(e^-) + B(e^+) = 0 + 0 = 0$$

Hence $B(\text{LHS}) \neq B(\text{RHS})$.
So baryon number B is not conserved. Reaction forbidden.
(vi) In $p + p \to p + p + \bar{n}$

$$B(\text{LHS}) = B(p) + B(p) = +1 + 1 = +2$$

$$B(\text{RHS}) = B(p) + B(p) + B(\bar{n}) = 1 + 1 - 1 = 1$$

Hence $B(\text{LHS}) \neq B(\text{RHS})$.
So baryon number B is not conserved. Reaction forbidden.
(vii) In $p + n \to p + p + n + \bar{p}$

$$B(\text{LHS}) = B(p) + B(n) = +1 + 1 = +2$$

$$B(\text{RHS}) = B(p) + B(p) + B(n) + B(\bar{p}) = 1 + 1 + 1 - 1 = +2$$

Hence $B(\text{LHS}) = B(\text{RHS})$.
So baryon number B is conserved. Reaction may occur.
(viii) In $p + n \to p + p + \bar{p}$

$$B(\text{LHS}) = B(p) + B(n) = +1 + 1 = +2$$

$$B(\text{RHS}) = B(p) + B(p) + B(\bar{p}) = 1 + 1 - 1 = +1$$

Hence $B(\text{LHS}) \neq B(\text{RHS})$.
So baryon number B is not conserved. Reaction forbidden.

Exercise 8.7 *Compare the effect of charge conjugation C parity and G parity on pions.*
$\boxed{\text{Ans}}$ Under charge conjugation

$$C \mid \pi^\pm > \; = \mid \pi^\mp >$$

$$C \mid \pi^0 > \; = \mid \pi^0 >$$

Under G parity operation

$$G \mid \pi^\pm > \; = - \mid \pi^\pm >$$

$$G \mid \pi^0 > \; = - \mid \pi^0 >$$

Clearly the G parity of pions is -1, i.e. $G(\pi) = -1$.

Exercise 8.8 *Verify the formula of equation (8.245) for a baryon octet.*
  $\boxed{\text{Ans}}$ We note that

$$m_N = 939.6 \, MeV, \, m_\Xi = 1321.3 \, MeV$$

Consider the term

$$\frac{1}{2}(m_N + m_\Xi) = \frac{1}{2}(939.6 \, MeV + 1321.3 \, MeV) = 1130.45 \, MeV \qquad (8.255)$$

We also note that $m_\Lambda = 1115.7 \, MeV$, $m_\Sigma = 1189.4 \, MeV$
Consider the term

$$\frac{1}{4}(3m_\Lambda + m_\Sigma) = \frac{1}{4}(3 \times 1115.7 \, MeV + 1189.4 \, MeV) = 1134.125 \, MeV \quad (8.256)$$

We see from equations (8.255) and (8.256) that the values are nearly equal and so the formula of equation (8.245)

$$\frac{1}{2}(m_N + m_\Xi) = \frac{1}{4}(3m_\Lambda + m_\Sigma)$$

is verified for a baryon octet.

Exercise 8.9 *Verify the formula of equation (8.254) for a meson octet.*
  $\boxed{\text{Ans}}$ We note that $m_{K^0} = 493 \, MeV$, $m_{\overline{K^0}} = 498 \, MeV$
Consider the term

$$\frac{1}{2}\left(m_{K^0}^2 + m_{\overline{K^0}}^2\right) = \frac{1}{2}\left[(493 \, MeV)^2 + (498 \, MeV)^2\right] = 245527 \, MeV^2 \quad (8.257)$$

We also note that $m_\eta = 550 \, MeV$, $m_{\pi^0} = 138 \, MeV$
Consider the term

$$\frac{1}{4}\left(3m_\eta^2 + m_{\pi^0}^2\right) = \frac{1}{4}\left[3(550 \, MeV)^2 + (138 \, MeV)^2\right] = 231\,636 \, MeV^2 \quad (8.258)$$

We see from equations (8.257) and (8.258) that the values are nearly equal (not exact) and so the formula of equation (8.254)

$$\frac{1}{2}(M_{K^0}^2 + M_{\overline{K^0}}^2) = \frac{1}{4}(3M_\eta^2 + M_{\pi^0}^2)$$

is verified. However, the formula holds only approximately for a meson octet.

Exercise 8.10 *Show that Gell-Mann–Okubo formula holds for a nucleonic isodoublet and pionic isotriplet but not for strange particles.*

[Ans] The charge Q and the 3rd component of isospin I_3 are connected by the Gell-Mann–Okubo relationship as

$$Q = I_3 + \frac{B}{2}$$

- For the nucleonic isodoublet $I = \frac{1}{2}$ we have

For p: $Q = +1$, $I_3 = \frac{1}{2}$, $B = 1$ and so $+1 = +\frac{1}{2} + \frac{1}{2}$, i.e. $Q = I_3 + \frac{B}{2}$ holds.

For n: $Q = 0$, $I_3 = -\frac{1}{2}$, $B = 1$ and so $0 = -\frac{1}{2} + \frac{1}{2}$, i.e. $Q = I_3 + \frac{B}{2}$ holds.

and the Gell-Mann–Okubo relationship $Q = I_3 + \frac{B}{2}$ holds.

- For the pionic isotriplet $I = 1$ we have

For π^+: $Q = +1$, $I_3 = +1$, $B = 0$ and so $+1 = +1 + \frac{0}{2}$, i.e. $Q = I_3 + \frac{B}{2}$ holds.

For π^0: $Q = 0$, $I_3 = 0$, $B = 0$ and so $0 = 0 + \frac{0}{2}$, i.e. $Q = I_3 + \frac{B}{2}$ holds.

For π^-: $Q = -1$, $I_3 = -1$, $B = 0$ and so $-1 = -1 + \frac{0}{2}$, i.e. $Q = I_3 + \frac{B}{2}$ holds.

So the Gell-Mann–Okubo relationship $Q = I_3 + \frac{B}{2}$ holds for a pionic isotriplet.

- For a strange particle say Ω^- we have $I = 0$

$Q = -1$, $I_3 = 0$, $B = 1$ and so $-1 \neq 0 + \frac{1}{2}$ i.e. $Q = I_3 + \frac{B}{2}$ does not hold.

Exercise 8.11 *Show that charged pion parity is odd.*

[Ans] We consider the case of a charged pion. Consider the reaction

$$\pi^- + d \rightarrow n + n \tag{8.259}$$

$$\pi^+ + d \rightarrow p + p \tag{8.260}$$

We invoke angular momentum conservation relation to equation (8.259)

$$\vec{J} = \vec{J}_{\pi^-} + \vec{J}_d = \vec{J}_{2n}$$

$$= \vec{0} + \vec{1} = \vec{l} + \overset{\scriptscriptstyle 1}{\frac{1}{2}} + \overset{\scriptscriptstyle 1}{\frac{1}{2}}$$

$$\vec{J} = \vec{1} = \vec{l} + \begin{cases} \vec{0} \text{ (singlet)} \\ \vec{1} \text{ (triplet)} \end{cases}$$

Thus $J = 1$ and $\vec{l} = \vec{1} - \begin{cases} \vec{0} \text{ (singlet)} \\ \vec{1} \text{ (triplet)} \end{cases} = \begin{cases} 1 \\ 2,\ 1,\ 0 \end{cases}$

Corresponding possible final states are

$J = 1$, $l = 0$, $s = 0$, spectral term $^1P_1$

$J = 1$, $l = 2$, $s = 1$, spectral term $^3D_1$, $J = 1$, $l = 1$, $s = 1$, spectral term $^3P_1$,

$J = 1$, $l = 1$, $s = 1$, spectral term $^3S_1$

Again n is a fermion and the nn system is fermion system and hence wave function is anti-symmetric, i.e.

$$\psi_{nn} = \psi_{\text{space}} \psi_{\text{spin}} = (-) \text{ (anti symmetric)}$$

$$(-) = (-)^l (-)^{1+s} \text{ i. e. } (-) = (-)^{l+1+s}$$

$$(-)^{l+s} = 1 \text{ i. e.}$$

$$l + s = \text{even}$$

The only possibility is $l = 1$, $s = 1$, spectral term ${}^3P_1$
Parity equation from equation (8.255) is

$$P(\pi^-)P(d) = P(nn)$$

Now $P(d) = P(np) = (+)(+) = +ve$ as n and p are by convention assigned $+ve$ parity and parity is multiplicative. Hence

$$P(\pi^-)(+) = (-)^{l=1}$$

$$P(\pi^-) = -ve$$

Similarly, from equation (8.256) we get $P(\pi^+) = -ve$
So both charged pions $\pi^\pm$ have intrinsic parity -1.

Exercise 8.12 *Show that neutral pion parity is odd.*
 Consider decay of a neutral pion

$$\pi^0 \to \gamma + \gamma \text{ i. e. } \pi^0 \to \gamma_1 + \gamma_2$$

In the centre-of-mass frame of reference

$$\vec{p}_{\pi^0} = 0, \quad \vec{p}_{\gamma 1} + \vec{p}_{\gamma 2} = 0$$

$$\vec{p}_{\gamma 1} = -\vec{p}_{\gamma 2}$$

$$\hbar \vec{k}_1 = -\hbar \vec{k}_2$$

$$\vec{k}_1 = -\vec{k}_2$$

Since $\vec{J} = 0$ the two photons must be in a state such that the total angular momentum is zero. To construct a state with zero angular momentum $J = 0$, the pion wave function is a scalar, i.e.
$\psi \xrightarrow{\text{parity}} \psi$, does not change sign, has even parity and then $J^\pi = 0^+$
Or the pion wave function is a pseudoscalar, i.e.
$\psi \xrightarrow{\text{parity}} -\psi$, changes sign, has odd parity and then $J^\pi = 0^-$.
Let $\vec{E}_1$ and $\vec{E}_2$ be the plane polarization vectors (electric field vectors) of the two photons.
Three possibilities arise.

1) Pion wave function $\psi(\pi^0)$ may be a mixture of scalar and pseudoscalar. This corresponds to no definite parity.

However, parity is conserved in strong interaction and electromagnetic interaction, i.e.

$$[P\ H] = 0$$

i.e. P is a good quantum number. So the pion wave function should be of definite parity. In view of this we discard this possibility.

2) $\psi(\pi^0)$ is a scalar, i.e. expressible as a dot product.

$$\vec{E_1}.\,\vec{E_2} \xrightarrow{\text{parity}} (-\vec{E_1}).\,(-\vec{E_2}) = \vec{E_1}.\,\vec{E_2} \Leftarrow \text{even parity}$$

with $\vec{E_1}$ not orthogonal to $\vec{E_2}$.

But experiment shows that the photon $\vec{E}$ vectors are orthogonal. So we discard this possibility.

3) $\psi(\pi^0)$ is a pseudoscalar, i.e. expressible as a scalar triple product.

$$\psi(\pi^0) \sim \vec{k}.\,\vec{E_1} \times \vec{E_2} \xrightarrow{\text{parity}} (-\vec{k}).\left[(-\vec{E_1}) \times (-\vec{E_2})\right] = -\vec{k}.\,\vec{E_1} \times \vec{E_2} \Leftarrow \text{odd parity}$$

with $\vec{E_1}$ not parallel to $\vec{E_2}$.

This possibility is acceptable. Also, under interchange of photons $\vec{E_1} \times \vec{E_2}$ becomes $\vec{E_2} \times \vec{E_1} = -\vec{E_1} \times \vec{E_2}$ (antisymmetric), but then $\vec{k} \to -\vec{k}$. So under interchange

$$\psi(\pi^0) \sim \vec{k}.\,\vec{E_1} \times \vec{E_2} \to (-\vec{k}).\,(-\vec{E_1} \times \vec{E_2}) = \vec{k}.\,\vec{E_1} \times \vec{E_2} \text{ (symmetric under interchange)}.$$

This corresponds to a bosonic system of a neutral pion.

Clearly neutral pion wave function is a pseudoscalar and has odd parity ($-$ve).

Exercise 8.13 *Establish that $\pi^0 \to 2\gamma$ is possible but $\pi^0 \to 3\gamma$ is not possible.*
[Ans.] For π^0 state $|\ \psi > = |\ \pi^0 >$

$$C\,|\ \pi^0 > \ = |\ \pi^0 > \ \text{(equation (8.118))}$$

π^0 has positive charge parity, i.e. C is even $+1$ for π^0. Again

$$C\,|\ \gamma > \ = -|\ \gamma > .$$

For a single photon charge parity is odd (-1).
✓Consider the reaction

$$\pi^0 \to \gamma + \gamma$$

Check conservation of C parity

$$C(\text{LHS}) = C(\pi^0) = +1$$

$$C(\text{RHS}) = C(\gamma + \gamma) \ = (-1)(-1) = +1$$

$$C(\text{LHS}) = C(\text{RHS}).$$

As C is conserved the reaction $\pi^0 \to \gamma + \gamma$ is possible.
✓Consider the reaction

$$\pi^0 \to \gamma + \gamma + \gamma$$

$$C(\text{LHS}) = C(\pi^0) = +1$$

$$C(\text{RHS}) = C(\gamma + \gamma + \gamma = (-1)(-1)(-1) = -1$$

$$C(\text{LHS}) \neq C(\text{RHS}).$$

As C is not conserved the reaction $\pi^0 \to \gamma + \gamma + \gamma$ is not possible.

Exercise 8.14 *With respect to the strangeness quantum number explain whether the following reactions can occur.*
   (i) $K^0 \to \pi^+ + \pi^-$
   (ii) $K^0 \to \gamma + \gamma$
   (iii) $\pi^+ + p \to K^+ + \Sigma^+$
   (iv) $K^+ \to \pi^+ + \pi^-$
   (v) $\Sigma^+ \to n + \pi^+$

$\boxed{\text{Ans.}}$ (*i*) In $K^0 \to \pi^+ + \pi^-$

$$S(\text{LHS}) = S(K^0) = +1$$

$$S(\text{RHS}) = S(\pi^+) + S(\pi^-) = 0 + 0 = 0$$

$$S(\text{LHS}) \neq S(\text{RHS}).$$

As strangeness quantum number is not conserved this reaction cannot occur through strong interaction.
(*ii*) $K^0 \to \gamma + \gamma$

$$S(\text{LHS}) = S(K^0) = +1$$

$$S(\text{RHS}) = S(\gamma) + S(\gamma) = 0 + 0 = 0$$

$$S(\text{LHS}) \neq S(\text{RHS}).$$

As strangeness quantum number is not conserved this reaction cannot occur through strong interaction.
(*iii*) $\pi^+ + p \to K^+ + \Sigma^+$

$$S(\text{LHS}) = S(\pi^+) + S(p) = 0 + 0 = 0$$

$$S(\text{RHS}) = S(K^+) + S(\Sigma^+) = 1 - 1 = 0$$

$$S(\text{LHS}) = S(\text{RHS}).$$

As strangeness quantum number is conserved this reaction occurs through strong interaction.

(iv) $K^+ \rightarrow \pi^+ + \pi^-$

$$S(\text{LHS}) = S(K^+) = +1$$

$$S(\text{RHS}) = S(\pi^+) + S(\pi^-) = 0 + 0 = 0$$

$$S(\text{LHS}) \neq S(\text{RHS}).$$

As strangeness quantum number is not conserved this reaction cannot occur through strong interaction. It can proceed through weak interaction.

(v) $\Sigma^+ \rightarrow n + \pi^+$

$$S(\text{LHS}) = S(\Sigma^+) = -1$$

$$S(\text{RHS}) = S(n) + S(\pi^+) = 0 + 0 = 0$$

$$S(\text{LHS}) \neq S(\text{RHS}).$$

As strangeness quantum number is not conserved this reaction cannot occur through strong interaction. It can proceed through weak interaction.

Exercise 8.15 *The reaction $K^- + p \rightarrow \Xi^- + K^+$ occurs through strong interaction. Find the strangeness of Ξ^-.*

Ans. Consider the reaction

$$K^- + p \rightarrow \Xi^- + K^+$$

It occurs through strong interaction in which strangeness is conserved.

$$S(\text{LHS}) = S(\text{RHS})$$

$$S(K^-) + S(p) = S(\Xi^-) + S(K^+)$$

$$-1 + 0 = S(\Xi^-) + 1$$

$$S(\Xi^-) = -1 - 1 = -2$$

Exercise 8.16 *Verify that $\pi^- + p \rightarrow K^0 + \Sigma^0$ is allowed through strong interaction but the reaction $\pi^- + p \rightarrow \eta^0 + \Sigma^0$ is not.*

Ans. Consider the reaction

$$\pi^- + p \rightarrow K^0 + \Sigma^0$$

✓ Charge conservation

$$Q(\text{LHS}) = Q(\pi^-) + Q(p) = -1 + 1 = 0$$

$$Q(\text{RHS}) = Q(K^0) + Q(\Sigma^0) = 0 + 0 = 0$$

$$Q(\text{LHS}) = Q(\text{RHS})$$

Charge is conserved.
✓ Angular momentum conservation

$$\vec{s}(\text{LHS}) = \vec{s}(\pi^-) + \vec{s}(p) = \vec{0} + \frac{\vec{1}}{2} = \frac{\vec{1}}{2}$$

$$\vec{s}(\text{RHS}) = \vec{s}(K^0) + \vec{s}(\Sigma^0) = \vec{0} + \frac{\vec{1}}{2} = \frac{\vec{1}}{2}$$

$$\vec{s}(\text{LHS}) = \vec{s}(\text{RHS})$$

Spin conserved.
✓ Lepton number conservation
Lepton numbers L_e, L_μ, L_τ are conserved since $L = 0$ for all π^-, p, K^0, Σ^0.
✓ Baryon number conservation

$$B(\text{LHS}) = B(\pi^-) + B(p) = 0 + 1 = 1$$

$$B(\text{RHS}) = B(K^0) + B(\Sigma^0) = 0 + 1 = 1$$

$$B(\text{LHS}) = B(\text{RHS})$$

Baryon number conserved.
✓ Strangeness conservation

$$S(\text{LHS}) = S(\pi^-) + S(p) = 0 + 0 = 0$$

$$S(\text{RHS}) = S(K^0) + S(\Sigma^0) = 1 - 1 = 0$$

$$S(\text{LHS}) = S(\text{RHS})$$

Strangeness conserved.
✓ Conservation of isospin projection

$$I_z(\text{LHS}) = I_z(\pi^-) + I_z(p) = -1 + \frac{1}{2} = -\frac{1}{2}$$

$$I_z(\text{RHS}) = I_z(K^0) + Q(\Sigma^0) = -\frac{1}{2} + 0 = -\frac{1}{2}$$

$$I_z(\text{LHS}) = I_z(\text{RHS})$$

Isospin projection conserved.

The reaction is thus allowed through strong interaction

✓ Consider the reaction $\pi^- + p \to \eta + \Sigma^0$

$$S(\text{LHS}) = S(\pi^-) + S(p) = 0 + 0 = 0$$

$$S(\text{RHS}) = S(\eta^0) + S(\Sigma^0) = 0 - 1 = -1$$

$$S(\text{LHS}) \neq S(\text{RHS})$$

Strangeness quantum number not conserved.

The reaction is not allowed through strong interaction.

Exercise 8.17 *Give examples of weak interactions in which*
(i) only hadrons are present;
(ii) both leptons and hadrons are present;
(iii) only leptons are present in the final state.

$\boxed{\text{Ans.}}$ (*i*) Examples of weak interactions in which only hadrons are present

$$\Lambda \to p + \pi^-$$

(ii) Examples of weak interactions in which both leptons and hadrons are present

$$n \to p + e^- + \bar{\nu}_e$$

$$\Lambda \to p + e^- + \bar{\nu}_e$$

$$\Sigma^- \to n + e^- + \bar{\nu}_e$$

$$\pi^+ \to \pi^0 + e^+ + \nu_e$$

$$K^+ \to \pi^0 + \mu^+ + \nu_\mu$$

(*iii*) Examples of weak interactions in which only leptons are present.

$$\mu^- \to e^- + \bar{\nu}_e + \nu_\mu$$

Exercise 8.18 *Give example of weak interactions that occur without neutrino. Check conservation of strangeness and isospin projection.*

$\boxed{\text{Ans.}}$ An example of weak interactions without neutrino is

$$✓K^0 \to \pi^+ + \pi^-$$

Conservation of strangeness

$$S(\text{LHS}) = S(K^0) = +1, \quad S(\text{RHS}) = S(\pi^+ + \pi^-) = 0 + 0 = 0$$

$$S(\text{LHS}) \neq S(\text{RHS})$$

$$I_z(\text{LHS}) = I_z(K^0) = -\frac{1}{2}, \quad I_z(\text{RHS}) = I_z(\pi^+ + \pi^-) = +1 - 1 = 0$$

$$I_z(\text{LHS}) \neq I_z(\text{RHS})$$

✓ $\Sigma^+ \rightarrow p + \pi^0$

$$S(\text{LHS}) = S(\Sigma^+) = -1, \quad S(\text{RHS}) = S(p + \pi^0) = 0 + 0 = 0$$

$$S(\text{LHS}) \neq S(\text{RHS})$$

$$I_z(\text{LHS}) = I_z(\Sigma^+) = +1, \quad I_z(\text{RHS}) = I_z(p + \pi^0) = +\frac{1}{2} + 0 = \frac{1}{2}$$

$$I_z(\text{LHS}) \neq I_z(\text{RHS})$$

✓ $\Xi^- \rightarrow \Lambda + \pi^-$

$$S(\text{LHS}) = S(\Xi^-) = -2, \quad S(\text{RHS}) = S(\Lambda + \pi^-) = -1 + 0 = -1$$

$$S(\text{LHS}) \neq S(\text{RHS})$$

$$I_z(\text{LHS}) = I_z(\Xi^-) = -\frac{1}{2}, \quad I_z(\text{RHS}) = I_z(\Lambda + \pi^-) = 0 - 1 = -1$$

$$I_z(\text{LHS}) \neq I_z(\text{RHS})$$

Clearly in weak interaction isospin and strangeness quantum numbers are not conserved.

Exercise 8.19 *Verify whether isospin, isospin projection and strangeness are conserved in the following electromagnetic reactions.*
    (i) $\pi^0 \rightarrow \gamma + \gamma$
    (ii) $\eta^0 \rightarrow \gamma + \gamma$
    (iii) $\Sigma^0 \rightarrow \Lambda + \gamma$

$\boxed{\text{Ans.}}$ (*i*) $\pi^0 \rightarrow \gamma + \gamma$
✓ Isospin conservation

$$\vec{I}(\text{LHS}) = \vec{I}(\pi^0) = \vec{1}$$

$$\vec{I}(\text{RHS}) = \vec{I}(\gamma + \gamma) = \text{ not defined.}$$

$$\vec{I}(\text{LHS}) \neq \vec{I}(\text{RHS})$$

$\vec{I}$ not conserved.

✓ Conservation of isospin projection

$$I_z(\text{LHS}) = I_z(\pi^0) = 0$$

$$I_z(\text{RHS}) = I_z(\gamma + \gamma) = \text{ not defined} \sim 0$$

$$I_z(\text{LHS}) = I_z(\text{RHS})$$

I_z is conserved.

✓ Conservation of strangeness

$$S(\text{LHS}) = S(\pi^0) = 0$$

$$S(\text{RHS}) = S(\gamma + \gamma) = \text{ not defined} \sim 0.$$

$$S(\text{LHS}) = S(\text{RHS})$$

Strangeness is conserved.

(*ii*) $\eta \to \gamma + \gamma$

✓ Isospin conservation

$$\vec{I}(\text{LHS}) = \vec{I}(\eta^0) = \vec{0}$$

$$\vec{I}(\text{RHS}) = \vec{I}(\gamma + \gamma) = \text{ not defined}.$$

$$\vec{I}(\text{LHS}) \neq \vec{I}(\text{RHS})$$

$\vec{I}$ not conserved.

✓ Conservation of isospin projection

$$I_z(\text{LHS}) = I_z(\eta^0) = 0$$

$$I_z(\text{RHS}) = I_z(\gamma + \gamma) = \text{ not defined} \sim 0.$$

$$I_z(\text{LHS}) = I_z(\text{RHS})$$

✓ Conservation of strangeness

$$S(\text{LHS}) = S(\eta^0) = 0$$

$$S(\text{RHS}) = S(\gamma + \gamma) = \text{ not defined} \sim 0.$$

$$S(\text{LHS}) = S(\text{RHS})$$

(*iii*) $\Sigma^0 \to \Lambda + \gamma$

✓ Isospin conservation

$$\vec{I}(\text{LHS}) = \vec{I}(\Sigma^0) = \vec{1}$$

$$\vec{I}(\text{RHS}) = \vec{I}(\Lambda + \gamma) = \vec{0} + \text{undefined} \sim 0$$

$$\vec{I}(\text{LHS}) \neq \vec{I}(\text{RHS})$$

✓ Conservation of isospin projection

$$I_z(\text{LHS}) = I_z(\Sigma^0) = 0$$

$$I_z(\text{RHS}) = I_z(\Lambda + \gamma) = 0 + \text{undefined} \sim 0$$

$$I_z(\text{LHS}) = I_z(\text{RHS})$$

✓ Conservation of strangeness

$$S(\text{LHS}) = S(\Sigma^0) = -1$$

$$S(\text{RHS}) = S(\Lambda + \gamma) = -1 + \text{undefined} \sim 0$$

$$S(\text{LHS}) = S(\text{RHS})$$

In electromagnetic interaction isospin is not conserved while isospin projection and strangeness are conserved.

Exercise 8.20 *Which of the following weak interactions are possible?*
   (i) $p \rightarrow n + e^+ + \nu_e$
   (ii) $\pi^- + p \rightarrow \pi^0 + \Lambda$

$\boxed{\text{Ans.}}$ (i) $p \rightarrow n + e^+ + \nu_e$

$$\text{Since } m_p = 938 \; MeV < m_n = 939 \; MeV$$

Energy conservation is violated and so the decay is forbidden.
However this decay, i.e. transformation of $p \rightarrow n$, is possible within the nucleus where the deficit energy is supplied by the nucleons of the nucleus. This occurs in β^+ decay.
   (ii) $\pi^- + p \rightarrow \pi^0 + \Lambda$
✓ Charge conservation

$$Q(\text{LHS}) = Q(\pi^- + p) = -1 + 1 = 0$$

$$Q(\text{RHS}) = Q(\pi^0 + \Lambda) = 0 + 0 = 0,$$

$$Q(\text{LHS}) = Q(\text{RHS})$$

Charge is conserved.
✓ Spin conservation

$$\vec{s}(\text{LHS}) = \vec{s}(\pi^- + p) = \vec{0} + \frac{\vec{1}}{2} = \frac{\vec{1}}{2}$$

$$\vec{s}(\text{RHS}) = \vec{s}(\pi^0 + \Lambda) = \vec{0} + \frac{\vec{1}}{2} = 0$$

$$\vec{s}(\text{LHS}) = \vec{s}(\text{RHS})$$

Spin is conserved.
✓ Lepton number conservation
$L(\text{LHS}) = L(\text{RHS})$ as non-leptons are involved.
L is conserved.
✓ Baryon number conservation

$$B(\text{LHS}) = B(\pi^- + p) = 0 + 1 = 1$$

$$B(\text{RHS}) = B(\pi^0 + \Lambda) = 0 + 1 = 1,$$

$$B(\text{LHS}) = B(\text{RHS})$$

B is conserved.
✓ Strangeness conservation

$$S(\text{LHS}) = S(\pi^- + p) = 0 + 0 = 0$$

$$S(\text{RHS}) = S(\pi^0 + \Lambda) = 0 - 1 = -1,$$

$$S(\text{LHS}) \neq S(\text{RHS})$$

Strangeness is not conserved.
✓ Conservation of isospin projection

$$I_3(\text{LHS}) = I_3(\pi^- + p) = -1 + \frac{1}{2} = -\frac{1}{2}$$

$$I_3(\text{RHS}) = I_3(\pi^0 + \Lambda) = 0 + 0 = 0,$$

$$I_3(\text{LHS}) \neq I_3(\text{RHS})$$

Isospin projection is not conserved.
This reaction is possible through weak interaction.
(*iii*) $K^- + p \rightarrow K^0 + n$
✓ Charge conservation

$$Q(\text{LHS}) = Q(K^- + p) = -1 + 1 = 0$$

$$Q(\text{RHS}) = Q(K^0 + n) = 0 + 0 = 0$$

$$Q(\text{LHS}) = Q(\text{RHS})$$

Charge is conserved.
✓ Spin conservation

$$\vec{s}(\text{LHS}) = \vec{s}(K^- + p) = \vec{0} + \frac{\vec{1}}{2} = \frac{\vec{1}}{2}$$

$$\vec{s}\,(\text{RHS}) = \vec{s}\,(K^0 + n) = \vec{0} + \frac{\vec{1}}{2} = \frac{\vec{1}}{2}$$

$$\vec{s}\,(\text{LHS}) = \vec{s}\,(\text{RHS})$$

spin conserved.
   ✓ Lepton number conservation
$L(\text{LHS}) = L(\text{RHS})$ as non-leptons are involved.
L is conserved.
   ✓ Baryon number conservation

$$B(\text{LHS}) = B(K^- + p) = 0 + 1 = 1$$

$$B(\text{RHS}) = B(K^0 + n) = 0 + 1 = 1$$

$$B(\text{LHS}) = B(\text{RHS})$$

B conserved.
   ✓ Strangeness conservation

$$S(\text{LHS}) = S(K^- + p) = -1 + 0 = -1$$

$$S(\text{RHS}) = S(K^0 + n) = +1 + 0 = 1,$$

$$S(\text{LHS}) \neq S(\text{RHS})$$

Strangeness quantum number not conserved.
   ✓ Conservation of isospin projection

$$I_3(\text{LHS}) = I_3(K^- + p) = -\frac{1}{2} + \frac{1}{2} = 0$$

$$I_3(\text{RHS}) = I_3(K^0 + n) = -\frac{1}{2} - \frac{1}{2} = -1,$$

$$I_3(\text{LHS}) \neq I_3(\text{RHS})$$

Isospin projection is not conserved.
This reaction is possible through weak interaction.

Exercise 8.21 *Which of the interactions shown in* figure 8.31 *is not allowed?*
   (a) $e^- e^+ \xrightarrow{\gamma} \mu^- \mu^+$
   (b) $e^- e^+ \xrightarrow{Z^0} \mu^- \mu^+$
   (c) $e^- e^+ \xrightarrow{Z^0} \nu_e \bar{\nu}_e$
   (d) $e^- e^+ \xrightarrow{\gamma} \nu_e \bar{\nu}_e$
   (e) $e^- e^+ \xrightarrow{Z^0} \nu_\mu \bar{\nu}_\mu$

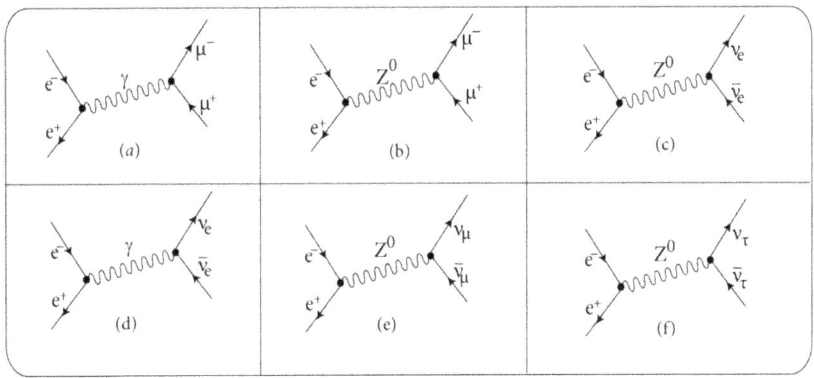

Figure 8.31. Pertaining to exercise 8.21.

(f) $e^- e^+ \xrightarrow{Z^0} \nu_\tau \bar{\nu}_\tau$

$\boxed{\text{Ans.}}$ (a) $e^- e^+ \xrightarrow{\gamma} \mu^- \mu^+$ is an electromagnetic interaction where e^- and e^+ interact through exchange of γ.

(b) $e^- e^+ \xrightarrow{Z^0} \mu^- \mu^+$ is a weak interaction where e^- and e^+ interact through exchange of weak gauge boson Z^0 which is the propagator of weak neutral current interaction. There is no charge change along the current line.

(c) $e^- e^+ \xrightarrow{Z^0} \nu_e \bar{\nu}_e$ is a weak interaction where e^- and e^+ interact through exchange of weak gauge boson Z^0

(d) $e^- e^+ \xrightarrow{\gamma} \nu_e \bar{\nu}_e$ is not possible since γ cannot directly interact with neutral fermions $\nu_e, \bar{\nu}_e$. A photon interacts with only charged particles. So it is not allowed.

(e) $e^- e^+ \xrightarrow{Z^0} \nu_\mu \bar{\nu}_\mu \bar{\nu}_e$ is a weak interaction where e^- and e^+ interact through exchange of weak gauge boson Z^0.

(f) $e^- e^+ \xrightarrow{Z^0} \nu_\tau \bar{\nu}_\tau \bar{\nu}_e$ is a weak interaction where e^- and e^+ interact through exchange of weak gauge boson Z^0.

Exercise 8.22 *Are the following reactions allowed?*
 (i) $\Sigma^0 \to \pi^0 + \gamma$
 (ii) $\Omega^- \to \Xi^- + \pi^0$
 (iii) $n \to e^+ + e^-$
 (iv) $p + \bar{p} \to p + p + \bar{p}$
 (v) $\pi^- \to \mu^- + \nu_\mu + \bar{\nu}_\mu$

$\boxed{\text{Ans.}}$ (*i*) Consider $\Sigma^0 \to \pi^0 + \gamma$
Check spin conservation

$$\vec{s}\,(\text{LHS}) = \vec{s}\,(\Sigma^0) = \frac{\vec{1}}{2}$$

$$\vec{s}\,(\text{RHS}) = \vec{s}\,(\pi^0 + \gamma) = \vec{0} + \vec{1} = 1$$

$$\vec{s}\,(\text{LHS}) \neq \vec{s}\,(\text{RHS})$$

Spin $\vec{s}$ is not conserved. Reaction forbidden.
(ii) Consider $\Omega^- \rightarrow \Xi^- + \pi^0$
Check spin conservation

$$\vec{s}\,(\text{LHS}) = \vec{s}\,(\Omega^-) = \frac{\vec{3}}{2}$$

$$\vec{s}\,(\text{RHS}) = \vec{s}\,(\Xi^- + \pi^0) = \frac{\vec{1}}{2} + \vec{0} = \frac{\vec{1}}{2}$$

$$\vec{s}\,(\text{LHS}) \neq \vec{s}\,(\text{RHS})$$

Spin $\vec{s}$ is not conserved. Reaction forbidden.
(*iii*) $n \rightarrow e^+ + e^-$
Check baryon number conservation

$$B(\text{LHS}) = B(n) = 1$$

$$B(\text{RHS}) = B(e^+ + e^-) = 0 + 0 = 0,$$

$$B(\text{LHS}) \neq B(\text{RHS})$$

Baryon number is not conserved. Reaction forbidden.
(iv) $p + \bar{p} \rightarrow p + p + \bar{p}$

$$B(\text{LHS}) = B(p + \bar{p}) = 1 - 1 = 0$$

$$B(\text{RHS}) = B(p + p + \bar{p}) = 1 + 1 - 1 = 1$$

$$B(\text{LHS}) \neq B(\text{RHS})$$

Baryon number is not conserved. Reaction forbidden.
(v) $\pi^- \rightarrow \mu^- + \nu_\mu + \bar{\nu}_\mu$
Check conservation of lepton number

$$L_\mu(\text{LHS}) = L_\mu(\pi^-) = 0$$

$$L_\mu(\text{RHS}) = L_\mu(\mu^- + \nu_\mu + \bar{\nu}_\mu) = +1 - 1 + 1 = 1$$

$$L_\mu(\text{LHS}) \neq L_\mu(\text{RHS}).$$

Baryon number is not conserved. Reaction forbidden.

Exercise 8.23 *Is the reaction $p + \bar{p} \to \pi^+ + \pi^- + \pi^0$ invariant under C operation?*

Ans. $C(p + \bar{p} \to \pi^+ + \pi^- + \pi^0) = \bar{p} + p \to \pi^- + \pi^+ + \pi^0$

Clearly under C operation the given reaction is invariant.

Exercise 8.24 *Check conservation of angular momentum in the reaction*

$$n \to p + e^- + \overline{\nu_e}$$

Ans. Consider the reaction $n \to p + e^- + \overline{\nu_e}$

$$\vec{s}\,(\text{LHS}) = \vec{s}\,(n) = \frac{\vec{1}}{2}$$

$$\vec{s}\,(\text{RHS}) = \vec{s}\,(p + e^- + \overline{\nu_e}) = \frac{\vec{1}}{2} + \frac{\vec{1}}{2} + \frac{\vec{1}}{2} = \frac{\vec{1}}{2} + \left\{ \begin{matrix} \vec{0} \\ \vec{1} \end{matrix} \right. = \left\{ \begin{matrix} \frac{1}{2} \\ \frac{1}{2}, \frac{3}{2} \end{matrix} \right.$$

$$\vec{s}\,(\text{LHS}) = \vec{s}(\text{RHS})$$

Spin $\vec{s}$ is conserved.

Exercise 8.25 *Verify if the following can proceed through strong reaction?*
   (i) $p + p \to p + n + K^+$
   (ii) $p + p \to \Lambda + K^0 + p + \pi^+$

Ans. (*i*) $p + p \to p + n + K^+$
Consider conservation of isospin projection

$$I_z(\text{LHS}) = I_z(p + p) = \frac{1}{2} + \frac{1}{2} = 1$$

$$I_z(\text{RHS}) = I_z(p + n + K^+) = \frac{1}{2} - \frac{1}{2} + \frac{1}{2} = \frac{1}{2}$$

$$I_z(\text{LHS}) \neq I_z(\text{RHS})$$

Reaction cannot proceed through strong interaction.
(*ii*) $p + p \to \Lambda + K^0 + p + \pi^+$
✓ Check charge conservation

$$Q(\text{LHS}) = Q(p + p) = 1 + {}^{\backprime}1 = 2$$

$$Q(\text{RHS}) = Q(\Lambda + K^0 + p + \pi^+) = 0 + 0 + 1 + 1 = 2$$

$$Q(\text{LHS}) = Q(\text{RHS})$$

Charge is conserved.

✓ Check spin conservation

$$\vec{s}(\text{LHS}) = \vec{s}(p + p) = \frac{\vec{1}}{2} + \frac{\vec{1}}{2} = 1, 0$$

$$\vec{s}(\text{RHS}) = \vec{s}(\Lambda + K^0 + p + \pi^+) = \frac{\vec{1}}{2} + \vec{0} + \frac{\vec{1}}{2} + \vec{0} = 1, 0$$

$$\vec{s}(\text{LHS}) = \vec{s}(\text{RHS})$$

Spin is conserved.
✓ Check Lepton number conservation
$L(\text{LHS}) = L(\text{RHS})$ as non-leptons are involved. L is conserved.
✓ Check baryon number conservation

$$B(\text{LHS}) = B(p + p) = 1 + 1 = 2$$

$$B(\text{RHS}) = B(\Lambda + K^0 + p + \pi^+) = 1 + 0 + 1 + 0 = 2$$

$$B(\text{LHS}) = B(\text{RHS})$$

B conserved.
✓ Check strangeness conservation

$$S(\text{LHS}) = S(p + p) = 0 + 0 = 0$$

$$S(\text{RHS}) = S(\Lambda + K^0 + p + \pi^+) = -1 + 1 + 0 + 0 = 0$$

$$S(\text{LHS}) = S(\text{RHS})$$

This reaction is possible through strong interaction.

Exercise 8.26 *Explain why a Λ hyperon does not decay into a $\pi^\pm$ meson.*
Ans. Consider the reaction

$$\Lambda \rightarrow \pi^+ + \pi^-$$

Check baryon number conservation.

$$B(\text{LHS}) = B(\Lambda) = 1$$

$$B(\text{RHS}) = B(\pi^+ + \pi^-) = 0 + 0 = 0$$

Baryon number is violated. So the reaction does not occur.

Exercise 8.27 *Identify the particle x in the given reactions.*
   (i) $\mu^- + p \rightarrow x + n$
   (ii) $K^- + p \rightarrow K^+ + x$

(iii) $\pi^- + n \rightarrow K^0 + x$

(iv) $\pi^- + p \rightarrow K^0 + x$

$\boxed{\text{Ans.}}$ (*i*) $\mu^- + p \rightarrow x + n$ (weak interaction)

✓ Charge is conserved.

$$Q(x) = Q(\mu^-) + Q(p) - Q(n) = -1 + 1 - 0 = 0 = \text{charge of } x$$

✓ Baryon number is conserved.

$$B(x) = B(\mu^-) + B(p) - B(n) = 0 + 1 - 1 = 0 = \text{baryon number of } x$$

✓ Spin is conserved.

$$\vec{s}(x) = \vec{s}(\mu^-) + \vec{s}(p) - \vec{s}(n) = \frac{\vec{1}}{2} + \frac{\vec{1}}{2} - \frac{\vec{1}}{2} = \frac{\vec{1}}{2} = \text{spin of } x$$

✓ Muon lepton number is conserved.

$$L_\mu(x) = L_\mu(\mu^-) + L_\mu(p) - L_\mu(n) = +1 + 0 - 0 = +1 = \text{muon lepton number of } x$$

We identify such a particle to be

$$x \equiv \nu_\mu \text{ having } Q = 0, \ B = 0, \ \vec{s} = \frac{\vec{1}}{2}, \ L_\mu = +1$$

(*ii*) $K^- + p \rightarrow K^+ + x$

✓ Charge is conserved.

$$Q(x) = Q(K^-) + Q(p) - Q(K^+) = -1 + 1 - 1 = -1 = \text{charge of } x$$

✓ Baryon number is conserved.

$$B(x) = B(K^-) + B(p) - B(K^+) = 0 + 1 - 0 = +1 = \text{baryon number of } x$$

✓ Spin is conserved.

$$\vec{s}(x) = \vec{s}(K^-) + \vec{s}(p) - \vec{s}(K^+) = \vec{0} + \frac{\vec{1}}{2} - \vec{0} = \frac{\vec{1}}{2} = \text{spin of } x$$

✓ Lepton number is conserved

$$L(x) = L(K^-) + L(p) - L(K^+) = 0 + 0 - 0 = 0 = \text{lepton number of } x$$

✓ Isospin projection is conserved.

$$I_3(x) = I_3(K^-) + I_3(p) - I_3(K^+) = -\frac{1}{2} + \frac{1}{2} - \frac{1}{2} = -\frac{1}{2} = \text{isospin projection of } x$$

Such a particle is

$$x \equiv \Xi^- \text{ having } Q = -1, \ B = +1, \ \vec{s} = \frac{\vec{1}}{2}, \ L = 0, \ I_3 = -\frac{1}{2}$$

(*iii*) $\pi^- + n \rightarrow K^0 + x$

✓ Charge is conserved

$$Q(x) = Q(\pi^-) + Q(n) - Q(K^0) = -1 + 0 - 0 = -1 = \text{charge of } x$$

✓ Baryon number is conserved.

$$B(x) = B(\pi^-) + B(n) - B(K^0) = 0 + 1 - 0 = +1 = \text{baryon number of } x$$

✓ Spin is conserved

$$\vec{s}(x) = \vec{s}(\pi^-) + \vec{s}(n) - \vec{s}(K^0) = \vec{0} + \frac{\vec{1}}{2} - \vec{0} = \frac{\vec{1}}{2} = \text{spin of } x$$

✓ Lepton number is conserved.

$$L(x) = L(\pi^-) + L(n) - L(K^0) = 0 + 0 - 0 = 0 = \text{lepton number of } x$$

✓ Isospin projection is conserved.

$$I_3(x) = I_3(\pi^-) + I_3(n) - I_3(K^0) = -1 - \frac{1}{2} - (-\frac{1}{2}) = -1 = \text{isospin projection of } x$$

Such a particle is

$$x \equiv \Sigma^- \text{ having } Q = -1, \ B = +1, \ \vec{s} = \frac{\vec{1}}{2}, \ L = 0, \ I_3 = -1$$

(iv) $\pi^- + p \rightarrow K^0 + x$

✓ Charge is conserved.

$$Q(x) = Q(\pi^-) + Q(p) - Q(K^0) = -1 + 1 - 0 = 0 = \text{charge of } x$$

✓ Baryon number is conserved.

$$B(x) = B(\pi^-) + B(p) - B(K^0) = 0 + 1 - 0 = +1 = \text{baryon number of } x$$

✓ Spin is conserved.

$$\vec{s}(x) = \vec{s}(\pi^-) + \vec{s}(p) - \vec{s}(K^0) = \vec{0} + \frac{\vec{1}}{2} - \vec{0} = \frac{\vec{1}}{2} = \text{spin of } x$$

✓ Lepton number is conserved.

$$L(x) = L(\pi^-) + L(p) - L(K^0) = 0 + 0 - 0 = 0 = \text{lepton number of } x$$

✓ isospin projection is conserved.

$$I_3(x) = I_3(\pi^-) + I_3(p) - I_3(K^0) = -1 + \frac{1}{2} - (-\frac{1}{2}) = 0 = \text{isospin projection of } x$$

Such a particle is

$$x \equiv \lambda \text{ or } \Sigma^0 \text{ having } Q = 0, \ B = +1, \ \vec{s} = \frac{\vec{1}}{2}, \ L = 0, \ I_3 = 0$$

Exercise 8.28 *A hadron has the quark content $u\bar{d}$. Find baryon number, charge, spin, strangeness. Identify the hadron.*

$\boxed{\text{Ans.}}$ $B(u\bar{d}) = B(u) + B(\bar{d}) = \dfrac{1}{3} - \dfrac{1}{3} = 0, \quad Q(u\bar{d}) = Q(u) + Q(\bar{d}) = \dfrac{2}{3} + \dfrac{1}{3} = 1,$

$\vec{s}(u\bar{d}) = \vec{s}(u) + \vec{s}(\bar{d}) = \dfrac{\vec{1}}{2} + \dfrac{\vec{1}}{2} = 1, 0, \quad S(u\bar{d}) = S(u) + S(\bar{d}) = 0 + 0 = 0$

We identify $u\bar{d} \equiv \pi^+$ (spin 0)

Exercise 8.29 *Establish that for a 2-pion system the $\pi^+\pi^+$ combination is equivalent to $\pi^-\pi^-$ combination so far as strong interaction is concerned.*

$\boxed{\text{Ans.}}$ A pion is an isotriplet having $I = 1$ and has isospin projections $1, 0, -1$.

$|I\ I_z> \ =|1\ 1> \ =|\ \pi^+> , \quad |I\ I_z> \ =|1\ 0> \ =|\ \pi^0> , \quad |I\ I_z> \ =|1-1> \ =|\ \pi^->$

Hence for $\pi^+\pi^+$ $I_z = 1 + 1 = 2$ and for $\pi^-\pi^-$ $I_z = -1 - 1 = -2$

The iso-projections $I_z = \pm 1$ follow from $I = 2$. This means that $\pi^+\pi^+$ and $\pi^-\pi^-$ both correspond to $I = 2$ and hence are equivalent. Actually

$|\ \pi^+\pi^+> \ = |\ I = 2\ I_z = 2>$ that can be rewritten as $|11 > |\ 11 > \ =|\ 22>$

$|\ \pi^-\pi^- = \ |\ I = 2\ I_z = -2>$ that can be rewritten as $|1-1>|1-1> \ =|2-2>.$

Exercise 8.30 *Analyse the reaction $\pi^- + p \to \Lambda + K^0$ in terms of quarks.*

$\boxed{\text{Ans.}}$ Consider the reaction $\pi^- + p \to \Lambda + K^0$. Let us rewrite it in terms of quarks as follows

$$\bar{u}d + uud \to uds + d\bar{s}$$

Both sides have one u quark and two d quarks in common. They do not change in the reaction. The residual quark reaction are

$$\bar{u} + u \to s + \bar{s}$$

Obviously the quark–anti-quark pair u and $\bar{u}$ annihilate each other into energy which creates the other quark–anti-quark pair s and $\bar{s}$.

Exercise 8.31 *Identify the correct decay scheme*

(a) $\mu^- \to e^- + \bar{\nu}_e$ (b) $\mu^- \to e^- + \bar{\nu}_\mu + \nu_e$

(c) $\mu^- \to e^- + \bar{\nu}_e + \nu_\mu$ (d) $\mu^- \to e^- + \bar{\nu}_\mu + \nu_\mu$

Exercise 8.32 *Identify the particle having zero baryon number.*

(a) *neutron*   (b) *proton*   (c) *pion*   (d) Δ^+.

Exercise 8.33 *A neutron at rest in free space decaying into a proton and an electron would violate which of the conservation laws?*

(a) *angular momentum*   (b) *energy*   (c) *charge*   (d) *linear momentum.*

Exercise 8.34 Consider the reaction

$$K^- + p \rightarrow \Sigma^- + \pi^+ + \pi^- + \pi^+.$$

The incident kaon has kinetic energy 1.63 *GeV*. Find average energy per product particle if masses are

$$K^-(0.4938\ GeV), \quad p(938.3\ MeV), \quad \Sigma^-(1197.3\ MeV), \quad \pi(139.6\ MeV).$$

$\boxed{\text{Ans.}}$ Consider the reaction

$$K^- + p \rightarrow \Sigma^- + \pi^+ + \pi^- + \pi^+$$

$$E(\text{LHS}) = M(K^-) + M(p) + T_i$$
$$= 0.4938\ GeV + 9383\ GeV + 1.63\ GeV = 3.0621\ GeV$$

$$E(\text{RHS}) = M(\Sigma^-) + M(\pi^+) + M(\pi^-) + M(\pi^+)$$
$$= 1.1973\ GeV + 0.1396\ GeV + 0.1396\ GeV + 0.1396\ GeV = 1.6161\ GeV$$

$$\text{Excess energy} = E(\text{LHS}) - E(\text{RHS}) = 3.0621\ GeV - 1.6161\ GeV = 1.446\ GeV$$

$$\text{Average energy per product particle} = \frac{1.446\ \text{GeV}}{4} = 0.3615\ GeV \equiv =361.5\ MeV$$

Exercise 8.35 *A pion decays from rest to give a muon of energy 4MeV. What is the energy of the accompanying neutrino?*

$\boxed{\text{Ans.}}$ Consider the reaction

$$\pi^+ \rightarrow \mu^+ + \nu_\mu$$

$$E_\pi = E_\mu + E_\nu$$

$$E_\nu = E_\pi - E_\mu = (m_\pi c^2 + T_\pi) - (m_\mu c^2 + T_\mu)$$
$$= (m_\pi c^2 + T_\pi) - (m_\mu c^2 + T_\nu)$$
$$= (273 m_e + 0) - (207 m_e + 4\ MeV)$$
$$= 66 m_e - 4\ MeV = 66 \times 0.51\ MeV - 4\ MeV$$

$$E_\nu = m_\nu c^2 + T_\nu = 33.7\ MeV - 4\ MeV = 29.7\ MeV$$

Exercise 8.36 *Obtain the third component of isospin of Ξ^- from the strong interaction given by*

$$\pi^+ + n \rightarrow \Xi^- + K^+ + K^-$$

Ans. Consider the interaction

$$\pi^+ + n \rightarrow \Xi^- + K^+ + K^+$$

$$I_3(\text{LHS}) = I_3(\pi^+) + I_3(n) = 1 - \frac{1}{2} = \frac{1}{2}$$

$$I_3(\text{RHS}) = I_3(\Xi^-) + I_3(K^+) + I_3(K^+) = I_3(\Xi^-) + \frac{1}{2} + \frac{1}{2} = I_3(\Xi^-) + 1$$

Since strong interaction conserves I_3 we have

$$I_3(\text{LHS}) = I_3(\text{RHS})$$

$$\frac{1}{2} = I_3(\Xi^-) + 1$$

$$I_3(\Xi^-) = \frac{1}{2} - 1 = -\frac{1}{2}$$

Answers to multiple choice questions
8.31*c*, 8.32*c*, 8.33 *a,b*.

8.61 Question bank

Q8.1 What is the unitary operator for space translation?

Q8.2 Show that the linear momentum operator is the generator for space translation.

Q8.3 What is the unitary operator for rotation?

Q8.4 Show that the z component of angular momentum operator is the generator for rotation.

Q8.5 What do we mean by the terms invariance and symmetry?

Q8.6 Give examples of continuous and discrete symmetry.

Q8.7 What is the conservation law for space-time invariance, rotation invariance?

Q8.8 Mention Noether's theorem.

Q8.9 How do we explain stability of the electron?

Q8.10 How do we explain stability of the proton?

Q8.11 How are hadrons classified?

Q8.12 What is the baryon number of anti-quarks?

Q8.13 An electron neutrino is always emitted with the emission of an electron. Explain this from conservation of the lepton number.

Q8.14 Explain isospin symmetry of proton and neutron.

Q8.15 Write down the Gell-Mann–Okubo formula. Show that it holds for nucleons, pions but not for strange particles.

Q8.16 Obtain quark structure for pions.

Q8.17 What are the strange properties of strange particles?

Q8.18 Show a meson octet in I_3 versus S plane.

Q8.19 Show a baryon octet in I_3 versus S plane.

Q8.20 Show a baryon decuplet in I_3 versus S plane

Q8.21 Define hypercharge. Write down the hypercharge for hadrons.

Q8.22 Write down the Gell-Mann–Nishijima formula. Show that it holds for nucleons, pions and strange particles.

Q8.23 How can charge be evaluated from the Gell-Mann–Nishijima formula?

Q8.24 What is CPT symmetry?

Q8.25 Show that charge conjugation operator is unitary, Hermitian.

Q8.26 What are the eigenvalues of charge conjugation operator?

Q8.27 What do we mean by a self-conjugate or truly neutral particle?

Q8.28 Define parity.

Q8.29 Find the eigenvalues of a parity operator.

Q8.30 How do you identify a scalar, pseudoscalar, vector and a pseudovector?

Q8.31 Obtain parity of spherically symmetric potential.

Q8.32 Explain intrinsic parity.

Q8.33 Explain helicity or handedness.

Q8.34 What is $\tau - \theta$ puzzle?

Q8.35 What was Lee and Yang's explanation regarding parity violation in weak decay?

Q8.36 What was the experimental evidence of parity violation in weak decay?

Q8.37 What do we mean by parity violation in weak decay?

Q8.38 Explain time reversal symmetry.

Q8.39 Obtain the condition of time reversal invariance of the Schrödinger equation.

Q8.40 Show that the time reversal operator is anti-linear and anti-unitary.

Q8.41 What is CP symmetry? How is it violated in kaon decay?

Q8.42 What is *G* parity?

Q8.43 Obtain *G* parity for an *n* pion system.

Q8.44 Write down the Gell-Mann–Okubo mass formula. Obtain mass formula for a baryon octet and a meson octet.

Q8.45 What do we mean by current–current interaction in the case of weak decay?

Q8.46 Which conservation laws are violated and conserved in strong, electromagnetic and weak interaction?

Further reading

[1] Perkins H 2000 *Introduction to High Energy Physics* (Cambridge: Cambridge University Press)

[2] Griffiths D 2008 *Introduction to Elementary Particles* (NewYork: Wiley-VCH)

[3] Tayal D C 2009 *Nuclear Physics* (Mumbai: Himalaya Publishing House)

[4] Prakash S 2005 *Nuclear Physics & Particle Physics* (New Delhi: Sultan Chand & Sons)

[5] Guha J 2019 *Quantum Mechanics: Theory, Problems & Solutions* 3rd edn (Kolkata: Books and Allied (P) Ltd)

[6] Lim Y K 2002 *Problems and Solutions on Atomic, Nuclear and Particle Physics* (Singapore: World Scientific)

IOP Publishing

Nuclear and Particle Physics with Cosmology, Volume 2
Particle physics and cosmology
Jyotirmoy Guha

Chapter 9

A brief introduction to group theory

In this chapter we define what is a group and elaborate on its characteristics, giving examples of discrete and continuous groups. Abelian and non-abelian properties are discussed. We discuss order, rank and generator of a group. We derive the generator of a space translation group, time translation operation and rotation. We discuss orthogonal matrix and unitary matrix. Symmetry representation of $SO(n)$ group and generator of $SO(2)$, $SO(3)$ are discussed. Symmetry representation of the $SU(n)$ group and generator of $SU(2)$ and $SU(3)$ are discussed. Gell-Mann matrices are derived. We introduce Young's diagram. Young tableaux are discussed in detail with several examples.

9.1 Introduction

Particle physics, and in particular the Standard Model, has quite a few conservation laws. Presence of conservation law means there is symmetry. In the Standard Model we represent these symmetries as groups. In other words, symmetry plays a crucial role in group theory.

9.2 Group

A group G is essentially a set of elements or members, say $\{A, B, C, D, \ldots\}$ and a group operation (denoted by $\otimes$).

The group operation has to be defined. It might be, for instance, operation by some binary operator like addition, subtraction, multiplication or division. So group G is the elements plus group operation, i.e.

$$G: \{A, B, C, D, \ldots\} + \otimes \tag{9.1}$$

Also, the following four conditions have to be satisfied.
    (1) Closure property
        A group is a closed group, i.e. satisfies closure property.

doi:10.1088/978-0-7503-5032-7ch9

If A and B are members of the group then the group operation $\otimes$ generates another element of the same group, i.e. if A operates upon B to generate C, i.e.

$$A \otimes B = C$$

then C is also a member of the same group.

(2) Associative property

Group elements obey associative property, i.e.

$$A \otimes (B \otimes C) = (A \otimes B) \otimes C \tag{9.2}$$

(3) Occurrence of identity element

A group has an identity element E such that for any member A of the group the following relation is obeyed

$$A \otimes E = E \otimes A = A \tag{9.3}$$

i.e. we get back the group element after group operation with identity element E performed on it.

(4) Occurrence of an inverse element

For each element A of the group there is an inverse element B in the group such that

$$A \otimes B = B \otimes A = E \tag{9.4}$$

where E is inverse element of the same group, i.e. A, B, E all are members of the same group. So identity element E can be generated by a group element through group operation with its inverse element.

Abelian or non-abelian is an optional property of a group that we discuss now.

9.3 Abelian property

If for all members A, B, ...of a group

$$A \otimes B = B \otimes A \tag{9.5}$$

i.e group multiplication is commutative, i.e. order independent, then the group is called abelian.

9.4 Non-abelian property

If for any one pair of elements we find that group operation is non-commutative, i.e.

$$A \otimes B \neq B \otimes A \tag{9.6}$$

then the group is called non-abelian.

9.5 Explanation of group properties

Consider a group formed by all the integers $-\infty$, ... , 0, ...∞ and suppose the group operation is that of addition, i.e. $\otimes \equiv +$.

Let us check the four conditions.

✓ Closure property

If we add any two integers it leads to another member of the group, e.g.

$$2 + 3 = 5$$

and 5 is also a member of the group. So closure property is obeyed.

✓ Associative property

$$1 + (2 + 3) = (1 + 2) + 3 \text{ (similar to equation (9.2))}$$

✓ Occurrence of identity element

Consider an element of group say 7. We note that

$$7 + 0 = 0 + 7 = 7$$

i.e. we get back 7 after performing the group operation of addition. Comparing with equation (9.3) viz. $A \otimes E = E \otimes A = A$ we identify $A \equiv 7$ and $E \equiv 0$ as the identity element for the integers under addition, i.e.

$$7 + E = E + 7 = 7 \text{ with } E \equiv 0$$

✓ Occurrence of inverse element

Consider an element of a group, say the positive integer 7. We note that

$$7 + (-7) = (-7) + 7 = 0$$

Comparing with equation (9.4) viz. $A \otimes B = B \otimes A = E$ we identify $A \equiv 7$ and $B \equiv -7$ which is the inverse of $A = 7$ since 7 and -7 can be combined through group operation of addition to generate identity element $E = 0$.

So the inverse of every positive integer is the same integer with negative sign and obviously there is an inverse for every element.

As the four conditions are satisfied, the integers form a group under addition.

✓ Abelian property

For any two group elements say 5 and 4 we have

$$5 + 4 = 4 + 5 \text{ (the group operation being } \otimes = +)$$

In other words, commutation relation is obeyed by the elements of the group and hence the group is an abelian group.

9.6 Order of a group

Order of a group is the number of independent parameters needed to specify its fundamental representation. In fact order of group is the number of elements in the group.

9.7 Discrete group and continuous group

All discrete and continuous groups can be represented by square matrices.

A discrete group is said to be of order n if the group possesses n elements.

A continuous group is said to be of order n if the elements of the group are defined by n independent parameters.

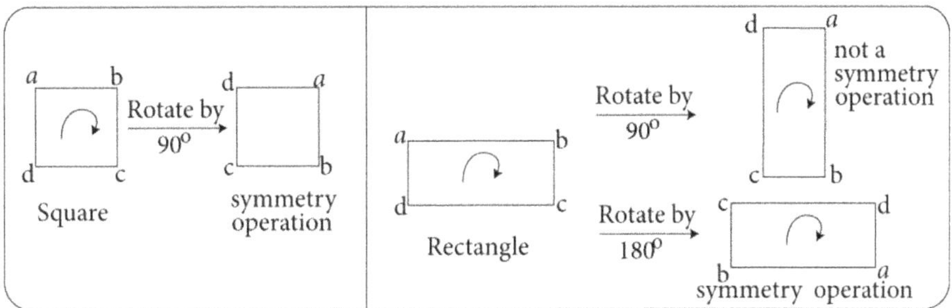

Figure 9.1. Rotating a square and rectangle by 90°, 180° about the centre of a square.

9.8 Example of a discrete group

Consider a square as shown in figure 9.1.

A rotation about the centre of a square by 90° keeps the square in its original position, i.e. it falls upon itself as if no rotation has been made. Clearly we cannot distinguish whether rotation by 90° has been made or not. So rotation by 90° $\equiv \frac{\pi}{2}$ is a symmetry operation. Let us denote rotation by $\frac{\pi}{2}$ as $R_{\pi/2}$.

Other symmetry operations of a square are rotations by $180° \equiv 2\pi$, $270° \equiv \frac{3\pi}{2}$, $360° \equiv 2\pi$ or 0° (this is identity operation denoted by E).

However, if we make a rotation by 89° or 91° we end up with a tilted square. Since now the square changes its position, rotation by 89° or 91° are not symmetry operations.

The symmetry operations for a square are the following four distinct operations of rotation

$$E \equiv R_{0_\circ \text{ or } 2\pi}, \; R_{\pi/2}, \; R_\pi, \; R_{3\pi/2} \tag{9.7}$$

We can represent rotation of angle θ by the rotation matrix

$$R_\theta = \begin{pmatrix} \cos\theta & \sin\theta \\ -\sin\theta & \cos\theta \end{pmatrix} \tag{9.8}$$

Hence

$$R_{\pi/2} = \begin{pmatrix} \cos\frac{\pi}{2} & \sin\frac{\pi}{2} \\ -\sin\frac{\pi}{2} & \cos\frac{\pi}{2} \end{pmatrix} = \begin{pmatrix} 0 & 1 \\ -1 & 0 \end{pmatrix}, \quad R_\pi = \begin{pmatrix} \cos\pi & \sin\pi \\ -\sin\pi & \cos\pi \end{pmatrix} = \begin{pmatrix} -1 & 0 \\ 0 & -1 \end{pmatrix}$$

$$R_{3\pi/2} = \begin{pmatrix} \cos\frac{3\pi}{2} & \sin\frac{3\pi}{2} \\ -\sin\frac{3\pi}{2} & \cos\frac{3\pi}{2} \end{pmatrix} = \begin{pmatrix} 0 & -1 \\ 1 & 0 \end{pmatrix}, \quad R_{2\pi} = \begin{pmatrix} \cos 2\pi & \sin 2\pi \\ -\sin 2\pi & \cos 2\pi \end{pmatrix} = \begin{pmatrix} 1 & 0 \\ 0 & 1 \end{pmatrix} = E$$

Let us check whether the four elements of equation (9.7) together form a group under the group operation $\otimes \equiv$ rotation about the centre of a square.

We have to check the following four essential properties.

✓ Closure property

$$R_{\pi/2}R_{\pi/2} = \begin{pmatrix} 0 & 1 \\ -1 & 0 \end{pmatrix}\begin{pmatrix} 0 & 1 \\ -1 & 0 \end{pmatrix} = \begin{pmatrix} -1 & 0 \\ 0 & -1 \end{pmatrix} = R_{\pi}$$

i.e. two rotations by $\frac{\pi}{2}$ is equivalent to one rotation by π.

$$R_{\pi/2}R_{\pi} = \begin{pmatrix} 0 & 1 \\ -1 & 0 \end{pmatrix}\begin{pmatrix} -1 & 0 \\ 0 & -1 \end{pmatrix} = \begin{pmatrix} 0 & -1 \\ 1 & 0 \end{pmatrix} = R_{3\pi/2}$$

i.e. two rotations by $\frac{\pi}{2}$ and π is equivalent to one rotation by $\frac{3\pi}{2}$.

So rotations can be combined to generate elements of the same set. This is closure property.

✓ Associative property

$$R_{\pi/2}(R_{\pi}\, R_{3\pi/2}) = (R_{\pi/2}\, R_{\pi})R_{3\pi/2} = \begin{pmatrix} -1 & 0 \\ 0 & -1 \end{pmatrix} \text{ (after matrix multiplication)}$$

✓ Occurrence of identity element

$$R_{\pi/2}E = E\, R_{\pi/2} = R_{\pi/2} = \begin{pmatrix} 0 & 1 \\ -1 & 0 \end{pmatrix} \text{ (after matrix multiplication)}$$

✓ Occurrence of inverse element

$$R_{\pi/2}\, R_{3\pi/2} = R_{3\pi/2}\, R_{\pi/2} = E \text{ (after matrix multiplication)}$$

$$R_{\pi}\, R_{\pi} = E \text{ (self inverse) (after matrix multiplication)}$$

✓ Abelian property

Since

$$R_{\pi/2}\, R_{\pi} = R_{\pi}\, R_{\pi/2} \text{ (elements commute)}$$

the group G: E, $R_{\pi/2}$, R_{π}, $R_{3\pi/2}$ is abelian.

We note that for a rectangle (figure 9.1) rotation by 180° is a symmetry operation (rotation by 90° is not a symmetry operation).

9.9 Example of a continuous group

Consider a circle as shown in figure 9.2.

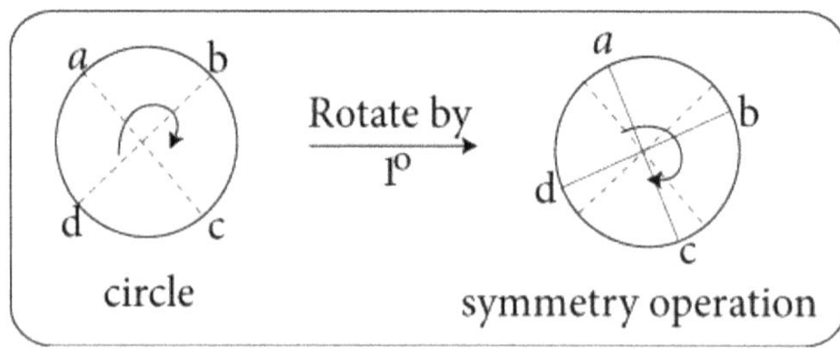

Figure 9.2. Rotation of a circle about an axis passing through the centre by any angle is a symmetry operation.

A rotation about the centre of a circle by angle θ, $\delta\theta$, $\theta + \delta\theta$ (with $0° \leqslant \theta \leqslant 2\pi$) keeps the circle in its original position, i.e. it falls upon itself as if no rotation has been made. Clearly we cannot distinguish whether rotation by any amount has been made or not. So rotation by θ, $\theta + \delta\theta$ are symmetry operations. The rotation angle θ varies continuously and forms a continuous group where rotation about the centre of the circle is the group operation. Since the operation of rotations by θ, $\theta + \delta\theta$ commutes, i.e.

$$R_\theta R_{\delta\theta} = R_{\delta\theta} R_\theta = R_{\theta+\delta\theta}$$

this group is abelian.

9.10 Generator of a group

The generator is an element with which we can generate all the elements of the group.

Consider for instance the group of symmetry operations on a square (equation (9.7)) viz. G: E, $R_{\pi/2}$, R_π, $R_{3\pi/2}$(group operation $\otimes \equiv$ rotation about the centre of a square). We note that

$$\left(R_{\frac{\pi}{2}}\right)^1 = R_{\frac{\pi}{2}}, \quad \left(R_{\frac{\pi}{2}}\right)^2 = R_\pi, \quad \left(R_{\frac{\pi}{2}}\right)^3 = R_{\frac{3\pi}{2}}, \quad \left(R_{\frac{\pi}{2}}\right)^4 = R_{2\pi} = E$$

(after matrix multiplication).

Since all the elements of the group viz.

$$E, R_{\pi/2}, R_\pi, R_{3\pi/2}$$

can be generated from $R_{\pi/2}$ it follows that the generator of this group is $R_{\pi/2}$. Since

$$(R_{\pi/2})^4 = R_{2\pi} = E \tag{9.9}$$

the order of the group is 4 (there are 4 elements in the group).

For an n-sided regular polygon, the generator is $R_{2\pi/n}$ and order of the generator element is n since

$$(R_{2\pi/n})^n = R_{2\pi} = E \tag{9.10}$$

9.11 Generator of space translation operation

Consider a 1D free particle of mass m described by a Hamiltonian H given by

$$H = \frac{p^2}{2m} \tag{9.11}$$

where $p = m\frac{\mathrm{d}x}{\mathrm{d}t}$ is momentum of a free particle.

Suppose the free particle suffers translation of amount a as

$$x \rightarrow x' = x + a \text{ where } a \text{ is constant } (x \text{ changes}) \tag{9.12}$$

$$p \rightarrow p' = m\frac{dx'}{dt} = m\frac{d}{dt}(x + a) = m\frac{dx}{dt} = p \ (p \text{ is invariant}) \tag{9.13}$$

$$H \rightarrow H' = \frac{p'^2}{2m} = \frac{p^2}{2m} = H \ (H \text{ is invariant}) \tag{9.14}$$

Clearly translation has no effect on Hamiltonian, i.e. H is not altered despite change in x. So the free particle has translational symmetry.

Let us find the explicit form of translation operator $T(a)$ which represents translation by amount a in 1D. This can be represented as

$$T(a)f(x) = f(x + a) \tag{9.15}$$

Taylor expansion of $f(x)$ about $x = x_0$ is

$$f(x) = f(x_0) + f^{(1)}(x_0)(x - x_0) + \frac{f^{(2)}(x_0)}{2!}(x - x_0)^2 + \ldots \tag{9.16}$$

Writing x instead of x_0 and $x + a$ instead of x in equation (9.16) gives

$$f(x + a) = f(x) + f^{(1)}(x)(x + a - x) + \frac{f^{(2)}(x)}{2!}(x + a - x)^2 + \ldots$$

$$= f(x) + \frac{a}{1!}f^{(1)}(x) + \frac{a^2}{2!}f^{(2)}(x) + \ldots$$

$$f(x + a) = f(x) + \frac{a}{1!}\frac{\partial}{\partial x}f(x) + \frac{a^2}{2!}\frac{\partial^2}{\partial x^2}f(x) + \ldots$$

$$f(x + a) = (1 + \frac{a}{1!}\frac{\partial}{\partial x} + \frac{a^2}{2!}\frac{\partial^2}{\partial x^2} + \ldots)f(x) \tag{9.17}$$

$$f(x + a) = e^{a\frac{\partial}{\partial x}}f(x) \tag{9.18}$$

Combining equations (9.15) and (9.18) we get

$$T(a)f(x) = f(x + a) = e^{a\frac{\partial}{\partial x}}f(x) \tag{9.19}$$

Hence

$$T(a) = e^{a\frac{\partial}{\partial x}} \tag{9.20}$$

Any finite translation of amount a can be obtained by adding small translations δa. For small translation δa, i.e. for

$$x \rightarrow x + \delta a \tag{9.21}$$

we have (from equations (9.19) and (9.20))

$$T(\delta a)f(x) = f(x + \delta a) = e^{\delta a \frac{\partial}{\partial x}} f(x) \tag{9.22}$$

$$= \left[1 + \delta a\frac{\partial}{\partial x} + O(\delta a)^2\right] f(x) \tag{9.23}$$

Ignoring higher order terms, i.e. putting $O(\delta a)^2 \to 0$ we get

$$T(\delta a)f(x) = f(x + \delta a) = \left(1 + \delta a\frac{\partial}{\partial x}\right)f(x) \tag{9.24}$$

$$T(\delta a) \cong 1 + \delta a\frac{\partial}{\partial x} \tag{9.25}$$

Here $\frac{\partial}{\partial x}$ is called the generator of translation and δa is called parameter (one generator and one parameter).

- We can derive equation (9.20) from equation (9.25) as follows.
    Suppose N small translations of amount δa produce finite translation a, i.e.

$$a = N\delta a \quad \text{and} \quad \delta a = \frac{a}{N} \tag{9.26}$$

We have from equation (9.24)

$$f(x + \delta a) = (1 + \frac{a}{N}\frac{\partial}{\partial x})f(x) \tag{9.27}$$

For finite translation a

$$f(x + a) = \underset{N \to \infty}{\text{Lt}} \left(1 + \frac{a}{N}\frac{\partial}{\partial x}\right)\left(1 + \frac{a}{N}\frac{\partial}{\partial x}\right) \cdots \left(1 + \frac{a}{N}\frac{\partial}{\partial x}\right)f(x)$$

$$= \underset{N \to \infty}{\text{Lt}} \left(1 + \frac{a}{N}\frac{\partial}{\partial x}\right)^N f(x) \tag{9.28}$$

$f(x + a) = e^{a\frac{\partial}{\partial x}}f(x)$ (this is equation (9.18))
- In the case of translation of a 3D free particle we can write (like in equation (9.12))

$$\vec{r} \to \vec{r} + \vec{a} \tag{9.29}$$

Analogous to equation (9.17) that holds in 1D we have in 3D

$$f(\vec{r} + \vec{a}) = \left(1 + \vec{a}.\vec{\nabla} + \frac{1}{2!}(\vec{a}.\vec{\nabla})^2 + \ldots\right)f(\vec{r}) \tag{9.30}$$

$$T(\vec{a})f(\vec{r}) = f(\vec{r} + \vec{a}) = e^{\vec{a}.\vec{\nabla}} f(\vec{r}) \text{ (like equation (9.18))} \tag{9.31}$$

Since

$$\vec{a} \cdot \vec{\nabla} = a_x \frac{\partial}{\partial x} + a_y \frac{\partial}{\partial y} + a_z \frac{\partial}{\partial z} \tag{9.32}$$

it follows that for translation in 3D there are three generators $\frac{\partial}{\partial x}, \frac{\partial}{\partial y}, \frac{\partial}{\partial z}$ and three parameters a_x, a_y, a_z.

We note that the number of generators and number of parameters are equal.

9.12 Quantum mechanical description of space translation operation

According to the postulate of quantum mechanics, observables will be operators. We find the operators for the generators describing 1D and 3D space translation.

Suppose wave function of a free particle has translational symmetry such that for 1D spatial translation of

$$x \rightarrow x' = x - a \text{ (compare with equation (9.12), we have } - a \text{ instead of } a) \tag{9.33}$$

we would have an equation similar to equation (9.15) viz.

$$T(a)\psi(x) = \psi(x - a) \tag{9.34}$$

And as in equation (9.20) we have

$$T(a) = e^{-a\frac{\partial}{\partial x}} \tag{9.35}$$

$$T(a)\psi(x) = \psi(x - a) = e^{-a\frac{\partial}{\partial x}} \psi(x) \tag{9.36}$$

As $\frac{\partial}{\partial x} \neq \frac{\partial}{\partial x}^{\dagger}$ i.e. $\frac{\partial}{\partial x}$ is not Hermitian (*exercise 9.1*) we cannot work with equation (9.35) in quantum mechanics, i.e. for a micro system.

Let us introduce in equation (9.36) the momentum operator which is a Hermitian operator viz.

$$p = -i\hbar \frac{\partial}{\partial x} \equiv p_x \tag{9.37}$$

as follows.

$$T(a)\psi(x) = e^{-a\frac{\partial}{\partial x}} \psi(x) = e^{-\frac{a}{-i\hbar}(-i\hbar\frac{\partial}{\partial x})}\psi(x) = e^{\frac{a}{i\hbar}p}\psi(x) = e^{-iap/\hbar}\psi(x) \tag{9.38}$$

$$T(a) = e^{-iap/\hbar} \tag{9.39}$$

Such translation is not going to affect system—it gives the same energy of particle. Here we note that the operator $p \equiv p_x$ is the generator for translation and $a \equiv a_x$ is the parameter.

For 3D translation (similar to equation (9.33))

$$\vec{r} \rightarrow \vec{r} - \vec{a} \tag{9.40}$$

we have as in equation (9.31)

$$\psi(\vec{r} - \vec{a}) = e^{-\vec{a}.\vec{\nabla}}\,\psi(\vec{r}) \tag{9.41}$$

With the 3D momentum operator $\vec{p} = (p_x, p_y, p_z) = -i\hbar\,\vec{\nabla}$ and $\vec{a} = (a_x, a_y, a_z)$ we get

$$T(\vec{a})\psi(\vec{r}) = \psi(\vec{r} - \vec{a}) = e^{-\frac{\vec{a}}{-i\hbar}(-i\hbar\vec{\nabla})}\psi(\vec{r}) = e^{-i\vec{a}.\vec{p}\,/\hbar}\psi(\vec{r}) \tag{9.42}$$

$$T(\vec{a}) = e^{-i\vec{a}.\vec{p}\,/\hbar} \tag{9.43}$$

In other words, in quantum mechanics the generator for 3D translation is linear momentum operator $\vec{p} = (p_x, p_y, p_z) = -i\hbar\,\vec{\nabla}$ and there are three parameters a_x, a_y, a_z to carry out the translation.

- Also, for a small translation (like equation (9.21))

$$x \rightarrow x - \delta a \tag{9.44}$$

we have, similar to equation (9.25),

$$T(\delta a) = 1 - \delta a\frac{\partial}{\partial x} = 1 - \frac{\delta a}{-i\hbar}\left(-i\hbar\frac{\partial}{\partial x}\right) = 1 - \frac{i\delta a}{\hbar}p \tag{9.45}$$

Hence

$$T(\delta a_x) = 1 - \frac{i\delta a_x}{\hbar}p_x, \quad T(\delta a_y) = 1 - \frac{i\delta a_y}{\hbar}p_y. \tag{9.46}$$

9.13 Abelian property of generators of translation

An important property of translation operator T is that it is abelian since order of operation is immaterial, as evident from the calculation of the following commutator bracket, namely

$$T(\delta a_x)T(\delta a_y) - T(\delta a_y)T(\delta a_x) \tag{9.47}$$

$$=(1 - \frac{i\delta a_x}{\hbar}p_x)(1 - \frac{i\delta a_y}{\hbar}p_y) - (1 - \frac{i\delta a_y}{\hbar}p_y)(1 - \frac{i\delta a_x}{\hbar}p_x) \text{ (using equation (9.46))}$$

$$=1 - \frac{i\delta a_y}{\hbar}p_y - \frac{i\delta a_x}{\hbar}p_x + \frac{i\delta a_x}{\hbar}p_x\frac{i\delta a_y}{\hbar}p_y$$

$$-[1 - \frac{i\delta a_x}{\hbar}p_x - \frac{i\delta a_y}{\hbar}p_y + \frac{i\delta a_y}{\hbar}p_y\frac{i\delta a_x}{\hbar}p_x]$$

$$=\frac{i\delta a_x}{\hbar}p_x\frac{i\delta a_y}{\hbar}p_y - \frac{i\delta a_y}{\hbar}p_y\frac{i\delta a_x}{\hbar}p_x$$

$$= \frac{i\delta a_x}{\hbar}\frac{i\delta a_y}{\hbar}(p_x p_y - p_y p_x)$$

$$= \frac{i\delta a_x}{\hbar}\frac{i\delta a_y}{\hbar}[p_x, p_y] = 0 \text{ (as } p_x \text{ and } p_y \text{ commute)} \tag{9.48}$$

Hence

$$T(\delta a_x)T(\delta a_y) = T(\delta a_y)T(\delta a_x) \tag{9.49}$$

Similarly,

$$T(a_x)T(a_y) = T(a_y)T(a_x) \tag{9.50}$$

$$\text{i. e. } [T(a_x)T(a_y)] = 0 \tag{9.51}$$

So the commutator of the generators of translation like $[p_x, p_y]$ is zero, i.e. generators of translation commute. In other words these generators of translation form an abelian group.

The commutator bracket of the generators of the group carries all information about the group and we can work with generators instead of the group elements.

Algebra of the generators of a group is essentially a study of the commutator bracket of the generators.

9.14 Time translation operation

In the case of time translation we consider (similar to equation (9.12))

$$t \rightarrow t + \varepsilon \tag{9.52}$$

Similar to equation (9.18) we have

$$T(\varepsilon)\psi(t) = \psi(t + \varepsilon) = e^{\varepsilon\frac{\partial}{\partial t}}\psi(t) \tag{9.53}$$

Let us make the operator a Hermitian operator.

$$T(\varepsilon)\psi(t) = e^{\frac{\varepsilon}{i\hbar}(i\hbar\frac{\partial}{\partial t})}\psi(t) = e^{-i\varepsilon H/\hbar}\psi(t) \tag{9.54}$$

From the time dependent Schrödinger equation we have $H\psi = i\hbar\frac{\partial}{\partial t}\psi$ and so replacing $i\hbar\frac{\partial}{\partial t}$ by the Hamiltonian operator H we get from equation (9.54)

$$T(\varepsilon) = e^{-i\varepsilon H/\hbar} \tag{9.55}$$

Clearly Hamiltonian operator H is the generator of time translation.

9.15 Space-time translation operation

A translation in 1D space ($x \rightarrow x - a$) is described by the operator $e^{-a\frac{\partial}{\partial x}}$ where a is the parameter and $\frac{\partial}{\partial x}$ is the generator. In quantum mechanics it is described by

$e^{-iap/\hbar}$ (equation (9.39)) where $p = -i\hbar\frac{\partial}{\partial x}$ is the linear momentum operator which is the generator of space translation in 1D and a is the parameter.

A translation in 3D space ($\vec{r} \to \vec{r} - \vec{a}$) is described by operator $e^{\vec{a}\cdot\vec{\nabla}}$ where a_x, a_y, a_z are parameters and $\frac{\partial}{\partial x}$, $\frac{\partial}{\partial y}$, $\frac{\partial}{\partial z}$ are generators. In quantum mechanics it is described by $e^{-i\vec{a}\cdot\vec{p}/\hbar}$ (equation (9.43)) where the three momentum operators $p_x = -i\hbar\frac{\partial}{\partial x}$, $p_y = -i\hbar\frac{\partial}{\partial y}$, $p_z = -i\hbar\frac{\partial}{\partial z}$ are the generators of space translation in 3D and a_x, a_y, a_z are the parameters.

Time translation is described by operator $e^{-\varepsilon\frac{\partial}{\partial t}}$ where the generator is $\frac{\partial}{\partial t}$ and in quantum mechanics it is described by $e^{-i\varepsilon H/\hbar}$ (equation (9.55)) where Hamiltonian H is the generator of time translation.

Combining space translation and time translation we can talk about space-time translation

$$x^{\mu} \to x^{\mu} + a^{\mu} \tag{9.56}$$

and build

$$T(a_{\mu})\psi(x^{\mu}) = \psi(x^{\mu} + a^{\mu}) = e^{ia_{\mu}O^{\mu}/\hbar}\psi(x^{\mu}) \tag{9.57}$$

where $x^0 = ct$, $x^1 = x$, $x^2 = y$, $x^3 = z$ and the generators O^{μ} are taken to be

$$O^0 = H, \quad O^1 = p_x, \quad O^2 = p_y, \quad O^3 = p_z \tag{9.58}$$

9.16 Rotation in physical space

Consider rotation operation about the Z-axis $\hat{k}$ by a small angle $\delta\theta$ carried out by rotation operator $R(\delta\theta, \hat{k})$ because of which $\vec{r}$ transforms to $\vec{r}'$ expressed as

$$\vec{r}' = \vec{r} + \delta\vec{r} = R(\delta\theta, \hat{z})\vec{r} \tag{9.59}$$

Rotation operation can be represented by a matrix called a rotation matrix of equation (9.8). Let us make a 3D representation of the rotation matrix and also replace θ by $\delta\theta$ to get

$$R(\delta\theta, \hat{z}) = \begin{pmatrix} \cos\delta\theta & \sin\delta\theta & 0 \\ -\sin\delta\theta & \cos\delta\theta & 0 \\ 0 & 0 & 1 \end{pmatrix} \cong \begin{pmatrix} 1 & \delta\theta & 0 \\ -\delta\theta & 1 & 0 \\ 0 & 0 & 1 \end{pmatrix} \tag{9.60}$$

since $\cos\delta\theta \cong 1$, $\sin\delta\theta \cong \delta\theta$ for infinitesimally small $\delta\theta$.

Writing equation (9.59) in matrix form we have

$$\begin{pmatrix} x' \\ y' \\ z' \end{pmatrix} = \begin{pmatrix} 1 & \delta\theta & 0 \\ -\delta\theta & 1 & 0 \\ 0 & 0 & 1 \end{pmatrix}\begin{pmatrix} x \\ y \\ z \end{pmatrix} = \begin{pmatrix} x + y\delta\theta \\ -x\delta\theta + y \\ z \end{pmatrix} = \begin{pmatrix} x \\ y \\ z \end{pmatrix} + \begin{pmatrix} y\delta\theta \\ -x\delta\theta \\ 0 \end{pmatrix} \tag{9.61}$$

$$\vec{r}\,' = \vec{r} + (\hat{i}\,y\delta\theta - \hat{j}\,x\delta\theta) \tag{9.62}$$

$$\vec{r}\,' = \vec{r} + \delta\theta(\hat{i}\,y - \hat{j}\,x) = \vec{r} + \delta\theta\,(\hat{j} \times \hat{k}\,y - \hat{k} \times \hat{i}\,x)$$

$$=\vec{r} + \delta\theta\,(-\hat{k} \times \hat{j}\,y - \hat{k} \times \hat{i}\,x) = \vec{r} - \delta\theta\hat{k} \times (\hat{j}\,y + \hat{i}\,x)$$

$$\vec{r}\,' = \vec{r} - \delta\theta\hat{k} \times (\hat{i}\,x + \hat{j}\,y + \hat{k}\,z)\ (\text{as } \hat{k} \times \hat{k} = 0)$$

$$\vec{r}\,' = \vec{r} - \delta\theta\hat{k} \times \vec{r} \tag{9.63}$$

9.17 Non-abelian nature of generators of rotation

In the case of rotation operation about an axis $\hat{n}$ we have to replace $\hat{k}$ by $\hat{n}$ to get

$$\vec{r}\,' = \vec{r} - \delta\theta\hat{n} \times \vec{r} = \vec{r} - \overrightarrow{\delta\theta} \times \vec{r} = \vec{r} + \delta\,\vec{r}\ (\text{from equation } (9.59)) \tag{9.64}$$

$$\delta\,\vec{r} = -\overrightarrow{\delta\theta} \times \vec{r} \tag{9.65}$$

$$\overrightarrow{\delta\theta} = \delta\theta\hat{n}$$

Now consider operation of rotation by a small angle $\delta\theta$ such that $\psi(\vec{r})$ is rotated about $\hat{n}$ slightly. This is represented as

$$R(\delta\theta, \hat{n})\psi(\vec{r}) = \psi(\vec{r} + \delta\,\vec{r})$$

Make a Taylor expansion

$$R(\delta\theta, \hat{n})\psi(\vec{r}) = \psi(\vec{r}) + \delta\vec{r}.\overrightarrow{\nabla}\,\psi(\vec{r})$$

$$=\psi(\vec{r}) - \overrightarrow{\delta\theta} \times \vec{r}.\overrightarrow{\nabla}\psi(\vec{r})\ (\text{using equation } (9.65))$$

$$=\psi(\vec{r}) - \frac{1}{i\hbar}\overrightarrow{\delta\theta} \times \vec{r}.\,i\hbar\,\overrightarrow{\nabla}\,\psi(\vec{r})$$

$$=\psi(\vec{r}) + \frac{1}{i\hbar}\overrightarrow{\delta\theta} \times \vec{r}.(-i\hbar\overrightarrow{\nabla})\psi(\vec{r})\ (\text{using } \overrightarrow{p} = -i\hbar\overrightarrow{\nabla})$$

$$=\psi(\vec{r}) + \frac{1}{i\hbar}\overrightarrow{\delta\theta} \times \vec{r}.\overrightarrow{p}\,\psi(\vec{r})$$

$$=\psi(\vec{r}) - \frac{i}{\hbar}\overrightarrow{\delta\theta} \times \vec{r}.\overrightarrow{p}\,\psi(\vec{r})$$

$$=\psi(\vec{r}) - \frac{i}{\hbar}\overrightarrow{\delta\theta}.\,\vec{r} \times \overrightarrow{p}\,\psi(\vec{r})$$

$$R(\delta\theta, \hat{n})\psi(\vec{r}) = \left(1 - \frac{i}{\hbar}\overrightarrow{\delta\theta}.\,\vec{l}\right)\psi(\vec{r}) \tag{9.66}$$

where $\vec{l} = \vec{r} \times \vec{p} =$ orbital angular momentum of particle. It is the generator for rotation in 3D space. It has three components l_x, l_y, l_z, i.e. there are three generators and there are three parameters, namely $\delta\theta_x$, $\delta\theta_y$, $\delta\theta_z$.

A finite rotation θ is reached by adding the small rotations $\delta\theta$ where

$$\delta\theta = \frac{\theta}{N} \tag{9.67}$$

$$R(\theta, \hat{n})\psi(\vec{r}) = \operatorname*{Lt}_{N\to\infty}\left(1 - \frac{i}{\hbar}\frac{1}{N}\vec{\theta}.\,\vec{l}\right)\left(1 - \frac{i}{\hbar}\frac{1}{N}\vec{\theta}.\,\vec{l}\right)...\left(1 - \frac{i}{\hbar}\frac{1}{N}\vec{\theta}.\,\vec{l}\right)\psi(\vec{r})$$

$$= \operatorname*{Lt}_{N\to\infty}\left(1 - \frac{i}{\hbar}\frac{1}{N}\vec{\theta}.\,\vec{l}\right)^{N}\psi(\vec{r})$$

$$R(\theta, \hat{n})\psi(\vec{r}) = e^{-i\vec{\theta}.\vec{l}/\hbar}\psi(\vec{r}) \tag{9.68}$$

$$R(\theta, \hat{n}) = e^{-i\vec{\theta}.\vec{l}/\hbar} \tag{9.69}$$

- The generators of rotation l_x, l_y, l_z satisfy the commutation relation (*exercise 9.28*)

$$[l_i l_j] = i\hbar\varepsilon_{ijk}l_k \tag{9.70}$$

Since angular momenta do not commute, it follows that the generators of rotation form a non-abelian group.
- In *exercise 9.27* we show that

$$T(\delta\theta, \hat{i})T(\delta\theta, \hat{j}) - T(\delta\theta, \hat{j})T(\delta\theta, \hat{i}) = T((\delta\theta)^2, \hat{k}) - I \neq 0 \tag{9.71}$$

- Symmetry requires that the generators of symmetry operation should commute with the Hamiltonian.

A free particle is described by the Hamiltonian

$$H = \frac{p^2}{2m} = \frac{\vec{p}\cdot\vec{p}}{2m} = \frac{p_i^2}{2m} \tag{9.72}$$

The generator of space translation is p_i and

$$[H, p_i] = 0 \text{ for } i = x, y, z \tag{9.73}$$

This means the free particle Hamiltonian has translation symmetry, i.e. under translation the Hamiltonian remains invariant.

The generator of space rotation is l_i and

$$[H, l_i] = 0 \text{ for } i = x, y, z \tag{9.74}$$

This means the free particle Hamiltonian has rotation symmetry, i.e. under rotation the Hamiltonian remains invariant.

9.18 Orthogonal matrix

Let us define an orthogonal matrix.

A real matrix is termed as an orthogonal matrix if its transpose is equal to its inverse, i.e.

$$R^T = R^{-1} \tag{9.75}$$

$$RR^T = I, \quad R^T R = I \ (I = \text{identity matrix}) \tag{9.76}$$

This is the condition for a matrix to be an orthogonal matrix.

9.19 Symmetry representation of the $SO(n)$ group

In general, rotations form a group because combination or succession of rotations is again a rotation.

We note that the square of the distance from origin to any point does not change upon rotation of coordinate axis about the fixed origin, i.e. distance square is invariant under rotation about the origin. From this relation we can obtain the nature of a rotation matrix as follows.

Consider coordinate x_i, $i = 1,2, \ldots , n$ in n dimensional case and $n = 3$ for 3D.

The rotation is a linear transformation and each coordinate in the new system is a linear combination of coordinates in the old coordinate system.

$$x_i' = R_{ij}x_j \tag{9.77}$$

where x_j, x_i' are coordinates in the old and new coordinate systems, respectively, and R_{ij} is the matrix element, R is the rotation matrix.

9.20 Rotation matrix

- Rotation matrix is an orthogonal matrix
  Under rotation in n dimensional space

$$x_i x_i = x_i' x_i' \text{ (distance square is invariant)} \tag{9.78}$$

Now

$$x_i x_i = x_j x_j = \delta_{jk} x_j x_k \tag{9.79}$$

Now

$$x_i' x_i' = R_{ij} x_j R_{ik} x_k \text{ (using equation (9.77))} \tag{9.80}$$

$$= R_{ij} R_{ik} x_j x_k$$

$$= R_{ji}^{T} R_{ik} x_j x_k$$

$$x_i' x_i' = (R^T R)_{jk} x_j x_k \tag{9.81}$$

Using equation (9.78) viz. $x_i x_i = x_i' x_i'$ in equation (9.81) we write

$$x_i x_i = (R^T R)_{jk} x_j x_k \tag{9.82}$$

Using equation (9.79) viz. $x_i x_i = x_j x_j = \delta_{jk} x_j x_k$ in equation (9.82)

$$\delta_{jk} x_j x_k = (R^T R)_{jk} x_j x_k$$

$$(R^T R)_{jk} = \delta_{jk} \text{ for all } j, k$$

$$R^T R = I \tag{9.83}$$

This is orthogonality condition, i.e. R is an orthogonal matrix. Clearly the rotation matrix is an orthogonal matrix.

- The set of $n \times n$ orthogonal matrices form a group, an element of the group is denoted by $O(n)$.

An important property of the orthogonal matrices follows if we take the determinant of both sides of equation (9.83).

$$\det(R^T R) = \det I$$

$$(\det R^T)(\det R) = \det I$$

$$(\det R)^2 = 1 (\text{as } \det R^T = \det R)$$

$$\det R = \pm 1 \tag{9.84}$$

✓ Orthogonal matrices with

$$\det R = +1$$

correspond to proper rotation (*exercise 9.15*). These matrices can be constructed by moving continuously from the identity matrix through a succession of infinitesimal transformations each of the matrices having $\det R = +1$. This set includes identity transformation matrix I and hence this set of matrices forms a group. This group of rotation matrices is called the $SO(n)$ group. The letter S denotes the special choice

$$\det R = +1 \tag{9.85}$$

The letter O denotes that the matrices are orthogonal matrices. And n denotes the fact that the corresponding matrices are of dimension $n \times n$.

✓ Orthogonal matrices with

$$\det R = -1 \tag{9.86}$$

correspond to improper rotation (representing reflection and 3D parity operation, which are discrete operations). These matrices represent a discontinuous or discrete transformation. This set does not include identity element I and hence this set of matrices does not form a group.

9.20.1 Construction of rotation matrix

Consider for example a circular disc in the XY plane and the Z-axis passes through its centre about which rotation by any angle can be performed.

2D rotation is completely specified by a single angle θ. Rotation by angle θ of a circular disc about the symmetry axis Z forms a continuous group (figure 9.3). Group elements are rotations through angle θ described by 2×2 matrices denoted by say $R(\theta)$. Number of elements is infinite as θ can vary from 0 to ∞. Since elements of the group are defined by one parameter θ the order of the continuous group is 1. This group is denoted by $SO(2)$.

   ✓ The letter S stands for the determinantal value of $+1$.
   ✓ The letter O indicates that the matrices describing rotations are orthogonal.
   ✓ The 2 of $SO(2)$ denotes that matrices describing this group are of dimension 2×2.

The disc is rotated by angle θ in the XY plane about a fixed Z-axis in the clockwise sense as a result of which the point $\vec{r}(x, y)$ is rotated to the point $\vec{r}'(x', y')$ as shown in figure 9.3.

We have proved in *exercise 9.10(ii)* that clockwise rotation of a vector keeping axes fixed is described by the matrix $R(\theta)$ as

$$\begin{pmatrix} x' \\ y' \end{pmatrix} = \begin{pmatrix} \cos\theta & \sin\theta \\ -\sin\theta & \cos\theta \end{pmatrix} \begin{pmatrix} x \\ y \end{pmatrix} = R\begin{pmatrix} x \\ y \end{pmatrix} \tag{9.87}$$

The matrix

$$R = \begin{pmatrix} \cos\theta & \sin\theta \\ -\sin\theta & \cos\theta \end{pmatrix} = R(\theta) \tag{9.88}$$

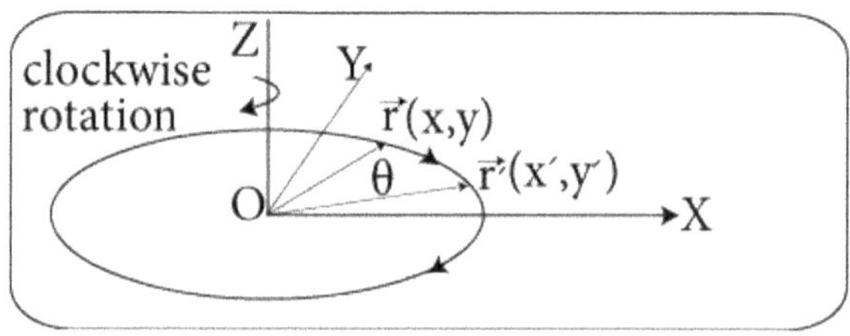

Figure 9.3. Disc is rotated in the XY plane about the Z-axis in the clockwise sense.

is called a transformation matrix and has determinantal value of $|R| = +1$ and such a matrix characterizes rotation. Clearly rotations in 2D space can be described by 2×2 orthogonal matrices with determinant $+1$.

- In *exercise 9.10(iv)* we have derived the rotation matrix for anti-clockwise rotation of a vector keeping axes fixed to be

$$R'(\theta) = \begin{pmatrix} \cos\theta & -\sin\theta \\ \sin\theta & \cos\theta \end{pmatrix}. \tag{9.89}$$

- In *exercise 9.10(i), (iii)* we have shown that $R(\theta)$ also describes anti-clockwise rotation of axes keeping the vector fixed while $R'(\theta)$ also describes clockwise rotation of axes keeping the vector fixed.
- In *exercise 9.17* we have shown that

$$R^T R = R R^T = I \tag{9.90}$$

$$R'^T R' = R' R'^T = I \tag{9.91}$$

So R, R' are orthogonal matrices.
- *Exercise 9.12* shows the properties of a rotation matrix.
- Let us find $R(0)$, $R(2\pi)$ which represent the identity element I.

$$R(0) = \begin{pmatrix} \cos\theta & \sin\theta \\ -\sin\theta & \cos\theta \end{pmatrix} = \begin{pmatrix} \cos\theta & \sin\theta \\ -\sin\theta & \cos\theta \end{pmatrix} = \begin{pmatrix} 1 & 0 \\ 0 & 1 \end{pmatrix} = I = \text{Identity element.}$$

$$R(2\pi) = \begin{pmatrix} \cos\theta & \sin\theta \\ -\sin\theta & \cos\theta \end{pmatrix} = \begin{pmatrix} \cos 2\pi & \sin 2\pi \\ -\sin 2\pi & \cos 2\pi \end{pmatrix} = \begin{pmatrix} 1 & 0 \\ 0 & 1 \end{pmatrix} = I$$

$$= \text{Identity element.}$$

- Let us find the inverse element

$$R(\theta) = \begin{pmatrix} \cos\theta & \sin\theta \\ -\sin\theta & \cos\theta \end{pmatrix}, \quad R(-\theta) = \begin{pmatrix} \cos(-\theta) & \sin(-\theta) \\ -\sin(-\theta) & \cos(-\theta) \end{pmatrix} = \begin{pmatrix} \cos\theta & -\sin\theta \\ \sin\theta & \cos\theta \end{pmatrix}$$

Now consider

$$R(\theta)R(\phi) = \begin{pmatrix} \cos\theta & \sin\theta \\ -\sin\theta & \cos\theta \end{pmatrix}\begin{pmatrix} \cos\phi & \sin\phi \\ -\sin\phi & \cos\phi \end{pmatrix}$$

$$= \begin{pmatrix} \cos\theta\cos\phi - \sin\theta\sin\phi & \cos\theta\sin\phi + \sin\theta\cos\phi \\ -(\sin\theta\cos\phi + \cos\theta\sin\phi) & \cos\theta\cos\phi - \sin\theta\sin\phi \end{pmatrix}$$

$$= \begin{pmatrix} \cos(\theta+\phi) & \sin(\theta+\phi) \\ -\sin(\theta+\phi) & \cos(\theta+\phi) \end{pmatrix} = R(\theta+\phi)$$

Hence for $\phi = -\theta$ we have

$$R(\theta)R(-\theta) = R(\theta - \theta) = R(0) = I \tag{9.92}$$

So the inverse of $R(\theta)$ is $R(-\theta)$.

- Let us find the determinant of R

$$|R| = \begin{vmatrix} \cos\theta & \sin\theta \\ -\sin\theta & \cos\theta \end{vmatrix} = \cos^2\theta + \sin^2\theta = 1 \tag{9.93}$$

R is an orthogonal matrix of order 2×2 with determinantal value 1. It is called special orthogonal matrix of order 2 due to this special choice of determinantal value and represented as $SO(2)$. (S for special and 2 denoting that the matrix is 2×2).

- The elements of rotation matrix $R(\theta)$ contain real numbers—it does not involve complex numbers.
- Clearly $SO(2)$ group consists of 2×2 orthogonal matrices and represents 2D rotation.
- $SO(2)$ is an abelian group since in a plane rotation by θ, θ' can be performed in any order

$$R(\theta)R(\theta') = R(\theta')R(\theta) = R(\theta + \theta'). \tag{9.94}$$

- If we have a system that never changes under 2D rotation, i.e. all laws of physics remain invariant under rotation in a plane, then it has $SO(2)$ symmetry. This is how a group is related to symmetry.
- We have given examples of matrices representing rotation in *exercise 9.15*.

9.20.2 Reflection

- If we include reflection then the determinant can be ± 1. This group is called $O(2)$. So $SO(2)$ is a special case of $O(2)$.
- We give examples of matrices representing reflection in *exercise 9.15*.

9.20.3 3D rotation

- Rotation in 3D space is described by 3×3 orthogonal matrices. Resulting groups are $O(3)$ and $SO(3)$.
- $SO(3)$ group has 3 angles as group parameters. We take these three parameters as the Euler angles. So order of $SO(3)$ is 3.
- We have given examples of 3×3 matrices representing rotation in *exercise 9.15*.
- If R is an $n \times n$ orthogonal matrix with determinantal value 1 then it is called a special orthogonal matrix and forms a group represented as $SO(n)$.
- Orthogonal matrices form a group because of the following.
    - ✓ Closure relation
        Let O_1, O_2... be orthogonal matrices. Then

$$O_1^T = O_1^{-1}$$

$$O_2^T = O_2^{-1}$$

$$(O_1 O_2)^T = O_2^T O_1^T = O_2^{-1} O_1^{-1} = (O_1 O_2)^{-1}$$

Hence $O_1 O_2$ is also orthogonal.

✓ Matrix multiplication is associative.

✓ The definition of an orthogonal matrix implies identity matrix element viz. $OO^T = I$.

✓ The definition of an orthogonal matrix implies inverse matrix element viz. $O^T = O^{-1}$.

✓ Since $OO^T = I$ and $\det O = \det O^T$ it follows that

$$(\det O)^2 = 1$$

$$\det O = \pm 1$$

If we choose $\det O = +1$ we have $SO(2)$ group.

- In *exercise 9.18* we show that an orthogonal matrix with $\det O = -1$ cannot form a group.

9.21 Lie group

- Orthogonal groups are called Lie groups. They depend on continuously varying parameters.
- The angles vary over a closed finite interval, say over 0 to 2π.
- Lie groups are continuous groups.
- Group operations are differentiable.
- Generators of the group.

  These groups can be studied in terms of generators which are a minimal set of quantities that could be used in a specific way to produce any element of the group.
- Order.

  Order is the number of linearly independent generators of a group.

$$\text{Order of } SO(n) \text{ is } \frac{n(n-1)}{2} \tag{9.95}$$

For $n = 2$, order of $SO(2)$ is $\frac{2(2-1)}{2} = 1$ (i.e. one angle θ is needed).

For $n = 3$, order of $SO(3)$ is $\frac{3(3-1)}{2} = 3$ (i.e. 3 Euler angles are needed).

$SO(n)$ is a special case of $O(n)$.

This is shown in figure 9.4.

9.22 Generator of the $SO(2)$ group

Consider the rotational group $SO(2)$ represented by rotation matrix (equation (9.88)) viz.

Order		Generators			
$SO(n) = \dfrac{n(n-1)}{2}$	$SU(n) = n^2 - 1$	$SO(2)$	σ_2	$SU(2)$	$\dfrac{\sigma_x}{2}, \dfrac{\sigma_y}{2}, \dfrac{\sigma_z}{2}$
$SO(2) = 1$	$SU(2) = 3$			$SU(3)$	$\dfrac{\lambda_1}{2}, \dfrac{\lambda_2}{2}, \dfrac{\lambda_3}{2}, \dfrac{\lambda_4}{2},$
$SO(3) = 3$	$SU(3) = 8$	$SO(3)$	S_x, S_y, S_z		$\dfrac{\lambda_5}{2}, \dfrac{\lambda_6}{2}, \dfrac{\lambda_7}{2}, \dfrac{\lambda_8}{2}$

Figure 9.4. Order of a group and generators of a group.

$$R(\theta) = \begin{pmatrix} \cos\theta & \sin\theta \\ -\sin\theta & \cos\theta \end{pmatrix}$$

Let us find the generator of the $SO(2)$ group.

Consider

$$-i\frac{d}{d\theta}R(\theta) = -i\frac{d}{d\theta}\begin{pmatrix} \cos\theta & \sin\theta \\ -\sin\theta & \cos\theta \end{pmatrix} = -i\begin{pmatrix} -\sin\theta & \cos\theta \\ -\cos\theta & -\sin\theta \end{pmatrix}$$

$$-i\frac{d}{d\theta}R(\theta)\Big|_{\theta=0} = -i\begin{pmatrix} -\sin\theta & \cos\theta \\ -\cos\theta & -\sin\theta \end{pmatrix}\Big|_{\theta=0} = -i\begin{pmatrix} 0 & 1 \\ -1 & 0 \end{pmatrix}$$

$$-i\frac{d}{d\theta}R(\theta) = \begin{pmatrix} 0 & -i \\ i & 0 \end{pmatrix} = \sigma_2 \tag{9.96}$$

The matrix σ_2 is obtained from the rotation matrix $R(\theta)$.

This σ_2 is the generator of the $SO(2)$ group, i.e. $SO(2)$ has one generator σ_2.

- Let us now express $R(\theta)$ in terms of σ_2. In other words we generate $R(\theta)$ from σ_2.

$$R(\theta) = \begin{pmatrix} \cos\theta & \sin\theta \\ -\sin\theta & \cos\theta \end{pmatrix} = \begin{pmatrix} \cos\theta & -i^2\sin\theta \\ i^2\sin\theta & \cos\theta \end{pmatrix}$$

$$= \begin{pmatrix} 1 & 0 \\ 0 & 1 \end{pmatrix}\cos\theta + i\begin{pmatrix} 0 & -i \\ i & 0 \end{pmatrix}\sin\theta$$

$$R(\theta) = I\cos\theta + i\sigma_2\sin\theta \tag{9.97}$$

- Consider the expansion of $e^{i\sigma_2\theta}$

$$e^{i\sigma_2\theta} = \sum_{n=0}^{\infty}\frac{(i\sigma_2\theta)^n}{n!} \tag{9.98}$$

$$= \frac{(i\sigma_2\theta)^0}{0!} + \frac{(i\sigma_2\theta)^1}{1!} + \frac{(i\sigma_2\theta)^2}{2!} + \frac{(i\sigma_2\theta)^3}{3!} + \frac{(i\sigma_2\theta)^4}{4!} + \frac{(i\sigma_2\theta)^5}{5!} + \frac{(i\sigma_2\theta)^6}{6!} + \dots$$

$$= 1 + i\sigma_2\theta + \frac{i^2\sigma_2^2\theta^2}{2!} + \frac{i^3\sigma_2^3\theta^3}{3!} + \frac{i^4\sigma_2^4\theta^4}{4!} + \frac{i^5\sigma_2^5\theta^5}{5!} + \frac{i^6\sigma_2^6\theta^6}{6!} + \cdots$$

$$e^{i\sigma_2\theta} = 1 + i\sigma_2\theta - \frac{\sigma_2^2\theta^2}{2!} - \frac{i\sigma_2\sigma_2^2\theta^3}{3!} + \frac{\sigma_2^4\theta^4}{4!} + \frac{i\sigma_2\sigma_2^4\theta^5}{5!} + \frac{\sigma_2^6\theta^6}{6!} + \cdots \quad (9.99)$$

Using $\sigma_i^2 = 1$, i.e. $\sigma^2 = 1$ we have

$$e^{i\sigma_2\theta} = 1 + i\sigma_2\theta - \frac{\theta^2}{2!} - \frac{i\sigma_2\theta^3}{3!} + \frac{\theta^4}{4!} + \frac{i\sigma_2\theta^5}{5!} - \frac{\theta^6}{6!} + \cdots$$

$$e^{i\sigma_2\theta} = \left(1 - \frac{\theta^2}{2!} + \frac{\theta^4}{4!} - \frac{\theta^6}{6!} + \cdots\right) + i\sigma_2\left(\theta - \frac{\theta^3}{3!} - \frac{\theta^4}{4!} + \frac{\theta^5}{5!} + \cdots\right) \quad (9.100)$$

Attaching a unit matrix I in the first term on the RHS of equation (9.100) we write

$$e^{i\sigma_2\theta} = I\cos\theta + i\sigma_2\sin\theta \quad (9.101)$$

$$= \begin{pmatrix} 1 & 0 \\ 0 & 1 \end{pmatrix}\cos\theta + i\begin{pmatrix} 0 & -i \\ i & 0 \end{pmatrix}\sin\theta$$

$$= \begin{pmatrix} \cos\theta & -i^2\sin\theta \\ i^2\sin\theta & \cos\theta \end{pmatrix}$$

$$e^{i\sigma_2\theta} = \begin{pmatrix} \cos\theta & \sin\theta \\ -\sin\theta & \cos\theta \end{pmatrix} = R(\theta) \quad (9.102)$$

Hence σ_2 generates $SO(2)$.

$$R(\theta) = I\cos\theta + i\sigma_2\sin\theta = e^{i\sigma_2\theta} \quad (9.103)$$

Hence σ_2 is the generator of the $SO(2)$ group. This is shown in figure 9.4.

9.22.1 3×3 rotation matrices

We include the axis of rotation in the rotation matrix. We then end up with 3×3 orthogonal rotation matrices in 3D space.

For rotation by angle θ about the Z-axis in the XY plane the rotation matrix is

$$R_z(\theta) = \begin{pmatrix} \cos\theta & \sin\theta & 0 \\ -\sin\theta & \cos\theta & 0 \\ 0 & 0 & 1 \end{pmatrix} \quad (9.104)$$

In this matrix the 3×3 element is 1.

9.23 Generator of $SO(3)$

Let us find the generator of the $SO(3)$ group.

Consider

$$-i\frac{d}{d\theta}R_z(\theta) = -i\frac{d}{d\theta}\begin{pmatrix} \cos\theta & \sin\theta & 0 \\ -\sin\theta & \cos\theta & 0 \\ 0 & 0 & 1 \end{pmatrix} = -i\begin{pmatrix} -\sin\theta & \cos\theta & 0 \\ -\cos\theta & -\sin\theta & 0 \\ 0 & 0 & 0 \end{pmatrix}$$

$$-i\frac{d}{d\theta}R_z(\theta)\Big|_{\theta=0} = -i\begin{pmatrix} -\sin\theta & \cos\theta & 0 \\ -\cos\theta & -\sin\theta & 0 \\ 0 & 0 & 0 \end{pmatrix}\Big|_{\theta=0} = -i\begin{pmatrix} 0 & 1 & 0 \\ -1 & 0 & 0 \\ 0 & 0 & 0 \end{pmatrix}$$

$$-i\frac{d}{d\theta}R_z(\theta)|_\theta = \begin{pmatrix} 0 & -i & 0 \\ i & 0 & 0 \\ 0 & 0 & 0 \end{pmatrix} = S_z \tag{9.105}$$

This is how $R_z(\theta)$ gives out S_z. In other words, S_z is a generator of the $SO(3)$ group.
- For an infinitesimal rotation by angle $\delta\theta$ we can write from equation (9.104)

$$R_z(\delta\theta) = \begin{pmatrix} \cos\delta\theta & \sin\delta\theta & 0 \\ -\sin\delta\theta & \cos\delta\theta & 0 \\ 0 & 0 & 1 \end{pmatrix} \tag{9.106}$$

For $\delta\theta \sim 0$, $\cos\delta\theta \approx 1$, $\sin\delta\theta \approx \delta\theta$ we can write

$$R_z(\delta\theta) = \begin{pmatrix} 1 & \delta\theta & 0 \\ -\delta\theta & 1 & 0 \\ 0 & 0 & 1 \end{pmatrix} = \begin{pmatrix} 1 & 0 & 0 \\ 0 & 1 & 0 \\ 0 & 0 & 1 \end{pmatrix} + \begin{pmatrix} 0 & \delta\theta & 0 \\ -\delta\theta & 0 & 0 \\ 0 & 0 & 0 \end{pmatrix}$$

$$R_z(\delta\theta) = \begin{pmatrix} 1 & 0 & 0 \\ 0 & 1 & 0 \\ 0 & 0 & 1 \end{pmatrix} + i\delta\theta\begin{pmatrix} 0 & -i & 0 \\ i & 0 & 0 \\ 0 & 0 & 0 \end{pmatrix}$$

$$R_z(\delta\theta) = I + i\delta\theta\, S_z \tag{9.107}$$

A finite rotation is superposition of successive infinitesimal rotations as

$$R_z(\delta\theta_1 + \delta\theta_2) = (I + i\delta\theta_1 S_z)(I + i\delta\theta_2 S_z)$$

As $\delta\theta_1 \sim \delta\theta_2 \sim \delta\theta$

$$R_z(\delta\theta_1 + \delta\theta_2) = (I + i\delta\theta S_z)^2 \tag{9.108}$$

For N such infinitesimal rotations leading to finite angle θ we get

$$R_z(\theta) = (I + i\delta\theta S_z)^N \tag{9.109}$$

If $N\delta\theta = \theta$ then $\delta\theta = \frac{\theta}{N}$ and hence

$$R_z(\theta) = \left(I + i\frac{\theta}{N}S_z\right)^N \tag{9.110}$$

With $x = i\theta S_z$ we get

$$R_z(\theta) = \mathop{\mathrm{Lt}}_{N \to \infty} \left(1 + \frac{x}{N}\right)^N \tag{9.111}$$

$$= \mathop{\mathrm{Lt}}_{N \to \infty} [1 + N.\frac{x}{N} + \frac{N(N+1)}{2!}\left(\frac{x}{N}\right)^2 + \frac{N(N+1)(N+3)}{3!}\left(\frac{x}{N}\right)^3 + ...]$$

$$= \mathop{\mathrm{Lt}}_{N \to \infty} [1 + x + \frac{N^2}{2!}\left(\frac{x}{N}\right)^2 + \frac{N^3}{3!}\left(\frac{x}{N}\right)^3 + ...] = 1 + x + \frac{x^2}{2!} + \frac{x^3}{3!} + ... = e^x$$

$$R_z(\theta) = e^{i\theta S_z} \tag{9.112}$$

In other words, we have

$$R_Z(\theta) = I \cos\theta + iS_z \sin\theta = e^{iS_z\theta} \tag{9.113}$$

which is analogous to equation (9.103).

✓ In an identical manner we can define the rotation matrix for rotation about the X-axis as

$$R_x(\phi) = \begin{pmatrix} 1 & 0 & 0 \\ 0 & \cos\phi & \sin\phi \\ 0 & -\sin\phi & \cos\phi \end{pmatrix} \tag{9.114}$$

(Here the 1×1 element of the matrix is 1)

The corresponding generator of rotation is

$$S_x = \begin{pmatrix} 0 & 0 & 0 \\ 0 & 0 & -i \\ 0 & i & 0 \end{pmatrix} \tag{9.115}$$

✓ Similarly, we can define the rotation matrix for rotation about the Y-axis as

$$R_y(\psi) = \begin{pmatrix} \cos\psi & 0 & \sin\psi \\ 0 & 1 & 0 \\ -\sin\psi & 0 & \cos\psi \end{pmatrix} \tag{9.116}$$

(Here the 2×2 element of the matrix is 1)

The corresponding generator of rotation is

$$S_y = \begin{pmatrix} 0 & 0 & i \\ 0 & 0 & 0 \\ -i & 0 & 0 \end{pmatrix} \tag{9.117}$$

Hence, S_x, S_y, S_z are generators of the $SO(3)$ group. This is shown in figure 9.4.

9.24 Relation between orbital angular momentum l_z about the Z-axis and rotation R_z about the Z-axis

Consider a function $\psi(x, y, z)$ with respect to a fixed coordinate system x, y, z.

Let $R_z(\delta\theta)$ be the rotation operator that affects rotation by angle $\delta\theta$ about the Z-axis and is given by

$$R_z(\delta\theta) = \begin{pmatrix} \cos\delta\theta & \sin\delta\theta & 0 \\ -\sin\delta\theta & \cos\delta\theta & 0 \\ 0 & 0 & 1 \end{pmatrix} \xrightarrow{\text{for } \delta\theta \to 0} \begin{pmatrix} 1 & \delta\theta & 0 \\ -\delta\theta & 1 & 0 \\ 0 & 0 & 1 \end{pmatrix} \tag{9.118}$$

Let $R_z(\delta\theta)$ operate on $\psi = \psi(x, y, z) = \begin{pmatrix} x \\ y \\ z \end{pmatrix}$ and we have

$$R_z(\delta\theta)\begin{pmatrix} x \\ y \\ z \end{pmatrix} = \begin{pmatrix} 1 & \delta\theta & 0 \\ -\delta\theta & 1 & 0 \\ 0 & 0 & 1 \end{pmatrix}\begin{pmatrix} x \\ y \\ z \end{pmatrix} = \begin{pmatrix} x + y\delta\theta \\ -x\delta\theta + y \\ z \end{pmatrix} \tag{9.119}$$

So we can write

$$R_z(\delta\theta)\psi(x, y, z) = \psi(x + y\delta\theta, y - x\delta\theta, z) \tag{9.120}$$

The formula for Taylor expansion is

$$f(x + \delta x, y + \delta y) = f(x, y) + f_x\,\delta x + f_y\,\delta y + \text{higher order terms} \tag{9.121}$$

where $f_x = \frac{\partial f}{\partial x}, f_y = \frac{\partial f}{\partial y}$

So making a Taylor expansion of the RHS of equation (9.120) we get

$$R_z(\delta\theta)\psi(x, y, z) = \psi(x, y, z) + \frac{\partial\psi}{\partial x}(y\delta\theta) + \frac{\partial\psi}{\partial y}(-x\delta\theta)$$

$$+\text{higher order terms in } \delta\theta.$$

Hence neglecting higher order terms in $\delta\theta$ we have

$$R_z(\delta\theta)\psi(x, y, z) = \psi(x, y, z) - \delta\theta(x\frac{\partial\psi}{\partial y} - y\frac{\partial\psi}{\partial x}) \tag{9.122}$$

Again, orbital angular momentum is given by

$$\vec{l} = \vec{r} \times \vec{p} = \begin{vmatrix} \hat{x} & \hat{y} & \hat{z} \\ x & y & z \\ p_x & p_y & p_z \end{vmatrix} = \hat{x}(yp_z - zp_y) + \hat{y}(zp_x - xp_z) + \hat{z}(xp_y - yp_x)$$

$$=\hat{x}l_x + \hat{y}l_y + \hat{z}l_z \tag{9.123}$$

Making operator replacement $p_y = -i\hbar\frac{\partial}{\partial y} = -i\frac{\partial}{\partial y}$ and $p_x = -i\hbar\frac{\partial}{\partial x} = -i\frac{\partial}{\partial x}$ in natural units $\hbar = 1$ we have

$$l_z = xp_y - yp_x = x\left(-i\frac{\partial}{\partial y}\right) - y\left(-i\frac{\partial}{\partial x}\right) = -i\left(x\frac{\partial}{\partial y} - y\frac{\partial}{\partial x}\right)$$

$$x\frac{\partial}{\partial y} - y\frac{\partial}{\partial x} = \frac{1}{-i}l_z = il_z \left(\text{as } i^2 = -1, \frac{1}{-i} = i\right)$$

Hence

$$x\frac{\partial \psi}{\partial y} - y\frac{\partial \psi}{\partial x} = \left(x\frac{\partial}{\partial y} - y\frac{\partial}{\partial x}\right)\psi = il_z\psi \tag{9.124}$$

Now from equation (9.122)

$$R_z(\delta\theta)\psi(x, y, z) = \psi - \delta\theta il_z\psi$$

$$R_z(\delta\theta)\psi = (1 - i\delta\theta l_z)\psi \tag{9.125}$$

$$R_z(\delta\theta) = (1 - i\delta\theta l_z) \tag{9.126}$$

Then rotation by angle $\theta + \delta\theta$ can be looked upon as rotation by θ and then by $\delta\theta$. Hence

$$R_z(\theta + \delta\theta)\psi = R_z(\delta\theta)R_z(\theta) = (1 - i\delta\theta l_z)R_z(\theta)\psi \equiv R_z\psi - i\delta\theta l_z R_z\psi$$

$$R_z(\theta + \delta\theta) - R_z = -i\delta\theta l_z R_z \tag{9.127}$$

$$dR_z = -\mathrm{i}d\theta l_z R_z \tag{9.128}$$

$$\frac{dR_z}{R_z} = -il_z d\theta$$

$$\mathrm{d}\ln R_z = -il_z d\theta$$

Integrate to get

$$R_z = e^{-il_z\theta} \text{ (overlooking a constant)} \tag{9.129}$$

This is the relation between orbital angular momentum l_z about the Z-axis and rotation R_z about the Z-axis.

- Commutation relation satisfied by angular momentum is (*exercise 9.28*)

$$[l_i l_j] = i\varepsilon_{ijk}l_k \text{ e. g } [l_x l_y] = il_z. \tag{9.130}$$

9.25 Unitary matrix

A unitary matrix U is defined as

$$U^\dagger U = UU^\dagger = I \tag{9.131}$$

$$(U^*)^T = U^\dagger = U^{-1}, \quad UU^{-1} = I \tag{9.132}$$

As complex conjugation is involved the elements of the matrix can be complex numbers.

9.26 Symmetry representation of the $SU(n)$ group

In particle physics, there are particular types of groups of unitary matrices that are useful in describing the symmetries of the Standard Model.

We first discuss what we mean by the $U(1)$ group.

9.27 $U(1)$ group

Consider the function

$$U(\theta) = e^{i\theta} \tag{9.133}$$

where θ is real.

Let us find if $U(\theta) = e^{i\theta}$ forms a group.

✓ Closure property

$$U(\theta_1)U(\theta_2) = e^{i\theta_1}e^{i\theta_2} = e^{i(\theta_1 + \theta_2)}$$

As θ_1, θ_2 are real, so $\theta_1 + \theta_2$ is also real. So $e^{i(\theta_1 + \theta_2)}$ is also an element of the same set.

✓ Associative property

Multiplication of complex numbers is associative.

✓ Identity element

We note that for $\theta = 0$, $U(\theta) = e^{i\theta} = e^0 = 1$ and also

$$e^{i\theta}. 1 = e^{i\theta}$$

So $\theta = 0$ is an identity element.

✓ Inverse element

$$U(\theta)U(-\theta) = e^{i\theta}. e^{-i\theta} = 1 \text{ which is the identity element.}$$

So $U(-\theta)$ is the inverse of $U(\theta)$. We do have an inverse for all elements of the set.

It therefore follows that since all the four properties or conditions are satisfied, $U(\theta) = e^{i\theta}$ forms a group called the $U(1)$ group.

U stands for unitary property of $U(\theta) = e^{i\theta}$ since

$$U^\dagger U = (e^{i\theta})^{*T} e^{i\theta} = e^{-i\theta}e^{i\theta} = 1$$

The 1 in $U(1)$ stands for the fact that the group has one parameter viz. θ. The group consists of 1×1 matrices. It corresponds to one dimension.

So $U(1)$ is a unitary group of 1×1 unitary matrices.

This $U(1)$ group represents symmetry associated with electromagnetism.

9.28 Group formed by unitary matrices

Unitary matrices form a group because of the following.

✓ Closure property

Let U_1, U_2. . be unitary matrices. Then

$$U_1^\dagger = U_1^{-1}$$

$$U_2^\dagger = U_2^{-1}$$

$$(U_1 U_2)^\dagger = U_2^\dagger U_1^\dagger = U_2^{-1} U_1^{-1} = (U_1 U_2)^{-1}$$

So $U_1 U_2$ is a unitary matrix formed from U_1 and U_2. So closure property is satisfied.

✓ Associative property

Matrix multiplication is associative

✓ Identity matrix element

The definition of unitary matrix implies identity matrix element viz. $UU^\dagger = I$.

✓ Inverse matrix element

The definition of a unitary matrix implies an inverse matrix element viz. $U^\dagger = U^{-1}$.

- The set of all 2×2 unitary matrices forms a group called the $U(2)$ group, U is for unitary and 2 suggesting that it involves 2×2 unitary matrices.
- The set of all 3×3 unitary matrices forms a group called the $U(3)$ group, U is for unitary and 3 suggesting that it involves 3×3 unitary matrices.
- The set of all $n \times n$ unitary matrices forms a group called the $U(n)$ group, U is for unitary and n suggesting that it involves $n \times n$ unitary matrices.

9.29 Unimodular unitary matrices

Let us make a special choice of the unitary matrices taking det $U = +1$ (unimodular). This special choice, i.e. unimodular property is denoted by the letter S. We thus have the following.

- 2×2 unitary matrices U with det $U = +1$ forms a group denoted by the $SU(2)$ group called a special unitary group of 2×2 unitary matrices.

  The $SU(2)$group represents symmetry associated with weak interaction.
- 3×3 unitary matrices U with det $U = +1$ form a group denoted by the $SU(3)$ group called a special unitary group of 3×3 unitary matrices.

  SU(3) group represents symmetry associated with strong interaction.
- $n \times n$ unitary matrices U with det $U = +1$ form a group denoted by the $SU(n)$ group called a special unitary group of $n \times n$ unitary matrices.

9.30 Order of $SU(n)$

Order of $SU(n)$ is $n^2 - 1$.

For $n = 2$ the order of $SU(2)$ is $2^2 - 1 = 3$ (three independent parameters). So there will be 3 generators for $SU(2)$.

For $n = 3$ the order of $SU(3)$ is $3^2 - 1 = 8$ (eight independent parameters). So there will be 8 generators for $SU(3)$.

This is shown in figure 9.4.

9.31 Generators of the $SU(n)$ group

Since for the $SU(n)$ group there are $n^2 - 1$ parameters we write the unitary matrices as

$$U(\alpha_1, \alpha_2, \ldots, \alpha_{n^2 - 1})$$

where α are the parameters.

If all the parameters are zero then the unitary matrix is an identity matrix I and if the parameters are infinitesimals then the unitary matrix will be infinitesimally deviated from identity. This infinitesimal deviation from identity is, say H, which is of the order of α, i.e. linear in α, plus there would be terms higher order in α. We thus can write

$$U(\alpha_1, \alpha_2, \ldots, \alpha_{n^2 - 1}) = I + iH + O(\alpha^2) \tag{9.134}$$

$$U^\dagger(\alpha_1, \alpha_2, \ldots, \alpha_{n^2 - 1}) = I - iH^\dagger + O(\alpha^2)$$

Neglecting higher order terms we have

$$U \approx I + iH \tag{9.135}$$

$$U^\dagger \approx I - iH^\dagger \tag{9.136}$$

Using $UU^\dagger = I$ (unitarity condition)

$$(I + iH)(I - iH^\dagger) = I \tag{9.137}$$

$$I + iH - iH^\dagger + HH^\dagger = I$$

We neglect $HH^\dagger \sim \alpha^2$. So we are left with

$$H = H^\dagger \tag{9.138}$$

H is $n \times n$ Hermitian matrix.

We can generate U by expressing it as a sum of a linear combination of Hermitian matrices as follows.

$$U = I + i\sum_{i=1}^{n^2-1} \alpha_i H_i \tag{9.139}$$

We have shown in exercise 9.32 that the special choice of $\det U = 1$ imposes the restriction that H is traceless. These Hermitian, traceless matrices H are called the generators of $SU(n)$.

For $SU(2)$ we require $2^2 - 1 = 3$ generators, i.e. 3 traceless Hermitian matrices.

For $SU(3)$ we require $3^2 - 1 = 8$ generators, i.e. 8 traceless Hermitian matrices. Let us construct the generators.

9.32 General form of 2×2 unitary unimodular matrix in $SU(2)$

- In exercise 9.21 we show that the general form of a 2×2 unitary unimodular matrix in $SU(2)$ is

$$U = \begin{pmatrix} a & b \\ -b^* & a^* \end{pmatrix} \tag{9.140}$$

where a, b, c, d are complex numbers in general.

9.33 General element of $SU(2)$

Let us first construct the most general element of $SU(2)$.

A spin up state can be represented by

$$|\uparrow\rangle = \begin{pmatrix} 1 \\ 0 \end{pmatrix}$$

and a spin down state can be represented by

$$|\downarrow\rangle = \begin{pmatrix} 0 \\ 1 \end{pmatrix}$$

Let us introduce a phase factor $e^{i\alpha}$ to $|\uparrow\rangle$ to get $\begin{pmatrix} e^{i\alpha} \\ 0 \end{pmatrix}$ and introduce a phase factor $e^{-i\alpha}$ to $|\downarrow\rangle$ to get $\begin{pmatrix} 0 \\ e^{-i\alpha} \end{pmatrix}$.

- We construct a matrix U_α with its columns as $\begin{pmatrix} e^{i\alpha} \\ 0 \end{pmatrix}$ and $\begin{pmatrix} 0 \\ e^{-i\alpha} \end{pmatrix}$ i.e.

$$U_\alpha = \begin{pmatrix} e^{i\alpha} & 0 \\ 0 & e^{-i\alpha} \end{pmatrix} \tag{9.141}$$

$$U_\alpha^\dagger = (U_\alpha^*)^T = \begin{pmatrix} e^{i\alpha} & 0 \\ 0 & e^{-i\alpha} \end{pmatrix}^{*T} = \begin{pmatrix} e^{-i\alpha} & 0 \\ 0 & e^{i\alpha} \end{pmatrix}^T = \begin{pmatrix} e^{-i\alpha} & 0 \\ 0 & e^{i\alpha} \end{pmatrix}$$

$$U_\alpha U_\alpha^\dagger = \begin{pmatrix} e^{i\alpha} & 0 \\ 0 & e^{-i\alpha} \end{pmatrix}\begin{pmatrix} e^{-i\alpha} & 0 \\ 0 & e^{i\alpha} \end{pmatrix} = \begin{pmatrix} 1 & 0 \\ 0 & 1 \end{pmatrix} = I$$

Also

$$\det U_\alpha = \begin{vmatrix} e^{i\alpha} & 0 \\ 0 & e^{-i\alpha} \end{vmatrix} = +1 \tag{9.142}$$

So U_α is 2×2 unitary unimodular matrix.

- To get the most general element construct the product

$$U = U_\alpha R_z(\eta) U_\beta = \begin{pmatrix} e^{i\alpha} & 0 \\ 0 & e^{-i\alpha} \end{pmatrix}\begin{pmatrix} \cos\eta & \sin\eta \\ -\sin\eta & \cos\eta \end{pmatrix}\begin{pmatrix} e^{i\beta} & 0 \\ 0 & e^{-i\beta} \end{pmatrix} \tag{9.143}$$

where U_α, U_β are unitary matrices and $R_z(\eta)$ is a rotation matrix.

$$U = \begin{pmatrix} e^{i\alpha}\cos\eta & e^{i\alpha}\sin\eta \\ -e^{-i\alpha}\sin\eta & e^{-i\alpha}\cos\eta \end{pmatrix}\begin{pmatrix} e^{i\beta} & 0 \\ 0 & e^{-i\beta} \end{pmatrix} = \begin{pmatrix} e^{i(\alpha+\beta)}\cos\eta & e^{i(\alpha-\beta)}\sin\eta \\ -e^{-i(\alpha-\beta)}\sin\eta & e^{-i(\alpha+\beta)}\cos\eta \end{pmatrix}$$

Let $\alpha + \beta = \xi$, $\alpha - \beta = \zeta$

$$U = \begin{pmatrix} e^{i\xi}\cos\eta & e^{i\zeta}\sin\eta \\ -e^{-i\zeta}\sin\eta & e^{-i\xi}\cos\eta \end{pmatrix} \tag{9.144}$$

$$U^\dagger = (U^*)^T = \begin{pmatrix} e^{i\xi}\cos\eta & e^{i\zeta}\sin\eta \\ -e^{-i\zeta}\sin\eta & e^{-i\xi}\cos\eta \end{pmatrix}^{*T} = \begin{pmatrix} e^{-i\xi}\cos\eta & e^{-i\zeta}\sin\eta \\ -e^{i\zeta}\sin\eta & e^{i\xi}\cos\eta \end{pmatrix}^{T}$$

$$= \begin{pmatrix} e^{-i\xi}\cos\eta & -e^{i\zeta}\sin\eta \\ e^{-i\zeta}\sin\eta & e^{i\xi}\cos\eta \end{pmatrix}$$

$$UU^\dagger = \begin{pmatrix} e^{i\xi}\cos\eta & e^{i\zeta}\sin\eta \\ -e^{-i\zeta}\sin\eta & e^{-i\xi}\cos\eta \end{pmatrix}\begin{pmatrix} e^{-i\xi}\cos\eta & -e^{i\zeta}\sin\eta \\ e^{-i\zeta}\sin\eta & e^{i\xi}\cos\eta \end{pmatrix} = \begin{pmatrix} 1 & 0 \\ 0 & 1 \end{pmatrix} = I$$

Also

$$\det U = \begin{vmatrix} e^{i\xi}\cos\eta & e^{i\zeta}\sin\eta \\ -e^{-i\zeta}\sin\eta & e^{-i\xi}\cos\eta \end{vmatrix} = \cos^2\eta + \sin^2\eta = 1 \tag{9.145}$$

U is the most general element of $SU(2)$.
We have re-derived it in *exercise 9.30*.

- We shall now establish how the Pauli matrices σ_1, σ_2, σ_3 are associated with $SU(2)$.

In fact Pauli matrices emerge out of U as follows.
Differentiating this matrix U we can obtain the Pauli spin matrices σ_1, σ_2, σ_3.
✓ Let us evaluate

$$-i\frac{\partial U}{\partial \xi}\Big|_{\xi=0,\ \eta=0} = -i\frac{\partial}{\partial \xi}\begin{pmatrix} e^{i\xi}\cos\eta & e^{i\zeta}\sin\eta \\ -e^{-i\zeta}\sin\eta & e^{-i\xi}\cos\eta \end{pmatrix}\Big|_{\xi=0,\ \eta=0}$$

$$= -i\begin{pmatrix} ie^{i\xi}\cos\eta & 0 \\ 0 & -ie^{-i\xi}\cos\eta \end{pmatrix}\Big|_{\xi=0,\ \eta=0} = -i\begin{pmatrix} i & 0 \\ 0 & -i \end{pmatrix} = \begin{pmatrix} 1 & 0 \\ 0 & -1 \end{pmatrix}$$

$$-i\frac{\partial U}{\partial \xi}\Big|_{\xi=0,\ \eta=0} = \begin{pmatrix} 1 & 0 \\ 0 & -1 \end{pmatrix} = \sigma_3 \qquad (9.146)$$

✓ Let us evaluate

$$-i\frac{\partial U}{\partial \eta}\Big|_{\eta=0,\ \zeta=0} = -i\frac{\partial}{\partial \eta}\begin{pmatrix} e^{i\xi}\cos\eta & e^{i\zeta}\sin\eta \\ -e^{-i\zeta}\sin\eta & e^{-i\xi}\cos\eta \end{pmatrix}\Big|_{\eta=0,\ \zeta=0}$$

$$=-i\begin{pmatrix} -e^{i\xi}\sin\eta & e^{i\zeta}\cos\eta \\ -e^{-i\zeta}\cos\eta & -e^{-i\xi}\sin\eta \end{pmatrix}\Big|_{\eta=0,\ \zeta=0} = -i\begin{pmatrix} 0 & 1 \\ -1 & 0 \end{pmatrix} = \begin{pmatrix} 0 & -i \\ i & 0 \end{pmatrix}$$

$$-i\frac{\partial U}{\partial \eta}\Big|_{\eta=0,\ \zeta=0} = \begin{pmatrix} 0 & -i \\ i & 0 \end{pmatrix} = \sigma_2 \qquad (9.147)$$

✓ Let us choose the general form as given in equation (9.140) viz.

$$U = \begin{pmatrix} a & b \\ -b^* & a^* \end{pmatrix}$$

With $b = i\beta$ (purely imaginary), $\beta=$ real

$$-b^*=-(i\beta)^*=-(-i)\beta = i\beta$$

We thus have the following form for U

$$U = \begin{pmatrix} a & i\beta \\ i\beta & a^* \end{pmatrix}$$

$$\det U = \begin{vmatrix} a & i\beta \\ i\beta & a^* \end{vmatrix} = aa^* - (i\beta)(i\beta) = |a|^2 + |\beta|^2 = 1$$

$$a^2 = 1 - \beta^2$$

$$a = \sqrt{1-\beta^2}$$

$$U = \begin{pmatrix} \sqrt{1-\beta^2} & i\beta \\ i\beta & \sqrt{1-\beta^2} \end{pmatrix}$$

Consider

$$-i\frac{\partial U}{\partial \beta}\Big|_{\beta=0} = -i\frac{\partial}{\partial \beta}\begin{pmatrix} \sqrt{1-\beta^2} & i\beta \\ i\beta & \sqrt{1-\beta^2} \end{pmatrix}\Big|_{\beta=0}$$

$$= -i \begin{pmatrix} \dfrac{-2\beta}{2\sqrt{1-\beta^2}} & i \\ i & \dfrac{-2\beta}{2\sqrt{1-\beta^2}} \end{pmatrix}\Big|_{\beta=0} = -i \begin{pmatrix} -\beta & i \\ i & -\beta \end{pmatrix}\Big|_{\beta=0} = -i \begin{pmatrix} 0 & i \\ i & 0 \end{pmatrix}$$

$$-i\frac{\partial U}{\partial \beta}\Big|_{\beta=0} = \begin{pmatrix} 0 & 1 \\ 1 & 0 \end{pmatrix} = \sigma_1 \tag{9.148}$$

Clearly Pauli matrices σ_1, σ_2, σ_3 are associated with $SU(2)$.

9.34 $SU(2)$ generators

We write down an arbitrary 2×2 Hermitian matrix comprising complex numbers

$$H = \begin{pmatrix} z_1 & z_2 \\ z_3 & z_4 \end{pmatrix} \tag{9.149}$$

$$H^\dagger = H^{*T} = \begin{pmatrix} z_1^* & z_2^* \\ z_3^* & z_4^* \end{pmatrix}^T = \begin{pmatrix} z_1^* & z_3^* \\ z_2^* & z_4^* \end{pmatrix} \tag{9.150}$$

Since H is Hermitian, i.e. $H = H^\dagger$ we have from equations (9.149) and (9.150)

$$\begin{pmatrix} z_1 & z_2 \\ z_3 & z_4 \end{pmatrix} = \begin{pmatrix} z_1^* & z_3^* \\ z_2^* & z_4^* \end{pmatrix}$$

$$z_1 = z_1^* \text{ i.e. } z_1 \text{ is real, say } z_1 = a$$

$$z_4 = z_4^* \text{ i.e. } z_4 \text{ is real, say } z_4 = b$$

$$z_2 = z_3^*$$

Let us write $z_3 = c + id$, $z_2 = z_3^* = c - id$. Then

$$H = \begin{pmatrix} z_1 & z_2 \\ z_3 & z_4 \end{pmatrix} = \begin{pmatrix} a & c - id \\ c + id & b \end{pmatrix}$$

H has to be traceless. So

$$a + b = 0$$

$$b = -a$$

We split and write as follows

$$H = \begin{pmatrix} z_1 & z_2 \\ z_3 & z_4 \end{pmatrix} = \begin{pmatrix} a & c - id \\ c + id & -a \end{pmatrix} \tag{9.151}$$

$$H = c\begin{pmatrix} 0 & 1 \\ 1 & 0 \end{pmatrix} + d\begin{pmatrix} 0 & -i \\ i & 0 \end{pmatrix} + a\begin{pmatrix} 1 & 0 \\ 0 & -1 \end{pmatrix} = c\sigma_1 + d\sigma_2 + a\sigma_3 \qquad (9.152)$$

where σ_1, σ_2, σ_3 are Pauli spin matrices. Only σ_3 is diagonal.

Pauli matrices are traceless, i.e. $Tr\sigma_i = 0$ (*exercise 9.22*), Hermitian i.e. $\sigma_i^\dagger = \sigma_i$ (*exercise 9.23*) and unitary, i.e. $\sigma_i^\dagger = \sigma_i^{-1}$ (*exercise 9.24*) and non-commuting satisfying the commutation relation (*exercise 9.25*).

$$[\sigma_i, \sigma_j] = 2i\varepsilon_{ijk}\sigma_k \qquad (9.153)$$

and having trace (exercise 9.35)

$$\mathrm{Tr}\left(\frac{\sigma_i}{2}\frac{\sigma_j}{2}\right) = \frac{1}{2}\delta_{ij} \qquad (9.154)$$

Similar to equation (9.103) viz. $R(\theta) = I\cos\theta + i\sigma_2\sin\theta = e^{i\sigma_2\theta}$ we take

$$U_1 = e^{i\frac{\sigma_1}{2}a_1} = e^{iS_1a_1} \qquad (9.155)$$

$$U_2 = e^{i\frac{\sigma_2}{2}a_2} = e^{iS_2a_2} \qquad (9.156)$$

$$U_3 = e^{i\frac{\sigma_3}{2}a_3} = e^{iS_3a_3} \qquad (9.157)$$

$$\text{i.e. } U_j = e^{i\frac{\sigma_j}{2}a_j} = e^{iS_ja_j} \qquad (9.158)$$

where

$$S_j = \frac{\sigma_j}{2} \qquad (9.159)$$

and a_j is a parameter.

The factor 2 in equation (9.158) is because of the relations shown in equations (9.153) and (9.154).

Hence $\frac{\sigma_j}{2}$ i.e. $\frac{\sigma_x}{2}, \frac{\sigma_y}{2}, \frac{\sigma_z}{2}$ are the generators of $SU(2)$ as indicated in figure 9.4.

The commutation relation between the $SU(2)$ generators $\frac{1}{2}\sigma_i$ is (*exercise 9.34*)

$$\left[\frac{\sigma_i}{2}, \frac{\sigma_j}{2}\right] = i\varepsilon_{ijk}\frac{\sigma_k}{2} \qquad (9.160)$$

- The reason for taking the scale factor $\frac{1}{2}$ in $S_i = \frac{\sigma_i}{2}$ is as follows.
  Commutation relation of the Pauli matrices are

$$[\sigma_i\sigma_j] = 2i\varepsilon_{ijk}\sigma_k \ (\textit{exercise 9.25}) \qquad (9.161)$$

  Clearly a factor of 2 is involved.
  Commutation relation of $S_i = \frac{\sigma_i}{2}$ will be

$$[S_i S_j] = S_i S_j - S_j S_i = \frac{\sigma_i}{2}\frac{\sigma_j}{2} - \frac{\sigma_j}{2}\frac{\sigma_i}{2} = \frac{1}{4}(\sigma_i \sigma_j - \sigma_j \sigma_i) = \frac{1}{4}[\sigma_i \sigma_j] = \frac{1}{4}2i\varepsilon_{ijk}\sigma_k$$

$$[S_i S_j] = i\varepsilon_{ijk}\frac{\sigma_k}{2} = i\varepsilon_{ijk} S_k \tag{9.162}$$

This matches the commutation relation for orbital angular momentum (*exercise 9.28*) viz.

$$[l_i l_j] = i\varepsilon_{ijk} l_k \tag{9.163}$$

- Similar to the relation of equation (9.103) viz. $R(\theta) = e^{i\sigma_2 \theta} = I \cos \theta + i\sigma_2 \sin \theta$ we can write from equation (9.158)

$$U_j = e^{i\frac{\sigma_j}{2}a_j} = e^{i\sigma_j \frac{a_j}{2}} = I \cos \frac{a_j}{2} + i\sigma_j \sin \frac{a_j}{2}. \tag{9.164}$$

9.35 Correspondence between $SU(2)$ and $SO(3)$

$SO(3)$ describes rotation in 3D ordinary space such that $x^2 + y^2 + z^2$ is invariant.
Determinantal value is $+1$.
There are 3 independent parameters for the $SO(3)$ group.
The elements of $SU(2)$ describe rotation in 2D complex space such that $|z_1|^2 + |z_2|^2$ is invariant.
Determinantal value is $+1$.
There are 3 independent parameters for the $SU(2)$ group.
Clearly $SO(3)$ and $SU(2)$ groups have the same number of parameters and the same rotational interpretation. This suggests that there is correspondence between $SO(3)$ and $SU(2)$. Let us develop the correspondence.
Operation of $SU(2)$ on a matrix M is given by the unitary transformation

$$M' = UMU^\dagger \tag{9.165}$$

where M is a 2×2 matrix of zero trace.
We have shown in *exercise 9.26* that any 2×2 matrix can be expressed as a linear combination of Pauli spin matrices and an identity matrix, i.e. in terms of $\sigma_1, \sigma_2, \sigma_3, I$.
Let us take

$$M = x\sigma_1 + y\sigma_2 + z\sigma_3 = x\begin{pmatrix} 0 & 1 \\ 1 & 0 \end{pmatrix} + y\begin{pmatrix} 0 & -i \\ i & 0 \end{pmatrix} + z\begin{pmatrix} 1 & 0 \\ 0 & -1 \end{pmatrix} = \begin{pmatrix} z & x - iy \\ x + iy & -z \end{pmatrix}$$

Trace is invariant under unitary transformation. So

$$M' = x'\sigma_1 + y'\sigma_2 + z'\sigma_3 = \begin{pmatrix} z' & x' - iy' \\ x' + iy' & -z' \end{pmatrix}$$

Determinantal value is also invariant under unitary transformation. So we can equate the determinants.

$$\det M = \det M'$$

$$\begin{vmatrix} z & x - iy \\ x + iy & -z \end{vmatrix} = \begin{vmatrix} z' & x' - iy' \\ x' + iy' & -z' \end{vmatrix}$$

$$(z)(-z) - (x - iy)(x + iy) = (z')(-z') - (x' - iy')(x' + iy')$$

$$-z^2 - (x^2 + y^2) = -z'^2 - (x'^2 + y'^2)$$

$$x^2 + y^2 + z^2 = x'^2 + y'^2 + z'^2 = \text{invariant}$$

This holds under $SO(3)$.

Similarly, for $SU(2)$ this quantity $x^2 + y^2 + z^2$ is also invariant.

So operation of $SU(2)$ on M produces rotation of coordinates in XYZ. Hence $SU(2)$ and $SO(3)$ have correspondence.

Let us develop further correspondence as follows.

Consider the general element of $SU(2)$ as given in equation (9.140) viz.

$$U = \begin{pmatrix} a & b \\ -b^* & a^* \end{pmatrix} \text{ with } a = e^{i\frac{\alpha}{2}}, \, b = 0 \text{ i. e.}$$

$$U = \begin{pmatrix} e^{i\frac{\alpha}{2}} & 0 \\ 0 & e^{-i\frac{\alpha}{2}} \end{pmatrix} = U\left(\frac{\alpha}{2}\right) \tag{9.166}$$

and so

$$U^\dagger = \begin{pmatrix} e^{-i\frac{\alpha}{2}} & 0 \\ 0 & e^{i\frac{\alpha}{2}} \end{pmatrix} \tag{9.167}$$

Consider unitary operation

$$M' = UMU^\dagger = U(x\sigma_1 + y\sigma_2 + z\sigma_3)U^\dagger = x'\sigma_1 + y'\sigma_2 + z'\sigma_3 \tag{9.168}$$

$$M' = Ux\sigma_1 U^\dagger + Uy\sigma_2 U^\dagger + Uz\sigma_3 U^\dagger$$

$$= \begin{pmatrix} e^{i\frac{\alpha}{2}} & 0 \\ 0 & e^{-i\frac{\alpha}{2}} \end{pmatrix} x \begin{pmatrix} 0 & 1 \\ 1 & 0 \end{pmatrix} \begin{pmatrix} e^{-i\frac{\alpha}{2}} & 0 \\ 0 & e^{i\frac{\alpha}{2}} \end{pmatrix} + \begin{pmatrix} e^{i\frac{\alpha}{2}} & 0 \\ 0 & e^{-i\frac{\alpha}{2}} \end{pmatrix} y \begin{pmatrix} 0 & -i \\ i & 0 \end{pmatrix} \begin{pmatrix} e^{-i\frac{\alpha}{2}} & 0 \\ 0 & e^{i\frac{\alpha}{2}} \end{pmatrix}$$

$$+ \begin{pmatrix} e^{i\frac{\alpha}{2}} & 0 \\ 0 & e^{-i\frac{\alpha}{2}} \end{pmatrix} z \begin{pmatrix} 1 & 0 \\ 0 & -1 \end{pmatrix} \begin{pmatrix} e^{-i\frac{\alpha}{2}} & 0 \\ 0 & e^{i\frac{\alpha}{2}} \end{pmatrix}$$

$$M' = \begin{pmatrix} e^{i\frac{\alpha}{2}} & 0 \\ 0 & e^{-i\frac{\alpha}{2}} \end{pmatrix} \begin{pmatrix} 0 & x \\ x & 0 \end{pmatrix} \begin{pmatrix} e^{-i\frac{\alpha}{2}} & 0 \\ 0 & e^{i\frac{\alpha}{2}} \end{pmatrix} + \begin{pmatrix} e^{i\frac{\alpha}{2}} & 0 \\ 0 & e^{-i\frac{\alpha}{2}} \end{pmatrix} \begin{pmatrix} 0 & -iy \\ iy & 0 \end{pmatrix} \begin{pmatrix} e^{-i\frac{\alpha}{2}} & 0 \\ 0 & e^{i\frac{\alpha}{2}} \end{pmatrix}$$

$$+\begin{pmatrix} e^{i\frac{\alpha}{2}} & 0 \\ 0 & e^{-i\frac{\alpha}{2}} \end{pmatrix}\begin{pmatrix} z & 0 \\ 0 & -z \end{pmatrix}\begin{pmatrix} e^{-i\frac{\alpha}{2}} & 0 \\ 0 & e^{i\frac{\alpha}{2}} \end{pmatrix}$$

$$M' = \begin{pmatrix} e^{i\frac{\alpha}{2}} & 0 \\ 0 & e^{-i\frac{\alpha}{2}} \end{pmatrix}\begin{pmatrix} 0 & xe^{i\frac{\alpha}{2}} \\ xe^{-i\frac{\alpha}{2}} & 0 \end{pmatrix} + \begin{pmatrix} e^{i\frac{\alpha}{2}} & 0 \\ 0 & e^{-i\frac{\alpha}{2}} \end{pmatrix}\begin{pmatrix} 0 & -\mathrm{i}ye^{i\frac{\alpha}{2}} \\ \mathrm{i}ye^{-i\frac{\alpha}{2}} & e^{i\frac{\alpha}{2}} \end{pmatrix}$$

$$+\begin{pmatrix} e^{i\frac{\alpha}{2}} & 0 \\ 0 & e^{-i\frac{\alpha}{2}} \end{pmatrix}\begin{pmatrix} ze^{-i\frac{\alpha}{2}} & 0 \\ 0 & -ze^{i\frac{\alpha}{2}} \end{pmatrix}$$

$$M' = \begin{pmatrix} 0 & xe^{i\alpha} \\ xe^{-i\alpha} & 0 \end{pmatrix} + \begin{pmatrix} 0 & -\mathrm{i}ye^{i\alpha} \\ \mathrm{i}ye^{-i\alpha} & 0 \end{pmatrix} + \begin{pmatrix} z & 0 \\ 0 & -z \end{pmatrix}$$

$$= x\begin{pmatrix} 0 & e^{i\alpha} \\ e^{-i\alpha} & 0 \end{pmatrix} + \mathrm{i}y\begin{pmatrix} 0 & -e^{i\alpha} \\ e^{-i\alpha} & 0 \end{pmatrix} + z\begin{pmatrix} 1 & 0 \\ 0 & -1 \end{pmatrix}$$

$$= x\begin{pmatrix} 0 & \cos\alpha + i\sin\alpha \\ \cos\alpha - i\sin\alpha & 0 \end{pmatrix} + \mathrm{i}y\begin{pmatrix} 0 & -(\cos\alpha + i\sin\alpha) \\ \cos\alpha - i\sin\alpha & 0 \end{pmatrix} + z\begin{pmatrix} 1 & 0 \\ 0 & -1 \end{pmatrix}$$

$$= x\cos\alpha\begin{pmatrix} 0 & 1 \\ 1 & 0 \end{pmatrix} - x\sin\alpha\begin{pmatrix} 0 & -i \\ i & 0 \end{pmatrix} + y\cos\alpha\begin{pmatrix} 0 & -i \\ i & 0 \end{pmatrix} + y\sin\alpha\begin{pmatrix} 0 & 1 \\ 1 & 0 \end{pmatrix} + z\begin{pmatrix} 1 & 0 \\ 0 & -1 \end{pmatrix}$$

$$M' = x\cos\alpha\,\sigma_1 - x\sin\alpha\,\sigma_2 + y\cos\alpha\,\sigma_2 + y\sin\alpha\,\sigma_1 + z\sigma_3$$

$$M' = (x\cos\alpha + y\sin\alpha)\sigma_1 + (-x\sin\alpha + y\cos\alpha)\sigma_2 + z\sigma_3$$

Comparing with equation (9.168) viz. $M' = x'\sigma_1 + y'\sigma_2 + z'\sigma_3$ we get

$$x' = x\cos\alpha + y\sin\alpha$$

$$y' = -x\sin\alpha + y\cos\alpha$$

$$z' = z$$

This suggests the transformation relation

$$\begin{pmatrix} x' \\ y' \\ z' \end{pmatrix} = \begin{pmatrix} \cos\alpha & \sin\alpha & 0 \\ -\sin\alpha & \cos\alpha & 0 \\ 0 & 0 & 1 \end{pmatrix}\begin{pmatrix} x \\ y \\ z \end{pmatrix} = R_z(\alpha)\begin{pmatrix} x \\ y \\ z \end{pmatrix} \tag{9.169}$$

where $R_z(\alpha)$ is the rotation matrix.

Clearly the 2×2 unitary transformation using $U\left(\frac{\alpha}{2}\right) \equiv U_3\left(\frac{\alpha}{2}\right)$ is equivalent to the rotational operation performed by $R_z(\alpha)$ of $SO(3)$.

Also, clear that the correspondence is not $1:1$ but $\frac{\alpha}{2}: \alpha = 1:2$ between SU(2) and $SO(3)$.

Clear that the element $U\left(\frac{\alpha}{2}\right)$ of $SU(2)$ corresponds to rotation through α about the z-axis of the group $SO(3)$, i.e.

$$U\left(\frac{\alpha}{2}\right) = \begin{pmatrix} e^{i\frac{\alpha}{2}} & 0 \\ 0 & e^{-i\frac{\alpha}{2}} \end{pmatrix} \xrightarrow{\text{corresponds to}} R_z(\alpha) = \begin{pmatrix} \cos\alpha & \sin\alpha & 0 \\ -\sin\alpha & \cos\alpha & 0 \\ 0 & 0 & 1 \end{pmatrix} \qquad (9.170)$$

✓ Let us now choose (equation (9.140))

$$U = \begin{pmatrix} a & b \\ -b^* & a^* \end{pmatrix} \text{ with } a = \cos\frac{\beta}{2}, \ b = -\sin\frac{\beta}{2} \text{ i. e.}$$

$$U = \begin{pmatrix} \cos\frac{\beta}{2} & -\sin\frac{\beta}{2} \\ \sin\frac{\beta}{2} & \cos\frac{\beta}{2} \end{pmatrix} = U\left(\frac{\beta}{2}\right) \text{ and so } U^\dagger = \begin{pmatrix} \cos\frac{\beta}{2} & -\sin\frac{\beta}{2} \\ \sin\frac{\beta}{2} & \cos\frac{\beta}{2} \end{pmatrix}$$

Consider unitary operation

$$M' = UMU^\dagger = U(x\sigma_1 + y\sigma_2 + z\sigma_3)U^\dagger = x'\sigma_1 + y'\sigma_2 + z'\sigma_3 \qquad (9.171)$$

$$M' = Ux\sigma_1 U^\dagger + Uy\sigma_2 U^\dagger + Uz\sigma_3 U^\dagger$$

$$= \begin{pmatrix} \cos\frac{\beta}{2} & -\sin\frac{\beta}{2} \\ \sin\frac{\beta}{2} & \cos\frac{\beta}{2} \end{pmatrix} x \begin{pmatrix} 0 & 1 \\ 1 & 0 \end{pmatrix} \begin{pmatrix} \cos\frac{\beta}{2} & \sin\frac{\beta}{2} \\ -\sin\frac{\beta}{2} & \cos\frac{\beta}{2} \end{pmatrix}$$

$$+ \begin{pmatrix} \cos\frac{\beta}{2} & -\sin\frac{\beta}{2} \\ \sin\frac{\beta}{2} & \cos\frac{\beta}{2} \end{pmatrix} y \begin{pmatrix} 0 & -i \\ i & 0 \end{pmatrix} \begin{pmatrix} \cos\frac{\beta}{2} & \sin\frac{\beta}{2} \\ -\sin\frac{\beta}{2} & \cos\frac{\beta}{2} \end{pmatrix}$$

$$+ \begin{pmatrix} \cos\frac{\beta}{2} & -\sin\frac{\beta}{2} \\ \sin\frac{\beta}{2} & \cos\frac{\beta}{2} \end{pmatrix} z \begin{pmatrix} 1 & 0 \\ 0 & -1 \end{pmatrix} \begin{pmatrix} \cos\frac{\beta}{2} & \sin\frac{\beta}{2} \\ -\sin\frac{\beta}{2} & \cos\frac{\beta}{2} \end{pmatrix}$$

$$M' = \begin{pmatrix} \cos\frac{\beta}{2} & -\sin\frac{\beta}{2} \\ \sin\frac{\beta}{2} & \cos\frac{\beta}{2} \end{pmatrix} \begin{pmatrix} 0 & x \\ x & 0 \end{pmatrix} \begin{pmatrix} \cos\frac{\beta}{2} & \sin\frac{\beta}{2} \\ -\sin\frac{\beta}{2} & \cos\frac{\beta}{2} \end{pmatrix}$$

$$+ \begin{pmatrix} \cos\frac{\beta}{2} & -\sin\frac{\beta}{2} \\ \sin\frac{\beta}{2} & \cos\frac{\beta}{2} \end{pmatrix} \begin{pmatrix} 0 & -iy \\ iy & 0 \end{pmatrix} \begin{pmatrix} \cos\frac{\beta}{2} & \sin\frac{\beta}{2} \\ -\sin\frac{\beta}{2} & \cos\frac{\beta}{2} \end{pmatrix}$$

$$+ \begin{pmatrix} \cos\frac{\beta}{2} & -\sin\frac{\beta}{2} \\ \sin\frac{\beta}{2} & \cos\frac{\beta}{2} \end{pmatrix} \begin{pmatrix} z & 0 \\ 0 & -z \end{pmatrix} \begin{pmatrix} \cos\frac{\beta}{2} & \sin\frac{\beta}{2} \\ -\sin\frac{\beta}{2} & \cos\frac{\beta}{2} \end{pmatrix}$$

$$= \begin{pmatrix} \cos\frac{\beta}{2} & -\sin\frac{\beta}{2} \\ \sin\frac{\beta}{2} & \cos\frac{\beta}{2} \end{pmatrix} \begin{pmatrix} -x\sin\frac{\beta}{2} & x\cos\frac{\beta}{2} \\ x\cos\frac{\beta}{2} & x\sin\frac{\beta}{2} \end{pmatrix} + \begin{pmatrix} \cos\frac{\beta}{2} & -\sin\frac{\beta}{2} \\ \sin\frac{\beta}{2} & \cos\frac{\beta}{2} \end{pmatrix} \begin{pmatrix} iy\sin\frac{\beta}{2} & -iy\cos\frac{\beta}{2} \\ iy\cos\frac{\beta}{2} & iy\sin\frac{\beta}{2} \end{pmatrix}$$

$$+\begin{pmatrix} \cos\frac{\beta}{2} & -\sin\frac{\beta}{2} \\ \sin\frac{\beta}{2} & \cos\frac{\beta}{2} \end{pmatrix}\begin{pmatrix} z\cos\frac{\beta}{2} & z\sin\frac{\beta}{2} \\ z\sin\frac{\beta}{2} & -z\cos\frac{\beta}{2} \end{pmatrix}$$

$$M' = \begin{pmatrix} -x2\cos\frac{\beta}{2}\sin\frac{\beta}{2} & x(\cos^2\frac{\beta}{2}-\sin^2\frac{\beta}{2}) \\ x(\cos^2\frac{\beta}{2}-\sin^2\frac{\beta}{2}) & x2\cos\frac{\beta}{2}\sin\frac{\beta}{2} \end{pmatrix} + \begin{pmatrix} 0 & -iy \\ iy & 0 \end{pmatrix}$$

$$+\begin{pmatrix} z(\cos^2\frac{\beta}{2}-\sin^2\frac{\beta}{2}) & z2\cos\frac{\beta}{2}\sin\frac{\beta}{2} \\ z2\cos\frac{\beta}{2}\sin\frac{\beta}{2} & -z(\cos^2\frac{\beta}{2}-\sin^2\frac{\beta}{2}) \end{pmatrix}$$

$$M' = \begin{pmatrix} -x\sin\beta & x\cos\beta \\ x\cos\beta & x\sin\beta \end{pmatrix} + \begin{pmatrix} 0 & -iy \\ iy & 0 \end{pmatrix} + \begin{pmatrix} z\cos\beta & z\sin\beta \\ z\sin\beta & -z\cos\beta \end{pmatrix}$$

$$M' = x\cos\beta\begin{pmatrix} 0 & 1 \\ 1 & 0 \end{pmatrix} - x\sin\beta\begin{pmatrix} 1 & 0 \\ 0 & -1 \end{pmatrix} + y\begin{pmatrix} 0 & -i \\ i & 0 \end{pmatrix} + z\cos\beta\begin{pmatrix} 1 & 0 \\ 0 & -1 \end{pmatrix} + z\sin\beta\begin{pmatrix} 0 & 1 \\ 1 & 0 \end{pmatrix}$$

$$M' = x\cos\beta\sigma_1 - x\sin\beta\sigma_3 + y\sigma_2 + z\cos\beta\sigma_3 + z\sin\beta\sigma_1$$

$$=(x\cos\beta + z\sin\beta)\sigma_1 + y\sigma_2 + (-x\sin\beta + z\cos\beta)\sigma_3$$

Comparison with equation (9.171) viz. $M' = x'\sigma_1 + y'\sigma_2 + z'\sigma_3$ gives

$$x' = x\cos\beta + z\sin\beta$$

$$y' = y$$

$$z' = -x\sin\beta + z\cos\beta$$

This suggests the transformation relation

$$\begin{pmatrix} x' \\ y' \\ z' \end{pmatrix} = \begin{pmatrix} \cos\beta & 0 & \sin\beta \\ 0 & 1 & 0 \\ -\sin\beta & 0 & \cos\beta \end{pmatrix}\begin{pmatrix} x \\ y \\ z \end{pmatrix} = R_Y(\beta)\begin{pmatrix} x \\ y \\ z \end{pmatrix} \tag{9.172}$$

where $R_Y(\beta)$ is the rotation matrix about Y axis.

Clear that the element $U(\frac{\beta}{2})$ of $SU(2)$ corresponds to rotation through β about Y axis of the group $SO(3)$, i.e.

$$U\left(\frac{\beta}{2}\right) = \begin{pmatrix} \cos\frac{\beta}{2} & -\sin\frac{\beta}{2} \\ \sin\frac{\beta}{2} & \cos\frac{\beta}{2} \end{pmatrix} \xrightarrow{\text{corresponds to}} R_Y(\beta) = \begin{pmatrix} \cos\beta & 0 & \sin\beta \\ 0 & 1 & 0 \\ -\sin\beta & 0 & \cos\beta \end{pmatrix} \tag{9.173}$$

✓ The general form of $SU(2)$ (*exercise 9.30*) is

$$U\left(\frac{\alpha}{2}, \frac{\beta}{2}, \frac{\gamma}{2}\right) = U_3(\frac{\gamma}{2})U_2(\frac{\beta}{2})U_3(\frac{\alpha}{2}) = \begin{pmatrix} \cos\frac{\beta}{2}e^{i\frac{(\alpha+\gamma)}{2}} & -\sin\frac{\beta}{2}e^{-i\frac{(\alpha-\gamma)}{2}} \\ \sin\frac{\beta}{2}e^{i\frac{(\alpha-\gamma)}{2}} & \cos\frac{\beta}{2}e^{-i\frac{(\alpha+\gamma)}{2}} \end{pmatrix} \quad (9.174)$$

and it corresponds to the general form of $SO(3)$ (*exercise 9.29*) which is

$$R_Z(\gamma)R_Y(\beta)R_Z(\alpha)$$

$$= \begin{pmatrix} \cos\alpha\cos\beta\cos\gamma - \sin\alpha\sin\gamma & \sin\alpha\cos\beta\cos\gamma + \cos\alpha\sin\gamma & \sin\beta\cos\gamma \\ -\cos\alpha\cos\beta\sin\gamma - \sin\alpha\cos\gamma & -\sin\alpha\cos\beta\sin\gamma + \cos\alpha\cos\gamma & -\sin\beta\sin\gamma \\ -\cos\alpha\sin\beta & -\cos\alpha\sin\beta & -\cos\beta \end{pmatrix} \quad (9.175)$$

i.e. $U_3\left(\frac{\gamma}{2}\right)U_2\left(\frac{\beta}{2}\right)U_3\left(\frac{\alpha}{2}\right) \xrightarrow{\text{corresponds to}} R_Z(\gamma)R_Y(\beta)R_Z(\alpha).$ \quad (9.176)

9.36 Generators of $SU(3)$, Gell-Mann matrices

We write down an arbitrary 3×3 Hermitian matrix comprising of complex numbers

$$H = \begin{pmatrix} z_1 & z_2 & z_3 \\ z_4 & z_5 & z_6 \\ z_7 & z_8 & z_9 \end{pmatrix} \quad (9.177)$$

$$H^\dagger = H^{*T} = \begin{pmatrix} z_1^* & z_2^* & z_3^* \\ z_4^* & z_5^* & z_6^* \\ z_7^* & z_8^* & z_9^* \end{pmatrix}^T = \begin{pmatrix} z_1^* & z_4^* & z_7^* \\ z_2^* & z_5^* & z_8^* \\ z_3^* & z_6^* & z_9^* \end{pmatrix}$$

Since H is Hermitian, i.e. $H = H^\dagger$ we have

$$\begin{pmatrix} z_1 & z_2 & z_3 \\ z_4 & z_5 & z_6 \\ z_7 & z_8 & z_9 \end{pmatrix} = \begin{pmatrix} z_1^* & z_4^* & z_7^* \\ z_2^* & z_5^* & z_8^* \\ z_3^* & z_6^* & z_9^* \end{pmatrix}$$

$$z_1 = z_1^* \text{ i. e. } z_1 \text{ is real say } z_1 = a$$

$$z_5 = z_5^* \text{ i. e. } z_5 \text{ is real say } z_5 = b$$

$$z_9 = z_9^* \text{ i. e. } z_9 \text{ is real say } z_9 = c$$

$$z_2 = z_4^*, \quad z_3 = z_7^*, \quad z_4 = z_2^*, \quad z_6 = z_8^*, \quad z_7 = z_3^*, \quad z_8 = z_6^*$$

Let us write

$$z_4 = d + ie, \quad z_2 = z_4^* = d - ie$$

$$z_7 = f + ig, \quad z_3 = z_7^* = f - ig$$

$$z_8 = h + ik, \quad z_6 = z_8^* = h - ik$$

$$H = \begin{pmatrix} z_1 & z_2 & z_3 \\ z_4 & z_5 & z_6 \\ z_7 & z_8 & z_9 \end{pmatrix} = \begin{pmatrix} a & d - ie & f - ig \\ d + ie & b & h - ik \\ f + ig & h + ik & c \end{pmatrix}$$

H has to be traceless. So

$$a + b + c = 0$$

$$c = -a - b$$

$$H = \begin{pmatrix} a & d - ie & f - ig \\ d + ie & b & h - ik \\ f + ig & h + ik & -a - b \end{pmatrix} \tag{9.178}$$

We split and write as follows

$$H = \begin{pmatrix} a & 0 & 0 \\ 0 & b & 0 \\ 0 & 0 & -a-b \end{pmatrix} + \begin{pmatrix} 0 & d - ie & 0 \\ d + ie & 0 & 0 \\ 0 & 0 & 0 \end{pmatrix}$$

$$+ \begin{pmatrix} 0 & 0 & f - ig \\ 0 & 0 & 0 \\ f + ig & 0 & 0 \end{pmatrix} + \begin{pmatrix} 0 & 0 & 0 \\ 0 & 0 & h - ik \\ 0 & h + ik & 0 \end{pmatrix}$$

$$= \begin{pmatrix} a & 0 & 0 \\ 0 & b & 0 \\ 0 & 0 & -a-b \end{pmatrix} + d \begin{pmatrix} 0 & 1 & 0 \\ 1 & 0 & 0 \\ 0 & 0 & 0 \end{pmatrix} + e \begin{pmatrix} 0 & -i & 0 \\ i & 0 & 0 \\ 0 & 0 & 0 \end{pmatrix}$$

$$+ f \begin{pmatrix} 0 & 0 & 1 \\ 0 & 0 & 0 \\ 1 & 0 & 0 \end{pmatrix} + g \begin{pmatrix} 0 & 0 & -i \\ 0 & 0 & 0 \\ i & 0 & 0 \end{pmatrix}$$

$$+ h \begin{pmatrix} 0 & 0 & 0 \\ 0 & 0 & 1 \\ 0 & 1 & 0 \end{pmatrix} + k \begin{pmatrix} 0 & 0 & 0 \\ 0 & 0 & -i \\ 0 & i & 0 \end{pmatrix}$$

$$H = \begin{pmatrix} a & 0 & 0 \\ 0 & b & 0 \\ 0 & 0 & -a-b \end{pmatrix} + d\lambda_1 + e\lambda_2 + f\lambda_4 + g\lambda_5 + h\lambda_6 + k\lambda_7 \tag{9.179}$$

Let us construct the remaining 2 matrices viz. λ_3,λ_8 as follows.

Consider the first matrix on the RHS of equation (9.179). To give the look of σ_3 we choose $a = 1$, $b = -1$ giving

$$\lambda_3 = \begin{pmatrix} 1 & 0 & 0 \\ 0 & -1 & 0 \\ 0 & 0 & 0 \end{pmatrix} \text{(traceless)} \tag{9.180}$$

And we can build another linearly independent matrix by choosing say $a = b = 1$, namely

$$\lambda_8 = k \begin{pmatrix} 1 & 0 & 0 \\ 0 & 1 & 0 \\ 0 & 0 & -2 \end{pmatrix} \tag{9.181}$$

where k is determined using the relation (*exercise 9.36*)

$$Tr(\lambda_i \lambda_j) = 2\delta_{ij} \tag{9.182}$$

For $i = 8, j = 8$

$$Tr(\lambda_8 \lambda_8) = 2$$

$$Trk^2 \begin{pmatrix} 1 & 0 & 0 \\ 0 & 1 & 0 \\ 0 & 0 & -2 \end{pmatrix} \begin{pmatrix} 1 & 0 & 0 \\ 0 & 1 & 0 \\ 0 & 0 & -2 \end{pmatrix} = 2$$

$$Trk^2 \begin{pmatrix} 1 & 0 & 0 \\ 0 & 1 & 0 \\ 0 & 0 & 4 \end{pmatrix} = 2$$

$$k^2(1 + 1 + 4) = 2$$

$k^2 = \frac{2}{6} = \frac{1}{3}$ and $k = \frac{1}{\sqrt{3}}$. Hence

$$\lambda_8 = \frac{1}{\sqrt{3}} \begin{pmatrix} 1 & 0 & 0 \\ 0 & 1 & 0 \\ 0 & 0 & -2 \end{pmatrix} \tag{9.183}$$

$$H = d\lambda_1 + e\lambda_2 + \lambda_3 + f\lambda_4 + g\lambda_5 + h\lambda_6 + k\lambda_7 + \lambda_8 \tag{9.184}$$

where λ_i with $i = 1$ to 8 are called the Gell-Mann matrices viz.

$$\lambda_1 = \begin{pmatrix} 0 & 1 & 0 \\ 1 & 0 & 0 \\ 0 & 0 & 0 \end{pmatrix}, \quad \lambda_2 = \begin{pmatrix} 0 & -i & 0 \\ i & 0 & 0 \\ 0 & 0 & 0 \end{pmatrix}, \quad \lambda_3 = \begin{pmatrix} 1 & 0 & 0 \\ 0 & -1 & 0 \\ 0 & 0 & 0 \end{pmatrix}, \quad \lambda_4 = \begin{pmatrix} 0 & 0 & 1 \\ 0 & 0 & 0 \\ 1 & 0 & 0 \end{pmatrix}$$

$$\lambda_5 = \begin{pmatrix} 0 & 0 & -i \\ 0 & 0 & 0 \\ i & 0 & 0 \end{pmatrix}, \quad \lambda_6 = \begin{pmatrix} 0 & 0 & 0 \\ 0 & 0 & 1 \\ 0 & 1 & 0 \end{pmatrix}, \quad \lambda_7 = \begin{pmatrix} 0 & 0 & 0 \\ 0 & 0 & -i \\ 0 & i & 0 \end{pmatrix}, \quad \lambda_8 = \frac{1}{\sqrt{3}} \begin{pmatrix} 1 & 0 & 0 \\ 0 & 1 & 0 \\ 0 & 0 & -2 \end{pmatrix}$$

$$\lambda_1 = \begin{pmatrix} \sigma_1 & 0 \\ 0 & 0 & 0 \end{pmatrix} \qquad \lambda_2 = \begin{pmatrix} \sigma_2 & 0 \\ 0 & 0 & 0 \end{pmatrix} \qquad \lambda_3 = \begin{pmatrix} \sigma_3 & 0 \\ 0 & 0 & 0 \end{pmatrix} \qquad \lambda_4 = \begin{pmatrix} 0 & 0 & 1 \\ 0 & 0 & 0 \\ 1 & 0 & 0 \end{pmatrix}$$

$$\lambda_5 = \begin{pmatrix} 0 & 0 & -i \\ 0 & 0 & 0 \\ -i & 0 & 0 \end{pmatrix} \qquad \lambda_6 = \begin{pmatrix} 0 & 0 & 0 \\ 0 & \sigma_1 \\ 0 \end{pmatrix} \qquad \lambda_7 = \begin{pmatrix} 0 & 0 & 0 \\ 0 & \sigma_2 \\ 0 \end{pmatrix} \qquad \lambda_8 = \frac{1}{\sqrt{3}} \begin{pmatrix} I & 0 \\ & 0 \\ 0 & 0 & -2 \end{pmatrix}$$

Figure 9.5. Generators λ_j, $j = 1$ to 8 of $SU(3)$ involving the Pauli spin matrices.

These are the basis for 8D vector space. A re-representation of the Gell-Mann matrices is shown in figure 9.5.

Among these, λ_3 and λ_8 are diagonal (similar to the z component of angular momentum J_z) and the other 6 correspond to the off-diagonal part of the Hermitian matrices (similar to J_x, J_y).

The commutation relation between λ_i is (as shown in *exercise 9.37*)

$$[\lambda_j, \lambda_k] = 2if_{jkl}\,\lambda_l \tag{9.185}$$

- In $SU(2)$ unitary unimodular 2×2 matrices can be represented as $U_j = e^{i\frac{\sigma_j}{2}a_j}$ where $\frac{\sigma_j}{2}$ are the generators of (2).

  In $S\overset{\cdot}{U}(3)$ we can represent unitary unimodular 3×3 matrices as

  $$U = e^{i\frac{\lambda_j}{2}b_j} \tag{9.186}$$

  $\frac{\lambda_j}{2}$ being the generators of $SU(3)$ where b_j is a parameter.

  The commutation relation between the $SU(3)$ generators $\frac{1}{2}\lambda_i$ is (*exercise 9.37*)

  $$\left[\frac{\lambda_j}{2}, \frac{\lambda_k}{2}\right] = if_{jkl}\frac{\lambda_l}{2} \tag{9.187}$$

  where f_{jkl} is called the structure constant, the values of which are given in *exercise 9.37*.
- These generator matrices can be chosen in many ways. However, it is a convention to choose matrices in the above manner.
- Matrices λ_3 and λ_8, being diagonal, commute, i.e.

  $$[\lambda_3\lambda_8] = 0 \tag{9.188}$$

  and also

  $$\left[\frac{\lambda_3}{2}, \frac{\lambda_8}{2}\right] = 0 \tag{9.189}$$

So they have simultaneous eigenstate.

- Proton and neutron are bound in the nucleus in the same way, though the proton is of positive charge and the neutron is of zero charge (i.e. they are not exactly identical). There is nothing to distinguish between them as far as their strong interaction is concerned. This is strong isospin symmetry. So in the case of strong interaction isospin represents symmetry between proton and neutron $\begin{pmatrix} p \\ n \end{pmatrix}$.

Now $p = uud$ and shifting $u \leftrightarrow d$ we get $n = ddu$. So there is symmetry between p and n, i.e. between u and d. So we can represent this isospin symmetry as

$$\begin{pmatrix} p \\ n \end{pmatrix} = \begin{pmatrix} u \\ d \end{pmatrix}.$$

The strong interaction conserves isospin and treats members of an iso-multiplet in an identical way. The corresponding symmetry is the $SU(2)$ isospin group. The relevant conserved quantum numbers that are analogues and generalization of l_z and l^2 from $SO(3)$ are I_3 and I^2 for isospin and Y for hypercharge.

- ✓ Gell-Mann suggested that the strong interaction is invariant under $SU(3)$ operation, i.e. strong interaction has $SU(3)$ flavour symmetry. The choice of $SU(3)$ was based upon two facts.
- ✓ Two quantum numbers I_3 and Y are conserved.
- ✓ The group has 8D representation to account for the 8 baryons. $SU(3)$ has $3^2 - 1 = 8$ generators. Hence the name eightfold way (figure 10.12).

9.37 Rank of group

$SU(n)$ has $n^2 - 1$ generators. The number of diagonal generators is called the rank of the group.

$SU(n)$ has $n - 1$ diagonal generators and hence the rank is $n - 1$.

- ✓ For $n = 2$ $SU(2)$ has rank $2 - 1 = 1$ (it has one diagonal generator, namely $\frac{\sigma_z}{2}$).
- ✓ For $n = 3$ $SU(3)$ has rank $3 - 1 = 2$ (it has two diagonal generators, namely $\frac{\lambda_3}{2}$, $\frac{\lambda_8}{2}$).

9.38 Young's diagram

9.38.1 In $SU(2)$

The fundamental or defining representation of $SU(2)$ is

$$| \uparrow \rangle = \begin{pmatrix} 1 \\ 0 \end{pmatrix} = |m = \tfrac{1}{2}\rangle(|\text{up isospin}\rangle \text{ or } |\tfrac{1}{2}\tfrac{1}{2}\rangle) \tag{9.190}$$

$$| \downarrow \rangle = \begin{pmatrix} 0 \\ 1 \end{pmatrix} = |m = -\tfrac{1}{2}\rangle(|\text{down isospin}\rangle \text{ or } |\tfrac{1}{2} - \tfrac{1}{2}\rangle) \tag{9.191}$$

So $m = |\tfrac{1}{2}$ is the eigenvalue of the diagonal generator S_z and we can denote the state as $\vec{m} \equiv |\tfrac{1}{2}m\rangle$.

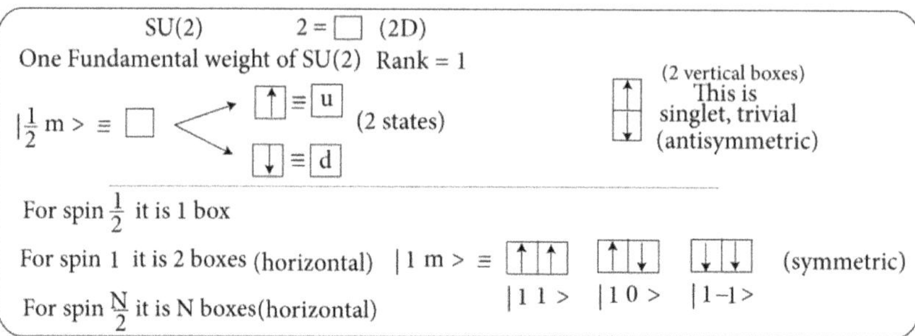

Figure 9.6. 2D Young's diagrammatic representation of $|\frac{1}{2}m\rangle$ and $|1m\rangle$ states.

There are two fundamental states and this 2-state system $|\frac{1}{2}m\rangle$ can be represented by a box □ referred to as one fundamental weight. So $SU(2)$ has one fundamental weight □.

We can put upspin ↑ in the single box as ↑ or we can put downspin ↓ in the single box as ↓ as shown in figure 9.6.

Figure 9.6 also shows how to represent $|1m\rangle$. This is the Young diagram in 2D.

There is only one fundamental weight in $SU(2)$ since number of diagonal generators is one, namely S_z and rank of $SU(2)$ is $n - 1 = 2 - 1 = 1$.

Actually, the number of fundamental weights is equal to the rank and also equal to the number of diagonal generators.

9.38.2 In $SU(3)$

The fundamental or defining representation of $SU(3)$ is

$$|u\rangle = \begin{pmatrix} 1 \\ 0 \\ 0 \end{pmatrix}, \quad |d\rangle = \begin{pmatrix} 0 \\ 1 \\ 0 \end{pmatrix}, \quad |s\rangle = \begin{pmatrix} 0 \\ 0 \\ 1 \end{pmatrix} \tag{9.192}$$

These are the simultaneous eigenstates of the two diagonal generators $\frac{\lambda_3}{2}, \frac{\lambda_8}{2}$.

- Consider

$$\frac{\lambda_3}{2}|u\rangle = \frac{1}{2}\begin{pmatrix} 1 & 0 & 0 \\ 0 & -1 & 0 \\ 0 & 0 & 0 \end{pmatrix}\begin{pmatrix} 1 \\ 0 \\ 0 \end{pmatrix} = \frac{1}{2}\begin{pmatrix} 1 \\ 0 \\ 0 \end{pmatrix} = \frac{1}{2}|u\rangle \text{ (eigenvalue is } \frac{1}{2}) \tag{9.193}$$

$$\frac{\lambda_8}{2}|u\rangle = \frac{1}{2}\frac{1}{\sqrt{3}}\begin{pmatrix} 1 & 0 & 0 \\ 0 & 1 & 0 \\ 0 & 0 & -2 \end{pmatrix}\begin{pmatrix} 1 \\ 0 \\ 0 \end{pmatrix} = \frac{1}{2\sqrt{3}}\begin{pmatrix} 1 \\ 0 \\ 0 \end{pmatrix}$$

$$= \frac{\sqrt{3}}{6}\begin{pmatrix} 1 \\ 0 \\ 0 \end{pmatrix} = \frac{\sqrt{3}}{6}|u\rangle \text{ (eigenvalue is } \frac{1}{2\sqrt{3}}) \tag{9.194}$$

Hence we denote

$$\vec{\mu}_1 \equiv |u\rangle = \left(\frac{1}{2}, \frac{\sqrt{3}}{6}\right) \text{ as one state.}$$

- Consider

$$\frac{\lambda_3}{2}|d\rangle = \frac{1}{2}\begin{pmatrix} 1 & 0 & 0 \\ 0 & -1 & 0 \\ 0 & 0 & 0 \end{pmatrix}\begin{pmatrix} 0 \\ 1 \\ 0 \end{pmatrix} = \frac{1}{2}\begin{pmatrix} 0 \\ 1 \\ 0 \end{pmatrix} = -\frac{1}{2}|d\rangle \text{ (eigenvalue is } -\frac{1}{2}\text{)} \tag{9.195}$$

$$\frac{\lambda_8}{2}|d\rangle = \frac{1}{2}\frac{1}{\sqrt{3}}\begin{pmatrix} 1 & 0 & 0 \\ 0 & 1 & 0 \\ 0 & 0 & -2 \end{pmatrix}\begin{pmatrix} 0 \\ 1 \\ 0 \end{pmatrix} = \frac{1}{2\sqrt{3}}\begin{pmatrix} 0 \\ 1 \\ 0 \end{pmatrix}$$

$$= \frac{\sqrt{3}}{6}\begin{pmatrix} 0 \\ 1 \\ 0 \end{pmatrix} = \frac{\sqrt{3}}{6}|d\rangle \text{ (eigenvalue is } \frac{\sqrt{3}}{6}\text{)} \tag{9.196}$$

Hence we denote

$$\vec{\mu}_2 \equiv |d\rangle = \left(-\frac{1}{2}, \frac{\sqrt{3}}{6}\right) \text{ as one state.}$$

- Consider

$$\frac{\lambda_3}{2}|s\rangle = \frac{1}{2}\begin{pmatrix} 1 & 0 & 0 \\ 0 & -1 & 0 \\ 0 & 0 & 0 \end{pmatrix}\begin{pmatrix} 0 \\ 0 \\ 1 \end{pmatrix} = \begin{pmatrix} 0 \\ 0 \\ 0 \end{pmatrix} = 0\begin{pmatrix} 0 \\ 0 \\ 1 \end{pmatrix} = 0|s\rangle \text{ (eigenvalue is 0)} \tag{9.197}$$

$$\frac{\lambda_8}{2}|s\rangle = \frac{1}{2}\frac{1}{\sqrt{3}}\begin{pmatrix} 1 & 0 & 0 \\ 0 & 1 & 0 \\ 0 & 0 & -2 \end{pmatrix}\begin{pmatrix} 0 \\ 0 \\ 1 \end{pmatrix} = \frac{1}{2\sqrt{3}}\begin{pmatrix} 0 \\ 0 \\ -2 \end{pmatrix}$$

$$= -\frac{1}{\sqrt{3}}\begin{pmatrix} 0 \\ 0 \\ 1 \end{pmatrix} = -\frac{\sqrt{3}}{3}|s\rangle \text{ (eigenvalue is } \frac{\sqrt{3}}{3}\text{)} \tag{9.198}$$

Hence we denote

$$\vec{\mu}_3 \equiv |s\rangle = \left(0, -\frac{\sqrt{3}}{3}\right) \text{ as one state.}$$

We have obtained the states $\vec{\mu}_1$, $\vec{\mu}_2, \vec{\mu}_3$ in $SU(3)$.

9.39 Fundamental weights

The diagram obtained by plotting $\frac{\lambda_3}{2}$ along the X-axis and $\frac{\lambda_8}{2}$ along the Y-axis is called a weight diagram.

We have shown in figure 9.7 the three states $(\frac{1}{2}, \frac{\sqrt{3}}{6}), (-\frac{1}{2}, \frac{\sqrt{3}}{6}), (0, -\frac{\sqrt{3}}{3})$ in the weight diagram.

In figure 9.7 we represent the states $\vec{\mu}_1, \vec{\mu}_2, \vec{\mu}_3$ corresponding to quarks in a weight diagram.

In figure 9.8 we represent states $\vec{\mu}'_1, \vec{\mu}'_2, \vec{\mu}'_3$ corresponding to anti-quarks in a weight diagram.

There are three fundamental states and this three-state system can be represented by two fundamental weights, as depicted in figure 9.9.

One is a single box and the other is a combination of two vertical boxes (which is an antisymmetric arrangement). This is the Young's diagram in 3D vector space.

There are two weights in $SU(3)$ since number of diagonal generators is two, namely $\frac{\lambda_3}{2}$, $\frac{\lambda_8}{2}$, and rank of $SU(3)$ is $n - 1 = 3 - 1 = 2$.

There are $n - 1$ fundamental weights in $SU(n)$ that has rank $n - 1$. This is schematically depicted in a Young's diagram in figure 9.9.

In a Young's diagram the entries in the horizontal boxes should be such that it is symmetric and entries in the vertical boxes should be such that it is antisymmetric.

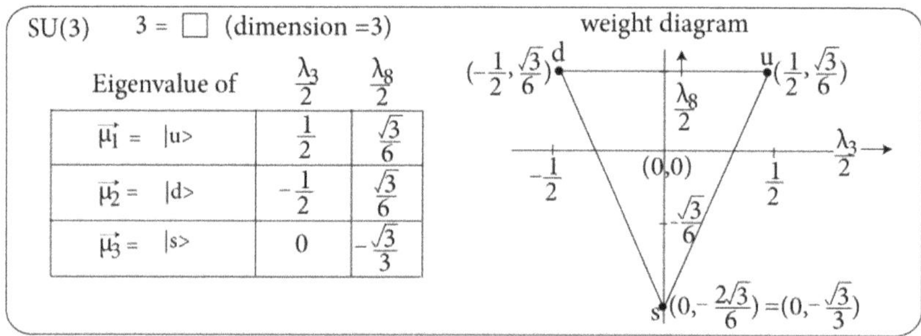

Figure 9.7. States $\vec{\mu}_1, \vec{\mu}_2, \vec{\mu}_3$ in weight diagram (see figure 10.5 also). It corresponds to quarks.

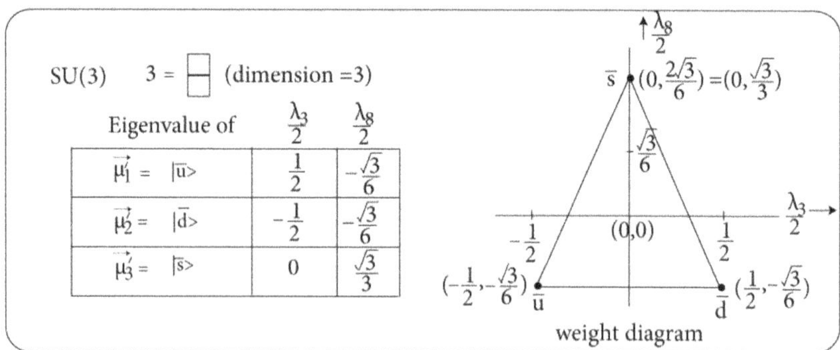

Figure 9.8. States $\vec{\mu}'_1, \vec{\mu}'_2, \vec{\mu}'_3$ in weight diagram (see figure 10.6 also) It corresponds to anti-quarks.

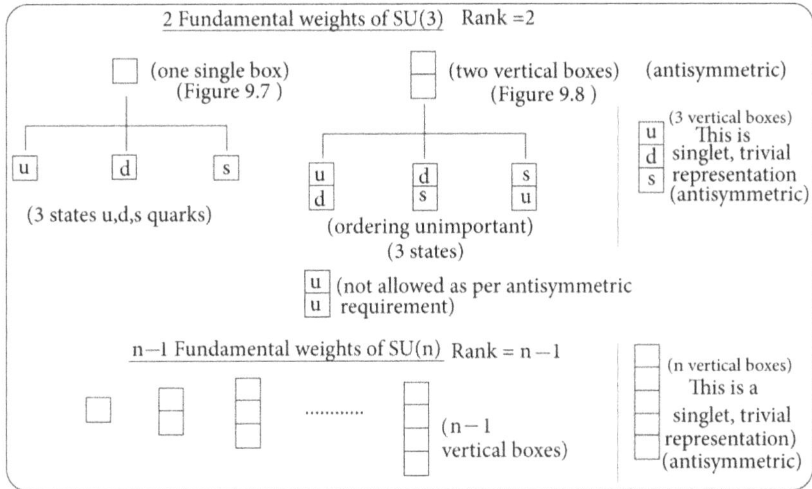

Figure 9.9. Fundamental weights of $SU(3)$ and $SU(n)$.

9.40 Young tableaux

With the help of Young tableaux we can construct higher dimensional representation of a given group using the lowest dimensional representation, i.e. using fundamental and its conjugate representation.

9.40.1 Thumb rule for $SU(n)$

Fundamental representation is denoted by a single box as shown in figure 9.10(a).

Conjugate of fundamental representation is denoted by $n - 1$ vertical boxes, as shown in figure 9.10(b).

Singlet representation is denoted by n vertical boxes, as shown in figure 9.10(c). One cannot have less than singlet representation as that is unphysical.

The general diagram is shown in figure 9.10(d). R_i represents the ith row and C_j represents the jth column.

Number of boxes in the upper row is greater than or equal to the number of boxes in the lower row. Hence figure 9.10(e) does not represent a valid diagram.

Number of boxes in any column should not be more than n for $SU(n)$.

Filling up of boxes in any column must start from C_1 (first column)

The boxes in a specific row are symmetric to each other while the boxes for a specific column are antisymmetric to each other.

Figure 9.10(f) shows how the boxes are filled up. Number in a box increases by one unit to the right. Number in a box decreases by one unit as we go down.

Define dimensionality (which gives the number of states) as

$$\text{Dimensionality} = \frac{\text{Numerator}}{\text{Denominator}} \tag{9.199}$$

Numerator = product of numbers shown in all the boxes

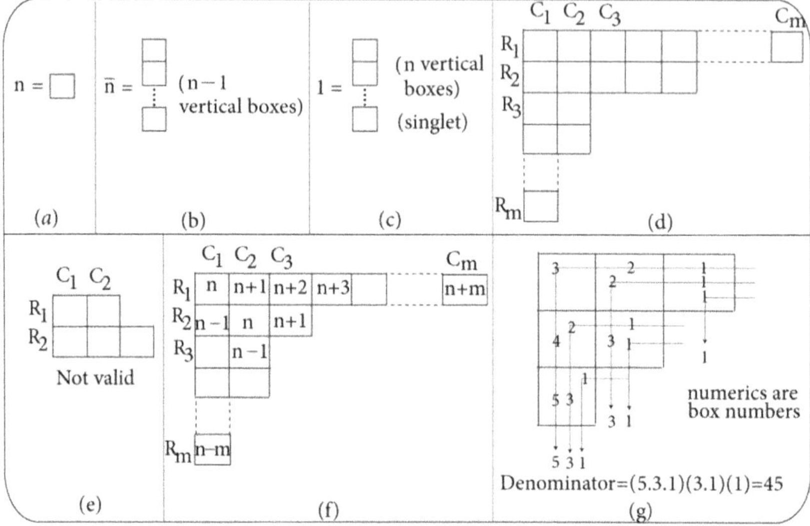

Figure 9.10. Young tableaux.

Denominator = product of numbers that are obtained by counting the boxes from right to left (horizontally) and then moving vertically down. This is called the Hook number.

This is illustrated in figures 9.10(g), 9.11 and 9.12.

9.41 Some examples of Young tableaux

We furnish some examples of Young Tableaux in $SU(2)$.

Consider $n = 2$ dimensional representation which is depicted as a single box and its conjugate representation $\bar{n} = \bar{2}$ depicted by $n - 1 = 2 - 1 = 1$ vertical box, i.e. another single box as shown in figure 9.13(a).

9.41.1 $2 \otimes \bar{2} = 3 \oplus 1$

In figure 9.13(b) we represent a tensor product of two boxes represented as

$$2 \otimes \bar{2} = \boxed{A} \otimes \boxed{B}$$

in figure 9.13(b) where we establish by evaluation of dimensionality that

$$2 \otimes \bar{2} = 3 \oplus 1 \tag{9.200}$$

- Equation (9.200) has resemblance to the addition of angular momenta $\vec{j_1} = \frac{\vec{1}}{2}$ and $\vec{j_2} = \frac{\vec{1}}{2}$.

Let us explain this in the following.

Consider $j_1 = \frac{1}{2}$, $m_1 = \pm\frac{1}{2}$. So multiplicity is $2j_1 + 1 = 2\frac{1}{2} + 1 = 2$. The two states are

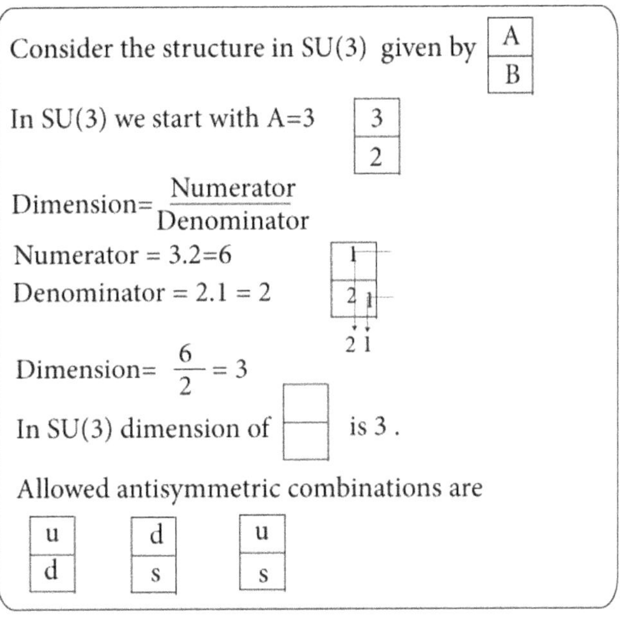

Figure 9.11. Finding dimension of a combination in $SU(3)$ and allowed states in a Young diagram.

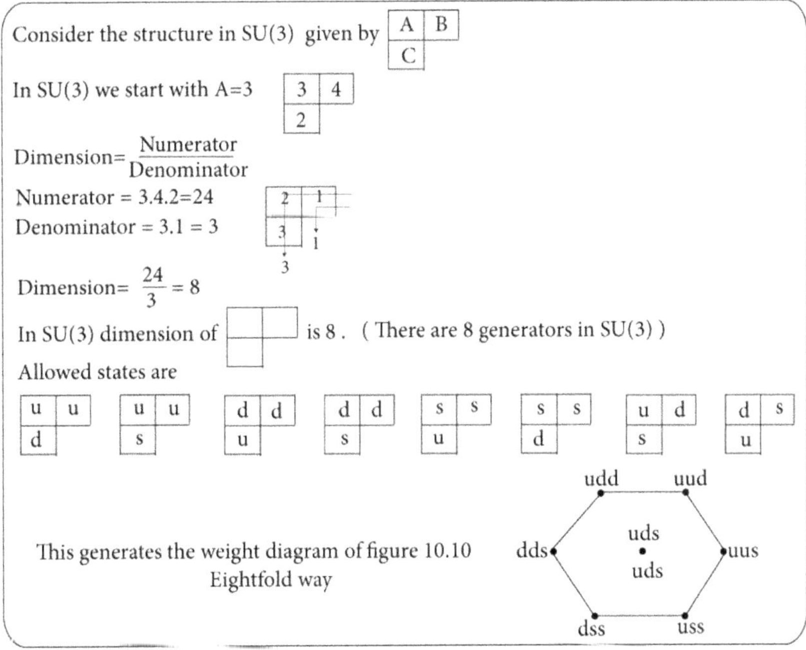

Figure 9.12. Finding dimension of a combination in $SU(3)$, the allowed states and their interpretation in a weight diagram.

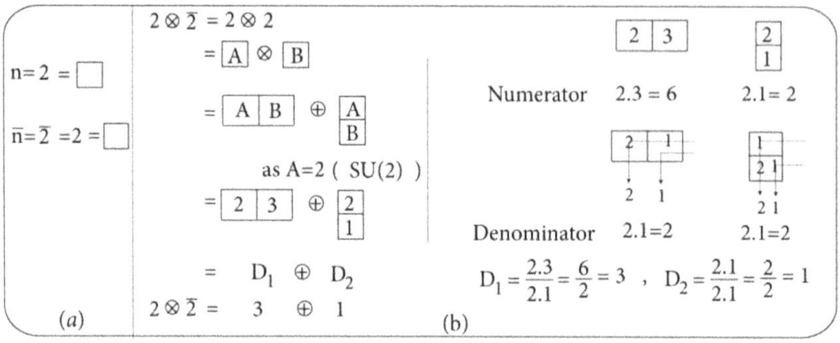

Figure 9.13. Young Tableaux showing $2 \otimes 2 = 3 \oplus 1$.

$$|j_1 m_1\rangle = |\tfrac{1}{2}\tfrac{1}{2}\rangle_1, \ |\tfrac{1}{2} - \tfrac{1}{2}\rangle_1 \ \text{(doublet as } 2j_1 + 1 = 2\tfrac{1}{2} + 1 = 2 \text{ states)}$$

Consider $j_2 = \tfrac{1}{2}, m_2 = \pm\tfrac{1}{2}$. So multiplicity is $2j_2 + 1 = 2\tfrac{1}{2} + 1 = 2$. The two states are

$$|j_2 m_2\rangle| = |\tfrac{1}{2}\tfrac{1}{2}\rangle_2, \ |\tfrac{1}{2} - \tfrac{1}{2}\rangle_2 \ \text{(doublet as } 2j_2 + 1 = 2\tfrac{1}{2} + 1 = 2 \text{ states)}$$

Clearly there are two states each and the product of multiplicities 2×2 is 4, which gives the total number of states to be 4.

So $2 \otimes 2$ represents 4 states.

Again, by angular momentum addition rule we have from $\vec{j_1} = \tfrac{1}{2}$ and $\vec{j_2} = \tfrac{1}{2}$ the j values to be

$$\frac{\vec{1}}{2} + \frac{\vec{1}}{2} = \begin{cases} \tfrac{1}{2} + \tfrac{1}{2} = 1 \\ \tfrac{1}{2} - \tfrac{1}{2} = 0 \end{cases}$$

The projection of $j = 1$ is $m = 1, 0, -1$. The corresponding three states are

$$|jm\rangle = |11\rangle, \ |10\rangle, \ |1 - 1\rangle \ \text{(triplet as } 2j + 1 = 2.1 + 1 = 3 \text{ states)}$$

The projection of $j = 0$ is $m = 0$. The corresponding one state is

$$|jm\rangle = |00\rangle \ \text{(singlet as } 2j + 1 = 2.0 + 1 = 1 \text{ state)}$$

Clearly there are 3 states for $j = 1$ and 1 state for $j = 0$ and the sum of multiplicities $3 + 1$ is 4, which gives the total number of states to be 4.

So $3 \oplus 1$ represents 4 states.

Obviously equation (9.200) viz. $2 \otimes \bar{2} = 3 \oplus 1$ is consistent with the addition of angular momenta

$$\vec{j_1} = \frac{\vec{1}}{2} \ \text{and} \ \vec{j_2} = \frac{\vec{1}}{2}.$$

9.41.2 $3 \otimes 3 = 5 \oplus 3 \oplus 1$

We have discussed the tensor product $3 \otimes 3$ in figure 9.14 through Young Tableaux leading to

$$3 \otimes 3 = 5 \oplus 3 \oplus 1 \tag{9.201}$$

This result $3 \otimes 3 = 5 \oplus 3 \oplus 1$ has resemblance to the addition of angular momenta

$$\vec{j_1} = \vec{1}, \ \vec{j_2} = \vec{1}.$$

Consider $j_1 = 1$, $m_1 = 1, 0, -1$. So multiplicity is $2j_1 + 1 = 2.1 + 1 = 3$. The 3 states are

$$|j_1 m_1\rangle = |11\rangle_1, \ |10\rangle_1, \ |1-1\rangle_1 \text{ (triplet as } 2j_1 + 1 = 2.1 + 1 = 3 \text{ states)}$$

Consider $j_2 = 1$, $m_2 = \pm 1$. So multiplicity is $2j_2 + 1 = 2.1 + 1 = 3$. The 3 states are

$$|j_2 m_2\rangle = |11\rangle_2, \ |10\rangle_2, \ |1-1\rangle_1 \text{ (triplet as } 2j_2 + 1 = 2.1 + 1 = 3 \text{ states)}$$

Clearly there are 3 states each and the product of multiplicities 3×3 is 9, which gives the total number of states to be 9.

So $3 \otimes 3$ represents 9 states.

Again, by angular momentum addition rule we have from $\vec{j_1} = \vec{1}$ and $\vec{j_2} = 1$ the j values to be

$$\vec{1} + \vec{1} = \begin{cases} 1 + 1 = 2 \\ 1 + 1 - 1 = 1 \\ 1 - 1 = 0 \end{cases}$$

The projection of $j = 2$ is $m = 2, 1, 0, -1, -2$. The corresponding 5 states are

$$|jm\rangle = |22\rangle, \ |21\rangle, \ |20\rangle, \ |2-1\rangle, \ |2-2\rangle \text{ (5-plet as } 2j + 1 = 2.2 + 1 = 5 \text{ states)}.$$

The projection of $j = 1$ is $m = 1, 0, -1$. The corresponding 3 states are

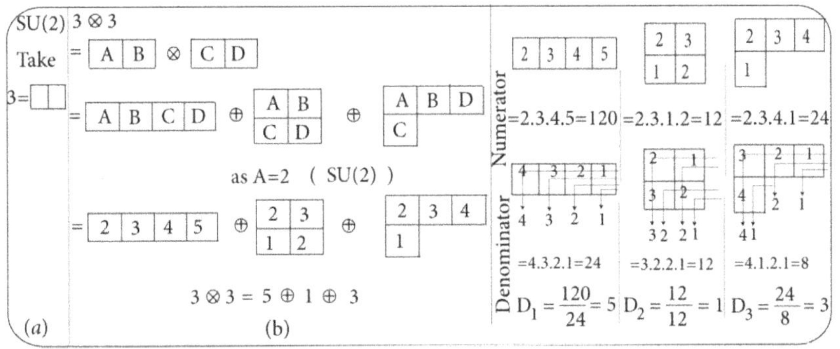

Figure 9.14. Young Tableaux showing $3 \otimes 3 = 5 \oplus 1 \oplus 3$.

$$|jm\rangle = |11\rangle, \ |10\rangle, |1-1\rangle \ \text{(triplet as } 2j+1 = 2.1+1 = 3 \text{ states)}.$$

The projection of $j = 0$ is $m = 0$. The corresponding 1 state is

$$|jm\rangle = |00\rangle \ \text{(singlet as } 2j+1 = 2.0+1 = 1 \text{ state)}$$

Clearly there are 5 states for $j = 2$; 3 states for $j = 1$ and 1 state for $j = 0$ and the sum of multiplicities $5 + 3 + 1$ is 9, which gives the total number of states to be 9.

So $5 \oplus 3 \oplus 1$ represents 9 states.

Obviously the equation (9.201) viz. $3 \otimes 3 = 5 \oplus 3 \oplus 1$ is consistent with the addition of angular momenta $\vec{j_1} = \vec{1}$ and $\vec{j_2} = \vec{1}$.

- The notion of angular momentum and their related algebra are applicable for $SU(2)$ only. The angular momentum addition method cannot be used to construct higher dimensional representations for other $SU(n)$ groups for $n \rangle 2$.

We furnish some examples of Young Tableaux in $SU(3)$.

9.41.3 $3 \otimes 3$ and $\bar{3} \otimes 3$

We have discussed the tensor product $3 \otimes 3$ in figure 9.15 through Young Tableaux leading to

$$3 \otimes 3 = 6 \oplus \bar{3} \tag{9.202}$$

We have discussed the tensor product $\bar{3} \otimes 3$ in figure 9.16 through Young Tableaux leading to

$$\bar{3} \otimes 3 = 1 \oplus 8 \tag{9.203}$$

In general

$$n \otimes \bar{n} = (n^2 - 1) \oplus 1 \tag{9.204}$$

Figure 9.7 (that corresponds to quarks) and figure 9.8 (that corresponds to anti-quarks) are two fundamental diagrams in $SU(3)$. They can be combined through tensor product to create various states, as shown in figure 9.17. In other words, quarks and anti-quarks can be combined in various ways to generate states.

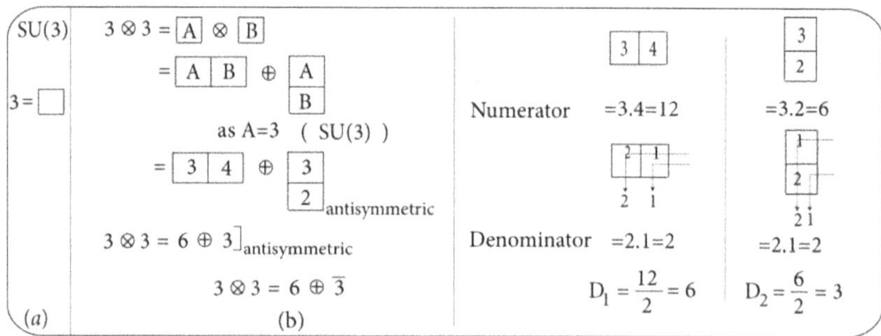

Figure 9.15. Young Tableaux $3 \otimes 3 = 6 \oplus \bar{3}$.

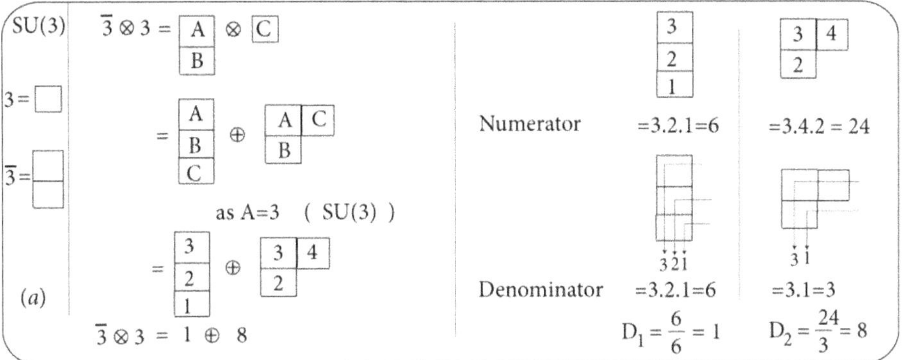

Figure 9.16. Young Tableaux $\bar{3} \otimes 3 = 1 \oplus 8$.

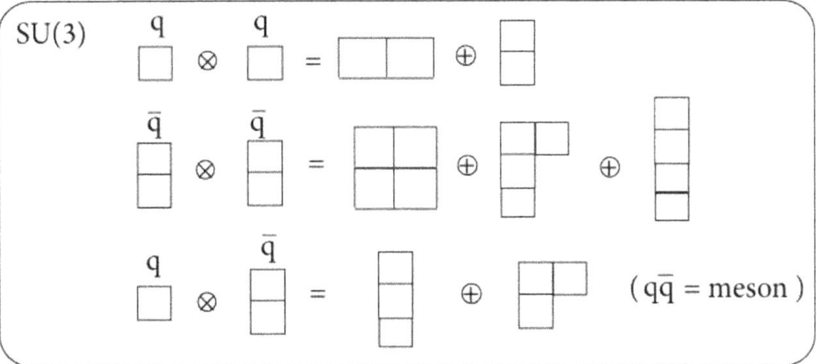

Figure 9.17. Combination of quarks and anti-quarks to generate various states.

9.42 Exercises

Exercise 9.1 *Check Hermiticity of the operators* $x, \frac{d}{dx}, i\frac{d}{dx}, p.$
[Ans.] Condition of Hermiticity is

$$<\psi \,|A\psi\rangle = <A\psi|\psi\rangle \text{ or } A = A^\dagger$$

- We discuss Hermiticity of x. Consider

$$<\psi|x\psi\rangle = \int_{-\infty}^{\infty} \psi^*(x\psi)\mathrm{d}x = \int_{-\infty}^{\infty} x\psi^*\psi\mathrm{d}x = \int_{-\infty}^{\infty} (x\psi)^*\psi\mathrm{d}x \text{ (as } x = x^*)$$

$$<\psi \,|x\psi\rangle = <x\psi|\psi\rangle$$

So x is Hermitian, i.e. $x = x^\dagger$.
- We discuss Hermiticity of $\frac{d}{dx}$. Consider

$$\langle\psi|\frac{d\psi}{dx}\rangle = \int_{-\infty}^{\infty} \psi^* \frac{d\psi}{dx} dx = \psi^*\psi \Big|_{-\infty}^{\infty} - \int_{-\infty}^{\infty} \frac{d\psi^*}{dx}\psi dx$$

$$=0 - \int_{-\infty}^{\infty}\left(\frac{d\psi}{dx}\right)^*\psi dx (\text{as } \psi \to 0 \text{ at } x \to \pm\infty)$$

$$= -\langle\frac{d\psi}{dx}|\psi\rangle$$

So $\frac{d}{dx}$ is anti-Hermitian i.e. $\frac{d}{dx}^\dagger = -\frac{d}{dx}$ (since $\langle\psi|A\psi\rangle = -\langle A\psi|\psi\rangle$ or $A = -A^\dagger$ is the condition of anti-Hermiticity)

- We discuss Hermiticity of $i\frac{d}{dx}$. Consider

$$\left(i\frac{d}{dx}\right)^\dagger = -i\left(\frac{d}{dx}\right)^\dagger = -i\left(-\frac{d}{dx}\right) = i\frac{d}{dx}$$

So $i\frac{d}{dx}$ is Hermitian.

- We discuss Hermiticity of momentum operator $p = -i\hbar\frac{\partial}{\partial x}$. Consider

$$p^\dagger = \left(-i\hbar\frac{\partial}{\partial x}\right)^\dagger = i\hbar\left(\frac{\partial}{\partial x}\right)^\dagger = i\hbar\left(-\frac{\partial}{\partial x}\right) = -i\hbar\frac{\partial}{\partial x} = p$$

So $p = -i\hbar\frac{\partial}{\partial x}$ is Hermitian.

Exercise 9.2 *Generators of space translation and time translation symmetries are which of the following operators?*

(a) *Linear momentum, Hamiltonian*

(b) *Hamiltonian, linear momentum*

(c) *Linear momentum, angular momentum*

(d) *Hamiltonian, Angular momentum.*

Hint. $\psi'(x) = e^{-iap/\hbar}\psi(x)$, $\psi'(t) = e^{i\varepsilon H/\hbar}\psi(t)$

Exercise 9.3 *Show that the elements* $1, -1$ *satisfy properties of a group? What is the order of this group?*

$\boxed{\text{Ans.}}$ ✓ Closure property
$1 \times (-1) = -1$ is a member of the group. This is closure property.
 ✓ Associative property
  Number multiplication is always associative.
 ✓ Occurrence of identity element E
  The relation $(-1) \times 1 = -1$ shows that $E = 1$ is the identity element.
 ✓ Occurrence of inverse element

$1 \times 1 = 1, -1 \times -1 = 1$ so 1 is the inverse of 1 and -1 is the inverse of -1.

Hence $1, -1$ satisfy the properties of a group.
The order of this group is 2 since this discrete group has 2 elements.

Exercise 9.4 *Consider the set of all complex numbers under the operation of addition. Then which of the following is the best description of this given set?*

(a) *Abelian group*   (b) *non-abelian group*   (c) *complex group*   (d) *does not form a group.*

Exercise 9.5 *Consider the set of all 2×2 matrices with complex numbers under the operation of matrix addition. Then which of the following is the best description of this given set?*

(a) *Abelian group*   (b) *non-abelian group*   (c) *complex group*   (d) *does not form a group.*

Exercise 9.6 *Consider the set of all 2×2 matrices with complex numbers under the operation of matrix multiplication. Then which of the following is the best description of this given set?*

(a) *Abelian group*   (b) *non-abelian group*   (c) *complex group*   (d) *does not form a group.*

Exercise 9.7 *The generator of a square symmetry group under rotation and its order will be*

(a) $R_{\pi/2}$, 4   (b) $R_{2\pi/4}$, 4   (c) $R_{2\pi/6}$, 6   d) $R_{2\pi}$,1

Hint. A square has four sides. The generator is $R_{2\pi/4} = R_{\pi/2}$.
Since $(R_{2\pi/4})^4 = (R_{\pi/2}) = I$ order of generator is 4. Also there are 4 elements in the group

$$(R_{2\pi/4},\ R_{4\pi/4},\ R_{6\pi/4},\ R_{8\pi/4}).$$

Exercise 9.8 *Generator of a hexagon symmetry group under rotation is*

(a) $R_{\pi/2}$, 2   (b) $R_{2\pi/4}$, 4   (c) $R_{2\pi/6}$,6   (d) $R_{2\pi}$, 1.

Hint. A hexagon has 6 sides. The generator is $R_{2\pi/6} = R_{\pi/3}$. Since $(R_{2\pi/6})^6 = I$ order of generator is 6. Also there are 6 elements in the group

$$(R_{2\pi/6},\ R_{4\pi/6},\ R_{6\pi/6},\ R_{8\pi/6},\ R_{10\pi/6},\ R_{12\pi/6}).$$

Exercise 9.9 *Construct a group with two generators a, b under the conditions $a^2 = I$, $b^2 = I$, $ab = ba$.*

Ans. The general group element is $a^m b^n$. So the group is $\{I,\ a,\ b,\ ab\}$. It has 4 elements.

Exercise 9.10

   (i) *Obtain transformation matrix for anti-clockwise rotation of axes keeping the vector fixed.*

   (ii) *Obtain transformation matrix for clockwise rotation of a vector keeping axes fixed.*

   (iii) *Obtain transformation matrix for clockwise rotation of axes keeping the vector fixed.*

   (iv) *Obtain transformation matrix for anti-clockwise rotation of a vector keeping axes fixed*

      (consider all rotations to be in the 2D XY plane and about Z-axis).

   (v) *Show that rotation of a vector in clockwise direction produces the same result as the rotation of axes in the anti-clockwise direction.*

   (vi) *Show that rotation of a vector in anti-clockwise direction produces the same result as the rotation of axes in the clockwise direction.*

Ans. We consider all rotations occurring in the 2D XY plane and about the Z-axis through the origin.

   (i) The transformation matrix for anti-clockwise rotation of axes keeping the vector fixed (figure 9.18(a)).

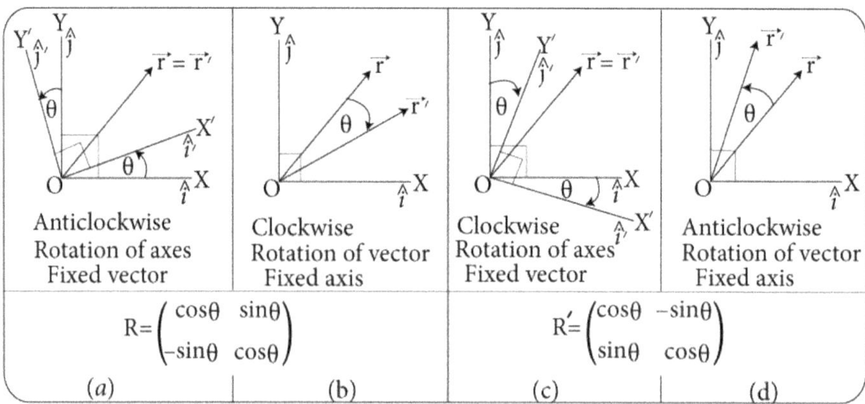

$$R = \begin{pmatrix} \cos\theta & \sin\theta \\ -\sin\theta & \cos\theta \end{pmatrix}$$

$$R' = \begin{pmatrix} \cos\theta & -\sin\theta \\ \sin\theta & \cos\theta \end{pmatrix}$$

Figure 9.18. (a) Anti-clockwise rotation of axes keeping vector fixed. (b) Clockwise rotation of vector keeping axes fixed. (c) Clockwise rotation of axes keeping vector fixed. (d) Anti-clockwise rotation of vector keeping axes fixed.

Vector fixed means

$$\vec{r}' = \vec{r} \tag{9.205}$$

$$x'\hat{i}' + y'\hat{j}' = x\hat{i} + y\hat{j} \tag{9.206}$$

Take dot product of equation (9.206) with $\hat{i}'$

$$x'\underbrace{(\hat{i}'.\,\hat{i}')}_{=1} + y'\underbrace{(\hat{j}'.\,\hat{i}')}_{=0} = x\underbrace{(\hat{i},\,\hat{i}')}_{=\cos\theta} + y\underbrace{(\hat{j}.\,\hat{i}')}_{=\cos(90°-\theta)=\sin\theta} \tag{9.207}$$

$$x' = x\cos\theta + y\sin\theta \tag{9.208}$$

Take dot product of equation (9.206) with $\hat{j}'$

$$x'\underbrace{(\hat{i}'.\,j')}_{=0} + y'\underbrace{(\hat{j}'.\,\hat{j}')}_{=1} = x\underbrace{(\hat{i},\,\hat{j}')}_{=\cos(90°+\theta)=-\sin\theta} + y\underbrace{(\hat{j}.\,\hat{j}')}_{=\cos\theta} \tag{9.209}$$

$$y' = -x\sin\theta + y\cos\theta \tag{9.210}$$

Combining equation (9.208) and equation (9.210) we can write

$$\begin{pmatrix} x' \\ y' \end{pmatrix} = \begin{pmatrix} \cos\theta & \sin\theta \\ -\sin\theta & \cos\theta \end{pmatrix} \begin{pmatrix} x \\ y \end{pmatrix} = R\begin{pmatrix} x \\ y \end{pmatrix} \text{ i. e. } \begin{pmatrix} x \\ y \end{pmatrix} \xrightarrow[\text{of axes by } \theta]{\text{anticlockwise rotation}} \begin{pmatrix} x' \\ y' \end{pmatrix} \tag{9.211}$$

where

$$R = \begin{pmatrix} \cos\theta & \sin\theta \\ -\sin\theta & \cos\theta \end{pmatrix} \tag{9.212}$$

is the rotation matrix or transformation matrix for anti-clockwise rotation of axes keeping the vector fixed.

(ii) Transformation matrix for clockwise rotation of vector keeping axes fixed (figure 9.18(b)).

Consider

$$\vec{r} \to re^{i\phi} = x + iy = (x, y) \tag{9.213}$$

$$= r(\cos\phi + i\sin\phi) = r\cos\phi + ir\sin\phi = x + iy \tag{9.214}$$

$$x = r\cos\phi,\ y = r\sin\phi \tag{9.215}$$

Multiply $\vec{r}$ by $e^{-i\theta}$ which is equivalent to clockwise rotation of $\vec{r}$ by θ, to get the rotated vector $\vec{r}'$.

$$\vec{r}' \to e^{-i\theta}\vec{r} = e^{-i\theta}(re^{i\phi}) = re^{i(\phi-\theta)} = (x', y') \tag{9.216}$$

$$= r[\cos(\phi - \theta) + ir\sin(\phi - \theta)]$$

$$= r\cos(\phi - \theta) + r\sin(\phi - \theta) = x' + iy' \tag{9.217}$$

$$x' = r \cos (\phi - \theta) \tag{9.218}$$

$$x' = r(\cos \theta \cos \phi + \sin \theta \sin \phi) = (r \cos \phi) \cos \theta + (r \sin \phi) \sin \theta \tag{9.219}$$

Using equation (9.215) we have

$$x' = x \cos \theta + y \sin \theta \tag{9.220}$$

$$y' = r \sin (\phi - \theta) = r(\sin \phi \cos \theta - \cos \phi \sin \theta)$$

$$y' = (r \sin \phi) \cos \theta - (r \cos \phi) \sin \theta \tag{9.221}$$

Using equation (9.215) we have

$$y' = -x \sin \theta + y \cos \theta \tag{9.222}$$

Combining (9.220) and (9.222) we get

$$\begin{pmatrix} x' \\ y' \end{pmatrix} = \begin{pmatrix} \cos \theta & \sin \theta \\ -\sin \theta & \cos \theta \end{pmatrix} \begin{pmatrix} x \\ y \end{pmatrix} = R \begin{pmatrix} x \\ y \end{pmatrix} \text{ i. e. } \begin{pmatrix} x \\ y \end{pmatrix} \xrightarrow[\text{of vector by } \theta]{\text{Clockwise rotation}} \begin{pmatrix} x' \\ y' \end{pmatrix} \text{(equation (9.87))} \tag{9.223}$$

where

$$R = \begin{pmatrix} \cos \theta & \sin \theta \\ -\sin \theta & \cos \theta \end{pmatrix} \text{(equation (9.88))} \tag{9.224}$$

is the rotation matrix or transformation matrix for clockwise rotation of the vector keeping axes fixed.

(iii) Transformation matrix for clockwise rotation of axes keeping the vector fixed (figure 9.18(c)).

Vector fixed means

$$\vec{r}' = \vec{r} \tag{9.225}$$

$$x'\hat{i}' + y'\hat{j}' = x\hat{i} + y\hat{j} \tag{9.226}$$

Take dot product of equation (9.226) with $\hat{i}'$

$$x'\underbrace{(\hat{i}'. \hat{i}')}_{=1} + y'\underbrace{(\hat{j}'. \hat{i}')}_{=0} = x\underbrace{(\hat{i}, \hat{i}')}_{=\cos \theta} + y\underbrace{(\hat{j}. \hat{i}')}_{=\cos (90°+\theta)=-\sin \theta} \tag{9.227}$$

$$x' = x \cos \theta - y \sin \theta \tag{9.228}$$

Take dot product of equation (9.226) with $\hat{j}'$

$$x'\underbrace{(\hat{i}'. \hat{j}')}_{=0} + y'\underbrace{(\hat{j}'. \hat{j}')}_{=1} = x\underbrace{(\hat{i}, \hat{j}')}_{=\cos (90°-\theta)=\sin \theta} + y\underbrace{(\hat{j}. \hat{j}')}_{=\cos \theta} \tag{9.229}$$

$$y' = x \sin \theta + y \cos \theta \tag{9.230}$$

Combining equation (9.228) and equation (9.230) we can write

$$\begin{pmatrix} x' \\ y' \end{pmatrix} = \begin{pmatrix} \cos\theta & -\sin\theta \\ \sin\theta & \cos\theta \end{pmatrix} \begin{pmatrix} x \\ y \end{pmatrix} = R' \begin{pmatrix} x \\ y \end{pmatrix} \text{ i.e. } \begin{pmatrix} x \\ y \end{pmatrix} \xrightarrow[\text{of axes by } \theta]{\text{Clockwise rotation}} \begin{pmatrix} x' \\ y' \end{pmatrix} \qquad (9.231)$$

where

$$R' = \begin{pmatrix} \cos\theta & -\sin\theta \\ \sin\theta & \cos\theta \end{pmatrix} \text{ (equation (9.89))} \qquad (9.232)$$

is the rotation matrix or transformation matrix for clockwise rotation of axes keeping the vector fixed.

(iv) Transformation matrix for anti-clockwise rotation of a vector keeping axes fixed (figure 9.18(d)).

Consider

$$\vec{r} \rightarrow re^{i\phi} = x + iy = (x,\ y) \qquad (9.233)$$

$$=r(\cos\phi + i\sin\phi) = r\cos\phi + ir\sin\phi = x + iy \qquad (9.234)$$

$$x = r\cos\phi,\ y = r\sin\phi \qquad (9.235)$$

Multiply $\vec{r}$ by $e^{i\theta}$ which is equivalent to anti-clockwise rotation of $\vec{r}$ by θ, to get the rotated vector $\vec{r}'$.

$$\vec{r}' \rightarrow e^{i\theta}\vec{r} = e^{i\theta}(re^{i\phi}) = re^{i(\theta+\phi)} = (x',\ y') \qquad (9.236)$$

$$=r[\cos(\theta + \phi) + ir\sin(\theta + \phi)]$$

$$=r\cos(\theta + \phi) + ir\sin(\theta + \phi) = x' + iy' \qquad (9.237)$$

$$x' = r\cos(\theta + \phi) \qquad (9.238)$$

$$x' = r(\cos\theta\cos\phi - \sin\theta\sin\phi) = (r\cos\phi)\cos\theta - (r\sin\phi)\sin\theta \qquad (9.239)$$

Using equation (9.235) we have

$$x' = x\cos\theta - y\sin\theta \qquad (9.240)$$

$$y' = r\sin(\theta + \phi) = r(\sin\theta\cos\phi + \sin\phi\cos\theta)$$

$$y' = (r\cos\phi)\sin\theta + (r\sin\phi)\cos\theta \qquad (9.241)$$

$$y' = x\sin\theta + y\cos\theta \qquad (9.242)$$

Combining (9.240) and (9.242) we get

$$\begin{pmatrix} x' \\ y' \end{pmatrix} = \begin{pmatrix} \cos\theta & -\sin\theta \\ \sin\theta & \cos\theta \end{pmatrix} \begin{pmatrix} x \\ y \end{pmatrix} = R' \begin{pmatrix} x \\ y \end{pmatrix} \text{ i.e. } \begin{pmatrix} x \\ y \end{pmatrix} \xrightarrow[\text{of vector by } \theta]{\text{anticlockwise rotation}} \begin{pmatrix} x' \\ y' \end{pmatrix} \qquad (9.243)$$

where

$$R' = \begin{pmatrix} \cos\theta & -\sin\theta \\ \sin\theta & \cos\theta \end{pmatrix} \text{(equation (9.89))} \qquad (9.244)$$

is the rotation matrix or transformation matrix for anti-clockwise rotation of a vector keeping axes fixed.

(v) Rotation of a vector in clockwise direction and the rotation of axes in the anti-clockwise direction are described by the same rotation matrix

$$R = \begin{pmatrix} \cos\theta & \sin\theta \\ -\sin\theta & \cos\theta \end{pmatrix} \text{(equation (9.88))} \qquad (9.245)$$

So the two operations are equivalent.

(vi) Rotation of a vector in anti-clockwise direction and the rotation of axes in the clockwise direction are described by the same rotation matrix

$$R' = \begin{pmatrix} \cos\theta & -\sin\theta \\ \sin\theta & \cos\theta \end{pmatrix} \text{(equation (9.89))} \qquad (9.246)$$

So the two operations are equivalent.

Exercise 9.11 *Find the inverse of a rotation matrix.*

Ans. • Consider the rotation matrix $R = \begin{pmatrix} \cos\theta & \sin\theta \\ -\sin\theta & \cos\theta \end{pmatrix}$

$R^{-1} = \frac{R^{cT}}{|R|}$, c for cofactor matrix, T for the transpose of the matrix.

$$|R| = \begin{vmatrix} \cos\theta & \sin\theta \\ -\sin\theta & \cos\theta \end{vmatrix} = \cos^2\theta + \sin^2\theta = 1$$

$$R^{cT} = \begin{pmatrix} \cos\theta & \sin\theta \\ -\sin\theta & \cos\theta \end{pmatrix}^{cT} = \begin{pmatrix} \cos\theta & -(-\sin\theta) \\ -\sin\theta & \cos\theta \end{pmatrix}^{T} = \begin{pmatrix} \cos\theta & -\sin\theta \\ \sin\theta & \cos\theta \end{pmatrix}$$

Hence

$$R^{-1} = \begin{pmatrix} \cos\theta & -\sin\theta \\ \sin\theta & \cos\theta \end{pmatrix}$$

• Consider the rotation matrix $R' = \begin{pmatrix} \cos\theta & -\sin\theta \\ \sin\theta & \cos\theta \end{pmatrix}$

$$R'^{-1} = \frac{R'^{cT}}{|R'|}$$

$$|R'| = \begin{vmatrix} \cos\theta & -\sin\theta \\ \sin\theta & \cos\theta \end{vmatrix} = \cos^2\theta + \sin^2\theta = 1$$

$$R'^{cT} = \begin{pmatrix} \cos\theta & -\sin\theta \\ \sin\theta & \cos\theta \end{pmatrix}^{cT} = \begin{pmatrix} \cos\theta & -\sin\theta \\ -(-\sin\theta) & \cos\theta \end{pmatrix}^{T} = \begin{pmatrix} \cos\theta & \sin\theta \\ -\sin\theta & \cos\theta \end{pmatrix}$$

Hence

$$R'^{-1} = \begin{pmatrix} \cos\theta & -\sin\theta \\ \sin\theta & \cos\theta \end{pmatrix}$$

Hence $' = \begin{pmatrix} \cos\theta & \sin\theta \\ -\sin\theta & \cos\theta \end{pmatrix}$, $R^{-1} = \begin{pmatrix} \cos\theta & -\sin\theta \\ \sin\theta & \cos\theta \end{pmatrix}$

Exercise 9.12 *What are the properties of a rotation matrix?*

Ans. Properties of rotation matrix are the following.

- $R = \begin{pmatrix} \cos\theta & \sin\theta \\ -\sin\theta & \cos\theta \end{pmatrix}$, $R^{-1} = \begin{pmatrix} \cos\theta & -\sin\theta \\ \sin\theta & \cos\theta \end{pmatrix} = R'$; $R'^{-1} = R$. R and R' are the inverse of one another.
- $R = R'^{T}$, $R' = R^{T}$
- $RR^{T} = R^{T}R = I$, i.e. R is an orthogonal matrix ($R^{-1} = R^{T}$).
  $R'R'^{T} = R'^{T}R' = I$, i.e. R' is an orthogonal matrix ($R^{-1} = R^{T}$).
- $|R| = 1, |R'| = 1$, i.e. determinant of the rotation matrix is $+1$.
- Rotation of a vector in clockwise direction and the rotation of axes in the anti-clockwise direction are described by the rotation matrix

$$R = \begin{pmatrix} \cos\theta & \sin\theta \\ -\sin\theta & \cos\theta \end{pmatrix} \text{(equation (9.88))}$$

Rotation of a vector in anti-clockwise direction and the rotation of axes in the clockwise direction are described by the rotation matrix

$$R' = \begin{pmatrix} \cos\theta & -\sin\theta \\ \sin\theta & \cos\theta \end{pmatrix} \text{(equation (9.89))}$$

Exercise 9.13 *What is orthogonal transformation? Mention its significance.*

Ans. If the matrix that performs transformation of coordinates $x, y \leftrightarrow x', y'$ is an orthogonal matrix then the transformation is called orthogonal transformation.

Significance

- As $R = \begin{pmatrix} \cos\theta & \sin\theta \\ -\sin\theta & \cos\theta \end{pmatrix}$ performs rotation of a vector in clockwise direction and also rotation of axes in the anti-clockwise direction such a transformation is an orthogonal transformation.

Also since $R' = \begin{pmatrix} \cos\theta & -\sin\theta \\ \sin\theta & \cos\theta \end{pmatrix}$ performs rotation of a vector in anti-clockwise direction and the rotation of axes in the clockwise direction such a transformation is an orthogonal transformation.

- R, R' are 2×2 orthogonal matrices with determinant $+1$ and correspond to rotation—hence they are called rotation matrices.
- Any 2×2 orthogonal matrix with determinant -1 corresponds to reflection through a line.
- Length invariance is a consequence of orthogonal transformation (*exercises 9.14 and 9.16*).

Exercise 9.14 *Show that length invariance is a consequence of orthogonal transformation.*

$\boxed{\text{Ans.}}$ • Consider the orthogonal transformation performed by R

$$\begin{pmatrix} x' \\ y' \end{pmatrix} = \begin{pmatrix} \cos\theta & \sin\theta \\ -\sin\theta & \cos\theta \end{pmatrix}\begin{pmatrix} x \\ y \end{pmatrix} \Rightarrow \begin{matrix} x' = x\cos\theta + y\sin\theta \\ y' = -x\sin\theta + y\cos\theta \end{matrix}$$

$$x'^2 + y'^2 = (x\cos\theta + y\sin\theta)^2 + (-x\sin\theta + y\cos\theta)^2$$

$$= x^2\cos^2\theta + y^2\sin^2\theta + 2xy\cos\theta\sin\theta + x^2\sin^2\theta + y^2\cos^2\theta - 2xy\cos\theta\sin\theta$$

$$= x^2(\cos^2\theta + \sin^2\theta) + y^2(\cos^2 + \sin^2\theta) = x^2 + y^2$$

Hence

$$\sqrt{x'^2 + y'^2} = \sqrt{x^2 + y^2}$$

i.e. $r' = r$, i.e. length invariance follows from orthogonal transformation.

- Consider the orthogonal transformation performed by R'

$$\begin{pmatrix} x' \\ y' \end{pmatrix} = \begin{pmatrix} \cos\theta & -\sin\theta \\ \sin\theta & \cos\theta \end{pmatrix}\begin{pmatrix} x \\ y \end{pmatrix} \Rightarrow \begin{matrix} x' = x\cos\theta - y\sin\theta \\ y' = x\sin\theta + y\cos\theta \end{matrix}$$

$$x'^2 + y'^2 = (x\cos\theta - y\sin\theta)^2 + (x\sin\theta + y\cos\theta)^2$$

$$= x^2\cos^2\theta + y^2\sin^2\theta - 2xy\cos\theta\sin\theta + x^2\sin^2\theta + y^2\cos^2\theta + 2xy\cos\theta\sin\theta$$

$$= x^2(\cos^2\theta + \sin^2\theta) + y^2(\cos^2 + \sin^2\theta) = x^2 + y^2$$

Hence

$$\sqrt{x'^2 + y'^2} = \sqrt{x^2 + y^2}$$

i.e. $r' = r$, i.e. length invariance follows from orthogonal transformation.

Exercise 9.15 *What do the following matrices represent?*

$$A = \begin{pmatrix} -1 & 0 \\ 0 & 1 \end{pmatrix}, \quad B = \begin{pmatrix} 0 & 1 \\ -1 & 0 \end{pmatrix}, \quad C = BA, \quad D = \begin{pmatrix} 1 & 0 \\ 0 & -1 \end{pmatrix}, \quad E = \frac{1}{\sqrt{2}} \begin{pmatrix} 1 & 1 \\ -1 & 1 \end{pmatrix},$$

$$F = \frac{1}{2} \begin{pmatrix} -1 & \sqrt{3} \\ -\sqrt{3} & -1 \end{pmatrix}$$

$$G = \begin{pmatrix} \cos\theta & -\sin\theta & 0 \\ \sin\theta & \cos\theta & 0 \\ 0 & 0 & 1 \end{pmatrix}, \quad H = \begin{pmatrix} \cos\theta & 0 & \sin\theta \\ 0 & 1 & 0 \\ -\sin\theta & 0 & \cos\theta \end{pmatrix}, \quad J = \begin{pmatrix} 1 & 0 & 0 \\ 0 & \cos\theta & \sin\theta \\ 0 & -\sin\theta & \cos\theta \end{pmatrix}$$

Find rotation of $\hat{i}$ by F
Find rotation of $\hat{j}$ by F.

$\boxed{\text{Ans.}}$ Let $x\hat{i} + y\hat{j} = \begin{pmatrix} x \\ y \end{pmatrix}$

- Consider $A\begin{pmatrix} x \\ y \end{pmatrix} = \begin{pmatrix} -1 & 0 \\ 0 & 1 \end{pmatrix}\begin{pmatrix} x \\ y \end{pmatrix} = \begin{pmatrix} -x \\ y \end{pmatrix}$ i.e. $(x, y) \overset{A}{\to} (-x, y)$

$$\text{Also } |A| = \begin{vmatrix} -1 & 0 \\ 0 & 1 \end{vmatrix} = -1$$

As shown in figure 9.19(*a*) matrix A represents reflection about the Y-axis, i.e. about $\hat{j}$.

- Consider $B\begin{pmatrix} x \\ y \end{pmatrix} = \begin{pmatrix} 0 & 1 \\ -1 & 0 \end{pmatrix}\begin{pmatrix} x \\ y \end{pmatrix} = \begin{pmatrix} y \\ -x \end{pmatrix}$ i.e. $(x, y) \overset{B}{\to} (y, -x)$

$$\text{Also } |B| = \begin{vmatrix} 0 & 1 \\ -1 & 0 \end{vmatrix} = 1$$

As shown in figure 9.19(b) matrix B represents clockwise rotation about the Z-axis by $90°$.

- Consider $C\begin{pmatrix} x \\ y \end{pmatrix} = BA\begin{pmatrix} x \\ y \end{pmatrix} = \begin{pmatrix} 0 & 1 \\ -1 & 0 \end{pmatrix}\begin{pmatrix} -1 & 0 \\ 0 & 1 \end{pmatrix}\begin{pmatrix} x \\ y \end{pmatrix} = \begin{pmatrix} 0 & 1 \\ 1 & 0 \end{pmatrix}\begin{pmatrix} x \\ y \end{pmatrix} = \begin{pmatrix} y \\ x \end{pmatrix}$

$$\text{i.e. } (x, y) \overset{C}{\to} (y, x)$$

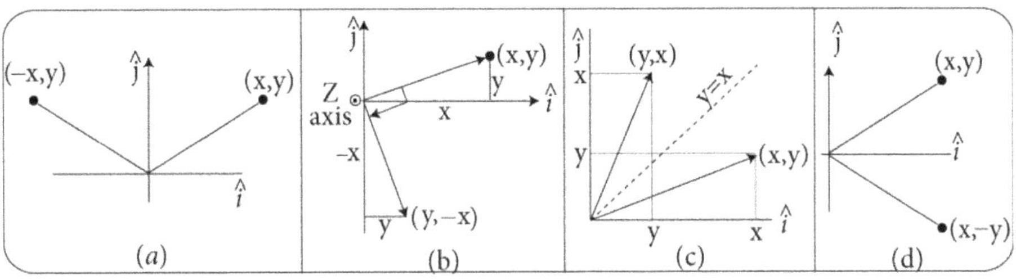

Figure 9.19. (*a*) Matrix A represents reflection. (b) Matrix B represents rotation. (c) Matrix C represents reflection. (d) Matrix D represents reflection.

$$|C| = \begin{vmatrix} 0 & 1 \\ 1 & 0 \end{vmatrix} = -1$$

As shown in figure 9.19(c), matrix C represents reflection about the $y = x$ line.

- Consider $D\begin{pmatrix} x \\ y \end{pmatrix} = \begin{pmatrix} 1 & 0 \\ 0 & -1 \end{pmatrix}\begin{pmatrix} x \\ y \end{pmatrix} = \begin{pmatrix} x \\ -y \end{pmatrix}$

$$\text{i.e. } (x, y) \xrightarrow{D} (x, -y)$$

$$\text{Also } |D| = \begin{vmatrix} 1 & 0 \\ 0 & -1 \end{vmatrix} = -1$$

As shown in figure 9.19(d) matrix D represents reflection about $\hat{\imath}$.

- $E = \frac{1}{\sqrt{2}}\begin{pmatrix} 1 & 1 \\ -1 & 1 \end{pmatrix} = \begin{pmatrix} 1/\sqrt{2} & 1/\sqrt{2} \\ -1/\sqrt{2} & 1/\sqrt{2} \end{pmatrix} = \begin{pmatrix} \cos\theta & -\sin\theta \\ \sin\theta & \cos\theta \end{pmatrix} = R'$

$$\cos\theta = \frac{1}{\sqrt{2}}, \quad \sin\theta = -\frac{1}{\sqrt{2}} \quad \theta = -45°$$

$$\text{Also } |E| = \begin{vmatrix} 1/\sqrt{2} & 1/\sqrt{2} \\ -1/\sqrt{2} & 1/\sqrt{2} \end{vmatrix} = 1$$

A represents rotation of $-45°$.

- $F = \frac{1}{2}\begin{pmatrix} -1 & \sqrt{3} \\ -\sqrt{3} & -1 \end{pmatrix} = \begin{pmatrix} \cos\theta & -\sin\theta \\ \sin\theta & \cos\theta \end{pmatrix} = R'$

$$\cos\theta = \frac{1}{2}, \quad \sin\theta = -\frac{\sqrt{3}}{2} \quad \text{i. e } \theta = 240° \text{ or } -120°$$

$$\text{Also } |F| = \begin{vmatrix} -1/2 & \sqrt{3}/2 \\ -\sqrt{3}/2 & -1/2 \end{vmatrix} = 1$$

A represents rotation by $240°$ or $-120°$

- $G = \begin{pmatrix} \cos\theta & -\sin\theta & 0 \\ \sin\theta & \cos\theta & 0 \\ 0 & 0 & 1 \end{pmatrix}, \quad |G| = 1,$

G represents rotation of vectors about the Z-axis by angle θ.

- $H = \begin{pmatrix} \cos\theta & 0 & \sin\theta \\ 0 & 1 & 0 \\ -\sin\theta & 0 & \cos\theta \end{pmatrix}, \quad |H| = 1,$

H represents rotation of vectors about the Y-axis by angle θ.

- $J = \begin{pmatrix} 1 & 0 & 0 \\ 0 & \cos\theta & \sin\theta \\ 0 & -\sin\theta & \cos\theta \end{pmatrix}$, $\quad |J| = 1$,

J represents rotation of vector about the X-axis by angle θ.

✓ Rotation of $\hat{i}$ by F.

$$\hat{i} = 1\hat{i} + 0\hat{j} = \begin{pmatrix} 1 \\ 0 \end{pmatrix}$$

Operate $\hat{i}$ by F

$$F\begin{pmatrix} 1 \\ 0 \end{pmatrix} = \frac{1}{2}\begin{pmatrix} -1 & \sqrt{3} \\ -\sqrt{3} & -1 \end{pmatrix}\begin{pmatrix} 1 \\ 0 \end{pmatrix} = \frac{1}{2}\begin{pmatrix} -1 \\ -\sqrt{3} \end{pmatrix} = \begin{pmatrix} -\frac{1}{2} \\ -\frac{\sqrt{3}}{2} \end{pmatrix}$$

$$= -\frac{1}{2}\hat{i} - \frac{\sqrt{3}}{2}\hat{j} = -\frac{1}{2}(\hat{i} + \hat{j}\sqrt{3}) = \hat{a}$$

Now $\hat{i}.\hat{a} = \hat{i}.[-\frac{1}{2}(\hat{i} + \hat{j}\sqrt{3})]$

$\cos\theta = -\frac{1}{2}$, i.e. $\theta = 240°$ or $-120°$ (shown in figure 9.20(a))

✓ Rotation of $\hat{j}$ by F.

$$\hat{j} = 0\hat{i} + 1\hat{j} = \begin{pmatrix} 0 \\ 1 \end{pmatrix}$$

$$F\begin{pmatrix} 0 \\ 1 \end{pmatrix} = \frac{1}{2}\begin{pmatrix} -1 & \sqrt{3} \\ -\sqrt{3} & -1 \end{pmatrix}\begin{pmatrix} 0 \\ 1 \end{pmatrix} = \frac{1}{2}\begin{pmatrix} \sqrt{3} \\ -1 \end{pmatrix}$$

$$= \begin{pmatrix} \sqrt{3}/2 \\ -1/2 \end{pmatrix} = \frac{\sqrt{3}}{2}\hat{i} - \frac{1}{2}\hat{j} = \frac{1}{2}(\sqrt{3}\hat{i} - \hat{j}) = \hat{b}$$

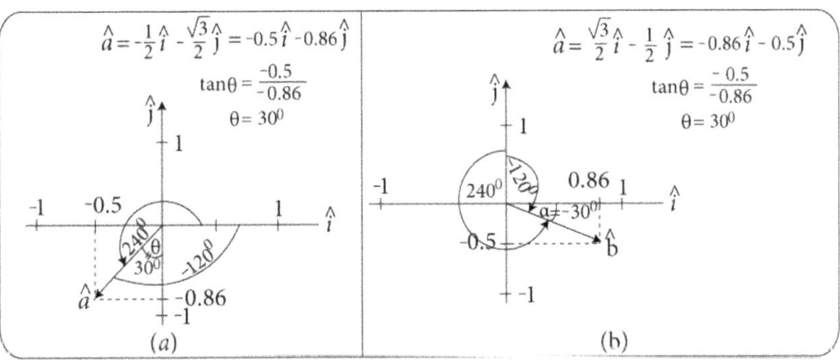

Figure 9.20. (a) Rotation of $\hat{i}$ by F. (b) Rotation of $\hat{j}$ by F.

Now $\hat{j}\cdot\hat{b} = \hat{j}\cdot[\frac{1}{2}(\sqrt{3}\,\hat{i} - \hat{j}\,)]$

$\cos\theta = -\frac{1}{2}$ i.e. $\theta = 240°$ or $-120°$ (shown in figure 9.20(b)).

Exercise 9.16 *Show that a linear transformation*

$$\begin{pmatrix} x' \\ y' \end{pmatrix} = \begin{pmatrix} a & b \\ c & d \end{pmatrix}\begin{pmatrix} x \\ y \end{pmatrix} = R\begin{pmatrix} x \\ y \end{pmatrix}$$

that preserves the length of vector is an orthogonal transformation. Predict and interpret determinantal value of R.

Ans. It follows from the given matrix relation that

$$x' = ax + by$$

$$y' = cx + dy$$

Preservation of length means that

$$x'^2 + y'^2 = x^2 + y^2$$

$$(ax + by)^2 + (cx + dy)^2 = x^2 + y^2$$

$$a^2x^2 + 2abxy + b^2y^2 + c^2x^2 + 2cdxy + d^2y^2 = x^2 + y^2$$

$$(a^2 + c^2)x^2 + (b^2 + d^2)y^2 + 2(ab + cd)xy = x^2 + y^2$$

Comparing, we have

$$a^2 + c^2 = 1$$

$$b^2 + d^2 = 1$$

$$ab + cd = 0$$

$$R^T R = \begin{pmatrix} a & b \\ c & d \end{pmatrix}^T \begin{pmatrix} a & b \\ c & d \end{pmatrix} = \begin{pmatrix} a & c \\ b & d \end{pmatrix}\begin{pmatrix} a & b \\ c & d \end{pmatrix}$$

$$= \begin{pmatrix} a^2 + c^2 & ab + cd \\ ab + cd & b^2 + d^2 \end{pmatrix} = \begin{pmatrix} 1 & 0 \\ 0 & 1 \end{pmatrix} = I$$

$$R^{-1} = R^T.$$

So R is an orthogonal matrix.
- Consider $R^T R = I$

$$det(R^T R) = detI$$

$$detR^T detR = detI$$

$$\begin{vmatrix} a & c \\ b & d \end{vmatrix} \begin{vmatrix} a & b \\ c & d \end{vmatrix} = \begin{vmatrix} 1 & 0 \\ 0 & 1 \end{vmatrix}$$

$$(ad - bc)(ad - bc) = 1$$

$$(ad - bc)^2 = 1$$

As $detR = ad - bc$ we can write

$$(detR)^2 = 1$$

$$detR = \pm 1$$

$$detR = +1 \text{ corresponds to rotation.}$$

$$detR = -1 \text{ corresponds to reflection.}$$

Exercise 9.17 *Check equations (9.90) and (9.91) i.e.* $RR^T = R^T R = I$ *and* $R'R'^T = R'^T R' = I$. *Comment on the nature of the matrix* R, R'.

$$\boxed{\text{Ans.}}\ RR^T = \begin{pmatrix} \cos\phi & \sin\phi \\ -\sin\phi & \cos\phi \end{pmatrix} \begin{pmatrix} \cos\phi & \sin\phi \\ -\sin\phi & \cos\phi \end{pmatrix}^T$$

$$= \begin{pmatrix} \cos\phi & \sin\phi \\ -\sin\phi & \cos\phi \end{pmatrix} \begin{pmatrix} \cos\phi & -\sin\phi \\ \sin\phi & \cos\phi \end{pmatrix}$$

$$= \begin{pmatrix} \cos^2\phi + \sin^2\phi & -\cos\phi\sin\phi + \sin\phi\cos\phi \\ -\sin\phi\cos\phi + \cos\phi\sin\phi & \sin^2\phi + \cos^2\phi \end{pmatrix}$$

$$= \begin{pmatrix} 1 & 0 \\ 0 & 1 \end{pmatrix} = I$$

Similarly $R^T R = I$

Hence $RR^T = R^T R = I$

So R is an orthogonal matrix.

Similarly, $R'R'^T = R'^T R' = I$ and R' is an orthogonal matrix.

Exercise 9.18 *Show that an orthogonal matrix with* $detO = -1$ *cannot form a group.*

$\boxed{\text{Ans.}}$ Consider the 2×2 orthogonal matrix

$$O = \begin{pmatrix} -1 & 0 \\ 0 & 1 \end{pmatrix} \text{ having det } O = \begin{vmatrix} -1 & 0 \\ 0 & 1 \end{vmatrix} = -1$$

3×3 orthogonal matrix

$$O = \begin{pmatrix} -1 & 0 & 0 \\ 0 & 1 & 0 \\ 0 & 0 & 1 \end{pmatrix} \text{ having } detO = \begin{vmatrix} -1 & 0 & 0 \\ 0 & 1 & 0 \\ 0 & 0 & 1 \end{vmatrix} = -1.$$

- Closure property

  Consider two matrices O_1, O_2 with determinantal values

$$detO_1 = -1, \quad detO_2 = -1.$$

Then their combination O_1O_2 has determinantal value

$$det(O_1O_2) = (detO_1)(detO_2) = (-1)(-1) = +1$$

So combination of two members does not lead to another member of the set. So closure property fails.

- Associative property

  Matrix multiplication is associative.

- Identity element

  Let us search for an identity matrix I such that

$$OI = O$$

and this requires that

$$I = \begin{pmatrix} 1 & 0 \\ 0 & 1 \end{pmatrix} \text{ for } 2 \times 2 \text{ matrix or } I = \begin{pmatrix} 1 & 0 & 0 \\ 0 & 1 & 0 \\ 0 & 0 & 1 \end{pmatrix} \text{ for } 3 \times 3 \text{ matrix.}$$

But $detI = +1$ which is not a member of the set. So we fail to find an identity matrix I that fits in the set.

- Inverse element

The requirement is

$$OO^{-1} = I = \text{Identity matrix.}$$

But as Identity matrix I does not exist in the set we cannot find inverse.

Since 3 requirements (out of 4) are not met we conclude that an orthogonal matrix with det $O = -1$ cannot form a group.

Exercise 9.19 *Find the determinantal value of O_1O_2 if O_1 and O_2 are members of the SO(2) group.*

$\boxed{\text{Ans.}}$ The $SO(2)$ group is a special unitary group and so determinantal value of its matrix elements will be $+1$ i.e.

$$detO_1 = +1, \quad detO_2 = +1.$$

Hence

$$det(O_1O_2) = detO_1detO_2 = (+1)(+1) = +1$$

Exercise 9.20 *How is S_z the generator of the SO(3) group connected to orbital angular momentum l_z about the Z-axis?*

[Ans.] Let us evaluate

$$
\begin{pmatrix} x & y & z \end{pmatrix} S_z \begin{pmatrix} \frac{\partial}{\partial x} \\ \frac{\partial}{\partial y} \\ \frac{\partial}{\partial z} \end{pmatrix} = \begin{pmatrix} x & y & z \end{pmatrix} \begin{pmatrix} 0 & -i & 0 \\ i & 0 & 0 \\ 0 & 0 & 0 \end{pmatrix} \begin{pmatrix} \frac{\partial}{\partial x} \\ \frac{\partial}{\partial y} \\ \frac{\partial}{\partial z} \end{pmatrix} = \begin{pmatrix} x & y & z \end{pmatrix} \begin{pmatrix} -i\frac{\partial}{\partial y} \\ i\frac{\partial}{\partial x} \\ 0 \end{pmatrix}
$$

$$
= -ix\frac{\partial}{\partial y} + iy\frac{\partial}{\partial x} = = -i\left(x\frac{\partial}{\partial y} - y\frac{\partial}{\partial x} \right) = l_z
$$

This is how S_z the generator of the $SO(3)$ group is connected to orbital angular momentum l_z about the Z-axis.

Exercise 9.21 *Show that the general form of a 2×2 unitary matrix in $SU(2)$ is $\begin{pmatrix} a & b \\ -b* & a* \end{pmatrix}$ where a, b, c, d are complex numbers in general.*

[Ans.] Consider a general matrix $U = \begin{pmatrix} a & b \\ c & d \end{pmatrix}$ where a, b, c, d are complex numbers in general.

$SU(2)$ involves a unimodular matrix and so

$$
\begin{vmatrix} a & b \\ c & d \end{vmatrix} = 1 = det\,U
$$

Condition of unitarity is

$$
U^\dagger = U^{-1} \tag{9.247}
$$

Now $U^\dagger = (U^*)^T = \begin{pmatrix} a & b \\ c & d \end{pmatrix}^{*T} = \begin{pmatrix} a* & b* \\ c* & d* \end{pmatrix}^T = \begin{pmatrix} a* & c* \\ b* & d* \end{pmatrix}$

$U^{-1} = \frac{(U^c)^T}{det\,U} = (U^c)^T$ as $det\,U = 1$ for the $SU(2)$ group.

$$
U^{-1} = (U^c)^T = \begin{pmatrix} a & b \\ c & d \end{pmatrix}^{cT} = \begin{pmatrix} d & -c \\ -b & a \end{pmatrix}^T = \begin{pmatrix} d & -b \\ -c & a \end{pmatrix}
$$

Equation (9.247) gives

$$
\begin{pmatrix} a* & c* \\ b* & d* \end{pmatrix} = \begin{pmatrix} d & -b \\ -c & a \end{pmatrix}
$$

$$
d - a*
$$

$$c = -b*$$

Hence the general form of a 2×2 unitary matrix in $SU(2)$ is

$$U = \begin{pmatrix} a & b \\ c & d \end{pmatrix} = \begin{pmatrix} a & b \\ -b* & a* \end{pmatrix}.$$

Exercise 9.22 *Show that Pauli spin matrices are traceless.*

[Ans.] $\mathrm{Tr}\,\sigma_x = \mathrm{Tr}\begin{pmatrix} 0 & 1 \\ 1 & 0 \end{pmatrix} = 0 + 0 = 0,$

$$Tr\,\sigma_y = Tr\begin{pmatrix} 0 & -i \\ i & 0 \end{pmatrix} = 0 + 0 = 0$$

$$Tr\,\sigma_z = Tr\begin{pmatrix} 1 & 0 \\ 0 & -1 \end{pmatrix} = 1 - 1 = 0$$

Hence $Tr\sigma_i = 0$.

Exercise 9.23 *Show that Pauli spin matrices are Hermitian.*

$$[Ans.]\ \sigma_x^\dagger = \begin{pmatrix} 0 & 1 \\ 1 & 0 \end{pmatrix}^\dagger = \begin{pmatrix} 0 & 1 \\ 1 & 0 \end{pmatrix} = \sigma_x$$

$$\sigma_y^\dagger = \begin{pmatrix} 0 & -i \\ i & 0 \end{pmatrix}^\dagger = \begin{pmatrix} 0 & -i \\ i & 0 \end{pmatrix} = \sigma_y$$

$$\sigma_z^\dagger = \begin{pmatrix} 1 & 0 \\ 0 & -1 \end{pmatrix}^\dagger = \begin{pmatrix} 1 & 0 \\ 0 & -1 \end{pmatrix} = \sigma_z$$

Hence $\sigma_i^\dagger = \sigma_i$
i.e. Pauli spin matrices are Hermitian.

Exercise 9.24 *Pauli spin matrices are unitary.*
[Ans.] Using $\sigma_i^\dagger = \sigma_i$ (Hermiticity of σ_i) we can write

$$\sigma_i^\dagger \sigma_i = \sigma_i \sigma_i^\dagger = \sigma_i^2 = I$$

So $\sigma_i^\dagger = \sigma_i^{-1}$
i.e. σ_i is unitary.

Exercise 9.25 *Establish the commutation relation for Pauli spin matrices.*
[Ans.] Consider

$$[\sigma_x \sigma_y] = \sigma_x \sigma_y - \sigma_y \sigma_x = \begin{pmatrix} 0 & 1 \\ 1 & 0 \end{pmatrix}\begin{pmatrix} 0 & -i \\ i & 0 \end{pmatrix} - \begin{pmatrix} 0 & -i \\ i & 0 \end{pmatrix}\begin{pmatrix} 0 & 1 \\ 1 & 0 \end{pmatrix}$$

$$[\sigma_x \sigma_y] = \begin{pmatrix} i & 0 \\ 0 & -i \end{pmatrix} - \begin{pmatrix} -i & 0 \\ 0 & i \end{pmatrix} = \begin{pmatrix} 2i & 0 \\ 0 & -2i \end{pmatrix} = 2i\begin{pmatrix} 1 & 0 \\ 0 & -1 \end{pmatrix} = 2i\sigma_z$$

and

$$[\sigma_y \sigma_x] = -2i\sigma_z$$

Similarly

$$[\sigma_y \sigma_z] = 2i\sigma_x \text{ and } [\sigma_z \sigma_y] = -2i\sigma_x$$

$$[\sigma_z \sigma_x] = 2i\sigma_y \text{ and } [\sigma_x \sigma_z] = -2i\sigma_y$$

Hence $[\sigma_i \sigma_j] = 2i\varepsilon_{ijk}\sigma_k$ is the commutation relation for Pauli spin matrices.

Exercise 9.26 *Can any 2×2 matrix be expressed as a linear combination of the Pauli matrices and an identity matrix?*

$\boxed{\text{Ans.}}$ Consider a 2×2 matrix with constant elements $a_{11}, a_{12}, a_{21}, a_{22}$ as

$$A = \begin{pmatrix} a_{11} & a_{12} \\ a_{21} & a_{22} \end{pmatrix} = \begin{pmatrix} \frac{(a_{11}+a_{22})+(a_{11}-a_{22})}{2} & \frac{(a_{12}+a_{21})+(a_{12}-a_{21})}{2} \\ \frac{(a_{12}+a_{21})-(a_{12}-a_{21})}{2} & \frac{(a_{11}+a_{22})-(a_{11}-a_{22})}{2} \end{pmatrix}$$

$$= \frac{1}{2}\begin{pmatrix} a_{11}+a_{22} & 0 \\ 0 & a_{11}+a_{22} \end{pmatrix} + \frac{1}{2}\begin{pmatrix} a_{11}-a_{22} & 0 \\ 0 & -(a_{11}-a_{22}) \end{pmatrix}$$

$$+ \frac{1}{2}\begin{pmatrix} 0 & a_{12}+a_{21} \\ a_{12}+a_{21} & 0 \end{pmatrix} + \frac{1}{2}i\begin{pmatrix} 0 & -i(a_{12}-a_{21}) \\ i(a_{12}-a_{21}) & 0 \end{pmatrix}$$

$$= \frac{1}{2}(a_{11}+a_{22})\begin{pmatrix} 1 & 0 \\ 0 & 1 \end{pmatrix} + \frac{1}{2}(a_{11}-a_{22})\begin{pmatrix} 1 & 0 \\ 0 & -1 \end{pmatrix}$$

$$+ \frac{1}{2}(a_{12}+a_{21})\begin{pmatrix} 0 & 1 \\ 1 & 0 \end{pmatrix} + \frac{1}{2}i(a_{12}-a_{21})\begin{pmatrix} 0 & -i \\ i & 0 \end{pmatrix}$$

$$A = C_0\begin{pmatrix} 1 & 0 \\ 0 & 1 \end{pmatrix} + C_z\begin{pmatrix} 1 & 0 \\ 0 & -1 \end{pmatrix} + C_2\begin{pmatrix} 0 & 1 \\ 1 & 0 \end{pmatrix} + C_4\begin{pmatrix} 0 & -i \\ i & 0 \end{pmatrix}$$

$$A = C_0 I + C_z \sigma_z + C_x \sigma_x + C_y \sigma_y$$

Exercise 9.27 *Show that $T(\delta\theta, \hat{i})T(\delta\theta, \hat{j}) - T(\delta\theta, \hat{j})T(\delta\theta, \hat{i}) \neq 0$*

$\boxed{\text{Ans.}}$ Let us find

$$T(\delta\theta, \hat{i})T(\delta\theta, \hat{j}) - T(\delta\theta, \hat{j})T(\delta\theta, \hat{i})$$

$$=(1 - \frac{i}{\hbar}\,\delta\theta\,\hat{i}.\,\vec{l})(1 - \frac{i}{\hbar}\,\delta\theta\,\hat{j}.\,\vec{l}) - (1 - \frac{i}{\hbar}\,\delta\theta\,\hat{j}.\,\vec{l})(1 - \frac{i}{\hbar}\,\delta\theta\,\hat{i}.\,\vec{l})$$

$$=(1 - \frac{i}{\hbar}\,\delta\theta\,l_x)(1 - \frac{i}{\hbar}\,\delta\theta\,l_y) - (1 - \frac{i}{\hbar}\,\delta\theta\,l_y)(1 - \frac{i}{\hbar}\,\delta\theta\,l_x)$$

$$=1 - \frac{i}{\hbar}\,\delta\theta\,l_y - \frac{i}{\hbar}\,\delta\theta\,l_x + (\frac{i}{\hbar}\,\delta\theta\,l_x)(\frac{i}{\hbar}\,\delta\theta\,l_y)$$

$$-(1 - \frac{i}{\hbar}\,\delta\theta\,l_x - \frac{i}{\hbar}\,\delta\theta\,l_y + (\frac{i}{\hbar}\,\delta\theta\,l_y)(\frac{i}{\hbar}\,\delta\theta\,l_x))$$

$$=-\frac{1}{\hbar^2}(\delta\theta)^2 l_x l_y + \frac{1}{\hbar^2}(\delta\theta)^2 l_y l_x$$

$$=-\frac{1}{\hbar^2}(\delta\theta)^2(l_x l_y - l_y l_x)$$

$$=-\frac{1}{\hbar^2}(\delta\theta)^2[l_x,l_y]$$

$$=-\frac{1}{\hbar^2}(\delta\theta)^2 i\hbar l_z$$

$$=-\frac{1}{\hbar^2}(\delta\theta)^2 i\hbar \hat{k}.\,\vec{l}$$

$$=\left(I - \frac{i}{\hbar}(\delta\theta)^2 \hat{k}.\,\vec{l}\right) - I$$

$$=T((\delta\theta)^2, \hat{k}) - I \neq 0$$

Exercise 9.28 *Prove that* $[l_i l_j] = i\hbar\varepsilon_{ijk} l_k$ *where* l_i *is orbital angular momentum.*
Ans. Consider the commutator bracket

$$[l_x l_y]\psi = (l_x l_y - l_y l_x)\psi$$

$$=\left[\left\{-i\hbar\left(y\frac{\partial}{\partial z} - z\frac{\partial}{\partial y}\right)\right\}\left\{-i\hbar\left(z\frac{\partial}{\partial x} - x\frac{\partial}{\partial z}\right)\right\} - \left\{-i\hbar\left(z\frac{\partial}{\partial x} - x\frac{\partial}{\partial z}\right)\right\}\left\{-i\hbar\left(y\frac{\partial}{\partial z} - z\frac{\partial}{\partial y}\right)\right\}\right]\psi$$

$$=(-i\hbar)^2\left[\left(y\frac{\partial}{\partial z} - z\frac{\partial}{\partial y}\right)\left(z\frac{\partial}{\partial x} - x\frac{\partial}{\partial z}\right) - \left(z\frac{\partial}{\partial x} - x\frac{\partial}{\partial z}\right)\left(y\frac{\partial}{\partial z} - z\frac{\partial}{\partial y}\right)\right]\psi$$

$$= (i\hbar)^2 \left[y\frac{\partial}{\partial z}\left(z\frac{\partial}{\partial x} - x\frac{\partial}{\partial z} \right) - z\frac{\partial}{\partial y}\left(z\frac{\partial}{\partial x} - x\frac{\partial}{\partial z} \right) - z\frac{\partial}{\partial x}\left(y\frac{\partial}{\partial z} - z\frac{\partial}{\partial y} \right) + x\frac{\partial}{\partial z}\left(y\frac{\partial}{\partial z} - z\frac{\partial}{\partial y} \right) \right]\psi$$

$$= (i\hbar)^2 [y\frac{\partial}{\partial z}\left(z\frac{\partial}{\partial x} \right) - yx\frac{\partial^2}{\partial z^2} - z^2\frac{\partial^2}{\partial y\partial x} - zx\frac{\partial^2}{\partial y\partial z}$$

$$- zy\frac{\partial^2}{\partial x\partial z} + z^2\frac{\partial^2}{\partial x\partial y} + xy\frac{\partial^2}{\partial z^2} - x\frac{\partial}{\partial z}\left(z\frac{\partial}{\partial y} \right)]\,\psi = 0$$

As 2nd and 7th, 3rd and 6th terms cancel out we have

$$[l_x l_y]\psi = (i\hbar)^2 \left[y\left(\frac{\partial}{\partial x} + z\frac{\partial^2}{\partial z\partial x} \right) + zx\frac{\partial^2}{\partial y\partial z} - zy\frac{\partial^2}{\partial x\partial z} - x\left(\frac{\partial}{\partial y} + z\frac{\partial^2}{\partial z\partial y} \right) \right]\psi$$

As 2nd and 4th, 3rd and 6th terms cancel out we have

$$[l_x l_y]\psi = (i\hbar)^2 \left(y\frac{\partial}{\partial x} - x\frac{\partial}{\partial y} \right)\psi = i\hbar \left[-i\hbar\left(x\frac{\partial}{\partial y} - y\frac{\partial}{\partial x} \right) \right]\psi = i\hbar l_z \psi$$

As ψ is arbitrary

$$[l_x l_y] = i\hbar l_z$$

Similarly $[l_y l_z] = i\hbar l_x$, $[l_z l_x] = i\hbar l_y$

$$\text{Also } [l_x l_z] = -i\hbar l_y, \quad [l_y l_x] = -i\hbar l_z, \quad [l_z l_y] = -i\hbar l_x$$

Combining the commutation relations we get $[l_i l_j] = i\hbar \varepsilon_{ijk} l_k$.

Exercise 9.29 *Find the general element of SO(3).*
    Ans. Let us express rotations in terms of Euler angles.
    Rotation through Euler angles consists of 3 successive rotations.
    (1) Rotation through α about the z-axis.
    (2) Rotation through β about the new Y-axis.
    (3) Rotation through γ about the new z-axis. Thus

$$R_Z(\gamma)R_Y(\beta)R_Z(\alpha)$$

$$= \begin{pmatrix} \cos\gamma & \sin\gamma & 0 \\ -\sin\gamma & \cos\gamma & 0 \\ 0 & 0 & 1 \end{pmatrix} \begin{pmatrix} \cos\beta & 0 & \sin\beta \\ 0 & 1 & 0 \\ -\sin\beta & 0 & \cos\beta \end{pmatrix} \begin{pmatrix} \cos\alpha & \sin\alpha & 0 \\ -\sin\alpha & \cos\alpha & 0 \\ 0 & 0 & 1 \end{pmatrix}$$

$$= \begin{pmatrix} \cos\gamma & \sin\gamma & 0 \\ -\sin\gamma & \cos\gamma & 0 \\ 0 & 0 & 1 \end{pmatrix} \begin{pmatrix} \cos\alpha\cos\beta & \sin\alpha\cos\beta & \sin\beta \\ -\sin\alpha & \cos\alpha & 0 \\ -\cos\alpha\sin\beta & -\sin\alpha\sin\beta & \cos\beta \end{pmatrix}$$

$$= \begin{pmatrix} \cos\alpha\cos\beta\cos\gamma - \sin\alpha\sin\gamma & \sin\alpha\cos\beta\cos\gamma + \cos\alpha\sin\gamma & \sin\beta\cos\gamma \\ -\cos\alpha\cos\beta\sin\gamma - \sin\alpha\cos\gamma & -\sin\alpha\cos\beta\sin\gamma + \cos\alpha\cos\gamma & -\sin\beta\sin\gamma \\ -\cos\alpha\sin\beta & -\sin\alpha\sin\beta & \cos\beta \end{pmatrix}$$

This is an orthogonal matrix with determinant $+1$ and represents the general element of the $SO(3)$ group.

Exercise 9.30 *Find the general element of $SU(2)$.*

Ans. Corresponding to the general element of $SO(3)$ group represented by $R_Z(\gamma)R_Y(\beta)R_Z(\alpha)$ which is 3 successive rotations by Euler angles through α about the z-axis, through β about the new Y-axis and through γ about the new Z-axis let us write down the $SU(2)$ elements

$$U\left(\frac{\alpha}{2}, \frac{\beta}{2}, \frac{\gamma}{2}\right) = U_3\left(\frac{\gamma}{2}\right)U_2\left(\frac{\beta}{2}\right)U_3\left(\frac{\alpha}{2}\right)$$

$$= \begin{pmatrix} e^{i\frac{\gamma}{2}} & 0 \\ 0 & e^{-i\frac{\gamma}{2}} \end{pmatrix} \begin{pmatrix} \cos\frac{\beta}{2} & \sin\frac{\beta}{2} \\ -\sin\frac{\beta}{2} & \cos\frac{\beta}{2} \end{pmatrix} \begin{pmatrix} e^{i\frac{\alpha}{2}} & 0 \\ 0 & e^{-i\frac{\alpha}{2}} \end{pmatrix}$$

$$= \begin{pmatrix} e^{i\frac{\gamma}{2}} & 0 \\ 0 & e^{-i\frac{\gamma}{2}} \end{pmatrix} \begin{pmatrix} \cos\frac{\beta}{2}e^{i\frac{\alpha}{2}} & \sin\frac{\beta}{2}e^{-i\frac{\alpha}{2}} \\ -\sin\frac{\beta}{2}e^{i\frac{\alpha}{2}} & \cos\frac{\beta}{2}e^{-i\frac{\alpha}{2}} \end{pmatrix}$$

$$U\left(\frac{\alpha}{2}, \frac{\beta}{2}, \frac{\gamma}{2}\right) = \begin{pmatrix} \cos\frac{\beta}{2}e^{i\frac{\alpha+\gamma}{2}} & \sin\frac{\beta}{2}e^{-i\frac{\alpha-\gamma}{2}} \\ -\sin\frac{\beta}{2}e^{i\frac{\alpha-\gamma}{2}} & \cos\frac{\beta}{2}e^{-i\frac{\alpha+\gamma}{2}} \end{pmatrix}$$

With $\eta = \frac{\beta}{2}$, $\xi = \frac{\alpha+\gamma}{2}$, $\zeta = -\frac{\alpha-\gamma}{2}$ the matrix $U(\frac{\alpha}{2}, \frac{\beta}{2}, \frac{\gamma}{2})$ reduces to the form of equation (9.144).

This is the most general form of $SU(2)$ and corresponds to $R_Z(\gamma)R_Y(\beta)R_Z(\alpha)$ of $SO(3)$ evaluated in *exercise 9.29*.

Exercise 9.31 *Show that 2×2 unitary matrices with $det\,1$ form a group.*

Ans. Closure property

✓ U_1, U_2. . are unitary matrices with det 1. Then

$$U_1^\dagger = U_1^{-1}$$

$$U_2^\dagger = U_2^{-1}$$

$$(U_1 U_2)^\dagger = U_2^\dagger U_1^\dagger = U_2^{-1}U_1^{-1} = (U_1 U_2)^{-1}$$

So $U_1 U_2$ is a unitary matrix formed from U_1 and U_2.

Also consider

$$det(U_1 U_2) = det U_1 det U_2 = (+1)(+1) = +1.$$

So the product matrix is unitary having determinant $=1$ and hence belongs to the same set.

So closure property is satisfied.

✓ Associative property

Matrix multiplication is associative.

✓ Identity element

The definition of unitary matrix implies identity matrix element viz. $UU^\dagger = I$.

✓ Inverse element

The definition of unitary matrix implies inverse matrix element viz. $U^\dagger = U^{-1}$.

The set of 2×2 unitary matrices with $det 1$ form a group called $U(2)$ group.

Exercise 9.32 *Consider a matrix written as $U = e^{iH}$ where H is Hermitian matrix. Show that U is unitary. Find the condition on H so that we can make a special choice of $det U = +1$.*

$\boxed{\text{Ans.}}$ Consider a matrix written as

$$U = e^{iH}$$

where $H = H^\dagger$, i.e. Hermitian.

Expand U into a Taylor series

$$U = \sum_{k=0}^{\infty} \frac{(iH)^k}{k!}$$

$$U^\dagger = \left[\sum_{k=0}^{\infty} \frac{(iH)^k}{k!} \right]^\dagger = \sum_{k=0}^{\infty} \frac{(-iH^\dagger)^k}{k!}$$

Using $H = H^\dagger$ we have

$$U^\dagger = \sum_{k=0}^{\infty} \frac{(-iH)^k}{k!} = e^{-iH}$$

Thus we have

$$UU^\dagger = e^{iH} e^{-iH} = I$$

So U is unitary if H is Hermitian.

We can write any matrix as the product of its eigenvalues. Let the eigenvalue of H be λ_j, $j = 1, 2, ..n$ with eigenvector v_j, i.e.

$$Hv_j = \lambda_j v_j$$

Let us operate U on v_j

$$Uv_j = e^{iH}v_j = \sum_{k=0}^{\infty} \frac{(iH)^k}{k!}v_j = \sum_{k=0}^{\infty} \frac{i^k}{k!}H^k v_j$$

Since $H^k v_j = H. \ldots\ldots Hv_j = \lambda_j \ldots \ldots \lambda_j v_j = \lambda_j^k v_j$

$$Uv_j = e^{iH}v_j = \sum_{k=0}^{\infty} \frac{i^k}{k!}\lambda_j^k v_j = \sum_{k=0}^{\infty} \frac{(i\lambda_j)^k}{k!}v_j = e^{i\lambda_j}v_j$$

So $U = e^{iH}$ has eigenvalues $e^{i\lambda_j}$.

The trace of any matrix is the sum of its eigenvalues. Hence for H we write

$$TrH = \sum_{j=1}^{n} \lambda_j$$

And the determinant is the product of eigenvalues. Hence for $U = e^{iH}$ we have

$$det\ U = \prod_{j=1}^{n} e^{i\lambda_j} = e^{i\lambda_1}e^{i\lambda_2} \ldots \ldots\ldots = e^{i(\lambda_1 + \lambda_2 + \ldots.)} = e^{i\sum_j \lambda_j}$$

$$det\ U = e^{iTrH}$$

If we set $det\ U = 1$ then it follows that $TrH = 0$.

So the special choice of $det\ U = 1$ restricts the generator Hermitian matrix H to be traceless.

Exercise 9.33 *Show that for a unitary matrix $det\ U$ is a phase factor.*

Ans. A unitary matrix is defined as

$$UU^{\dagger} = I \text{ (where } I = \text{ unit matrix)}$$

$$det(UU^{\dagger}) = detI$$

$$det\ U det\ U^{\dagger} = detI \tag{9.248}$$

For $n \times n$ matrix

$$detI = 1$$

And

$$det\ U^{\dagger} = det\ U^{T}*$$

Hence from equation (9.248)

$$(det\ U)(det\ U^{T})* = 1$$

Since $det\ U = det\ U^{T}$ it follows that

$$det\ U\ (det\ U)* = 1$$

$$|det\ U|^2 = 1$$

This means $det\,U$ is a phase, say $e^{i\alpha}$ i.e.

$det\,U = e^{i\alpha}$ (α is a parameter) which is a number of unit magnitude.

This is a property of any unitary matrix.

If we choose $\alpha = 0$ then $det\,U = 1$ which is a property of $SU(n)$.

Exercise 9.34 *Establish the commutation relation between the SU (2) generators $\frac{1}{2}\sigma_i$ s.*

$\boxed{\text{Ans.}}$ ✓ Consider

$$[\sigma_1, \sigma_2] = \sigma_1\sigma_2 - \sigma_2\sigma_1$$

$$= \begin{pmatrix} 0 & 1 \\ 1 & 0 \end{pmatrix}\begin{pmatrix} 0 & -i \\ i & 0 \end{pmatrix} - \begin{pmatrix} 0 & -i \\ i & 0 \end{pmatrix}\begin{pmatrix} 0 & 1 \\ 1 & 0 \end{pmatrix} = 2i\begin{pmatrix} 1 & 0 \\ 0 & -1 \end{pmatrix}$$

$$[\sigma_1, \sigma_2] = 2i\sigma_3$$

✓ Consider

$$[\sigma_2, \sigma_3] = \sigma_2\sigma_3 - \sigma_3\sigma_2$$

$$= \begin{pmatrix} 0 & -i \\ i & 0 \end{pmatrix}\begin{pmatrix} 1 & 0 \\ 0 & -1 \end{pmatrix} - \begin{pmatrix} 1 & 0 \\ 0 & -1 \end{pmatrix}\begin{pmatrix} 0 & -i \\ i & 0 \end{pmatrix} = 2i\begin{pmatrix} 0 & 1 \\ 1 & 0 \end{pmatrix}$$

$$[\sigma_2, \sigma_3] = 2i\sigma_1$$

✓ Consider

$$[\sigma_3, \sigma_1] = \sigma_3\sigma_1 - \sigma_1\sigma_3$$

$$= \begin{pmatrix} 1 & 0 \\ 0 & -1 \end{pmatrix}\begin{pmatrix} 0 & 1 \\ 1 & 0 \end{pmatrix} - \begin{pmatrix} 0 & 1 \\ 1 & 0 \end{pmatrix}\begin{pmatrix} 1 & 0 \\ 0 & -1 \end{pmatrix} = -2\begin{pmatrix} 0 & -1 \\ 1 & 0 \end{pmatrix}$$

$$= -2\frac{i}{i}\begin{pmatrix} 0 & -1 \\ 1 & 0 \end{pmatrix} = -\frac{2}{i}\begin{pmatrix} 0 & -i \\ i & 0 \end{pmatrix}$$

$$[\sigma_3, \sigma_1] = 2i\sigma_2$$

We combine the results to get

$$[\sigma_i, \sigma_j] = 2i\varepsilon_{ijk}\sigma_k$$

where ε_{ijk} is the Levi-Civita symbol.

Hence the commutation relation between the $SU(2)$ generators $\frac{1}{2}\sigma_i$ s will be

$$\left[\frac{\sigma_i}{2}, \frac{\sigma_j}{2}\right] = \frac{\sigma_i}{2}\frac{\sigma_j}{2} - \frac{\sigma_j}{2}\frac{\sigma_i}{2} = \frac{1}{4}(\sigma_i\sigma_j - \sigma_j\sigma_i) = \frac{1}{4}[\sigma_i, \sigma_j] = \frac{1}{4}2i\varepsilon_{ijk}\sigma_k = i\varepsilon_{ijk}\frac{\sigma_k}{2}$$

Exercise 9.35 *Establish the relation* $Tr(\frac{\sigma_i}{2}\frac{\sigma_j}{2}) = \frac{1}{2}\delta_{ij}$.

$$\boxed{\text{Ans.}}\ Tr\left(\frac{\sigma_i}{2}\frac{\sigma_j}{2}\right) = Tr\,\frac{1}{4}\sigma_i\sigma_j$$

Let us take $i = 1$, $j = 2$

$$Tr\left(\frac{\sigma_i}{2}\frac{\sigma_j}{2}\right) = Tr\frac{1}{4}\sigma_1\sigma_2 = Tr\frac{1}{4}\begin{pmatrix} 0 & 1 \\ 1 & 0 \end{pmatrix}\begin{pmatrix} 0 & -i \\ i & 0 \end{pmatrix} = Tr\frac{1}{4}\begin{pmatrix} i & 0 \\ 0 & -i \end{pmatrix} = \frac{1}{4}(i - i) = 0$$

Let us take $i = 1$, $j = 1$

$$Tr\left(\frac{\sigma_i}{2}\frac{\sigma_i}{2}\right) = Tr\frac{1}{4}\sigma_i^2 = Tr\frac{1}{4}\sigma_1^2 = Tr\frac{1}{4}\begin{pmatrix} 0 & 1 \\ 1 & 0 \end{pmatrix}\begin{pmatrix} 0 & 1 \\ 1 & 0 \end{pmatrix}$$

$$= Tr\frac{1}{4}\begin{pmatrix} 1 & 0 \\ 0 & 1 \end{pmatrix} = \frac{1}{4}(1 + 1) = \frac{1}{2}$$

Similarly, for other combinations of i, j.

Hence for $i \neq j$, $Tr(\frac{\sigma_i}{2}\frac{\sigma_j}{2}) = 0$ and for $i = j$, $Tr(\frac{\sigma_i}{2}\frac{\sigma_j}{2}) = \frac{1}{2}$

Exercise 9.36 *Establish* $Tr(\lambda_i\lambda_j) = 2\delta_{ij}$ *where* λ_i *are the generators of* $SU(3)$.

$\boxed{\text{Ans.}}$ Consider $Tr(\lambda_i\lambda_j)$

For $i = 1$, $j = 8$

$$Tr(\lambda_1\lambda_8) = Tr\begin{pmatrix} 0 & 1 & 0 \\ 1 & 0 & 0 \\ 0 & 0 & 0 \end{pmatrix}\frac{1}{\sqrt{3}}\begin{pmatrix} 1 & 0 & 0 \\ 0 & 1 & 0 \\ 0 & 0 & -2 \end{pmatrix} = Tr\begin{pmatrix} 0 & 0 & 0 \\ 1 & 0 & 0 \\ 0 & 0 & 0 \end{pmatrix} = 0 = 2\delta_{18}$$

For $i = 1, j = 1$

$$Tr(\lambda_1\lambda_1) = Tr\begin{pmatrix} 0 & 1 & 0 \\ 1 & 0 & 0 \\ 0 & 0 & 0 \end{pmatrix}\begin{pmatrix} 0 & 1 & 0 \\ 1 & 0 & 0 \\ 0 & 0 & 0 \end{pmatrix} = Tr\begin{pmatrix} 1 & 0 & 0 \\ 0 & 1 & 0 \\ 0 & 0 & 0 \end{pmatrix} = 2 = 2\delta_{11}$$

Similarly for others. Hence

$$Tr(\lambda_i\lambda_j) = 2\delta_{ij}.$$

Exercise 9.37 *Establish the commutation relation between the* $SU(3)$ *generators* $\frac{1}{2}\lambda_i$.

$\boxed{\text{Ans.}}$ Consider

$$[\lambda_1, \lambda_2] = \lambda_1\lambda_2 - \lambda_2\lambda_1$$

$$= \begin{pmatrix} 0 & 1 & 0 \\ 1 & 0 & 0 \\ 0 & 0 & 0 \end{pmatrix}\begin{pmatrix} 0 & -i & 0 \\ i & 0 & 0 \\ 0 & 0 & 0 \end{pmatrix} - \begin{pmatrix} 0 & -i & 0 \\ i & 0 & 0 \\ 0 & 0 & 0 \end{pmatrix}\begin{pmatrix} 0 & 1 & 0 \\ 1 & 0 & 0 \\ 0 & 0 & 0 \end{pmatrix} = 2i\begin{pmatrix} 1 & 0 & 0 \\ 0 & -1 & 0 \\ 0 & 0 & 0 \end{pmatrix}$$

$$[\lambda_1, \lambda_2] = 2i\lambda_3 = 2if_{123}\,\lambda_3 \ (f_{123} = 1)$$

Consider

$$[\lambda_1, \lambda_4] = \lambda_1\lambda_4 - \lambda_4\lambda_1$$

$$= \begin{pmatrix} 0 & 1 & 0 \\ 1 & 0 & 0 \\ 0 & 0 & 0 \end{pmatrix}\begin{pmatrix} 0 & 0 & 1 \\ 0 & 0 & 0 \\ 1 & 0 & 0 \end{pmatrix} - \begin{pmatrix} 0 & 0 & 1 \\ 0 & 0 & 0 \\ 1 & 0 & 0 \end{pmatrix}\begin{pmatrix} 0 & 1 & 0 \\ 1 & 0 & 0 \\ 0 & 0 & 0 \end{pmatrix}$$

$$= \begin{pmatrix} 0 & 0 & 0 \\ 0 & 0 & 1 \\ 0 & -1 & 0 \end{pmatrix} = i\begin{pmatrix} 0 & 0 & 0 \\ 0 & 0 & -i \\ 0 & i & 0 \end{pmatrix}$$

$$[\lambda_1, \lambda_4] = i\lambda_7 = 2i\frac{1}{2}\lambda_7 = 2if_{147}\,\lambda_7 \quad (f_{147} = \frac{1}{2})$$

Similarly

$$[\lambda_5, \lambda_1] = 2if_{516}\,\lambda_6$$

$$[\lambda_2, \lambda_4] = 2if_{246}\,\lambda_6$$

$$[\lambda_2, \lambda_5] = 2if_{257}\,\lambda_7$$

$$[\lambda_3, \lambda_4] = 2if_{345}\,\lambda_5,$$

$$[\lambda_6, \lambda_3] = 2if_{637}\,\lambda_7$$

with $f_{516} = f_{246} = f_{257} = f_{345} = f_{637} = \frac{1}{2}$

Consider

$$[\lambda_4, \lambda_5] = \lambda_4\lambda_5 - \lambda_5\lambda_4$$

$$= \begin{pmatrix} 0 & 0 & 1 \\ 0 & 0 & 0 \\ 1 & 0 & 0 \end{pmatrix}\begin{pmatrix} 0 & 0 & -i \\ 0 & 0 & 0 \\ i & 0 & 0 \end{pmatrix} - \begin{pmatrix} 0 & 0 & -i \\ 0 & 0 & 0 \\ i & 0 & 0 \end{pmatrix}\begin{pmatrix} 0 & 0 & 1 \\ 0 & 0 & 0 \\ 1 & 0 & 0 \end{pmatrix} = \begin{pmatrix} 2i & 0 & 0 \\ 0 & 0 & 1 \\ 0 & 0 & -2i \end{pmatrix}$$

$$= \begin{pmatrix} i & 0 & 0 \\ 0 & i & 0 \\ 0 & 0 & -2i \end{pmatrix} + \begin{pmatrix} i & 0 & 0 \\ 0 & -i & 1 \\ 0 & 0 & 0 \end{pmatrix} = i\begin{pmatrix} 1 & 0 & 0 \\ 0 & 1 & 0 \\ 0 & 0 & -2 \end{pmatrix} + i\begin{pmatrix} 1 & 0 & 0 \\ 0 & -1 & 0 \\ 0 & 0 & 0 \end{pmatrix}$$

$$[\lambda_4, \lambda_5] = 2i\frac{\sqrt{3}}{2}\cdot\frac{1}{\sqrt{3}}\begin{pmatrix} 1 & 0 & 0 \\ 0 & 1 & 0 \\ 0 & 0 & -2 \end{pmatrix} + 2i\frac{1}{2}\begin{pmatrix} 1 & 0 & 0 \\ 0 & -1 & 0 \\ 0 & 0 & 0 \end{pmatrix}$$

$$= 2if_{458}\,\lambda_8 + 2if_{453}\,\lambda_3$$

$$[\lambda_4, \lambda_5] = 2if_{45l}\,\lambda_l$$

where $f_{458} = \frac{\sqrt{3}}{2}$, $f_{453} = \frac{1}{2}$, other combinations of $f_{45l} = 0$

Consider

$$[\lambda_6, \lambda_7] = \lambda_6\lambda_7 - \lambda_7\lambda_6$$

$$= \begin{pmatrix} 0 & 0 & 0 \\ 0 & 0 & 1 \\ 0 & 1 & 0 \end{pmatrix}\begin{pmatrix} 0 & 0 & 0 \\ 0 & 0 & -i \\ 0 & i & 0 \end{pmatrix} - \begin{pmatrix} 0 & 0 & 0 \\ 0 & 0 & -i \\ 0 & i & 0 \end{pmatrix}\begin{pmatrix} 0 & 0 & 0 \\ 0 & 0 & 1 \\ 0 & 1 & 0 \end{pmatrix} = \begin{pmatrix} 0 & 0 & 0 \\ 0 & 2i & 1 \\ 0 & 0 & -2i \end{pmatrix}$$

$$= \begin{pmatrix} i & 0 & 0 \\ 0 & i & 0 \\ 0 & 0 & -2i \end{pmatrix} - \begin{pmatrix} i & 0 & 0 \\ 0 & -i & 0 \\ 0 & 0 & 0 \end{pmatrix} = i\begin{pmatrix} 1 & 0 & 0 \\ 0 & 1 & 0 \\ 0 & 0 & -2 \end{pmatrix} - i\begin{pmatrix} 1 & 0 & 0 \\ 0 & -1 & 0 \\ 0 & 0 & 0 \end{pmatrix}$$

$$[\lambda_6, \lambda_7] = 2i\frac{\sqrt{3}}{2}\cdot\frac{1}{\sqrt{3}}\begin{pmatrix} 1 & 0 & 0 \\ 0 & 1 & 0 \\ 0 & 0 & -2 \end{pmatrix} - 2i\frac{1}{2}\begin{pmatrix} 1 & 0 & 0 \\ 0 & -1 & 0 \\ 0 & 0 & 0 \end{pmatrix}$$

$$[\lambda_6, \lambda_7] = 2if_{678}\,\lambda_8 + 2if_{673}\,\lambda_3 = 2if_{67l}\,\lambda_l$$

where $f_{678} = \frac{\sqrt{3}}{2}$, $f_{673} = -\frac{1}{2}$, other combinations of $f_{67l} = 0$

We combine the results to get $[\lambda_j,\ \lambda_k] = 2if_{jkl}\,\lambda_l$ (sum is over the index l) where non-vanishing f_{ijk} are

$$f_{123} = 1,\ \ f_{147} = f_{516} = f_{246} = f_{257} = f_{345} = f_{637} = \frac{1}{2}, f_{458} = f_{678} = \frac{\sqrt{3}}{2}$$

value is positive for cyclic permutations, f_{ijk} is antisymmetric under interchange of any two indices.

Hence the commutation relation between the $SU(3)$ generators $\frac{1}{2}\lambda_i$ will be

$$\left[\frac{\lambda_j}{2}, \frac{\lambda_k}{2}\right] = \frac{\lambda_j}{2}\frac{\lambda_k}{2} - \frac{\lambda_k}{2}\frac{\lambda_j}{2} = \frac{1}{4}(\lambda_j\lambda_k - \lambda_k\lambda_j) = \frac{1}{4}[\lambda_j, \lambda_k] = \frac{1}{4}2if_{jkl}\,\lambda_l = if_{jkl}\frac{\lambda_l}{2}.$$

Exercise 9.38 *Select the correct answer for rank, order, number of generators and number of diagonal generators of the group $SU(4)$.*

(a) 3, 8, 8, 2   (b) 4,15, 8, 3   (c) 3, 4, 15, 3   (d) 3,15, 15, 3.

Ans. $SU(n)$ has rank $n - 1$, order $n^2 - 1$, number of generators $n^2 - 1$, number of diagonal generators $n - 1$.

Exercise 9.39 *Justify the following:*

(i) *A meson with strangeness $+1$ and electric charge -1 cannot exist.*

(ii) *A baryon with strangeness -2 and electric charge $+1$ cannot exist.*

Ans.

(i) The structure of meson is $q\bar{q}$. To achieve strangeness $+1$ an $\bar{s}$ quark has to be present. So the possible structure is $q\bar{s}$. Total charge of the said meson is

$$Q(q\bar{s}) = -1$$

$$Q(q) + Q(\bar{s}) = -1$$

$$Q(q) = -1 - Q(\bar{s}) = -1 - \left(+\frac{1}{3}\right) = -\frac{4}{3}$$

Such a quark with electric charge $\frac{4}{3}$ does not exist. So a meson with strangeness $+1$ and electric charge -1 cannot exist.

(ii) The structure of a baryon is qqq. To achieve strangeness -2 two s quarks have to be present. So the possible structure is qss. Total charge of the said meson is

$$Q(qss) = +1$$

$$Q(q) + Q(s) + Q(s) = +1$$

$$Q(q) = +1 - 2Q(s) = 1 - 2\left(-\frac{1}{3}\right) = \frac{5}{3}$$

Such a quark with electric charge $\frac{5}{3}$ does not exist. So a baryon with strangeness -1 and electric charge $+1$ cannot exist.

Answers to multiple choice questions

9.2*a*, 9.4*a*, 9.5*a*, 9.6*b*, 9.7*b*, 9.8*c*, 9.38*d*.

9.43 Question bank

Q9.1 What is the criterion for a set of elements to form a group?
Q9.2 What do we mean by abelian property of a group?
Q9.3 Give an example of a discrete group.
Q9.4 Give an example of a continuous group.
Q9.5 What do we mean by order of a group?

Q9.6 What do we mean by generator of a group?
Q9.7 Obtain the generator of space translation operator.
Q9.8 What are the properties of an orthogonal matrix?
Q9.9 What is a Lie group?
Q9.10 Obtain the generator of $SO(2)$.
Q9.11 Obtain the generator of $SO(3)$.
Q9.12 What are the properties of a unitary matrix?
Q9.13 Obtain the generator of $SU(2)$.
Q9.14 Obtain the generator of $SU(3)$.
Q9.15 What are Gell-Mann matrices?
Q9.16 What do we mean by rank of a group?
Q9.17 Why is Young's diagram used?
Q9.18 Why are Young tableaux used?

Further reading

[1] Perkins H 2000 *Introduction to High Energy Physics* (Cambridge: Cambridge University Press)
[2] Griffiths D 2008 *Introduction to Elementary Particles* (New York: Wiley-VCH)
[3] Tayal D C 2009 *Nuclear Physics* (Mumbai: Himalaya Publishing House)
[4] Satya P 2005 *Nuclear Physics & Particle Physics* (New Delhi: Sultan Chand & Sons)
[5] Guha J 2019 *Quantum Mechanics: Theory, Problems & Solutions* 3rd edn (Kolkata: Books and Allied (P) Ltd))
[6] Yung K L 2002 *Problems and Solutions on Atomic, Nuclear and Particle Physics* (Singapore: World Scientific)

IOP Publishing

Nuclear and Particle Physics with Cosmology, Volume 2
Particle physics and cosmology
Jyotirmoy Guha

Chapter 10

Physics of quarks

In this chapter we study the quark structure of hadrons, i.e. mesons and baryons. In a weight diagram we represent quarks and antiquarks. We represent pseudoscalar mesons and vector mesons in a weight diagram. We also represent baryons and resonance baryons in a weight diagram. Representation of antiparticle in $SU(2)$ is discussed. We discuss asymptotic freedom and quark confinement. Also, the characteristic feature of gluons is discussed. We show how the color octet for gluon states corresponds to Gell-Mann matrices.

10.1 Introduction

Quark model gives some understanding of the different compound particles like proton, neutron, pion, kaon etc.

Hadron is a composite subatomic particle made up of two or more quarks held together by strong interaction, as shown in figures 7.10 and 10.1.

Figure 10.2 shows the mass, spin parity J^π, I, I_3, S and lifetime of various hadrons. Pion is the lightest of all strongly interacting particles.

$SU(3)$ scheme helps us to understand hadron spectrum.

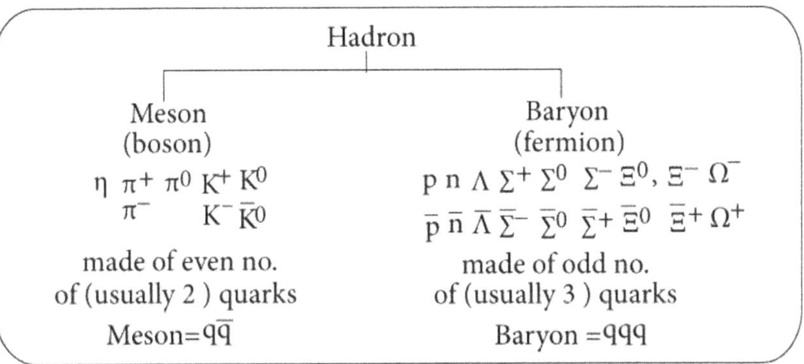

Figure 10.1. Hadron structure.

doi:10.1088/978-0-7503-5032-7ch10

particle symbol		mass	spin parity J^P	Isospin I	I_3	Strangeness S	lifetime
N	p	938.3 MeV	$\frac{1}{2}^+$	$\frac{1}{2}$	$\frac{1}{2}$	0	$> 10^{34}$ yrs
	n	939.6 MeV			$-\frac{1}{2}$		887 s ($n \rightarrow p + e^+ + \bar{\nu}_e$)
Λ		1115.7 MeV	$\frac{1}{2}^+$	0	0	-1	10^{-10}s
Σ	Σ^-	1197.4 MeV			1		
	Σ^0	1192.5 MeV	$\frac{1}{2}^+$	1	0	-1	10^{-10}s
	Σ^+	1189.4 MeV			-1		
Ξ	Ξ^-	1321.3 MeV	$\frac{1}{2}^+$	$\frac{1}{2}$	$\frac{1}{2}$	-2	10^{-10}s
	Ξ^0	1314.9 MeV			$-\frac{1}{2}$		
Ω^-		1670 MeV	$\frac{3}{2}^+$	0	0	-3	10^{-10}s
π	$\pi^\pm$	140 MeV	0^-	1	± 1	0	10^{-8}s
	π^0	138 MeV			0		
K	K^+	493 MeV	0^-	$\frac{1}{2}$	$\frac{1}{2}$	1	K_S^0 10^{-11}s
	K^0	493 MeV			$-\frac{1}{2}$	1	K_L^0 10^{-8}s
$\bar{K}$	$\bar{K}^0$	498 MeV	0^-	$\frac{1}{2}$	$\frac{1}{2}$	-1	
	K^-	498 MeV			$-\frac{1}{2}$	-1	
η		550 MeV	0^-	0	0	0	10^{-19}s
η'		957 MeV	0^-	0	0	0	10^{-21}s

Figure 10.2. Properties of hadrons (baryons and mesons).

Gell-Mann and Zweig introduced triplet of quarks. We have discussed $SU(3)$ in chapter 9.

10.2 About quarks and anti-quarks

Quark is a type of elementary particle and a fundamental constituent of matter.

Quarks combine to form a composite particle called hadron.

There are six flavours of quarks and their anti-particles exist.

Quantum numbers associated with quarks u, d, s, c, t, b are shown in figure 10.3. The quantum numbers defined to characterize various quarks are as follows.

I = Isospin

I_3 = Isospin projection

B = baryon quantum number

C = charm quantum number or charmness

S = strangeness quantum number

T = top quantum number or topness

B' = bottom quantum number or bottomness

Y = hypercharge

Q = electric charge (in units of $|e|$, the electronic charge magnitude)

These quantities are inter-related as

$$Y = B + C + S + T + B' \tag{10.1}$$

Quantum numbers associated to quarks										
	Symbol	I I_3	B	c	S	T	B'	$B+C+S+T+B'=Y$	Q $=I_3+\frac{Y}{2}$
1st generation										
up quark	u	$\frac{1}{2}$ $\frac{1}{2}$	$\frac{1}{3}$	0	0	0	0	$\frac{1}{3}+0+0+0+0=\frac{1}{3}$	$\frac{2}{3}$
down quark	d	$\frac{1}{2}$ $-\frac{1}{2}$	$\frac{1}{3}$	0	0	0	0	$\frac{1}{3}+0+0+0+0=\frac{1}{3}$	$-\frac{1}{3}$
2nd generation										
charm quark	c	0 0	$\frac{1}{3}$	1	0	0	0	$\frac{1}{3}+1+0+0+0=\frac{4}{3}$	$\frac{2}{3}$
strange quark	s	0 0	$\frac{1}{3}$	0	-1	0	0	$\frac{1}{3}+0-1+0+0=-\frac{2}{3}$	$-\frac{1}{3}$
3rd generation										
top quark	t	0 0	$\frac{1}{3}$	0	0	1	0	$\frac{1}{3}+0+0+1+0=\frac{4}{3}$	$\frac{2}{3}$
bottom quark	b	0 0	$\frac{1}{3}$	0	0	0	-1	$\frac{1}{3}+0+0+0-1=-\frac{2}{3}$	$-\frac{1}{3}$

Figure 10.3. Quantum numbers associated with quarks.

Quantum numbers associated to antiquarks										
	Symbol	I I_3	B	C	S	T	B'	$B+C+S+T+B'=Y$	Q $=I_3+\frac{Y}{2}$
1st generation										
up antiquark	$\bar{u}$	$\frac{1}{2}$ $-\frac{1}{2}$	$-\frac{1}{3}$	0	0	0	0	$-\frac{1}{3}+0+0+0+0=-\frac{1}{3}$	$-\frac{2}{3}$
down antiquark	$\bar{d}$	$\frac{1}{2}$ $\frac{1}{2}$	$-\frac{1}{3}$	0	0	0	0	$-\frac{1}{3}+0+0+0+0=-\frac{1}{3}$	$\frac{1}{3}$
2nd generation										
charm antiquark	$\bar{c}$	0 0	$-\frac{1}{3}$	-1	0	0	0	$-\frac{1}{3}-1+0+0+0=-\frac{2}{3}$	$-\frac{2}{3}$
strange antiquark	$\bar{s}$	0 0	$-\frac{1}{3}$	0	1	0	0	$-\frac{1}{3}+0+1+0+0=\frac{2}{3}$	$\frac{1}{3}$
3rd generation										
top antiquark	$\bar{t}$	0 0	$-\frac{1}{3}$	0	0	-1	0	$-\frac{1}{3}+0+0-1+0=-\frac{4}{3}$	$-\frac{2}{3}$
bottom antiquark	$\bar{b}$	0 0	$-\frac{1}{3}$	0	0	0	1	$-\frac{1}{3}+0+0+0+1=\frac{2}{3}$	$\frac{1}{3}$

Figure 10.4. Quantum numbers associated with anti-quarks.

These quantum numbers are not all independent but are inter-related through Gell-Mann–Nishijima relation

$$Q = I_3 + \frac{Y}{2} \tag{10.2}$$

Quantum numbers associated with quarks u, d, s, c, t, b are shown in figure 10.4. Figure 10.4 also shows the Gell-Mann–Nishijima relation for quarks.

Quantum numbers associated to anti-quarks $\bar{u}$, $\bar{d}$, $\bar{c}$, $\bar{s}$, $\bar{t}$, $\bar{b}$ are shown in figure 10.4. Figure 10.4 also shows the Gell-Mann–Nishijima relation for anti-quarks.

We can extend this Gell-Mann–Nishijima relation to hadrons, as done in chapter 8.

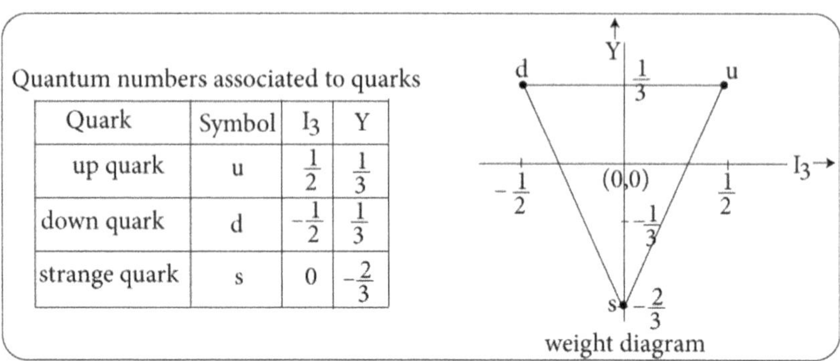

Figure 10.5. Weight diagram for quarks u, d, s.

10.3 Weight diagram of quarks

In matrix form let us define field operators for quarks.

$$q = \begin{pmatrix} u \\ d \\ s \end{pmatrix} \text{ is a column matrix, } \bar{q} = \begin{pmatrix} \bar{u} & \bar{d} & \bar{s} \end{pmatrix} \text{ is a row matrix} \tag{10.3}$$

The quark states in the $|I_3\ Y\rangle$ representation are

$$|u\rangle = |\tfrac{1}{2}\ \tfrac{1}{3}\rangle, \ \ |d\rangle = |-\tfrac{1}{2}\ \tfrac{1}{3}\rangle, \ \ |s\rangle = |0\ -\tfrac{2}{3}\rangle \tag{10.4}$$

where I_3 is isospin projection and Y is hypercharge.

The plot of Y against I_3 is called a weight diagram.

Quarks u, d, s belong to triplet representation of $SU(3)$ and can be represented in the weight diagram. We have shown the quantum numbers I_3 and Y in figure 10.5 along with the corresponding weight diagram for quarks u, d, s. It is an inverted triangle with the vertices or corners having coordinates $(I_3,\ Y)$ representing the quarks as

$$u\left(\frac{1}{2},\ \frac{1}{3}\right), \ \ d\left(-\frac{1}{2},\ \frac{1}{3}\right), \ \ s\left(0,\ -\frac{2}{3}\right) \tag{10.5}$$

The anti-quark states in the $|I_3\ Y\rangle$ representation are

$$|\bar{u}\rangle = |-\frac{1}{2}\ -\frac{1}{3}\rangle, \ \ |\bar{d}\rangle = |\frac{1}{2}\ -\frac{1}{3}\rangle, \ \ |\bar{s}\rangle = |0\ \frac{2}{3}\rangle \tag{10.6}$$

Anti-quarks $\bar{u}$, $\bar{d}$, $\bar{s}$ belong to triplet representation of $SU(3)$ and can be represented in the weight diagram. We have shown the quantum numbers I_3 and Y in figure 10.6 along with the corresponding weight diagram for anti-quarks $\bar{u}$, $\bar{d}$, $\bar{s}$. It is an erect triangle with the vertices or corners having coordinates $(I_3,\ Y)$ representing the anti-quarks as

$$\bar{u}\left(-\frac{1}{2},\ -\frac{1}{3}\right), \ \ \bar{d}\left(\frac{1}{2},\ -\frac{1}{3}\right), \ \ \bar{s}\left(0\ \frac{2}{3}\right) \tag{10.7}$$

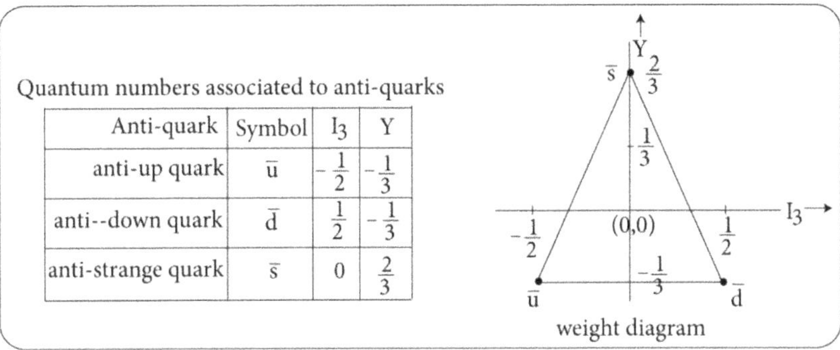

Figure 10.6. Weight diagram for anti-quarks $\bar{u}$, $\bar{d}$, $\bar{s}$.

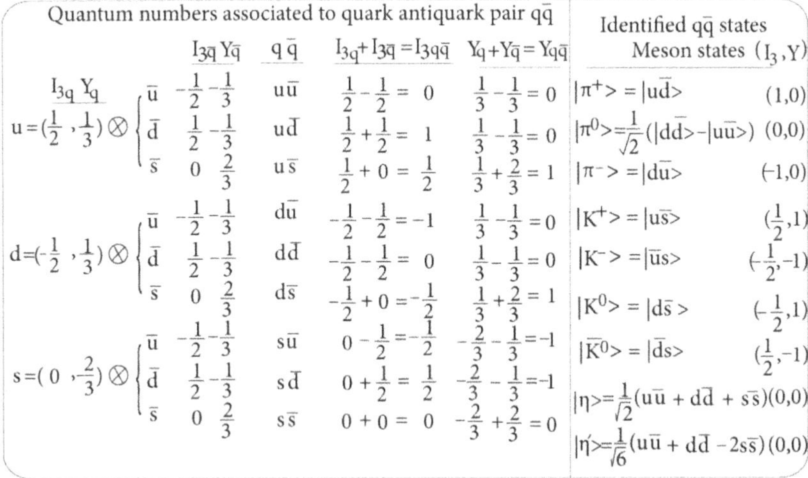

Figure 10.7. Evaluation of quantum numbers of $q\bar{q}$.

10.4 Nonet and octet representation of mesons

We now evaluate the quantum numbers that a quark–anti-quark pair can assume.

The quantum numbers of quark–anti-quark $q\bar{q}$ combination are evaluated in figure 10.7.

We combine the quark states $q \equiv u(\frac{1}{2}, \frac{1}{3})$, $d(-\frac{1}{2}, \frac{1}{3})$, $s(0, -\frac{2}{3})$ with the anti-quark states $\bar{q} \equiv \bar{u}(-\frac{1}{2}, -\frac{1}{3})$, $\bar{d}(\frac{1}{2}, -\frac{1}{3})$, $\bar{s}(0, \frac{2}{3})$ i.e. perform the operation $q \otimes \bar{q}$ in figure 10.7. This actually leads to the meson states or the quark–anti-quark $q\bar{q}$ states identified in figure 8.6 and re-mentioned in figure 10.7.

These states have been shown in the I_3 versus Y weight diagram in figure 10.8 occupying positions having coordinates (I_3 Y) as follows.

$$\pi^+(1, 0), \quad K^+\left(\frac{1}{2}, 1\right), \quad K^0\left(-\frac{1}{2}, 1\right), \quad \pi^-(-1, 0),$$

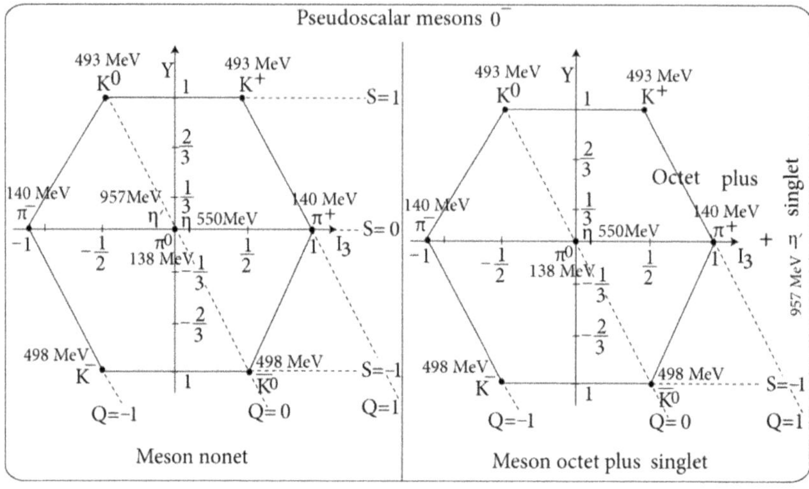

Figure 10.8. Weight diagram showing nonet and octet representation of mesons.

$$K^-\left(-\frac{1}{2}, -1\right), \quad \overline{K^0}\left(\frac{1}{2}, -1\right), \quad \pi^0(0, 0), \quad \eta(0, 0), \quad \eta'(0, 0).$$

So there are nine meson states in the weight diagram.

$$|\pi^+\rangle = |u\overline{d}\rangle = |1\ 0\rangle$$

$$|K^+\rangle = |u\overline{s}\rangle = |\tfrac{1}{2}\ 1\rangle$$

$$|\pi^-\rangle = |d\overline{u}\rangle = |-1\ 0\rangle$$

$$|K^0\rangle = |d\overline{s}\rangle = |-\tfrac{1}{2}\ 1\rangle$$

$$|K^-\rangle = |s\overline{u}\rangle = |-\tfrac{1}{2}\ -1\rangle$$

$$|\overline{K^0}\rangle = |s\overline{d}\rangle = |\tfrac{1}{2}\ -1\rangle$$

$$|\pi^0\rangle = \frac{1}{\sqrt{2}}\left(|u\overline{u}\rangle - |d\overline{d}\rangle\right) = |0\ 0\rangle$$

$$|\eta^0\rangle \equiv |\eta^1\rangle \equiv |\eta\rangle = \frac{1}{3}\left(|u\overline{u}\rangle + |d\overline{d}\rangle + |s\overline{s}\rangle\right) = |0\ 0\rangle$$

$$|\eta^8\rangle \equiv |\eta'\rangle = \frac{1}{\sqrt{6}}\left(|u\overline{u}\rangle + |d\overline{d}\rangle - 2|s\overline{s}\rangle\right) = |0\ 0\rangle$$

Particle Meson $J^P{=}1^-$	quark structure $q\bar{q}$	Mass in MeV	Q	I	I_3	B	S	$Y{=}B{+}S$
K^{*0}	$\lvert d\bar{s}\rangle$	890	0	$\frac{1}{2}$	$-\frac{1}{2}$	0	1	1
K^{*+}	$\lvert u\bar{s}\rangle$	890	+1	$\frac{1}{2}$	$\frac{1}{2}$	0	1	1
$\bar{K}^{*-}$	$\lvert s\bar{u}\rangle$	890	-1	$\frac{1}{2}$	$-\frac{1}{2}$	0	-1	-1
$\bar{K}^{*0}$	$\lvert s\bar{d}\rangle$	890	0	$\frac{1}{2}$	$\frac{1}{2}$	0	-1	-1
ρ^+	$\lvert u\bar{d}\rangle$	770	+1	1	1	0	0	0
ρ^0	$\lvert u\bar{u}-d\bar{d}\rangle$	770	0	1	0	0	0	0
ρ^-	$\lvert d\bar{u}\rangle$	770	-1	1	-1	0	0	0
ω	$\lvert u\bar{s}\rangle$	780	0	0	0	0	0	0
ϕ	$\lvert s\bar{s}\rangle$	1020	0	0	0	0	0	0

Figure 10.9. Properties of vector mesons $J^P = 1^-$.

The nine mesons $\lvert\pi^+\rangle$, $\lvert K^+\rangle$, $\lvert\pi^-\rangle$, $\lvert K^0\rangle$, $\lvert K^-\rangle$, $\lvert\overline{K^0}\rangle$, $\lvert\pi^0\rangle$, $\lvert\eta^0\rangle$, $\lvert\eta'\rangle$ can be represented as a meson nonet, as shown in figure 10.8(a).

Sometimes $\lvert\eta'\rangle$ is taken outside and shown separately and then we have a meson octet plus singlet η' as depicted in figure 10.8(b).

These are all mesons whose total spin is $J = 0$ and parity is –ve, i.e. $J^\pi = 0^-$. Therefore these are pseudoscalar mesons.

We show in figure 10.9 the properties of vector mesons $J^\pi = 1^-$. The characterizing quantum numbers that define these vector mesons have been shown, namely the values of Q, I, I_3, B, S, Y. We also write down the quark structure of these mesons as follows.

$$\lvert K^{*0}\rangle = \lvert d\bar{s}\rangle$$

$$\lvert K^{*+}\rangle = \lvert u\bar{s}\rangle$$

$$\lvert K^{*-}\rangle = \lvert s\bar{u}\rangle$$

$$\lvert\overline{K}^{*0}\rangle = \lvert s\bar{d}\rangle$$

$$\lvert\rho^+\rangle = \lvert u\bar{d}\rangle$$

$$\lvert\rho^0\rangle = \lvert u\bar{u} - d\bar{d}\rangle$$

$$\lvert\rho^-\rangle = \lvert d\bar{u}\rangle$$

$$\lvert\omega\rangle = \lvert u\bar{s}\rangle$$

$$\lvert\phi\rangle = \lvert s\bar{s}\rangle$$

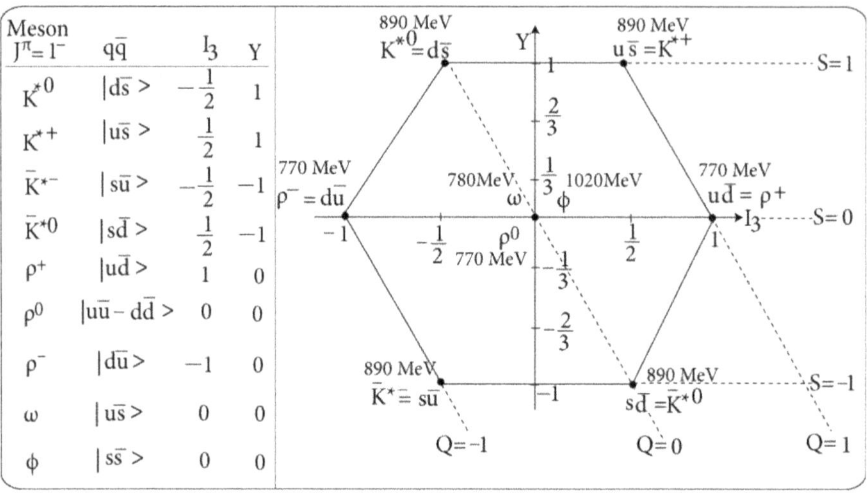

Figure 10.10. Nonet formed by vector mesons $J^\pi = 1^-$ in a weight diagram.

The vector mesons form a nonet in the weight diagram shown in figure 10.10 where we also show the corresponding I_3 and Y values. These vector mesons occupy the corners and centre of the nonet having the coordinates $(I_3\ Y)$ as follows.

$$K^{*0}\left(-\frac{1}{2},\ 1\right),\quad K^{*+}\left(\frac{1}{2},\ 1\right),\quad \rho^+(1,\ 0),\quad \overline{K}^{*0}\left(\frac{1}{2},\ -1\right),$$

$$\overline{K}^{*-}\left(-\frac{1}{2},\ -1\right),\quad \rho^-(-1,\ 0),\quad \rho^0(0,\ 0),\quad \omega(0,\ 0),\quad \phi(0,\ 0).$$

10.5 Baryon octet, the eightfold way

The strong interaction conserves isospin and treats members of the iso-multiplet in an identical way. The corresponding symmetry is the $SU(2)$ isospin group. The relevant conserved quantum numbers are I_3 and I^2 for isospin and Y for hypercharge.

The quantum numbers, namely isospin values I, I_3, baryon number B, strangeness S and hypercharge $Y = B + S$ of baryons, are listed in figure 10.11.

These baryons form an octet, as shown in figure 10.12, and the corresponding symmetry is called the eightfold way.

Six baryons occupy corners and two baryons occupy the centre of the hexagon having the following coordinates $(I_3,\ Y)$.

$$n(-\frac{1}{2},\ 1)$$

$$p(\frac{1}{2},\ 1)$$

Particle	Family	Mass in MeV	I	I_3	B	S	Y=B+S
Ξ^-		1321.32	$\frac{1}{2}$	$-\frac{1}{2}$	1	-2	-1
Ξ^0	Ξ	1314.9	$\frac{1}{2}$	$\frac{1}{2}$	1	-2	-1
Σ^-		1197.43	1	-1	1	-1	0
Σ^0	Σ	1192.55	1	0	1	-1	0
Σ^+		1189.37	1	1	1	-1	0
Λ	Λ	1115.63	0	0	1	-1	0
n		939.566	$\frac{1}{2}$	$-\frac{1}{2}$	1	0	1
p	N	938.272	$\frac{1}{2}$	$\frac{1}{2}$	1	0	1

Figure 10.11. Masses of eight baryons that form baryon octet in Y versus I_3 weight diagram called the eightfold way.

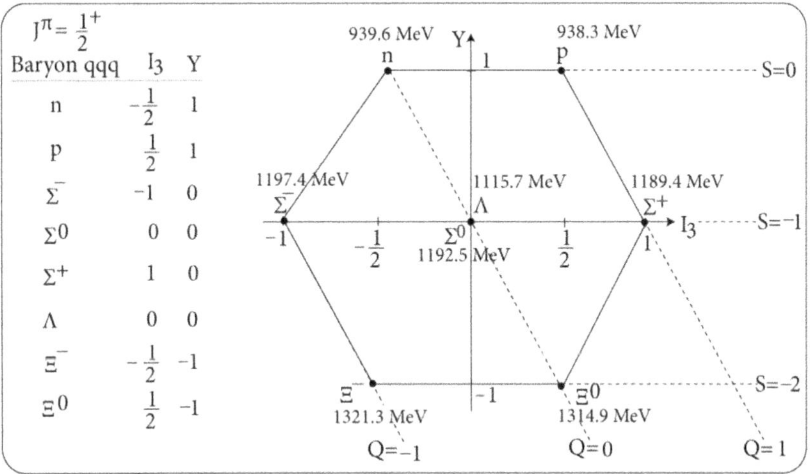

Figure 10.12. Hexagon representing the eightfold way. The eight baryons form an octet in the Y versus I_3 weight diagram.

$$\Sigma^-(-1, 0)$$

$$\Sigma^0(0, 0)$$

$$\Sigma^+(1, 0)$$

$$\Lambda(0, 0)$$

$$\Xi^-\left(-\frac{1}{2}, -1\right)$$

$$\Xi^0\left(\frac{1}{2}, -1\right).$$

Clearly the eight baryons form an octet in the Y versus I_3 weight diagram.
The quark structure is indicated in figure 10.17.
The Hamiltonian for eight baryons is composed of three parts

$$H = H_{\text{strong}} + H_{\text{medium}} + H_{\text{em}}$$

The first part H_{strong} has $SU(3)$ symmetry and leads to the eightfold degeneracy.
In figure 10.13 we show the splitting caused by H_{medium} and H_{em}.
Introduction of the symmetry breaking term H_{medium} removes part of degeneracy giving the four isospin multiplets with different masses viz.
(1) Ξ (isodoublet)
(2) Σ (isotriplet)
(3) Λ (isosinglet)
(4) N (isodoublet)

Finally, the presence of charge dependent forces splits the isospin multiplets and removes degeneracy.

(1) Ξ^- (2) Ξ^0 (3) Σ^- (4) Σ^0 (5) Σ^+ (6) Λ (7) p (8) n

This explains the baryon octet shown in figure 10.12.
The octet representation is not the simplest $SU(3)$ representation.
The simplest representation of $SU(3)$ is the triangular ones of figure 10.5 of u, d, s and figure 10.6 for $\bar{u}, \bar{d}, \bar{s}$ from which all other structures are generated by generalized coupling.

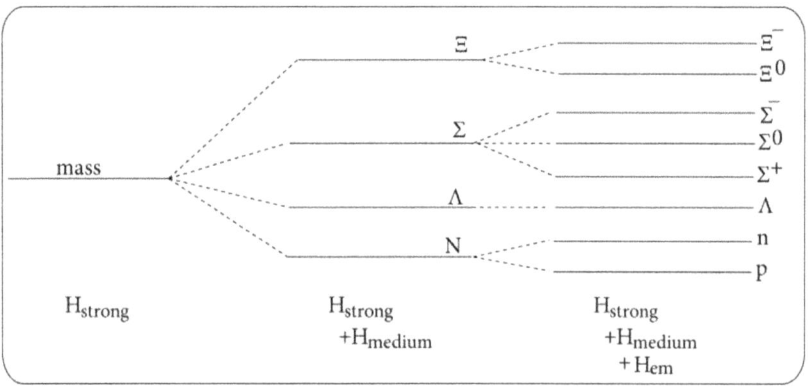

Figure 10.13. Mass splitting.

Group theory identifies and formalizes symmetries. It classifies and sometimes predicts particles.

10.6 Resonance particles

Whenever two particles collide, sometimes cross-section becomes large, indicating a strong or vigorous reaction, and we say that there is resonance. As large energy is involved at resonance it is associated with particle production. So resonance is identified as particles. Resonance particles are hadrons.

In 1950, Fermi and Anderson collided π^+ and p in a cyclotron

$$\pi^+ + p \rightarrow \pi^+ + p$$

And resonance was found to occur at energy $\sim 1234\ MeV$ identified as a delta particle (figure 10.14), the resonance width being $\Gamma \sim 120\ MeV$.

$$\Delta(1234\ MeV) \text{ has } I = \frac{3}{2}, \quad J^\pi = \frac{3^+}{2}.$$

We note that mass of $(\pi^+ + p)$ is $\sim(140 + 938)\ MeV = 1078\ MeV$.

Clearly the Δ that is produced momentarily can decay through strong interaction as

$$\Delta \rightarrow \pi^+ + p$$

with lifetime $10^{-23}\ s$.

Lifetime can be calculated using the Heisenberg uncertainty principle

$$\Delta E\ \Delta t = h$$

$$\Delta t = \frac{h}{\Delta E} = \frac{h}{\Gamma} = \frac{6.626 \times 10^{-34}\ J.\ s}{120 \times 10^6 \times 1.6 \times 10^{-19}\ J} = 5.5 \times 10^{-23}\ s$$

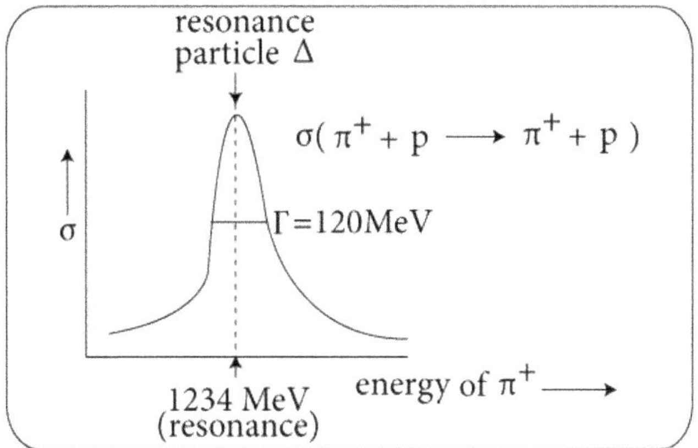

Figure 10.14. Resonance reaction producing resonance particle $\Delta(1234\ MeV)$ in the reaction $\pi^+ + p \rightarrow \pi^+ + p$ having resonance width $\Gamma \propto \frac{1}{\tau}$, $\tau = 10^{-23}s$, $\Gamma \sim 120\ MeV$.

Resonance is treated as a particle called a resonance particle. These particles come into existence and vanish so quickly, in about 10^{-23} s, that they cannot even cross an atom. So the resonance particles do not produce any track and cannot trigger a counter either. So they cannot be detected by any direct measurement.

Resonance particles are recognized by occurrence of resonance peaks in the normal energy spectrum. Resonance particle manifests as the width of resonance $\Gamma \propto \frac{1}{\tau}$ (figure 10.14). Their existence can be inferred from studies of the longer-lived products of their decay.

There are a large number of resonance particles. Some of these are shown in figure 10.14.

10.7 Baryon decuplet

We show the various properties of baryon resonances, i.e. resonance particles which are baryons, in figure 10.15. Their quantum numbers like isospin I, isospin projection I_3, baryon number B, strangeness S, hypercharge $Y = B + S$ are specified.

These baryon resonance particles form a triangular structure in the Y versus I_3 weight diagram, as shown in figue 10.16, called baryon decuplet.

The $(I_3,\ Y)$ coordinates of these particles occupy corners and various points on the triangle as indicated in the following.

$$\Delta^{++}(\frac{3}{2}, 1)$$

$J^\pi = \frac{3^+}{2}$ Baryon resonances	I	I_3	B	S	$Y=B+S$	mass in MeV
Δ^{++}	$\frac{3}{2}$	$\frac{3}{2}$	1	0	1	1232
Δ^{+}	$\frac{3}{2}$	$\frac{1}{2}$	1	0	1	1232
Δ^{0}	$\frac{3}{2}$	$-\frac{1}{2}$	1	0	1	1232
Δ^{-}	$\frac{3}{2}$	$-\frac{3}{2}$	1	0	1	1232
Σ^{*+}	1	1	1	-1	0	1385
Σ^{*0}	1	0	1	-1	0	1385
Σ^{*-}	1	-1	1	-1	0	1385
Ξ^{*0}	$\frac{1}{2}$	$\frac{1}{2}$	1	-2	-1	1520
Ξ^{*-}	$\frac{1}{2}$	$-\frac{1}{2}$	1	-2	-1	1520
Ω^{-}	0	0	1	-3	-2	1670

Figure 10.15. Properties of baryon resonances.

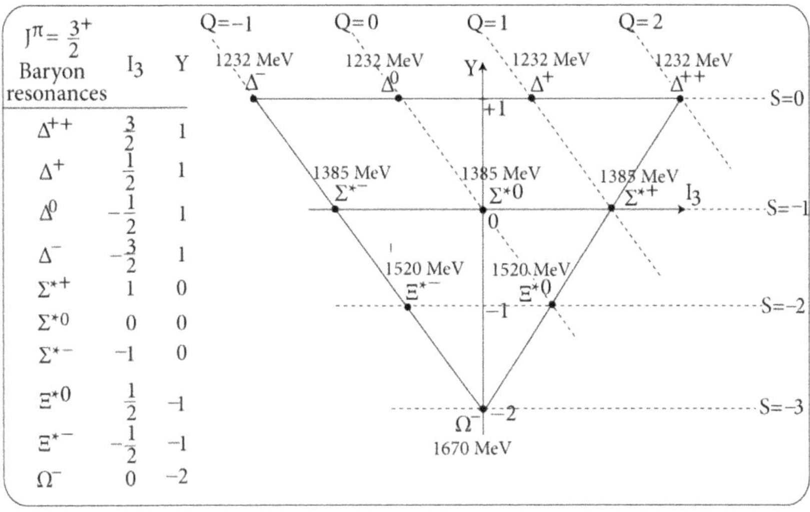

Figure 10.16. Baryon decuplet in weight diagram.

$$\Delta^+(\tfrac{1}{2}, 1)$$

$$\Delta^0(-\tfrac{1}{2}, 1)$$

$$\Delta^-(-\tfrac{3}{2}, 1)$$

$$\Sigma^{*+}(1, 0)$$

$$\Sigma^{*0}(0, 0)$$

$$\Sigma^{*-}(-1, 0)$$

$$\Xi^{*0}(\tfrac{1}{;2}, -1)$$

$$\Xi^{*-}(-\tfrac{1}{2}, -1)$$

$$\Omega(0, -2)$$

- We note the following pattern in the masses of baryon resonances, i.e. in the baryon decuplet

$$m_\Sigma - m_\Delta = 1385\ MeV - 1232\ MeV = 153\ MeV$$

$$m_\Xi - m_\Sigma = 1520\ MeV - 1385\ MeV = 135\ MeV$$

$$m_\Omega - m_\Xi = 1670\ MeV - 1520\ MeV = 150\ MeV$$

Clearly m_Δ, m_Σ, m_Ξ, m_Ω are nearly equally spaced in mass.

Again, the corresponding hypercharge values are also equispaced since

$$Y_\Sigma - Y_\Delta = 0 - 1 = -1$$

$$Y_\Xi - Y_\Sigma = -1 - 0 = -1$$

$$Y_\Omega - Y_\Xi = -2 - (-1) = -1$$

So mass varies linearly with hypercharge and we can write

$$\Delta M = a + bY \tag{10.8}$$

where ΔM is mass difference, a, b are constants.

Equation (10.8) is referred to as the equal spacing rule for a decuplet.

The resonance particles Δ, Σ^*, Ξ^* do not leave any track and exhibit strong decay as

$$\Delta \to N\pi$$

$$\Sigma^* \to \Lambda\pi$$

$$\Xi^* \to \Sigma\overline{K}$$

However Ω^- leaves a track in a bubble chamber and exhibits weak decay ($\sim 10^{-10}\ s$) as

$$\Omega^- \to \Xi^-\pi^0$$

For Ω^- to exhibit strong decay the following strangeness conserving interaction should have occurred.

$$\underset{S=-3}{\Omega^-} \to \underset{S=-2}{\Xi^-} + \underset{S=-1}{\overline{K}}\quad (S_{\mathrm{LHS}} = S_{\mathrm{RHS}})$$

But then let us check mass–energy conservation

$$m_{\mathrm{RHS}} = m_{\Xi^-} + m_{\overline{K}} = 1520\ MeV + 493\ MeV = 2013\ MeV\ \text{while}$$

$$m_{\mathrm{LHS}} = m_{\Omega^-} = 1670\ MeV$$

$$\text{Clearly } m_{\mathrm{LHS}} \ne m_{\mathrm{RHS}}.$$

So Ω^- cannot decay via strong interaction.

Hence Ω^- decays via weak interaction in which strangeness is not conserved since

$$\underset{S=-3}{\Omega^-} \to \underset{S=-2}{\Xi^-} + \underset{S=0}{\pi^0}\quad (S_{\mathrm{LHS}} \ne S_{\mathrm{RHS}})$$

10.8 Purpose of $SU(2)$, $SU(3)$

Numerous particles and resonances (which are also identified with particles) were discovered and to handle such a multitude of particles efficiently, i.e. to categorize or order them one uses the help of symmetries.

Heisenberg brought some order by considering $SU(2)$ which is mathematically the same as the theory of angular momentum but it refers to internal physically unobserved space and is the symmetry between two objects (proton and neutron, which have a very small mass difference). Heisenberg saw that p and n have similar properties attributing

$$I = \frac{1}{2}, I_z = \frac{1}{2} \text{ for } p \text{ and } I = \frac{1}{2}, I_z = -\frac{1}{2} \text{ for } n.$$

This situation is like $s = \frac{1}{2}$, $s_z = \frac{1}{2}$ for spin up e^- and $s = \frac{1}{2}$, $s_z = -\frac{1}{2}$ for spin down e^-.

So instead of considering proton and neutron as separate particles they can be made to collapse into one particle called the nucleon $N = \begin{pmatrix} p \\ n \end{pmatrix}$.

Instead of considering π^+, π^0, π^- as separate particles they can be viewed as the components of one particle π with isospin $I = 1$ since

$$|\pi^+\rangle = |I = 1\ I_z = 1\rangle, \ |\pi^0\rangle = |I = 1\ I_z = 0\rangle, \ |\pi^-\rangle = |I = 1\ I_z = -1\rangle$$

So

$$\pi = \begin{pmatrix} \pi^+ \\ \pi^0 \\ \pi^- \end{pmatrix}$$

Clearly isospin symmetry itself brought some order. Isospin symmetry is only used to classify particles.

$SU(2)$ transformation rotates u, d quarks in 2D.

$SU(3)$ goes one step further and enlarges the symmetry. $SU(3)$ is an extension of $SU(2)$ and refers to the symmetry between three objects (the three quarks u, d, s, which have small mass difference).

In $SU(3)$ the number 3 refers to the three quarks u, d, s. They are supposed to be elementary particles of which all other particles are composed.

$SU(3)$ transformation rotates u, d, s quarks in 3D.

$SU(3)$ includes $SU(2)$. All $SU(2)$ multiplets, i.e. isospin multiplets (or iso-multiplets) are put together into a bigger multiplet like an octet or decuplet.

It was observed that the 8 particles (baryons) p, n, Λ, Σ^+, Σ^0, Σ^-, Ξ^0, Ξ^- have a similar property to some extent as they all possess $J^\pi = \frac{1}{2}^+$. It is possible to arrange them into an octet (which is a hexagon with 6 particles at the 6 corners and 2 particles at the centre, as shown in figure 10.12) in $SU(3)$ representation called the eightfold way. So all these 8 particles collapse into a single particle, i.e. the 8 particles can be looked upon as different states of a single particle.

Also the 8 particles (mesons) K^+, K^0, π^+, π^0, π^-, K^-, $\overline{K}^0$, η have similar properties to some extent as they all possess $J^\pi = 0^-$. It is possible to arrange them into an octet (which is a hexagon with 6 particles at the 6 corners and 2 particles at the centre, as shown in figure 10.8) in $SU(3)$ representation. So all these 8 particles collapse into a single particle, i.e. the 8 particles can be looked upon as different states of a single particle.

Again, 10 resonance particles (baryons) Δ^{++}, Δ^{+}, Δ^{0}, Δ^{-}, Σ^{*+}, Σ^{*0}, Σ^{*-}, Ξ^{*0}, Ξ^{*-}, Ω^{-} have a similar property to some extent as they all possess $J^{\pi} = \frac{3}{2}^{+}$. It is possible to arrange them into a decuplet (which is a triangle, as shown in figure 10.16) in $SU(3)$ representation. So all these 10 particles collapse into a single particle, i.e. the 10 particles can be looked upon as different states of a single particle.

So in making I_3 versus Y plot and arranging particles into multiplets (octets, decuplets) some regularity among the hadrons was established.

However, regularity is broken when we consider the masses, since the masses of particles go on changing.

For instance in $SU(2)$ multiplet masses are not same—they are slightly different as

$$m_p = 938.3\ MeV,\ m_n = 939.6\ MeV \text{ in nucleon isodoublet,}$$

$$m_{\pi^\pm} =\ 140\ MeV,\ m_{\pi^0} = 138\ MeV \text{ in isotriplet.}$$

However, in $SU(3)$ multiplets the mass differences are large and so $SU(3)$ is a badly broken symmetry. For instance

✓ In the baryon eightfold way shown in figure 10.12 we have.

$$m_p\ = 938.3\ MeV,\ \ m_n = 939.6\ MeV,\ \ m_{\Sigma^+} = 1189.4\ MeV,$$

$$m_{\Sigma^0} = 1192.5\ MeV,\ \ m_{\Sigma^-} = 1197.4\ MeV,\ \ m_{\Xi^0} = 131.9\ MeV,$$

$$m_{\Xi^-} = 1321.3\ MeV,\ \ m_\Lambda = 1115.7\ MeV.$$

✓ In a meson octet, as shown in figure 10.8, we have

$$m_{K^+} = m_{K^0} = 493\ MeV,\ \ m_{\pi^\pm} = 140\ MeV,\ \ m_{\pi^0} = 138\ MeV,$$

$$m_{\overline{K}^0} = m_{K^-} = 498\ MeV.$$

✓ In the baryon decuplet, as shown in figure 10.16, we have

$$m_\Delta = 1232\ MeV,\ \ m_{\Sigma^*} = 1385\ MeV,\ \ m_{\Xi^*} = 1520\ MeV,\ \ m_{\Omega^-} = 1670\ MeV$$

The above data show that the $SU(2)$ and $SU(3)$ are both broken symmetries.

We shall construct all the hadrons (baryons and mesons, strange and non-strange) from quarks.

$$\text{Baryon } B \sim qqq$$

Meson $M \sim q\bar{q}$

Resonance baryon $B^* \sim$ qqq (27 wave functions $3 \otimes 3 \otimes 3$)

This is shown in figure 10.17.

Quark interaction does not depend on the spin (up or down) of the quark. So the symmetry is larger than $SU(3)$ and is called $SU(6)$ symmetry (proposed in 1965 by Gursey, Radicati and Sakita). Forces are spin independent. So we can construct the wave function

$$\psi_i = \begin{pmatrix} u\uparrow \\ u\downarrow \\ d\uparrow \\ d\downarrow \\ s\uparrow \\ s\downarrow \end{pmatrix}$$

u, d, s represent quarks, $\uparrow$ represents upspin, $\downarrow$ represents downspin, $i = 1$ to 6.

$SU(3)$ represents symmetry of unitary transformation of three objects.

$SU(6)$ represents 6-dimensinal unitary transformation. We can classify the hadrons according to $SU(6)$ and then particles collapse more and further simplification can be achieved.

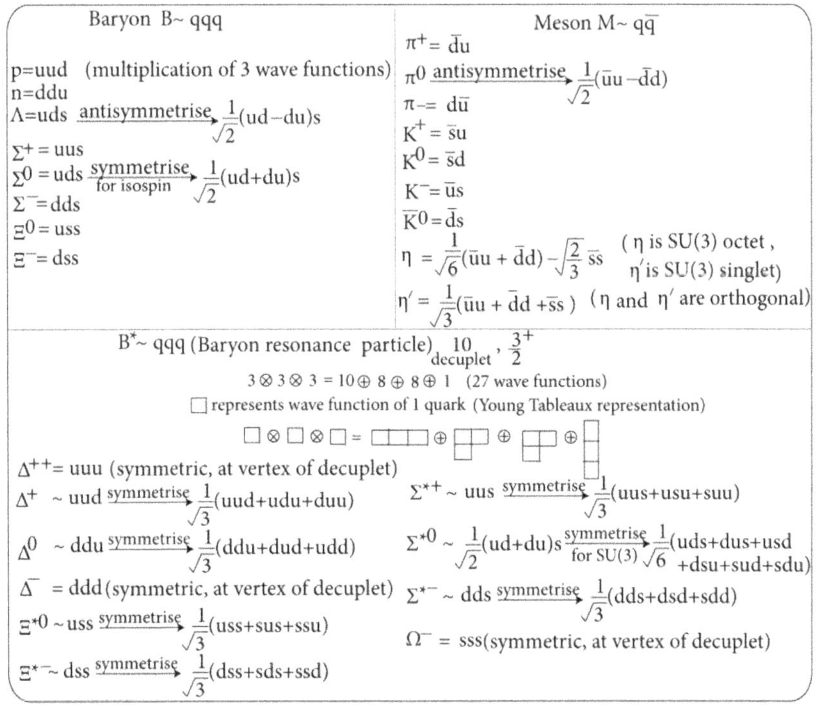

Figure 10.17. Quark structure of hadrons (baryon, resonance baryon, meson).

However $SU(6)$ is an approximate symmetry—more badly broken than $SU(3)$.

First we classified particles w.r.t. isospin, then w.r.t. $SU(3)$ as octet, or decuplet etc. Now we consider classification w.r.t. $SU(6)$ which corresponds to a 56-dimensional multiplet

$$56 = (8, 2) \oplus (10, 4) \text{ under } (3) \otimes SU(2).$$

This leads to $8 \times 2 + 10 \times 4 = 16 + 40 = 56$ particles, $SU(3)$ being internal symmetry and $SU(2)$ being isospin and spins $\frac{1}{2}^+$ and $\frac{3}{2}^+$ being combined.

10.9 Discovery of colour quantum number from wave function of Δ^{++}

Consider the Δ^{++} wave function expressed as

$$\Delta^{++}_{\mathrm{spin}\frac{3}{2}} = \mathrm{uuu} \uparrow \uparrow \uparrow \psi(\vec{r_1}, \vec{r_2}, \vec{r_3})$$

where the quark combination is uuu, spin combination is $\uparrow \uparrow \uparrow$ and spatial wave function is $\psi(\vec{r_1}, \vec{r_2}, \vec{r_3})$.

In ground state the relative orbital angular momentum is zero leading to

$$\psi(\vec{r_1}, \vec{r_2}, \vec{r_3}) = \text{symmetric.}$$

Also, the uuu part is symmetric and the $\uparrow \uparrow \uparrow$ part is symmetric.

Hence wave function $\Delta^{++}_{3/2}$ is symmetric under interchange of quarks, i.e. $SU(3)$, spin and space coordinates. However, quarks are fermions and so a three-fermion system $\Delta^{++}_{3/2}$ has to be antisymmetric obeying Fermi–Dirac statistics.

For this essential requirement we define another quantum number called colour quantum number that takes three values red (R), blue (B) and green(G). Accordingly, the colour part, say

$$\chi_{\mathrm{colour}}(R, B, G)$$

has to be incorporated into wave function $\Delta^{++}_{3/2}$ i.e.

$$\Delta^{++}_{\mathrm{spin}\frac{3}{2}} = \mathrm{uuu} \uparrow \uparrow \uparrow \psi(\vec{r_1}, \vec{r_2}, \vec{r_3}) \chi_{\mathrm{colour}}(R, B, G)$$

This makes $\Delta^{++}_{3/2}$ antisymmetric w.r.t. interchange of quarks, i.e. (3), spin, space and colour. This was how colour quantum number was discovered. With gauged $SU(3)$ this subsequently gave rise to quantum chromodynamics or QCD.

10.10 Comparison between $SU(2)$ and $SU(3)$

$SU(2)$, $SU(3)$ are called Lie groups.

In $SU(2)$ let us take the constituents as u and d and consider the column vector $\begin{pmatrix} u \\ d \end{pmatrix}$. Consider the unitary transformation

$$\begin{pmatrix} u' \\ d' \end{pmatrix} = U \begin{pmatrix} u \\ d \end{pmatrix}$$

where U is 2×2 unitary unimodular matrix with

$$U^\dagger U = I, \ \det U = |U| = 1$$

Take the most general unitary unimodular matrix to have the form (equation (9.164))

$$U = e^{i\frac{\tau_i}{2}\theta_i}, \ i = 1, 2, 3$$

$\frac{\tau_i}{2}$ is the generator and

$$\tau_1 = \begin{pmatrix} 0 & 1 \\ 1 & 0 \end{pmatrix}, \ \tau_2 = \begin{pmatrix} 0 & -i \\ i & 0 \end{pmatrix}, \ \tau_3 = \begin{pmatrix} 1 & 0 \\ 0 & -1 \end{pmatrix}$$

$$\vec{\tau} = (\tau_1, \tau_2, \tau_3) = \ \text{isospin vector}$$

In the case of angular momentum we use Pauli matrices σ_i instead of τ_i. The three matrices are identical for σ_i and τ_i. Also (see equation (9.160))

$$\left[\frac{\tau_i}{2}, \frac{\tau_j}{2} \right] = i\varepsilon_{ijk} \frac{\tau_k}{2} \text{ and } \left\{ \frac{\tau_i}{2}, \frac{\tau_i}{2} \right\} = 0 \text{ (Lie algebra)}$$

Multiplets of $SU(2)$ are 2,3,4,5 which have multiplicities corresponding to $\tau = \frac{1}{2}, 1, \frac{3}{2}, 2$.

- $\vec{\frac{1}{2}} + \vec{\frac{1}{2}} = 0, 1$. We can rewrite this relation in terms of multiplicities as

$$\underbrace{\left(2.\frac{1}{2} + 1 \right)}_{\text{doublet}} \otimes \underbrace{\left(2.\frac{1}{2} + 1 \right)}_{\text{doublet}} = (2.0 + 1) \otimes (2.1 + 1) \text{ i. e.}$$
$$\qquad\qquad\qquad\qquad\qquad\qquad\ \underset{\text{singlet}}{\ } \qquad \underset{\text{triplet}}{\ }$$

$$2 \otimes 2 = 1 \oplus 3$$

as the number of states are $2 \times 2 = 4$ or $1 + 3 = 4$.

2 is referred to as the fundamental representation.
- Denote $|I \ I_3\rangle = |\frac{1}{2} \ \frac{1}{2}\rangle = \alpha, |I \ I_3\rangle = |\frac{1}{2} \ -\frac{1}{2}\rangle = \beta$.

Upon addition of $\vec{I_1} = \vec{\frac{1}{2}}$ and $\vec{I_2} = \vec{\frac{1}{2}}$ we get

$$|I \ I_3\rangle = |0 \ 0\rangle \text{ and}$$

$$|I \ I_3\rangle = |1 \ 1\rangle, |1 \ 0\rangle, |1 \ -1\rangle.$$

Normalization and proper symmetrization gives

$$|0 \ 0\rangle = \frac{1}{\sqrt{2}} [\alpha(1)\beta(2) - \alpha(2)\beta(1)]$$

$$|1 0\rangle - \alpha(1)\alpha(2)$$

$$|1\ 0\rangle = \frac{1}{\sqrt{2}}\,[\alpha(1)\beta(2) + \alpha(2)\beta(1)]$$

$$|1 - 1\rangle = \beta(1)\beta(2)$$

10.11 Representation of anti-particle under $SU(2)$

Let

$$\psi = \begin{pmatrix} u \\ d \end{pmatrix}$$

be an isodoublet in $SU(2)$. Anti-particle is represented by the complex conjugate

$$\psi^* = \begin{pmatrix} u^* \\ d^* \end{pmatrix}$$

For anti-particle, I_3 has to change by a negative sign, i.e. a flip is needed. This is achieved by multiplying ψ^* by $i\tau_2$ to get

$$i\tau_2\psi = i\begin{pmatrix} 0 & -i \\ i & 0 \end{pmatrix}\begin{pmatrix} u^* \\ d^* \end{pmatrix} = \begin{pmatrix} 0 & 1 \\ -1 & 0 \end{pmatrix}\begin{pmatrix} u^* \\ d^* \end{pmatrix} = \begin{pmatrix} d^* \\ -u^* \end{pmatrix}$$

which represents anti-particle.

- Multiplets of $SU(3)$

In $SU(3)$ let us take the constituents as u, d, s and consider the column vector $\begin{pmatrix} u \\ d \\ s \end{pmatrix}$.

Consider the unitary transformation

$$\begin{pmatrix} u' \\ d' \\ s' \end{pmatrix} = U\begin{pmatrix} u \\ d \\ s \end{pmatrix}.$$

where U is a 3×3 unitary unimodular matrix with

$$U^\dagger U = I, \ \det U = |U| = 1$$

Take the most general unitary unimodular matrix to have the form (equation (9.186))

$$U = e^{i\frac{\lambda_i}{2}\theta_i}, \ i = 1, 2, 3, \dots 8$$

where $\frac{\lambda_i}{2}$ is the generator satisfying

$$\left[\frac{\lambda_i}{2}, \frac{\lambda_j}{2}\right] = if_{ijk}\frac{\lambda_k}{2} \ \text{(obeys Lie algebra)}$$

$$\lambda_1 = \begin{pmatrix} 0 & 1 & 0 \\ 1 & 0 & 0 \\ 0 & 0 & 0 \end{pmatrix} = \begin{pmatrix} \tau_1 & & 0 \\ & & 0 \\ 0 & 0 & 0 \end{pmatrix} \quad \lambda_2 = \begin{pmatrix} 0 & -i & 0 \\ i & 0 & 0 \\ 0 & 0 & 0 \end{pmatrix} = \begin{pmatrix} \tau_2 & & 0 \\ & & 0 \\ 0 & 0 & 0 \end{pmatrix} \quad \lambda_3 = \begin{pmatrix} 1 & 0 & 0 \\ 0 & -1 & 0 \\ 0 & 0 & 0 \end{pmatrix} = \begin{pmatrix} \sigma_3 & & 0 \\ & & 0 \\ 0 & 0 & 0 \end{pmatrix} \quad \lambda_4 = \begin{pmatrix} 0 & 0 & 1 \\ 0 & 0 & 0 \\ 1 & 0 & 0 \end{pmatrix}$$

$$\lambda_5 = \begin{pmatrix} 0 & 0 & -i \\ 0 & 0 & 0 \\ -i & 0 & 0 \end{pmatrix} \quad \lambda_6 = \begin{pmatrix} 0 & 0 & 0 \\ 0 & 0 & 1 \\ 0 & 1 & 0 \end{pmatrix} = \begin{pmatrix} 0 & 0 & 0 \\ 0 & & \\ 0 & \tau_1 & \end{pmatrix} \quad \lambda_7 = \begin{pmatrix} 0 & 0 & 0 \\ 0 & 0 & -i \\ 0 & i & 0 \end{pmatrix} = \begin{pmatrix} 0 & 0 & 0 \\ 0 & & \\ 0 & \tau_2 & \end{pmatrix} \quad \lambda_8 = \begin{pmatrix} 1 & 0 & 0 \\ 0 & 1 & 0 \\ 0 & 0 & -2 \end{pmatrix} = \frac{1}{\sqrt{3}} \begin{pmatrix} I & & 0 \\ & & 0 \\ 0 & 0 & -2 \end{pmatrix}$$

Figure 10.18. Gell-Mann matrices where $\frac{\lambda_i}{2}$ are generators of $SU(3)$.

where the value of f_{ijk} is given in exercise 9.37 and λ_i is given in figure 10.18.

The fundamental representation of $SU(3)$ is 3×3, i.e. the quark triplets. But with it we can generate various multiplets.

The $SU(3)$ multiplets are 3, 3*, 6, 8, 10, 27. This tallies with the multiplet structures obtained with particles (figure 10.8 representing an octet with mesons, figure 10.12 representing the eightfold way, figure 10.16 representing the baryon decuplet with resonance particles).

- In $SU(2)$ we do not have 2, 2* since when we take the complex conjugate of u, d doublet it also behaves like a doublet and so 2* is not different from 2. This is not the case for higher groups such as $SU(3)$. 3* (corresponding to anti-quarks) is different from 3 and hence it has been written separately.
- There is one anti-commutation relation that the $SU(2)$ generators $\frac{\tau_i}{2}$ have to obey. But in $SU(3)$ there is no anti-commutation that generators $\frac{\lambda_i}{2}$ have to satisfy.
- The non-commuting nature of angular momenta components I_1, I_2, I_3 means that I^2 and I_3 are diagonal. So in $SU(2)$ there is one diagonal generator matrix $\frac{\tau_3}{2}$. Members of a multiplet are distinguished by only one quantum number, namely I_3 values.

  In $SU(3)$ there are two independent 3×3 diagonal generator matrices viz. $\frac{\lambda_3}{2}$, $\frac{\lambda_8}{2}$. And so we need two quantum numbers, namely I_3 and Y to distinguish members of a multiplet (in octet, decuplet representation of particles). In other words, I_3 and Y correspond to $\frac{\lambda_3}{2}$, $\frac{\lambda_8}{2}$.
- $SU(2)$ has 3 generators while $SU(3)$ has 8 generators and $SU(n)$ has $n^2 - 1$ generators.

10.12 Evidence supporting the quark model

The evidence supporting the quark model is as follows:
- ✓ The deep inelastic scattering data of electrons on nucleons indicate that the nucleon has a substructure.
- ✓ The $SU(3)$ symmetry of hadrons can be explained by the quark model.
- ✓ The quark model gives the correct cross-section relationship of hadronic reactions.
- ✓ The quark model explains anomalous magnetic moments of nucleons.

10.13 Approximate $SU(3)$ symmetry of strong interaction

If $SU(3)$ symmetry were perfect, particles of the same super multiplet should have the same mass. However, the masses of particles in the same super multiplet differ appreciably, which means the symmetry is only approximate. The greater the difference the greater is the breaking of symmetry.

For instance, though π^0, K^0 are members of the same meson supermultiplet 0^-, their masses differ as

$$m(\pi^0) = 138 \ MeV, \ m(K^0) = 493 \ MeV.$$

10.14 Heavy quarks

Light quarks are

$$u(m_u = 4 \ MeV)$$

$$d(m_d = 8 \ MeV)$$

$$s(m_s = 30 \ MeV)$$

Heavy quarks are

$$c(m_c = 1.5 \ GeV, \text{ discovered in 1974})$$

$$b \ (m_b = 5 \ GeV, \text{ discovered in 1977})$$

$$t \ (m_t = 174 \ GeV, \text{ discovered in 1995})$$

10.14.1 c quark

The hadron J/ψ was discovered in 1974 by Richer and Ting.

The c quark was first seen in the J/ψ meson particle. The structure of J/ψ was identified as a bound state $J/\psi = c\bar{c}$.

Mass is $m_{J/\psi} = 3.1 \ GeV$. Again $m_c = m_{\bar{c}} = 1. \ 6 \ GeV$.

So the difference of masses of $c\bar{c}$ and J/ψ is the binding energy. It was seen as an extremely narrow peak in an e^+–e^- collider (figure 10.19).

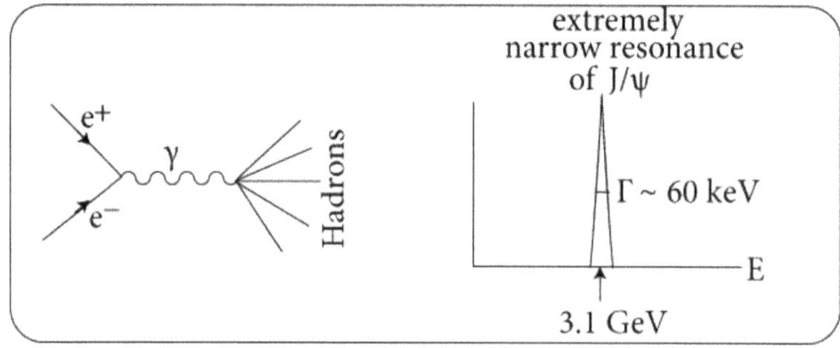

Figure 10.19. Narrow resonance observed in an electron–positron collider at 3.1 GeV.

Particle with c quark	Mass	Particle with b quark	Mass
Meson		Meson	
$J/\psi = c\bar{c}$	3 GeV	$\Upsilon = b\bar{b}$	9.5 GeV
$D^+ = c\bar{d}$, $D^- = \bar{c}d$	1.87 GeV	$B^+ = u\bar{b}$, $B^- = \bar{u}b$	5.3 GeV
$D^0 = c\bar{u}$, $\bar{D}^0 = \bar{c}u$	1.86 GeV	$B^0 = d\bar{b}$, $\bar{B}^0 = \bar{d}b$	5.3 GeV
$D_s^+ = c\bar{s}$, $D_s^- = \bar{c}s$	1.97 GeV	$B_s^0 = s\bar{b}$, $\bar{B}_s^0 = b\bar{s}$	5.4 GeV
Baryon		Baryon	
$\Lambda_c^+ = udc$	2.28 GeV	$\Lambda_b^0 = udb$	5.6 GeV
$\Sigma_c^{++} = uuc$	2.455 GeV	$\Sigma_b^+ = uub$	5.8 GeV
$\Sigma_c^+ = udc$	2.455 GeV	$\Sigma_b^0 = udb$	—
$\Sigma_c^0 = ddc$	2.455 GeV	$\Sigma_b^- = ddb$	5.8 GeV

Figure 10.20. Mesons and baryons with c quark and b quark as constituents.

A few mesons with c quarks and a few baryons with c quarks as their constituent are shown in figure 10.20.

10.14.2 b quark

The hadron upsilon denoted by Υ was discovered in 1977 by Ledermann.

The b quark was first seen in the upsilon Υ meson particle. The structure of Υ was identified as the bound state $\Upsilon = b\bar{b}$.

Mass is $m_\Upsilon = 10\ GeV$ Again $m_b = m_{\bar{b}} \sim 5\ GeV$.

The difference of masses of $b\bar{b}$ and Υ is the binding energy.

A few mesons with b quark and a few baryons with b quark as its constituent are shown in figure 10.20.

10.14.3 t quark

Bound states with the t quark do not exist because of asymptotic freedom. Because of its high mass $m_t = 174\ GeV$ strong coupling becomes so weak that it is not able to bind. It takes a long time for the Bohr orbit to collapse into a bound state and by that time the t quark decays through weak interaction, namely (figure 10.21)

$$t \rightarrow b, e^-, \bar{\nu}_e$$

$$t \rightarrow b, \mu^-, \bar{\nu}_\mu$$

In other words, the top quarks produced in high energy scattering experiments decay before they can form bound states, i.e. before they can hadronize. All other five quarks u, d, s, c, b hadronise, i.e. get confined to form bound states of different composition.

The quarks usually pick up other quarks from a vacuum and with the available quarks they get confined into bound states—they hadronize either into a paired state of quark–anti-quark (meson) or into a bound state of three quarks (baryon).

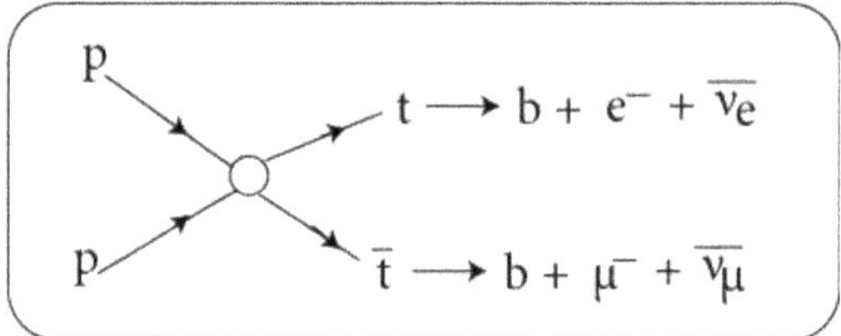

Figure 10.21. t quarks do not form a bound state as they decay before formation.

A top quark when produced also looks for other quarks in an effort to hadronize, but before the hadronization process gets complete it decays. So a bound state with top quark t is not formed.

10.15 Asymptotic freedom

Asymptotic freedom was discovered in an experiment on deep inelastic scattering performed by Friedman, Kendall and Taylor in 1969. An electron is accelerated to very high energies towards a target (stationary protons) and there is electromagnetic interaction. There is very large momentum transfer to a hadron, as shown in figure 10.22.

A study was made of the cross-section of a reaction as a function of q^2 where

$$q = |\vec{p}_f - \vec{p}_i| = \text{momentum transfer}$$

$\vec{p}_i$ is initial momentum of the electron and $\vec{p}_f$ is final momentum of the electron.

The whole cross-section looks as if it is a superposition of elementary cross-sections corresponding to multiple free-point particles existing within the proton. In other words, as observed through high q^2 probes, the constituents of a proton behaved like free particles. So a particle behaves as a free particle at asymptotically high energy. This is referred to as asymptotic freedom.

This was analogous to the discovery of one point nucleus within atom in Rutheford's α scattering experiment. Rutherford probed the atom and in deep inelastic scattering the nucleon (proton) was being probed. In Rutherford's experiment one point nucleus was discovered but here multiple point particles (within the proton) were discovered.

A theory describing asymptotic freedom was given by Gross, Politzer and Wilczek in 1972 as a non-abelian gauge theory (Yang–Mills theory).

Non-abelian gauge theory is the only quantum field theory that has asymptotic freedom. Self-interaction (figure 10.23) in the Yang–Mills gauge field makes it have asymptotic freedom.

In contrast, photons do not have self -interaction.

Distance is inversely connected to momentum. Large momentum corresponds to short distance and small momentum corresponds to large distance.

Figure 10.23 shows the coupling constant as a function of momentum in non-abelian gauge theory. At large distance, i.e. for small momentum, coupling is large and then it decreases with momentum increase and at small distance, i.e. for large

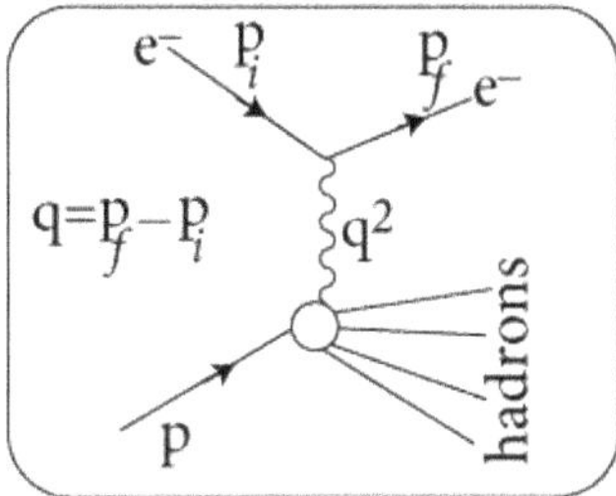

Figure 10.22. Deep inelastic scattering of electron from proton target leading to discovery of asymptotic freedom.

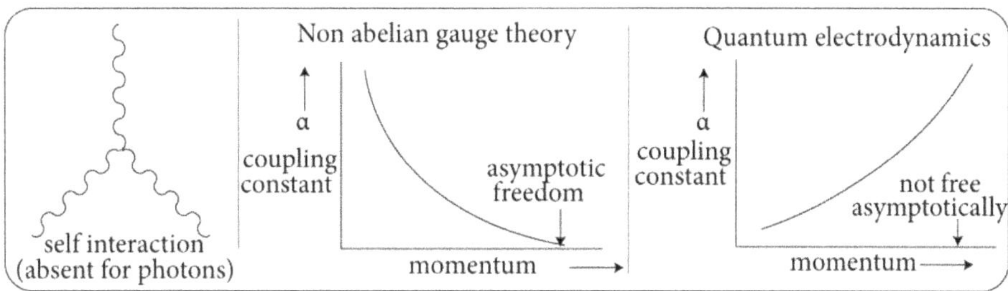

Figure 10.23. Non-abelian gauge theory and asymptotic freedom.

(asymptotic) momentum, the coupling constant is zero. This is called asymptotic freedom.

Because of this we say that non-abelian gauge theory is a fit theory for strong interaction. It explains why quarks are almost free at large momentum transfer.

In quantum electrodynamics (QED) or photon theory the behaviour of coupling constant is opposite since it is shown in the figure that coupling constant is very large asymptotically.

10.15.1 6 quarks and 6 leptons

We write the quarks and leptons as doublets—6 quarks form 3 doublets and 6 leptons form 3 doublets. They are

$$\binom{u}{d},\ \binom{c}{s},\ \binom{t}{b},\ \binom{\nu_e}{e},\ \binom{\nu_\mu}{\mu},\ \binom{\nu_\tau}{\tau}$$

Actually there are $6 \times 3 = 18$ quarks taking colour into consideration, namely

$$u_\alpha,\ d_\alpha,\ s_\alpha,\ c_\alpha,\ b_\alpha,\ t_\alpha$$

where $\alpha = R,\ B,\ G$ and 6 leptons $e,\ \mu,\ \tau,\ \nu_e,\ \nu_\mu,\ \nu_\tau$.

Let us rewrite the doublets in terms of their masses in MeV

$$\binom{4}{8},\ \binom{1600}{30},\ \binom{174\,000}{5000}\binom{<\,1\text{ eV}}{0.51},\ \binom{<\,1\text{ eV}}{107},\ \binom{<\,1\text{ eV}}{1700}$$

The proton has quark structure $p = uud$. Now

$$m_p = 938 \ MeV \text{ and } m_u = 4 \ MeV, \ m_d = 8 \ MeV.$$

So there is a great mismatch between m_p and m_{uud}. Actually, most of the mass of p and n comes from the intense glue field or the gluonic field that surrounds the quarks within the proton. When the glue field energy is taken care of there is no mismatch.

Actually, most of the mass of all objects comes from nuclei and most of the mass of the nucleus comes from the glue field.

10.16 Heavy lepton

The tauon was discovered by Perl in 1971 at SLAC.

The scheme of the corresponding interaction is shown in figure 10.24. It was an interaction between e^+ and e^-. The collision gives off a virtual photon that materializes into two particles $\tau^\pm$. The τ^+ decays into e^+, ν_e and $\bar{\nu}_\tau$ (figures 10.24 (a) and (c)) or into μ^+, ν_μ and $\bar{\nu}_\tau$ (figures 10.24(b) and (d)) while τ^- decays into μ^-, $\bar{\nu}_\mu$, ν_τ (figures 10.24(a) and (d)) or into e^-, $\bar{\nu}_e$ and ν_τ (figures 10.24(b) and (c)).

Figure 10.25 shows the quark structure of particles and anti-particles (hadrons).

We note that the quark content of π^+ is $u\bar{d}$ and the quark content of π^- is obtained by replacing a quark by an anti-quark and vice versa to get $\pi^- = \bar{u}d$. Any of the pairs π^+, π^- can be treated as a particle and the others as an anti-particle.

$$\text{Similarly } K^+ = u\bar{s} \text{ and } K^- = \bar{u}s.$$

Let us note that we do not have normal matter made of anti-particles.

10.17 Gluon colour

Gluons are massless quanta that mediate or exchange strong interaction between quarks. They are vector bosons of spin $s = 1$ and negative intrinsic parity.

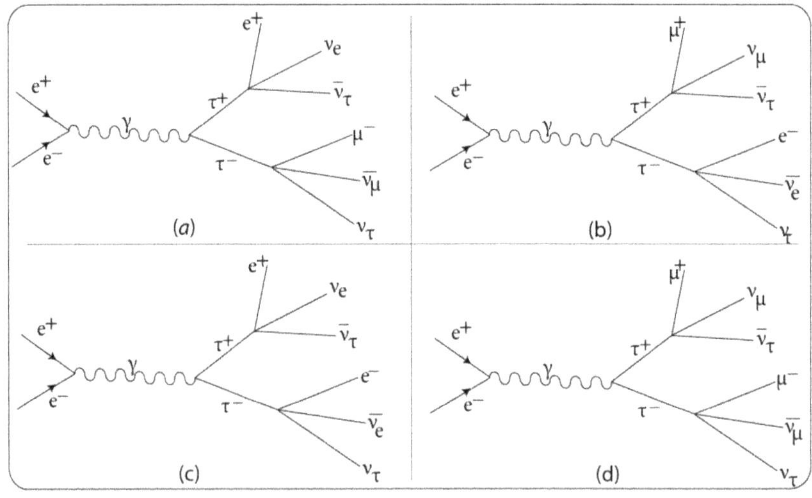

Figure 10.24. Discovery of heavy lepton τ by Perl in 1971.

Meson		Baryon			
Particle	Anti-particle	Particle	Anti-particle	Particle	Anti-particle
$\pi^+ = ud$	$\pi^- = \bar{u}d$	$p = uud$	$\bar{p}$	$\Lambda_c^+ = udc$	Λ_c^-
$K^+ = u\bar{s}$	$K^- = \bar{u}s$	$n = udd$	$\bar{n}$	$\Sigma_c^{++} = uuc$	$\bar{\Sigma}_c^{--}$
$K^0 = \bar{s}d$	$\bar{K}^0 = d\bar{s}$	$\Lambda = \frac{1}{\sqrt{2}}(ud-du)s$	$\bar{\Lambda}$	$\Sigma_c^+ = udc$	$\bar{\Sigma}_c^-$
$D^+ = c\bar{d}$	$D^- = \bar{c}d$	$\Lambda^{++} = uuu$	$\bar{\Lambda}^{--}$	$\Sigma_c^0 = ddc$	$\bar{\Sigma}_c^0$
$D^0 = c\bar{u}$	$\bar{D}^0 = \bar{c}u$	$\Sigma^- = dds$	$\bar{\Sigma}^+$	$\Lambda_b^0 = udb$	$\bar{\Lambda}_b^0$
$D_s^+ = c\bar{s}$	$D_s^- = \bar{c}s$	$\Sigma^+ = uus$	$\bar{\Sigma}^-$	$\Sigma_b^+ = uub$	$\bar{\Sigma}_b^-$
$B^+ = u\bar{b}$	$B^- = \bar{u}b$	$\Sigma^0 = \frac{1}{\sqrt{2}}(ud+du)s$	$\bar{\Sigma}^0$	$\Sigma_b^0 = udb$	$\bar{\Sigma}_b^0$
$B^0 = d\bar{b}$	$\bar{B}^0 = \bar{d}b$	$\Xi^0 = uss$	$\bar{\Xi}^0$	$\Sigma_b^- = ddb$	$\bar{\Sigma}_b^+$
$B_s^0 = s\bar{b}$	$\bar{B}_s^0 = b\bar{s}$	$\Xi^- = dss$	Ξ^+		
$\pi^0 = \frac{1}{\sqrt{2}}(\bar{u}u-\bar{d}d)$	π^0	$\Omega^- = sss$	Ω^+	$\Xi^{*0} = \frac{1}{\sqrt{3}}(uss+sus+ssu)$	$\bar{\Xi}^{*0}$
$\Upsilon = b\bar{b}$	Υ	$\Delta^{++} = uuu$	Δ^{--}	$\Xi^{*-} = \frac{1}{\sqrt{3}}(dss+sds+ssd)$	$\bar{\Xi}^{*+}$
$J/\psi = c\bar{c}$	J/ψ	$\Delta^+ = \frac{1}{\sqrt{3}}(uud+udu+duu)$	$\bar{\Delta}^-$	$\Sigma^{*+} = \frac{1}{\sqrt{3}}(uus+usu+suu)$	$\bar{\Sigma}^{*-}$
$\eta = \frac{1}{\sqrt{6}}(\bar{u}u+\bar{d}d)-\sqrt{\frac{2}{3}}\,\bar{s}s$	η	$\Delta^0 = \frac{1}{\sqrt{3}}(ddu+dud+udd)$	$\bar{\Delta}^0 / \bar{\Delta}^+$	$\Sigma^{*0} = \frac{1}{\sqrt{3}}(uds+dus+usd+dsu+sud+sdu)$	$\bar{\Sigma}^{*0}$
		$\Delta^- = ddd$		$\Sigma^{*-} = \frac{1}{\sqrt{3}}(dds+dsd+sdd)$	$\bar{\Sigma}^{*+}$

Figure 10.25. Quark structure of particles and anti-particles.

According to the theory of QCD, gluons bind or glue quarks into groups forming hadrons like protons, neutrons etc.

There are 8 independent types of gluons in QCD.

Gluons carry colour charge.

Quarks carry three types of colour charge (R, B, G) and anti-quarks carry three types of anti-colour charges ($\bar{R}$, $\bar{B}$, $\bar{G}$). So gluons that act as an adhesive between quarks and anti-quarks are a combination of colour and anti-colour. Accordingly there can be nine possible combinations. They are

✓ $R\bar{R}$ (Red–anti-Red), $R\bar{B}$ (Red–anti-Blue), $R\bar{G}$ (Red–anti-Green)

✓ $B\bar{R}$ (Blue–anti-Red), $B\bar{B}$ (Blue–anti-Blue), $B\bar{G}$ (Blue–anti-Green)

✓ $G\bar{R}$ (Green–anti-Red), $G\bar{B}$ (Green–anti-Blue), $G\bar{G}$ (Green–anti-Green).

These effective states can be suitably combined or mixed according to the mathematics of colour charge to get 8 independent colour states (colour octet) which correspond to the 8 types of gluons. They are

$$\frac{1}{\sqrt{2}}(R\bar{B} + B\bar{R}) \text{ (corresponding to } \lambda_1)$$

$$-\frac{i}{\sqrt{2}}(R\bar{B} - B\bar{R}) \text{ (corresponding to } \lambda_2)$$

$$\frac{1}{\sqrt{2}}(R\bar{R} - B\bar{B}) \text{ (corresponding to } \lambda_3)$$

$$\frac{1}{\sqrt{2}}(R\overline{G} + G\overline{R}) \text{ (corresponding to } \lambda_4)$$

$$-\frac{i}{\sqrt{2}}(R\overline{G} - G\overline{R}) \text{ (corresponding to } \lambda_5)$$

$$\frac{1}{\sqrt{2}}(B\overline{G} + G\overline{B}) \text{ (corresponding to } \lambda_6)$$

$$-\frac{i}{\sqrt{2}}(B\overline{G} - G\overline{B}) \text{ (corresponding to } \lambda_7)$$

$$\frac{1}{\sqrt{6}}(R\overline{R} + B\overline{B} - 2G\overline{G}) \text{ (corresponding to } \lambda_8) \tag{10.9}$$

These correspond to the Gell-Mann matrices λ_i, $i = 1$ to 8 as evident from figure 10.26. These 8 states are linearly independent.

We mention that long-range gluon interaction does not exist.

Stable strongly interacting particles, i.e. hadrons observed in nature are colourless, also called the colour singlet state.

We can also build a the colour singlet state as

$$\frac{1}{\sqrt{3}}(R\overline{R} + B\overline{B} + G\overline{G}) \tag{10.10}$$

But gluons in the singlet state (equation (10.10)) do not exist. This is also evident since we cannot derive this forbidden colour singlet state (equation (10.10)) starting from the states of the colour octet (equation (10.9)).

10.18 Weak interaction

In the list of elementary particles we have the following. Let us write down the left-handed and right-handed ones.

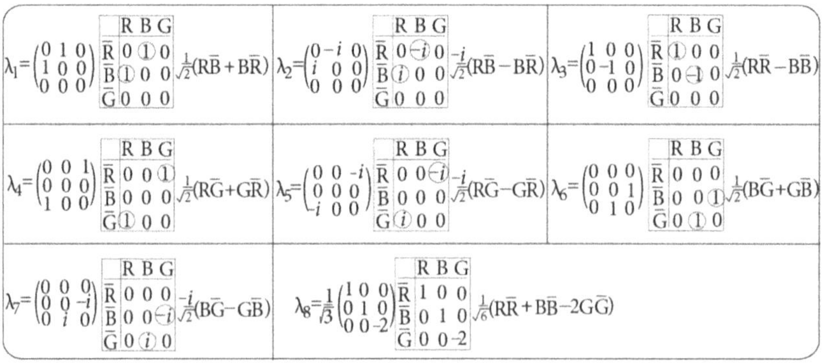

$$\lambda_1 = \begin{pmatrix} 0 & 1 & 0 \\ 1 & 0 & 0 \\ 0 & 0 & 0 \end{pmatrix} \quad \begin{array}{c|ccc} & R & B & G \\ \hline \overline{R} & 0 & 1 & 0 \\ \overline{B} & 1 & 0 & 0 \\ \overline{G} & 0 & 0 & 0 \end{array} \quad \frac{1}{\sqrt{2}}(R\overline{B}+B\overline{R})$$

$$\lambda_2 = \begin{pmatrix} 0 & -i & 0 \\ i & 0 & 0 \\ 0 & 0 & 0 \end{pmatrix} \quad \begin{array}{c|ccc} & R & B & G \\ \hline \overline{R} & 0 & -i & 0 \\ \overline{B} & i & 0 & 0 \\ \overline{G} & 0 & 0 & 0 \end{array} \quad \frac{-i}{\sqrt{2}}(R\overline{B}-B\overline{R})$$

$$\lambda_3 = \begin{pmatrix} 1 & 0 & 0 \\ 0 & -1 & 0 \\ 0 & 0 & 0 \end{pmatrix} \quad \begin{array}{c|ccc} & R & B & G \\ \hline \overline{R} & 1 & 0 & 0 \\ \overline{B} & 0 & -1 & 0 \\ \overline{G} & 0 & 0 & 0 \end{array} \quad \frac{1}{\sqrt{2}}(R\overline{R}-B\overline{B})$$

$$\lambda_4 = \begin{pmatrix} 0 & 0 & 1 \\ 0 & 0 & 0 \\ 1 & 0 & 0 \end{pmatrix} \quad \begin{array}{c|ccc} & R & B & G \\ \hline \overline{R} & 0 & 0 & 1 \\ \overline{B} & 0 & 0 & 0 \\ \overline{G} & 1 & 0 & 0 \end{array} \quad \frac{1}{\sqrt{2}}(R\overline{G}+G\overline{R})$$

$$\lambda_5 = \begin{pmatrix} 0 & 0 & -i \\ 0 & 0 & 0 \\ -i & 0 & 0 \end{pmatrix} \quad \begin{array}{c|ccc} & R & B & G \\ \hline \overline{R} & 0 & 0 & -i \\ \overline{B} & 0 & 0 & 0 \\ \overline{G} & i & 0 & 0 \end{array} \quad \frac{-i}{\sqrt{2}}(R\overline{G}-G\overline{R})$$

$$\lambda_6 = \begin{pmatrix} 0 & 0 & 0 \\ 0 & 0 & 1 \\ 0 & 1 & 0 \end{pmatrix} \quad \begin{array}{c|ccc} & R & B & G \\ \hline \overline{R} & 0 & 0 & 0 \\ \overline{B} & 0 & 0 & 1 \\ \overline{G} & 0 & 1 & 0 \end{array} \quad \frac{1}{\sqrt{2}}(B\overline{G}+G\overline{B})$$

$$\lambda_7 = \begin{pmatrix} 0 & 0 & 0 \\ 0 & 0 & -i \\ 0 & i & 0 \end{pmatrix} \quad \begin{array}{c|ccc} & R & B & G \\ \hline \overline{R} & 0 & 0 & 0 \\ \overline{B} & 0 & 0 & -i \\ \overline{G} & 0 & i & 0 \end{array} \quad \frac{-i}{\sqrt{2}}(B\overline{G}-G\overline{B})$$

$$\lambda_8 = \frac{1}{\sqrt{3}}\begin{pmatrix} 1 & 0 & 0 \\ 0 & 1 & 0 \\ 0 & 0 & -2 \end{pmatrix} \quad \begin{array}{c|ccc} & R & B & G \\ \hline \overline{R} & 1 & 0 & 0 \\ \overline{B} & 0 & 1 & 0 \\ \overline{G} & 0 & 0 & -2 \end{array} \quad \frac{1}{\sqrt{6}}(R\overline{R}+B\overline{B}-2G\overline{G})$$

Figure 10.26. Colour octet for gluon states correspond to the 8 Gell-Mann matrices.

We have left-handed electron written as e_L and corresponding to it we have left-handed electron neutrino written as ν_{eL}.

We have left-handed muon written as μ_L and corresponding to it we have left-handed muon neutrino written as $\nu_{\mu L}$.

We have left-handed tauon written as τ_L and corresponding to it we have left-handed taon neutrino written as $\nu_{\tau L}$.

We have right-handed electron written as e_R and corresponding to it we do not have a right-handed electron neutrino, i.e. we do not have ν_{eR}.

We have right-handed muon written as μ_R and corresponding to it we do not have a right-handed muon neutrino, i.e. we do not have $\nu_{\mu R}$.

This is summarized in figure 10.27.

Right-handed neutrinos are not experimentally found. In other words, they do not exist in nature. Only left-handed neutrinos ν_{eL}, $\nu_{\mu L}$, $\nu_{\tau L}$ exist in nature.

Only left-handed neutrinos ν_{eL}, $\nu_{\mu L}$, $\nu_{\tau L}$ and right-handed anti-neutrinos $\bar{\nu}_{eR}$, $\bar{\nu}_{\mu R}$, $\bar{\nu}_{\tau R}$ take part in beta decay and weak interactions.

Let us now discuss the same regarding quarks.

Up quark, down quark, charm quark, strange quark, top quark and bottom quark may all be left-handed as well as right-handed, i.e. u_L, d_L, c_L, s_L, t_L, b_L (L for left-handed quarks), u_R, d_R, c_R, s_R, t_R, b_R (R for right-handed quarks). This is also indicated in figure 10.27.

Experimental observation is that only left-handed particles take part in weak interactions.

We consider weak interaction and the gauge transformation corresponding to the weak interaction. The gauge group is identified as $SU(2)_L$ gauge group. Suffix L denotes that left-handed particles take part in weak interactions.

We can group the particles together as follows.

Let $\psi_{\nu_{eL}}$, ψ_{eL} be the wave function corresponding to left-handed electron neutrino and left-handed electron, respectively. They can be grouped together and considered as a doublet $\begin{pmatrix} \psi_{\nu_{eL}} \\ \psi_{eL} \end{pmatrix}$ under $SU(2)_L$.

In contrast the right-handed field, i.e. the wave function corresponding to the right-handed electron has no such partner and so (ψ_{eR}) can be treated as a singlet under $SU(2)_L$. This singlet means that it is invariant and does not take part in the interaction corresponding to $SU(2)_L$ which is weak interaction.

The situation is similar to the flavour isospin (for strong interaction) where we saw that the proton and neutron were treated as different projections of the same

e_L , μ_L , τ_L , e_R , μ_R , τ_R	u_L , c_L , t_L , u_R , c_R , t_R
ν_{eL} , $\nu_{\mu L}$, $\nu_{\tau L}$, No right handed neutrinos	d_L , s_L , b_L , d_R , s_R , b_R

Figure 10.27. There are no right-handed neutrinos.

entity called a nucleon. The isospin of the nucleon is $\frac{1}{2}$ and the proton is the state with $+\frac{1}{2}$ projection and the neutron is the state with $-\frac{1}{2}$ projection.

In exactly the same way we consider the $SU(2)_L$ isospin or the weak isospin.

10.19 Weak isospin or $SU(2)_L$ isospin

Weak isospin or $SU(2)_L$ isospin is completely different from the isospin that was considered in the case of strong interaction.

Under weak isospin we can categorize these fields as follows.

- $\begin{pmatrix} \psi_{\nu_{eL}} \\ \psi_{eL} \end{pmatrix}$ transform as a doublet. This means that its weak isospin has the quantum numbers as follows.

$$I\begin{pmatrix} \nu_{eL} \\ e_L \end{pmatrix} = \frac{1}{2} \text{ and } I_3(\nu_{eL}) = +\frac{1}{2}, \ I_3(e_L) = -\frac{1}{2}$$

- (ψ_{eR}) transforms as a singlet. This means that its weak isospin will have the quantum numbers $I(e_R) = 0, \ I_3(e_R) = 0$.

Similarly, the other fields can be grouped and their weak isospin values attached. Dropping the ψ notation let us rewrite the fields for brevity

$$\begin{pmatrix} \nu_{eL} \\ e_L \end{pmatrix}, \ \begin{pmatrix} \nu_{\mu L} \\ \mu_L \end{pmatrix}, \ \begin{pmatrix} \nu_{\tau L} \\ \tau_L \end{pmatrix} \text{ are doublets} \left(I = \frac{1}{2}, \ I_3 = \pm\frac{1}{2}\right) \text{ and } (e_R),$$

$$(\mu_R), \ (\tau_R) \text{ are singlets } (I = 0, \ I_3 = 0).$$

Similarly, we can group the quarks as follows

$$\begin{pmatrix} u_L \\ d_L \end{pmatrix}, \ \begin{pmatrix} c_L \\ s_L \end{pmatrix}, \ \begin{pmatrix} t_L \\ b_L \end{pmatrix} \text{ are doublets} \left(I = \frac{1}{2}, \ I_3 = \pm\frac{1}{2}\right) \text{ and } (u_R),$$

$$(d_R), \ (c_R), \ (s_R), \ (t_R), \ (b_R) \text{ are singlets } (I = 0, \ I_3 = 0).$$

10.20 Electroweak interaction

Analysis of weak interaction has been done using gauge symmetry similar to QCD for strong interactions and QED for electromagnetic interactions. Analysis shows that we have to consider the electromagnetic interaction and the weak interaction together for consistency. Let us call it electroweak interaction which is a unification of electromagnetic interaction with weak interaction. The gauge group we are to consider is not just $SU(2)$ but a direct product of $SU(2)_L$ and $U(1)_Y$ i.e.

$$SU(2)_L \otimes U(1)_Y \text{ (electroweak interaction)}$$

where $SU(2)$ refers to weak isospin under electroweak interaction, L denotes that only the left-handed particles are going to be affected by transformations under

$SU(2)$ and Y stands for hypercharge. So there is a mix-up between $SU(2)$ isospin and hypercharge in both electromagnetic interaction as well as weak interaction.

10.20.1 A comparison

Electrostatics involves electric field produced by a stationary point charge while magnetostatics involves a magnetic induction field produced by stationary current. These can be treated as separate subjects. But at high energy or for a moving charged particle proceeding with high velocity we cannot neglect the electric effects and magnetic effects. We then have to consider electromagnetic theory or electromagnetic interactions between charged particles and this is described by Maxwell's equations. A changing electric field induces magnetic field and a changing magnetic field induces electric field. So we have to speak of electromagnetic field and we cannot speak of electric field and magnetic field separately.

Unification of weak and electromagnetic interactions is a similar step.

We can think of weak and electromagnetic interactions as unified, i.e. consider them as a unified theory at sufficiently large energies.

10.20.2 Assignment of hypercharge

We can assign hypercharge through a consistent way. Electroweak interaction includes electrodynamics and hence involves a description of photons. So a photon is a mixture of $SU(2)_L$ and $U(1)_Y$ gauge fields. The corresponding charges are related as

$$Q = I_3 + \frac{Y}{2}$$

$$Y = 2(Q - I_3) \tag{10.11}$$

where Q is the charge of electromagnetic interaction, I_3 is the third projection of isospin which is the charge corresponding to $SU(2)$

Consider the doublet formed by left-handed electron neutrino and left-handed electron $\begin{pmatrix} \nu_{eL} \\ e_L \end{pmatrix}$.

$$\text{For } \nu_{eL}\colon \ Q = 0, \ I_3 = +\frac{1}{2}, \ Y = 2(0 - \frac{1}{2}) = -1$$

$$\text{For } e_L\colon \ Q = -1, \ I_3 = -\frac{1}{2}, \ Y = 2(-1 + \frac{1}{2}) = -1$$

The result $Y(\nu_{eL}) = Y(e_L) = -1$ is consistent with the fact that all the $SU(2)$ multiplets have the same hypercharge. To ensure such a result we write ν_{eL} as the first component and e_L as the second component of the doublet $\begin{pmatrix} \nu_{eL} \\ e_L \end{pmatrix}$.

Also, since I_3 changes by one unit when we go from one component to another component in the same multiplet, i.e. move from ν_{eL} to e_L the charge Q also changes by one unit.

For right-handed electron e_R : $Q = -1$, $I_3 = 0$, $Y = 2(-1 - 0) = -2$ i.e. $Y(e_R) = -2$.

Clearly $Y(e_L) \neq Y(e_R)$, i.e. there is a difference in the hypercharge of the left-handed electron and the right-handed electron.

10.20.3 Distinction between e_L and e_R

There are two charges that allow us to distinguish between left-handed electron e_L and right-handed electron e_R. They are

(i) Hypercharge

$Y(e_L) = -1$, $Y(e_R) = -2$ and so $(e_L) \neq Y(e_R)$. So hypercharge differs.

(ii) Isospin

$I_3(e_L) = -\dfrac{1}{2}$, $I_3(e_R) = 0$ and so $I_3(e_L) \neq I_3(e_R)$. So isospin projection differs.

Clearly hypercharge and weak isospin quantum numbers distinguish the left-handed and right-handed electrons from one another. This is the reason why they interact differently under weak interaction, which is a combination of $SU(2)_L$ and $(1)_Y$. They are distinguishable under weak interaction.

The electromagnetic charge or the electric charge of electron is -1 irrespective of whether it is right-handed or left-handed. This is the reason why we cannot distinguish between the left-handed and right-handed electrons under electromagnetic interaction.

Similar is the case for other leptons like muon, tauon and their corresponding neutrinos.

Let us consider quarks now.

Consider the doublet formed by left-handed u quark and left-handed d quark $\begin{pmatrix} u_L \\ d_L \end{pmatrix}$.

$$\text{For left-handed } u \text{ quark } u_L: \quad Q = \frac{2}{3}, \ I_3 = +\frac{1}{2}, \ Y = 2\left(\frac{2}{3} - \frac{1}{2}\right) = \frac{1}{3}$$

$$\text{For left-handed } d \text{ quark } d_L: \quad Q = -\frac{1}{3}, \ I_3 = -\frac{1}{2}, \ Y = 2\left(-\frac{1}{3} + \frac{1}{2}\right) = \frac{1}{3}$$

Consider the doublet formed by left-handed c quark and left-handed s quark $\begin{pmatrix} c_L \\ s_L \end{pmatrix}$.

$$\text{For left-handed } c \text{ quark } c_L: \quad Q = \frac{2}{3}, \ I_3 = +\frac{1}{2}, \ Y = 2\left(\frac{2}{3} - \frac{1}{2}\right) = \frac{1}{3}$$

$$\text{For left-handed } s \text{ quark } s_L: \quad Q = -\frac{1}{3}, \ I_3 = -\frac{1}{2}, \ Y = 2\left(-\frac{1}{3} + \frac{1}{2}\right) = \frac{1}{3}$$

Consider the doublet formed by left-handed t quark and left-handed b quark $\begin{pmatrix} t_L \\ b_L \end{pmatrix}$.

For left-handed t quark t_L: $Q = \frac{2}{3}, I_3 = +\frac{1}{2}, Y = 2(\frac{2}{3} - \frac{1}{2}) = \frac{1}{3}$

For left-handed b quark b_L: $Q = -\frac{1}{3}, I_3 = -\frac{1}{2}, Y = 2(-\frac{1}{3} + \frac{1}{2}) = \frac{1}{3}$

It follows that

$$Y(u_L) = \frac{1}{3}, \quad Y(d_L) = \frac{1}{3}, \quad Y(c_L) = \frac{1}{3}, \quad Y(s_L) = \frac{1}{3}, \quad Y(t_L) = \frac{1}{3}, \quad Y(b_L) = \frac{1}{3}.$$

So all left-handed quarks $(u_L, d_L, c_L, s_L, t_L, b_L)$ have $Y = \frac{1}{3}$.

For right-handed u quark u_R : $Q = \frac{2}{3}, I_3 = 0, Y = 2(\frac{2}{3} - 0) = \frac{4}{3}$

Clearly $Y(u_L) = \frac{1}{3}, \quad Y(u_R) = \frac{4}{3}$ i.e. $Y(u_L) \neq Y(u_R)$

For right-handed d quark d_R : $Q = -\frac{1}{3}, I_3 = 0, Y = 2(-\frac{1}{3} - 0) = -\frac{2}{3}$

Clearly $Y(d_L) = \frac{1}{3}, \quad Y(d_R) = -\frac{2}{3}$ i.e. $Y(d_L) \neq Y(d_R)$

For right-handed c quark c_R : $Q = \frac{2}{3}, I_3 = 0, Y = 2(\frac{2}{3} - 0) = \frac{4}{3}$

Clearly $Y(c_L) = \frac{1}{3}, Y(c_R) = \frac{4}{3}$ i.e. $Y(c_L) \neq Y(c_R)$

For right-handed s quark s_R : $Q = -\frac{1}{3}, I_3 = 0, Y = 2(-\frac{1}{3} - 0) = -\frac{2}{3}$

Clearly $Y(s_L) = \frac{1}{3}, Y(s_R) = -\frac{2}{3}$ i.e. $Y(c_L) \neq Y(c_R)$

For right-handed t quark t_R : $Q = \frac{2}{3}, I_3 = 0, Y = 2(\frac{2}{3} - 0) = \frac{4}{3}$

Clearly $Y(t_L) = \frac{1}{3}, Y(t_R) = \frac{4}{3}$ i.e. $Y(t_L) \neq Y(t_R)$

For right-handed b quark b_R : $Q = -\frac{1}{3}, I_3 = 0, Y = 2(-\frac{1}{3} - 0) = -\frac{2}{3}$

Clearly $Y(b_L) = \frac{1}{3}, Y(b_R) = -\frac{2}{3}$ i.e. $Y(b_L) \neq Y(b_R)$

Thus

$$Y \text{ (all left-handed quarks } u_L, \ d_L, \ c_L, \ s_L, \ t_L, \ b_L) = \frac{1}{3}$$

$$Y(u_R, \ c_R, \ t_R) = \frac{4}{3}, \quad Y(d_R, \ s_R, \ b_R) = -\frac{2}{3}.$$

10.21 Exercises

Exercise 10.1 *Why is baryon number of quarks $\frac{1}{3}$?*

[Ans] Baryons are characterized by a quantum number called baryon quantum number which is integral. Since three quarks form a baryon, the baryon quantum number of quarks has to be fractional and equal to $\frac{1}{3}$.

Exercise 10.2 *Deuteron is a stable nucleus consisting of a neutron which decays. How do you interpret this?*

Ans. The deuteron nucleus is made up of a neutron and a proton and is stable. It is built up of neutron and proton. The deuteron nucleus has binding energy 2. 225 *MeV* that prevents the neutron from decaying. A free neutron can decay.

Exercise 10.3 *Electrons, neutrinos and photons are observed but quarks are not visible. Why?*

Ans A very special property of quarks is that they possess colour quantum number.

Since a detectable entity has to be colourless, quarks are not visible and are confined within hadrons. This phenomenon is called quark confinement.

In other words, confinement of quarks can be explained through colour quantum number of quarks.

Electrons, neutrinos and photons are observed since they do not possess colour quantum number.

Exercise 10.4 *Check Gell-Mann–Nishijima relation for members of the meson octet and baryon octet and baryon decuplet.*

Ans The Gell-Mann–Nishijima relation for members of the meson octet is shown in figure 10.28.

The Gell-Mann–Nishijima relation for members of the baryon octet is shown in figure 10.29.

The Gell-Mann–Nishijima relation for members of the baryon decuplet is shown in figure 10.30.

Meson octet : $B=0$, $C=0$, $B'=0$, $T=0$, $Y = B+C+S+T+B' = S$

Members	I	I_3	$S=Y$	$Q = I_3 + \dfrac{Y}{2} = I_3 + \dfrac{S}{2}$
K^+	$\frac{1}{2}$	$\frac{1}{2}$	1	1
K^0	$\frac{1}{2}$	$-\frac{1}{2}$	1	0
π^+	1	1	0	1
π^0	1	0	0	0
π^-	1	-1	0	-1
$\bar{K}^0$	$\frac{1}{2}$	$\frac{1}{2}$	-1	0
K^-	$\frac{1}{2}$	$-\frac{1}{2}$	-1	-1
η	0	0	0	0

Figure 10.28. Gell-Mann–Nishijima relation for members of the meson octet.

Baryon octet : $C=0$, $B=0$, $T'=0$, $Y=B+C+S+T+B'=B+S$

Members	I	I_3	B	S	$Y=B+S$	$Q=I_3+\dfrac{Y}{2}$
Ξ^-	$\dfrac{1}{2}$	$-\dfrac{1}{2}$	1	-2	-1	-1
Ξ^0	$\dfrac{1}{2}$	$\dfrac{1}{2}$	1	-2	-1	0
Σ^-	1	-1	1	-1	0	-1
Σ^0	1	0	1	-1	0	0
Σ^+	1	1	1	-1	0	$+1$
Λ	0	0	1	-1	0	0
n	$\dfrac{1}{2}$	$-\dfrac{1}{2}$	1	0	1	0
p	$\dfrac{1}{2}$	$\dfrac{1}{2}$	1	0	1	$+1$

Figure 10.29. Gell-Mann–Nishijima relation for members of the baryon octet.

Baryon decuplet $C=0$, $B=0$, $T'=0$, $Y=B+C+S+T+B'=B+S$

Members	I	I_3	B	S	$Y=B+S$	$Q=I_3+\dfrac{Y}{2}$
Δ^{++}	$\dfrac{3}{2}$	$\dfrac{3}{2}$	1	0	1	$+2$
Δ^+	$\dfrac{3}{2}$	$\dfrac{1}{2}$	1	0	1	$+1$
Δ^0	$\dfrac{3}{2}$	$-\dfrac{1}{2}$	1	0	1	0
Δ^-	$\dfrac{3}{2}$	$-\dfrac{3}{2}$	1	0	1	-1
Σ^{*+}	1	1	1	-1	0	$+1$
Σ^{*0}	1	0	1	-1	0	0
Σ^{*-}	1	-1	1	-1	0	-1
Ξ^{*0}	$\dfrac{1}{2}$	$\dfrac{1}{2}$	1	-2	-1	0
Ξ^{*-}	$\dfrac{1}{2}$	$-\dfrac{1}{2}$	1	-2	-1	-1
Ω^-	0	0	1	-3	-2	-1

Figure 10.30. Gell-Mann–Nishijima relation for members of the baryon decuplet.

To understand quark structure, i.e. the quark level picture of proton, neutron and other hadrons, we rely on scattering experiments.

Rutherford's experiment says that there are substructures in the nucleus like protons and neutrons.

Exercise 10.5 *Why can baryons not be constructed with two quarks?*
[Ans] Quarks are spin-half particles. With two spin-half particles we cannot make a baryon as a baryon is a spin-half particle. We require three quarks to build a baryon. And hence quarks have baryon number $B = \frac{1}{3}$.

Exercise 10.6 *If quark structure of D^+ meson is $c\bar{d}$ find the quark structure of D^-.*
[Ans] Quark structure of meson $D^+ = c\bar{d}$.
Replacing quarks by corresponding anti-quarks and anti-quarks by quarks we get the quark structure of $D^- = \bar{c}d$.

Exercise 10.7 *Weight diagrams for quarks u, d, s and $\bar{u}$, $\bar{d}$, $\bar{s}$ are*

(a) inverted triangle, inverted triangle   *(b) triangle, inverted triangle*
(c) triangle, triangle   *(d) inverted triangle, triangle.*

Exercise 10.8 *Eightfold way in a weight diagram represents which of he following?*

(a) triangle   *(b) square*   *(c) hexagon*   *(d) octagon.*

Answers to multiple choice questions

$10.7d$, $10.8c$.

10.22 Question bank

Q10.1 What do we mean by a hadron? Classify hadrons.
Q10.2 Write down the quantum numbers associated with quarks.
Q10.3 Write down the quantum numbers associated with anti-quarks.
Q10.4 Draw a weight diagram for quarks u, d, s.
Q10.5 Draw a weight diagram for anti-quarks $\bar{u}$, $\bar{d}$, $\bar{s}$.
Q10.6 Show how pseudoscalar mesons can be represented by a nonet.
Q10.7 Show how pseudoscalar mesons can be represented by an octet plus a singlet.
Q10.8 Show how vector mesons can be represented by a nonet.
Q10.9 Show how baryons can be represented by an octet.
Q10.10 Show how resonance baryons can be represented by a decuplet.

Q10.11 What do we mean by resonance particle?

Q10.12 What do we mean by baryon decuplet? How could it predict Ω^-?

Q10.13 Write down the quark structure of various hadrons.

Q10.14 How could study of Δ^{++} predict colour quantum number?

Q10.15 How can anti-particles be represented in $SU(2)$?

Q10.16 What is asymptotic freedom?

Q10.17 What is weak isospin?

Q10.18 Show that the colour octet for gluon states corresponds to the 8 Gell-Mann matrices.

Further reading

[1] Perkins H 2000 *Introduction to High Energy Physics* (Cambridge: Cambridge University Press)

[2] Griffiths D 2008 *Introduction to Elementary Particles* (New York: Wiley-VCH)

[3] Tayal D C 2009 *Nuclear Physics* (Mumbai: Himalaya Publishing House)

[4] Satya P 2005 *Nuclear Physics & Particle Physics* (New Delhi: Sultan Chand & Sons)

[5] Guha J 2019 *Quantum Mechanics: Theory, Problems & Solutions* 3rd edn (Kolkata: Books and Allied (P) Ltd)

[6] Yung K L 2002 *Problems and Solutions on Atomic, Nuclear and Particle Physics* (Singapore: World Scientific)

IOP Publishing

Nuclear and Particle Physics with Cosmology, Volume 2
Particle physics and cosmology
Jyotirmoy Guha

Chapter 11

Relativistic wave equation for particles

In this chapter we underscore the need for relativistic quantum mechanics in dealing with particles. Various notations used in relativity are defined and explained. We derive the Klein–Gordon equation for a free particle as well as for a particle in an electromagnetic field. We obtain a solution and interpret it. We derive the DIrac equation for a free particle as well as for a particle in an electromagnetic field. The plane wave solution to the Dirac equation is obtained. We show how the spin of an electron, the idea of the antiparticle and the phenomena of pair production and pair annihilation follow from the Dirac equation.

11.1 Relativistic quantum mechanics

In particle physics we deal with tiny particles moving with lots of energy and with high velocity $v \sim c$. We need to use both special relativity and quantum mechanics. We can combine them as was done by Dirac to obtain fundamental equations that describe high energy particles.

When we consider scattering of high energy electrons or protons or other particles we need to consider relativistic effects. It is not enough to consider the Schrödinger picture and then make corrections to the Schrödinger equation and other mathematical formalism by considering some aspects of relativistic effect. Rather, it is necessary to find out a formalism suitable to the relativistic systems.

Einstein has given two theories of relativity, one is the special theory of relativity in 1905 and its generalization, namely the general theory of relativity in 1915.

The special theory of relativity deals with inertial frame of reference (that moves with constant velocity). It is also called the Lorentz frame of reference because this frame is invariant under Lorentz transformation.

The general theory of relativity deals with non-inertial frame of reference (that is, moving with acceleration).

The special theory of relativity deals with plane or flat geometry—there being no curvature in space-time. We focus our attention on the special theory of relativity and

doi:10.1088/978-0-7503-5032-7ch11   11-1

try to couple it with quantum mechanics. But the general theory of relativity deals with curved geometry since due to heavy mass space-time is curved. Again, the laws of quantum mechanics are not designed for such curved space-time. For this reason the general theory of relativity cannot be coupled or combined with quantum mechanics.

Consider the particle for which wave properties are dominant. It is described by quantum mechanics. Suppose the particle is moving with high velocity $v \sim c$ and so the laws of special theory of relativity are applicable. So for such a particle we apply relativistic quantum mechanics. Such particles have very large energy $\sim GeV$ or TeV.

One example of this is particles like protons that are accelerated in the LHC and are described by relativistic quantum mechanics.

11.2 Notations used

In the special theory of relativity we consider a 4D orthogonal space called Minkowski space with 3 spatial axes X-axis, Y-axis and Z-axis and one time axis which is taken as the ct-axis (ct being length) so that space and time are treated on an equal footing. This space is, so to say, an extension of Euclidean space with time incorporated.

In this Minkowski space any physical quantity will have 4 components referred to as 4-vectors. Again, 4-vectors can be expressed in two ways:

Contravariant 4-vector A^μ, $\mu = 0$, $i = 0, 1, 2, 3$, i.e.

$$A^\mu = (A^0, A^i) = (A^0, A^1, A^2, A^3) = \left(A^0, A_x, A_y, A_z\right) = \left(A^0, \vec{A}\right) \qquad (11.1)$$

Covariant 4-vector A_μ, $\mu = 0$, $i = 0, 1, 2, 3$, i.e.

$$A_\mu = \left(A_0, A_i\right) = (A_0, A_1, A_2, A_3) = \left(A_0, -A_x, -A_y, -A_z\right) = \left(A_0, -\vec{A}\right) \quad (11.2)$$

Clearly

$$A^0 = A_0, \quad A^i = -A_i = \vec{A} = \left(A_x, A_y, A_z\right) \qquad (11.3)$$

Example
Position 4-vector is

$$x^\mu = (x^0, x^i) = (x^0, x^1, x^2, x^3) = (ct, \vec{r}) = (ct, x, y, z) \qquad (11.4)$$

$$x_\mu = (x_0, x_i) = (x_0, x_1, x_2, x_3) = (ct, -\vec{r}) = \left(ct, -x, -y, -z\right) \qquad (11.5)$$

where

$$x^0 = x_0 = ct, \quad x^i = -x_i = \vec{r} = (x, y, z) \; i.\, e.$$

$$x^1 = -x_1 = x, \quad x^2 = -x_2 = y, \quad x^3 = -x_3 = z \qquad (11.6)$$

The covariant and contravariant vectors are related as follows

$$x_\mu = \eta_{\mu\nu} x^\nu \qquad (11.7)$$

$$x^\mu = \eta^{\mu\nu} x_\nu \qquad (11.8)$$

where η is the metric tensor given by the following 4×4 transformation matrix

$$\eta = \begin{pmatrix} 1 & 0 & 0 & 0 \\ 0 & -1 & 0 & 0 \\ 0 & 0 & -1 & 0 \\ 0 & 0 & 0 & -1 \end{pmatrix} = \mathrm{diag}(1 - 1 - 1 - 1) \tag{11.9}$$

that transforms between covariant and contravariant forms.

Clearly

$$\eta^{00} = -\eta^{11} = -\eta^{22} = -\eta^{33} = 1. \text{ All other } \eta^{\mu\nu} = 0 \tag{11.10}$$

$$\eta_{00} = -\eta_{11} = -\eta_{22} = -\eta_{33} = 1. \text{ All other } \eta_{\mu\nu} = 0 \tag{11.11}$$

11.3 Differential operators

11.3.1 First-order differential operators

$$\text{In 3D gradient is } \vec{\nabla} = \left(\frac{\partial}{\partial x}, \frac{\partial}{\partial y}, \frac{\partial}{\partial z} \right)$$

In 4D we have the following contravariant derivative

$$\frac{\partial}{\partial x_\mu} \equiv \partial^\mu = (\partial^0, \partial^i) = \left(\frac{\partial}{\partial x_0}, \frac{\partial}{\partial x_i} \right) = \left(\frac{\partial}{\partial ct}, \frac{\partial}{\partial x_i} \right) = \left(\frac{1}{c}\frac{\partial}{\partial t}, \frac{\partial}{\partial x_i} \right)$$

$$= \left(\frac{1}{c}\frac{\partial}{\partial t}, \frac{\partial}{\partial x_1}, \frac{\partial}{\partial x_2}, \frac{\partial}{\partial x_3} \right) = \left(\frac{1}{c}\frac{\partial}{\partial t}, \frac{\partial}{\partial(-x)}, \frac{\partial}{\partial(-y)}, \frac{\partial}{\partial(-z)} \right) \text{ (using equation (11.6))}$$

$$= \left(\frac{1}{c}\frac{\partial}{\partial t}, -\frac{\partial}{\partial x}, -\frac{\partial}{\partial y}, -\frac{\partial}{\partial z} \right) = \left(\frac{1}{c}\frac{\partial}{\partial t}, -\vec{\nabla} \right) \tag{11.12}$$

In 4D we have the following covariant derivative

$$\frac{\partial}{\partial x^\mu} \equiv \partial_\mu = (\partial_0, \partial_i) = \left(\frac{\partial}{\partial x^0}, \frac{\partial}{\partial x^i} \right) = \left(\frac{\partial}{\partial ct}, \frac{\partial}{\partial x^i} \right) = \left(\frac{1}{c}\frac{\partial}{\partial t}, \frac{\partial}{\partial x^i} \right)$$

$$= \left(\frac{1}{c}\frac{\partial}{\partial t}, \frac{\partial}{\partial x^1}, \frac{\partial}{\partial x^2}, \frac{\partial}{\partial x^3} \right) = \left(\frac{1}{c}\frac{\partial}{\partial t}, \frac{\partial}{\partial x}, \frac{\partial}{\partial y}, \frac{\partial}{\partial z} \right) \text{ (using equation (11.6))}$$

$$= \left(\frac{1}{c}\frac{\partial}{\partial t}, \vec{\nabla} \right) \tag{11.13}$$

Also

$$\partial^\mu = \eta^{\mu\nu}\partial_\nu = \begin{pmatrix} 1 & 0 & 0 & 0 \\ 0 & -1 & 0 & 0 \\ 0 & 0 & -1 & 0 \\ 0 & 0 & 0 & -1 \end{pmatrix} \begin{pmatrix} \frac{1}{c}\partial/\partial t \\ \partial/\partial x \\ \partial/\partial y \\ \partial/\partial z \end{pmatrix} = \begin{pmatrix} \frac{1}{c}\partial/\partial t \\ -\partial/\partial x \\ -\partial/\partial y \\ -\partial/\partial z \end{pmatrix} \tag{11.14}$$

11.3.2 Second-order differential operator

$$\partial^2 = \partial^\mu \partial_\mu = \partial^0 \partial_0 + \partial^i \partial_i$$

$$= \frac{1}{c}\frac{\partial}{\partial t}\frac{1}{c}\frac{\partial}{\partial t} + (-\vec{\nabla} \cdot \vec{\nabla}) \text{ (using equations (11.12) and (11.13))}$$

$$= \frac{1}{c^2}\frac{\partial^2}{\partial t^2} - \nabla^2 = \square \tag{11.15}$$

This operator $\square$ is called the d'Alembertian operator.

11.4 Scalar product or dot product

Scalar product or dot product of 2 four vectors A_μ, B_μ is

$$A_\mu B^\mu = \eta_{\mu\nu} A^\nu \eta^{\mu\nu} B_\nu = \eta_{\mu\nu}\eta^{\mu\nu} A^\nu B_\nu = A^\nu B_\nu \tag{11.16}$$

Also

$$A_\mu B^\mu = A_0 B^0 + A_i B^i \tag{11.17}$$

If $A = B$ then

$$A_\mu A^\mu = A_0 A^0 + A_i A^i \tag{11.18}$$

Using

$$A_0 = A^0, \quad A^i = -A_i = \vec{A} \tag{11.19}$$

we have from equations (11.18) and (11.19)

$$A_\mu A^\mu = (A^0)^2 + (-\vec{A}) \cdot (\vec{A}) = (A^0)^2 - (\vec{A})^2 \tag{11.20}$$

Also

$$A^\mu A_\mu = A^0 A_0 + A^i A_i = (A^0)^2 + (\vec{A}) \cdot (-\vec{A}) = (A^0)^2 - (\vec{A})^2 \tag{11.21}$$

Combining equations (11.20) and (11.21) we have

$$A_\mu A^\mu = A^\mu A_\mu = (A^0)^2 - (\vec{A})^2 \tag{11.22}$$

11.4.1 Displacement four-vector

Analogous to the relation viz.

$$\vec{x} \cdot \vec{x} = x^2 + y^2 + z^2 = \text{invariant under rotation in 3 space}$$

we have

$$x^\mu x_\mu = x_\mu x^\mu = \text{invariant quantity in 4 space} \tag{11.23}$$

Again summing over repeated indices we have

$$x^\mu x_\mu = (ct, x, y, z) \cdot (ct, -x, -y, -z)$$

$$= c^2 t^2 - x^2 - y^2 - z^2. \tag{11.24}$$

11.4.2 Momentum four-vector

$$p^\mu = \left(\frac{E}{c}, \vec{p}\right) = \left(p^0, p^i\right) = \left(p^0, p^1, p^2, p^3\right) = \left(\frac{E}{c}, \vec{p}\right) = \left(\frac{E}{c}, p_x, p_y, p_z\right) \quad (11.25)$$

$$\begin{aligned} p_\mu &= \left(\frac{E}{c}, -\vec{p}\right) = \left(p_0, p_i\right) = (p_0, p_1, p_2, p_3) \\ &= \left(\frac{E}{c}, -\vec{p}\right) = \left(\frac{E}{c}, -p_x, -p_y, -p_z\right) \end{aligned} \quad (11.26)$$

$$p_\mu p^\mu = \left(\frac{E}{c}, -\vec{p}\right)\left(\frac{E}{c}, \vec{p}\right) = \left(\frac{E}{c}\right)^2 + \left(-\vec{p}\right)\cdot\vec{p} = \frac{E^2}{c^2} - p^2 = \frac{E^2 - p^2 c^2}{c^2} = \frac{m^2 c^4}{c^2}$$

$$p_\mu p^\mu = m^2 c^2. \quad (11.27)$$

11.5 Metric

Conversion from contravariant to covariant or covariant to contravariant vector is possible by introducing the metric $\eta^{\mu\nu}$ or $\eta_{\mu\nu}$. We can write

$$x^\mu = \eta^{\mu\nu}x_\nu = \eta^{\mu 0}x_0 + \eta^{\mu 1}x_1 + \eta^{\mu 2}x_2 + \eta^{\mu 3}x_3 \quad (11.28)$$

$$x^0 = x_0 \Rightarrow \eta^{00} = 1, \ \eta^{01} = \eta^{02} = \eta^{03} = 0$$

$$x^1 = -x_1 \Rightarrow \eta^{11} = -1, \ \eta^{10} = \eta^{12} = \eta^{13} = 0$$

$$x^2 = -x_2 \Rightarrow \eta^{22} = -1, \ \eta^{20} = \eta^{21} = \eta^{23} = 0$$

$$x^3 = -x_3 \Rightarrow \eta^{33} = -1, \ \eta^{30} = \eta^{31} = \eta^{32} = 0$$

$$x_\mu = \eta_{\mu\nu}x^\nu = \eta_{\mu 0}x^0 + \eta_{\mu 1}x^1 + \eta_{\mu 2}x^2 + \eta_{\mu 3}x^3 \quad (11.29)$$

$$x_0 = x^0 \Rightarrow \eta_{00} = 1, \ \ \eta_{01} = \eta_{02} = \eta_{03} = 0$$

$$x_1 = -x^1 \Rightarrow \eta_{11} = -1, \ \ \eta_{10} = \eta_{12} = \eta_{13} = 0$$

$$x_2 = -x^2 \Rightarrow \eta_{22} = -1, \ \ \eta_{20} = \eta_{21} = \eta_{23} = 0$$

$$x_3 = -x^3 \Rightarrow \eta_{33} = -1, \ \ \eta_{30} = \eta_{31} = \eta_{32} = 0$$

$$\eta_{\mu\nu} = \begin{pmatrix} \eta_{00} & \eta_{01} & \eta_{02} & \eta_{03} \\ \eta_{10} & \eta_{11} & \eta_{12} & \eta_{13} \\ \eta_{20} & \eta_{21} & \eta_{22} & \eta_{23} \\ \eta_{30} & \eta_{31} & \eta_{32} & \eta_{33} \end{pmatrix} = \begin{pmatrix} \eta^{00} & \eta^{01} & \eta^{02} & \eta^{03} \\ \eta^{10} & \eta^{11} & \eta^{12} & \eta^{13} \\ \eta^{20} & \eta^{21} & \eta^{22} & \eta^{23} \\ \eta^{30} & \eta^{31} & \eta^{32} & \eta^{33} \end{pmatrix} = \begin{pmatrix} 1 & 0 & 0 & 0 \\ 0 & -1 & 0 & 0 \\ 0 & 0 & -1 & 0 \\ 0 & 0 & 0 & -1 \end{pmatrix} = \eta^{\mu\nu}$$

$$=\text{diag}(1 - 1 - 1 - 1) \tag{11.30}$$

$$\eta^{00} = \eta_{00} = 1, \quad \eta^{11} = \eta_{11} = -1, \quad \eta^{22} = \eta_{22} = -1, \quad \eta^{33} = \eta_{33} = -1 \tag{11.31}$$

$$\eta^{\mu\nu} = \eta_{\mu\nu} = 0 \text{ for } \mu \neq \nu. \tag{11.32}$$

11.6 Coordinate transformation

Consider rotation about the Z-axis in 2D so that transformation of axes occurs as follows

$$(\hat{x}, \hat{y}) \rightarrow (\hat{x}', \hat{y}') \tag{11.33}$$

as shown in figure 11.1. The position vector is given by

$$\vec{r}' = x'\hat{x}' + y'\hat{y}' \tag{11.34}$$

in primed frame and

$$\vec{r} = x\hat{x} + y\hat{y} \tag{11.35}$$

in unprimed frame. And

$$\vec{r}' = \vec{r} \tag{11.36}$$

$$x' = x \cos\theta + y \sin\theta \text{ (equation (9.208))} \tag{11.37}$$

$$y' = -x \sin\theta + y \cos\theta \text{ (equation (9.210))} \tag{11.38}$$

In matrix form we have

$$\begin{pmatrix} x' \\ y' \end{pmatrix} = \begin{pmatrix} \cos\theta & \sin\theta \\ -\sin\theta & \cos\theta \end{pmatrix} \begin{pmatrix} x \\ y \end{pmatrix} = \begin{pmatrix} a_1^1 & a_2^1 \\ a_1^2 & a_2^2 \end{pmatrix} \begin{pmatrix} x \\ y \end{pmatrix} \tag{11.39}$$

where

$$a_1^1 = a_2^2 = \cos\theta, \quad a_2^1 = -a_1^2 = \sin\theta$$

and so we write equations (11.37) and (11.38) as

$$x' = a_1^1 x + a_2^1 y \tag{11.40}$$

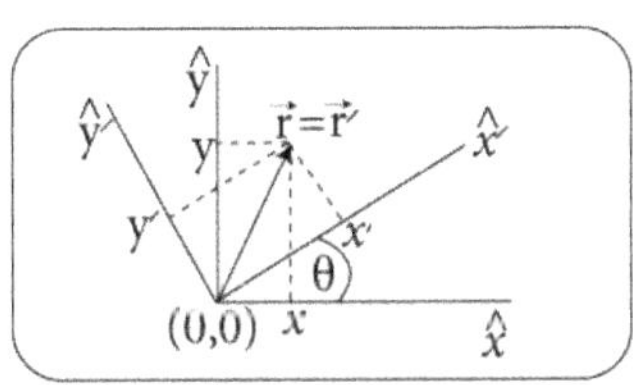

Figure 11.1. Rotation of axes $(\hat{x}, \hat{y}) \rightarrow (\hat{x}', \hat{y}')$ in 2D about the Z-axis.

$$y' = a_1^2 x + a_2^2 y \tag{11.41}$$

Defining

$$x = x^1, \quad y = x^2; \; x' = x'^1, \quad y' = x'^2 \tag{11.42}$$

we rewite equations (11.40) and (11.41) as

$$x'^1 = a_1^1 x^1 + a_2^1 x^2 = a_\beta^1 x^\beta$$

$$x'^2 = a_1^2 x^1 + a_2^2 x^2 = a_\beta^2 x^\beta$$

We can thus inter-relate primed and unprimed quantities as follows:

$$x'^\alpha = a_\beta^\alpha x^\beta \tag{11.43}$$

where $\alpha, \; \beta = 1, 2$

Generalizing to 4D we can write

$$x'^\mu = a_\nu^\mu x^\nu \; \mu, \nu \to 0, 1, 2, 3 \; e.\, g.$$

$$x'^0 = a_0^0 x^0 + a_1^0 x^1 + a_2^0 x^2 + a_3^0 x^3 \tag{11.44}$$

and also

$$x'_\mu = a_\mu^\lambda x_\lambda \tag{11.45}$$

Any 4-vector transforms in the same way. For instance

$$A'^\mu = a_\nu^\mu A^\nu \tag{11.46}$$

$$A'_\mu = a_\mu^\lambda A_\lambda \tag{11.47}$$

We note that rotation in 2D preserves length since

$$\vec{r}'^2 = x'^2 + y'^2 \tag{11.48}$$

Using equations (11.37) and (11.38) we get

$$\vec{r}'^2 = (x \cos\theta + y \sin\theta)^2 + (-x \sin\theta + y \cos\theta)^2 = x^2 + y^2 = \vec{r}^2 \tag{11.49}$$

In 4D

$$x'^\mu x'_\mu = a_\nu^\mu x^\nu a_\mu^\lambda x_\lambda = a_\nu^\mu a_\mu^\lambda x^\nu x_\lambda = \delta_\nu^\lambda x^\nu x_\lambda = x^\nu x_\nu \tag{11.50}$$

where δ_ν^λ is Kroneckar delta symbol given by

$$\delta_\nu^\lambda = 1 \; \text{for} \; \nu = \lambda \; \text{and} \; \delta_\nu^\lambda = 0 \; \text{for} \; \nu \neq \lambda$$

Again from equation (11.50)

$$a_\nu^\mu a_\mu^\lambda = \delta_\nu^\lambda \tag{11.51}$$

For $\nu = \lambda$ we write $a_\nu^\mu a_\mu^\nu = \delta_\nu^\nu = 1$ i.e.

$$a_\nu^\mu a_\mu^\nu = 1 \tag{11.52}$$

$$\frac{1}{a_\nu^\mu} = a_\mu^\nu \quad \text{and} \quad \frac{1}{a_\mu^\nu} = a_\nu^\mu \tag{11.53}$$

- Transformation of $\frac{\partial}{\partial x^\mu}$

$$\text{As } x^\mu \to x'^\mu = a_\nu^\mu x^\nu$$

$$\frac{\partial}{\partial x^\mu} \to \frac{\partial}{\partial x'^\mu} = \frac{\partial}{\partial(a_\nu^\mu x^\nu)} = \frac{1}{a_\nu^\mu} \frac{\partial}{\partial x^\nu} = a_\mu^\nu \frac{\partial}{\partial x^\nu} \quad \text{(using equation (11.53))} \tag{11.54}$$

Similarly as $x_\mu \to x'_\mu = a_\mu^\nu x_\nu$

$$\frac{\partial}{\partial x_\mu} \to \frac{\partial}{\partial x'_\mu} = \frac{\partial}{\partial\left(a_\mu^\nu x_\nu\right)} = \frac{1}{a_\mu^\nu} \frac{\partial}{\partial x_\nu} = a_\nu^\mu \frac{\partial}{\partial x_\nu} \quad \text{(using equation (11.53))}. \tag{11.55}$$

11.7 Non-relativistic Schrödinger equation for a free particle

A free particle means no interaction and so no potential energy, i.e. $V = 0$. The particle has only kinetic energy.

The classical non-relativistic energy–momentum relation for the free particle moving with low speed is

$$\frac{\vec{p}^2}{2m} = E \tag{11.56}$$

where m is particle mass, $\vec{p}$ is particle momentum, E is particle energy.

We make operator replacement in equation (11.56) as is usual in quantum mechanics viz.

$$\vec{p} \to -i\hbar\,\vec{\nabla}\,, \ E \to i\hbar\frac{\partial}{\partial t}$$

$$\frac{\left(-i\hbar\,\vec{\nabla}\,\right)\left(-i\hbar\,\vec{\nabla}\,\right)}{2m} = i\hbar\frac{\partial}{\partial t}$$

$$-\frac{\hbar^2}{2m}\,\nabla^2 = i\hbar\frac{\partial}{\partial t}$$

This operator equation will have meaning if it operates on a wave function.

Let $\psi = \psi(\vec{r}, t)$ be the wave function describing the state of a non-relativistic free particle. The corresponding Schrödinger equation describing motion of the non-relativistic free particle is

$$-\frac{\hbar^2}{2m}\,\nabla^2\,\psi = i\hbar\frac{\partial}{\partial t}\psi \tag{11.57}$$

We note that in this non-relativistic Schrödinger equation time derivative is of first order and the space derivative is of second order. Clearly in this equation space and time are not on equal footing as seen in the special theory of relativity. Clearly this equation does not fit into the fabric of the special theory of relativity and is a non-relativistic equation.

11.8 Relativistic wave equation for a free particle

In the special theory of relativity, space and time are treated on an equal footing. Then the equation involving space and time should have equal order of differentiation. Two cases arise.

✓ Both space and time derivatives are of second order. Such an arrangement gives rise to an equation which is called the Klein–Gordon equation.

✓ Both space and time derivatives are of first order. Such an arrangement gives rise to an equation which is called the Dirac equation.

11.9 Klein–Gordon equation for a free particle

The relativistic energy–momentum relation for a free particle moving with relativistic speed is given by

$$E^2 = p^2 c^2 + m^2 c^4 \tag{11.58}$$

where m is rest mass of particle, p its momentum, E is energy, c is speed of light in vacuum.

We make operator replacement in equation (11.58) as follows

$$\vec{p} \to -i\hbar\,\vec{\nabla}\,, \quad p^2 = (-i\hbar\,\vec{\nabla}\,).\,(-i\hbar\,\vec{\nabla}\,) = -\hbar^2\nabla^2$$

$$E \to i\hbar\frac{\partial}{\partial t}, \quad E^2 = \left(i\hbar\frac{\partial}{\partial t}\right)\left(i\hbar\frac{\partial}{\partial t}\right) = -\hbar^2\frac{\partial^2}{\partial t^2}$$

Let the state of the relativistic free particle be described by the wave function $\psi = \psi(\vec{r},\,t)$. Then the equation describing its motion will be obtained from equation (11.58) by operator replacement as

$$-\hbar^2\frac{\partial^2}{\partial t^2}\psi = -\hbar^2\,\nabla^2\,\psi\,c^2 + m^2 c^4\psi$$

$$\hbar^2 c^2 \nabla^2\psi - \hbar^2\frac{\partial^2}{\partial t^2}\psi = m^2 c^4\psi \tag{11.59}$$

$$\nabla^2\psi - \frac{1}{c^2}\frac{\partial^2}{\partial t^2}\psi = \frac{m^2 c^2}{\hbar^2}\psi$$

$$\left(\nabla^2 - \frac{1}{c^2}\frac{\partial^2}{\partial t^2}\right)\psi = \frac{m^2 c^2}{\hbar^2}\psi \qquad (11.60)$$

This is Klein–Gordon equation

With D'Alembertian operator $\Box = \frac{1}{c^2}\frac{\partial^2}{\partial t^2} - \nabla^2$ (equation (11.15)) we can rewrite Klein–Gordon equation (11.60) as

$$-\Box\psi = \frac{m^2 c^2}{\hbar^2}\psi$$

$$\left(\Box + \frac{m^2 c^2}{\hbar^2}\right)\psi = 0 \qquad (11.61)$$

Using equation (11.15) viz. $\partial^\mu \partial_\mu = \frac{1}{c^2}\frac{\partial^2}{\partial t^2} - \nabla^2 = \Box$ in equation (11.60) we have the Klein–Gordon equation

$$\left(\partial^\mu \partial_\mu + \frac{m^2 c^2}{\hbar^2}\right)\psi = 0 \qquad (11.62)$$

In natural units $\hbar = c = 1$ and so the Klein–Gordon equation (11.60) has the form

$$\left(\nabla^2 - \frac{\partial^2}{\partial t^2}\right)\psi = m^2\psi \qquad (11.63)$$

$$(\Box + m^2)\psi = 0 \qquad (11.64)$$

And from equation (11.62) we have the Klein–Gordon equation in natural units

$$(\partial^\mu \partial_\mu + m^2)\psi = 0 \qquad (11.65)$$

11.10 Klein–Gordon equation for massless particle (photon)

For a massless particle (like a photon) Klein–Gordon equations (11.63) and (11.64) reduce to

$$\left(\nabla^2 - \frac{1}{c^2}\frac{\partial^2}{\partial t^2}\right)\psi = 0 \qquad (11.66)$$

$$\Box\psi = 0 \qquad (11.67)$$

From equation equation (11.66) we have

$$\nabla^2\psi = \frac{1}{c^2}\frac{\partial^2}{\partial t^2}\psi \qquad (11.68)$$

which is the wave equation of Maxwell. So Maxwell's wave equation follows from the Klein–Gordon equation.

- We have derived the non-relativistic limit of the Klein–Gordon equation in *exercise 11.4* which is obviously the Schrödinger equation.

11.11 Covariant form of the Klein–Gordon equation

We derived in equation (11.60) the Klein–Gordon equation for a free particle to be

$$\left(\nabla^2 - \frac{1}{c^2}\frac{\partial^2}{\partial t^2} \right)\psi = \frac{m^2 c^2}{\hbar^2}\psi$$

Using equation (11.15) viz. $\partial^\mu \partial_\mu = \frac{1}{c^2}\frac{\partial^2}{\partial t^2} - \nabla^2$ we have

$$-\partial^\mu \partial_\mu \psi = \frac{m^2 c^2}{\hbar^2}\psi$$

$$\left(\partial^\mu \partial_\mu + \frac{m^2 c^2}{\hbar^2} \right)\psi = 0 \tag{11.69}$$

This is the covariant form of the Klein–Gordon equation.

11.12 Solution of the Klein–Gordon equation for a free particle

Since we have considered a free particle let us seek a plane wave solution and take

$$\psi(\vec{r},\, t) \sim e^{i(\vec{k}.\vec{r}-wt)} \tag{11.70}$$

considering the phase factor only, where $\vec{k}$ is wave vector, $\vec{r}$ is position vector, w is frequency.

Substituting equation (11.70) in equation (11.60) we get

$$\left(\nabla^2 - \frac{1}{c^2}\frac{\partial^2}{\partial t^2} \right)e^{i(\vec{k}.\vec{r}-wt)} = \frac{m^2 c^2}{\hbar^2}e^{i(\vec{k}.\vec{r}-wt)}$$

$$(i\vec{k})^2 e^{i(\vec{k}.\vec{r}-wt)} - \frac{1}{c^2}(-iw)^2 e^{i(\vec{k}.\vec{r}-wt)} = \frac{m^2 c^2}{\hbar^2}e^{i(\vec{k}.\vec{r}-wt)}$$

$$-k^2 + \frac{w^2}{c^2} = \frac{m^2 c^2}{\hbar^2}$$

$$\hbar^2 w^2 = \hbar^2 k^2 c^2 + m^2 c^4 \tag{11.71}$$

Using the expression of energy

$$E = h\nu = \hbar\frac{\nu}{2\pi} = \hbar w \tag{11.72}$$

$$E^2 = \hbar^2 k^2 c^2 + m^2 c^4 \tag{11.73}$$

$$E = \pm\sqrt{\hbar^2 k^2 c^2 + m^2 c^4} \tag{11.74}$$

Also $p = \hbar k$ and so

$$E^2 = p^2 c^2 + m^2 c^4 \tag{11.75}$$

$$E = \pm\sqrt{p^2 c^2 + m^2 c^4} = \pm c\sqrt{p^2 + m^2 c^2} \tag{11.76}$$

Both positive and negative values of energy E are allowed in the relativistic case.

But in the non-relativistic case a free particle can possess a positive kinetic energy only, namely $E = \frac{p^2}{2m}$.

The negative energies are physically connected with anti-particles.

11.13 Interpretation of the solution of the Klein–Gordon

Solution should be Lorentz invariant.

The continuity equation gives the law of conservation of probability and is Lorentz invariant. It is given by

$$\frac{\partial}{\partial t}\rho(\vec{r},\, t) + \vec{\nabla}\,.\,\vec{J}(\vec{r},\, t) = 0 \tag{11.77}$$

where $\rho(\vec{r},\, t)$ is the position probability density and $\vec{J}\,(\vec{r},\, t)$ is the probability current density.

The Klein–Gordon equation is given by equation (11.59)

$$\hbar^2 c^2 \nabla^2 \psi - \hbar^2 \frac{\partial^2 \psi}{\partial t^2} = m^2 c^4 \psi \tag{11.78}$$

and its complex conjugate is

$$\hbar^2 c^2 \nabla^2 \psi^* - \hbar^2 \frac{\partial^2 \psi^*}{\partial t^2} = m^2 c^4 \psi^* \tag{11.79}$$

Multiplying equation (11.78) from the left by ψ^* we get

$$\hbar^2 c^2 \psi \nabla^2 \psi - \hbar^2 \psi^* \frac{\partial^2 \psi}{\partial t^2} = m^2 c^4 \psi^* \psi \tag{11.80}$$

Multiplying equation (11.79) from the left by ψ we get

$$\hbar^2 c^2 \psi \nabla^2 \psi^* - \hbar^2 \psi \frac{\partial^2 \psi^*}{\partial t^2} = m^2 c^4 \psi \psi^* \tag{11.81}$$

Subtracting equation (11.81) from equation (11.80) we get

$$\hbar^2 c^2 \psi^* \nabla^2 \psi - \hbar^2 c^2 \psi \nabla^2 \psi^* - \left(\hbar^2 \psi^* \frac{\partial^2 \psi}{\partial t^2} - \hbar^2 \psi \frac{\partial^2 \psi^*}{\partial t^2}\right) = 0$$

$$c^2(\psi^*\nabla^2\psi - \psi\nabla^2\psi^*) - \left(\psi^*\frac{\partial^2\psi}{\partial t^2} - \psi\frac{\partial^2\psi^*}{\partial t^2}\right) = 0 \tag{11.82}$$

$$c^2\vec{\nabla}\cdot(\psi^*\vec{\nabla}\psi - \psi\vec{\nabla}\psi^*) - \frac{\partial}{\partial t}\left(\psi^*\frac{\partial\psi}{\partial t} - \psi\frac{\partial\psi^*}{\partial t}\right) = 0 \tag{11.83}$$

Multiplying equation (11.83) by $-\frac{i\hbar}{2m}$ we have

$$-\frac{i\hbar}{2m}c^2\vec{\nabla}\cdot(\psi^*\vec{\nabla}\psi - \psi\vec{\nabla}\psi^*) + \frac{i\hbar}{2m}\frac{\partial}{\partial t}\left(\psi^*\frac{\partial\psi}{\partial t} - \psi\frac{\partial\psi^*}{\partial t}\right) = 0$$

$$\frac{\partial}{\partial t}\frac{i\hbar}{2mc^2}\left(\psi^*\frac{\partial\psi}{\partial t} - \psi\frac{\partial\psi^*}{\partial t}\right) - \vec{\nabla}\cdot\frac{i\hbar}{2m}\left(\psi^*\vec{\nabla}\psi - \psi\vec{\nabla}\psi^*\right) = 0$$

$$\frac{\partial\rho}{\partial t} + \vec{\nabla}\cdot\vec{J} = 0 \tag{11.84}$$

where

$$\rho(\vec{r},t) = \frac{i\hbar}{2mc^2}\left(\psi^*\frac{\partial\psi}{\partial t} - \psi\frac{\partial\psi^*}{\partial t}\right) \tag{11.85}$$

$$\vec{J}(\vec{r},t) = \frac{i\hbar}{2m}\left(\psi^*\vec{\nabla}\psi - \psi\vec{\nabla}\psi^*\right) \tag{11.86}$$

The non-relativistic position probability density is

$$\rho(\vec{r},t)|_{\text{non-relativistic}} = \psi^*(\vec{r},t)\psi(\vec{r},t) \tag{11.87}$$

And non-relativistic current density is

$$\vec{J}(\vec{r},t)|_{\text{non-relativistic}} = \vec{J}(\vec{r},t) \tag{11.88}$$

We note that if $\psi(\vec{r},t)$ is real then $\psi^*=\psi$ and equation (11.84) gives

$$\rho(\vec{r},t) = \frac{i\hbar}{2mc^2}\left(\psi^*\frac{\partial\psi}{\partial t} - \psi\frac{\partial\psi^*}{\partial t}\right) = \frac{i\hbar}{2mc^2}\left(\psi\frac{\partial\psi}{\partial t} - \psi\frac{\partial\psi}{\partial t}\right) = 0 \tag{11.89}$$

Suppose ψ is complex. Then from equation (11.84) we have

$$\rho(\vec{r},t) = \frac{1}{2mc^2}\left[\psi^* i\hbar\frac{\partial}{\partial t}\psi - \psi i\hbar\frac{\partial\psi^*}{\partial t}\right]$$

$$= \frac{1}{2mc^2}\left[\psi^* i\hbar\frac{\partial}{\partial t}\psi + \left(i\hbar\frac{\partial\psi}{\partial t}\psi^*\right)^\dagger\right]$$

Using the energy operator $\hat{E} \equiv i\hbar\frac{\partial}{\partial t}$ and $\hat{E}\psi = i\hbar\frac{\partial\psi}{\partial t}$ we rewrite

$$\rho(\vec{r},\, t) = \frac{1}{2mc^2}[\psi^* \hat{E}\psi + ((\hat{E}\psi)\psi^*)^\dagger]$$

Writing $\hat{E}\psi = E\psi$ ($\hat{E}$ = energy operator, E = energy eigenvalue)

$$\rho(\vec{r},\, t) = \frac{1}{2mc^2}[\psi^* E\psi + ((E\psi)\psi^*)^\dagger]$$

$$\rho(\vec{r},\, t) = \frac{E}{2mc^2}[\psi^*\psi + \psi^*\psi] = \frac{2E}{2mc^2}\psi^*\psi = \frac{E}{mc^2}\psi^*\psi$$

$$\rho(\vec{r},\, t) = \frac{E}{mc^2}|\psi(\vec{r},\, t)| \tag{11.90}$$

By equation (11.74) viz. $E = \pm\sqrt{\hbar^2 k^2 c^2 + m^2 c^4}$ energy can be both positive as well as negative. So $\rho(\vec{r},\, t)$ can also be both positive as well as negative. Again, negative position probability density is meaningless. So we make a redefinition by multiplying $\rho(\vec{r},\, t)$ by electric charge magnitude $|e|$ and calling $|e|\rho(\vec{r},\, t)$ the electric charge probability density and $|e|\vec{J}(\vec{r},\, t)$ the electrical current density.

Since the Klein–Gordon equation shows two types of charge probability densities it does not correspond to a system of similarly charged particles.

The Klein–Gordon equation thus describes a multi-particle system of particles and anti-particles. Positively charged particles will correspond to positive electric charge density and anti-particles will correspond to negative electric charge density. A many-particle system of particles and anti-particles is known as a field. So it is a relativistic field theory.

In *exercise 11.5* we show that the continuity equation that follows from the Klein–Gordon equation has the form $\partial_\mu J^\mu = 0$.

11.14 Spin state to which the Klein–Gordon equation corresponds

The Klein–Gordon equation (11.60) is given by

$$\left(\nabla^2 - \frac{1}{c^2}\frac{\partial^2}{\partial t^2} \right)\psi = \frac{m^2 c^2}{\hbar^2}\psi$$

This equation has only one component, i.e. ψ is a single component.

For a particle with spin s the total number of components (multiplicity) is $2s + 1$.

For $s = 0$, there are $2s + 1 = 2 . 0 + 1 = 1$ components only.

So the Klein–Gordon equation is sufficient to describe the behaviour of a spin zero particle.

Also, we note that the Klein–Gordon equation can be written as (*exercise 11.3*)

$$p^\mu p_\mu \psi - m^2 c^2 \psi = 0$$

$$p^\mu p_\mu \psi = m^2 c^2 \psi \tag{11.91}$$

Since $p^\mu p_\mu = m^2 c^2$ (equation (11.27)) is a scalar or invariant Klein–Gordon equation it corresponds to spin zero particles.

11.15 Klein–Gordon equation for a charged particle in an electromagnetic field

If a spinless particle has electric charge q and is placed in an external electromagnetic field described by a vector potential $\vec{A}$ and a scalar potential ϕ we make the following replacements

$$\vec{p} \rightarrow \vec{p} - q\vec{A} \tag{11.92}$$

$$E \rightarrow E - q\phi \tag{11.93}$$

in the relativistic Hamiltonian $H = E = (c^2 p^2 + m^2 m^4)^{1/2}$ the square of which is given by equation (11.75) to be

$$E^2 = p^2 c^2 + m^2 c^4$$

Hence

$$(E - q\phi)^2 = c^2(\vec{p} - q\vec{A})^2 + m^2 c^4 \tag{11.94}$$

The Klein–Gordon equation for an electromagnetic field is obtained by making the operator replacements

$$E \rightarrow i\hbar \frac{\partial}{\partial t}$$

$$\vec{p} \rightarrow -i\hbar \vec{\nabla}$$

$$\left(i\hbar \frac{\partial}{\partial t} - q\phi\right)^2 \psi = c^2(-i\hbar \vec{\nabla} - q\vec{A})^2 \psi + m^2 c^4 \psi$$

$$\left(i\hbar \frac{\partial}{\partial t} - q\phi\right)\left(i\hbar \frac{\partial}{\partial t} - q\phi\right)\psi = c^2(-i\hbar \vec{\nabla} - q\vec{A})(-i\hbar \vec{\nabla} - q\vec{A})\psi + m^2 c^4 \psi$$

$$-\hbar^2 \frac{\partial^2 \psi}{\partial t^2} + q^2 \phi^2 \psi - i\hbar q \frac{\partial}{\partial t}(\phi\psi) - q\phi i\hbar \frac{\partial \psi}{\partial t}$$

$$= c^2[-\hbar^2 \nabla^2 \psi + q^2 A^2 \psi + i\hbar q \vec{\nabla} . (\vec{A} \psi) + i\hbar q \vec{A} . \vec{\nabla} \psi] + m^2 c^4 \psi$$

$$-\hbar^2 \frac{\partial^2 \psi}{\partial t^2} + q^2 \phi^2 \psi - i\hbar q \left(\frac{\partial \phi}{\partial t}\psi - \phi\frac{\partial \psi}{\partial t}\right) - i\hbar q \phi \frac{\partial \psi}{\partial t}$$

$$= -\hbar^2 c^2 \nabla^2 \psi + q^2 c^2 A^2 \psi + i\hbar q c^2[(\vec{\nabla} . \vec{A})\psi + \vec{A} . \vec{\nabla} \psi] + i\hbar q c^2 \vec{A} . \vec{\nabla} \psi + m^2 c^4 \psi$$

$$-\hbar^2\frac{\partial^2\psi}{\partial t^2} + q^2\phi^2\psi - i\hbar q\frac{\partial\phi}{\partial t}\psi - 2i\hbar q\phi\frac{\partial\psi}{\partial t}$$

$$=-\hbar^2c^2\,\nabla^2\,\psi + q^2c^2A^2\psi + i\hbar qc^2(\vec{\nabla}.\,\vec{A})\psi + 2i\hbar qc^2\vec{A}.\,\vec{\nabla}\,\psi + m^2c^4\psi \quad (11.95)$$

The complex conjugate of equation (11.95) is

$$-\hbar^2\frac{\partial^2\psi^*}{\partial t^2} + q^2\phi^2\psi^* + i\hbar q\frac{\partial\phi}{\partial t}\psi^* + 2i\hbar q\phi\frac{\partial\psi^*}{\partial t}$$

$$=-\hbar^2c^2\,\nabla^2\,\psi^* + q^2c^2A^2\psi^* - i\hbar qc^2(\vec{\nabla}.\,\vec{A})\psi^* - 2i\hbar qc^2\vec{A}.\,\vec{\nabla}\,\psi^* + m^2c^4\psi^* \quad (11.96)$$

Let us consider $\psi^* \times$ equation (11.95)

$$-\hbar^2\psi^*\frac{\partial^2\psi}{\partial t^2} + q^2\phi^2\psi^*\psi - i\hbar q\frac{\partial\phi}{\partial t}\psi^*\psi - 2i\hbar q\phi\psi^*\frac{\partial\psi}{\partial t}$$

$$=-\hbar^2c^2\psi^*\,\nabla^2\,\psi + q^2c^2A^2\psi^*\psi + i\hbar qc^2(\vec{\nabla}.\,\vec{A})\psi^*\psi + 2i\hbar qc^2\psi^*\vec{A}.\,\vec{\nabla}\,\psi + m^2c^4\psi^*\psi \quad (11.97)$$

Let us consider equation (11.96) $\times \psi$

$$-\hbar^2\frac{\partial^2\psi^*}{\partial t^2}\psi + q^2\phi^2\psi^*\psi + i\hbar q\frac{\partial\phi}{\partial t}\psi^*\psi + 2i\hbar q\phi\frac{\partial\psi^*}{\partial t}\psi$$

$$=-\hbar^2c^2(\nabla^2\psi^*)\psi + q^2c^2A^2\psi^*\psi - i\hbar qc^2(\vec{\nabla}.\,\vec{A})\psi^*\psi - 2i\hbar qc^2(\vec{A}.\,\vec{\nabla}\,\psi^*)\psi + m^2c^4\psi^*\psi \quad (11.98)$$

Equation (11.97)–(11.98) gives

$$-\hbar^2\psi^*\frac{\partial^2\psi}{\partial t^2} + q^2\phi^2\psi^*\psi - i\hbar q\frac{\partial\phi}{\partial t}\psi^*\psi - 2i\hbar q\phi\psi^*\frac{\partial\psi}{\partial t}$$

$$-\left[-\hbar^2\frac{\partial^2\psi^*}{\partial t^2}\psi + q^2\phi^2\psi^*\psi + i\hbar q\frac{\partial\phi}{\partial t}\psi^*\psi + 2i\hbar q\phi\frac{\partial\psi^*}{\partial t}\psi\right]$$

$$=[-\hbar^2c^2\psi^*\,\nabla^2\,\psi + q^2c^2A^2\psi^*\psi + i\hbar qc^2(\vec{\nabla}.\,\vec{A})\psi^*\psi + 2i\hbar qc^2\psi^*\vec{A}.\,\vec{\nabla}\,\psi + m^2c^4\psi^*\psi]$$

$$-[-\hbar^2c^2(\nabla^2\psi^*)\psi + q^2c^2A^2\psi^*\psi - i\hbar qc^2(\vec{\nabla}.\,\vec{A})\psi^*\psi - 2i\hbar qc^2(\vec{A}.\,\vec{\nabla}\,\psi^*)\psi + m^2c^4\psi^*\psi] \quad (11.99)$$

In equation (11.99) the 2nd term and 6th terms cancel out in the LHS, 2nd term and 7th terms cancel out on the RHS, 5th and 10th terms cancel out on the RHS. We are left with

$$-\hbar^2\psi^*\frac{\partial^2\psi}{\partial t^2} - i\hbar q\frac{\partial\phi}{\partial t}\psi^*\psi - 2i\hbar q\phi\psi^*\frac{\partial\psi}{\partial t} - \left[-\hbar^2\frac{\partial^2\psi^*}{\partial t^2}\psi + i\hbar q\frac{\partial\phi}{\partial t}\psi^*\psi + 2i\hbar q\phi\frac{\partial\psi^*}{\partial t}\psi\right]$$

$$=[-\hbar^2c^2\psi^*\,\nabla^2\,\psi + i\hbar qc^2(\vec{\nabla}.\,\vec{A})\psi^*\psi + 2i\hbar qc^2\psi^*\vec{A}.\,\vec{\nabla}\,\psi]$$

$$-[-\hbar^2c^2(\nabla^2\psi^*)\psi - i\hbar qc^2(\vec{\nabla}.\,\vec{A})\psi^*\psi - 2i\hbar qc^2(\vec{A}.\,\vec{\nabla}\,\psi^*)\psi]$$

$$-\hbar^2\left(\psi^*\frac{\partial^2\psi}{\partial t^2} - \frac{\partial^2\psi^*}{\partial t^2}\psi\right) - 2i\hbar q\phi\left(\psi^*\frac{\partial\psi}{\partial t} + \frac{\partial\psi^*}{\partial t}\psi\right) - i\hbar q\frac{\partial\phi}{\partial t}2\psi^*\psi$$

$$=-\hbar^2 c^2(\psi^* \nabla^2 \psi - (\nabla^2\psi^*)\psi) + i\hbar qc^2(\vec{\nabla}.\vec{A})(2\psi^*\psi) + 2i\hbar qc^2\vec{A}.(\psi^* \vec{\nabla} \psi + (\vec{\nabla}\psi^*)\psi]$$

$$-\hbar^2\frac{\partial}{\partial t}\left(\psi^*\frac{\partial\psi}{\partial t} - \frac{\partial\psi^*}{\partial t}\psi\right) - 2i\hbar q\frac{\partial}{\partial t}\left(\psi^*\psi\right) - 2i\hbar q\frac{\partial\phi}{\partial t}\left(\psi^*\psi\right)$$

$$=-\hbar^2 c^2 \vec{\nabla}.[\psi^* \vec{\nabla} \psi - (\vec{\nabla}\psi^*)\psi] + 2i\hbar qc^2(\vec{\nabla}.\vec{A})(\psi^*\psi) + 2i\hbar qc^2\vec{A}.\vec{\nabla}(\psi^*\psi)$$

$$-\hbar^2\frac{\partial}{\partial t}\left(\psi^*\frac{\partial\psi}{\partial t} - \frac{\partial\psi^*}{\partial t}\psi\right) - 2i\hbar q\frac{\partial}{\partial t}\left(\phi\psi^*\psi\right)$$

$$=-\hbar^2 c^2 \vec{\nabla}.[\psi^* \vec{\nabla} \psi - (\vec{\nabla}\psi^*)\psi] + 2i\hbar qc^2 \vec{\nabla}.(\vec{A}\psi^*\psi)$$

$$\frac{\partial}{\partial t}\left[-\hbar^2\left(\psi^*\frac{\partial\psi}{\partial t} - \frac{\partial\psi^*}{\partial t}\psi\right) - 2i\hbar q\left(\phi\psi^*\psi\right)\right]$$

$$=- \vec{\nabla}.[\hbar^2 c^2(\psi^* \vec{\nabla} \psi - (\vec{\nabla}\psi^*)\psi) - 2i\hbar qc^2\vec{A}\psi^*\psi] \qquad (11.100)$$

Multiply equation (11.100) throughout by $\frac{q}{2im\hbar c^2}$ to get

$$\frac{\partial}{\partial t}\left[-\frac{q\hbar}{2imc^2}\left(\psi^*\frac{\partial\psi}{\partial t} - \frac{\partial\psi^*}{\partial t}\psi\right) - \frac{q^2}{mc^2}\left(\phi\psi^*\psi\right)\right]$$

$$=- \vec{\nabla}.\left[\frac{q\hbar}{2im}(\psi^* \vec{\nabla} \psi - (\vec{\nabla}\psi^*)\psi) - \frac{q^2}{m}\vec{A}\psi^*\psi\right]$$

$$\frac{\partial}{\partial t}\left[-\frac{qi\hbar}{2mc^2}\left(\psi^*\frac{\partial\psi}{\partial t} - \frac{\partial\psi^*}{\partial t}\psi\right) + \frac{q^2}{mc^2}\left(\phi\psi^*\psi\right)\right]$$

$$= \vec{\nabla}.\left[\frac{q\hbar}{2im}(\psi^* \vec{\nabla} \psi - (\vec{\nabla}\psi^*)\psi) - \frac{q^2}{m}\vec{A}\psi^*\psi\right]$$

$$\vec{\nabla}.\left[\frac{q\hbar}{2im}(\psi^* \vec{\nabla} \psi - (\vec{\nabla}\psi^*)\psi) - \frac{q^2}{m}\vec{A}\psi^*\psi\right] + \frac{\partial}{\partial t}\left[\frac{qi\hbar}{2mc^2}\left(\psi^*\frac{\partial\psi}{\partial t} - \frac{\partial\psi^*}{\partial t}\psi\right) - \frac{q^2}{mc^2}\left(\phi\psi^*\psi\right)\right] = 0$$

$$\vec{\nabla}.\vec{J} + \frac{\partial\rho}{\partial t} = 0 \qquad (11.101)$$

with

$$\vec{J} = \frac{q\hbar}{2im}(\psi^* \vec{\nabla} \psi - (\vec{\nabla}\psi^*)\psi) - \frac{q^2}{m}\vec{A}\psi^*\psi = \vec{J}_{\text{free}} - \frac{q^2}{m}\vec{A}\psi^*\psi \qquad (11.102)$$

is the electrical current density and

$$\rho = \frac{qi\hbar}{2mc^2}\left(\psi^*\frac{\partial\psi}{\partial t} - \frac{\partial\psi^*}{\partial t}\psi\right) - \frac{q^2}{mc^2}(\phi\psi^*\psi) = \rho_{\text{free}} - \frac{q^2\phi}{mc^2}\psi^*\psi \tag{11.103}$$

is the electrical charge density

The equation of continuity is Lorentz invariant and is not affected by the presence of electromagnetic potential.

For the stationary state

$$\psi = \psi(\vec{r}, t) = \psi(\vec{r})e^{-iEt/\hbar} \tag{11.104}$$

we have from equation (11.103)

$$\rho = \frac{qi\hbar}{2mc^2}\left[\psi^*\left(-\frac{iE}{\hbar}\right)\psi - \left(\frac{iE}{\hbar}\right)\psi^*\psi\right] - \frac{q^2\phi}{mc^2}\psi^*\psi$$

$$= \frac{qi\hbar}{2mc^2}\left(\frac{2iE}{\hbar}\right)\psi^*\psi - \frac{q^2\phi}{mc^2}\psi^*\psi \tag{11.105}$$

$$\rho = \frac{q^2}{mc^2}(E - q\phi)\psi^*(\vec{r})\psi(\vec{r}) \tag{11.106}$$

This is the form of ρ in the stationary states of a particle in an electromagnetic field.

When $E > q\phi$ the sign of the charge density is the same as the sign of the charge of the particle.

When $E < q\phi$ the sign of the charge density is opposite to that of the charge of the particle.

We cannot thus retain the single-particle interpretation in the region of very strong field ϕ.

The physical meaning of the change of sign of ρ in strong fields can only be understood from a theory describing the behaviour of systems with a variable number of particles and taking the creation and annihilation of particles with positive and negative charges into account.

- Consequence of the new interpretation of the continuity equation

✓ Position probability P is replaced by electrical charge density

$$\rho = qP \tag{11.107}$$

Positional current density $\vec{J}$ is replaced by electrical current density

$$\vec{j} = q\vec{J} \tag{11.108}$$

Then the equation of continuity expresses charge conservation rather than probability conservation.

✓ We note that

$$E^2 = p^2c^2 + m^2c^4$$

Quantity (Definition)	Charge	Energy state	Charge density $\rho = \dfrac{qE}{mc^2}$
Particle	$q = \lvert q \rvert$	+ve , E=\lvert E \rvert	ρ = +ve
Anti-particle	$q = -\lvert q \rvert$	$-$ve , E=$-\lvert E \rvert$	ρ = +ve
Particle	$q = \lvert q \rvert$	$-$ve , E=$-\lvert E \rvert$	ρ = $-$ve
Anti-particle	$q = -\lvert q \rvert$	+ve , E=\lvert E \rvert	ρ = $-$ve

Figure 11.2. A particle in the negative energy state is equivalent to its anti-particle in the positive energy state.

$$E = \pm\sqrt{p^2 c^2 + m^2 c^4} = \pm\lvert E \rvert$$

Energy of a particle may be either positive or negative.

We have established in figure 11.2 the following:

ρ = positive for a positively charged particle in positive energy state

ρ = positive for a negatively charged particle in negative energy state

ρ = negative for a positively charged particle in negative energy state

ρ = negative for a negatively charged particle in positive energy state.

In view of the above we conclude that a particle in the negative energy state is equivalent to its anti-particle in the positive energy state.

✓ $\int_\tau P d\tau = \int_\tau \psi^* \psi \, d\tau$ is non-relativistically conserved in time (Schrödinger case) which permits a positive definite probability density. But in the Klein–Gordon case $\int_\tau \rho_{KG} \, d\tau$ is not conserved in time.

In fact, at relativistic velocities the principle of mass energy equivalence $E = mc^2$ hints at the possibility of creation of new particles at these energies. Obviously thus conservation of particle number cannot be a fundamental feature of relativistic quantum mechanics.

On the other hand, the electrical charge is strictly a conserved quantity. The equation of continuity of a relativistic wave equation should thus refer to the conservation of electric charge rather than to the conservation of particle number. So

$$\int \rho(\vec{r}, t)d\tau = \int qP(\vec{r}, t)d\tau = q \int P_{KG}(\vec{r}, t)d\tau = q$$

refers to charge conservation. Such normalization is conserved in time. It is invariant under Lorentz transformation.

11.16 Derivation of Dirac equation for a free particle

In 1928 Dirac formulated an equation that was first order both in time and space derivative and designed to avoid difficulties of negative probability density of the

Klein–Gordon equation that arose because it was first order in time and second order in space.

The Hamiltonian for a free particle is

$$H = \frac{p^2}{2m} = \frac{(-i\hbar \, \vec{\nabla} \,)^2}{2m} = -\frac{\hbar^2 \nabla^2}{2m} \tag{11.109}$$

and the Schrödinger equation is

$$-\frac{\hbar^2 \nabla^2}{2m}\psi = i\hbar \frac{\partial \psi}{\partial t} \tag{11.110}$$

This equation is linear in time derivative but quadratic in space derivative.

To keep Hamiltonian linear in space derivative also, Dirac modified the Hamiltonian of the free particle by expressing it as follows

$$H = c \, \vec{\alpha}.\vec{p} \, + \beta \, mc^2 \tag{11.111}$$

$$H = c \left(\alpha_x p_x + \alpha_y p_y + \alpha_z p_z \right) + \beta \, mc^2 \tag{11.112}$$

where α_x, α_y, α_z, β are matrices called Dirac matrices. This is called the Dirac Hamiltonian for a free particle.

With momentum operator $\vec{p} = -i\hbar \, \vec{\nabla}$ we can write the Dirac Hamiltonian of equation (11.111) as

$$H = c \, \vec{\alpha}. \, (-i\hbar \, \vec{\nabla} \,) + \beta \, mc^2 = -i\hbar c \, \vec{\alpha}. \, \vec{\nabla} + \beta \, mc^2$$

$$H = -i\hbar c \left(\alpha_x \frac{\partial}{\partial x} + \alpha_y \frac{\partial}{\partial y} + \alpha_z \frac{\partial}{\partial z} \right) + \beta \, mc^2 \tag{11.113}$$

Equations (11.111), (11.112) and (11.113) are Dirac's relativistic Hamiltonian which is linear in momentum, linear in space and time.

Dirac's relativistic equation for a free particle would thus be

$$H\psi(\vec{r},\, t) = i\hbar \frac{\partial}{\partial t}\psi(\vec{r},\, t)$$

$$(c \, \vec{\alpha}.\vec{p} \, + \beta \, mc^2)\psi = i\hbar \frac{\partial}{\partial t}\psi \tag{11.114}$$

$$-i\hbar c(\vec{\alpha}. \, \vec{\nabla})\psi + \beta \, mc^2\psi = i\hbar \frac{\partial}{\partial t}\psi \tag{11.115}$$

Such an equation would be free of the difficulties experienced in dealing with the Klein–Gordon equation since it is first order in space and time. In other words, this equation (11.115) explicitly preserves the symmetry between space and time.

In natural units $\hbar = c = 1$ and Dirac's relativistic equation (11.114) for a free particle becomes

$$(\vec{\alpha}.\vec{p} \; + \; \beta\, m)\psi = i\frac{\partial}{\partial t}\psi \tag{11.116}$$

11.17 Dirac matrices α, β and their properties

The relation between total energy and momentum of a relativistic free particle according to the special theory of relativity is

$$E^2 = p^2 c^2 + m^2 c^4 \tag{11.117}$$

Taking square root we have

$$E = \pm\sqrt{p^2 c^2 + m^2 c^4}$$

$$E = \pm c\sqrt{p^2 + m^2 c^2} \tag{11.118}$$

To proceed further we replace the physical observables E, $\vec{p}$ by their corresponding operators $\to i\hbar\frac{\partial}{\partial t}$, $\vec{p} \to -i\hbar\,\vec{\nabla}$, $p^2 \to -\hbar^2\nabla^2$.

The RHS of equation (11.118) involves the square root of the p^2 operator which is difficult to handle. Again, in quantum mechanics we replace operators by matrices and so Dirac introduced matrices α, β are suitable for the problem. In other words, Dirac expressed the quantity within the root of the RHS of equation (11.118), i.e. $p^2 + m^2 c^2$ as a square making the following choice

$$p^2 + m^2 c^2 = (\vec{\alpha}.\vec{p} \; + \; \beta mc)^2 \tag{11.119}$$

Writing in component form

$$p_x^2 + p_y^2 + p_z^2 + m^2 c^2 = \; (\vec{\alpha}.\vec{p} \; + \; \beta mc)(\vec{\alpha}.\vec{p} \; + \; \beta mc) \tag{11.120}$$

We are to find a suitable representation for the quantities $\vec{\alpha} \equiv (\alpha_x, \; \alpha_y, \; \alpha_z)$, β.

Expanding the RHS of equation (11.118) we have

$$p_x^2 + p_y^2 + p_z^2 + m^2 c^2 = \; (\alpha_x p_x + \alpha_y p_y + \alpha_z p_z + \beta mc)(\alpha_x p_x + \alpha_y p_y + \alpha_z p_z + \beta mc)$$

$$= \alpha_x p_x \alpha_x p_x + \alpha_x p_x \alpha_y p_y + \alpha_x p_x \alpha_z p_z + \alpha_x p_x \beta mc$$

$$+ \alpha_y p_y \alpha_x p_x + \alpha_y p_y \alpha_y p_y + \alpha_y p_y \alpha_z p_z + \alpha_y p_y \beta mc$$

$$+ \alpha_z p_z \alpha_x p_x + \alpha_z p_z \alpha_y p_y + \alpha_z p_z \alpha_z p_z + \alpha_z p_z \beta mc$$

$$+ \beta mc\, \alpha_x p_x + \beta mc\, \alpha_y p_y + \beta mc\, \alpha_z p_z + \beta mc\, \beta mc$$

We note that

$$[p_i \; p_j] = 0 \tag{11.121}$$

- i.e. p_i commutes and also assume that α and p_i commute, i.e.

$$[\alpha \; p_i] = 0 \tag{11.122}$$

and β and p_i commute, i.e.

$$[\beta \; p_i] = 0 \tag{11.123}$$

Hence

$$p_x^2 + p_y^2 + p_z^2 + m^2c^2 = \alpha_x^2 p_x^2 + \alpha_x \alpha_y p_x p_y + \alpha_x \alpha_z p_x p_z + \alpha_x \beta p_x mc$$

$$+ \alpha_y \alpha_x p_y p_x + \alpha_y^2 p_y^2 + \alpha_y \alpha_z p_y p_z + \alpha_y \beta p_y mc$$

$$+ \alpha_z \alpha_x p_z p_x + \alpha_z \alpha_y p_z p_y + \alpha_z^2 p_z^2 + \alpha_z \beta p_z mc$$

$$+ mc\beta \alpha_x p_x + mc\beta \alpha_y p_y + mc\beta \alpha_z p_z + \beta^2 m^2 c^2 \tag{11.124}$$

Comparison of LHS and RHS shows that

$$\alpha_x^2 = 1, \quad \alpha_y^2 = 1, \quad \alpha_z^2 = 1, \quad \beta^2 = 1 \tag{11.125}$$

- $$\alpha_i^2 = 1, \quad \beta^2 = 1 \;\text{ with }\; i = 1, 2, 3 \;\text{ or }\; x, y, z \tag{11.126}$$

- α_i and β have eigenvalues ± 1. This follows from equation (11.126).
With this we have from equation (11.124)

$$0 = \alpha_x \alpha_y p_x p_y + \alpha_x \alpha_z p_x p_z + \alpha_x \beta p_x mc + \alpha_y \alpha_x p_y p_x + \alpha_y \alpha_z p_y p_z + \alpha_y \beta p_y mc$$

$$+ \alpha_z \alpha_x p_z p_x + \alpha_z \alpha_y p_z p_y + \alpha_z \beta p_z mc + mc\beta \alpha_x p_x + mc\beta \alpha_y p_y + mc\beta \alpha_z p_z$$

$$0 = \left(\alpha_x \alpha_y + \alpha_y \alpha_x\right) p_x p_y + (\alpha_x \alpha_z + \alpha_z \alpha_x) p_x p_z + (\alpha_x \beta - \beta \alpha_x) p_x mc$$

$$+ (\alpha_y \alpha_z - \alpha_z \alpha_y) p_y p_z + (\alpha_y \beta - \beta \alpha_y) p_y mc + \left(\alpha_z \beta - \beta \alpha_z\right) p_z mc$$

$$0 = \{\alpha_x \; \alpha_y\} p_x p_y + \{\alpha_x \; \alpha_z\} p_x p_z + \{\alpha_x \; \beta\} p_x mc$$

$$+ \{\alpha_y \; \alpha_z\} p_y p_z + \{\alpha_y \; \beta\} p_y mc + \left\{\alpha_z \; \beta\right\} p_z mc \tag{11.127}$$

Comparison of LHS and RHS gives from equation (11.127)

$$\{\alpha_x \; \alpha_y\} = 0, \; \{\alpha_x \; \alpha_z\} = 0, \; \{\alpha_y \; \alpha_z\} = 0, \; \{\alpha_x \; \beta\} = 0, \; \{\alpha_y \; \beta\} = 0, \; \{\alpha_z \; \beta\} = 0 \tag{11.128}$$

- $$\{\alpha_i \; \alpha_j\} = 0, \quad \{\alpha_i \; \beta\} \;\text{ with }\; i = x, y, z \;\text{ or }\; 1, 2, 3 \tag{11.129}$$

Since α_i s and β anti-commute it follows that they cannot be numbers because numbers commute. So α_i and β are matrices.

- Since the Dirac Hamiltonian $H = c\,\vec{\alpha}.\vec{p} + \beta\,mc^2$ is Hermitian, i.e.

$$H^\dagger = H$$

it follows from equation (11.111) that

$$(c\,\vec{\alpha}.\vec{p} + \beta\,mc^2)^\dagger = c\,\vec{\alpha}.\vec{p} + \beta\,mc^2$$

$$c\,\vec{\alpha}^\dagger.\vec{p} + \beta^\dagger mc^2 = c\,\vec{\alpha}.\vec{p} + \beta\,mc^2 \quad (p \text{ is Hermitian } i.\,e.\ p^\dagger = p)$$

This gives

$$\alpha^\dagger = \alpha, \quad \beta^\dagger = \beta \tag{11.130}$$

i.e. α and β are Hermitian.

- Hermitian means α, β are square matrices, i.e. their number of rows and number of columns are equal.
- Since

$$\alpha_i^2 = I \quad i.\,e.\ \alpha_i\alpha_i = 1 \tag{11.131}$$

$$\alpha_i = \alpha_i^{-1} \tag{11.132}$$

Since

$$\beta^2 = I \quad i.\,e.\ \beta\beta = 1 \tag{11.133}$$

$$\beta = \beta^{-1} \tag{11.134}$$

This means the inverse of the Dirac matrices α_i, β exists, i.e. Dirac matrices α_i, β are non-singular, i.e. they have a non-vanishing determinant.

- Consider $\alpha_i^2 = I$ (equation (11.131))

$$\alpha_i\alpha_i = I$$

As α_i is Hermitian $\alpha_i = \alpha_i^\dagger$ and so

$$\alpha_i^\dagger\alpha_i = I \tag{11.135}$$

So α_i is unitary.

Consider $\beta^2 = I$ (equation (11.133))

$$\beta\beta = I$$

As β is Hermitian $\beta = \beta^\dagger$

$$\beta^\dagger\beta = I \tag{11.136}$$

So β is unitary.

- Consider the α matrices

$$\alpha_i = \alpha_i\beta^2 \text{ as } \beta^2 = I$$

$$= \alpha_i \beta\beta \tag{11.137}$$

Since α_i and β anti-commute $\{\alpha_i \; \beta\} = 0$, i.e. $\alpha_i \beta = -\beta \; \alpha_i$ we have from equation (11.137)

$$\alpha_i = -\beta \; \alpha_i \beta \tag{11.138}$$

Since β is Hermitian, i.e. $\beta^\dagger = \beta$

$$\alpha_i = -\beta^\dagger \alpha_i \beta \tag{11.139}$$

- Consider the β matrix

$$\beta = \beta\alpha_i^2 \; as \; \alpha_i^2 = I$$

$$= \beta\alpha_i\alpha_i \tag{11.140}$$

Since α_i and β anti-commute $\{\alpha_i \; \beta\} = 0 \Rightarrow \beta \; \alpha_i = -\; \alpha_i\beta$ we have from equation (11.140)

$$\beta = -\alpha_i\beta\alpha_i \tag{11.141}$$

Since α_i is Hermitian, i.e. $\alpha_i^\dagger = \alpha_i$

$$\beta = -\alpha_i^\dagger \beta\alpha_i \tag{11.142}$$

These are the transformations between the α, β matrices (unitary transformation as the Dirac matrices are unitary).
- Eigenvalue of unity of the Dirac matrices imply that these matrices consist of numbers and do not depend upon space and time.
- Let us calculate the trace of the Dirac matrices

$$Tr \; \alpha_i = Tr(-\beta^\dagger \alpha_i \beta) \; \text{(using equation (11.139))}$$

$$= -Tr \; (\beta^\dagger \alpha_i \beta) = -Tr(\alpha_i \beta\beta^\dagger) \; \text{since} \; Tr \; (ABC) = Tr(BCA)$$

$$= -Tr \; \alpha_i \; \text{as} \; \beta\beta^\dagger = \beta\beta = \beta^2 = I \; \text{as} \; \beta \; \text{is Hermitian} \; \beta^\dagger = \beta$$

$$2Tr\alpha_i = 0$$

$$Tr\alpha_i = 0 \tag{11.143}$$

$$Tr \; \beta = Tr\left(-\alpha_i^\dagger \beta\alpha_i\right) \; \text{(using equation (11.142))}$$

$$= -Tr \left(\alpha_i^\dagger \beta\alpha_i\right) = -Tr(\beta\alpha_i\alpha_i^\dagger) \; \text{since} \; Tr \; (ABC) = Tr(BCA)$$

$$= -Tr\beta \; \text{as} \; \alpha_i\alpha_i^\dagger = \alpha_i\alpha_i = \alpha_i^2 = I \; \text{as} \; \alpha_i \; \text{is Hermitian} \; \alpha_i^\dagger = \alpha_i$$

$$2Tr\beta = 0$$

$$Tr\beta = 0 \qquad (11.144)$$

Dirac matrices α_i, β are traceless.
- Dirac matrices $\alpha_{i,}$, β are linearly independent.
- Since Dirac matrices α_i, β are traceless it follows that they are even order matrices, i.e. number of rows and number of columns are equal, i.e. the Dirac matrices can be 2×2, 4×4 etc (*exercise 11.8*).

Invoke a 2×2 set of matrices—the Pauli spin matrices

$$\sigma_i = \sigma_x = \begin{pmatrix} 0 & 1 \\ 1 & 0 \end{pmatrix}, \quad \sigma_y = \begin{pmatrix} 0 & -i \\ i & 0 \end{pmatrix}, \quad \sigma_z = \begin{pmatrix} 1 & 0 \\ 0 & -1 \end{pmatrix} \qquad (11.145)$$

The Pauli spin matrices σ_i satisfy properties like

$$\sigma_i^2 = I \quad (\sigma_i s \text{ have eigenvalues } \pm 1)$$

$$\{\sigma_i \, \sigma_j\} = 0 \, (\sigma_i \, s \text{ anti-commute})$$

$$Tr\sigma_i = 0 \, (\sigma_i \, s \text{ are traceless}).$$

It thus follows that we can build the Dirac matrices α_i, β in terms of the Pauli spin matrices. But cannot use the 2×2 Pauli matrices as the Dirac matrices (*exercise 11.9*).

In fact, 2×2 representation does not allow us to construct 4 linearly independent matrices that satisfy the anti-commutation property that is required in the Dirac Hamiltonian.

Let us use a 4×4 representation of the Dirac matrices α_i, β in terms of Pauli spin matrices. This is called Pauli–Dirac representation. We construct

$$\alpha_x = \begin{pmatrix} 0 & \sigma_x \\ \sigma_x & 0 \end{pmatrix} = \begin{pmatrix} 0 & 0 & 0 & 1 \\ 0 & 0 & 1 & 0 \\ 0 & 1 & 0 & 0 \\ 1 & 0 & 0 & 0 \end{pmatrix}, \quad \alpha_y = \begin{pmatrix} 0 & \sigma_y \\ \sigma_y & 0 \end{pmatrix} = \begin{pmatrix} 0 & 0 & 0 & -i \\ 0 & 0 & i & 0 \\ 0 & -i & 0 & 0 \\ i & 0 & 0 & 0 \end{pmatrix},$$

$$\alpha_z = \begin{pmatrix} 0 & \sigma_z \\ \sigma_z & 0 \end{pmatrix} = \begin{pmatrix} 0 & 0 & 1 & 0 \\ 0 & 0 & 0 & -1 \\ 1 & 0 & 0 & 0 \\ 0 & -1 & 0 & 0 \end{pmatrix}, \quad \beta = \begin{pmatrix} I & 0 \\ 0 & -I \end{pmatrix} = \begin{pmatrix} 1 & 0 & 0 & 0 \\ 0 & 1 & 0 & 0 \\ 0 & 0 & -1 & 0 \\ 0 & 0 & 0 & -1 \end{pmatrix}$$

Hence in short

$$\alpha_i = \begin{pmatrix} 0 & \sigma_i \\ \sigma_i & 0 \end{pmatrix}, \quad \beta = \begin{pmatrix} I & 0 \\ 0 & -I \end{pmatrix} \qquad (11.146)$$

As α_i, β are 4×4 matrices it follows from the Hamiltonian H of equation (11.111) that the Hamiltonian contains 4×4 operator matrices and they operate on the wave function $\psi(\vec{r}, t)$. So the wave function $\psi(\vec{r}, t)$ will be a 4×1 column matrix as

$$|\psi(\vec{r},\,t)\rangle \;=\; \begin{pmatrix} \psi_1 \\ \psi_2 \\ \psi_3 \\ \psi_4 \end{pmatrix} \tag{11.147}$$

This is called a Dirac spinor and represents spin $\frac{1}{2}$ particles, say electron, positron. The Dirac equation (11.111) is invariant under Lorentz transformation only when the state vector $|\psi(\vec{r},\,t)\rangle$ has this form. This state vector $|\psi(\vec{r},\,t)\rangle$ does not transform like a 4-vector in Minkowski space but corresponds to a Dirac spinor.

11.18 Covariant form of the Dirac equation through introduction of γ matrices

Multiply the Dirac equation (11.115) viz. $-i\hbar c(\vec{\alpha}.\,\vec{\nabla})\psi + \beta\,mc^2\psi = i\hbar\frac{\partial}{\partial t}\psi$ by $\frac{\beta}{c\hbar}$ to get

$$\frac{\beta}{c\hbar}(-)i\hbar c(\vec{\alpha}.\,\vec{\nabla})\psi + \frac{\beta}{c\hbar}\beta\,mc^2\psi = \frac{\beta}{c\hbar}i\hbar\frac{\partial}{\partial t}\psi$$

$$\frac{i\beta}{c}\frac{\partial}{\partial t}\psi + i\beta(\vec{\alpha}.\,\vec{\nabla})\psi - \beta^2\frac{mc}{\hbar}\,\psi = 0$$

Using $\beta^2 = 1$ we have

$$i\beta\frac{\partial}{\partial ct}\psi + i\beta(\vec{\alpha}.\,\vec{\nabla})\psi - \frac{mc}{\hbar}\,\psi = 0 \tag{11.148}$$

Using equation (11.28)

$$x^0 = x_0 = ct,\ x^1 = -x_1 = x,\ x^2 = -x_2 = y,\ x^3 = -x_3 = z$$

$$\alpha_x = -\alpha_1 = \alpha^1,\ \alpha_y = -\alpha_2 = \alpha^2,\ \alpha_z = -\alpha_3 = \alpha^3 \ i.\ e.\ \vec{\alpha} = -\alpha_k = \alpha^k$$

$$\vec{\nabla} = -\partial^k = \partial_k,\ \frac{\partial}{\partial ct}\ = \partial^0 = \partial_0 \ \text{(equations (11.12) and (11.13))}$$

we rewrite equation (11.148) as

$$i\beta\partial^0\psi + i\beta(-\,\alpha_k)(-\partial^k)\psi - \frac{mc}{\hbar}\,\psi = 0$$

$$i\beta\partial^0\psi + i\beta\alpha_k\partial^k\psi - \frac{mc}{\hbar}\,\psi = 0 \tag{11.149}$$

Define a matrix called gamma matrix as

$$\gamma_0 = \beta,\ \gamma_k = \beta\alpha_k = \gamma_0\alpha_k \tag{11.150}$$

From equation (11.149) we have

$$i\gamma_0 \partial^0 \psi + i\gamma_k \partial^k \psi - \frac{mc}{\hbar}\,\psi = 0$$

$$i\gamma_\mu \partial^\mu \psi - \frac{mc}{\hbar}\,\psi = 0$$

$$\left(i\gamma_\mu \partial^\mu - \frac{mc}{\hbar}\right)\psi = 0 \tag{11.151}$$

Multiplying by i

$$\left(i^2 \gamma_\mu \partial^\mu - \frac{imc}{\hbar}\right)\psi = 0$$

$$\left(-\gamma_\mu \partial^\mu - \frac{imc}{\hbar}\right)\psi = 0$$

$$\left(\gamma_\mu \partial^\mu + \frac{imc}{\hbar}\right)\psi = 0 \tag{11.152}$$

Equations (11.151) and (11.152) are covariant forms of the Dirac relativistic equation.

In natural units $\hbar = c = 1$ we have from equation (11.152)

$$\left(\gamma_\mu \partial^\mu + im\right)\psi = 0 \tag{11.153}$$

To make this Dirac equation relativistically invariant, ψ is a 4×1 column vector called a spinor.

We can also recast the Dirac equation as follows (*exercise 11.10*)

$$(i\gamma^\mu \partial_\mu - m)\psi = 0$$

$$(\gamma^\mu \partial_\mu + im\ )\psi = 0 \tag{11.154}$$

where

$$\gamma^0 = \gamma_0 = \beta, \ \ \gamma^k = -\gamma_k = -\beta\alpha_k = -\gamma_0 \alpha_k \tag{11.155}$$

The momentum operator is $-i\hbar\,\vec{\nabla}$. Using $\vec{\nabla} = -\partial^i = \partial_i$ (equations (11.12) and (11.13)) we can write

$$-i\hbar\,\vec{\nabla} = -i\hbar\partial_i = p_i \tag{11.156}$$

$$-i\hbar\,\vec{\nabla} = -i\hbar(-\partial^i) = i\hbar\partial^i = p^i \tag{11.157}$$

$$\frac{E}{c} = \frac{1}{c}i\hbar\frac{\partial}{\partial t} = i\hbar\frac{\partial}{\partial ct} = i\hbar\frac{\partial}{\partial x^0} = i\hbar\frac{\partial}{\partial x_0} = i\hbar\partial^0 = i\hbar\partial_0 \tag{11.158}$$

Also,

$$p^{\mu} = \left(\frac{E}{c}, \; \vec{p}\right) = (p^0, p^i) = (i\hbar\partial^0, i\hbar\partial^i) = i\hbar\partial^{\mu} \tag{11.159}$$

With this we can rewrite equation (11.152) viz. $\left(\gamma_{\mu}\partial^{\mu} + \frac{imc}{\hbar}\right)\psi = 0$ as

$$\left(\gamma_{\mu}i\hbar\partial^{\mu} + \frac{(i\hbar)imc}{\hbar}\right)\psi = 0 \;\; \text{(multiplying by } i\hbar)$$

$$\left(\gamma_{\mu}p^{\mu} - mc\right)\psi = 0. \tag{11.160}$$

11.19 Properties of Dirac's γ matrices

The Dirac γ matrices are γ^0, γ^1, γ^2, γ^3 related to the Dirac matrices α_x, α_y, α_z, β as

$$\gamma^0 = \beta, \;\; \gamma^k = \beta\alpha_k \tag{11.161}$$

$$\gamma_0 = \beta, \;\; \gamma_k = -\beta\alpha_k = -\gamma_0\alpha_k \tag{11.162}$$

And so the structure of the γ matrices are as follows

$$\gamma^0 = \beta = \begin{pmatrix} I & 0 \\ 0 & -I \end{pmatrix} = \begin{pmatrix} 1 & 0 & 0 & 0 \\ 0 & 1 & 0 & 0 \\ 0 & 0 & -1 & 0 \\ 0 & 0 & 0 & -1 \end{pmatrix} \tag{11.163}$$

$$\gamma^1 = -\gamma_1 = \beta\alpha_1 = \gamma_0\alpha_x = \begin{pmatrix} I & 0 \\ 0 & -I \end{pmatrix}\begin{pmatrix} 0 & \sigma_x \\ \sigma_x & 0 \end{pmatrix} = \begin{pmatrix} 0 & \sigma_x \\ -\sigma_x & 0 \end{pmatrix} = \begin{pmatrix} 0 & 0 & 0 & 1 \\ 0 & 0 & 1 & 0 \\ 0 & -1 & 0 & 0 \\ -1 & 0 & 0 & 0 \end{pmatrix}$$

$$\gamma^2 = -\gamma_2 = \beta\alpha_2 = \gamma_0\alpha_y = \begin{pmatrix} I & 0 \\ 0 & -I \end{pmatrix}\begin{pmatrix} 0 & \sigma_y \\ \sigma_y & 0 \end{pmatrix} = \begin{pmatrix} 0 & \sigma_y \\ -\sigma_y & 0 \end{pmatrix} = \begin{pmatrix} 0 & 0 & 0 & -i \\ 0 & 0 & i & 0 \\ 0 & i & 0 & 0 \\ -i & 0 & 0 & 0 \end{pmatrix}$$

$$\gamma^3 = -\gamma_3 = \beta\alpha_3 = \gamma_0\alpha_z = \begin{pmatrix} I & 0 \\ 0 & -I \end{pmatrix}\begin{pmatrix} 0 & \sigma_z \\ \sigma_z & 0 \end{pmatrix} = \begin{pmatrix} 0 & \sigma_z \\ -\sigma_z & 0 \end{pmatrix} = \begin{pmatrix} 0 & 0 & 1 & 0 \\ 0 & 0 & 0 & -1 \\ -1 & 0 & 0 & 0 \\ 0 & 1 & 0 & 0 \end{pmatrix}$$

i.e.

$$\gamma^k = -\gamma_k = \beta\alpha_k = \gamma_0\alpha_k = \begin{pmatrix} I & 0 \\ 0 & -I \end{pmatrix}\begin{pmatrix} 0 & \sigma_k \\ \sigma_k & 0 \end{pmatrix} = \begin{pmatrix} 0 & \sigma_k \\ -\sigma_k & 0 \end{pmatrix}, \;\; k = x, y, z \tag{11.164}$$

- Hermiticity of γ matrices

✓ Consider $\gamma^0 = \beta$,

$$\gamma^{0\dagger} = \beta^\dagger = \beta = \gamma^0 \tag{11.165}$$

γ^0 is a Hermitian matrix.

✓ Consider $\gamma^k = -\gamma_k = \beta\alpha_k = \gamma_0\alpha_k$

$$\gamma^{k\dagger} = (\beta\alpha_k)^\dagger = \alpha_k^\dagger\beta^\dagger = \alpha_k\beta \text{ (as } \alpha_k, \ \beta \text{ are Hermitian)}$$

$$= -\beta\alpha_k \text{ as } \alpha_k, \ \beta \text{ anti-commutes.}$$

$$\gamma^{k\dagger} = -\gamma^k \tag{11.166}$$

So γ^k s are anti-Hermitian.

- The γ matrices are traceless, i.e.

$$Tr \ \gamma^\mu = 0 \tag{11.167}$$

- Square of γ matrices

$$\gamma^{02} = \gamma^0\gamma^0 = \beta\beta = \beta^2 = I \tag{11.168}$$

$$\gamma^{k2} = \gamma^k\gamma^k = (\beta\alpha_k)(\beta\alpha_k) = \beta\alpha_k\beta\alpha_k = -\beta\beta\alpha_k\alpha_k = -\beta^2\alpha_k^2 = -I \tag{11.169}$$

- Product of the γ matrices

$$\gamma^0\gamma^1\gamma^2\gamma^3 = \beta(\beta\alpha_x)(\beta\alpha_y)(\beta\alpha_z)$$

$$= \beta\beta\alpha_x\beta\alpha_y\beta\alpha_z = \beta^2\alpha_x\beta\alpha_y\beta\alpha_z = I\alpha_x\beta\alpha_y\beta\alpha_z$$

$$= \alpha_x\beta\alpha_y\beta\alpha_z = -\alpha_x\alpha_y\beta\beta\alpha_z = -\alpha_x\alpha_y\beta^2\alpha_z = -\alpha_x\alpha_y\alpha_z$$

$$= -\begin{pmatrix} 0 & \sigma_x \\ \sigma_x & 0 \end{pmatrix}\begin{pmatrix} 0 & \sigma_y \\ \sigma_y & 0 \end{pmatrix}\begin{pmatrix} 0 & \sigma_z \\ \sigma_z & 0 \end{pmatrix} = -\begin{pmatrix} 0 & \sigma_x \\ \sigma_x & 0 \end{pmatrix}\begin{pmatrix} \sigma_y\sigma_z & 0 \\ 0 & \sigma_y\sigma_z \end{pmatrix}$$

$$\gamma^0\gamma^1\gamma^2\gamma^3 = -\begin{pmatrix} 0 & \sigma_x\sigma_y\sigma_z \\ \sigma_x\sigma_y\sigma_z & 0 \end{pmatrix} \tag{11.170}$$

Use $\sigma_x\sigma_y\sigma_z = \begin{pmatrix} 0 & 1 \\ 1 & 0 \end{pmatrix}\begin{pmatrix} 0 & -i \\ i & 0 \end{pmatrix}\begin{pmatrix} 1 & 0 \\ 0 & -1 \end{pmatrix} = \begin{pmatrix} 0 & 1 \\ 1 & 0 \end{pmatrix}\begin{pmatrix} 0 & i \\ i & 0 \end{pmatrix} = \begin{pmatrix} i & 0 \\ 0 & i \end{pmatrix} = iI \tag{11.171}$

$$\gamma^0\gamma^1\gamma^2\gamma^3 = -\begin{pmatrix} 0 & iI \\ iI & 0 \end{pmatrix} = -i\begin{pmatrix} 0 & I \\ I & 0 \end{pmatrix} = -i\begin{pmatrix} 0 & 0 & 1 & 0 \\ 0 & 0 & 0 & 1 \\ 1 & 0 & 0 & 0 \\ 0 & 1 & 0 & 0 \end{pmatrix}$$

$$i\gamma^0\gamma^1\gamma^2\gamma^3 = \begin{pmatrix} 0 & I \\ I & 0 \end{pmatrix} = \begin{pmatrix} 0 & 0 & 1 & 0 \\ 0 & 0 & 0 & 1 \\ 1 & 0 & 0 & 0 \\ 0 & 1 & 0 & 0 \end{pmatrix} = \gamma^5 \text{ (call it)} \tag{11.172}$$

- Anti-commutator of the γ matrices $\{\gamma^\mu \;\; \gamma^\nu\}$
  Let us find

$$\{\gamma^0 \;\; \gamma^0\} = \gamma^0\gamma^0 + \gamma^0\gamma^0 = 2\gamma^0\gamma^0 = 2\beta\beta = 2\beta^2 = 2I \tag{11.173}$$

$$\{\gamma^0 \;\; \gamma^k\}_{k\neq 0} = \gamma^0\gamma^k + \gamma^k\gamma^0 = \beta(\beta\alpha_k) + (\beta\alpha_k)\beta = (\beta^2\alpha_k - \beta^2\alpha_k) \;\; = 0 \tag{11.174}$$

$$\{\gamma^k \;\; \gamma^k\} = \gamma^k\gamma^k + \gamma^k\gamma^k = 2\gamma^k\gamma^k = 2(\beta\alpha_k)(\beta\alpha_k)$$

$$=2\beta\alpha_k\beta\alpha_k = -2\beta\beta\alpha_k\alpha_k = -2\beta^2\alpha_k^2 = -2I \tag{11.175}$$

$$\{\gamma^k \;\; \gamma^j\}|_{k\neq 0,\, j\neq 0,\, k\neq j} = \gamma^k\gamma^j + \gamma^j\gamma^k = (\beta\alpha_k)(\beta\alpha_j) + \left(\beta\alpha_j\right)(\beta\alpha_k)$$

$$=\beta\alpha_k\beta\alpha_j + \beta\alpha_j\beta\alpha_k = -\beta\beta\alpha_k\alpha_j - \beta\beta\alpha_j\alpha_k = -\beta^2\alpha_k\alpha_j - \beta^2\alpha_j\alpha_k$$

$$=-(\alpha_k\alpha_j + \alpha_j\alpha_k) = -\left\{\alpha_k \;\; \alpha_j \;\right\} = 0 \tag{11.176}$$

We have defined the matrix $\eta = \mathrm{diag}(1, -1, -1, -1)$ in equation (11.30) such that

$$\eta^{00} = 1,\, \eta^{11} = -1,\;\; \eta^{22} = -1,\;\; \eta^{33} = -1,\;\; \eta^{\mu\nu}_{\mu\neq\nu} = 0 \tag{11.177}$$

We rewrite the above results as follows

$$\{\gamma^0 \;\; \gamma^0\} = 2I = 2\eta^{00}I \tag{11.178}$$

$$\{\gamma^0 \;\; \gamma^k\}_{k\neq 0} = 0 = 2\eta^{0k}I \tag{11.179}$$

$$\{\gamma^k \;\; \gamma^k\} = -2I = 2\eta^{kk}I \tag{11.180}$$

$$\{\gamma^k \;\; \gamma^j\}_{k\neq j} = 0 = 2\eta^{0k}I \tag{11.181}$$

Combining the results we have

$$\{\gamma^\mu \;\; \gamma^\nu\} = 2\eta^{\mu\nu}I \tag{11.182}$$

- We can combine and write γ^μ, γ_ν, $\eta_{\mu\nu} = \eta^{\mu\nu}$ as follows (Einstein summation convention of summing over repeated indices used)

$$\gamma_0 = \gamma^0 = \eta_{00}\gamma^0 = \eta_{0\nu}\gamma^\nu \ \left(\eta_{00} = 1, \ \ \eta_{0\nu}|_{\nu\neq 0} = 0\right) \tag{11.183}$$

$$\gamma_k|_{k\neq 0} = -\gamma^k = \eta_{kk}\gamma^k = \eta_{kj}\gamma^j \ \left(\eta_{kk}|_{k\neq 0} = -1, \ \ \eta_{kj}|_{k\neq 0,\,j\neq 0,\,k\neq j} = 0\right) \tag{11.184}$$

Combining

$$\gamma_\mu = \eta_{\mu\nu}\gamma^\nu \ \left(\text{since } (\gamma_0, \gamma_1, \gamma_2, \gamma_3) = (\gamma^0, \ -\gamma^1, -\gamma^2, \ -\gamma^3)\right) \tag{11.185}$$

-

$$\gamma^{5\dagger} = (i\gamma^0\gamma^1 \ \gamma^2\gamma^3)^{\ \dagger} = -i\gamma^{3\dagger}\gamma^{2\dagger}\gamma^{1\dagger}\gamma^{0\dagger} = -i(-\gamma^3)(-\gamma^2)(-\gamma^1)(\gamma^0)$$

$$= i\gamma^3\gamma^2\gamma^1\gamma^0 = -i\gamma^3\gamma^2\gamma^0\gamma^1 = i\gamma^3\gamma^0\gamma^2\gamma^1 = -i\gamma^0\gamma^3\gamma^2\gamma^1$$

$$= i\gamma^0\gamma^3\gamma^1\gamma^2 = -i\gamma^0\gamma^1\gamma^3\gamma^2 = i\gamma^0\gamma^1\gamma^2\gamma^3$$

$$\gamma^{5\dagger} = \gamma^5 \tag{11.186}$$

γ^5 is Hermitian.

-

$$(\gamma^5)^2 = \gamma^5\gamma^5 = \begin{pmatrix} 0 & I \\ I & 0 \end{pmatrix}\begin{pmatrix} 0 & I \\ I & 0 \end{pmatrix} = \begin{pmatrix} I & 0 \\ 0 & I \end{pmatrix} = I \tag{11.187}$$

Eigenvalues of γ^5 are ± 1

-

$$\gamma^\mu\gamma_\mu = \gamma^0\gamma_0 + \gamma^k\gamma_k = \beta\beta + (\beta\alpha_k)(-\beta\alpha_k) = \beta^2 - \beta\alpha_k\beta\alpha_k$$

$$= \beta^2 + \beta\beta\alpha_k\alpha_k = \beta^2 + \beta^2\alpha_k^2 = \ I + I\alpha_k^2$$

$$= I + \alpha_1^2 + \alpha_2^2 + \alpha_3^2 = I + I + I + I$$

$$\gamma^\mu\gamma_\mu = 4I \tag{11.188}$$

- Dirac γ matrices are not 4-vectors. They are constant matrices which remain invariant under a Lorentz transformation.
- The Dirac equation is Lorentz covariant.

11.20 Probability density from the Dirac equation

Equation (11.115) is

$$i\hbar\frac{\partial\psi}{\partial t} = -ic\hbar\,\vec{\alpha}.\overrightarrow{\nabla}\,\psi + \beta mc^2\psi$$

$$i\hbar\frac{\partial\psi}{\partial t} = -ic\hbar\left(\alpha_x\frac{\partial\psi}{\partial x} + \alpha_y\frac{\partial\psi}{\partial y} + \alpha_z\frac{\partial\psi}{\partial z}\right) + mc^2\beta\psi \tag{11.189}$$

The complex conjugate of this equation is

$$-i\hbar\frac{\partial\psi^\dagger}{\partial t} = ic\hbar\left(\alpha_x^\dagger\frac{\partial\psi^\dagger}{\partial x} + \alpha_y^\dagger\frac{\partial\psi^\dagger}{\partial y} + \alpha_z^\dagger\frac{\partial\psi^\dagger}{\partial z}\right) + mc^2\psi^\dagger\beta^\dagger$$

$$-i\hbar\frac{\partial\psi^\dagger}{\partial t} = ic\hbar\left(\alpha_x\frac{\partial\psi^\dagger}{\partial x} + \alpha_y\frac{\partial\psi^\dagger}{\partial y} + \alpha_z\frac{\partial\psi^\dagger}{\partial z}\right) + mc^2\psi^\dagger\beta \tag{11.190}$$

Multiply equation (11.189) by $\psi^\dagger$ from the left to get

$$i\hbar\psi^\dagger\frac{\partial\psi}{\partial t} = -ic\hbar\left(\psi^\dagger\,\alpha_x\frac{\partial\psi}{\partial x} + \psi^\dagger\alpha_y\frac{\partial\psi}{\partial y} + \psi^\dagger\alpha_z\frac{\partial\psi}{\partial z}\right) + mc^2\psi^\dagger\beta\psi \tag{11.191}$$

Multiply equation (11.190) by ψ from the right to get

$$-i\hbar\frac{\partial\psi^\dagger}{\partial t}\psi = ic\hbar\left(\alpha_x\frac{\partial\psi^\dagger}{\partial x}\psi + \alpha_y\frac{\partial\psi^\dagger}{\partial y}\psi + \alpha_z\frac{\partial\psi^\dagger}{\partial z}\psi\right) + mc^2\psi^\dagger\beta\psi \tag{11.192}$$

Subtracting equation (11.192) from equation (11.190)

$$i\hbar\left(\psi^\dagger\frac{\partial\psi}{\partial t} + \frac{\partial\psi^\dagger}{\partial t}\psi\right) = -ic\hbar\left(\psi^\dagger\,\alpha_x\frac{\partial\psi}{\partial x} + \psi^\dagger\alpha_y\frac{\partial\psi}{\partial y} + \psi^\dagger\alpha_z\frac{\partial\psi}{\partial z}\right) + mc^2\psi^\dagger\beta\psi$$

$$-\left[\,ic\hbar\left(\alpha_x\frac{\partial\psi^\dagger}{\partial x}\psi + \alpha_y\frac{\partial\psi^\dagger}{\partial y}\psi + \alpha_z\frac{\partial\psi^\dagger}{\partial z}\psi\right) + mc^2\psi^\dagger\beta\psi\,\right]$$

$$\frac{\partial}{\partial t}(\psi^\dagger\psi) = -c\left[\left(\psi^\dagger\,\alpha_x\frac{\partial\psi}{\partial x} + \alpha_x\frac{\partial\psi^\dagger}{\partial x}\psi\right) + \left(\psi^\dagger\alpha_y\frac{\partial\psi}{\partial y} + \alpha_y\frac{\partial\psi^\dagger}{\partial y}\psi\right)\right.$$

$$\left. + \left(\psi^\dagger\alpha_z\frac{\partial\psi}{\partial z} + \alpha_z\frac{\partial\psi^\dagger}{\partial z}\psi\right)\right]$$

$$\frac{\partial}{\partial t}(\psi^\dagger\psi) = -c\left[\frac{\partial}{\partial x}\left(\psi^\dagger\,\alpha_x\psi\right) + \frac{\partial}{\partial y}\left(\psi^\dagger\,\alpha_y\psi\right) + \frac{\partial}{\partial z}(\psi^\dagger\,\alpha_z\psi)\right]$$

$$\frac{\partial}{\partial t}(\psi^\dagger\psi) + \frac{\partial}{\partial k}(\psi^\dagger\,c\alpha_k\psi) = 0$$

$$\frac{\partial}{\partial t}(\psi^\dagger\psi) + \vec{\nabla}\cdot(\psi^\dagger\,c\vec{\alpha}\,\psi) = 0$$

$$\frac{\partial\rho}{\partial t} + \vec{\nabla}\cdot\vec{J} = 0 \tag{11.193}$$

where

$$\rho = \psi^\dagger \psi = \begin{pmatrix} \psi_1^* & \psi_2^* & \psi_3^* & \psi_4^* \end{pmatrix} \begin{pmatrix} \psi_1 \\ \psi_2 \\ \psi_3 \\ \psi_4 \end{pmatrix} = \psi_1^*\psi_1 + \psi_2^*\psi_2 + \psi_3^*\psi_3 + \psi_4^*\psi_4 \tag{11.194}$$

$$= \text{positive definite number called charge density}$$

$$\vec{J} = \psi^\dagger \, c\vec{\alpha} \, \psi \tag{11.195}$$
$$= \text{current density}$$

We mention that the problem of negative charge density of the Klein–Gordon equation is not there in the Dirac equation.

We define a 4-vector current as

$$J^\mu = (\rho c, \ \vec{J}) \tag{11.196}$$

where

$$J^0 = J_0 = \rho c, \ J^k = -J_k = J_x, \ k = 1, 2, 3 \text{ or } x, y, z \tag{11.197}$$

We can express the continuity equation in covariant form also as follows.

Equation (11.193) is

$$\frac{\partial \rho c}{\partial ct} + \frac{\partial J_x}{\partial x} + \frac{\partial J_y}{\partial y} + \frac{\partial J_z}{\partial z} = 0$$

$$\frac{\partial J^0}{\partial x^0} + \frac{\partial J^1}{\partial x^1} + \frac{\partial J^2}{\partial x^2} + \frac{\partial J^3}{\partial x^3} = 0$$

$$\frac{\partial J^\mu}{\partial x^\mu} = 0$$

$$\partial_\mu J^\mu = 0 \tag{11.198}$$

11.21 Plane wave solution of the Dirac equation

The Dirac equation

$$i\hbar \frac{\partial \psi}{\partial t} = -ic\hbar \, \vec{\alpha}. \, \overrightarrow{\nabla} \, \psi + \beta mc^2 \psi \tag{11.199}$$

contains the Dirac matrices α_k, β and these are 4×4 matrices. As they operate upon ψ it follows that ψ has to be a 4×1 column matrix, i.e. ψ has the form

$$\psi(\vec{r},\, t) = \begin{pmatrix} \psi_1(\vec{r},\, t) \\ \psi_2(\vec{r},\, t) \\ \psi_3(\vec{r},\, t) \\ \psi_4(\vec{r},\, t) \end{pmatrix} \tag{11.200}$$

Taking the free particle plane wave function as

$$\psi(\vec{r},\, t) = u e^{i(\vec{k}.\vec{r} - wt)} \tag{11.201}$$

where the space and time dependence enters through the exponential factor and u has 4 components and the form of a 4×1 column matrix and is called a spinor

$$u = \begin{pmatrix} u_1 \\ u_2 \\ u_3 \\ u_4 \end{pmatrix} = \begin{pmatrix} f \\ g \end{pmatrix}, \quad = \begin{pmatrix} u_1 \\ u_2 \end{pmatrix}, \quad g = \begin{pmatrix} u_3 \\ u_4 \end{pmatrix} \tag{11.202}$$

We have from the Dirac equation using equation (11.201)

$$i\hbar \frac{\partial}{\partial t}\, u e^{i(\vec{k}.\vec{r}-wt)} = -ic\hbar\, \vec{\alpha}.\, \overrightarrow{\nabla}\, u e^{i(\vec{k}.\vec{r}-wt)} + \beta mc^2 u e^{i(\vec{k}.\vec{r}-wt)}$$

$$i\hbar(-iw)\, u e^{i(\vec{k}.\vec{r}-wt)} = -ic\hbar\, \vec{\alpha}.\, (i\vec{k})\, u e^{i(\vec{k}.\vec{r}-wt)} + \beta mc^2 u e^{i(\vec{k}.\vec{r}-wt)} \tag{11.203}$$

With

$$E = \hbar w, \quad \vec{p} = \hbar \vec{k} \tag{11.204}$$

we rewrite

$$E\, u = c\, \vec{\alpha}.\vec{p}\, u + \beta mc^2 u \tag{11.205}$$

We thus get four simultaneous first order partial differential equations that are linear and homogeneous in u_j where the index $j - 1, 2, 3, 4$ represents the 4 components of wave function. Using equation (11.202) in equation (11.205) we get

$$E \begin{pmatrix} f \\ g \end{pmatrix} = c \begin{pmatrix} 0 & \vec{\sigma} \\ \vec{\sigma} & 0 \end{pmatrix}.\vec{p} \begin{pmatrix} f \\ g \end{pmatrix} + \begin{pmatrix} I & 0 \\ 0 & -I \end{pmatrix} mc^2 \begin{pmatrix} f \\ g \end{pmatrix}$$

$$\begin{pmatrix} Ef \\ Eg \end{pmatrix} = c \begin{pmatrix} 0 & \vec{\sigma}.\vec{p} \\ \vec{\sigma}.\vec{p} & 0 \end{pmatrix} \begin{pmatrix} f \\ g \end{pmatrix} + mc^2 \begin{pmatrix} fI \\ -gI \end{pmatrix}$$

$$\begin{pmatrix} Ef \\ Eg \end{pmatrix} = \begin{pmatrix} c\vec{\sigma}.\vec{p}\, g \\ c\vec{\sigma}.\vec{p}\, f \end{pmatrix} + \begin{pmatrix} mc^2 f \\ -mc^2 g \end{pmatrix} \tag{11.206}$$

This leads to two equations, namely

$$Ef = c\vec{\sigma}.\vec{p}\, g + mc^2 f \tag{11.207}$$

$$Eg = c\vec{\sigma}.\vec{p}f - mc^2 g \tag{11.208}$$

This gives

$$(E - mc^2)f = c\vec{\sigma}.\vec{p}\,g \tag{11.209}$$

$$(E + mc^2)g = c\vec{\sigma}.\vec{p}\,f \tag{11.210}$$

Multiply equation (11.209) by $(E + mc^2)$ to get

$$(E + mc^2)(E - mc^2)f = (E + mc^2)c\vec{\sigma}.\vec{p}\,g$$

$$(E^2 - m^2 c^4)f = c\vec{\sigma}.\vec{p}\,(E + mc^2)g$$

Using equation (11.210) we have

$$(E^2 - m^2 c^4)f = (c\vec{\sigma}.\vec{p})(c\vec{\sigma}.\vec{p}f)$$

$$(E^2 - m^2 c^4)f = c^2 (\vec{\sigma}.\vec{p})^2 f \tag{11.211}$$

Again

$$\vec{\sigma}.\vec{p} = \sigma_x p_x + \sigma_y p_y + \sigma_z p_z = \begin{pmatrix} 0 & 1 \\ 1 & 0 \end{pmatrix} p_x + \begin{pmatrix} 0 & -i \\ i & 0 \end{pmatrix} p_y + \begin{pmatrix} 1 & 0 \\ 0 & -1 \end{pmatrix} p_z$$

$$= \begin{pmatrix} 0 & p_x \\ p_x & 0 \end{pmatrix} + \begin{pmatrix} 0 & -ip_y \\ ip_y & 0 \end{pmatrix} + \begin{pmatrix} p_z & 0 \\ 0 & -p_z \end{pmatrix}$$

$$\vec{\sigma}.\vec{p} = \begin{pmatrix} p_z & p_x - ip_y \\ p_x + ip_y & -p_z \end{pmatrix} \tag{11.212}$$

$$(\vec{\sigma}.\vec{p})^2 = \begin{pmatrix} p_z & p_x - ip_y \\ p_x + ip_y & -p_z \end{pmatrix}\begin{pmatrix} p_z & p_x - ip_y \\ p_x + ip_y & -p_z \end{pmatrix} = \begin{pmatrix} p^2 & 0 \\ 0 & p^2 \end{pmatrix} = p^2 \begin{pmatrix} 1 & 0 \\ 0 & 1 \end{pmatrix} = p^2$$

$$(\vec{\sigma}.\vec{p})^2 = p^2 \tag{11.213}$$

Hence from equations (11.213) and (11.211)

$$(E^2 - m^2 c^4)f = c^2 p^2 f$$

$$E^2 = c^2 p^2 + m^2 c^4$$

$$E = \pm\sqrt{c^2 p^2 + m^2 c^4} \tag{11.214}$$

This is the energy eigenvalue and it can be positive

$$E_+ = +\sqrt{c^2 p^2 + m^2 c^4} \tag{11.215}$$

as well as negative

$$E_- = -\sqrt{c^2\, p^2 + m^2 c^4} \tag{11.216}$$

relativistically.

11.22 Energy eigenfunction

From equation (11.209) viz. $(E - mc^2)f = c\vec{\sigma}.\vec{p}\, g$ we have

$$f = \frac{c\vec{\sigma}.\vec{p}}{E - mc^2} g \tag{11.217}$$

In the non-relativistic situation which is a low momentum case

$$pc \ll <mc^2 \tag{11.218}$$

and so

$$E_+ = +\sqrt{c^2\, p^2 + m^2 c^4} \rightarrow +mc^2 \tag{11.219}$$

$$E_- = -\sqrt{c^2\, p^2 + m^2 c^4} \rightarrow -mc^2 \tag{11.220}$$

Again if we set $E = E_+$ in f (equation (11.217)) we have for $p \rightarrow 0$

$$f\big|_{p\rightarrow 0} = \frac{c\vec{\sigma}.\vec{p}}{(E_+ - mc^2)} g$$

Using equation (11.219) we have

$$f\big|_{p\rightarrow 0} = \frac{c\vec{\sigma}.\vec{p}}{(mc^2 - mc^2)} g \rightarrow \infty$$

Such divergence is not acceptable. So E_+ does not correspond to v. With $E = E_-$ we have from equation (11.217)

$$f = \frac{c\vec{\sigma}.\vec{p}}{(E_- - mc^2)} g \tag{11.221}$$

So f corresponds to $E = E_-$.

Similarly, from equation (11.210) viz. $(E + mc^2)g = c\vec{\sigma}.\vec{p}\, f$ we have

$$g = \frac{c\vec{\sigma}.\vec{p}}{E_- + mc^2} f \tag{11.222}$$

and for $E = E_-$ we have for $p \rightarrow 0$ from equation (11.220)

$$g\big|_{p\rightarrow 0} = \frac{c\vec{\sigma}.\vec{p}}{(-mc^2 + mc^2)} f \rightarrow \infty$$

So g is divergent for $E = E_- = -mc^2$. So g does not correspond to $E = E_-$.

$$g = \frac{c\vec{\sigma}\cdot\vec{p}}{E_+ + mc^2}f \tag{11.223}$$

Clearly g corresponds to $E = E_+$

- Assume $f = \begin{pmatrix} 1 \\ 0 \end{pmatrix}$ in equation (11.223) and using equation (11.212) we have

$$g = \frac{c\vec{\sigma}\cdot\vec{p}}{E_+ + mc^2}\begin{pmatrix} 1 \\ 0 \end{pmatrix} = \frac{c}{E_+ + mc^2}\begin{pmatrix} p_z & p_x - ip_y \\ p_x + ip_y & -p_z \end{pmatrix}\begin{pmatrix} 1 \\ 0 \end{pmatrix} = \frac{c}{E_+ + mc^2}\begin{pmatrix} p_z \\ p_x + ip_y \end{pmatrix}$$

$$g = \begin{pmatrix} \dfrac{cp_z}{E_+ + mc^2} \\[2ex] \dfrac{c(p_x + ip_y)}{E_+ + mc^2} \end{pmatrix} \tag{11.224}$$

From equation (11.202) we write

$$u = \begin{pmatrix} f \\ g \end{pmatrix} = \begin{pmatrix} 1 \\ 0 \\ \dfrac{cp_z}{E_+ + mc^2} \\[2ex] \dfrac{c(p_x + ip_y)}{E_+ + mc^2} \end{pmatrix} = u^{(1)} \text{ (one solution)} \tag{11.225}$$

- Assume $f = \begin{pmatrix} 0 \\ 1 \end{pmatrix}$ in equation (11.223) and using equation (11.212) we have

$$g = \frac{c\vec{\sigma}\cdot\vec{p}}{E_+ + mc^2}\begin{pmatrix} 0 \\ 1 \end{pmatrix} = \frac{c}{E_+ + mc^2}\begin{pmatrix} p_z & p_x - ip_y \\ p_x + ip_y & -p_z \end{pmatrix}\begin{pmatrix} 0 \\ 1 \end{pmatrix} = \frac{c}{E_+ + mc^2}\begin{pmatrix} p_x - ip_y \\ -p_z \end{pmatrix}$$

$$g = \begin{pmatrix} \dfrac{c(p_x - ip_y)}{E_+ + mc^2} \\[2ex] -\dfrac{cp_z}{E_+ + mc^2} \end{pmatrix} \tag{11.226}$$

From equation (11.202) we write

$$u = \begin{pmatrix} f \\ g \end{pmatrix} = \begin{pmatrix} 0 \\ 1 \\ \dfrac{c(p_x - ip_y)}{E_+ + mc^2} \\[2ex] -\dfrac{cp_z}{E_+ + mc^2} \end{pmatrix} = u^{(2)} \text{ (one solution)} \tag{11.227}$$

- Assume $g = \begin{pmatrix} 1 \\ 0 \end{pmatrix}$ in equation (11.221) to have

$$f = \frac{c\vec{\sigma}\cdot\vec{p}}{E_- - mc^2}\begin{pmatrix} 1 \\ 0 \end{pmatrix} = \frac{c}{E_- - mc^2}\begin{pmatrix} p_z & p_x - ip_y \\ p_x + ip_y & -p_z \end{pmatrix}\begin{pmatrix} 1 \\ 0 \end{pmatrix} = \frac{c}{E_- - mc^2}\begin{pmatrix} p_z \\ p_x + ip_y \end{pmatrix}$$

$$f = \begin{pmatrix} \dfrac{cp_z}{E_- - mc^2} \\[2mm] \dfrac{c(p_x + ip_y)}{E_- - mc^2} \end{pmatrix} \tag{11.228}$$

From equation (11.202) we write

$$u = \begin{pmatrix} f \\ g \end{pmatrix} = \begin{pmatrix} \dfrac{cp_z}{E_- - mc^2} \\[2mm] \dfrac{c(p_x + ip_y)}{E_- - mc^2} \\[2mm] 1 \\ 0 \end{pmatrix} = u^{(3)} \text{ (one solution)} \tag{11.229}$$

- Assume $g = \begin{pmatrix} 0 \\ 1 \end{pmatrix}$ in equation (11.221 to have)

$$f = \frac{c\vec{\sigma}\cdot\vec{p}}{E_- - mc^2}\begin{pmatrix} 0 \\ 1 \end{pmatrix} = \frac{c}{E_- - mc^2}\begin{pmatrix} p_z & p_x - ip_y \\ p_x + ip_y & -p_z \end{pmatrix}\begin{pmatrix} 0 \\ 1 \end{pmatrix} = \frac{c}{E_- - mc^2}\begin{pmatrix} p_x - ip_y \\ -p_z \end{pmatrix}$$

$$g = \begin{pmatrix} \dfrac{c(p_x - ip_y)}{E_- - mc^2} \\[2mm] -\dfrac{cp_z}{E_- - mc^2} \end{pmatrix} \tag{11.230}$$

From equation (11.202) we write

$$u = \begin{pmatrix} f \\ g \end{pmatrix} = \begin{pmatrix} \dfrac{c(p_x - ip_y)}{E_- - mc^2} \\[2mm] -\dfrac{cp_z}{E_- - mc^2} \\[2mm] 0 \\ 1 \end{pmatrix} = u^{(4)} \text{ (one solution)} \tag{11.231}$$

Let us write down the 4 planc wave solutions of the Dirac equation

$$\psi(\vec{r}, t) = u e^{i(\vec{k}\cdot\vec{r} - wt)} \tag{11.232}$$

where

$$
u^{(1)} = \begin{pmatrix} 1 \\ 0 \\ \dfrac{cp_z}{E_+ + mc^2} \\ \dfrac{c(p_x + ip_y)}{E_+ + mc^2} \end{pmatrix}, \quad
u^{(2)} = \begin{pmatrix} 0 \\ 1 \\ \dfrac{c(p_x - ip_y)}{E_+ + mc^2} \\ -\dfrac{cp_z}{E_+ + mc^2} \end{pmatrix},
$$

$$
u^{(3)} = \begin{pmatrix} \dfrac{cp_z}{E_- - mc^2} \\ \dfrac{c(p_x + ip_y)}{E_- - mc^2} \\ 1 \\ 0 \end{pmatrix}, \quad
u^{(4)} = \begin{pmatrix} \dfrac{c(p_x - ip_y)}{E_- - mc^2} \\ -\dfrac{cp_z}{E_- - mc^2} \\ 0 \\ 1 \end{pmatrix}
\tag{11.233}
$$

We have normalized these wave functions as in *exercise 11.14* and the completeness condition is satisfied.

In the non-relativistic limit

$$pc \ll mc^2 \text{ and so } E_- = -mc^2, \ E_+ = +mc^2$$

the quantities or components involving $\dfrac{1}{E_- - mc^2} = \dfrac{1}{-2mc^2} \to 0, \dfrac{1}{E_+ + mc^2} = \dfrac{1}{2mc^2} \to 0.$

In view of this the non-relativistic limit of the plane wave solutions of the Dirac equation given by equation (11.233) will be

$$
u^{(1)} = \begin{pmatrix} 1 \\ 0 \\ 0 \\ 0 \end{pmatrix}, \quad
u^{(2)} = \begin{pmatrix} 0 \\ 1 \\ 0 \\ 0 \end{pmatrix}, \quad
u^{(3)} = \begin{pmatrix} 0 \\ 0 \\ 1 \\ 0 \end{pmatrix}, \quad
u^{(4)} = \begin{pmatrix} 0 \\ 0 \\ 0 \\ 1 \end{pmatrix}
\tag{11.234}
$$

We note that the components $\begin{pmatrix} 1 \\ 0 \end{pmatrix}, \begin{pmatrix} 0 \\ 1 \end{pmatrix}$ do not change their values even in the non-relativistic limit and so are referred to as large components.

The other components that reduce to $\begin{pmatrix} 0 \\ 0 \end{pmatrix}$ in the non-relativistic limit are called small components.

We also note that the wave functions include the column matrices $\begin{pmatrix} 1 \\ 0 \end{pmatrix}, \begin{pmatrix} 0 \\ 1 \end{pmatrix}$ which are identical to the Pauli spin wave functions

$$|\text{spin up} > \ \equiv \alpha \equiv \begin{pmatrix} 1 \\ 0 \end{pmatrix} \equiv |\uparrow>, \ \text{spin down} > \ \equiv \beta \equiv \begin{pmatrix} 0 \\ 1 \end{pmatrix} \equiv |\downarrow> \tag{11.235}$$

Accordingly, the solutions $u^{(1)}$, $u^{(3)}$ are spin ups and $u^{(2)}$, $u^{(4)}$ are spin downs.

Clearly the total wave functions, say $\psi^{(1)}$, $\psi^{(2)}$, $\psi^{(3)}$, $\psi^{(4)}$, incorporate spin automatically and are called spinors. Clearly spin fits naturally into the Dirac equation.

The non-relativistic limit of the Dirac equation has been obtained in *exercise 11.16* and it reduces to the Schrödinger equation.

11.23 Interpretation of positive and negative energy states of a free particle and the idea of an anti-particle from Dirac's theory

Energy eigenvalues are

$$E_+ = +\sqrt{c^2\,p^2 + m^2 c^4} \xrightarrow{\text{non–relativistic limit}} + mc^2 \tag{11.236}$$

$$E_- = -\sqrt{c^2\,p^2 + m^2 c^4} \xrightarrow{\text{non–relativistic limit}} - mc^2 \tag{11.237}$$

Clearly the energy spectrum is divided into two parts.

In the non-relativistic limit (for instance for a particle at rest or moving with small momentum) the two parts are the positive rest energy $E_+ = mc^2$ and the negative rest energy $E_- = -mc^2$. Obviously they are separated by a gap of $\Delta E = 2mc^2$ as shown in figure 11.3.

This gap has no levels, i.e. particles are forbidden to occupy any state falling in the gap.

If the particle moves with more and more momentum we get positive energy states extending from mc^2 to ∞ and negative energy states extending from $-mc^2$ to $-\infty$. This is indicated in figure 11.3.

For a non-relativistic electron at rest or moving with small momentum, the rest energy is $mc^2 = 0.511\,MeV$ and hence the gap between positive and negative energy states is

$$\Delta E = 2mc^2 = 1.02\,MeV \tag{11.238}$$

It is clear that the Dirac equation describes particles which have negative energy $E_- < -mc^2$.

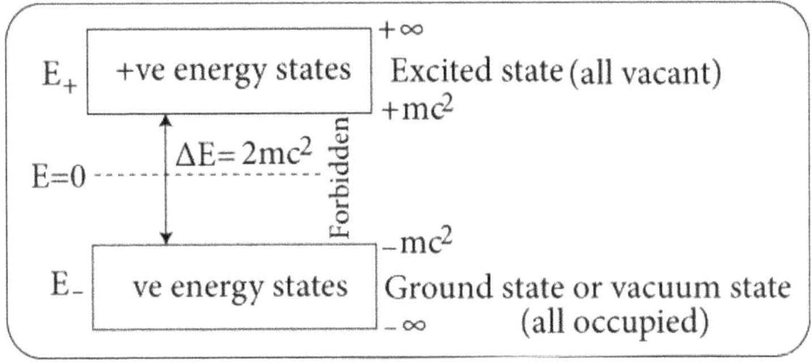

Figure 11.3. Positive and negative energy eigenvalues of a Dirac Hamiltonian. Ground state and excited states of a Dirac electron.

- The idea of a particle having negative energy and negative mass leads to the following consequences.

An attractive force acting on such a particle would repel it. No such particle was observed.

The Dirac electron has ground state (also called vacuum state) defined by all the states with energy $E \leqslant -mc^2$ and the states with $E \geqslant mc^2$ represent the excited states (figure 11.4).

In the ground state all the excited states (which are the positive energy states) are empty while all the negative energy states are occupied. The negative energy states are collectively called negative energy sea of particles say electrons (*exercise 11.11*).

- The system is quantum mechanical in nature. States are defined in the Hilbert space w.r.t. a basis. We have to incorporate all the linearly independent states as basis and they correspond to positive as well as negative energy states in Dirac theory. In other words, the negative energy states cannot be ignored and are part of Dirac theory.
- The Dirac vacuum consists of a large number (∞) of negative energy states $E = -mc^2$ to $E = -\infty$. A large number (∞) of particles fill up the negative energy states and there is a large number (∞) of degrees of freedom. This corresponds to field. So the concept of field which is avacuum with negative energy particles follows from the Dirac theory.

The vacuum state is not observable (since it is completely filled—it has no physically observable effect. Since electrons obey F–D statistics Pauli exclusion principle is followed. So these occupied states cannot accommodate any more electrons. Thus transition to negative energy states is prevented. A positive energy electron cannot make the transition to negative energy state and hence stability of positive energy electron is ensured and radiative collapse prevented. Only deviation from normal vacuum can be observed. We cannot measure vacuum properties, e.g. charge, field etc. In other words, the negative energy particles are not observable.

Particles with positive energy states are observables.

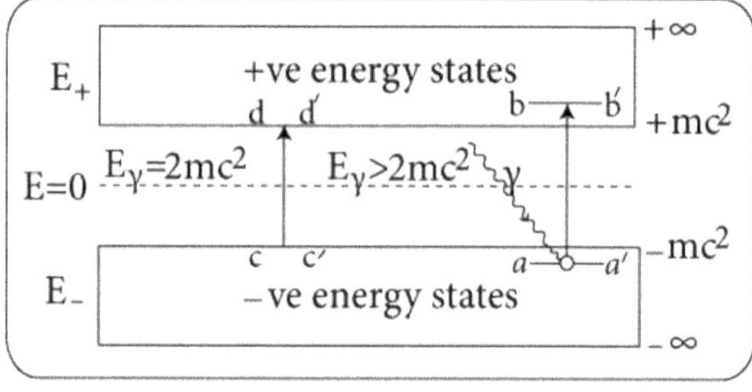

Figure 11.4. Dirac hole.

11.24 Anti-particle from Dirac theory

If a photon of energy $E_\gamma = 2mc^2$ interacts with a negative energy particle in state cc' say an electron (which is a negatively charged particle) it lifts the electron from the negative energy state cc' to the positive energy state dd' (figure 11.4).

If a photon of energy greater than $2mc^2$ interacts with the infinite negative energy sea of particles, say with an electron (which is a negatively charged particle) at state a', it lifts the electron from the negative energy state aa' to a positive energy state bb' (figure 11.4). And such transformation creates a vacant position or hole in the sea of negative energy states called the Dirac vacuum along with the simultaneous creation of an electron with positive energy. Since the hole corresponds to the vacancy of the electron this hole must have positive charge and also carries quantum numbers that are opposite to that of the particle (electron). This positively charged hole is called an anti-particle (positron). It has rest mass m (the same as the rest mass of the electron) and charge $-|e|$ (opposite to the electronic charge). This positron is the anti-particle of the electron. Clearly the idea of an anti-particle emerged from Dirac theory.

11.25 Pair production from Dirac theory

Suppose a hole that corresponds to an anti-particle is created in the negative energy sea of states. This hole represents a deviation from normal and so is manifested as a positively charged positron with positive energy. In other words, removal of a certain amount of negative energy and negative charge from the sea which is equivalent to creation of a hole is also equivalent to creation of an equal amount of positive energy and charge.

The whole process can be described as the disappearance of the quantum of electromagnetic radiation and associated appearance of a pair of observable particles—the electron e^- and positron e^+.

- Whenever there is transition from negative energy state to positive energy state there is creation of a particle–anti-particle pair. This is pair production and this concept follows from Dirac theory.

11.26 Pair annihilation from Dirac theory

A newly created hole in the vacuum will not last long, as an electron from the positive energy state will soon fall into it giving up the excess energy $\geqslant 2mc^2$ in the form of electromagnetic radiation. In the process both the electron and hole disappear. This phenomenon is the reverse of pair creation and is known as pair annihilation (figure 11.5).

- So hole theory very simply describes pair production and pair annihilation.
- Hole theory was vindicated by:
    1. the discovery of e^+
    2. the phenomenon of pair production
    3. the phenomenon of pair annihilation
    4. vacuum polarization.

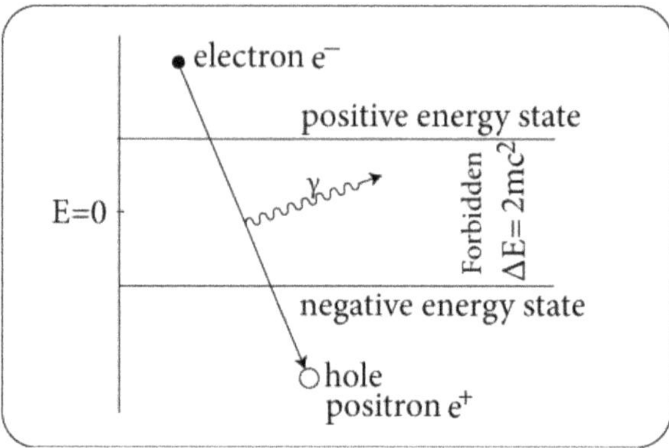

Figure 11.5. Pair annihilation.

11.27 Spin of a Dirac particle (electron, positron)

Presence of Pauli spin operators in the Dirac Hamiltonian indicates that the concept of spin is an inherent feature of the Dirac equation. This can be shown as follows.

The law of conservation says that if a physical quantity A commutes with the Hamiltonian H of a system then it is conserved and this follows from the relation

$$[A \ H] = i\hbar\frac{dA}{dt} \tag{11.239}$$

If $[A \ H] = 0$ then

$$\frac{dA}{dt} = 0$$

and A is conserved.

Rotational symmetry of a system, say of a free particle, requires that its total angular momentum be a constant of motion. So the total angular momentum must commute with the Dirac Hamiltonian H.

Let us check if Dirac Hamiltonian H conserves the orbital angular momentum $\vec{l} = \left(l_x, \ l_y, \ l_z\right)$ of the particle (say an electron).

We rewrite equation (11.239) with $A = l_x$

$$[\ l_x \ H \ ] = i\hbar\frac{dl_x}{dt} \tag{11.240}$$

Now $\vec{l} = \vec{r} \times \vec{p}$ and so

$$l_x = yp_z - zp_y$$

$$i\hbar\frac{dl_x}{dt} = [\ l_x \ H \ ] = [\ yp_z - zp_y, \ \ c(\vec{\alpha}.\,\vec{p}) + \beta mc^2 \ ]$$

$$= [\ yp_z - zp_y, \ \ c(\alpha_x p_x + \alpha_y p_y + \alpha_z p_z) + \beta mc^2 \ ]$$

$$i\hbar\frac{dl_x}{dt} = [yp_z - zp_y, \; c(\alpha_x p_x + \alpha_y p_y + \alpha_z p_z)] + [\, yp_z - zp_y, \; \beta mc^2 \,]$$

Since α, β do not depend upon space coordinate, it follows that α, β commute with $yp_x - xp_y$. Hence

$$i\hbar\frac{dl_x}{dt} = [\, yp_z - zp_y, \; c\alpha_x p_x \,] + [\, yp_z - zp_y, \; c\alpha_y p_y \,] + [\, yp_z - zp_y, \; c\alpha_z p_z]$$

$$i\hbar\frac{dl_x}{dt} = [yp_z, c\alpha_x p_x\,] - [\, zp_y, c\alpha_x p_x \,] + [\, yp_z, \; c\alpha_y p_y \,] - [\, zp_y, \; c\alpha_y p_y \,]$$

$$+[\, yp_z, \; c\alpha_z p_z \,] - [\, zp_y, \; c\alpha_z p_z \,]$$

$$i\hbar\frac{dl_x}{dt} = [\, yp_z, \; c\alpha_y p_y \,] - [zp_y, \; c\alpha_z p_z \,]$$

$$i\hbar\frac{dl_x}{dt} = (yp_z\, c\alpha_y p_y \; - c\alpha_y p_y\, yp_z\,) - (zp_y\, c\alpha_z p_z - c\alpha_z p_z\, zp_y)$$

$$i\hbar\frac{dl_x}{dt} = c\alpha_y p_z\,(yp_y - p_y y\,) - c\alpha_z p_y\,(zp_z - p_z z)$$

$$i\hbar\frac{dl_x}{dt} = c\alpha_y p_z[\, y\; p_y\,] - c\alpha_z p_y[\, z\; p_z\,]$$

$$i\hbar\frac{dl_x}{dt} = c\alpha_y p_z\,(i\hbar) - c\alpha_z p_y\,(i\hbar)$$

$$i\hbar\frac{dl_x}{dt} = i\hbar c(\alpha_y p_z - \alpha_z p_y) = i\hbar c(\vec{\alpha}\times\vec{p})_x \neq 0 \tag{11.241}$$

So l_x (and similarly l_y, l_z) are not a conserved quantity. And

$$i\hbar\frac{d\,\vec{l}}{dt} = i\hbar c\vec{\alpha}\times\vec{p} \neq 0 \tag{11.242}$$

Hence $\vec{l}$ is not a conserved quantity.

Let us construct a quantity

$$\vec{s} = (s_x, s_y, s_z) = \frac{\hbar}{2}\begin{pmatrix} \vec{\sigma} & 0 \\ 0 & \vec{\sigma} \end{pmatrix} \tag{11.243}$$

with

$$s_x = \frac{\hbar}{2}\begin{pmatrix} \sigma_x & 0 \\ 0 & \sigma_x \end{pmatrix}, \; s_y = \frac{\hbar}{2}\begin{pmatrix} \sigma_y & 0 \\ 0 & \sigma_y \end{pmatrix}, \; s_z - \frac{\hbar}{2}\begin{pmatrix} \sigma_z & 0 \\ 0 & \sigma_z \end{pmatrix} \tag{11.244}$$

and check whether s is conserved. Consider

$$i\hbar\frac{ds_x}{dt} = [\ s_x\ \ H\ \ ] = [\ s_x,\ \ c(\vec{\alpha}\cdot\vec{p}) + \beta mc^2] = [\ s_x,\ \ c(\vec{\alpha}\cdot\vec{p})] + [s_x,\ \ \beta mc^2] \quad (11.245)$$

$$= [\ s_x,\ c(\alpha_x p_x + \alpha_y p_y + \alpha_z p_z)\ ] + \left(s_x\beta mc^2 - \beta mc^2 s_x\right)$$

$$= [\ s_x,\ c\alpha_x p_x\ ] + [\ s_x,\ \ c\alpha_y p_y] + [\ s_x,\ \ c\alpha_z p_z] + mc^2\left(s_x\beta - \beta s_x\right)$$

$$i\hbar\frac{ds_x}{dt} = (s_x c\alpha_x p_x - c\alpha_x p_x s_x) + (s_x c\alpha_y p_y - c\alpha_y p_y s_x)$$

$$+ \left(s_x c\alpha_z p_z - c\alpha_z p_z s_x\right) + mc^2\left(s_x\beta - \beta s_x\right)$$

$$i\hbar\frac{ds_x}{dt} = cp_x(s_x\alpha_x - \alpha_x s_x) + cp_y(s_x\alpha_y - \alpha_y s_x)$$

$$+ cp_z(s_x\alpha_z - \alpha_z s_x)\ + mc^2\left(s_x\beta - \beta s_x\right) \quad (11.246)$$

$$i\hbar\frac{ds_x}{dt} = \text{1st term} + \text{2nd term} + \text{3rd term} + \text{4th term} \quad (11.247)$$

$$\text{1st term} = cp_x(s_x\alpha_x - \alpha_x s_x) = cp_x\left(\frac{\hbar}{2}\begin{pmatrix}\sigma_x & 0 \\ 0 & \sigma_x\end{pmatrix}\begin{pmatrix}0 & \sigma_x \\ \sigma_x & 0\end{pmatrix} - \begin{pmatrix}0 & \sigma_x \\ \sigma_x & 0\end{pmatrix}\frac{\hbar}{2}\begin{pmatrix}\sigma_x & 0 \\ 0 & \sigma_x\end{pmatrix}\right)$$

$$= \frac{\hbar}{2}cp_x\left(\begin{pmatrix}0 & \sigma_x^2 \\ \sigma_x^2 & 0\end{pmatrix} - \begin{pmatrix}0 & \sigma_x^2 \\ \sigma_x^2 & 0\end{pmatrix}\right) = 0 \quad (11.248)$$

$$\text{2nd term} = cp_y(s_x\alpha_y - \alpha_y s_x)$$

$$= cp_y\left(\frac{\hbar}{2}\begin{pmatrix}\sigma_x & 0 \\ 0 & \sigma_x\end{pmatrix}\begin{pmatrix}0 & \sigma_y \\ \sigma_y & 0\end{pmatrix} - \begin{pmatrix}0 & \sigma_y \\ \sigma_y & 0\end{pmatrix}\frac{\hbar}{2}\begin{pmatrix}\sigma_x & 0 \\ 0 & \sigma_x\end{pmatrix}\right)$$

$$= \frac{\hbar}{2}cp_y\left(\begin{pmatrix}0 & \sigma_x\sigma_y \\ \sigma_x\sigma_y & 0\end{pmatrix} - \begin{pmatrix}0 & \sigma_y\sigma_x \\ \sigma_y\sigma_x & 0\end{pmatrix}\right) = \frac{\hbar}{2}cp_y\begin{pmatrix}0 & [\sigma_x\sigma_y] \\ [\sigma_x\sigma_y] & 0\end{pmatrix}$$

$$= \frac{\hbar}{2}cp_y\begin{pmatrix}0 & 2i\sigma_z \\ 2i\sigma_z & 0\end{pmatrix} = \frac{\hbar}{2}(2icp_y\alpha_z) = i\hbar c\alpha_z p_y \quad (11.249)$$

$$\text{3rd term} = cp_z(s_x\alpha_z - \alpha_z s_x) = cp_z\left(\frac{\hbar}{2}\begin{pmatrix}\sigma_x & 0 \\ 0 & \sigma_x\end{pmatrix}\begin{pmatrix}0 & \sigma_z \\ \sigma_z & 0\end{pmatrix} - \begin{pmatrix}0 & \sigma_z \\ \sigma_z & 0\end{pmatrix}\frac{\hbar}{2}\begin{pmatrix}\sigma_x & 0 \\ 0 & \sigma_x\end{pmatrix}\right)$$

$$=\frac{\hbar}{2}cp_z\left(\begin{pmatrix}0 & \sigma_x\sigma_z\\ \sigma_x\sigma_z & 0\end{pmatrix}-\begin{pmatrix}0 & \sigma_z\sigma_x\\ \sigma_z\sigma_x & 0\end{pmatrix}\right)=\frac{\hbar}{2}cp_z\begin{pmatrix}0 & [\sigma_x\sigma_z]\\ [\sigma_x\sigma_z] & 0\end{pmatrix}$$

$$=\frac{\hbar}{2}cp_z\begin{pmatrix}0 & -2i\sigma_y\\ -2i\sigma_y & 0\end{pmatrix}=-\frac{\hbar}{2}(2icp_z\,\alpha_y)=-i\hbar c\alpha_y p_z \tag{11.250}$$

4th term $=mc^2(s_x\beta-\beta s_x)=mc^2\left(\frac{\hbar}{2}\begin{pmatrix}\sigma_x & 0\\ 0 & \sigma_x\end{pmatrix}\begin{pmatrix}I & 0\\ 0 & -I\end{pmatrix}-\begin{pmatrix}I & 0\\ 0 & -I\end{pmatrix}\frac{\hbar}{2}\begin{pmatrix}\sigma_x & 0\\ 0 & \sigma_x\end{pmatrix}\right)$

$$=\frac{\hbar}{2}mc^2\left(\begin{pmatrix}\sigma_x I & 0\\ 0 & -I\sigma_x\end{pmatrix}-\begin{pmatrix}\sigma_x I & 0\\ 0 & -\sigma_x I\end{pmatrix}\right)=0 \tag{11.251}$$

From equation (11.247)

$$i\hbar\frac{ds_x}{dt}=0+i\hbar c\alpha_z p_y-i\hbar c\alpha_y p_z+0=i\hbar c(\alpha_z p_y-\alpha_y p_z)=-i\hbar c(\vec{\alpha}\times\vec{p})_x \tag{11.252}$$

Similarly

$$i\hbar\frac{d\,\vec{s}}{dt}=-i\hbar c\vec{\alpha}\times\vec{p}\neq 0 \tag{11.253}$$

Adding equations (11.241) and (11.252)

$$i\hbar\frac{dl_x}{dt}+i\hbar\frac{ds_x}{dt}=i\hbar c(\alpha_y p_z-\alpha_z p_y)+i\hbar c(\alpha_z p_y-\alpha_y p_z)=0$$

$$i\hbar\frac{d}{dt}(l_x+s_x)=0 \tag{11.254}$$

Similarly

$$i\hbar\frac{d}{dt}(l_y+s_y)=0,\quad i\hbar\frac{d}{dt}(l_z+s_z)=0 \tag{11.255}$$

Adding

$$i\hbar\frac{d}{dt}(\vec{l}+\vec{s})=0 \tag{11.256}$$

Define

$$\vec{j}=\vec{l}+\vec{s} \tag{11.257}$$

Thus

$$\frac{d}{dt}\,\vec{j}=0 \tag{11.258}$$

$$\vec{j} = \text{constant of motion} \tag{11.259}$$

Clearly orbital angular momentum $\vec{l}$ is not a constant of motion but when we add to it $\vec{s} = \frac{\hbar}{2}\begin{pmatrix} \vec{\sigma} & 0 \\ 0 & \vec{\sigma} \end{pmatrix}$ where $\vec{s}$ is spin angular momentum then $\vec{j} = \vec{l} + \vec{s}$ becomes a constant of motion.

This means $\vec{j}$ is the total angular momentum.

$$\text{Though } [\vec{l} \ H] \neq 0, \ [\vec{s} \ H] \neq 0, \ [\vec{j} \ H] = 0, \ [\vec{p} \ H] = 0 \tag{11.260}$$

The concept of spin was hypothesized by Uhlenbach and Goudsmidt and experimentally demonstrated by Stern and Gerlach. Dirac theoretically showed that the electron has an intrinsic angular momentum called spin angular momentum $\vec{s}$ since it does not depend upon space $\vec{r}$ or the linear momentum $\vec{p}$ of the particle as the orbital angular momentum does (since $\vec{l} = \vec{r} \times \vec{p}$). So $\vec{s}$ is an intrinsic property of the system. Though orbital angular momentum $\vec{l}$ depends on the motion of the particle, spin angular momentum $\vec{s}$ does not arise from orbital motion. In fact $\vec{s}$ is an intrinsic or inherent property of the system or particle as there is no connection between spin and motion of the particle. Spin, mass electric charge etc are all intrinsic properties of the particle. It has no classical analogue. It does not depend on the state of motion.

The relation $\vec{j} = \vec{l} + \vec{s}$ shows that $\vec{s}$ is angular momentum and obeys angular momentum algebra.

11.28 Value of spin s of Dirac particle electron

In the non-relativistic Schrödinger theory the Schrödinger Hamiltonian for a free particle is

$$H_{\text{Sch}} = \frac{p^2}{2m} \tag{11.261}$$

The orbital angular momentum $\vec{l} = \vec{r} \times \vec{p}$ commutes with H_{Sch} and l^2 also commutes with H_{Sch}, i.e.

$$[\vec{l} \ H] = 0, \ [l^2 \ H] = 0 \tag{11.262}$$

So l^2 and $\vec{l}$ are conserved. In the Schrödinger theory simultaneous eigenstates of the operators H, l^2, l_z exist as $|n \ lm_l\rangle$ with eigenvalues E_n, $l(l + 1)\hbar^2$, $m_l\hbar$, respectively.

Let us now move over to Dirac theory.

The Dirac Hamiltonian for a free particle (say a free electron) is

$$H = c\vec{\alpha}.\vec{p} + \beta mc^2 \tag{11.263}$$

and

$$i\hbar\frac{d \ \vec{l}}{dt} = i\hbar c\vec{\alpha} \times \vec{p} \neq 0 \tag{11.264}$$

$$i\hbar \frac{d\,\vec{s}}{dt} = -i\hbar c\vec{\alpha} \times \vec{p} \neq 0 \tag{11.265}$$

$$i\hbar \frac{d\,\vec{j}}{dt} = 0 \tag{11.266}$$

So $\vec{l}, \vec{s}$ are not conserved but $\vec{j} = \vec{l} + \vec{s}$ is conserved.

So j^2 and $\vec{j}$ are conserved. Simultaneous eigenstates of the operators H, j^2, j_z exist as $|n\,jm_j\rangle$ with eigenvalues $E_n, j(j+1)\hbar^2, m_j\hbar$, respectively.

We note that in the non-relativistic limit, i.e. when linear momentum is small, i.e.

$$p \ll mc$$

$\vec{\alpha}$ reduces to the velocity operator and hence becomes parallel to $\vec{p}$ and so in this limit $\vec{\alpha} \times \vec{p} = 0$ and so $i\hbar \frac{d\,\vec{l}}{dt} = 0$, i.e. $\vec{l}$ is conserved.

It is clear that the concept of spin naturally fits into the Dirac theory.

$\vec{s}$ obeys the basic commutation rules of angular momentum viz.

$$[\,s_x,\ s_y] = i\hbar s_z \text{ and } [\,s_i,\ s_j] = i\hbar \varepsilon_{ijk} s_k \tag{11.267}$$

which is the same as the commutation relation satisfied by $\vec{l}$, i.e.

$$[\,l_x,\ l_y] = i\hbar l_z \text{ and } [\,l_i,\ l_j] = i\hbar \varepsilon_{ijk} l_k \tag{11.268}$$

and by $\vec{j}$, i.e.

$$[\,j_x,\ j_y] = i\hbar j_z \text{ and } [\,j_i,\ j_j] = i\hbar \varepsilon_{ijk} j_k \tag{11.269}$$

In fact the eigenvalue of s^2 is $s(s+1)\hbar^2$ where $s = 0, 1, 2, 3, ...$i.e. integral for a bosonic system and $s = \frac{1}{2}, \frac{3}{2}, \ ...$ i.e. half-integral for a fermionic system.

Also, components of $\vec{s}$ should commute with the components of $\vec{l}$ so that the usual rules of angular momentum addition can be used to obtain $\vec{j}$ from $\vec{l}$ and $\vec{s}$. Now

$$s^2 = ss = \frac{\hbar}{2}\begin{pmatrix} \vec{\sigma} & 0 \\ 0 & \vec{\sigma} \end{pmatrix}\frac{\hbar}{2}\begin{pmatrix} \vec{\sigma} & 0 \\ 0 & \vec{\sigma} \end{pmatrix} = \frac{\hbar^2}{4}\begin{pmatrix} \sigma^2 & 0 \\ 0 & \sigma^2 \end{pmatrix} = \frac{\hbar^2}{4}\sigma^2\begin{pmatrix} I & 0 \\ 0 & I \end{pmatrix} = \frac{\hbar^2}{4}\sigma^2 I_{4\times 4} = \frac{\hbar^2}{4}\sigma^2$$

$$s^2 = \frac{\hbar^2}{4}\left(\sigma_x^2 + \sigma_y^2 + \sigma_z^2\right) = \frac{\hbar^2}{4}(I + I + I\,) = \frac{3}{4}\hbar^2 = \frac{1}{2}\left(\frac{1}{2} + 1\right)\hbar^2 = s(s+1)\hbar^2$$

Comparison of the eigenvalue gives

$$s = \frac{1}{2} \tag{11.270}$$

So the Dirac equation describes spin $\frac{1}{2}$ particles such as electrons (called Dirac particles).

Also

$$s_z^2 = \frac{\hbar}{2}\begin{pmatrix} \sigma_z & 0 \\ 0 & \sigma_z \end{pmatrix}\frac{\hbar}{2}\begin{pmatrix} \sigma_z & 0 \\ 0 & \sigma_z \end{pmatrix} = \frac{\hbar^2}{4}\begin{pmatrix} \sigma_z^2 & 0 \\ 0 & \sigma_z^2 \end{pmatrix} = \frac{\hbar^2}{4}\sigma_z^2\begin{pmatrix} I & 0 \\ 0 & I \end{pmatrix} = \frac{\hbar^2}{4}\sigma_z^2 I_{4\times4} = \frac{\hbar^2}{4}\sigma_z^2$$

As $\sigma_z^2 = I$

$$s_z^2 = \frac{\hbar^2}{4}I = \frac{\hbar^2}{4}$$

$$\text{So } s_z = \pm\frac{\hbar}{2} = m_s\hbar \tag{11.271}$$

Comparison of eigenvalue gives

$$m_s = \pm\frac{1}{2} \tag{11.272}$$

11.28.1 Multiplicity

Since $s = \frac{1}{2}$ multiplicity is

$$2s + 1 = 2.\frac{1}{2} + 1 = 2,$$

So there are 2 independent spin orientations for a given energy and momentum viz.

$\alpha = \begin{pmatrix} 1 \\ 0 \end{pmatrix} = |\uparrow\rangle = \chi_{\frac{1}{2}\frac{1}{2}} \equiv \chi_{sm_s}$ (spin up state, spin parallel to momentum)

$\beta = \begin{pmatrix} 0 \\ 1 \end{pmatrix} = |\downarrow\rangle = \chi_{\frac{1}{2}-\frac{1}{2}} \equiv \chi_{sm_s}$ (spin down state, spin anti-parallel to momentum)

- For $s = \frac{1}{2}$ the eigenvector is a 4-component spinor (which is a vector in spin space).
- Spin is clearly a relativistic phenomenon. No classical analogue of spin can be found.

11.29 Helicity

Helicity is a quantum number that is used to label and identify the state of a particle or system moving with high energy and relativistic speed. We define helicity denoted by the symbol $\mathcal{H}$ as the projection of spin in the direction of linear momentum (figure 11.6).

$$\mathcal{H} = \frac{\vec{s}\cdot\vec{p}}{p} = \vec{s}\cdot\hat{p} = \frac{\hbar}{2}\begin{pmatrix} \vec{\sigma} & 0 \\ 0 & \vec{\sigma} \end{pmatrix}\cdot\hat{p} = \frac{\hbar}{2}\begin{pmatrix} \vec{\sigma}\cdot\hat{p} & 0 \\ 0 & \vec{\sigma}\cdot\hat{p} \end{pmatrix} \tag{11.273}$$

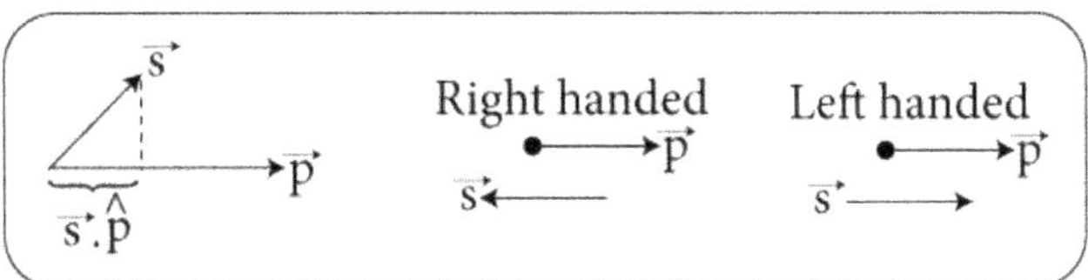

Figure 11.6. Projection of spin in the direction of linear momentum is helicity.

Since helicity commutes with the Hamiltonian, i.e. since $[h\ \mathcal{H}] = 0$ (*exercise 11.12*) it follows that helicity is conserved (it is an integral of motion).

Helicity is a good quantum number and both H and $\mathcal{H}$ have simultaneous eigenvectors or eigenstates $\psi(\vec{r}, t) = ue^{i(\vec{k}\cdot\vec{r} - wt)}$.

Let us find the square of helicity

$$\mathcal{H}^2 = \mathcal{H}\mathcal{H} = \frac{\hbar}{2}\begin{pmatrix} \vec{\sigma}\cdot\hat{p} & 0 \\ 0 & \vec{\sigma}\cdot\hat{p} \end{pmatrix}\frac{\hbar}{2}\begin{pmatrix} \vec{\sigma}\cdot\hat{p} & 0 \\ 0 & \vec{\sigma}\cdot\hat{p} \end{pmatrix} \text{ (from equation (11.273))}$$

$$=\frac{\hbar^2}{4p^2}\begin{pmatrix} \vec{\sigma}\cdot\vec{p} & 0 \\ 0 & \vec{\sigma}\cdot\vec{p} \end{pmatrix}\frac{\hbar}{2}\begin{pmatrix} \vec{\sigma}\cdot\vec{p} & 0 \\ 0 & \vec{\sigma}\cdot\vec{p} \end{pmatrix} = \frac{\hbar^2}{4p^2}\begin{pmatrix} (\vec{\sigma}\cdot\vec{p})^2 & 0 \\ 0 & (\vec{\sigma}\cdot\vec{p})^2 \end{pmatrix}$$

$$=\frac{\hbar^2}{4p^2}\begin{pmatrix} p^2 & 0 \\ 0 & p^2 \end{pmatrix}\left(\text{using equation (11.213)}\right)$$

$$\mathcal{H}^2 = \frac{\hbar^2}{4}\begin{pmatrix} I & 0 \\ 0 & I \end{pmatrix} = \frac{\hbar^2}{4}I_{4\times4} = \frac{\hbar^2}{4} \tag{11.274}$$

It is clear that the eigenvalue of helicity operator $\mathcal{H}$ is $\pm\frac{\hbar}{2}$. This is same as the eigenvalue of spin operator s_z which is also $\pm\frac{\hbar}{2}$.

We have thus 2 helicity states.

(1) Right-handed helicity in which spin is in the direction of motion, i.e. along linear momentum. Helicity eigenvalue is

$$\mathcal{H} = +1 \tag{11.275}$$

(2) Left-handed helicity in which spin is in the opposite direction to the motion, i.e. opposite to the linear momentum. Helicity eigenvalue is

$$\mathcal{H} = -1 \tag{11.276}$$

- Since $[\ \vec{s}\ H] \neq 0$, $[\mathcal{H}H] = 0$ it follows that to designate the state of a particle or system we can use helicity instead of spin since although spin is not a good quantum number, the helicity is a good quantum number.

Helicity is not a Lorentz invariant quantity.

11.30 Chirality

We define a Lorentz invariant quantity called chirality as

$$P_L = \frac{1}{2}(1 - \gamma_5) \text{ (for left-handed chirality)} \tag{11.277}$$

$$P_R = \frac{1}{2}(1 + \gamma_5) \text{ (for right-handed chirality)} \tag{11.278}$$

11.31 Sixteen independent elements from four Dirac matrices

With the help of the 4 Dirac matrices γ_μ we can construct 16 independent elements which are actually 4×4 independent matrices

   ✓ I (unit matrix, commutes with all) (one matrix)
   ✓ $\gamma_1,\ \gamma_2,\ \gamma_3,\ \gamma_4$ (4 matrices)
   ✓ $\gamma_1\gamma_2,\ \gamma_1\gamma_3,\ \gamma_1\gamma_4,\ \gamma_2\gamma_3,\ \gamma_2\gamma_4,\ \gamma_3\gamma_4$ (6 matrices)
   ✓ $\gamma_1\gamma_2\gamma_3,\ \gamma_2\gamma_3\gamma_4,\ \gamma_3\gamma_4\gamma_1,\ \gamma_4\gamma_1\gamma_2$ (4 matrices)
   ✓ $\gamma_1\gamma_2\gamma_3\gamma_4$ (one matrix).

In Dirac theory we deal with 4×4 matrices containing 16 elements. So we require 16 linearly independent matrices as the basis for expressing any 4×4 matrix.

11.32 Vacuum polarization

A positive energy electron repels the negative energy electron (of the vacuum) in its neighbourhood. This leads to vacuum polarization and will lead to a shift of energy levels. This has been experimentally verified.

11.33 Dirac equation for a particle of charge q in an electromagnetic field $(\vec{A}, \phi)$

The Dirac equation in its covariant form is given by equation (11.160)

$$\left(\gamma_\mu p^\mu - mc\right)\psi = 0$$

Replacing p^μ by $p^\mu - qA^\mu$ we get the covariant form of the Dirac equation for a charged particle in an electromagnetic field to be

$$(\gamma_\mu(p^\mu - qA^\mu) - mc\,)\psi = 0 \tag{11.279}$$

We have derived the same relation from equation (11.19) with the replacement of

$$E \to i\hbar\frac{\partial}{\partial t},\ \vec{p} \to -i\hbar\,\vec{\nabla}\ \text{ and } p^\mu \to i\hbar\partial^\mu$$

We have obtained the continuity equation from the covariant form of the Dirac equation for a particle of charge q in an electromagnetic field $(\vec{A}, \phi)$ in *exercise 11.20*.

11.34 Exercises

Exercise 11.1 *Which of the following relativity theories can be coupled with the laws of quantum mechanics?*

(a) special theory   *(b) general theory*   *(c) none*   *(d) both.*

Exercise 11.2 *The Schrödinger equation* $-\frac{\hbar^2}{2m}\nabla^2\psi = i\hbar\frac{\partial}{\partial t}\psi$ *does not describe a high speed free particle because:*

(a) it is not a Klein–Gordon equation   *(b) it is not a Dirac equation*
(c) space-time equivalence is not respected   *(d) it is a non-relativistic equation.*

Exercise 11.3 *Write down time independent Schrödinger equation for a free particle. Establish that* $p^\mu p_\mu = -\hbar^2\square$ *where* $\square = \frac{1}{c^2}\frac{\partial^2}{\partial t^2} - \nabla^2$ *is the d'Alembertian operator and make a relativistic extension to obtain the Klein–Gordon equation.*

   $\boxed{\text{Ans.}}$ The time independent Schrödinger equation for a free particle is

$$H\psi(\vec{r}, t) = E\psi(\vec{r}, t)$$

$$\frac{\vec{p}^{\,2}}{2m}\psi(\vec{r}, t) = i\hbar\frac{\partial}{\partial t}\psi(\vec{r}, t)$$

With $\vec{p} = -i\hbar\vec{\nabla}$ we have

$$-\frac{\hbar^2}{2m}\nabla^2\psi(\vec{r}, t) = i\hbar\frac{\partial}{\partial t}\psi(\vec{r}, t)$$

The 4-momentum vector is

$$p^\mu = (p^0, p^1, p^2, p^3) = \left(\frac{E}{c},\; p_x, p_y, p_z\right) = \left(\frac{E}{c},\; \vec{p}\right) \quad \text{(equation (11.25))}$$

$$p_\mu = (p_0, p_1, p_2, p_3) = \left(\frac{E}{c},\; -p_x, -p_y, -p_z\right) = \left(\frac{E}{c},\; -\vec{p}\right) \quad \text{(equation (11.26))}$$

In operator form

$$p^\mu = i\hbar\frac{\partial}{\partial x_\mu} = \left(i\hbar\frac{\partial}{\partial ct},\; -i\hbar\frac{\partial}{\partial x},\; i\hbar\frac{\partial}{\partial y},\; -i\hbar\frac{\partial}{\partial z}\right)$$

$$= i\hbar\left(\frac{\partial}{\partial ct}, \ -\frac{\partial}{\partial x}, \ -\frac{\partial}{\partial y}, \ -\frac{\partial}{\partial z}\right) = i\hbar\left(\frac{\partial}{\partial ct}, \ -\vec{\nabla}\right) \tag{11.280}$$

$$p_\mu = i\hbar\frac{\partial}{\partial x^\mu} = i\hbar\left(\frac{\partial}{\partial ct}, \ \vec{\nabla}\right) \tag{11.281}$$

$$p^\mu p_\mu = i\hbar\frac{\partial}{\partial x_\mu}\, i\hbar\frac{\partial}{\partial x^\mu} = -\hbar^2\frac{\partial}{\partial x_\mu}\frac{\partial}{\partial x^\mu} = -\hbar^2\left(\frac{\partial}{\partial x_0}\frac{\partial}{\partial x^0} + \frac{\partial}{\partial x_1}\frac{\partial}{\partial x^1} + \frac{\partial}{\partial x_2}\frac{\partial}{\partial x^2} + \frac{\partial}{\partial x_3}\frac{\partial}{\partial x^3}\right)$$

$$= -\hbar^2\left(\frac{\partial}{\partial ct}\frac{\partial}{\partial ct} - \frac{\partial}{\partial x}\frac{\partial}{\partial x} - \frac{\partial}{\partial y}\frac{\partial}{\partial y} - \frac{\partial}{\partial z}\frac{\partial}{\partial z}\right) \tag{11.282}$$

Using

$$x^0 = x_0 = ct, \ \ x^1 = -x_1 = x, \ \ x^2 = -x_2 = y, \ \ x^3 = -x_3 = z \tag{11.283}$$

$$p^\mu p_\mu = -\hbar^2\left(\frac{1}{c^2}\frac{\partial^2}{\partial t^2} - \frac{\partial^2}{\partial x^2} - \frac{\partial^2}{\partial y^2} - \frac{\partial^2}{\partial z^2}\right)$$

$$p^\mu p_\mu = -\hbar^2\left(\frac{1}{c^2}\frac{\partial^2}{\partial t^2} - \nabla^2\right) = -\hbar^2\square \tag{11.284}$$

where $\square = \frac{1}{c^2}\frac{\partial^2}{\partial t^2} - \nabla^2$ is the d'Alembertian operator

Now using

$$p^0 = p_0 = \frac{E}{c}, \ \ p^1 = -p_1 = p_x, \ \ p^2 = -p_2 = p_y, \ \ p^3 = -p_3 = p_z \tag{11.285}$$

$$p^\mu p_\mu = p^0 p_0 + p^1 p_1 + p^2 p_2 + p^3 p_3 \tag{11.286}$$

$$= \frac{E}{c}\frac{E}{c} - p_x p_x - p_y p_y - p_z p_z = \frac{E^2}{c^2} - \left(p_x^2 + p_y^2 + p_z^2\right) = \frac{E^2}{c^2} - p^2$$

$$p^\mu p_\mu = \frac{E^2}{c^2} - p^2 = \frac{E^2 - p^2 c^2}{c^2} = \frac{m^2 c^4}{c^2} = m^2 c^2$$

$$p^\mu p_\mu = m^2 c^2 \tag{11.287}$$

Making operator replacement and operating on wave function ψ

$$p^\mu p_\mu \psi = m^2 c^2 \psi$$

$$p^\mu p_\mu \psi - m^2 c^2 \psi = 0$$

$$-\hbar^2 \square \psi = m^2 c^2 \psi \ \text{(using equation (11.284))}$$

$$\hbar^2 \Box \psi + m^2 c^2 \psi = 0 \tag{11.288}$$

$$\left(\Box + \frac{m^2 c^2}{\hbar^2} \right) \psi = 0 \tag{11.289}$$

$$\frac{1}{c^2} \frac{\partial^2 \psi}{\partial t^2} - \nabla^2 \psi + \frac{m^2 c^2}{\hbar^2} \psi = 0 \tag{11.290}$$

This relativistic formulation of the Schrödinger equation is called the Klein–Gordon equation.

Exercise 11.4 *Obtain the non-relativistic limit of the Klein–Gordon equation and show that it is the Schrödinger equation.*

$\boxed{\text{Ans.}}$ The Klein–Gordon equation (relativistic equation) is

$$\left(\nabla^2 - \frac{1}{c^2} \frac{\partial^2}{\partial t^2} \right) \psi = \frac{m^2 c^2}{\hbar^2} \psi \quad \left(\text{equation (11.60)} \right)$$

$$\nabla^2 \psi - \frac{1}{c^2} \frac{\partial^2}{\partial t^2} \psi - \frac{m^2 c^2}{\hbar^2} \psi = 0 \tag{11.291}$$

where m is rest mass of the free particle wave function $\psi = \psi(\vec{r}, t)$.

Let us write the total wave function ψ as a product

$$\psi(\vec{r}, t) = \phi(\vec{r}, t) e^{-imc^2 t / \hbar} \tag{11.292}$$

where $\phi(\vec{r}, t)$ is the relativistic part and $e^{-imc^2 t / \hbar}$ is the non-relativistic part of wave function $\psi(\vec{r}, t)$. While the relativistic part $\phi(\vec{r}, t)$ is a function of space and time the non-relativistic part $e^{-imc^2 t / \hbar}$ is space independent.

Let us calculate $\frac{\partial^2}{\partial t^2} \psi$ of equation (11.291) using equation (11.292).

We first find $\frac{\partial}{\partial t} \psi$ where ψ is given by equation (11.292).

$$\frac{\partial}{\partial t} \psi = \frac{\partial}{\partial t} [\phi(\vec{r}, t) e^{-imc^2 t / \hbar}] = \phi \frac{\partial}{\partial t} e^{-imc^2 t / \hbar} + e^{-imc^2 t / \hbar} \frac{\partial}{\partial t} \phi$$

$$= -\frac{imc^2}{\hbar} \phi e^{-\frac{imc^2 t}{\hbar}} + e^{-\frac{imc^2 t}{\hbar}} \frac{\partial}{\partial t} \phi$$

$$\frac{\partial}{\partial t} \psi = \left(-\frac{imc^2}{\hbar} \phi + \frac{\partial \phi}{\partial t} \right) e^{-imc^2 t / \hbar} \tag{11.293}$$

Let us calculate $\frac{\partial^2}{\partial t^2} \psi$ using equation (11.293).

$$\frac{\partial^2}{\partial t^2} \psi = \frac{\partial}{\partial t} \frac{\partial}{\partial t} \psi = \frac{\partial}{\partial t} \left(-\frac{imc^2}{\hbar} \phi + \frac{\partial \phi}{\partial t} \right) e^{-imc^2 t / \hbar}$$

$$=\left(-\frac{imc^2}{\hbar}\phi + \frac{\partial\phi}{\partial t}\right)\frac{\partial}{\partial t}e^{-imc^2t/\hbar} + e^{-imc^2t/\hbar}\frac{\partial}{\partial t}\left(-\frac{imc^2}{\hbar}\phi + \frac{\partial\phi}{\partial t}\right)$$

$$=\left(-\frac{imc^2}{\hbar}\phi + \frac{\partial\phi}{\partial t}\right)\left(-\frac{imc^2}{\hbar}\right)e^{-imc^2t/\hbar} + e^{-imc^2t/\hbar}\left(-\frac{imc^2}{\hbar}\frac{\partial}{\partial t}\phi + \frac{\partial^2}{\partial t^2}\phi\right)$$

$$=\left[\left(\frac{imc^2}{\hbar}\cdot\frac{imc^2}{\hbar}\phi - \frac{imc^2}{\hbar}\frac{\partial\phi}{\partial t}\right) + \left(-\frac{imc^2}{\hbar}\frac{\partial}{\partial t}\phi + \frac{\partial^2}{\partial t^2}\phi\right)\right]e^{-imc^2t/\hbar}$$

$$=\left[-\frac{m^2c^4}{\hbar^2}\phi - \frac{imc^2}{\hbar}\frac{\partial\phi}{\partial t} + -\frac{imc^2}{\hbar}\frac{\partial}{\partial t}\phi + \frac{\partial^2}{\partial t^2}\phi\right]e^{-imc^2t/\hbar}$$

$$\frac{\partial^2}{\partial t^2}\psi = \left[-\frac{m^2c^4}{\hbar^2}\phi - \frac{2imc^2}{\hbar}\frac{\partial\phi}{\partial t} + \frac{\partial^2}{\partial t^2}\phi\right]e^{-imc^2t/\hbar}$$

$$\frac{\partial^2}{\partial t^2}\psi = -\frac{m^2c^4}{\hbar^2}\phi e^{-imc^2t/\hbar} - \frac{2imc^2}{\hbar}\frac{\partial\phi}{\partial t}e^{-imc^2t/\hbar} + \frac{\partial^2\phi}{\partial t^2}e^{-imc^2t/\hbar}$$

Using equation (11.292) we write

$$\frac{\partial^2}{\partial t^2}\phi e^{-imc^2t/\hbar} = -\frac{m^2c^4}{\hbar^2}\psi - \frac{2imc^2}{\hbar}\frac{\partial\phi}{\partial t}e^{-\frac{imc^2t}{\hbar}} + \frac{\partial^2\phi}{\partial t^2}e^{-imc^2t/\hbar} \qquad (11.294)$$

Putting equations (11.292) and (11.294) in the Klein–Gordon equation (11.291) we have

$$\nabla^2\phi e^{-imc^2t/\hbar} - \frac{1}{c^2}\left[-\frac{m^2c^4}{\hbar^2}\psi - \frac{2imc^2}{\hbar}\frac{\partial\phi}{\partial t}e^{-\frac{imc^2t}{\hbar}} + \frac{\partial^2\phi}{\partial t^2}e^{-imc^2t/\hbar}\right] - \frac{m^2c^2}{\hbar^2}\psi = 0$$

$$e^{-imc^2t/\hbar}\nabla^2\phi + \frac{2im}{\hbar}\frac{\partial\phi}{\partial t}e^{-\frac{imc^2t}{\hbar}} - \frac{1}{c^2}\frac{\partial^2\phi}{\partial t^2}e^{-imc^2t/\hbar} = 0$$

$$\nabla^2\phi + \frac{2im}{\hbar}\frac{\partial\phi}{\partial t} - \frac{1}{c^2}\frac{\partial^2\phi}{\partial t^2} = 0 \qquad (11.295)$$

In the non-relativistic limit the difference of total energy E and rest mass energy mc^2 is small, i.e.

$$E - mc^2 = E'$$

$E' \ll mc^2$ in the non-relativistic limit. We can set $E' \approx 0$.

Let us recall the time independent Schrödinger equation

$$i\hbar\frac{\partial}{\partial t}\phi = E'\phi \qquad (11.296)$$

$$i\hbar \frac{\partial}{\partial t} i\hbar \frac{\partial}{\partial t} \phi = E' i\hbar \frac{\partial}{\partial t} \phi$$

$$-\hbar^2 \frac{\partial^2}{\partial t^2}\phi = i\hbar E' \frac{\partial}{\partial t}\phi = i\hbar E' \frac{E'\phi}{i\hbar} = E'^2\phi$$

$$\frac{\partial^2}{\partial t^2}\phi = -\frac{E'^2\phi}{\hbar^2} \approx 0 \ \ (\text{since } E' \approx 0) \tag{11.297}$$

Using equations (11.296) and (11.297) in equation (11.295) we have, ignoring the term $-\frac{1}{c^2}\frac{\partial^2 \phi}{\partial t^2}$

$$\nabla^2\phi + \frac{2im}{\hbar}\frac{\partial\phi}{\partial t} = 0 \tag{11.298}$$

Multiplying by $-\frac{\hbar^2}{2m}$ we get

$$-\frac{\hbar^2}{2m}\nabla^2\phi - \frac{\hbar^2}{2m}\frac{2im}{\hbar}\frac{\partial\phi}{\partial t} = 0$$

$$-\frac{\hbar^2}{2m}\nabla^2\phi = i\hbar\frac{\partial\phi}{\partial t} \tag{11.299}$$

This is the Schrödinger equation for the non-relativistic free particle case which is the non-relativistic limit of the Klein–Gordon equation. In other words, the Schrödinger equation has been derived from the Klein–Gordon equation.

Exercise 11.5 *Show that the continuity equation that follows from the Klein–Gordon equation has the form $\partial_\mu J^\mu = 0$.*

$\boxed{\text{Ans.}}$ The continuity equation (11.84) follows from the Klein–Gordon equation

$$\frac{\partial\rho}{\partial t} + \vec{\nabla}\cdot\vec{J} = 0$$

We can recast this equation in the following form

$$\frac{\partial\rho c}{\partial ct} + \frac{\partial J_x}{\partial x} + \frac{\partial J_y}{\partial y} + \frac{\partial J_z}{\partial z} = 0$$

Using the notation

$$x^0 = x_0 = ct, \ \ x^1 = -x_1 = x, \ \ x^2 = -x_2 = y, \ \ x^3 = -x_3 = z$$

$$J^0 = J_0 = \rho c, \ \ J^1 = -J_1 = J_x, \ \ J^2 = -J_2 = J_y, \ \ J^3 = -J_3 = J_z$$

$J^\mu = (\rho c, \ \vec{J})$ and

$$\partial_\mu = \left(\frac{\partial}{\partial ct}, \ \vec{\nabla}\right)$$

$$\frac{\partial J^0}{\partial x^0} + \frac{\partial J^1}{\partial x^1} + \frac{\partial J^2}{\partial x^2} + \frac{\partial J^3}{\partial x^3} = 0$$

$$\frac{\partial J^0}{\partial x^0} + \frac{\partial J^i}{\partial x^i} = 0$$

$$\frac{\partial J^\mu}{\partial x^\mu} = 0$$

$$\partial_\mu J^\mu = 0$$

This is the continuity equation that follows from the Klein–Gordon equation.

Exercise 11.6 *Which of the following equations describe spin 0 and spin half particles?*
   (a) *Klein–Gordon equation, Schrödinger equation*
   (b) *Dirac equation, Klein–Gordon equation*
   (c) *Klein–Gordon equation, Dirac equation*
   (d) *Schrödinger equation, Dirac equation.*

Exercise 11.7 *Why should the α, β occurring in the Dirac equation have to be matrices and not numbers?*
   Ans. Because the α, β anti-commutes viz $\{\alpha_i \ \alpha_j\} = 0$, $\{\alpha_i \ \beta\} = 0$. Again numbers cannot anti-commute. Matrices can anti-commute. So the α, β have to be matrices.

Exercise 11.8 *Show that the dimension of Dirac matrices follows from their anti-commutation relation.*
   Ans. Dirac matrices $\alpha_i \ \beta$ anti-commute. Consider the anti-commutation relation

$$\{\alpha_i \ \beta\} = \alpha_i \beta + \beta \alpha_i = 0$$

$$\alpha_i \beta = -\beta \alpha_i$$

Taking the determinant of both sides we have

$$det \ (\alpha_i \beta) = det \ (-1) det \ (\alpha_i \beta)$$

$$det \ (\alpha_i) det \ (\beta) = det \ (-1) det \ (\alpha_i) det \ (\beta)$$

$$1 = det \ (-1)$$

$$1 = (-1)^N \ \text{(where } N = \text{ dimension of Dirac matrices)}$$

$$N = \text{even} = 2, 4, 6 \ \text{...} \ .$$

Exercise 11.9 *Can you use Pauli matrices as the Dirac matrices?*

Ans. In the Dirac equation there are 4 matrices α_x, α_y, α_z, β and they should be linearly independent. And these matrices anti-commute, i.e.

$$\{\ \alpha_i \quad \alpha_j\} = 0, \{\ \alpha_i\ \beta\} = 0.$$

Again, the 3 Pauli matrices σ_x, σ_y, σ_z and unit matrix I (all 2×2 matrices) form a linearly independent set. But the unit matrix I commutes with σ_is while we need 4 anti-commuting matrices.

So the set of Pauli matrices cannot be chosen as Dirac matrices. But since Pauli matrices σ_is satisfy all the properties expected of Dirac matrices we build the Dirac matrices α_i, β in terms of Pauli matrices σ_is.

Exercise 11.10 *Show that the covariant form of the Dirac equation can be recast as*

$$(i\gamma^\mu\partial_\mu - m\ )\psi = 0 \text{ or } (\ \gamma^\mu\partial_\mu + im\ )\psi = 0$$

Ans. The Dirac equation in covariant form was obtained in equation (11.151) to be

$$\left(i\gamma_\mu\partial^\mu - \frac{mc}{\hbar}\ \right)\psi = 0$$

$$\left(i\gamma_0\partial^0 + i\gamma_k\partial^k - \frac{mc}{\hbar}\ \right)\psi = 0$$

$$\left(i\gamma_0\frac{\partial}{\partial x_0} + i\gamma_k\frac{\partial}{\partial x_k} - \frac{mc}{\hbar}\ \right)\psi = 0$$

We rewrite using equation (11.155) $\gamma^0 = \gamma_0 = \beta$, $\gamma^k = -\gamma_k = -\beta\alpha_k = -\gamma_0\alpha_k$ and

$$x^0 = x_0 = ct,\ x^1 = -x_1 = x,\ x^2 = -x_2 = y,\ x^3 = -x_3 = z\ i.\,e.\ x^k = -x_k$$

$$\left(i\gamma^0\frac{\partial}{\partial x^0} + i\left(-\gamma^k\right)\frac{\partial}{\partial(-x^k)} - \frac{mc}{\hbar}\ \right)\psi = 0$$

$$\left(i\gamma^0\frac{\partial}{\partial x^0} + i\gamma^k\frac{\partial}{\partial x^k} - \frac{mc}{\hbar}\ \right)\psi = 0$$

$$\left(i\gamma^\mu\frac{\partial}{\partial x^\mu} - \frac{mc}{\hbar}\ \right)\psi = 0$$

$$\left(i\gamma^\mu\partial_\mu - \frac{mc}{\hbar}\ \right)\psi = 0 \tag{11.300}$$

In natural units $\hbar = c = 1$

$$(i\gamma^\mu \partial_\mu - m\ )\psi = 0$$

Multiplying equation f1 by i we get

$$\left(i^2\ \gamma^\mu \partial_\mu - \frac{imc}{\hbar}\ \right)\psi = 0$$

$$\left(-\ \gamma^\mu \partial_\mu - \frac{imc}{\hbar}\ \right)\psi = 0$$

$$\left(\gamma^\mu \partial_\mu + \frac{imc}{\hbar}\ \right)\psi = 0$$

In natural units we have

$$(\ \gamma^\mu \partial_\mu + im\ )\psi = 0.$$

Exercise 11.11 *How is stability of a system ensured by the vacuum state of Dirac theory.*

Ans. The infinite negative energy states $E = -mc^2$ to $E = -\infty$ of Dirac theory (figure 11.2) are completely filled by electrons as per the Pauli exclusion principle. And this multitude of filled negative energy states is called ground state and defined as the vacuum state by Dirac. There being no vacancy, the system is stable and there is no interaction. There is no possibility of an electron to go down and make the system collapse. In a nutshell, automatic collapse of a system is prevented by Pauli exclusion principle. This again indicates that the particles under discussion cannot be bosons which do not obey Pauli exclusion principle. They have to be fermions like electrons.

Exercise 11.12 *Show that helicity operator h commutes with the Dirac Hamiltonian H.*

Ans. $[\, h\ H] = hH - Hh = h(c\ \vec{\alpha}.\vec{p} + \beta mc^2) - (c\ \vec{\alpha}.\vec{p} + \beta mc^2\ )h$

$$= \frac{\hbar}{2}\begin{pmatrix} \vec{\sigma}.\hat{p} & 0 \\ 0 & \vec{\sigma}.\hat{p} \end{pmatrix}\left(c\begin{pmatrix} 0 & \vec{\sigma} \\ \vec{\sigma} & 0 \end{pmatrix}.\vec{p} + \begin{pmatrix} I & 0 \\ 0 & -I \end{pmatrix}mc^2 \right)$$

$$-\left(c\begin{pmatrix} 0 & \vec{\sigma} \\ \vec{\sigma} & 0 \end{pmatrix}.\vec{p} + \begin{pmatrix} I & 0 \\ 0 & -I \end{pmatrix}mc^2 \right)\frac{\hbar}{2}\begin{pmatrix} \vec{\sigma}.\hat{p} & 0 \\ 0 & \vec{\sigma}.\hat{p} \end{pmatrix}$$

$$= \frac{\hbar c}{2p}\begin{pmatrix} \vec{\sigma}.\vec{p} & 0 \\ 0 & \vec{\sigma}.\vec{p} \end{pmatrix}\begin{pmatrix} 0 & \vec{\sigma}.\vec{p} \\ \vec{\sigma}.\vec{p} & 0 \end{pmatrix} + mc^2\frac{\hbar}{2p}\begin{pmatrix} \vec{\sigma}.\vec{p} & 0 \\ 0 & \vec{\sigma}.\vec{p} \end{pmatrix}\begin{pmatrix} I & 0 \\ 0 & -I \end{pmatrix}$$

$$- \frac{\hbar c}{2p}\begin{pmatrix} 0 & \vec{\sigma}.\vec{p} \\ \vec{\sigma}.\vec{p} & 0 \end{pmatrix}\begin{pmatrix} \vec{\sigma}.\vec{p} & 0 \\ 0 & \vec{\sigma}.\vec{p} \end{pmatrix} - mc^2\frac{\hbar}{2p}\begin{pmatrix} I & 0 \\ 0 & -I \end{pmatrix}\begin{pmatrix} \vec{\sigma}.\vec{p} & 0 \\ 0 & \vec{\sigma}.\vec{p} \end{pmatrix}$$

$$= \frac{\hbar c}{2p}\begin{pmatrix} 0 & (\vec{\sigma}.\vec{p})^2 \\ (\vec{\sigma}.\vec{p})^2 & 0 \end{pmatrix} + mc^2 \frac{\hbar}{2p}\begin{pmatrix} \vec{\sigma}.\vec{p}\ I & 0 \\ 0 & -\vec{\sigma}.\vec{p}\ I \end{pmatrix}$$

$$- \frac{\hbar c}{2p}\begin{pmatrix} 0 & (\vec{\sigma}.\vec{p})^2 \\ (\vec{\sigma}.\vec{p})^2 & 0 \end{pmatrix} - mc^2 \frac{\hbar}{2p}\begin{pmatrix} I\vec{\sigma}.\vec{p} & 0 \\ 0 & -I\vec{\sigma}.\vec{p} \end{pmatrix}$$

$$[h\ H] = 0$$

Helicity commutes with the Dirac Hamiltonian and is a good quantum number and has simultaneous eigenvectors.

Exercise 11.13 *Show that* $(\vec{\sigma}.\vec{p})(\vec{\sigma}.\vec{p}) = p^2$

   Ans. Consider

$$(\vec{\sigma}.\vec{p})(\vec{\sigma}.\vec{p}) = \sigma_i p_i \sigma_j p_j = \sigma_i \sigma_j p_i p_j \tag{11.301}$$

as $\vec{\sigma}$ and $\vec{p}$ commute with each other as $\vec{p}$ is a differential operator in coordinate representation having no role in spin space. Interchanging the dummy indices we have

$$(\vec{\sigma}.\vec{p})(\vec{\sigma}.\vec{p}) = \sigma_j \sigma_i p_j p_i = \sigma_j \sigma_i p_i p_j \text{ since } p_i p_j = p_j p_i \tag{11.302}$$

Adding equations (11.301) and (11.302) we have

$$2(\vec{\sigma}.\vec{p})(\vec{\sigma}.\vec{p}) = \sigma_i \sigma_j p_i p_j + \sigma_j \sigma_i p_i p_j$$

$$(\vec{\sigma}.\vec{p})(\vec{\sigma}.\vec{p}) = \frac{1}{2}(\sigma_i \sigma_j p_i p_j + \sigma_j \sigma_i p_i p_j) = \frac{1}{2}(\sigma_i \sigma_j + \sigma_j \sigma_i)p_i p_j = \frac{1}{2}\{\sigma_i\ \sigma_j\}p_i p_j$$

Using $\{\sigma_i\ \sigma_j\} = 2\delta_{ij}$

$$(\vec{\sigma}.\vec{p})(\vec{\sigma}.\vec{p}) = \frac{1}{2}2\delta_{ij}p_i p_j = \delta_{ij}p_i p_j = p_i p_i = p_i^2 = \vec{p}^2 = p^2$$

In fact, $(\vec{\sigma}.\vec{p})(\vec{\sigma}.\vec{p}) = p^2$ holds for any vector $\vec{p}$ which commutes with the spin matrices $(\sigma_j p_i = p_i \sigma_j)$ and also the components of which commute among themselves, i.e.,

$$p_i p_j = p_j p_i.$$

Exercise 11.14 *Normalize the wave functions of equation (11.233).*

   Consider solutions for a Dirac particle with momentum $\vec{p} = p\hat{z}$. Identify the solutions with positive and negative energy, up and down spin, positive and negative helicity.

Show that positive energy solutions and negative energy solutions do not separately satisfy completeness condition but jointly they do.

Show that negative energy solutions have no role in the non-relativistic limit of Dirac equation.

Ans. Let us define $E = \pm\sqrt{p^2 c^2 + m^2 c^4} = E_\pm = \pm|E|$

With this we can rewrite the wave functions of equation (11.233) as follows

$$u^{(1)} = N \begin{pmatrix} 1 \\ 0 \\ \dfrac{cp_z}{|E| + mc^2} \\ \dfrac{c(p_x + ip_y)}{|E| + mc^2} \end{pmatrix}, \quad u^{(2)} = N \begin{pmatrix} 0 \\ 1 \\ \dfrac{c(p_x - ip_y)}{|E| + mc^2} \\ -\dfrac{cp_z}{|E| + mc^2} \end{pmatrix},$$

$$u^{(3)} = N \begin{pmatrix} -\dfrac{cp_z}{|E| + mc^2} \\ -\dfrac{c(p_x + ip_y)}{|E| + mc^2} \\ 1 \\ 0 \end{pmatrix}, \quad u^{(4)} = N \begin{pmatrix} -\dfrac{c(p_x - ip_y)}{|E| + mc^2} \\ \dfrac{cp_z}{|E| + mc^2} \\ 0 \\ 1 \end{pmatrix}$$

where N is the normalization constant and

$$|u^{(1)}> = N \begin{pmatrix} 1 \\ 0 \\ \dfrac{cp_z}{|E| + mc^2} \\ \dfrac{c(p_x + ip_y)}{|E| + mc^2} \end{pmatrix}, \quad <u^{(1)}|u^{(1)}> = 1$$

$$N^*N \begin{pmatrix} 1 & 0 & \dfrac{cp_z}{|E| + mc^2} & \dfrac{c(p_x - ip_y)}{|E| + mc^2} \end{pmatrix} \begin{pmatrix} 1 \\ 0 \\ \dfrac{cp_z}{|E| + mc^2} \\ \dfrac{c(p_x + ip_y)}{|E| + mc^2} \end{pmatrix} = 1$$

$$|N|^2 \left(1 + 0 + \frac{c^2 p_z^2}{(|E| + mc^2)^2} + \frac{c^2\left(p_x^2 + p_y^2\right)}{(|E| + mc^2)^2} \right) = 1$$

$$|N|^2 \left(1 + \frac{c^2 p^2}{(|E| + mc^2)^2} \right) = 1$$

$$N = \left(1 + \frac{c^2 p^2}{(|E| + mc^2)^2} \right)^{-1/2} = \left(1 + \frac{|E|^2 - m^2 c^4}{(|E| + mc^2)^2} \right)^{-1/2} \quad \left(as \ |E|^2 = p^2 c^2 + m^2 c^4 \right)$$

$$= \left(1 + \frac{(|E| + mc^2)(|E| - mc^2)}{(|E| + mc^2)^2}\right)^{-\frac{1}{2}} = \left(1 + \frac{|E| - mc^2}{|E| + mc^2}\right)^{-\frac{1}{2}}$$

$$= \left(\frac{|E| + mc^2 + |E| - mc^2}{|E| + mc^2}\right)^{-1/2}$$

$$N = \left(\frac{2|E|}{|E| + mc^2}\right)^{-1/2} = \sqrt{\frac{|E| + mc^2}{2|E|}}$$

To simplify we choose

$$\vec{p} = p\hat{z}$$

we have

$$p_x = 0, \ p_y = 0, \ p_z = p.$$

Also, let us rename

$$N = \sqrt{\frac{|E| + mc^2}{2|E|}} = a = \text{real quantity and}$$

$$b = \sqrt{\frac{|E| + mc^2}{2|E|}} \ \frac{cp_z}{|E| + mc^2} = \sqrt{\frac{c^2 p^2}{2|E|(|E| + mc^2)}} = \sqrt{\frac{|E|^2 - m^2 c^4}{2|E|(|E| + mc^2)}} = \sqrt{\frac{(|E| + mc^2)(|E| - mc^2)}{2|E|(|E| + mc^2)}}$$

$$b = \sqrt{\frac{(|E| - mc^2)}{2|E|}} = \text{real quantity}$$

Hence we get the basis

$$|u^{(1)}> = \begin{pmatrix} a \\ 0 \\ b \\ 0 \end{pmatrix}, \ |u^{(2)}> = \begin{pmatrix} 0 \\ a \\ 0 \\ -b \end{pmatrix}, \ |u^{(3)}> = \begin{pmatrix} -b \\ 0 \\ a \\ 0 \end{pmatrix}, \ u^{(4)}> = \begin{pmatrix} 0 \\ b \\ 0 \\ a \end{pmatrix} \qquad (11.303)$$

The energy eigen equations are

$$Hu^{(1)} = |E|u^{(1)}, \ Hu^{(2)} = |E|u^{(2)}, \ Hu^{(3)} = -|E|u^{(3)}, \ Hu^{(4)} = -|E|u^{(4)}$$

$$S_z = \frac{\hbar}{2}\begin{pmatrix} \sigma_z & 0 \\ 0 & \sigma_z \end{pmatrix} = \frac{\hbar}{2}\begin{pmatrix} 1 & 0 & 0 & 0 \\ 0 & -1 & 0 & 0 \\ 0 & 0 & 1 & 0 \\ 0 & 0 & 0 & -1 \end{pmatrix}$$

u	$u^{(1)} = \begin{pmatrix} a \\ 0 \\ b \\ 0 \end{pmatrix}$	$u^{(2)} = \begin{pmatrix} 0 \\ a \\ 0 \\ -b \end{pmatrix}$	$u^{(3)} = \begin{pmatrix} -b \\ 0 \\ a \\ 0 \end{pmatrix}$	$u^{(4)} = \begin{pmatrix} 0 \\ b \\ 0 \\ a \end{pmatrix}$
H	$|E|$	$|E|$	$-|E|$	$-|E|$
s_z	$+\dfrac{\hbar}{2}$	$-\dfrac{\hbar}{2}$	$+\dfrac{\hbar}{2}$	$-\dfrac{\hbar}{2}$
h	$+1$	-1	$+1$	-1
Nature	+ve energy spin up particle	+ve energy spin down particle	−ve energy spin up antiparticle	−ve energy spin down antiparticle

Figure 11.7. Classification of plane wave solution according to sign of energy, spin state and helicity (under relativistic condition).

The eigen equations for the operator s_z are

$$s_z u^{(1)} = \frac{\hbar}{2} u^{(1)}, \quad s_z u^{(2)} = -\frac{\hbar}{2} u^{(2)}, \quad s_z u^{(3)} = \frac{\hbar}{2} u^{(3)}, \quad s_z u^{(4)} = -\frac{\hbar}{2} u^{(4)}$$

The eigen equations for the helicity operator h are

$$h u^{(1)} = +u^{(1)}, \quad h u^{(2)} = -u^{(2)}, \quad h u^{(3)} = +u^{(3)}, \quad h u^{(4)} = -u^{(4)}$$

Results have been summarized in figure 11.7.

Completeness condition in the positive energy basis $|u^{(1)}\rangle$ and $|u^{(2)}\rangle$

$$|u^{(1)} > \quad <u^{(1)}| + |u^{(1)} > \quad <u^{(1)}|$$

$$= \begin{pmatrix} a \\ 0 \\ b \\ 0 \end{pmatrix}(a^*\,0\,b^*\,0) + \begin{pmatrix} a \\ 0 \\ b \\ 0 \end{pmatrix}(a^*\,0\,b^*\,0) = \begin{pmatrix} |a|^2 & 0 & ab^* & 0 \\ 0 & |a|^2 & 0 & -ab^* \\ ba^* & 0 & |b|^2 & 0 \\ 0 & -ba^* & 0 & |b|^2 \end{pmatrix} \neq 1$$

Positive energy solutions do not form a complete set, i.e. do not span the entire Hilbert space.

Completeness condition in the negative energy basis $|u^{(3)}\rangle$ and $|u^{(4)}\rangle$

$$|u^{(3)} > \quad <u^{(3)}| + |u^{(4)} > \quad <u^{(4)}|$$

$$= \begin{pmatrix} -b \\ 0 \\ a \\ 0 \end{pmatrix}(-b^*\,0\,a^*\,0) + \begin{pmatrix} 0 \\ b \\ 0 \\ a \end{pmatrix}(0\,b^*\,0\,a^*) = \begin{pmatrix} |b|^2 & 0 & -ba^* & 0 \\ 0 & |b|^2 & 0 & ba^* \\ -ab^* & 0 & |a|^2 & 0 \\ 0 & ab^* & 0 & |a|^2 \end{pmatrix} \neq 1$$

Negative energy solutions do not form a complete set, i.e. do not span the entire Hilbert space.

However, when we consider the complete basis of positive and negative energy states $|u^{(i)}\rangle$, $i = 1, 2, 3, 4$ they form a complete set. They form a closed Hilbert space.

$$|u^{(1)}> \quad <u^{(1)}|+|u^{(1)}> \quad <u^{(1)}| + |u^{(3)}> \quad <u^{(3)}|+|u^{(4)}> \quad <u^{(4)}| = 1$$

since

$$a = a^*, \ b = b^*, \ ab^* - ba^* = 0, \ |a|^2 + |b|^2 = \frac{|E| + mc^2}{2|E|} + \frac{(|E| - mc^2)}{2|E|} = 1$$

Exercise 11.15 *Find the non-relativistic limit of the solution of Dirac equation for a free particle.*

Ans. *The Dirac equation is*

$$(c\ \vec{\alpha}.\vec{p} + \beta mc^2)\psi = i\hbar\frac{\partial\psi}{\partial t}$$

$$\psi(\vec{r},\ t) = ue^{\frac{i(\vec{p}.\vec{r} - Et)}{\hbar}}, \quad u = \begin{pmatrix} f \\ g \end{pmatrix}, \ f = \begin{pmatrix} u_1 \\ u_2 \end{pmatrix}, \ g = \begin{pmatrix} u_3 \\ u_4 \end{pmatrix}$$

(solution of Dirac equation)

f, g are 2-component spinors and ψ is a 4-component spinor. The explicit solutions are given by equation (11.303)

To switch over to the non-relativistic limit let us define

$$E = E' + mc^2, \ |E'| \ll mc^2$$

where E' is kinetic energy

$$a = \sqrt{\frac{|E| + mc^2}{2|E|}} = \sqrt{\frac{1}{2}\left(1 + \frac{mc^2}{|E|}\right)} = \sqrt{\frac{1}{2}\left(1 + \frac{mc^2}{|E'| + mc^2}\right)} \approx \sqrt{\frac{1}{2}\left(1 + \frac{mc^2}{mc^2}\right)} = 1$$

$$b = \sqrt{\frac{(|E| - mc^2)}{2|E|}} = \sqrt{\frac{1}{2}\left(1 - \frac{mc^2}{|E|}\right)} = \sqrt{\frac{1}{2}\left(1 - \frac{mc^2}{|E'| + mc^2}\right)} \approx \sqrt{\frac{1}{2}\left(1 - \frac{mc^2}{mc^2}\right)} = 0$$

Hence with $a = 1$, $b = 0$ the solutions are from equation (11.303)

$$|u^{(1)}> = \begin{pmatrix} 1 \\ 0 \\ 0 \\ 0 \end{pmatrix}, \ |u^{(2)}> = \begin{pmatrix} 0 \\ 1 \\ 0 \\ 0 \end{pmatrix}, \ |u^{(3)}> = \begin{pmatrix} 0 \\ 0 \\ 1 \\ 0 \end{pmatrix}, \ u^{(4)}> = \begin{pmatrix} 0 \\ 0 \\ 0 \\ 1 \end{pmatrix} \tag{11.304}$$

For the positive energy case we put $=E_+ = E' + mc^2$, $|E'| \ll mc^2$ in equations (11.225) and (11.227). Now we note that

$$\frac{1}{E_+ + mc^2} = \frac{1}{E' + mc^2 + mc^2} = \frac{1}{E' + 2mc^2} \sim \frac{1}{2mc^2} \to 0$$

Hence from equations (11.225) and (11.227) we have

$$u^{(1)} = \begin{pmatrix} 1 \\ 0 \\ 0 \\ 0 \end{pmatrix} \to \begin{pmatrix} 1 \\ 0 \end{pmatrix}, \quad u^{(2)} = \begin{pmatrix} 0 \\ 1 \\ 0 \\ 0 \end{pmatrix} \to \begin{pmatrix} 0 \\ 1 \end{pmatrix}$$

Hence in the non-relativistic limit we can proceed with the positive energy solutions

$$|u^{(1)} > = \begin{pmatrix} 1 \\ 0 \end{pmatrix}, \quad | u^{(2)} > = \begin{pmatrix} 0 \\ 1 \end{pmatrix}.$$

which are 2-component spinors.

Let us check if they form a complete set.

$$|u^{(1)} > \quad <u^{(1)} | + |u^{(1)} > \quad <u^{(1)}|$$

$$= \begin{pmatrix} 1 \\ 0 \end{pmatrix}(1 \;\; 0) + \begin{pmatrix} 0 \\ 1 \end{pmatrix}(0 \;\; 1) = \begin{pmatrix} 1 & 0 \\ 0 & 0 \end{pmatrix} + \begin{pmatrix} 0 & 0 \\ 0 & 1 \end{pmatrix} = \begin{pmatrix} 1 & 0 \\ 0 & 1 \end{pmatrix} = I$$

Clearly the positive energy solutions are complete and they span the entire Hilbert space and so we do not need negative energy solutions to be a part of theory.

Exercise 11.16 *Show that the Dirac equation reduces to the Schrödinger equation in the non-relativistic limit.*

$\boxed{\text{Ans.}}$ The Dirac equation is

$$(c \; \vec{\alpha}. \vec{p} + \beta mc^2)\psi = i\hbar\frac{\partial \psi}{\partial t}$$

$$\psi(\vec{r},\, t) = ue^{i(\vec{p}.\vec{r} - Et)/\hbar}, \quad u = \begin{pmatrix} f \\ g \end{pmatrix}, \quad f = \begin{pmatrix} u_1 \\ u_2 \end{pmatrix}, \quad g = \begin{pmatrix} u_3 \\ u_4 \end{pmatrix}$$

We obtained from the Dirac equation equation (11.211) viz.

$$(E^2 - m^2 c^4)f = c^2 (\vec{\sigma}. \vec{p})^2 f$$

In the non-relativistic limit for positive energy $u \sim f = \begin{pmatrix} u_1 \\ u_2 \end{pmatrix}$, $g = \begin{pmatrix} 0 \\ 0 \end{pmatrix}$. Putting $E = E' + mc^2$ and using $(\vec{\sigma}. \vec{p})^2 = p^2$ from *exercise 11.13*.

$$[(E' + mc^2)^2 - m^2c^4]f = c^2p^2f$$

$$[E'^2 + m^2c^4 + 2E'mc^2 - m^2c^4]f = c^2p^2f$$

$$E'(E' + 2mc^2)f = c^2p^2f$$

$$E'2mc^2f = c^2p^2f$$

$$\frac{p^2}{2m}f = E'f$$

This is Schrödinger equation to which the Dirac equation reduces.

Exercise 11.17 *Prove that* $Tr(\vec{\alpha}.\,\vec{B})(\vec{\alpha}.\,\vec{C}) = 4\,\vec{B}.\,\vec{C}$ *where* $\vec{B}$ *and* $\vec{C}$ *commutes with* $\vec{\sigma}$.

$\boxed{\text{Ans.}}$ Consider

$$(\vec{\alpha}.\,\vec{B})(\vec{\alpha}.\,\vec{C}) = (\alpha_x B_x + \alpha_y B_y + \alpha_z B_z)\,(\alpha_x C_x + \alpha_y C_y + \alpha_z C_z)$$

$$= \alpha_x B_x \alpha_x C_x + \alpha_x B_x \alpha_y C_y + \alpha_x B_x \alpha_z C_z + \alpha_y B_y \alpha_x C_x + \alpha_y B_y \alpha_y C_y + \alpha_y B_y \alpha_z C_z$$

$$+ \alpha_z B_z \alpha_x C_x + \alpha_z B_z \alpha_y C_y + \alpha_z B_z \alpha_z C_z$$

$$= \alpha_x \alpha_x B_x C_x + \alpha_x \alpha_y B_x C_y + \alpha_x \alpha_z B_x C_z + \alpha_y \alpha_x B_y C_x + \alpha_y \alpha_y B_y C_y + \alpha_y \alpha_z B_y C_z$$

$$+ \alpha_z \alpha_x B_z C_x + \alpha_z \alpha_y B_z C_y + \alpha_z \alpha_z B_z C_z$$

$$(\vec{\alpha}.\,\vec{B})(\vec{\alpha}.\,\vec{C}) = \alpha_x^2 B_x C_x + \alpha_x \alpha_y B_x C_y + \alpha_x \alpha_z B_x C_z + \alpha_y \alpha_x B_y C_x$$

$$+ \alpha_y^2 B_y C_y + \alpha_y \alpha_z B_y C_z + \alpha_z \alpha_x B_z C_x + \alpha_z \alpha_y B_z C_y + \alpha_z^2 B_z C_z$$

Using $\alpha_x^2 = I$, $\alpha_y^2 = I$, $\alpha_z^2 = I$, $\sigma_x \sigma_y = i\sigma_z$, $\sigma_y \sigma_x = -i\sigma_z$ etc.

$$\alpha_x \alpha_y = \begin{pmatrix} \sigma_x & 0 \\ 0 & \sigma_x \end{pmatrix}\begin{pmatrix} \sigma_y & 0 \\ 0 & \sigma_y \end{pmatrix} = \begin{pmatrix} \sigma_x \sigma_y & 0 \\ 0 & \sigma_x \sigma_y \end{pmatrix} = \begin{pmatrix} i\sigma_z & 0 \\ 0 & i\sigma_z \end{pmatrix} = i\begin{pmatrix} \sigma_z & 0 \\ 0 & \sigma_z \end{pmatrix} = \frac{i}{\hbar} s_z$$

Similarly $\alpha_y \alpha_x = -\frac{i}{\hbar}s_z$, $\alpha_y \alpha_z = \frac{i}{\hbar}s_x$ etc hence we get

$$(\vec{\alpha}.\,\vec{B})(\vec{\alpha}.\,\vec{C}) = IB_x C_x + \frac{i}{\hbar}s_z B_x C_y - \frac{i}{\hbar}y B_x C_z - \frac{i}{\hbar}s_z B_y C_x$$

$$+ IB_y C_y + \frac{i}{\hbar}s_x B_y C_z \pm \frac{i}{\hbar}s_y B_z C_x - \frac{i}{\hbar}s_x B_z C_y + IB_z C_z$$

$$=I(B_x C_x + B_y C_y + B_z C_z) + \frac{i}{\hbar}s_z(B_x C_y - B_y C_x) + \frac{i}{\hbar}s_y(B_z C_x - B_x C_z)$$

$$+\frac{i}{\hbar}s_x\left(B_y C_z - B_z C_y\right)$$

$$=I(\vec{B}.\,\vec{C}) + \frac{i}{\hbar}s_z(\vec{B}\times\vec{C})_z + \frac{i}{\hbar}s_y(\vec{B}\times\vec{C})_y + \frac{i}{\hbar}s_x(\vec{B}\times\vec{C})_x$$

$$=I(\vec{B}.\,\vec{C}) + \frac{i}{\hbar}s_x(\vec{B}\times\vec{C})_x + \frac{i}{\hbar}s_y(\vec{B}\times\vec{C})_y + \frac{i}{\hbar}s_z(\vec{B}\times\vec{C})_z$$

$$=\begin{pmatrix} I & 0 \\ 0 & I \end{pmatrix}(\vec{B}.\,\vec{C}) + \begin{pmatrix} \sigma_x & 0 \\ 0 & \sigma_x \end{pmatrix}\frac{i}{\hbar}(\vec{B}\times\vec{C})_x + \begin{pmatrix} \sigma_y & 0 \\ 0 & \sigma_y \end{pmatrix}\frac{i}{\hbar}(\vec{B}\times\vec{C})_y + \begin{pmatrix} \sigma_z & 0 \\ 0 & \sigma_z \end{pmatrix}\frac{i}{\hbar}(\vec{B}\times\vec{C})_z$$

$$=\begin{pmatrix} 1 & 0 & 0 & 0 \\ 0 & 1 & 0 & 0 \\ 0 & 0 & 1 & 0 \\ 0 & 0 & 0 & 1 \end{pmatrix}(\vec{B}.\,\vec{C}) + \begin{pmatrix} 0 & 1 & 0 & 0 \\ 1 & 0 & 0 & 0 \\ 0 & 0 & 0 & 1 \\ 0 & 0 & 1 & 0 \end{pmatrix}\frac{i}{\hbar}(\vec{B}\times\vec{C})_x$$

$$+\begin{pmatrix} 0 & -i & 0 & 0 \\ i & 0 & 0 & 0 \\ 0 & 0 & 0 & -i \\ 0 & 0 & i & 0 \end{pmatrix}\frac{i}{\hbar}(\vec{B}\times\vec{C})_y + \begin{pmatrix} 1 & 0 & 0 & 0 \\ 0 & -1 & 0 & 0 \\ 0 & 0 & 1 & 0 \\ 0 & 0 & 0 & -1 \end{pmatrix}\frac{i}{\hbar}(\vec{B}\times\vec{C})_z$$

$$Tr(\vec{\alpha}.\,\vec{B})(\vec{\alpha}.\,\vec{C}) = 4\,\vec{B}.\,\vec{C} + 0 + 0 + 0$$

$$Tr(\vec{\alpha}.\,\vec{B})(\vec{\alpha}.\,\vec{C}) = 4\,\vec{B}.\,\vec{C}$$

Exercise 11.18 *Mention the distinguishing properties of spin 0, spin 1, spin 2, spin $\frac{1}{2}$ particles.*

$\boxed{\text{Ans.}}$ Spin 0 particle is called scalar. It is represented by a dot ($\bullet$). It looks the same from all directions.

Spin 1 particle is called vector. It is represented by an arrow ($\rightarrow$). It looks different from different directions. Only if it is turned around by 360° does the particle look the same.

Spin 2 particle is called tensor. It is represented by a double headed arrow ($\leftrightarrow$). It looks the same if turned through 180°.

Spin $\frac{1}{2}$ particle is called spinor. A spinor is a vector in spin space. The space of the Dirac matrices is spin space. A spinor is distinguished from an ordinary vector by its peculiar transformation properties under rotation. A rotation of 4π ($= 2$ complete revolutions) is required to restore a Dirac spinor to its original value.

Exercise 11.19 *Derive the covariant form of the Dirac equation for a particle of charge q in an electromagnetic field* $(\vec{A}, \phi)$.

$\boxed{\text{Ans.}}$ The Dirac wave equation (11.114) can be rewritten as

$$E\psi = c\ \vec{\alpha}.\vec{p}\,\psi + \beta mc^2\psi$$

$$(E - c\ \vec{\alpha}.\vec{p} - \beta mc^2)\psi = 0$$

To switch over to the electromagnetic field we make the usual replacements viz.

$$\vec{p} \to \vec{p} - q\vec{A}, \quad E \to E - q\phi$$

to get

$$((E - q\phi) - c\ \vec{\alpha}.(\vec{p} - q\,\vec{A}) - \beta mc^2))\psi = 0$$

With $E \to i\hbar\frac{\partial}{\partial t}$, $\vec{p} \to -i\hbar\,\vec{\nabla}$

$$\left(\left(i\hbar\frac{\partial}{\partial t} - q\phi\right) - c\ \vec{\alpha}.\left(-i\hbar\,\vec{\nabla} - q\,\vec{A}\right) - \beta mc^2\right)\psi = 0$$

$$i\hbar\frac{\partial\psi}{\partial t} = [c\ \vec{\alpha}.(-i\hbar\,\vec{\nabla} - q\,\vec{A}) + \beta mc^2 + q\phi]\psi \tag{11.305}$$

Multiply by $\frac{\beta}{c\hbar}$ to get

$$\frac{\beta}{c\hbar}i\hbar\frac{\partial\psi}{\partial t} = \left[\frac{\beta}{c\hbar}c\ \vec{\alpha}.(-i\hbar\,\vec{\nabla}) - \frac{\beta}{c\hbar}c\ \vec{\alpha}.q\,\vec{A} + \frac{\beta}{c\hbar}\beta mc^2 + \frac{\beta}{c\hbar}q\phi\right]\psi$$

$$\left[-i\beta\frac{\partial}{\partial ct} - i\beta\ \vec{\alpha}.\vec{\nabla} - \frac{\beta}{\hbar}\ \vec{\alpha}.q\,\vec{A} + \beta^2\frac{mc}{\hbar} + \frac{\beta}{c\hbar}q\phi\right]\psi = 0$$

$$\left[-i\beta\frac{\partial}{\partial ct} - i\beta\ \vec{\alpha}.\vec{\nabla} - \frac{\beta}{\hbar}\ \vec{\alpha}.q\,\vec{A} + \beta^2\frac{mc}{\hbar} + \frac{\beta}{c\hbar}q\phi\right]\psi = 0$$

Use $\dfrac{\partial}{\partial ct} = \partial^0$, $\vec{\nabla} = -\partial^k$, $\vec{\alpha} = -\alpha_k = \alpha^k$, $\vec{A} = -A_k = A^k$, $\dfrac{\phi}{c} = A^0 = A_0$, $\beta^2 = 1$

$$\left[-i\beta\partial^0 - i\beta\ (-\alpha_k)(-\partial^k) - \frac{\beta}{\hbar}\ (-\alpha_k)q\ (A^k) + \frac{mc}{\hbar} + \frac{\beta}{\hbar}qA^0\right]\psi = 0$$

Define $\beta = \gamma_0$, $\beta\alpha_k = \gamma_k$

$$\left[-i\beta\partial^0 - i\gamma_k\ \partial^k + \frac{q}{\hbar}\ \gamma_k A^k + \frac{mc}{\hbar} + \frac{q}{\hbar}\gamma_0 A^0\right]\psi = 0$$

$$\left[-i\gamma_\mu \partial^\mu + \frac{q}{\hbar}\,\gamma_\mu A^\mu + \frac{mc}{\hbar} \right]\psi = 0 \tag{11.306}$$

Multiplying by i we get

$$\left[\gamma_\mu \partial^\mu + \frac{iq}{\hbar}\,\gamma_\mu A^\mu + \frac{imc}{\hbar} \right]\psi = 0 \tag{11.307}$$

Multiplying equation (11.306) by $\hbar$ we get

$$[-i\hbar\gamma_\mu \partial^\mu + q\,\gamma_\mu A^\mu + mc\,]\psi = 0$$

Since $p^\mu = i\hbar p^\mu$ we get

$$[-\gamma_\mu p^\mu + q\gamma_\mu A^\mu + mc\,]\psi = 0$$

$$[\gamma_\mu\!\left(p^\mu - qA^\mu\right) - mc\,]\psi = 0 \tag{11.308}$$

This equation was referred to as equation (11.279).

Exercise 11.20 *Obtain the continuity equation from the covariant form of the Dirac equation for a particle of charge q in an electromagnetic field $(\vec{A},\ \phi)$.*

 Ans. The covariant form of the Dirac equation for a particle of charge q in an electromagnetic field $(\vec{A},\ \phi)$ is

$$[-i\hbar\gamma_\mu \partial^\mu + q\,\gamma_\mu A^\mu + mc\,]\psi = 0$$

$$-i\hbar\gamma_\mu\!\left(\partial^\mu\psi\right) + q\,\gamma_\mu A^\mu\psi + mc\,\psi = 0 \tag{11.309}$$

The Hermitian conjugate of equation (11.309) is

$$i\hbar(\partial^\mu\psi)^\dagger\gamma_\mu^\dagger + q(A^\mu\psi)^\dagger\,\gamma_\mu^\dagger + mc\,\psi^\dagger = 0$$

$$i\hbar(\partial^0\psi)^\dagger\gamma_0^\dagger + i\hbar(\partial^i\psi)^\dagger\gamma_i^\dagger + q(A^0\psi)^\dagger\,\gamma_0^\dagger + q(A^i\psi)^\dagger\,\gamma_i^\dagger + mc\,\psi^\dagger = 0$$

$$\text{As } \gamma_0^\dagger = \gamma_0,\ \gamma_i^\dagger = -\gamma_i$$

$$i\hbar\partial^0\psi^\dagger\gamma_0 + i\hbar\partial^i\psi^\dagger(-\gamma_i) + qA^0\psi^\dagger\,\gamma_0 + qA^i\psi^\dagger\,(-\gamma_i) + mc\,\psi^\dagger = 0$$

$$\text{As } \gamma_0 = \beta,\ \gamma_i = \beta\alpha_i$$

$$i\hbar\partial^0\psi^\dagger\beta - i\hbar\partial^i\psi^\dagger\beta\alpha_i + qA^0\psi^\dagger\,\beta - qA^i\psi^\dagger\beta\alpha_i + mc\,\psi^\dagger = 0 \tag{11.310}$$

Multiplying (11.310) by β from right we get

$$i\hbar\partial^0\psi^\dagger\beta\beta - i\hbar\partial^i\psi^\dagger\beta\alpha_i\beta + qA^0\psi^\dagger\,\beta\beta - qA^i\psi^\dagger\beta\alpha_i\beta + mc\,\psi^\dagger\beta = 0$$

$$\text{As } \alpha_i\beta = -\beta\alpha_i$$

$$i\hbar\partial^0\psi^\dagger\beta\beta + i\hbar\partial^i\psi^\dagger\beta\beta\alpha_i + qA^0\psi^\dagger\,\beta\beta + qA^i\psi^\dagger\beta\beta\alpha_i + mc\,\psi^\dagger\beta = 0$$

With $\overline{\psi} = \psi^\dagger\beta$

$$i\hbar\partial^0\overline{\psi}\beta + i\hbar\partial^i\,\overline{\psi}\beta\alpha_i + qA^0\overline{\psi}\,\beta + qA^i\,\overline{\psi}\beta\alpha_i + mc\,\overline{\psi} = 0$$

With $\gamma_0 = \beta$, $\gamma_i = \beta\alpha_i$

$$i\hbar\partial^0\overline{\psi}\gamma_0 + i\hbar\partial^i\,\overline{\psi}\gamma_i + qA^0\overline{\psi}\,\gamma_0 + qA^i\,\overline{\psi}\gamma_i + mc\,\overline{\psi} = 0$$

$$i\hbar\left(\partial^\mu\overline{\psi}\gamma_\mu\right) + qA^\mu\overline{\psi}\,\gamma_\mu + mc\,\overline{\psi} = 0 \tag{11.311}$$

$\overline{\psi}\times$ equation (11.309) + equation (11.311) $\times\,\psi$

$$\overline{\psi}\left(-i\hbar\gamma_\mu\left(\partial^\mu\psi\right) + q\;\gamma_\mu A^\mu\psi + mc\;\psi\right) - \left(i\hbar\left(\partial^\mu\overline{\psi}\gamma_\mu\right) + qA^\mu\overline{\psi}\;\gamma_\mu + mc\;\overline{\psi}\right)\psi = 0$$

$$\left(-i\hbar\overline{\psi}\gamma_\mu\left(\partial^\mu\psi\right) + q\;\overline{\psi}\gamma_\mu A^\mu\psi + mc\;\overline{\psi}\psi\right) - \left(i\hbar\left(\partial^\mu\overline{\psi}\gamma_\mu\right)\psi + qA^\mu\overline{\psi}\;\gamma_\mu\psi + mc\;\overline{\psi}\psi\right) = 0$$

$$-i\hbar\overline{\psi}\gamma_\mu\left(\partial^\mu\psi\right) - i\hbar\left(\partial^\mu\overline{\psi}\gamma_\mu\right)\psi = 0$$

$$\overline{\psi}\gamma_\mu\left(\partial^\mu\psi\right) + \left(\partial^\mu\overline{\psi}\gamma_\mu\right)\psi = 0$$

$$\partial^\mu(\overline{\psi}\gamma_\mu\psi) = 0$$

$$\partial^0(\overline{\psi}\gamma_0\psi) + \partial^i(\overline{\psi}\gamma_i\psi) = 0$$

$$\partial^0(\psi^\dagger\gamma_0\gamma_0\psi) + \partial^i(\psi^\dagger\gamma_0\gamma_i\psi) = 0$$

Using $\gamma_0^2 = 1$, $\gamma_0\gamma_i = \beta(\beta\alpha_i) = \alpha_i$

$$\partial^0(\psi^\dagger\psi) + \partial^i(\psi^\dagger\alpha_i\psi) = 0$$

Multiply by c to get

$$\partial^0(c\psi^\dagger\psi) + \partial^i(\psi^\dagger c\alpha_i\psi) = 0$$

Defining $J_0 = c\psi^\dagger\psi$, $J_i = \psi^\dagger c\alpha_i\psi$ we have

$$\partial^0 J_0 + \partial^i J_i = 0$$

$$\partial^\mu J_\mu = 0$$

This is the continuity equation $\frac{\partial\rho}{\partial t} + \vec{\nabla}\cdot\vec{J} = 0$.

Exercise 11.21 *What is the significance of the operator $c\vec{\alpha}$? What is Zitterbewegung?*
[Ans.] Consider

$$\frac{dx_1}{dt} = \frac{1}{i\hbar}[\ x_1,\ H\ ] = \frac{1}{i\hbar}[\ x_1,\ H\ ] = \frac{1}{i\hbar}[\ x_1,\ c\vec{\alpha}.\vec{p} + \beta mc^2\ ]$$

$$\frac{dx_1}{dt} = \frac{1}{i\hbar}[\ x_1,\ c\vec{\alpha}.\vec{p}\ ] + \frac{1}{i\hbar}\underbrace{[\ x_1,\ \beta mc^2\ ]}_{=0} = \frac{1}{i\hbar}[\ x_1,\ c\vec{\alpha}.\vec{p}\ ]$$

$$= \frac{c}{i\hbar}[\ x_1,\ \alpha_1 p_1 + \alpha_2 p_2 + \alpha_3 p_3\ ] = \frac{c}{i\hbar}[\ x_1,\ \alpha_1 p_1\ ] + \frac{c}{i\hbar}[\ x_1,\ \alpha_2 p_2\ ] + \frac{c}{i\hbar}[\ x_1,\ \alpha_3 p_3\ ]$$

$$= \frac{c\alpha_1}{i\hbar}\underbrace{[\ x_1,\ p_1\ ]}_{=i\hbar} + \frac{c\alpha_2}{i\hbar}\underbrace{[\ x_1,\ p_2\ ]}_{=0} + \frac{c\alpha_3}{i\hbar}\underbrace{[\ x_1,\ p_3\ ]}_{=0}$$

$$\frac{dx_1}{dt} = \frac{c\alpha_1}{i\hbar}i\hbar = c\alpha_1$$

Similarly

$$\frac{dx_2}{dt} = c\alpha_2,\quad \frac{dx_3}{dt} = c\alpha_3$$

$$\frac{d\vec{r}}{dt} = \frac{1}{i\hbar}[\ \vec{r},\ H\ ] = c\vec{\alpha}$$

The result $\frac{d\vec{r}}{dt} = c\vec{\alpha}$ shows that $c\vec{\alpha}$ acts like a velocity operator though it is not actually so, except in the case of non-relativistic limit. This is because of the following reason.

Components of α do not commute though components of velocity should commute.

Further eigenvalues are $\pm c$, which implies that the particle of mass $m \neq 0$ travels with the velocity of light which is impossible.

So the eigenvector of $c\vec{\alpha}$ and hence also the eigenvalues do not correspond to the physical states of a free Dirac particle.

We hope to get a sensible expression for the velocity of the particle only in a representation in which positive and negative energies are separated.

Actually, Dirac theory involves both positive and negative energy solutions. $\vec{\alpha}$ mixes the positive and negative energy states. Such mixing leads to rapid fluctuations in the coordinate of the electron over a distance of the order of Compton wavelength $(\frac{\hbar}{mc})$. This phenomenon is called Zitterbewegung or jittery motion.

This jittery motion or Zitterbewegung of the particle, due to spin orbit coupling is what makes the acceleration non-zero (despite the particle being free) and leads to $\pm c$ as eigenvalues for velocity.

Zitterbewegung can be interpreted by the uncertainty principle. A precise measurement of instantaneous velocity requires the accurate measurement of the position of the particle at two slightly different times. Such accurate position measurement implies that the momentum of the particle is completely unknown so that all momentum values are equally probable. Then very large momenta are more likely to result than small momenta and these correspond to velocity components close to the speed of light.

Exercise 11.22 *The special theory of relativity deals with which frame of reference?*

(a) Inertial *(b) non-inertial* *(c) both* *(d) neither.*

Exercise 11.23 *The general theory of relativity deals with which frame of reference?*

(a) Inertial *(b) non-inertial* *(c) both* *(d) neither.*

Exercise 11.24 *Frames moving with non-zero acceleration are referred to as*

(a) Inertial *(b) non-inertial* *(c) pseudo frame* *(d) rotating frame.*

Exercise 11.25 *Special theory of relativity deals with*

(a) Flat geometry *(b) curved geometry* *(c) simple geometry* *(d) complex geometry.*

Exercise 11.26 *General theory of relativity deals with*

(a) Flat geometry *(b) curved geometry* *(c) simple geometry* *(d) complex geometry.*

Exercise 11.27 *Identify the correct relation*

(a) $x_0 = x^0,\ x^i = -x_i = \vec{r} = (x, y, z)$ *(b)* $x^0 = x_0,\ x_i = -x^i = \vec{r} = (x, y, z)$
(c) $p^\mu = (\frac{E}{c}, \vec{p})$ *(d)* $p^\mu = (\frac{E}{c}, -\vec{p})$

Exercise 11.28 *Identify the correct relation*

(a) $\partial_\mu = (\frac{1}{c}\frac{\partial}{\partial t}, \vec{\nabla})$ *(b)* $\partial^\mu = (\frac{1}{c}\frac{\partial}{\partial t}, -\vec{\nabla})$
(c) $\partial^\mu = (\frac{1}{c}\frac{\partial}{\partial t}, -\vec{\nabla})$ *(d)* $\partial_\mu = (\frac{1}{c}\frac{\partial}{\partial t}, \vec{\nabla})$

$$\boxed{\text{Answers to multiple choice questions}}$$

11.1*a*, 11.2*c*, 11.6*c*, 11.22*a*, 11.23*b*, 11.24*b*, 11.25*a*, 11.26*b*, 11.27*a,c,d*, 11.28*a,b*.

11.35 Question bank

Q11.1 Obtain the Klein–Gordon equation for a free particle.

Q11.2 Obtain the Klein–Gordon equation for amassless particle.

Q11.3 Obtain the covariant form of the Klein–Gordon equation.

Q11.4 Obtain the solution of the Klein–Gordon equation for a free particle.

Q11.5 Obtain the Klein–Gordon equation for a charged particle in an electromagnetic field.

Q11.6 Find the spin state to which the Klein–Gordon equation corresponds.

Q11.7 Show that the equation of continuity that follows from the Klein–Gordon equation expresses charge conservation rather than probability conservation.

Q11.8 What was the problem of interpretation of the solution of the Klein–Gordon equation.

Q11.9 Derive the Dirac equation for a free particle.

Q11.10 Why did Dirac introduce the α, β and γ matrices ?

Q11.11 What are the properties of Dirac matrices?

Q11.12 Obtain the probability density from the Dirac equation.

Q11.13 Obtain the plane wave solution of the Dirac equation.

Q11.14 How does the idea of the anti-particle follow from Dirac theory?

Q11.15 How do we explain pair production from the Dirac theory?

Q11.16 How do we explain pair annihilation from the Dirac theory?

Q11.17 How can we show that Dirac theory corresponds to spin half particle.

Q11.18 What do we mean by multiplicity, helicity, chirality?

Q11.19 Obtain the Dirac equation for a charged particle in an electromagnetic field.

Further reading

[1] Perkins H 2000 *Introduction to High Energy Physics* (Cambridge: Cambridge University Press)

[2] Griffiths D 2008 *Introduction to Elementary Particles* (New York: Wiley-VCH)

[3] Tayal D C 2009 *Nuclear Physics* (Mumbai: Himalaya Publishing House)

[4] Satya P 2005 *Nuclear Physics & Particle Physics* (New Delhi: Sultan Chand & Sons)

[5] Guha J 2019 *Quantum Mechanics: Theory, Problems & Solutions* 3rd edn (Kolkata: Books and Allied (P) Ltd))

[6] Yung K L 2002 *Problems and Solutions on Atomic, Nuclear and Particle Physics* (Singapore: World Scientific)

IOP Publishing

Nuclear and Particle Physics with Cosmology, Volume 2
Particle physics and cosmology
Jyotirmoy Guha

Chapter 12

Introduction to quantum field theory, gauge theory and analysis of relativistic collisions

In this chapter we discuss first and second quantization. We obtain classical field equation in terms of Lagrangian density as well as in Hamiltonian formalism. Quantization of Schrödinger non- relativistic field, Klein–Gordon field, Dirac field and electromagnetic field is done. Maxwell field equations are expressed in terms of potentials and we explain gauge transformation and gauge invariance. Global and local gauge transformation is explained. Maxwell equations are derived from the Euler–Lagrange equation. We discuss gauge invariance of the Dirac equation. We also discuss spontaneous decay of particle, analyse relativistic collisions through Mandelstam variables, Möller and Bhava scattering, two-body scattering in center of mass frame and laboratory frame and the relation between scattering angles.

12.1 First quantization

Quantum mechanics, developed between 1925 and 1928 by Schrödinger, Heisenberg, Born, Pauli, Dirac and others is called first quantization.

In quantum mechanics, a physical system or system of particles is treated using quantum wave functions. The allowed values of momentum and energy for the matter particles (say electrons) are constrained by the Heisenberg uncertainty principle of quantum mechanics (viz. $\Delta p \geqslant \frac{\hbar}{2}$, $\Delta E \, \Delta t \geqslant \frac{\hbar}{2}$). Any value is not permitted—only certain discrete values are allowed to the bound system. So we say that the matter particles (say electrons) are quantized.

But the surrounding environment, e.g. the potential well or the field (electromagnetic field or gravitational field) the particles are subjected to is treated classically.

So a classical field interacts with quantized matter or a quantized system. In other words, the physical system is subjected to semi-classical treatment of quantum mechanics. This is called first quantization. Clearly there is asymmetry in the manner in which quantum mechanics (i.e. the first quantization theory) treats matter and the field. While matter is a system of quanta, field is a classical system. So the interaction occurs unevenly between matter and field.

The above discussion points to the fact that the field needs to be quantized. This is required for a valid quantum theory. This is done in the theory of second quantization where field is also quantized.

A true quantum mechanical description should include a quantized field.

12.2 Difference between first quantization and second quantization

✓First quantization is quantization of material particles.

Second quantization is quantization of fields.

✓In first quantization the phase space variables are trajectories.

In second quantization the phase space variables are fields.

✓Both the first and second quantization theories involve the process of quantization and hence are not fundamentally different.

✓In the case of first quantization there is a fixed and finite number of degrees of freedom.

In the case of second quantization there is an arbitrarily large but finite number of degrees of freedom or infinite degrees of freedom provided the limit exists.

✓In first quantization we deal with a problem (involving a system of material particles) in terms of the wave function associated with the particle. This wave function has value in a space called Hilbert space. So first quantization applies in Hilbert space. The number of particles is constant in this case.

Second quantization applies in Fock space. The number of particles is not constant in this case. A creation operator is responsible for the creation of particles leading to an increase in the number of particles. The annihilation operator is responsible for annihilation of particles leading to the decrease in the number of particles. Clearly the creation and annihilation operators have values in Fock space (which is an extended version of Hilbert space).

12.3 Hilbert space and Fock space

Hilbert space is a vector space. Each location in this Hilbert space has a vector attached to it. The bases of the space are eigenstates of a Hermitian operator (say the Hamiltonian). The base eigenstates represent systems that are constrained to have a fixed number of particles.

If we relax this constraint and use eigenstates that have a variable number of particles as the basis, the corresponding space is called Fock space or extended Hilbert space. This means it corresponds to creation of particles through creation operator as well as destruction of particles through annihilation operator. The action of creation operator and annihilation operator may be such as not to conserve the particle number.

• In other words, Hilbert space in which particle number is not conserved because of creation and destruction of particles is called Fock space.

12.4 Classical field equation

A classical mechanical system can be addressed using two approaches.

One is the Lagrangian approach or Lagrangian formulation of classical mechanics. The other approach is the Hamiltonian approach or Hamiltonian formulation of classical mechanics.

In classical field theory there are also two approaches. One is Lagrangian formalism and the other is Hamiltonian formalism.

A mechanical system has a finite number of degrees of freedom. In contrast, a field has a very large or infinite number of degrees of freedom.

As the field has an infinite number of degrees of freedom it can be treated as being analogous to a system consisting of an infinite number of particles.

A system of particles is specified by generalized coordinates q_i where $i = 1$ to f where f is the number of degrees of freedom of the system which is finite. So a mechanical system is specified by the f independent variables called generalized coordinates.

A field is specified by its amplitudes denoted by $\psi(\vec{r}, t)$ defined at all points $\vec{r}$ of the space. These amplitudes are independent of each other. The amplitudes of the field viz. $\psi(\vec{r}, t)$ play the same role as the generalized coordinates $q_i(t)$ in a mechanical system.

The amplitudes $\psi(\vec{r}, t)$ are also known as field functions.

The variable $\vec{r}$ is continuous and has to be considered as a continuous index.

In analogy with any mechanical system the field is the carrier of energy, momentum, angular momentum and other observable dynamical quantities.

We wish to derive and discuss the classical field equation. A field equation refers to the relation between $\psi(\vec{r}, t)$ and their derivatives. Field equations are analogous to the equation of motion in the case of a mechanical system.

A field propagates in accordance to the field equation and carries with it the dynamical variables or quantities.

12.5 Lagrangian formalism

Interaction of particles can be studied from a symmetry point of view instead of studying their equations of motion. In other words, gauge symmetry is a way of understanding interaction of particles.

12.5.1 Free particle case

The Lagrangian for a free particle is given by

$$L = T - V \tag{12.1}$$

where T is kinetic energy and V is potential energy. For a free particle moving in 1D

$$T = \frac{1}{2}m\dot{x}^2, \quad V = 0 \tag{12.2}$$

Hence

$$L = \frac{1}{2}m\dot{x}^2 \tag{12.3}$$

The Euler–Lagrange equation of motion is

$$\frac{d}{dt}\left(\frac{\partial L}{\partial \dot{x}}\right) - \frac{\partial L}{\partial x} = 0 \tag{12.4}$$

$$\frac{d}{dt}\left(\frac{\partial}{\partial \dot{x}}\,\frac{1}{2}m\dot{x}^2\right) - \frac{\partial}{\partial x}\frac{1}{2}m\dot{x}^2 = 0 \text{ i. e. } \ddot{x} = 0 \tag{12.5}$$

The Euler–Lagrange equation can be generalized to more than one coordinate, and for a continuous system of particles where there are an infinite number of particles the degrees of freedom is infinite. To describe the configuration we can use a function ψ of position coordinate and time, i.e. $\psi = \psi(\vec{x}, t) = \psi(x^\mu)$. (We denoted $\vec{r}$ by $\vec{x}$.)

12.6 Lagrangian density

We wish to derive and discuss the classical field equation using the Lagrangian approach. The Lagrangian of the field is

$$L = L(\psi, \dot{\psi}, t) \tag{12.6}$$

We define a quantity called the Lagrangian density $\mathcal{L}$.

The value of Lagrangian L per unit volume is called Lagrangian density $\mathcal{L}$.

In other words, the Lagrangian divided by volume is Lagrangian density. It is a function of ψ, space derivative of ψ, i.e. $\vec{\nabla}\psi = \hat{i}\frac{\partial \psi}{\partial x} + \hat{j}\frac{\partial \psi}{\partial y} + \hat{k}\frac{\partial \psi}{\partial z}$ and time derivative of ψ i.e. $\dot{\psi} = \frac{\partial \psi}{\partial t}$. Hence

$$\mathcal{L} = \mathcal{L}(\psi, \vec{\nabla}\psi, \dot{\psi}, t) \tag{12.7}$$

$$\mathcal{L} = \frac{dL}{dV} \text{ and so } dL = \mathcal{L}dV. \tag{12.8}$$

12.7 Classical field equation in terms of Lagrangian density $\mathcal{L}$

Let us first obtain the classical field equation in terms of Lagrangian density $\mathcal{L}$.

We start with the Hamilton's variational principle of classical mechanics

$$\delta \int_{t_1}^{t_2} L\,dt = 0, \ \delta q_i(t_i) = \delta q_i(t_2) = 0 \tag{12.9}$$

where $L = L(\psi, \dot{\psi}, t)$ and δ is variation.

Consider a region of space enclosed by a surface S and the volume enclosed is V, the elementary volume being $dV \equiv d^3\vec{r}$. Lagrangian density at dV is $\mathcal{L}$. So the relation between L and $\mathcal{L}$ is obtained by integration of $\mathcal{L}dV$ over the entire volume V as

$$L = \int_V \mathcal{L}dV \equiv \int_V \mathcal{L}d^3\vec{r} \tag{12.10}$$

Considering variation

$$\delta L = \delta \int_V \mathcal{L}dV = \int_V \delta\mathcal{L}dV \tag{12.11}$$

With this we have from equation (12.9) upon using equation (12.11)

$$\delta \int_{t_1}^{t_2} L \, dt = \int_{t_1}^{t_2} \delta L \, dt = 0 \tag{12.12}$$

$$\int_{t_1}^{t_2} dt \int_V \delta \mathcal{L} \, dV = 0 \tag{12.13}$$

The end point conditions of identical positions are

$$\delta\psi(\vec{r},\, t_1) = \delta\psi(\vec{r},\, t_2) = 0 \tag{12.14}$$

From $\mathcal{L} = \mathcal{L}(\psi,\ \text{grad}\ \psi,\ \dot{\psi},\ t)$ we take the variation

$$\delta\mathcal{L} = \frac{\partial\mathcal{L}}{\partial\psi}\delta\psi + \frac{\partial\mathcal{L}}{\partial(\text{grad}\ \psi)}\delta(\text{grad}\ \psi) + \frac{\partial\mathcal{L}}{\partial\dot{\psi}}\delta\dot{\psi} + \frac{\partial\mathcal{L}}{\partial t}\delta t \tag{12.15}$$

Since $\delta t = 0$

$$\delta\mathcal{L} = \frac{\partial\mathcal{L}}{\partial\psi}\delta\psi + \frac{\partial\mathcal{L}}{\partial(\text{grad}\ \psi)}\delta(\text{grad}\ \psi) + \frac{\partial\mathcal{L}}{\partial\dot{\psi}}\delta\dot{\psi} \tag{12.16}$$

Interchanging the grad and variation operations we get

$$\delta\mathcal{L} = \frac{\partial\mathcal{L}}{\partial\psi}\delta\psi + \frac{\partial\mathcal{L}}{\partial(\text{grad}\ \psi)}\ \text{grad}\ \delta\psi + \frac{\partial\mathcal{L}}{\partial\dot{\psi}}\delta\dot{\psi} \tag{12.17}$$

Using $\delta\dot{\psi} = \delta\frac{\partial\psi}{\partial t} = \frac{\partial}{\partial t}(\delta\psi)$

$$\delta\mathcal{L} = \frac{\partial\mathcal{L}}{\partial\psi}\delta\psi + \frac{\partial\mathcal{L}}{\partial(\text{grad}\ \psi)}\ \text{grad}\ \delta\psi + \frac{\partial\mathcal{L}}{\partial\dot{\psi}}\frac{\partial}{\partial t}(\delta\psi) \tag{12.18}$$

From equation (12.13) we get

$$\int_{t_1}^{t_2} dt \int_V \left[\frac{\partial\mathcal{L}}{\partial\psi}\delta\psi + \frac{\partial\mathcal{L}}{\partial(\text{grad}\ \psi)}\ \text{grad}\ \delta\psi + \frac{\partial\mathcal{L}}{\partial\dot{\psi}}\frac{\partial}{\partial t}(\delta\psi) \right] dV = 0 \tag{12.19}$$

$$\int_{t_1}^{t_2} dt \int_V \frac{\partial\mathcal{L}}{\partial\psi}\delta\psi \, dV + \int_{t_1}^{t_2} dt \int_V \frac{\partial\mathcal{L}}{\partial(\text{grad}\ \psi)}\ \text{grad}\ \delta\psi \, dV + \int_{t_1}^{t_2} dt \int_V \frac{\partial\mathcal{L}}{\partial\dot{\psi}}\frac{\partial}{\partial t}(\delta\psi) \, dV = 0 \tag{12.20}$$

Consider the volume integral portion of the second term on the RHS viz.

$$\int_V \frac{\partial\mathcal{L}}{\partial(\text{grad}\ \psi)}\ \text{grad}\ \delta\psi \, dV = \sum_{x,y,z} \int\int \left[\int \frac{\partial\mathcal{L}}{\partial(\text{grad}\ \psi)_x}\frac{\partial}{\partial x}\delta\psi \, dx \right] dy \, dz \tag{12.21}$$

where $dV = dx\,dy\,dz$

Integrating by parts the bracketed integral we have [$\frac{\partial\mathcal{L}}{\partial(\text{grad}\ \psi)_x} = $ 1st function, $\frac{\partial}{\partial x}\delta\psi -$ 2nd function]

$$\int_V \frac{\partial \mathcal{L}}{\partial(\mathrm{grad}\,\psi)}\,\mathrm{grad}\,\delta\psi dV =$$

$$= \sum_{x,y,z} \iint \left[\frac{\partial \mathcal{L}}{\partial(\mathrm{grad}\,\psi)_x} \int \frac{\partial}{\partial x}\delta\psi\,dx - \int \frac{\partial}{\partial x}\frac{\partial \mathcal{L}}{\partial(\mathrm{grad}\,\psi)_x}\int \frac{\partial}{\partial x}\delta\psi\,dx \right] dydz$$

$$= \sum_{x,y,z} \iint \left[\frac{\partial \mathcal{L}}{\partial(\mathrm{grad}\,\psi)_x}\delta\psi - \int \frac{\partial}{\partial x}\frac{\partial \mathcal{L}}{\partial(\mathrm{grad}\,\psi)_x}\delta\psi\,dx \right] dydz$$

$$= \sum_{x,y,z} \iint \left[\frac{\partial \mathcal{L}}{\partial(\mathrm{grad}\,\psi)_x}\delta\psi dydz - \int \frac{\partial}{\partial x}\frac{\partial \mathcal{L}}{\partial(\mathrm{grad}\,\psi)_x}\delta\psi\,dydz \right]$$

$$\int_V \frac{\partial \mathcal{L}}{\partial(\mathrm{grad}\,\psi)}\,\mathrm{grad}\,\delta\psi dV = \sum_{x,y,z} \int_S \frac{\partial \mathcal{L}}{\partial(\mathrm{grad}\,\psi)_x}\delta\psi\,dS - \sum_{x,y,z}\int_V \frac{\partial}{\partial x}\frac{\partial \mathcal{L}}{\partial(\mathrm{grad}\,\psi)_x}\delta\psi\,dV \tag{12.22}$$

The first term is a surface integral and it vanishes as ψ vanishes at x, y, $z \to \infty$ or satisfies periodic boundary conditions. Hence

$$\int_V \frac{\partial \mathcal{L}}{\partial(\mathrm{grad}\,\psi)}\,\mathrm{grad}\,\delta\psi dV = -\int_V \mathrm{div}\,\frac{\partial \mathcal{L}}{\partial(\mathrm{grad}\,\psi)}\delta\psi\,dV \tag{12.23}$$

From equations (12.20) and (12.23) we have

$$\int_{t_1}^{t_2} dt \int_V \frac{\partial \mathcal{L}}{\partial\psi}\delta\psi dV - \int_{t_1}^{t_2} dt \int_V \mathrm{div}\,\frac{\partial \mathcal{L}}{\partial(\mathrm{grad}\,\psi)}\delta\psi\,dV + \int_{t_1}^{t_2} dt \int_V \frac{\partial \mathcal{L}}{\partial\dot\psi}\frac{\partial}{\partial t}(\delta\psi)dV = 0$$

$$\int_{t_1}^{t_2} dt \int_V \frac{\partial \mathcal{L}}{\partial\psi}\delta\psi dV - \int_{t_1}^{t_2} dt \int_V \mathrm{div}\,\frac{\partial \mathcal{L}}{\partial(\mathrm{grad}\,\psi)}\delta\psi\,dV + \int_V dV \int_{t_1}^{t_2}\frac{\partial \mathcal{L}}{\partial\dot\psi}\frac{\partial}{\partial t}(\delta\psi)dt = 0 \tag{12.24}$$

Consider the time integral portion of the 3rd term on the RHS viz.

$$\int_{t_1}^{t_2} \frac{\partial \mathcal{L}}{\partial\dot\psi}\frac{\partial}{\partial t}(\delta\psi)dt$$

We integrate by parts taking $\frac{\partial \mathcal{L}}{\partial\dot\psi}$ as first function and $\frac{\partial}{\partial t}(\delta\psi)$ as second function.

$$\int_{t_1}^{t_2} \frac{\partial \mathcal{L}}{\partial\dot\psi}\frac{\partial}{\partial t}(\delta\psi)dt = \frac{\partial \mathcal{L}}{\partial\dot\psi}\int \frac{\partial}{\partial t}(\delta\psi)dt\,\Big|_{t_1}^{t_2} - \int_{t_1}^{t_2} \frac{\partial}{\partial t}\frac{\partial \mathcal{L}}{\partial\dot\psi}\int \frac{\partial}{\partial t}(\delta\psi)dt$$

$$= \frac{\partial \mathcal{L}}{\partial\dot\psi}\delta\psi\,\Big|_{t_1}^{t_2} - \int_{t_1}^{t_2}\frac{\partial}{\partial t}\frac{\partial \mathcal{L}}{\partial\dot\psi}\delta\psi dt \tag{12.25}$$

Using the end point conditions of equation (12.14) viz. $\delta\psi(\vec{r},\,t_1) = \delta\psi(\vec{r},\,t_2) = 0$ the first term vanishes. We are thus left with

$$\int_{t_1}^{t_2} \frac{\partial \mathcal{L}}{\partial\dot\psi}\frac{\partial}{\partial t}(\delta\psi)dt = -\int_{t_1}^{t_2}\frac{\partial}{\partial t}\frac{\partial \mathcal{L}}{\partial\dot\psi}\delta\psi dt \tag{12.26}$$

With equation (12.26) in equation (12.24) we get

$$\int_{t_1}^{t_2} dt \int_V \frac{\partial \mathcal{L}}{\partial \psi} \delta\psi\, dV - \int_{t_1}^{t_2} dt \int_V \text{div}\, \frac{\partial \mathcal{L}}{\partial(\text{grad}\,\psi)} \delta\psi\, dV - \int_V dV \int_{t_1}^{t_2} \frac{\partial}{\partial t} \frac{\partial \mathcal{L}}{\partial \dot\psi} \delta\psi\, dt = 0$$

$$\int_{t_1}^{t_2} dt \int_V \frac{\partial \mathcal{L}}{\partial \psi} \delta\psi\, dV - \int_{t_1}^{t_2} dt \int_V \text{div}\, \frac{\partial \mathcal{L}}{\partial(\text{grad}\,\psi)} \delta\psi\, dV - \int_{t_1}^{t_2} dt \int_V \frac{\partial}{\partial t} \frac{\partial \mathcal{L}}{\partial \dot\psi} \delta\psi\, dV = 0 \quad (12.27)$$

$$\int_{t_1}^{t_2} dt \int_V dV \left[\frac{\partial \mathcal{L}}{\partial \psi} - \text{div}\, \frac{\partial \mathcal{L}}{\partial(\text{grad}\,\psi)} - \frac{\partial}{\partial t} \frac{\partial \mathcal{L}}{\partial \dot\psi} \right] \delta\psi = 0 \quad (12.28)$$

Since $\delta\psi$ is arbitrary and mutually independent it follows that the integrand should be zero for validity of equation (12.28). Hence

$$\frac{\partial \mathcal{L}}{\partial \psi} - \text{div}\, \frac{\partial \mathcal{L}}{\partial(\text{grad}\,\psi)} - \frac{\partial}{\partial t} \frac{\partial \mathcal{L}}{\partial \dot\psi} = 0 \quad (12.29)$$

This is the classical field equation in terms of Lagrangian density $\mathcal{L}$.

This classical field equation is the counterpart of the Euler–Lagrange equation of classical mechanics, namely

$$\frac{d}{dt}\left(\frac{\partial L}{\partial \dot q_i} \right) - \frac{\partial L}{\partial q_i} = 0 \quad (12.30)$$

where $q_i = q_1, q_2, q_3 \dots q_f$ and $\dot q_i = \frac{\partial q_i}{\partial t}$.

The similarity of equation (12.29) to the Euler–Lagrange equation will be more apparent if we express equation (12.29) in terms of the Lagrangian of the field. This is achieved through the concept of functional derivatives.

12.8 Classical field equation in terms of Lagrangian L

Let us obtain the classical field equation in terms of Lagrangian L.

The amplitude of field is ψ.

The Lagrangian density is

$$\mathcal{L} = \frac{dL}{dV}\text{(Lagrangian per unit volume)} \quad (12.31)$$

And the Lagrangian is

$$L = \int \mathcal{L}\, dV \text{ (volume integral of Lagrangian density)} \quad (12.32)$$

It depends upon ψ and its space and time derivatives as

$$\mathcal{L} = \mathcal{L}(\psi,\ \text{grad}\,\psi,\ \dot\psi,\ t) \quad (12.33)$$

where grad $\psi = \hat{i}\dfrac{\partial \psi}{\partial x} + \hat{j}\dfrac{\partial \psi}{\partial y} + \hat{k}\dfrac{\partial \psi}{\partial z}$ is the derivative of ψ w.r.t. space coordinates x, y, z and $\dot{\psi} = \dfrac{\partial \psi}{\partial t}$ is the derivative w.r.t. time coordinate. The first bracket in the relation

$$\mathcal{L} = \mathcal{L}(\psi,\ \text{grad } \psi,\ \dot{\psi},\ t)$$

indicates that $\mathcal{L}$ is a function of ψ and $\dot{\psi}$.

We say that $\mathcal{L}$ is a function of ψ, $\dot{\psi}$ since $\mathcal{L}$ depends on the value of ψ, $\dot{\psi}$ at point $\vec{r}$ (not over volume V).

Again, Lagrangian L depends on ψ, $\dot{\psi}$ as

$$L = L[\psi,\ \dot{\psi},\ t] \tag{12.34}$$

We say that the L is a functional of ψ and $\dot{\psi}$ since L depends on the value of ψ, $\dot{\psi}$ over the entire volume V (not only at point $\vec{r}$) and write this as

$$L = L[\psi(\vec{r},\ t),\ \dot{\psi}(\vec{r},\ t),\ t]$$

the square bracket indicating that L is a functional of ψ and $\dot{\psi}$.

A functional can be regarded as the continuum limit of a function of discrete variables. Let us divide the volume into a large number of small portions ($i = 1$ to f) called cells and in the ith cell of volume δV_i the values of field and its derivative are ψ_i, $\dot{\psi}_i$, respectively. In other words, ψ_i is the value of $\psi(\vec{r},\ t)$ in the ith cell of volume δV_i. In the discrete case L would be a function of the discrete variables ψ_i, $\dot{\psi}_i$ i.e.

$$L = L(\psi_i,\ \dot{\psi}_i\ ,\ t) \tag{12.35}$$

$$\delta L = \sum_i \left(\frac{\partial L}{\partial \psi_i}\delta \psi_i + \frac{\partial L}{\partial \dot{\psi}_i}\delta \dot{\psi}_i \right) \tag{12.36}$$

In the continuum limit $\delta V_i \to 0$ and so this equation becomes

$$\delta L = \sum_i \underset{\delta V_i \to 0}{Lt} \frac{1}{\delta V_i}\left(\frac{\partial L}{\partial \psi_i}\delta \psi_i + \frac{\partial L}{\partial \dot{\psi}_i}\delta \dot{\psi}_i \right)\delta V_i$$

$$\delta L = \sum_i \underset{\delta V_i \to 0}{Lt} \left(\frac{1}{\delta V_i}\frac{\partial L}{\partial \psi_i}\delta \psi_i + \frac{1}{\delta V_i}\frac{\partial L}{\partial \dot{\psi}_i}\delta \dot{\psi}_i \right)\delta V_i \tag{12.37}$$

In the limit summation becomes integral and defining the partial functional derivatives as

$$\underset{\delta V_i \to 0}{Lt} \frac{1}{\delta V_i}\frac{\partial L}{\partial \psi_i}\delta \psi_i = \frac{\partial L}{\partial \psi} = \text{functional partial derivative of } L \text{ w. r. t. } \psi = \frac{\delta L}{\delta \psi} \tag{12.38}$$

$$\underset{\delta V_i \to 0}{Lt} \frac{1}{\delta V_i}\frac{\partial L}{\partial \dot{\psi}_i}\delta \dot{\psi}_i = \frac{\partial L}{\partial \dot{\psi}} = \text{functional partial derivative of } L \text{ w. r. t. } \dot{\psi} = \frac{\delta L}{\delta \dot{\psi}} \tag{12.39}$$

Thus we have the variation of the Lagrangian of the field in terms of the functional derivatives of the Lagrangian of the field given by

$$\delta L = \int_V \left(\frac{\partial L}{\partial \psi} \delta\psi + \frac{\partial L}{\partial \dot\psi} \delta\dot\psi \right) dV \tag{12.40}$$

From equation (12.11)

$$\delta L = \delta \int_V \mathcal{L} dV = \int_V \delta\mathcal{L} dV \tag{12.41}$$

With equation (12.18) for $\delta\mathcal{L}$

$$\delta L = \int_V \left(\frac{\partial \mathcal{L}}{\partial \psi} \delta\psi + \frac{\partial \mathcal{L}}{\partial(\mathrm{grad}\,\psi)} \,\mathrm{grad}\,\delta\psi + \frac{\partial \mathcal{L}}{\partial \dot\psi} \frac{\partial}{\partial t}(\delta\psi) \right) dV \tag{12.42}$$

Using equation (12.23) for the 2nd term we get

$$\delta L = \int_V \left(\frac{\partial \mathcal{L}}{\partial \psi} \delta\psi - \mathrm{div}\,\frac{\partial \mathcal{L}}{\partial(\mathrm{grad}\,\psi)} \delta\psi + \frac{\partial \mathcal{L}}{\partial \dot\psi} \delta\dot\psi \right) dV$$

$$\delta L = \int_V \left(\left(\frac{\partial \mathcal{L}}{\partial \psi} - \mathrm{div}\,\frac{\partial \mathcal{L}}{\partial(\mathrm{grad}\,\psi)} \right) \delta\psi + \frac{\partial \mathcal{L}}{\partial \dot\psi} \delta\dot\psi \right) dV \tag{12.43}$$

Comparing equations (12.40) and (12.43) we have

$$\frac{\partial L}{\partial \psi} = \frac{\partial \mathcal{L}}{\partial \psi} - \mathrm{div}\,\frac{\partial \mathcal{L}}{\partial(\mathrm{grad}\,\psi)} \tag{12.44}$$

$$\frac{\partial L}{\partial \dot\psi} = \frac{\partial \mathcal{L}}{\partial \dot\psi} \tag{12.45}$$

Substituting these in equation (12.29) which is the classical field equation in terms of Lagrangian density $\mathcal{L}$ we get

$$\frac{\partial L}{\partial \psi} - \frac{\partial}{\partial t} \frac{\partial L}{\partial \dot\psi} = 0 \tag{12.46}$$

This equation is analogous to the Euler–Lagrange equation (12.30).

So when we express the Lagrangian of the field in terms of the functional derivative of the Lagrangian, the field equation will be analogous to the Euler–Lagrange equation of classical mechanics.

12.9 Classical field equation in Hamiltonian formalism

In classical mechanics the Hamiltonian function is defined as

$$H = H(q, p, t) = \sum_i p_i \dot{q}_i - L \tag{12.47}$$

where $L = L(\psi, \dot{\psi}, t)$ is the Lagrangian of the field, p_i is momentum conjugate to the generalized coordinate q_i given by

$$p_i = \frac{\partial L}{\partial \dot{q}_i} \qquad (12.48)$$

Equations (12.47) and (12.48) hold for a particle moving in 3D space.

In the case of field the wave function ψ_i behaves as the coordinate q_i and the momentum conjugate to the canonical coordinate ψ_i is (in similarity to p_i of equation (12.48))

$$P_i = \frac{\partial L}{\partial \dot{\psi}_i} \qquad (12.49)$$

We can write it in terms of Lagrangian field density $\mathcal{L}$ using the relation

$$L = \int_V \mathcal{L}dV \qquad (12.50)$$

as

$$P_i = \frac{\partial \mathcal{L}}{\partial \dot{\psi}_i}dV_i = \pi dV_i \qquad (12.51)$$

where

$$\pi = \frac{\partial \mathcal{L}}{\partial \dot{\psi}_i} \qquad (12.52)$$

For infinitesimal cells we can move over to the continuum and so define

$$\pi = \frac{\partial \mathcal{L}}{\partial \dot{\psi}} = \frac{\partial L}{\partial \dot{\psi}} \qquad (12.53)$$

This π is referred to as the conjugate field. It is the partial derivative of the Lagrangian density w.r.t. $\dot{\psi}$ or equivalently the functional partial derivative of L w.r.t. $\dot{\psi}$.

The Lagrangian density can be replaced by the Lagrangian by replacing the partial derivatives ∂ by functional partial derivatives $\eth$.

We define the Hamiltonian of the field as (in similarity to equation (12.47))

$$H = \sum_i P_i \dot{\psi}_i - L \qquad (12.54)$$

Replacing $\sum_i$ by $\int dV$, $P = \pi dV$ and using $L = \int_V \mathcal{L}dV$ we get

$$H = \int_V dV \pi dV \dot{\psi} - \int_V \mathcal{L}dV = \int_V dV[\pi\dot{\psi} - \mathcal{L}]$$

$$H = \int_V dV \left[\pi(\vec{r}, t)\dot{\psi}(\vec{r}, t) - \mathcal{L} \right] \tag{12.55}$$

$$H = \int_V \mathcal{H} dV \tag{12.56}$$

where $\mathcal{H}$ is the Hamiltonian density given by

$$\mathcal{H} = \pi\dot{\psi} - \mathcal{L} \tag{12.57}$$

Hamiltonian density is the Hamiltonian per unit volume.
The Lagrangian classical field equation is given by equation (12.46) as

$$\frac{\partial L}{\partial \psi} - \frac{\partial}{\partial t}\frac{\partial L}{\partial \dot{\psi}} = 0 \tag{12.58}$$

Using equation (12.53) we replace $\frac{\partial L}{\partial \dot{\psi}}$ by π, the conjugate field to get

$$\frac{\partial L}{\partial \psi} = \frac{\partial}{\partial t}\frac{\partial L}{\partial \dot{\psi}} = \frac{\partial \pi}{\partial t} = \dot{\pi} \tag{12.59}$$

$$\dot{\pi} = \frac{\partial L}{\partial \psi} \tag{12.60}$$

From equations (12.56) and (12.57)

$$H = \int_V \mathcal{H} dV = \int_V \left[\pi\dot{\psi} - \mathcal{L} \right] dV \tag{12.61}$$

Variation in H gives

$$\delta H = \int_V \delta \left[\pi\dot{\psi} - \mathcal{L} \right] dV \tag{12.62}$$

$$= \int_V [(\delta\pi)\dot{\psi} + \pi(\delta\dot{\psi})] dV - \delta \int_V \mathcal{L} dV$$

$$\delta H = \int_V [(\delta\pi)\dot{\psi} + \pi(\delta\dot{\psi})] dV - \delta L \tag{12.63}$$

Using equation (12.40) i.e. $\delta L = \int_V (\frac{\partial L}{\partial \psi}\delta\psi + \frac{\partial L}{\partial \dot{\psi}}\delta\dot{\psi}) dV$ we get

$$\delta H = \int_V [(\delta\pi)\dot{\psi} + \pi(\delta\dot{\psi})] dV - \int_V \left(\frac{\partial L}{\partial \psi}\delta\psi + \frac{\partial L}{\partial \dot{\psi}}\delta\dot{\psi} \right) dV$$

$$\delta H = \int_V \left[(\delta\pi)\dot{\psi} + \pi(\delta\dot{\psi}) - \left(\frac{\partial L}{\partial \psi}\delta\psi + \frac{\partial L}{\partial \dot{\psi}}\delta\dot{\psi} \right) \right] dV \tag{12.64}$$

Using equation (12.59) viz. $\frac{\partial L}{\partial \psi} = \frac{\partial \pi}{\partial t} = \dot{\pi}$ and equation (12.53) viz. $\pi = \frac{\partial \mathcal{L}}{\partial \dot{\psi}} = \frac{\partial L}{\partial \dot{\psi}}$

$$\delta H = \int_V [\ (\delta\pi)\dot{\psi} + \pi(\delta\dot{\psi}) - \dot{\pi}(\delta\psi) - \pi(\delta\dot{\psi})\ ]dV$$

$$\delta H = \int_V [\ (\delta\pi)\dot{\psi} - \dot{\pi}(\delta\psi)\ ]dV \tag{12.65}$$

Again, H is a functional of π and ψ, i.e.

$$H = H[\ \psi,\ \pi\ ]\ \text{(square bracket used to denote functional)} \tag{12.66}$$

Hence variation of H will be an analogous relation as in equation (12.40), namely

$$\delta H = \int_V \left(\frac{\partial H}{\partial \psi}\delta\psi + \frac{\partial H}{\partial \pi}\delta\pi \right)dV \tag{12.67}$$

with

$$\frac{\partial H}{\partial \psi} = \frac{\partial \mathcal{H}}{\partial \psi} - \text{div}\ \frac{\partial \mathcal{H}}{\partial(\text{grad }\psi)}\text{(similar to equation (12.44))} \tag{12.68}$$

$$\frac{\partial H}{\partial \pi} = \frac{\partial \mathcal{H}}{\partial \pi} - \text{div}\ \frac{\partial \mathcal{H}}{\partial(\text{grad }\pi)} \tag{12.69}$$

Comparison of the δ variations in equations (12.65) and (12.67) gives ($\psi,\ \pi$ are the independent variables)

$$\dot{\psi} = \frac{\partial H}{\partial \pi},\ -\dot{\pi} = \frac{\partial H}{\partial \psi} \tag{12.70}$$

These equations are called classical field equations in Hamiltonian form. These are analogous to Hamilton's canonical equations of motion in classical mechanics viz. $\dot{q} = \frac{\partial H}{\partial p}, -\dot{p} = \frac{\partial H}{\partial q}$.

In place of partial derivatives we have functional partial derivatives. Instead of $\dot{q}$ there is $\dot{\psi}$ and in place of $\dot{p}$ there is $\dot{\pi}$. In place of the Hamiltonian of classical mechanics there is the Hamiltonian of the field.

If F is an arbitrary functional of ψ and π then

$$\dot{F} = \frac{\partial F}{\partial t} + \int_V \left(\frac{\partial F}{\partial \psi}\dot{\psi} + \frac{\partial F}{\partial \pi}\dot{\pi}\ \right)dV \tag{12.71}$$

The Poisson bracket of F with the Hamiltonian is given by

$$[F, H]_{PB} = \int_V \left(\frac{\partial F}{\partial \psi}\dot{\psi} + \frac{\partial F}{\partial \pi}\dot{\pi}\ \right)dV\text{(functional derivatives are involved)} \tag{12.72}$$

$$= \int_V \left(\frac{\partial F}{\partial \psi} \frac{\partial H}{\partial \pi} - \frac{\partial F}{\partial \pi} \frac{\partial H}{\partial \psi} \right) dV \tag{12.73}$$

Hence

$$\dot{F} = \frac{\partial F}{\partial t} + [F, \ H]_{PB} \tag{12.74}$$

12.10 Second quantization of non-relativistic field, classical field and Schrödinger field

The Schrödinger equation is a non-relativistic equation derivable from the classical wave equation. It describes motion of a non-relativistic particle of mass m moving in a potential V.

The Schrödinger time dependent equation is obtained from the classical Hamiltonian by replacing the canonical variables q and p by their corresponding operators and using the commutation relation

$$[q_i \ p_j] = i\hbar \delta_{ij} \tag{12.75}$$

to be

$$\left(-\frac{\hbar^2}{2m} \nabla^2 + V \right)\psi = i\hbar \frac{\partial \psi}{\partial t} \tag{12.76}$$

$$i\hbar \frac{\partial \psi}{\partial t} + \frac{\hbar^2}{2m} \nabla^2 \psi - V\psi = 0 \tag{12.77}$$

where $\psi = \psi(\vec{r}, t)$ is Schrödinger field and is a non-relativistic field. This process of quantization in quantum mechanics is called first quantization.

Though this equation is a quantized equation of motion the wave function ψ is treated as a classical field. So we look upon this equation as a classical field equation and we have to quantize it. Clearly it is the second time the equation is being quantized and so this is referred to as second quantization.

We have to obtain the Schrödinger equation (12.77) starting from the classical field equation (12.29) through a proper choice of Lagrangian density $\mathcal{L}_s$ say

$$\mathcal{L}_s = i\hbar\psi^*\dot{\psi} - \frac{\hbar^2}{2m} \text{grad } \psi^*. \ \text{grad } \psi - V\psi^*\psi \tag{12.78}$$

Substituting in equation (12.29) we get

$$\frac{\partial \mathcal{L}_s}{\partial \psi} - div \frac{\partial \mathcal{L}_s}{\partial (\text{grad } \psi)} - \frac{\partial}{\partial t} \frac{\partial \mathcal{L}_s}{\partial \dot{\psi}} = 0 \tag{12.79}$$

$$-V\psi^* + \frac{\hbar^2}{2m}\, \vec{\nabla} \cdot \frac{\partial}{\partial \vec{\nabla}\psi}\, \vec{\nabla}\psi^*. \ \vec{\nabla}\psi - \frac{\partial}{\partial t} i\hbar\psi^* = 0 \tag{12.80}$$

$$-V\psi^* + \frac{\hbar^2}{2m}\, \vec{\nabla} \cdot \ \vec{\nabla}\psi^* - \frac{\partial}{\partial t} i\hbar\psi^* = 0$$

$$-V\psi^* + \frac{\hbar^2}{2m}\, \nabla^2 \psi^* - i\hbar\frac{\partial}{\partial t}\psi^* = 0 \tag{12.81}$$

Taking the complex conjugate of equation (12.81) we get

$$-V\psi + \frac{\hbar^2}{2m}\, \nabla^2 \psi + i\hbar\frac{\partial}{\partial t}\psi = 0 \tag{12.82}$$

which is the Schrödinger equation (12.77).

The canonically conjugate field corresponding to ψ is

$$\pi = \frac{\partial \mathcal{L}_s}{\partial \dot{\psi}} \tag{12.83}$$

It is also called canonically conjugate momentum density.
From equation (12.78) we have

$$\pi = \frac{\partial}{\partial \dot{\psi}}\left(i\hbar\psi^*\dot{\psi} - \frac{\hbar^2}{2m}\, \mathrm{grad}\, \psi^*. \ \mathrm{grad}\, \psi - V\psi^*\psi \right) = \ i\hbar\psi^* \tag{12.84}$$

$$\pi = i\hbar\psi^* \tag{12.85}$$

From equation (12.57) we write with $\mathcal{L}_S$

$$\mathcal{H} = \pi\dot{\psi} - \mathcal{L}_S \tag{12.86}$$

With equations (12.78) and (12.85)

$$\mathcal{H} = i\hbar\psi^*\dot{\psi} - \left(i\hbar\psi^*\dot{\psi} - \frac{\hbar^2}{2m}\, \mathrm{grad}\, \psi^*. \ \mathrm{grad}\, \psi - V\psi^*\psi \right) \tag{12.87}$$

$$\mathcal{H} = \frac{\hbar^2}{2m}\, \mathrm{grad}\, \psi^*. \ \mathrm{grad}\, \psi + V\psi^*\psi \tag{12.88}$$

From equation (12.85)

$$\psi^* = \frac{\pi}{i\hbar} \tag{12.89}$$

$$\mathcal{H} = \frac{\hbar^2}{2m}\, \mathrm{grad}\, \frac{\pi}{i\hbar}. \ \mathrm{grad}\, \psi + V\frac{\pi}{i\hbar}\, \psi \tag{12.90}$$

$$\mathcal{H} = -\frac{i\hbar}{2m}\,\text{grad}\,\pi.\ \text{grad}\,\psi - \frac{i}{\hbar}V\pi\,\psi \equiv -\frac{i\hbar}{2m}\,\vec{\nabla}\,\pi.\ \vec{\nabla}\,\psi - \frac{i}{\hbar}V\pi\,\psi \quad (12.91)$$

The Hamiltonian is obtained by taking volume integral of the Hamiltonian density $\mathcal{H}$

$$H = \int_V \mathcal{H}dV \quad (12.92)$$

Using equation (12.88) we have the field Hamiltonian to be

$$H = \int_V \left(\frac{\hbar^2}{2m}\,\vec{\nabla}\,\psi^*.\ \vec{\nabla}\,\psi + V\psi^*\psi\right)dV \quad (12.93)$$

12.11 Quantization of Schrödinger field

In field theory we treat wave function $\psi(\vec{r},\,t)$ that represents the coordinate of the field as an operator. Let us expand $\psi(\vec{r},\,t)$ in terms of a complete orthonormal set of functions $\{u_k(\vec{r})\}$ which carry all the space dependence of $\psi(\vec{r},\,t)$. The operator properties of $\psi(\vec{r},\,t)$ are expressed through the expansion coefficients which depend on time.

$$\psi(\vec{r},\,t) = \sum_k a_k(t)u_k(\vec{r}) \quad (12.94)$$

$$\psi^\dagger(\vec{r},\,t) = \sum_k a_k^\dagger(t)u_k^*(\vec{r}) \quad (12.95)$$

Using equation (12.89) $\psi^* = \frac{\pi}{i\hbar}$ we have

$$\psi^\dagger\,(\vec{r},\,t) = \psi^*(\vec{r},\,t) = \frac{1}{i\hbar}\pi(\vec{r},\,t) = \sum_k a_k^\dagger(t)u_k^*(\vec{r}) \quad (12.96)$$

The orthonormal set of functions $\{u_k(\vec{r})\}$ are chosen to be the single-particle energy eigenfunctions of the Hamiltonian of a field, namely

$$H_p = -\frac{\hbar^2}{2m}\,\nabla^2 + V(\vec{r}) \quad (12.97)$$

so that

$$H_p u_k(\vec{r}) = E_k u_k(\vec{r}) \quad (12.98)$$

where V has been chosen to be independent of t.

Quantization is done by postulating suitable algebraic relations for the operators $a_k(t)$ and $a_k^\dagger(t)$.

12.12 Non-relativistic field: system of bosons

Particles having either zero or integral spin are called bosons. They obey Bose–Einstein statistics. There is no exclusion principle for bosons and so any number of particles can be placed in the same quantum state.

According to Jordan and Klein, such a condition can be achieved if the operators $a_k(t)$ and $a_k^{\dagger}(t)$ satisfy the following commutation relations

$$[a_k, \ a_l^{\dagger}] = \delta_{kl} \tag{12.99}$$

$$[a_k, \ a_l] = [a_k^{\dagger}, \ a_l^{\dagger}] = 0 \tag{12.100}$$

where a_k, a_l refer to the same time t.

We define another Hermitian operator called number operator

$$N_k = a_k^{\dagger} a_k \tag{12.101}$$

If ϕ_k is the normalized eigenfunction then the eigenvalue equation of $N_k \equiv \hat{N}_k$ operator is

$$\hat{N}_k \phi_k = N_k \phi_k \tag{12.102}$$

or in the ket symbol

$$\hat{N}_k |k\rangle = N_k |k\rangle \tag{12.103}$$

where N_k is the eigenvalue of $\hat{N}_k$.

Multiplying equation (12.103) from the left by $<k|$ we get

$$\langle k|\hat{N}_k |k\rangle = \langle k|N_k|k\rangle = N_k\langle k|k\rangle = \ N_k \tag{12.104}$$

since

$$\langle k|k\rangle = \int \phi_k^* \ \phi_k \, d\tau = 1 \tag{12.105}$$

Using equation (12.101) we have from equation (12.104)

$$\langle k|a_k^{\dagger} a_k|k\rangle = \ N_k \tag{12.106}$$

Using $(\ a_k^{\dagger}\ )^{\dagger} = a_k$ we rewrite equation (12.106) as (pushing $a_k^{\dagger}$ into bra where it becomes a_k)

$$\langle ka_k |a_k k\rangle = \ N_k \tag{12.107}$$

$$|a_k|k\rangle|^2 = N_k \tag{12.108}$$

$$N_k \geqslant 0 \tag{12.109}$$

Let us operate $a_k^{\dagger}$ from left on equation (12.103) to get

$$a_k^\dagger \hat{N}_k \left| k \right\rangle = a_k^\dagger N_k \left| k \right\rangle = N_k a_k^\dagger |k\rangle \tag{12.110}$$

From equation (12.99) viz. $[a_k, a_k^\dagger] = 1$ we have

$$a_k a_k^\dagger - a_k^\dagger a_k = 1 \tag{12.111}$$

$$a_k^\dagger a_k = a_k a_k^\dagger - 1 \tag{12.112}$$

$$\hat{N}_k = a_k a_k^\dagger - 1 \tag{12.113}$$

Equation (12.110) gives

$$a_k^\dagger (a_k a_k^\dagger - 1)|k\rangle = N_k a_k^\dagger |k\rangle \tag{12.114}$$

$$a_k^\dagger a_k a_k^\dagger |k\rangle - a_k^\dagger |k\rangle = N_k a_k^\dagger |k\rangle$$

$$a_k^\dagger a_k (a_k^\dagger |k\rangle) - (a_k^\dagger |k\rangle) = N_k (a_k^\dagger |k\rangle)$$

$$a_k^\dagger a_k \left(a_k^\dagger |k\rangle \right) = (N_k + 1)\left(a_k^\dagger |k\rangle \right)$$

$$\hat{N}_k \left(a_k^\dagger |k\rangle \right) = (N_k + 1)\left(a_k^\dagger |k\rangle \right) \; \left(\text{using } \hat{N}_k = a_k^\dagger a_k \right) \tag{12.115}$$

This result implies that if N_k is the eigenvalue of the operator $\hat{N}_k$ in the state $|k\rangle$ then $N_k + 1$ is the eigenvalue of the operator $\hat{N}_k$ in the state $a_k^\dagger |k\rangle$.

Apply a_k from the left on equation (12.103) to get

$$a_k \hat{N}_k \left| k \right\rangle = a_k N_k |k\rangle \tag{12.116}$$

$$a_k a_k^\dagger a_k \left| k \right\rangle = N_k a_k |k\rangle \; (\text{using } \hat{N}_k = a_k^\dagger a_k)$$

$$a_k a_k^\dagger (a_k \left| k \right\rangle) = N_k (a_k |k\rangle) \tag{12.117}$$

According to equation (12.113), i.e. $\hat{N}_k = a_k a_k^\dagger - 1$ we have

$$a_k a_k^\dagger = \hat{N}_k + 1 \tag{12.118}$$

Putting this in equation (12.117)

$$(\hat{N}_k + 1)(a_k \left| k \right\rangle) = N_k (a_k |k\rangle)$$

$$\hat{N}_k (a_k |k\rangle) + (a_k |k\rangle) = N_k (a_k |k\rangle)$$

$$\hat{N}_k (a_k |k\rangle) = (N_k - 1)(a_k |k\rangle) \tag{12.119}$$

This result implies that if N_k is the eigenvalue of the operator $\hat{N}_k$ in the state $|k\rangle$ then $N_k - 1$ is the eigenvalue of the operator $\hat{N}_k$ in the state $a_k |k\rangle$.

It is clear from equation (12.115) that the operator $a_k^\dagger$ increases the eigenvalue of $\hat{N}_k$ by 1 and from equation (12.118) that the operator a_k decreases the eigenvalue of $\hat{N}_k$ by 1. Hence the operators $a_k^\dagger$ and a_k are termed as rising operator (or creation operator) and lowering operator (or annihilation or destruction operator), respectively.

Clearly the eigenvalue spectrum of the Hermitian operator $\hat{N}_k = N_k = a_k^\dagger a_k$ is a set of non-negative integers $N_k = 0, 1, 2, \ldots, +\infty$

Hence the operator $\hat{N}_k$ is called number operator implying that $[\ \hat{N}_k\ \hat{N}_l\ ] = 0$ (as numbers commute) and since number operators commute a set of number operators can be diagonalized simultaneously.

This set of Hermitian operators forms a complete set which spans all the Hilbert space completely. Let this set have eigenvalues $N_1,\ N_2,\ N_3, \ldots,\ N_k,\ \ldots$ so that one such eigenfunction will be $\phi_{N_1,\ N_2,\ ..N_k,\ ..}$ or equivalently the eigenket will be

$$|N_1,\ N_2,\ \ldots N_k,\ \ldots\rangle = |\text{general state}\rangle$$

The most general eigenfunction ψ is obtained by linear superposition of these eigenfunctions or eigenkets. The eigenfunction for which all $N_k = 0$ is called the ground state or vacuum state ϕ_0 or $|0\rangle$ and $a_k|0\rangle = 0$, $\hat{N}_k|0\rangle = 0$ for all k.

To generate the eigenfunction $\phi_{N_1,\ N_2,\ ..N_k,\ ..}$ or the general state vector of the field we have to repeatedly apply the operator $a_k^\dagger$.

$$|N_1,\ N_2,\ ..N_k,\ ..\rangle = C\left(a_1^\dagger\ \right)^{N_1}\left(a_2^\dagger\ \right)^{N_2} \ldots \left(a_k^\dagger\ \right)^{N_k} \ldots |0\rangle \tag{12.120}$$

where C is normalization factor. Also we have

$$a_k^\dagger\ |N_1,\ N_2,\ ..N_k,\ \ldots\rangle = \alpha\ (N_k)|N_1,\ N_2,\ \ldots N_k + 1,\ \ldots\rangle \tag{12.121}$$

$$a_k\ |N_1,\ N_2,\ ..N_k,\ \ldots\rangle = \beta\ (N_k)|N_1,\ N_2,\ \ldots N_k - 1,\ \ldots\rangle \tag{12.122}$$

Operate equation (12.122) by $a_k^\dagger$ to get

$$a_k^\dagger a_k\ \big|\ N_1,\ N_2,\ \ldots N_k,\ \ldots\rangle = a_k^\dagger\beta\ (N_k)\big|\ N_1,\ N_2,\ \ldots N_k - 1,\ \ldots\rangle = \beta\ (N_k)a_k^\dagger|N_1,\ N_2,\ \ldots N_k - 1,\ \ldots\rangle$$

$$a_k^\dagger a_k|N_1,\ N_2,\ \ldots N_k,\ \ldots\rangle = \beta\ (N_k)\alpha\ (N_k - 1)|N_1,\ N_2,\ \ldots N_k,\ \ldots\rangle$$

As $\quad \hat{N}_k = a_k^\dagger a_k \quad$ and $\quad a_k^\dagger a_k\ |N_1,\ N_2,\ \ldots N_k,\ \ldots\rangle = \hat{N}_k|N_1,\ N_2,\ \ldots N_k,\ \ldots\rangle = N_k|N_1,\ N_2,\ \ldots N_k,\ \ldots\rangle$ we have

$$N_k|N_1,\ N_2,\ \ldots N_k,\ \ldots\rangle = \beta\ (N_k)\alpha\ (N_k - 1)|N_1,\ N_2,\ \ldots N_k,\ \ldots\rangle \tag{12.123}$$

Comparison gives

$$\beta\ (N_k)\alpha\ (N_k - 1) = N_k \tag{12.124}$$

Replacing the index N_k by $N_k + 1$

$$\beta\ (N_k + 1)\alpha\ (N_k) = N_k + 1 \tag{12.125}$$

We note that if we start with initial state $|N_1, N_2, ...N_k, ...\rangle$ then

$$a_k^\dagger \, |N_1, N_2, ...N_k, ...\rangle = \alpha \, (N_k)|N_1, N_2, ...N_k + 1, ...\rangle \qquad (12.126)$$

$$a_k \, |N_1, N_2, ...N_k + 1, ...\rangle = \beta \, (N_k + 1)|N_1, N_2, ...N_k, ...\rangle \ \text{(initial state restored)} \quad (12.127)$$

So we consider $\alpha \, (N_k)$ and $\beta \, (N_k + 1)$ to be equivalent.
Hence from equation (12.125)

$$\alpha \, (N_k)\alpha \, (N_k) = N_k + 1$$

$$\alpha(N_k) = \sqrt{N_k + 1} \qquad (12.128)$$

Also, if we start with initial state $|N_1, N_2, ...N_k, ...\rangle$ then

$$a_k \, |N_1, N_2, ...N_k, ...\rangle = \beta \, (N_k)|N_1, N_2, ...N_k - 1, ...\rangle \qquad (12.129)$$

$$a_k^\dagger \, |N_1, N_2, ...N_k - 1, ...\rangle = \alpha \, (N_k - 1)|N_1, N_2, ...N_k, ...\rangle \ \text{(initial state restored)} \quad (12.130)$$

So we consider $\beta(N_k)$ and $\alpha \, (N_k - 1)$ to be equivalent.
Hence from equation (12.124)

$$\beta \, (N_k)\beta(N_k) = N_k$$

$$\beta(N_k) = \sqrt{N_k} \qquad (12.131)$$

Let us rewrite equations (12.121) and (12.122)

$$a_k^\dagger \, |N_1, N_2, ...N_k, ...\rangle = \sqrt{N_k + 1} \, |N_1, N_2, ...N_k + 1, ...\rangle \qquad (12.132)$$

$$a_k \, |N_1, N_2, ...N_k, ...\rangle = \sqrt{N_k} \, |N_1, N_2, ...N_k - 1, ...\rangle \qquad (12.133)$$

$$\hat{N}_k \, |N_1, N_2, ...N_k, ...\rangle = N_k|N_1, N_2, ...N_k, ...\rangle \qquad (12.134)$$

From equation (12.120)

$$|N_1, N_2, ...N_k, ...\rangle = C\left(a_1^\dagger\right)^{N_1}\left(a_2^\dagger\right)^{N_2} ... \left(a_k^\dagger\right)^{N_k} ... |0\rangle \qquad (12.135)$$

$$= C\sqrt{N_1!N_2! \ ... \ .N_k!...} \, |N_1, N_2, ...N_k, ...\rangle \qquad (12.136)$$

since

$$a_k^\dagger \, |0\rangle = \sqrt{0 + 1} \, |1\rangle = \sqrt{1} \, |1\rangle \qquad (12.137)$$

$$\left(a_k^\dagger\right)^2 \, |0\rangle = \sqrt{1} \, a_k^\dagger \, |1\rangle = \sqrt{1} \, \sqrt{1 + 1} \, |2\rangle = \sqrt{2!} \, |2\rangle \qquad (12.138)$$

$$\left(a_k^\dagger\right)^{N_k} |0\rangle = \sqrt{N_k!} \, |N_K\rangle \qquad (12.139)$$

$$C\sqrt{N_1!N_2! \ \dots \ .N_k!\dots} = 1$$

$$C = \frac{1}{\sqrt{N_1!N_2! \ \dots \ N_k!\dots}} \text{(the normalization constant)} \tag{12.140}$$

$$|N_1, N_2, \dots N_k, \dots\rangle = \frac{1}{\sqrt{N_1!N_2! \ \dots \ .N_k!\dots}}\left(a_1^{\dagger}\right)^{N_1}\left(a_2^{\dagger}\right)^{N_2} \dots \left(a_k^{\dagger}\right)^{N_k} \dots |0\rangle \tag{12.141}$$

Some other important relations are

$$(a_k)^r\left(a_l^{\dagger}\right)^N |0\rangle = \frac{N!}{(N-r)!}\left(a_k^{\dagger}\right)^{N-r}|0\rangle\delta_{kl} \tag{12.142}$$

With $r = N$

$$(a_k)^N\left(a_l^{\dagger}\right)^N |0\rangle = \frac{N!}{(N-N)!}\left(a_k^{\dagger}\right)^{N-N}|0\rangle\delta_{kl} = N!\delta_{kl}|0\rangle \tag{12.143}$$

The condition of orthonormality of the base vectors is

$$\langle N_1', N_2', \dots N_k' \dots |N_1, N_2, \dots N_k \dots \rangle = \delta_{N_1'N_1}\delta_{N_2'N_2} \dots \delta_{N_k'N_k}\dots \tag{12.144}$$

$$\langle N_1', N_2', \dots N_k' \dots |\hat{N}_k|N_1, N_2, \dots N_k, \dots\rangle = N_k\delta_{N_1'N_1}\delta_{N_2'N_2} \dots \delta_{N_k'N_k}\dots \tag{12.145}$$

This equation (12.145) shows that $\hat{N}_k$ are diagonal operators in this representation with the diagonal matrix elements N_k.

The kets $|N_1, N_2, \dots N_k \dots \rangle$ constitute particle number representation for the operators a_k, $a_k^{\dagger}$, $\hat{N}_k$.

The quantum field is given by equation (12.94), i.e. $\psi(\vec{r}, t) = \sum_k a_k(t)u_k(\vec{r})$ with

$$\langle u_k|u_l\rangle = \int u_k^*(\vec{r})u_l(\vec{r}\,')d\tau = \delta_{kl} \tag{12.146}$$

$$\sum_k u_k(\vec{r})u_k^*(\vec{r}\,') = \delta(\vec{r} - \vec{r}\,') \tag{12.147}$$

And the single-particle Hamiltonian is given by equation (12.97) viz. $H_p = -\frac{\hbar^2}{2m}\nabla^2 + V(\vec{r})$ and the corresponding eigenvalue equation is given by equation (12.98), namely $H_p u_k(\vec{r}) = E_k u_k(\vec{r})$.

For a system of particles

$$H = \sum_k \hat{N}_k E_k \tag{12.148}$$

So the eigenvalues of the Hamiltonian for the quantized field are

$$E = \sum_k N_k E_k \tag{12.149}$$

The corresponding eigenvectors are $|N_1, N_2, \ldots N_k \ldots \rangle$.

This relation shows that the matter field can be thought of as an assembly of non-interacting individual particles each of which has energy E_k. The total energy is the sum of energies of individual particles. The total number of particles in the field is

$$N = N_1 + N_2 + \ldots + N_k + \ldots = \sum_k N_k \tag{12.150}$$

We have obtained this eigenvalue of system without any restriction on the distribution of particles in different quantum states. So there is no restriction on the number of particles in a particular quantum state. Any particle state can be occupied by any number of particles N_k. This is the basic feature of Bose–Einstein statistics. Thus the field represents an assembly of bosons.

The particle number operator $\hat{N}_k$ commutes with the Hamiltonian H i.e.

$$[\hat{N}_k \ H] = [\hat{N}_k \ \sum_l \hat{N}_l] = \sum_l [\hat{N}_k \ \hat{N}_l] = 0 \tag{12.151}$$

This implies that when potential function V does not depend on time the total number of particles in the field is a constant of motion. As the operator $a_k^\dagger$ increases the particle number by unity and a_k decreases the particle number by unity they are called creation and annihilation operators, respectively.

12.13 Non-relativistic field: system of fermions

Particles having half integral spin are called fermions. They obey Fermi–Dirac statistics. They follow the Pauli exclusion principle and so only one fermion can be placed in one quantum state. So any state can occupy either 0 or 1 fermion, i.e. $N_k = 0$ or 1 (in contrast to eigenvalues $N_k = 0, 1, 2 \ldots$. for bosons, since any number of bosons can be in the same quantum state).

According to Jordan and Winger, such a condition can be achieved if the operators a_k and $a_k^\dagger$ satisfy the following anti-commutation relations

$$\{a_k, \ a_l^\dagger\} = \{a_l^\dagger, \ a_k\} = a_k a_l^\dagger + a_l^\dagger a_k = \delta_{kl} \tag{12.152}$$

$$\{a_k, \ a_l\} = \{a_k^\dagger, \ a_l^\dagger\} = 0 \tag{12.153}$$

$$a_k a_k = 0 \text{ (exercise 12.1)} \tag{12.154}$$

$$a_k^\dagger a_k^\dagger = 0 \text{ (exercise 12.1)} \tag{12.155}$$

All the operators in these relations refer to the same time. Also

$$\hat{N}_k^2 = \left(a_k^\dagger a_k\right)\left(a_k^\dagger a_k\right) = a_k^\dagger a_k a_k^\dagger a_k \tag{12.156}$$

From equation (12.152) for $k = l$ we have

$$a_k a_k^\dagger + a_k^\dagger a_k = \delta_{kk} = 1 \tag{12.157}$$

$$a_k a_k^\dagger = 1 - a_k^\dagger a_k \qquad (12.158)$$

Hence

$$\hat{N}_k^2 = a_k^\dagger a_k a_k^\dagger a_k = a_k^\dagger \left(1 - a_k^\dagger a_k\right) a_k \qquad (12.159)$$

$$= a_k^\dagger a_k - a_k^\dagger a_k^\dagger a_k a_k$$

$$= \hat{N}_k - 0 \quad \text{(using equations (12.154) and (12.155))}$$

$$\hat{N}_k^2 - \hat{N}_k = 0$$

$$\hat{N}_k(\hat{N}_k - 1) = 0 \text{ i. e. } \hat{N}_k = 0 \text{ or } \hat{N}_k - 1 = 0 \qquad (12.160)$$

So for any arbitrary state $|k\rangle$

$$\hat{N}_k|k\rangle = 0 \text{ or } (\hat{N}_k - 1)|k\rangle = 0 \qquad (12.161)$$

This gives the eigenvalue spectrum of $\hat{N}_k$ as

$$N_k = 0 \text{ or } N_k = 1 \text{ for any } k \qquad (12.162)$$

$$\text{i. e. } N_1, \quad N_2, \quad N_3, \quad \ldots N_k \ldots . = 0 \text{ or } 1 \qquad (12.163)$$

Also

$$[\ \hat{N}_k \ \hat{N}_l \ ] = 0 \qquad (12.164)$$

So the base vectors can be characterized as simultaneous eigenfunctions of the number operator $\hat{N}_k$ with $N_k! = 1$ as $N_k = 0$ or 1.

To generate the general state vector of the field we have to repeatedly apply the operator $a_k^\dagger$ and then we get

$$|N_1, N_2, \ldots N_k, \ldots\rangle = \left(a_1^\dagger\right)^{N_1}\left(a_2^\dagger\right)^{N_2} \ldots \left(a_k^\dagger\right)^{N_k} \ldots |0\rangle \qquad (12.165)$$

As the operators anti-commute, i.e $\{\ a_k^\dagger, a_l^\dagger\} = 0$ (equation(12.153)) it follows that an undetermined phase factor ± 1 occurs in this equation unless we specify the order in which these operators operate. Let us put a $+$ sign when the creation operators are arranged in equation (12.165) in order of increasing superscript. With this convention and using the relation

$$(a_k)^r\left(a_l^\dagger\right)^N |0\rangle = \frac{N!}{(N - r)!}\left(a_k^\dagger\right)^{N-r}|0\rangle \delta_{kl} \qquad (12.166)$$

we arrive at

$$a_k|N_1, N_2, \ldots N_k, \ldots\rangle = (-1)^{N_1 + N_2 + \ldots + N_k - 1}\left(a_1^\dagger\right)^{N_1}\left(a_2^\dagger\right)^{N_2} \ldots \left(a_k^\dagger\right)^{N_k} \ldots |0\rangle \qquad (12.167)$$

$$= (-1)^{S_k} N_k \ |N_1, N_2, \ldots N_k - 1, \ldots\rangle$$

where $S_k = \sum\limits_{r=1}^{k-1} N_r$

Similarly we have

$$a_k^\dagger \, |N_1, N_2, \ldots N_k, \ldots\rangle = (-1)^{S_k}(1 - N_k) \, |N_1, N_2, \ldots N_k + 1, \ldots\rangle \qquad (12.168)$$

These equations show that in the case of fermions, a_k and $a_k^\dagger$ can also be interpreted as annihilation and creation operators. The factors N_k and $1 - N_k$ imply that an empty state ($N_k = 0$) cannot be further emptied while an occupied state ($N_k = 1$) cannot be further filled.

Let us combine equations (12.167) and (12.168). We operate equation (12.167) by $a_k^\dagger$ to get

$$a_k^\dagger a_k |N_1, N_2, \ldots N_k, \ldots\rangle = (-1)^{S_k} N_k \, a_k^\dagger |N_1, N_2, \ldots N_k - 1, \ldots\rangle \qquad (12.169)$$

Using equation (12.168) and $\hat{N}_k = a_k^\dagger a_k$ we get

$$\hat{N}_k |N_1, N_2, \ldots N_k, \ldots\rangle = (-1)^{S_k} N_k \, (-1)^{S_k}(1 - N_k)|N_1, N_2, \ldots N_k, \ldots\rangle$$

$$\hat{N}_k \, |N_1, N_2, \ldots N_k, \ldots\rangle = N_k(1 - N_k) \, |N_1, N_2, \ldots N_k, \ldots\rangle \qquad (12.170)$$

Also

$$\langle N_1', N_2', \ldots N_k' \ldots |\hat{N}_k|N_1, N_2, \ldots N_k, \ldots\rangle = N_k(1 - N_k) \, \delta_{N_1' N_1} \delta_{N_2' N_2} \cdots \delta_{N_k' N_k} \cdots \qquad (12.171)$$

12.14 Relativistic field

We have discussed quantization of the non-relativistic Schrödinger equation describing a system of bosons or a system of fermions.

Now we will consider relativistic fields. For instance the Klein–Gordon field corresponding to spin 0 scalar field, Dirac field corresponding to spin $\frac{1}{2}$ spinor field and the electromagnetic field corresponding to spin 1 vector field.

We have to write the relativistic field equations in covariant form.

For a system of bosons (0 or integral spin case) obeying Bose–Einstein statistics we have to quantize using proper commutation relations for the field operators.

For a system of fermions (spin $\frac{1}{2}$ case) obeying Fermi–Dirac statistics we have to quantize using proper anti-commutation relations.

12.15 Essence of quantum field theory

The relation between energy and mass of a moving particle, i.e. $E^2 = p^2 c^2 + m^2 c^4$ and $E = mc^2$ hints at the fact that energy and mass are equivalent and particles can be created and destroyed all the time. Clearly thus, be it the Klein–Gordon equation or the Dirac equation we run into problem if we focus on a single-particle approach. We have to use a multi-particle framework since a single-particle framework will not work. Wave function describes a single particle and now we call $\psi(\vec{r}, t)$ the field. The primary element of quantum field theory is the field $\psi(\vec{r}, t)$ that permeates the entire space.

Figure 12.1 shows a one-dimensional abstract situation where field $\psi(x, t)$ permeates the entire 1D space. Excitation of the field is looked upon as a particle. So the field naturally allows particles to come up and disappear. In other words, creation and destruction of particles occur and we have a multi-particle system. In other words, in field theory we allow exchange of energy and mass. For instance, the excitation of the electromagnetic field is called the photon. The field thus excites particles with a particular spin.

12.15.1 Occupation number representation

Bosons are described by totally symmetric states and fermions by totally anti-symmetric states. We can represent a system of identical quantum particles using occupation number representation that forms the basis of second quantization. In this representation we are concerned with how many particles are in which state.

Consider identical particles populating the states, as shown in figure 12.2. If we assign artificial labels to the particles then any exchange of particles leads to a different representation though such an exchange of identical particles does not change the configuration. We have to construct a symmetrized or anti-symmetrized state considering all possible permutations of particles.

The occupation number representation should have the inherent feature that if we exchange the identical particles the configuration does not change. This is done by stating the number of particles in a state and not addressing which particle is in which state. We thus do not have to associate each particle with a single-particle state, i.e. we do not have to assign labels to identical particles.

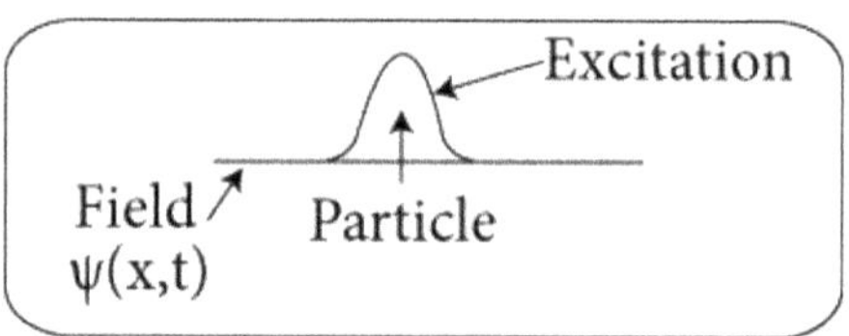

Figure 12.1. Particle is an excitation of the field.

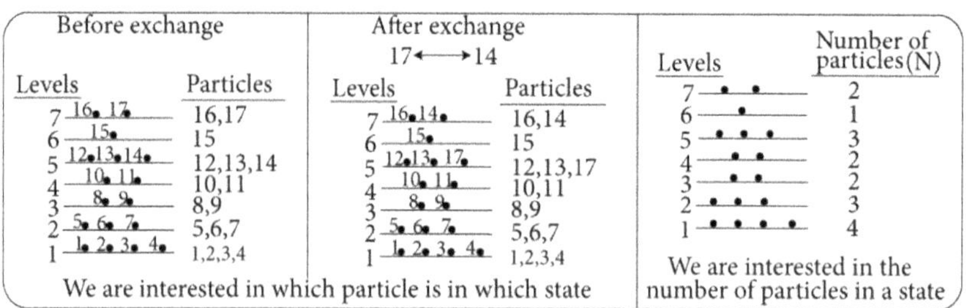

Figure 12.2. Identical particles populating the states.

12.16 Lagrangian density for the Klein–Gordon field

In the Klein–Gordon equation the field is real, i.e. $\psi = \psi^*$.

The field behaves as a scalar under Lorentz transformation, i.e. $\psi'(\vec{r}') = \psi(\vec{r})$

Also, $\underset{r \to 0}{Lt}\ \psi(\vec{r}) = 0$, i.e. at large distance there is little or no influence of the field.

We have derived the Euler–Lagrange equation in terms of Lagrangian density from the principle of least action in *exercise 12.2* viz.

$$\frac{\partial \mathcal{L}}{\partial \psi} - \partial_\mu \frac{\partial \mathcal{L}}{\partial(\partial_\mu \psi)} = 0 \tag{12.172}$$

Let us construct the Lagrangian density (in natural units)

$$\mathcal{L} = \frac{1}{2}\partial_\mu \psi \partial^\mu \psi - \frac{1}{2}m^2\psi^2 \tag{12.173}$$

Now

$$\frac{\partial \mathcal{L}}{\partial \psi} = \frac{\partial}{\partial \psi}\left(\frac{1}{2}\partial_\mu \psi \partial^\mu \psi - \frac{1}{2}m^2\psi^2\right) = -m^2\psi \tag{12.174}$$

$$\frac{\partial}{\partial(\partial_\mu \psi)}\left(\frac{1}{2}\partial_\mu \psi \partial^\mu \psi - \frac{1}{2}m^2\psi^2\right) = \frac{1}{2}\frac{\partial}{\partial(\partial_\mu \psi)}\partial_\mu \psi \partial^\mu \psi = \frac{1}{2}\frac{\partial}{\partial(\partial_\mu \psi)}\partial_\mu \psi \eta^{\mu\nu}\partial_\nu \psi \tag{12.175}$$

where

$$\partial^\mu \psi = \eta^{\mu\nu}\partial_\nu \psi \text{ and } \eta = \text{diag}(1 - 1 - 1 - 1) \tag{12.176}$$

$$\frac{\partial \mathcal{L}}{\partial \psi} = \frac{1}{2}\eta^{\mu\nu}\frac{\partial}{\partial(\partial_\mu \psi)}\partial_\mu \psi \partial_\nu \psi \tag{12.177}$$

$$\frac{\partial \mathcal{L}}{\partial \psi} = \frac{1}{2}\eta^{\mu\nu}\left[\left(\frac{\partial}{\partial(\partial_\mu \psi)}\partial_\mu \psi\right)\partial_\nu \psi + \partial_\mu \psi\left(\frac{\partial}{\partial(\partial_\mu \psi)}\partial_\nu \psi\right)\right] \tag{12.178}$$

$$\frac{\partial \mathcal{L}}{\partial \psi} = \frac{1}{2}\eta^{\mu\nu}[\partial_\nu \psi + \partial_\mu \psi\ \delta^\mu_\nu]\ (\delta^\mu_\nu \text{ is Kroneckar delta function}) \tag{12.179}$$

$$\frac{\partial \mathcal{L}}{\partial \psi} = \frac{1}{2}\eta^{\mu\nu}[\partial_\nu \psi + \partial_\nu \psi\ ] = \frac{1}{2}2\ \eta^{\mu\nu}\partial_\nu \psi \tag{12.180}$$

$$\frac{\partial \mathcal{L}}{\partial \psi} = \eta^{\mu\nu}\partial_\nu \psi = \partial^\mu \psi \tag{12.181}$$

From the Euler–Lagrange equation we get

$$\frac{\partial \mathcal{L}}{\partial \psi} - \partial_\mu \frac{\partial \mathcal{L}}{\partial(\partial_\mu \psi)} = 0 \tag{12.182}$$

$$-m^2\psi - \partial_\mu\partial^\mu\psi = 0 \tag{12.183}$$

$$(\partial_\mu\partial^\mu + m^2)\psi = 0 \tag{12.184}$$

which is the Klein–Gordon equation that we have obtained from Lagrangian density of equation (12.172).

12.17 Quantization of the Klein–Gordon field as harmonic oscillators

In classical mechanics the single-particle Lagrangian is

$$L = L(x, \dot{x}) \tag{12.185}$$

and the conjugate momentum is

$$p = \frac{\partial L}{\partial \dot{x}} \tag{12.186}$$

The Hamiltonian is

$$H = p\dot{x} - L \tag{12.187}$$

The Poisson brackets are

$$\{x\ x\} = 0, \ \{p\ p\} = 0, \ \{x\ p\} = 1 \tag{12.188}$$

and the equation of motion of a single particle is given by

$$\{x\ H\} = \dot{x}, \quad \{p\ H\} = \dot{p} \tag{12.189}$$

In classical field theory we deal with Lagrangian density

$$\mathcal{L} = \mathcal{L}\big(\psi(x), \ \partial_\mu\psi(x)\big) \tag{12.190}$$

The canonical momentum is

$$\pi = \frac{\partial \mathcal{L}}{\partial \dot{\psi}} \tag{12.191}$$

The Hamiltonian density is

$$\mathcal{H} = \pi\dot{\psi} - \mathcal{L} \tag{12.192}$$

The Hamiltonian is

$$H = \int d^3x \ \mathcal{H} = \int d^3x \ \left(\pi\dot{\psi} - \mathcal{L}\right) \tag{12.193}$$

The corresponding Poisson brackets are (evaluated at equal time $x^0 = y^0$)

$$\{\psi(x)\ \psi(y)\}_{x^0 = y^0} = 0, \ \{\pi(x)\ \pi(y)\}_{x^0 = y^0} = 0, \ \{\psi(x)\ \pi(y)\}_{x^0 = y^0} = \delta^3(x - y) \tag{12.194}$$

We move over to the quantum case now.
The Lagrangian density for the Klein–Gordon field is

$$\mathcal{L} = \frac{1}{2}\partial_\mu\psi\partial^\mu\psi - \frac{1}{2}m^2\psi^2 = \frac{1}{2}\partial_0\psi\partial^0\psi + \frac{1}{2}\partial_i\psi\partial^i\psi - \frac{1}{2}m^2\psi^2 \qquad (12.195)$$

$$= \frac{1}{2}\dot{\psi}\dot{\psi} + \frac{1}{2}\left(-\partial^i\psi\right)\partial^i\psi - \frac{1}{2}m^2\psi^2$$

$$\mathcal{L} = \frac{1}{2}\dot{\psi}^2 - \frac{1}{2}\ \vec{\nabla}\psi\,\vec{\nabla}\psi - \frac{1}{2}m^2\psi^2 \qquad (12.196)$$

The canonical momentum is

$$\pi = \frac{\partial\mathcal{L}}{\partial\dot{\psi}} = \frac{\partial}{\partial\dot{\psi}}\left(\frac{1}{2}\dot{\psi}^2 - \frac{1}{2}\ \vec{\nabla}\psi\,\vec{\nabla}\psi - \frac{1}{2}m^2\psi^2\right) = \dot{\psi} \qquad (12.197)$$

From equation (12.191)

$$\mathcal{H} = \pi\pi - \left(\frac{1}{2}\dot{\psi}^2 - \frac{1}{2}\ \vec{\nabla}\psi\,\vec{\nabla}\psi - \frac{1}{2}m^2\psi^2\right) = \pi^2 - \left(\frac{1}{2}\pi^2 - \frac{1}{2}\ \vec{\nabla}\psi\,\vec{\nabla}\psi - \frac{1}{2}m^2\psi^2\right) \quad (12.198)$$

$$\mathcal{H} = \frac{1}{2}\pi^2 + \frac{1}{2}\ \vec{\nabla}\psi\,\vec{\nabla}\psi + \frac{1}{2}m^2\psi^2 \qquad (12.199)$$

$$H = \int d^3x\ \left(\frac{1}{2}\pi^2 + \frac{1}{2}\ \vec{\nabla}\psi\,\vec{\nabla}\psi + \frac{1}{2}m^2\psi^2\right) \qquad (12.200)$$

We now quantize the field by switching over from dynamical field variables to field operators. The Poisson brackets are replaced by commutation brackets.

$$[\ \psi(x)\,\psi(y)] = 0, \quad [\ \pi(x)\,\pi(y)] = 0, \quad [\ \psi(x)\,\pi(y)] = i\delta^3(x - y) \qquad (12.201)$$

The last relation is analogous to the fundamental commutator bracket of quantization viz.

$$[x\ p] = i\hbar \rightarrow i \text{ in natural units} \qquad (12.202)$$

So we have quantized the fields through these commutation relations. The solution of field $\psi(x)$ in Fourier space is

$$\psi(x) = \frac{1}{(2\pi)^{3/2}} \int d^4k\ e^{-ikx}\ \tilde{\psi}(k) \qquad (12.203)$$

The Klein–Gordon equation is

$$\left(\partial_\mu\partial^\mu + m^2\right)\psi(x) = 0 \qquad (12.204)$$

$$(\partial_\mu \partial^\mu + m^2)\frac{1}{(2\pi)^{3/2}} \int d^4k \; e^{-ikx} \; \widetilde{\psi}(k) = 0 \tag{12.205}$$

Using $\partial_\mu \partial^\mu = \frac{\partial^2}{\partial t^2} - \nabla^2$ (equation (11.15))

$$\frac{1}{(2\pi)^{3/2}} \int d^4k \left(\frac{\partial^2}{\partial t^2} - \nabla^2 + m^2 \right) e^{-ikx} \; \widetilde{\psi}(k) = 0 \tag{12.206}$$

Using

$$- \nabla^2 \, e^{-ikx} = -(-ik)^2 e^{-ikx} = -i^2 k^2 e^{-ikx} = k^2 e^{-ikx} \xrightarrow{p=\hbar k=k} p^2 e^{-ikx} \tag{12.207}$$

$$\frac{1}{(2\pi)^{3/2}} \int d^4k \left(\frac{\partial^2}{\partial t^2} + p^2 + m^2 \right) e^{-ikx} \; \widetilde{\psi}(k) = 0 \tag{12.208}$$

Defining

$$w_p^2 = p^2 + m^2 \tag{12.209}$$

where $w_p=$ angular frequency.
So we get

$$\left(\frac{\partial^2}{\partial t^2} + w_p^2 \right) \widetilde{\psi}(k) = 0 \tag{12.210}$$

This is the equation of the simple harmonic oscillator.

$$\ddot{\widetilde{\psi}}(k) + \; w_p^2 \widetilde{\psi}(k) = 0 \tag{12.211}$$

So the equation of motion satisfied is that of the harmonic oscillator. Excitation of a field leads to a particle. Field consists of an infinite number of harmonic oscillators.

The Hamiltonian of a simple harmonic oscillator is

$$H = \frac{p^2}{2m} + \frac{1}{2}mw_p^2 x^2 \tag{12.212}$$

We can express p and x in terms of ladder operators

$$p = i\sqrt{\frac{mw}{2}}(a^\dagger - a) \tag{12.213}$$

$$x = \sqrt{\frac{1}{2mw}}(a + a^\dagger) \tag{12.214}$$

Using the relation

$$[\, x, \ p] = i \tag{12.215}$$

we get

$$\left[\sqrt{\frac{1}{2mw}}\,(a + a^\dagger), \ i\sqrt{\frac{mw}{2}}\,(a^\dagger - a)\right] = i \tag{12.216}$$

$$i\sqrt{\frac{mw}{2}}\,\sqrt{\frac{1}{2mw}}\,[\,(a + a^\dagger), \ (a^\dagger - a)\,] = i \tag{12.217}$$

$$\frac{i}{2}[\,(a + a^\dagger), \ (a^\dagger - a)\,] = i \tag{12.218}$$

$$\frac{1}{2}\,(\,[a \ a^\dagger] - [a \ a] + [a^\dagger \ a^\dagger] - [a^\dagger \ a]) = 1$$

$$[a \ a^\dagger] - [a^\dagger \ a] = 2$$

$$[a \ a^\dagger] + [a \ a^\dagger] = 2 \ \text{i. e.} \ \ 2[a \ a^\dagger] = 2$$

$$[a \ a^\dagger] = 1 \tag{12.219}$$

$$H = \frac{p^2}{2m} + \frac{1}{2}mw_p^2 x^2 = \frac{1}{2m}\left(i\sqrt{\frac{mw_p}{2}}\,(a^\dagger - a)\ \right)^2 + \frac{1}{2}mw_p^2\left(\sqrt{\frac{1}{2mw_p}}\,(a + a^\dagger)\ \right)^2 \tag{12.220}$$

$$= \frac{w_p}{4}[-(a^\dagger - a)^2] + \frac{w_p}{4}(a + a^\dagger)^2$$

$$= \frac{w_p}{4}\left[-(a^\dagger - a)(a^\dagger - a) + (a + a^\dagger)(a + a^\dagger)\right]$$

$$= \frac{w_p}{4}\left[-(a^\dagger a^\dagger - a^\dagger a - aa^\dagger + aa) + (aa + aa^\dagger + a^\dagger a + a^\dagger a^\dagger)\right]$$

$$H = \frac{w_p}{2}(aa^\dagger + a^\dagger a) = \frac{w_p}{2}(aa^\dagger - a^\dagger a + 2a^\dagger a)$$

$$H = \frac{w_p}{2}(\,[\,a \ a^\dagger] + 2a^\dagger a\,) \tag{12.221}$$

Using $[a \ a^\dagger] = 1$

$$H = \frac{w_p}{2}(\,1 + 2a^\dagger a\,)$$

$$H = w_p\left(a^\dagger a + \frac{1}{2}\right) \tag{12.222}$$

This is the Hamiltonian for a harmonic oscillator in terms of ladder operators. Since there are multitudes of harmonic oscillators we have to carry out summation.

Ground state is defined as $a|0\rangle = 0$ (since we cannot go further down, the ground state annihilation operator acting on $|0\rangle$ gives zero). Now

$$H|0\rangle = w_p\left(a^\dagger a + \frac{1}{2}\right)|0\rangle = w_p a^\dagger a|0\rangle + \frac{w_p}{2}|0\rangle = \frac{w_p}{2}|0\rangle \tag{12.223}$$

Clearly $\frac{w_p}{2}$ is the energy of ground state.

We calculate energy of first excited state considering $H(a^\dagger|0\rangle)$. We first consider

$$[\ H\ a^\dagger] = Ha^\dagger - a^\dagger H \tag{12.224}$$

$$= w_p\left(a^\dagger a + \frac{1}{2}\right)a^\dagger - a^\dagger w_p\left(a^\dagger a + \frac{1}{2}\right)$$

$$= w_p\left[a^\dagger a a^\dagger - a^\dagger a^\dagger a\right]$$

$$= w_p a^\dagger\left(aa^\dagger - a^\dagger a\right) = w_p a^\dagger[\ a\ a^\dagger] \tag{12.225}$$

$$[\ H\ a^\dagger] = w_p a^\dagger \tag{12.226}$$

$$H(a^\dagger|0\rangle) = Ha^\dagger|0\rangle = ([\ H\ a^\dagger] + a^\dagger H)\,|\,0\rangle \ \text{(using equation (12.224))}$$

$$= \left(w_p a^\dagger + a^\dagger H\ \right)|\,0\rangle\,\text{(using equation (12.226))}$$

$$= w_p a^\dagger|0\rangle + a^\dagger H|0\rangle$$

$$= w_p a^\dagger|0\rangle + a^\dagger \frac{w_p}{2}|0\rangle \ \text{(using equation (12.223))}$$

$$H(a^\dagger|0\rangle) = w_p\left(1 + \frac{1}{2}\right)(a^\dagger|0\rangle) = \frac{3}{2}w_p(a^\dagger|0\rangle) \tag{12.227}$$

Similarly

$$H|n\rangle = H((a^\dagger)^n|0\rangle) \tag{12.228}$$

$$= w_p\left(n + \frac{1}{2}\right)((a^\dagger)^n|0\rangle) = w_p\left(n + \frac{1}{2}\right)|n\rangle = E_n|n\rangle$$

$$E_n = w_p\left(n + \frac{1}{2}\right), \ n = 1, 2, 3, 4, \ .... \tag{12.229}$$

This is how we can quantize the Klein–Gordon field.

12.18 Quantization of the electromagnetic field

Let us express the electromagnetic field in terms of the vector potential $\vec{A}$ as follows.

Electric field is $\vec{E} = -\frac{\partial \vec{A}}{\partial t}$

Magnetic field is $\vec{B} = \vec{\nabla} \times \vec{A}$

Maxwell's equations of electrodynamics in a vacuum are

$$\vec{\nabla}.\vec{E} = 0, \ \ \vec{\nabla}.\vec{B} = 0, \ \ \vec{\nabla} \times \vec{E} = -\frac{\partial \vec{B}}{\partial t}, \ \ \vec{\nabla} \times \vec{B} = \mu_0 \varepsilon_0 \frac{\partial \vec{E}}{\partial t} \qquad (12.230)$$

Put $\vec{B} = \vec{\nabla} \times \vec{A}$, $\vec{E} = -\frac{\partial \vec{A}}{\partial t}$ in fourth Maxwell equation $\vec{\nabla} \times \vec{B} = \mu_0 \varepsilon_0 \frac{\partial \vec{E}}{\partial t}$ to get

$$\vec{\nabla} \times \vec{\nabla} \times \vec{A} = \mu_0 \varepsilon_0 \frac{\partial}{\partial t}\left(-\frac{\partial \vec{A}}{\partial t}\right)$$

$$-\nabla^2 \vec{A} + \vec{\nabla} \ (\vec{\nabla}.\vec{A}) = -\frac{1}{c^2}\frac{\partial^2 \vec{A}}{\partial t^2} \qquad (12.231)$$

Let us use the condition

$$\vec{\nabla}.\vec{A} = 0 \qquad (12.232)$$

which is called the Coulomb gauge. This gives

$$\nabla^2 \vec{A} - \frac{1}{c^2}\frac{\partial^2 \vec{A}}{\partial t^2} = 0 \qquad (12.233)$$

This is the classical wave equation.

Consider a box of length L, volume $V = L^3$. Applying periodic boundary condition we get a series of electromagnetic waves (figure 12.3).

We can decompose the vector potential $\vec{A}$ into waves of angular frequency $w_k = c|k|$

$$\vec{A} = \sum_k \left[\vec{A}_k^+ e^{-i(\vec{k}.\vec{r}-w_k t)} + \vec{A}_k^- e^{i(\vec{k}.\vec{r}-w_k t)}\right] \qquad (12.234)$$

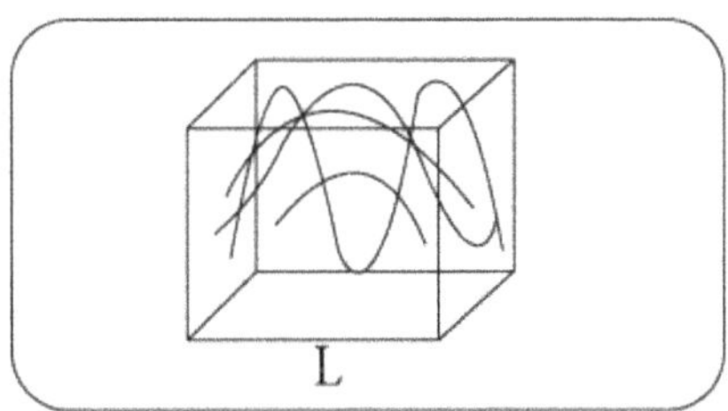

Figure 12.3. Set of electromagnetic waves of an electromagnetic field.

Vector potential $\vec{A}$ has to be real, i.e.

$$\vec{A} = \vec{A}^*$$
(12.235)

$$\vec{A}_k^+ e^{-i(\vec{k}.\vec{r}-w_k t)} + \vec{A}_k^- e^{i(\vec{k}.\vec{r}-w_k t)} = \left(\vec{A}_k^+\right)^* e^{i(\vec{k}.\vec{r}-w_k t)} + \left(\vec{A}_k^-\right)^* e^{-i(\vec{k}.\vec{r}-w_k t)}$$

$$\left(\vec{A}_k^+\right)^* = \vec{A}_k^-, \quad (\vec{A}_k^-)^* = \vec{A}_k^+ \quad \text{(left and right moving waves are orthogonal)} \quad (12.236)$$

Hence we have from equations (12.234) and (12.236)

$$\vec{A} = \sum_k \left[\left(\vec{A}_k^-\right)^* e^{-i(\vec{k}.\vec{r}-w_k t)} + \vec{A}_k^- e^{i(\vec{k}.\vec{r}-w_k t)} \right]$$
(12.237)

The first and second terms of the RHS are complex conjugates of one another.

The first term stands for waves travelling to the left and the second term stands for waves travelling to the right.

An electromagnetic wave can have two different types of polarizations. It can be polarized perpendicular or parallel to the plane and so we attach the index $\lambda = 1, 2$ to indicate them. Hence we rewrite equation (12.234) redefining the coefficients as

$$\left(\vec{A}_k^-\right)^* = A_{k,\lambda}^* \hat{\epsilon}_\lambda, \quad \vec{A}_k^- = A_{k,\lambda}\hat{\epsilon}_\lambda$$
(12.238)

$$\vec{A} = \sum_{k,\lambda} \left[A_{k,\lambda}^* \, e^{-i(\vec{k}.\vec{r}-w_k t)} + A_{k,\lambda} e^{i(\vec{k}.\vec{r}-w_k t)} \right] \hat{\epsilon}_\lambda$$
(12.239)

So we have decomposed the vector potential into waves characterized by wave vector $\vec{k}$ and polarization $\lambda = 1, 2$.

Because these are plane waves and the set $\{\hat{k}, \ \hat{\epsilon}_1, \ \hat{\epsilon}_2\}$ forms an orthonormal basis. The classical Hamiltonian of the electromagnetic field is

$$H = \frac{1}{2} \int \left(\varepsilon_0 \, |\vec{E}|^2 + \frac{1}{\mu_0} |\vec{B}|^2 \right) d^3\vec{r}$$
(12.240)

$$= \frac{1}{2} \int \left(\varepsilon_0 \left| -\frac{\partial \vec{A}}{\partial t} \right|^2 + \frac{1}{\mu_0} |\vec{\nabla} \times \vec{A}|^2 \right) d^3\vec{r}$$
(12.241)

$$H = \int \mathcal{H} \, d^3\vec{r}$$
(12.242)

where $\mathcal{H}$ is the Hamiltonian density given by

$$\mathcal{H} = \frac{1}{2} \left(\varepsilon_0 \left| \frac{\partial \vec{A}}{\partial t} \right|^2 + \frac{1}{\mu_0} |\vec{\nabla} \times \vec{A}|^2 \right)$$
(12.243)

Using Parseval's theorem we can switch over to Fourier space and calculation shows that

$$\int \left| \frac{\partial \vec{A}}{\partial t} \right|^2 d^3\vec{r} = 2 \sum_{k,\lambda} w_k^2 \, |A_{k,\lambda}|^2 \tag{12.244}$$

$$\int |\vec{\nabla} \times \vec{A}|^2 d^3\vec{r} = 2 \sum_{k,\lambda} \frac{w_k^2}{c^2} \, |A_{k,\lambda}|^2 \tag{12.245}$$

$$H = \frac{1}{2}\left(\varepsilon_0 2 \sum_{k,\lambda} w_k^2 \, |A_{k,\lambda}|^2 + \frac{1}{\mu_0} 2 \sum_{k,\lambda} \frac{w_k^2}{c^2} \, |A_{k,\lambda}|^2 \right) \tag{12.246}$$

$$H = \left(\varepsilon_0 \sum_{k,\lambda} w_k^2 \, |A_{k,\lambda}|^2 + \varepsilon_0 \sum_{k,\lambda} w_k^2 \, |A_{k,\lambda}|^2 \right) \tag{12.247}$$

$$H = 2\varepsilon_0 \sum_{k,\lambda} w_k^2 \, |A_{k,\lambda}|^2 \tag{12.247}$$

This is the classical Hamiltonian.

Let us dissolve it into real and imaginary parts as

$$A_{k,\lambda} = A_{k,\lambda}^R + iA_{k,\lambda}^I \tag{12.248}$$

$$H = 2\varepsilon_0 \sum_{k,\lambda} w_k^2 \left(\left| A_{k,\lambda}^R \right|^2 + \left| A_{k,\lambda}^I \right|^2 \right) \tag{12.249}$$

We can define 2 sets of conjugate variables

$$Q_{k,\lambda} = 2\sqrt{\varepsilon_0} \; A_{k,\lambda}^R \tag{12.250}$$

$$P_{k,\lambda} = 2w_k \sqrt{\varepsilon_0} \, A_{k,\lambda}^I \tag{12.251}$$

$$H = \frac{1}{2} \sum_{k,\lambda} (P_{k,\lambda}^2 + w_k^2 Q_{k,\lambda}^2) \tag{12.252}$$

such that

$$\frac{\partial H}{\partial Q_{k,\lambda}} = -\dot{P}_{k,\lambda}, \quad \frac{\partial H}{\partial P_{k,\lambda}} = \dot{Q}_{k,\lambda} \tag{12.253}$$

Equation (12.252) represents the Hamiltonian of the harmonic oscillator that we have arrived at starting from the field Hamiltonian of equation (12.240). We can now quantize it and switch over to the quantum harmonic oscillator through replacement of the canonically conjugate variables by their corresponding operators as

$$P_{k,\lambda} \to \hat{p}_{k,\lambda}, \quad Q_{k,\lambda} \to \hat{q}_{k,\lambda} \tag{12.254}$$

satisfying the commutation relation

$$[\, \hat{q}_{k,\lambda} \; \hat{p}_{k',\lambda'}] = i\hbar \delta_{kk'} \delta_{\lambda\lambda'} \tag{12.255}$$

Hence the quantized field Hamiltonian operator is from equation (12.252)

$$\hat{H} = \frac{1}{2} \sum_{k,\lambda} \left(\hat{p}_{k,\lambda}^2 + w_k^2 \hat{q}_{k,\lambda}^2 \right) \tag{12.256}$$

Introducing the ladder operators

$$a_{k,\lambda} = \sqrt{\frac{1}{2\hbar w_k}} \, (w_k \hat{q}_{k,\lambda} + i\hat{p}_{k,\lambda}) \tag{12.257}$$

$$a_{k,\lambda}^\dagger = \sqrt{\frac{1}{2\hbar w_k}} \, (w_k \hat{q}_{k,\lambda} - i\hat{p}_{k,\lambda}) \tag{12.258}$$

leading to

$$\hat{q}_{k,\lambda} = \sqrt{\frac{\hbar}{2w_k}} \, (a_{k,\lambda}^\dagger + a_{k,\lambda}) \tag{12.259}$$

$$\hat{p}_{k,\lambda} = i\sqrt{\frac{\hbar w_k}{2}} \, (a_{k,\lambda}^\dagger - a_{k,\lambda}) \tag{12.260}$$

With this the Hamiltonian of equation (12.256) becomes

$$
\begin{aligned}
\hat{H} &= \frac{1}{2} \sum_{k,\lambda} \left[\left(i\sqrt{\frac{\hbar w_k}{2}} \left(a_{k,\lambda}^\dagger - a_{k,\lambda} \right) \right)^2 + w_k^2 \left(\sqrt{\frac{\hbar}{2w_k}} \left(a_{k,\lambda}^\dagger + a_{k,\lambda} \right) \right)^2 \right] \\
&= \frac{1}{2} \sum_{k,\lambda} \left[-\frac{\hbar w_k}{2} \left(a_{k,\lambda}^\dagger a_{k,\lambda}^\dagger + a_{k,\lambda} a_{k,\lambda} - a_{k,\lambda}^\dagger a_{k,\lambda} - a_{k,\lambda} a_{k,\lambda}^\dagger \right) \right. \\
&\qquad \left. + w_k^2 \frac{\hbar}{2w_k} \left(a_{k,\lambda}^\dagger a_{k,\lambda}^\dagger + a_{k,\lambda} a_{k,\lambda} + a_{k,\lambda}^\dagger a_{k,\lambda} + a_{k,\lambda} a_{k,\lambda}^\dagger \right) \right]
\end{aligned}
\tag{12.261}
$$

$$\hat{H} = \frac{1}{2} \sum_{k,\lambda} \frac{\hbar w_k}{2} \, 2(a_{k,\lambda}^\dagger a_{k,\lambda} + a_{k,\lambda} a_{k,\lambda}^\dagger)$$

$$\hat{H} = \sum_{k,\lambda} \frac{\hbar w_k}{2} \left(a_{k,\lambda} a^{\dagger}_{k,\lambda} - a^{\dagger}_{k,\lambda} a_{k,\lambda} + 2a^{\dagger}_{k,\lambda} a_{k,\lambda} \right)$$

$$\hat{H} = \sum_{k,\lambda} \frac{\hbar w_k}{2} \left([a_{k,\lambda}\, a^{\dagger}_{k,\lambda}] + 2a^{\dagger}_{k,\lambda} a_{k,\lambda} \right) \tag{12.262}$$

$$\text{Using } \left[a_{k,\lambda}\, a^{\dagger}_{k,\lambda} \right] = 1$$

$$\hat{H} = \sum_{k,\lambda} \frac{\hbar w_k}{2} \left(1 + 2a^{\dagger}_{k,\lambda} a_{k,\lambda} \right)$$

$$\hat{H} = \sum_{k,\lambda} \hbar w_k \left(a^{\dagger}_{k,\lambda} a_{k,\lambda} + \frac{1}{2} \right) \tag{12.263}$$

So we have quantized the electromagnetic field to harmonic oscillators.

12.19 Dirac field

Let us construct the Lagrangian density (in natural units)

$$\mathcal{L} = i\overline{\psi}\gamma^{\mu}\partial_{\mu}\psi - m\overline{\psi}\psi \tag{12.264}$$

where

$$\overline{\psi} = \psi^{\dagger}\gamma^0 \tag{12.265}$$

From the Euler–Lagrange equation in $\overline{\psi}$ viz.

$$\frac{\partial \mathcal{L}}{\partial \overline{\psi}} - \partial_{\mu} \frac{\partial \mathcal{L}}{\partial(\partial_{\mu}\overline{\psi})} = 0 \tag{12.266}$$

we get

$$\frac{\partial}{\partial \overline{\psi}}(i\overline{\psi}\gamma^{\mu}\partial_{\mu}\psi - m\overline{\psi}\psi) - \partial_{\mu}\frac{\partial}{\partial(\partial_{\mu}\overline{\psi})}(i\overline{\psi}\gamma^{\mu}\partial_{\mu}\psi - m\overline{\psi}\psi) = 0 \tag{12.267}$$

$$\left(i\gamma^{\mu}\partial_{\mu}\psi - m\psi \right) - \partial_{\mu}(0) = 0 \tag{12.268}$$

$$\left(i\gamma^{\mu}\partial_{\mu} - m \right)\psi = 0 \tag{12.269}$$

This is the Dirac equation.

Quantizing of the field involves expressing the plane wave field solutions in terms of quantum harmonic oscillators. One can make Fourier expansions as follows

$$\psi(\vec{x}) = \int d^3p \; \frac{1}{(2\pi)^3} \frac{1}{\sqrt{2E_p}} \sum_s \left(a_p^s u^s(\vec{p}) e^{i\,\vec{p}.\vec{x}} + b_p^{s\dagger} v^s(\vec{p}) e^{-i\vec{p}.\vec{x}} \right) \qquad (12.270)$$

$$\overline{\psi}(\vec{x}) = \int d^3p \; \frac{1}{(2\pi)^3} \frac{1}{\sqrt{2E_p}} \sum_s \left(a_p^{s\dagger} \; \overline{u}^s(\vec{p})^\dagger e^{-i\vec{p}.\vec{x}} + b_p^s \; \overline{v}^s(\vec{p})^\dagger e^{i\vec{p}.\vec{x}} \right) \qquad (12.271)$$

where $a_p^{s\dagger}$ is the creation operator, i.e. it creates particles, a_p^s is the annihilation operator, i.e. it destroys particles corresponding to spinor $u^s(\vec{p})$, and $b_p^{s\dagger}$ is the creation operator, i.e. it creates particles, b_p^s is the annihilation operator, i.e. it destroys particles corresponding to spinor $v^s(\vec{p})$.

The fields obey the anti-commutation relation

$$\left\{ \psi_\alpha(x), \; \psi_\beta(y) \right\} = \left\{ \psi_\alpha^\dagger(x), \; \psi_\beta^\dagger(y) \right\} = 0 \qquad (12.272)$$

$$\left\{ \psi_\alpha(x)\psi_\beta^\dagger(y) \right\} = \delta^3(x-y)\delta_{\alpha\beta}. \qquad (12.273)$$

12.20 Quantum field theory is the language of particle physics

Whatever was done in particle physics was very qualitative. To make precise calculations we need quantum field theory. Actually, the Standard Model is the correct theory of particle physics. And the Standard Model is written in the language of quantum field theory. So quantum field theory is the only correct language of particle physics.

The main point is that the field is quantized.

In quantum mechanics we speak of particles (like electrons) whose motion is quantized. But in quantum field theory the classical field (say electromagnetic field) is quantized. The motion of a particle is described by three coordinates q_α, $\alpha = 1, 2, 3$ and corresponding momenta p_α. In the case of a field the field at every point is a dynamical variable. This means there are an infinite number of variables and they need to be quantized.

Dirac electron wave function will be re-interpreted as a field. Quantization of electromagnetic field leads to photon. Quantization of Dirac field leads to two types of particles, namely electron and positron.

Any dynamical system (field or particle) is quantized by using the Heisenberg commutation relation

$$[q_i \; p_j] = i\hbar\delta_{ij} \qquad (12.274)$$

In classical mechanics the qs and ps are numbers but in quantum mechanics qs and ps are operators. q_i is coordinate and $p_i = \frac{\partial \mathcal{L}}{\partial q_i}$ is momentum where $\mathcal{L}$ is Lagrangian density.

Every system is defined by specifying its Lagrangian.

The qs are coordinates and ps are momenta in the simple case. In the field case the qs are the field variables at every point in space (i.e. the field becomes a coordinate) and the ps are the canonical momentum.

Lagrangian density is a function of coordinate and velocity, i.e. $\mathcal{L} = \mathcal{L}(q_i, \dot{q}_i)$

12.21 Maxwell's field equations

Maxwell's field equations is

$$\vec{\nabla}.\vec{E} = \frac{\rho}{\varepsilon_0} \tag{12.275}$$

This is the Gauss equation. It gives $\vec{E}$ due to charge density ρ in the system.

$$\vec{\nabla}.\vec{B} = 0 \tag{12.276}$$

This is the Gauss equation. It tells about the non-existence of magnetic poles.

$$\vec{\nabla} \times \vec{E} = -\frac{\partial \vec{B}}{\partial t} \tag{12.277}$$

This is Faraday's law. It tells that time varying $\vec{B}$ generates $\vec{E}$.

$$\vec{\nabla} \times \vec{B} = \mu_0\, \vec{J} + \mu_0\varepsilon_0\frac{\partial \vec{E}}{\partial t} \tag{12.278}$$

This is Ampère's law with Maxwell's correction. It gives $\vec{B}$, a part of which is due to current density $\vec{J}$ and the other part is due to time varying $\vec{E}$.

Clearly these are coupled equations.

It is convenient to introduce potentials to reduce the number of second order differential equations. We then will not have to deal with the fields $\vec{E}$ and $\vec{B}$ directly but instead deal with potentials directly.

Consider the vector identity

$$\vec{\nabla}.\vec{\nabla} \times \vec{A} = 0 \tag{12.279}$$

Comparing equations (12.276) and (12.279) we have

$$\vec{B} = \vec{\nabla} \times \vec{A} \tag{12.280}$$

where $\vec{A}$ is defined to be a vector potential. So $\vec{B}$ can be derived from $\vec{A}$.

Putting $\vec{B} = \vec{\nabla} \times \vec{A}$ in equation (12.277) we get

$$\vec{\nabla} \times \vec{E} = -\frac{\partial}{\partial t}\vec{\nabla} \times \vec{A} = -\vec{\nabla} \times \frac{\partial}{\partial t}\vec{A}$$

$$\vec{\nabla} \times \left(\vec{E} + \frac{\partial}{\partial t}\vec{A} \right) = 0 \tag{12.281}$$

Consider the vector identity

$$\vec{\nabla} \times (-\vec{\nabla}\phi) = 0 \tag{12.282}$$

Compare equations (12.281) and (12.282), we have

$$\vec{E} + \frac{\partial}{\partial t}\vec{A} = -\vec{\nabla}\phi$$

$$\vec{E} = -\frac{\partial}{\partial t}\vec{A} - \vec{\nabla}\phi \tag{12.283}$$

where ϕ is called the scalar potential. So $\vec{E}$ can be derived from ϕ.

However, the above definition of potentials $\vec{A}$ and ϕ is not unique but arbitrary to some extent as there are other potentials $\vec{A}'$ and ϕ' that give the same $\vec{E}$ and $\vec{B}$, i.e.

$$\vec{B} = \vec{\nabla} \times \vec{A}, \quad \vec{B}' = \vec{\nabla} \times \vec{A}' \tag{12.284}$$

$$\vec{E} = -\frac{\partial}{\partial t}\vec{A} - \vec{\nabla}\phi, \quad \vec{E}' = -\frac{\partial}{\partial t}\vec{A}' - \vec{\nabla}\phi' \tag{12.285}$$

$\vec{A}'$, ϕ' and $\vec{A}$, ϕ differ by what is called gauge transformation.

In other words, the arbitrariness in the definition of the potentials $\vec{A}$ and ϕ is exploited by the gauge transformation and is employed to decouple the coupled equations.

12.22 Field equations in terms of potentials

Let us find the field equations in terms of potentials $\vec{A}$, ϕ

Putting $\vec{E} = -\frac{\partial}{\partial t}\vec{A} - \vec{\nabla}\phi$ in equation (12.275) we get

$$\vec{\nabla} \cdot \left(-\frac{\partial}{\partial t}\vec{A} - \vec{\nabla}\phi\right) = \frac{\rho}{\varepsilon_0}$$

$$-\frac{\partial}{\partial t}\vec{\nabla} \cdot \vec{A} - \vec{\nabla} \cdot \vec{\nabla}\phi = \frac{\rho}{\varepsilon_0}$$

$$\frac{\partial}{\partial t}\vec{\nabla} \cdot \vec{A} + \nabla^2\phi = -\frac{\rho}{\varepsilon_0} \tag{12.286}$$

Putting $\vec{B} = \vec{\nabla} \times \vec{A}$ in equation (12.278) we get

$$\vec{\nabla} \times \vec{\nabla} \times \vec{A} = \mu_0\vec{J} + \mu_0\varepsilon_0\frac{\partial}{\partial t}\left(-\frac{\partial}{\partial t}\vec{A} - \vec{\nabla}\phi\right)$$

$$-\nabla^2\vec{A} + \vec{\nabla}(\vec{\nabla} \cdot \vec{A}) = \mu_0\vec{J} - \mu_0\varepsilon_0\frac{\partial^2\vec{A}}{\partial t^2} - \mu_0\varepsilon_0\vec{\nabla}\frac{\partial\phi}{\partial t}$$

$$- \nabla^2 \; \vec{A} + \vec{\nabla} \left[\vec{\nabla}.\, \vec{A} + \mu_0 \varepsilon_0 \frac{\partial \phi}{\partial t} \right] = \mu_0 \; \vec{J} \; - \mu_0 \varepsilon_0 \frac{\partial^2 \vec{A}}{\partial t^2}$$

$$\left(\nabla^2 - \mu_0 \varepsilon_0 \frac{\partial^2}{\partial t^2} \right) \vec{A} = -\mu_0 \; \vec{J} + \vec{\nabla} \left[\vec{\nabla}.\, \vec{A} + \mu_0 \varepsilon_0 \frac{\partial \phi}{\partial t} \right] \qquad (12.287)$$

Equations (12.286) and (12.287) represent the set of Maxwell's equations in terms of potential $\vec{A}$ and ϕ. Clearly by use of potentials $\vec{A}$ and ϕ we have been able to reduce the 4 Maxwell equations to 2 equations. But equations (12.286) and (12.287) are still coupled equations.

Decoupling is achieved by the use of gauge transformation.

12.23 Gauge transformation, gauge invariance

Let us exploit the arbitrariness in the definition of $\vec{A}$ and ϕ. Consider equation (12.280) viz.

$$\vec{B} = \vec{\nabla} \times \vec{A}$$

Since

$$\vec{\nabla} \times (\vec{\nabla} \, \Gamma) = 0 \qquad (12.288)$$

we can write, for a scalar function $\Gamma = \Gamma(\vec{r}, t)$

$$\vec{B} = \vec{\nabla} \times (\vec{A} + \vec{\nabla} \, \Gamma) = \vec{\nabla} \times \vec{A}' = \vec{B}' \qquad (12.289)$$

Hence

$$\vec{A}' = \vec{A} + \vec{\nabla} \, \Gamma \qquad (12.290)$$

So replacement of $\vec{A}$ by $\vec{A}' = \vec{A} + \vec{\nabla} \, \Gamma$ still gives $\vec{B}$, i.e. $\vec{B}$ is left unchanged, i.e. invariant despite the transformation $\vec{A} \to \vec{A}' = \vec{A} + \vec{\nabla} \, \Gamma$.

Again, under the transformation $\vec{A} \to \vec{A}' = \vec{A} + \vec{\nabla} \, \Gamma$, $\phi \to \phi' = \phi + \phi'$ we demand that $\vec{E}$ is invariant and so we get from equation (12.283)

$$\vec{E}' = -\frac{\partial}{\partial t} \vec{A}' - \vec{\nabla} \, \phi' = -\frac{\partial}{\partial t}(\vec{A} + \vec{\nabla} \, \Gamma) - \vec{\nabla} \, (\phi + \phi') = \vec{E} = -\frac{\partial}{\partial t} \vec{A} - \vec{\nabla} \, \phi \quad (12.291)$$

Hence

$$-\frac{\partial}{\partial t} \vec{A} - \vec{\nabla} \; \frac{\partial}{\partial t} \Gamma - \vec{\nabla} \, \phi - \vec{\nabla} \, \phi' = -\frac{\partial}{\partial t} \vec{A} - \vec{\nabla} \, \phi$$

$$\vec{\nabla} \, \phi' = - \vec{\nabla} \, \frac{\partial}{\partial t} \Gamma$$

$$\phi' = -\frac{\partial}{\partial t} \Gamma \qquad (12.292)$$

So under the gauge transformation

$$\vec{A} \rightarrow \vec{A}' = \vec{A} + \vec{\nabla}\,\Gamma \tag{12.293}$$

$$\phi \rightarrow \phi' = \phi - \frac{\partial}{\partial t}\Gamma \tag{12.294}$$

where $\Gamma = \Gamma(\vec{r}, t)$ is an arbitrary function of $\vec{r}, t$ the fields $\vec{E}, \vec{B}$ are invariant, i.e.

$$\vec{E} = \vec{E}', \quad \vec{B} = \vec{B}'. \tag{12.295}$$

This invariance is called gauge invariance.

Any transformation of the potentials $(\phi, \vec{A}) \rightarrow (\phi', \vec{A}')$ is called gauge transformation. $\vec{E}, \vec{B}$ are invariant under gauge transformation. This is a very important aspect in the dynamics of particles in quantum field theory.

We note that $\vec{A}$ and ϕ are defined in such a way that they are not unique but ambiguous to the extent determined by the scalar function Γ.

It is the fields that have physical meaning and the electromagnetic potentials have been introduced for convenience. Because of the non-uniqueness of electromagnetic potentials $\vec{A}$ and ϕ we are free to impose an additional condition on $\vec{A}$. Alternatively, to specify a vector uniquely we must specify both divergence and curl. Here we take $\vec{\nabla} \times \vec{A} = \vec{B}$ but $\vec{\nabla}.\vec{A}$ is arbitrary. So $\vec{\nabla}.\vec{A}$ may be chosen conveniently to simplify the problem.

We are free to specify an extra relation between $\vec{A}$ and ϕ or conditions on $\vec{A}$ and ϕ (as long as $\vec{E}$ and $\vec{B}$ remain unchanged). This is called choosing a gauge.

12.23.1 Lorentz gauge

One useful choice, called Lorentz gauge, is

$$\vec{\nabla}.\vec{A} + \frac{1}{c^2}\frac{\partial \phi}{\partial t} = 0 \tag{12.296}$$

With this we have from equation (12.287)

$$\left(\nabla^2 - \mu_0\varepsilon_0\frac{\partial^2}{\partial t^2}\right)\vec{A} = -\mu_0\,\vec{J} \tag{12.297}$$

$$\text{i.e. } \Box\vec{A} = -\mu_0\,\vec{J} \tag{12.298}$$

Putting

$$\vec{A} \rightarrow \vec{A}' = \vec{A} + \vec{\nabla}\,\Gamma, \quad \phi \rightarrow \phi' = \phi - \frac{\partial}{\partial t}\Gamma \tag{12.299}$$

in Lorentz gauge condition we get

$$\vec{\nabla}.\,\vec{A}' + \frac{1}{c^2}\frac{\partial\phi'}{\partial t} = 0 \tag{12.300}$$

$$\vec{\nabla}.\left(\vec{A} + \vec{\nabla}\,\Gamma\right) + \frac{1}{c^2}\frac{\partial}{\partial t}\left(\phi - \frac{\partial}{\partial t}\Gamma\right) = 0$$

$$\vec{\nabla}.\,\vec{A} + \frac{1}{c^2}\frac{\partial\phi}{\partial t} + \vec{\nabla}.\,\vec{\nabla}\,\Gamma - \frac{1}{c^2}\frac{\partial^2}{\partial t^2}\Gamma = 0$$

$$\nabla^2\Gamma - \frac{1}{c^2}\frac{\partial^2}{\partial t^2}\Gamma = 0 \tag{12.301}$$

$$\Box\,\Gamma = 0 \tag{12.302}$$

Another useful choice in magnetostatics called Coulomb gauge is

$$\vec{\nabla}.\,\vec{A} = 0 \tag{12.303}$$

12.24 Global and local gauge transformation

Equations (12.293) and (12.294) can be grouped as follows. Using

$$\vec{A} = -A_i = A^i, \quad \vec{A}' = -A_i' = A^{i\prime}, \quad \frac{\phi}{c} = A^0 = A_0, \quad \frac{\phi'}{c} = A^{0\prime} = A_0'$$

and

$$\vec{\nabla} = \partial_i = -\partial^i, \quad \frac{\partial}{\partial ct} = \partial_0 = \partial^0$$

We have from these equations

$$A^{i\prime} = A^i - \partial^i\,\Gamma \tag{12.304}$$

$$\frac{\phi'}{c} = \frac{\phi}{c} - \frac{\partial}{\partial ct}\Gamma, \quad A^{0\prime} = A^0 - \partial^0\,\Gamma \tag{12.305}$$

Combining equations (12.304) and (12.305)

$$A^{\mu\prime} = A^\mu - \partial^\mu\,\Gamma \tag{12.306}$$

• If

$$\Gamma \neq \Gamma(\vec{r},\,t), \quad \partial^\mu\,\Gamma = 0 \tag{12.307}$$

i.e. if Γ is independent of space and time then we call it global gauge transformation.
• If

$$\Gamma = \Gamma(\vec{r},\,t), \quad \partial^\mu\,\Gamma \neq 0 \tag{12.308}$$

i.e. if Γ is dependent on space and time then we call it local gauge transformation.

In quantum mechanics we have to use potentials to write down Schrödinger equation.

12.25 Gauge freedom

We have to remove the unwanted degrees of freedom in A^μ so that there is no arbitrariness when we assign physical meaning to A^μ.

Consider the vector potential

$$A^\mu = \left(\frac{\phi}{c}, A^i\right) = \left(\frac{\phi}{c}, \quad A_x, \quad A_y, A_z\right) \tag{12.309}$$

These are not uniquely fixed. So let us think about some relation to fix this arbitrariness. Let us consider 4-divergence to fix the arbitrariness. Let us start with A^μ whose 4-divergence is not zero, i.e.

$$\partial_\mu A^\mu \neq 0 \tag{12.310}$$

and then make the transformation

$$A^\mu \to A'^\mu = A^\mu + \partial^\mu \chi \tag{12.311}$$

4-divergence of A'^μ gives us

$$\partial_\mu A'^\mu = \partial_\mu A^\mu + \partial_\mu \partial^\mu \chi \tag{12.312}$$

We can choose χ so that

$$\partial_\mu \partial^\mu \chi = -\partial_\mu A^\mu \tag{12.313}$$

This gives

$$\partial_\mu A'^\mu = \partial_\mu A^\mu - \partial_\mu A^\mu = 0 \tag{12.314}$$

Clearly if A^μ is not divergenceless ($\partial_\mu A^\mu \neq 0$) then we can always consider a scalar field which will change the A^μ to A'^μ that will be divergenceless ($\partial_\mu A'^\mu = 0$). So a divergenceless A^μ can always be achieved by a suitable choice of χ.

There is another arbitrariness also. Consider

$$\partial_\mu A'^\mu = 0$$

If we take

$$A'^\mu \to A''^\mu = A'^\mu + \partial^\mu \Lambda \tag{12.315}$$

and consider

$$\partial_\mu A''^\mu = \partial_\mu (A'^\mu + \partial^\mu \Lambda) = \partial_\mu A'^\mu + \partial_\mu \partial^\mu \Lambda = 0 + \partial_\mu \partial^\mu \Lambda$$

If Λ be such that

$$\partial_\mu \partial^\mu \Lambda = 0 \tag{12.316}$$

then

$$\partial_\mu A''^\mu = 0 \qquad (12.317)$$

Clearly even if 4-divergence of A^μ is zero there is still some arbitrariness in A^μ which is choosing $\partial_\mu \partial^\mu \Lambda = 0$. Such a choice keeps $\vec{E}$, $\vec{B}$ invariant.

Clearly there is freedom to choose the potentials without affecting $\vec{E}$, $\vec{B}$. In other words, the relations $\vec{E} = -\vec{\nabla}\phi - \frac{\partial \vec{A}}{\partial t}$ and $\vec{B} = \vec{\nabla} \times \vec{A}$ are satisfied by the following choices

$$A^\mu$$

$$A'^\mu = A^\mu + \partial^\mu \chi \text{ for any scalar function } \chi$$

$$A''^\mu = A'^\mu + \partial^\mu \Lambda \text{ where } \partial_\mu \partial^\mu \Lambda = 0$$

We mention that electromagnetic field is a collection of photons. Again, the photon which is the quantum of an electromagnetic field is more meaningful physically and the field associated with that is the potential A^μ.

12.26 Electromagnetic field tensor

$\vec{E}$ and $\vec{B}$ are 3-vectors. They are amalgamated to generate a 2nd rank anti-symmetric tensor in 4D called the electromagnetic field tensor $F_{\mu\nu}$ developed as follows.

Relation between fields and potentials is

$$\vec{E} = -\frac{\partial}{\partial t}\vec{A} - \vec{\nabla}\phi, \ \vec{B} = \vec{\nabla} \times \vec{A} \qquad (12.318)$$

We define

$$A_\mu = (A_0, A_i) = \left(\frac{\phi}{c}, -\vec{A}\right), \quad A^\mu = (A^0, A^i) = \left(\frac{\phi}{c}, \vec{A}\right)$$

$$A^0 = A_0 = \frac{\phi}{c}, \quad A^i = -A_i = \vec{A} \qquad (12.319)$$

$$\partial_\mu = (\partial_0, \ \partial_i) = \left(\frac{1}{c}\frac{\partial}{\partial t}, \vec{\nabla}\right), \quad \partial^\mu = (\partial^0, \partial^i) = \left(\frac{1}{c}\frac{\partial}{\partial t}, -\vec{\nabla}\right)$$

$$\frac{1}{c}\frac{\partial}{\partial t} = \partial^0 = \partial_0, \ \frac{\partial}{\partial x} = -\partial^1 = \partial_1, \ \frac{\partial}{\partial y} = -\partial^2 = \partial_2, \ \frac{\partial}{\partial z} = -\partial^3 = \partial_3 \ (12.320)$$

Consider the relation

$$\vec{E} = -\frac{\partial}{\partial t}\vec{A} - \vec{\nabla}\phi \qquad (12.321)$$

x component is

$$E_x = -\frac{\partial A_x}{\partial t} - \frac{\partial \phi}{\partial x} \tag{12.322}$$

$$\frac{E_1}{c} = -\frac{\partial}{\partial ct}A_x - \frac{\partial}{\partial x}\frac{\phi}{c} \quad \text{(denoting } E_x \text{ by } E_1\text{)} \tag{12.323}$$

Using $\frac{\partial}{\partial x} = -\partial^1$, $\frac{\partial}{\partial ct} = \partial^0$, $A_x = A^1$, $\frac{\phi}{c} = A^0$ we write equation (12.323) as

$$\frac{E_1}{c} = -\partial^0 A^1 - (-\partial^1)A^0$$

$$-\frac{E_1}{c} = \partial^0 A^1 - \partial^1 A^0 = F^{01} \quad \text{(define)} \tag{12.324}$$

$$F^{10} = \partial^1 A^0 - \partial^0 A^1 = \frac{E_1}{c} \tag{12.325}$$

Again, using $\frac{\partial}{\partial x} = \partial_1$, $\frac{\partial}{\partial ct} = \partial_0$, $A_x = -A_1$, $\frac{\phi}{c} = A_0$ we write equation (12.323) as

$$\frac{E_1}{c} = -\partial_0(-A_1) - \partial_1 A_0$$

$$\frac{E_1}{c} = \partial_0 A_1 - \partial_1 A_0 = F_{01} \quad \text{(define)} \tag{12.326}$$

$$F_{10} = \partial_1 A_0 - \partial_0 A_1 = -\frac{E_1}{c} \tag{12.327}$$

Similarly

$$F^{02} = \partial^0 A^2 - \partial^2 A^0 = -\frac{E_2}{c} = -F^{20}$$

$$F^{03} = \partial^0 A^3 - \partial^3 A^0 = -\frac{E_3}{c} = -F^{30}$$

$$F_{02} = \partial_0 A_2 - \partial_2 A_0 = \frac{E_2}{c} = -F_{20}$$

$$F_{03} = \partial_0 A_3 - \partial_3 A_0 = \frac{E_3}{c} = -F_{30}$$

$$F^{0i} = \partial^0 A^i - \partial^i A^0 = -\frac{E_i}{c} = -F^{i0} \tag{12.328}$$

$$F_{0i} = \partial_0 A_i - \partial_i A_0 = \frac{E_i}{c} = -F_{i0} \tag{12.329}$$

Consider $\vec{B} = \vec{\nabla} \times \vec{A}$
x component is

$$B_x = (\vec{\nabla} \times \vec{A})_x$$

$$B_1 = \frac{\partial A_z}{\partial y} - \frac{\partial A_y}{\partial z} \quad \text{(denoting } B_x \text{ by } B_1) \tag{12.330}$$

Using $\frac{\partial}{\partial y} = -\partial^2$, $\frac{\partial}{\partial z} = -\partial^3$, $A_z = A^3$, $A_y = A^2$ we write equation (12.330) as

$$B_1 = -\partial^2 A^3 - (-\partial^3)A^2 = \partial^3 A^2 - \partial^2 A^3$$

$$-B_1 = \partial^2 A^3 - \partial^3 A^2 = F^{23} \quad \text{(define)} \tag{12.331}$$

$$F^{32} = \partial^3 A^2 - \partial^2 A^3 = B_1 \tag{12.332}$$

Again using $\frac{\partial}{\partial y} = \partial_2$, $\frac{\partial}{\partial z} = \partial_3$, $A_z = -A_3$, $A_y = -A_2$ we write equation (12.330) as

$$B_1 = \partial_2(-A_3) - \partial_3(-A_2)$$

$$-B_1 = \partial_2 A_3 - \partial_3 A_2 = F_{23} \quad \text{(define)} \tag{12.333}$$

$$F_{32} = \partial_3 A_2 - \partial_2 A_3 = B_1 \tag{12.334}$$

$$\text{Hence } F^{23} = F_{23} = -B_1, \quad F^{32} = F_{32} = B_1 \tag{12.335}$$

$$\text{Similarly } F^{31} = F_{31} = -B_2, \quad F^{13} = F_{13} = B_2 \tag{12.336}$$

$$F^{12} = F_{12} = -B_3, \quad F^{21} = F_{21} = B_3 \tag{12.337}$$

In general

$$F^{ij} = F_{ij} = -\varepsilon_{ijk} B_k \tag{12.338}$$

$$(\text{e. g. } \quad F^{31} = F_{31} = -\varepsilon_{312} B_2 = -B_2, \quad F^{13} = F_{13} = -\varepsilon_{132} B_2 = B_2)$$

$$F_{\mu\nu} = \partial_\mu A_\nu - \partial_\nu A_\mu = \frac{\partial A_\nu}{\partial x^\mu} - \frac{\partial A_\mu}{\partial x^\nu} \tag{12.339}$$

$$F_{\mu\nu} = -F_{\nu\mu} \quad (\text{antisymmetry property of } F_{\mu\nu}) \tag{12.340}$$

$$F^{\mu\nu} = \partial^\mu A^\nu - \partial^\nu A^\mu = \frac{\partial A_\nu}{\partial x_\mu} - \frac{\partial A_\mu}{\partial x_\nu} \tag{12.341}$$

$$F_{\mu\mu} = \partial_\mu A_\mu - \partial_\mu A_\mu = 0 \tag{12.342}$$

Clearly $F_{\mu\nu}$ has $4^2 = 16$ components. And can be expressed as a matrix given by

$$F_{\mu\nu} = \begin{pmatrix} F_{00} & F_{01} & F_{02} & F_{03} \\ F_{10} & F_{11} & F_{12} & F_{13} \\ F_{20} & F_{21} & F_{22} & F_{23} \\ F_{30} & F_{31} & F_{32} & F_{33} \end{pmatrix} = \begin{pmatrix} 0 & E_1/c & E_2/c & E_3/c \\ -E_1/c & 0 & -B_3 & B_2 \\ -E_2/c & B_3 & 0 & -B_1 \\ -E_3/c & -B_2 & B_1 & 0 \end{pmatrix} \tag{12.343}$$

$$F^{\mu\nu} = \begin{pmatrix} F^{00} & F^{01} & F^{02} & F^{03} \\ F^{10} & F^{11} & F^{12} & F^{13} \\ F^{20} & F^{21} & F^{22} & F^{23} \\ F^{30} & F^{31} & F^{32} & F^{33} \end{pmatrix} = \begin{pmatrix} 0 & -E_1/c & -E_2/c & -E_3/c \\ E_1/c & 0 & -B_3 & B_2 \\ E_2/c & B_3 & 0 & -B_1 \\ E_3/c & -B_2 & B_1 & 0 \end{pmatrix} \tag{12.344}$$

- Relation between $F^{\mu\nu}$ and $F_{\mu\nu}$ is

$$F_{\mu\nu} = g_{\mu\rho} g_{\nu\sigma} F^{\rho\sigma} \tag{12.345}$$

$$\bullet F^{\mu\nu} = -F^{\nu\mu} \tag{12.346}$$

- Between contravariant and covariant electromagnetic field tensor the electric field components change sign and magnetic field components remain the same.

12.27 Lagrangian density of an electromagnetic field

Based on the requirement of Lorentz invariance and gauge invariance the Lagrangian density of an electromagnetic field is taken as (for zero current density)

$$\mathcal{L} = -\frac{1}{4\mu_0} F_{\mu\nu} F^{\mu\nu} \tag{12.347}$$

where $F_{\mu\nu} = \partial_\mu A_\nu - \partial_\nu A_\mu$
A_μ being the dynamical variable.

The Lagrangian density of equation (12.347) can also be written in terms of the Lorentz invariant quantity $c^2 B^2 - E^2$.

$$\mathcal{L} = \frac{1}{2}\left(\varepsilon_0 E^2 - \frac{1}{\mu_0} B^2\right) = \frac{1}{2\mu_0}\left(\frac{E^2}{c^2} - B^2\right) = -\frac{1}{2\mu_0}(c^2 B^2 - E^2) \tag{12.348}$$

We shall derive the Euler–Lagrange equation from $\mathcal{L}$ of equation (12.347) and check if that leads to the 4 Maxwell's equations.

12.28 Maxwell equations from the Euler–Lagrange equation

The Euler–Lagrange equation is

$$\partial^\mu \left(\frac{\partial \mathcal{L}}{\partial(\partial_\mu A_\nu)} \right) - \frac{\partial \mathcal{L}}{\partial A_\nu} = 0 \tag{12.349}$$

$\mathcal{L}$ does not depend on A_μ and hence $\frac{\partial \mathcal{L}}{\partial A_\nu} = 0$.

$$\partial^\mu \left(\frac{\partial \mathcal{L}}{\partial(\partial_\mu A_\nu)} \right) = 0$$

Using equation (12.347) viz. $\mathcal{L} = -\frac{1}{4\mu_0} F_{\alpha\beta} F^{\alpha\beta}$ we get

$$\partial^\mu \left(\frac{\partial}{\partial(\partial_\mu A_\nu)} \ F_{\alpha\beta} F^{\alpha\beta} \right) = 0$$

As $F_{\alpha\beta} F^{\alpha\beta} \sim F_{\alpha\beta}^2$

$$\partial^\mu \left(2 F_{\alpha\beta} \frac{\partial}{\partial(\partial_\mu A_\nu)} \ F_{\alpha\beta} \right) = 0 \tag{12.350}$$

As $F_{\alpha\beta} = \partial_\alpha A_\beta - \partial_\beta A_\alpha$

$$\partial^\mu \left(F_{\alpha\beta} \frac{\partial}{\partial(\partial_\mu A_\nu)} \left(\partial_\alpha A_\beta - \partial_\beta A_\alpha \right) \right) = 0$$

$$\partial^\mu \left(F_{\alpha\beta} \frac{\partial}{\partial(\partial_\mu A_\nu)} \partial_\alpha A_\beta - F_{\alpha\beta} \frac{\partial}{\partial(\partial_\mu A_\nu)} \partial_\beta A_\alpha \right) = 0$$

$$\partial^\mu \left(F_{\alpha\beta} \ \delta^\mu_\alpha \delta^\nu_\beta - F_{\alpha\beta} \delta^\mu_\beta \delta^\nu_\alpha \right) = 0$$

$$\partial^\mu (F_{\mu\nu} - F_{\nu\mu}) = 0$$

$$\partial^\mu (F_{\mu\nu} + F_{\mu\nu}) = 0$$

$$\partial^\mu F_{\mu\nu} = 0 \tag{12.351}$$

This is Maxwell's equation.

Let us now obtain the 4 Maxwell's equations in its familiar form from equation (12.351).

- Let us consider $\nu = 0$

$$\partial^\mu F_{\mu 0} - 0$$

$$\partial^0 F_{00} + \partial^i F_{i0} = 0$$

Since all diagonal elements of $F_{\mu\nu}$ are zero, i.e. $F_{\mu\mu} = 0$ we have $F_{00} = 0$

$$\partial^i F_{i0} = 0$$

$$\partial^i(-E_i/c) = 0$$

$$-\partial^i E_i = 0 \text{ i. e. } \partial_i E_i = 0 \; \partial_1 E_1 + \partial_2 \; E_2 + \partial_3 E_3 = 0$$

$$\frac{\partial E_1}{\partial x} + \frac{\partial E_2}{\partial y} + \frac{\partial E_3}{\partial z} = 0$$

$$\vec{\nabla}. \vec{E} = 0$$

This is Maxwell's first equation or Gauss's equation for $\vec{E}$ in vacuum (i.e. for $\rho = 0$).

• Let us consider $\nu = 1$

$$\partial^\mu F_{\mu 1} = 0$$

$$\partial^0 F_{01} + \partial^1 F_{11} + \partial^2 F_{21} + \partial^3 F_{31} = 0 \tag{12.352}$$

Using $\partial^0 = \frac{\partial}{\partial ct}, \quad \partial^1 = -\frac{\partial}{\partial x}, \quad \partial^2 = -\frac{\partial}{\partial y}, \quad \partial^3 = -\frac{\partial}{\partial z}$

$$F_{01} = \frac{E_1}{c} = \frac{E_x}{c}, \quad F_{11} = 0, \quad F_{21} = B_3 = B_z, \quad F_{31} = -B_2 = -B_y$$

Putting in equation (12.352) we get

$$\frac{\partial}{\partial ct} \frac{E_x}{c} - \frac{\partial}{\partial x}(0) - \frac{\partial}{\partial y}B_z - \frac{\partial}{\partial z}(-B_y) = 0$$

$$\frac{1}{c^2}\frac{\partial}{\partial t}E_x - \frac{\partial}{\partial y}B_z + \frac{\partial}{\partial z}B_y = 0 \Rightarrow \frac{\partial}{\partial y}B_z - \frac{\partial}{\partial z}B_y = \frac{1}{c^2}\frac{\partial}{\partial t}E_x \Rightarrow (\vec{\nabla} \times \vec{B})_x = \frac{1}{c^2}\frac{\partial}{\partial t}E_x$$

Similarly for other components. Combining we get

$$\vec{\nabla} \times \vec{B} = \frac{1}{c^2}\frac{\partial \vec{E}}{\partial t}$$

This is Maxwell's 4th equation or Ampère's law with Maxwell's correction for $\vec{J} = 0$.

• Let us operate by ∂^μ on $F_{\nu\lambda} = \partial_\nu A_\lambda - \partial_\lambda A_\nu$

$$\partial^\mu F_{\nu\lambda} = \partial^\mu \partial_\nu A_\lambda - \partial^\mu \partial_\lambda A_\nu \tag{12.353}$$

Let us operate by ∂^ν on $F_{\lambda\nu} = \partial_\lambda A_\nu - \partial_\nu A_\lambda$

$$\partial^\nu F_{\lambda\mu} = \partial^\nu \partial_\lambda A_\mu - \partial^\nu \partial_\mu A_\lambda \tag{12.354}$$

Let us operate by ∂^λ on $F_{\mu\nu} = \partial_\mu A_\nu - \partial_\nu A_\nu$

$$\partial^\lambda F_{\mu\nu} = \partial^\lambda \partial_\mu A_\nu - \partial^\lambda \partial_\nu A_\mu \tag{12.355}$$

Adding equations (12.353), (12.354), and (12.355) we get

$$\partial^\mu F_{\nu\lambda} + \partial^\nu F_{\lambda\mu} + \partial^\lambda F_{\mu\nu} = \partial^\mu \partial_\nu A_\lambda - \partial^\mu \partial_\lambda A_\nu + \partial^\nu \partial_\lambda A_\mu - \partial^\nu \partial_\mu A_\lambda + \partial^\lambda \partial_\mu A_\nu - \partial^\lambda \partial_\nu A_\mu$$

As the derivatives commute

$$\partial^\mu F_{\nu\lambda} + \partial^\nu F_{\lambda\mu} + \partial^\lambda F_{\mu\nu} = 0 \ \text{ (This is called Bianchi identity)} \tag{12.356}$$

Take $\mu = 1, \ \nu = 2, \ \lambda = 3$ to get

$$\partial^1 F_{23} + \partial^2 F_{31} + \partial^3 F_{12} = 0$$

$$\partial^1 B_1 + \partial^2 B_2 + \partial^3 B_3 = 0$$

$$\vec{\nabla} \cdot \vec{B} = 0$$

This is Maxwell's second equation or Gauss's equation for $\vec{B}$.
• With $\mu = 0, \ \nu = 1, \ \lambda = 2$ the Bianchi identity becomes

$$\partial^0 F_{12} + \partial^1 F_{20} + \partial^2 F_{01} = 0$$

Using $F_{12} = B_3, \ F_{20} = -E_2/c, \ F_{01} = E_1/c$ we have

$$\partial^0 B_3 - \partial^1 E_2/c + \partial^2 E_1/c = 0$$

$$c\frac{\partial}{\partial x_0} B_3 - \frac{\partial}{\partial x_1} E_2 + \frac{\partial}{\partial x_2} E_1 = 0$$

$$c\frac{\partial}{\partial ct} B_3 + \frac{\partial}{\partial x} E_2 - \frac{\partial}{\partial y} E_1 = 0 \ \text{ (as } x_0 = ct, \ x_1 = -x, \ x_2 = -y)$$

$$\frac{\partial}{\partial t} B_z + \frac{\partial}{\partial x} E_y - \frac{\partial}{\partial y} E_x = 0$$

$$\frac{\partial}{\partial x} E_y - \frac{\partial}{\partial y} E_x = -\frac{\partial}{\partial t} B_z$$

$$(\vec{\nabla} \times \vec{E})_z = -\frac{\partial}{\partial t} B_z$$

Similarly for the other components and hence combining we have

$$\vec{\nabla} \times \vec{E} = -\frac{\partial B}{\partial t}$$

This is Maxwell's 3rd equation or Faraday's equation.

In an electromagnetic field we recognize the vector potential defined at every point of space and time A_μ is the dynamical variable and so we replace it by an operator. Since A_μ is an operator, fields E_i and B_i will also be operators.

We thus have derived the 4 famaliar Maxwell's equations of electrodynamics starting from Lagrangian density $\mathcal{L} = -\frac{1}{4\mu_0}F_{\mu\nu}F^{\mu\nu}$ and also showed that these 4 equations collapse into a single equation $\partial^\mu F_{\mu\nu} = 0$.

12.29 Hamiltonian

Generalized momentum is given by

$$p_k = \frac{\partial L}{\partial v_k} = \frac{\partial}{\partial v_k}\left(\frac{1}{2}mv_k^2 - q\phi + qv_k A_k\right) \tag{12.357}$$

$$p_k = mv_k + qA_k \tag{12.358}$$

$$v_k = \frac{p_k - qA_k}{m} \tag{12.359}$$

By definition the Hamiltonian is given by

$$H = \sum_k p_k v_k - L = \sum_k (mv_k + qA_k)v_k - L = \sum_k mv_k^2 + \sum_k qv_k A_k - L$$

$$H = \sum_k mv_k^2 + \sum_k qv_k A_k - \left(\sum_k \frac{1}{2}mv_k^2 - q\phi + q\sum_k v_k A_k\right)$$

$$H = \sum_k \frac{1}{2}mv_k^2 + q\phi = \sum_k \frac{1}{2}m\left(\frac{p_k - qA_k}{m}\right)^2 + q\phi = \sum_k \frac{(p_k - qA_k)^2}{2m} + q\phi$$

$$H = \frac{1}{2m}(\vec{p} - q\vec{A})^2 + q\phi \tag{12.360}$$

In contrast to equation (12.360) a free particle has Hamiltonian $H_{\text{free}} = E = \frac{p^2}{2m}$. This means that to switch over to an electromagnetic field we have to make the following replacements

$$\vec{p} \rightarrow \vec{p} - q\vec{A}, \quad E \rightarrow E - q\phi \tag{12.361}$$

We have derived the Lagrangian and Hamiltonian of an lectromagnetic field in *exercise 12.4.*

12.30 Gauge invariance of Dirac equation

Covariant form of Dirac equation for a particle of charge q in an electromagnetic field $(\vec{A}, \phi)$ is by equation (11.279)

$$[\gamma_\mu(p^\mu - qA^\mu) - mc\ ]\psi = 0 \qquad (12.362)$$

Perform a unitary transformation (change of phase)

$$\psi = \psi' e^{-iq\Gamma/\hbar} \qquad (12.363)$$

Perform gauge transformation

$$A^\mu = A^{\mu\prime} + \partial^\mu\Gamma \qquad (12.364)$$

$$[\gamma_\mu(p^\mu - q(A^{\mu\prime} + \partial^\mu\Gamma)\,) - mc\ ]\psi' e^{-iq\Gamma/\hbar} = 0$$

$$\gamma_\mu p^\mu \psi' e^{-iq\Gamma/\hbar} - \gamma_\mu qA^{\mu\prime}\psi' e^{-iq\Gamma/\hbar} - \gamma_\mu q(\partial^\mu\Gamma)\,\psi' e^{-iq\Gamma/\hbar} - mc\,\psi' e^{-iq\Gamma/\hbar} = 0$$

With $p^\mu \to i\hbar\partial^\mu$

$$\gamma_\mu i\hbar\partial^\mu \psi' e^{-iq\Gamma/\hbar} - \gamma_\mu qA^{\mu\prime}\psi' e^{-iq\Gamma/\hbar} - \gamma_\mu q(\partial^\mu\Gamma)\,\psi' e^{-iq\Gamma/\hbar} - mc\,\psi' e^{-iq\Gamma/\hbar} = 0$$

$$i\hbar\gamma_\mu((\partial^\mu\psi')e^{-iq\Gamma/\hbar} + i\hbar\gamma_\mu\psi'(\partial^\mu\,e^{-iq\Gamma/\hbar})$$

$$-\gamma_\mu qA^{\mu\prime}\psi' e^{-iq\Gamma/\hbar} - \gamma_\mu q(\partial^\mu\Gamma)\,\psi' e^{-iq\Gamma/\hbar} - mc\,\psi' e^{-iq\Gamma/\hbar} = 0$$

$$i\hbar\gamma_\mu(\partial^\mu\psi')e^{-iq\Gamma/\hbar} + i\hbar\gamma_\mu\psi'\left(-\frac{iq}{\hbar}\right)e^{-iq\Gamma/\hbar}(\partial^\mu\Gamma)$$

$$-\gamma_\mu qA^{\mu\prime}\psi' e^{-iq\Gamma/\hbar} - \gamma_\mu q(\partial^\mu\Gamma)\,\psi' e^{-iq\Gamma/\hbar} - mc\,\psi' e^{-iq\Gamma/\hbar} = 0$$

$$\gamma_\mu p^\mu \psi' + \gamma_\mu q(\partial^\mu\Gamma)\psi' - \gamma_\mu qA^{\mu\prime}\psi' - \gamma_\mu q(\partial^\mu\Gamma)\,\psi' - mc\,\psi' = 0$$

$$\gamma_\mu p^\mu \psi' - \gamma_\mu qA^{\mu\prime}\psi' - mc\,\psi' = 0$$

$$(\gamma_\mu(p^\mu - qA^{\mu\prime}) - mc)\,\psi' = 0 \qquad (12.365)$$

We have thus shown that the Dirac equation is invariant under local gauge transformation.

ψ and ψ' have a relative phase shift and A^μ and $A^{\mu\prime}$ are related by local gauge transformation.

12.31 Global gauge transformation

Let us consider the Dirac Lagrangian density given by equation (12.264)

$$\mathcal{L} = i\bar{\psi}\gamma^\mu\partial_\mu\psi - m\bar{\psi}\psi$$

Consider a transformation in which we add a phase factor to ψ as

$$\psi \to \psi' - e^{i\alpha}\psi \qquad (12.366)$$

where α is a constant. This transformation is called global gauge transformation. We are not changing the space-time coordinate but changing the form of the wave field. Now

$$\overline{\psi} \to \overline{\psi'} = e^{-i\alpha}\overline{\psi} \tag{12.367}$$

$$\partial_\mu\psi \to \partial_\mu\psi' = \partial_\mu(e^{i\alpha}\psi) = e^{i\alpha}\partial_\mu\psi \tag{12.368}$$

Under the global gauge transformation the Lagrangian becomes

$$\mathcal{L} \to \mathcal{L}' = i\overline{\psi'}\gamma^\mu\partial_\mu\psi' - m\overline{\psi'}\psi' = i\left(e^{-i\alpha}\overline{\psi}\right)\gamma^\mu\partial_\mu\left(e^{i\alpha}\psi\right) - m\left(e^{-i\alpha}\overline{\psi}\right)\left(e^{i\alpha}\psi\right)$$

$$= i\overline{\psi}\gamma^\mu\partial_\mu\psi - m\overline{\psi}\psi = \mathcal{L} \tag{12.369}$$

So the Lagrangian is invariant under global gauge transformation $\to\psi' = e^{i\alpha}\psi$.

12.32 Local gauge transformation

Let us consider the Dirac Lagrangian density given by equation (12.264)

$$\mathcal{L} = i\overline{\psi}\gamma^\mu\partial_\mu\psi - m\overline{\psi}\psi \tag{12.370}$$

Consider a transformation in which we add a phase factor to ψ as

$$\psi \to \psi' = e^{i\alpha(x)}\psi \tag{12.371}$$

where α is a function of space-time denoted by x. This transformation is called local gauge transformation. It is called local because α depends on local points, i.e. it can have one value at x_1 and another value at x_2. It is not globally defined, i.e. does not have same value for all points. Now

$$\overline{\psi} \to \overline{\psi'} = e^{-i\alpha(x)}\overline{\psi} \tag{12.372}$$

$$\partial_\mu\psi \to \partial_\mu\psi' = \partial_\mu(e^{i\alpha(x)}\psi) = i(\partial_\mu\alpha)e^{i\alpha}\psi + e^{i\alpha}\partial_\mu\psi = e^{i\alpha}\left[i(\partial_\mu\alpha) + \partial_\mu\right]\psi \tag{12.373}$$

$$\mathcal{L} \to \mathcal{L}' = i\overline{\psi'}\gamma^\mu\partial_\mu\psi' - m\overline{\psi'}\psi'$$

$$= i(e^{-i\alpha}\overline{\psi})\gamma^\mu(e^{i\alpha}[\,i(\partial_\mu\alpha) + \partial_\mu]\psi) - m\left(e^{-i\alpha}\overline{\psi}\right)\left(e^{i\alpha}\psi\right)$$

$$= i\overline{\psi}\gamma^\mu\left[i(\partial_\mu\alpha) + \partial_\mu\right]\psi - m\overline{\psi}\psi$$

$$\mathcal{L}' = i\overline{\psi}\gamma^\mu\partial_\mu\psi - m\overline{\psi}\psi - \overline{\psi}\gamma^\mu(\partial_\mu\alpha)\psi \tag{12.374}$$

$$\mathcal{L}' = \mathcal{L} - \overline{\psi}\gamma^\mu(\partial_\mu\alpha)\psi \neq \mathcal{L} \tag{12.375}$$

This Lagrangian $\mathcal{L}$ is not invariant under local gauge transformation due to the presence of an additional term viz. $-\overline{\psi}\gamma^\mu(\partial_\mu\alpha)\psi$ in it.

We can define a new derivative as

$$D_\mu = \partial_\mu + eA_\mu \tag{12.376}$$

With this we get

$$D_\mu \psi \to D'_\mu \psi' = \left(\partial_\mu + eA'_\mu\right)\left(e^{i\alpha(x)}\psi\right)$$

$$= \partial_\mu(e^{i\alpha}\psi) + eA'_\mu e^{i\alpha}\psi$$

$$D'_\mu \psi' = e^{i\alpha}\left(\partial_\mu\psi + i(\partial_\mu\alpha)\psi + eA'_\mu\psi\right) \tag{12.377}$$

Take

$$A'_\mu = A_\mu + \delta A_\mu \tag{12.378}$$

This gives

$$D'_\mu \psi' = e^{i\alpha}\Big[(\partial_\mu\psi + i(\partial_\mu\alpha)\psi + eA_\mu\psi + e\delta A_\mu\psi)$$

$$= e^{i\alpha}(\partial_\mu\psi + eA_\mu\psi) + e^{i\alpha}i(\partial_\mu\alpha)\psi + e^{i\alpha}(e\delta A_\mu)\psi$$

$$= e^{i\alpha}(\partial_\mu + eA_\mu)\psi + e^{i\alpha}e\left[\frac{i}{e}\partial_\mu\alpha + \delta A_\mu\right]\psi$$

$$D'_\mu \psi' = e^{i\alpha}D_\mu\psi + e^{i\alpha}e\left(\delta A_\mu + \frac{i}{e}\partial_\mu\alpha\right)\psi \tag{12.379}$$

Choose

$$\delta A_\mu = -\frac{i}{e}\partial_\mu\alpha \tag{12.380}$$

and so

$$D'_\mu \psi' = e^{i\alpha}D_\mu\psi \tag{12.381}$$

$$A'_\mu = A_\mu + \delta A_\mu = A_\mu - \frac{i}{e}\partial_\mu\alpha \tag{12.382}$$

Hence

$$\mathcal{L} = i\overline{\psi}\gamma^\mu D_\mu\psi - m\overline{\psi}\psi \tag{12.383}$$

$$\mathcal{L} \to \mathcal{L}' = i\overline{\psi'}\gamma^\mu D'_\mu\psi' - m\overline{\psi'}\psi'$$

$$= i(e^{-i\alpha}\overline{\psi})\gamma^\mu(e^{i\alpha}D_\mu\psi) - m(e^{-i\alpha}\overline{\psi})(e^{i\alpha}\psi)$$

$$\mathcal{L}' = i\overline{\psi}\gamma^\mu(D_\mu\psi) - m\overline{\psi}\psi = \mathcal{L} \tag{12.384}$$

So the Lagrangian is invariant under the local gauge transformation

$$\psi \to \psi' = e^{i\alpha(x)}\psi$$

$$A'_\mu = A_\mu - \frac{i}{e}\partial_\mu\alpha. \tag{12.385}$$

Clearly we have a new Lagrangian $\mathcal{L} = i\overline{\psi}\gamma^\mu(D_\mu\psi) - m\overline{\psi}\psi$ with $D_\mu = \partial_\mu + eA_\mu$ which has an additional term compared to the old Lagrangian $\mathcal{L} = i\overline{\psi}\gamma^\mu\partial_\mu\psi - m\overline{\psi}\psi$. But this new Lagrangian $\mathcal{L} = i\overline{\psi}\gamma^\mu(D_\mu\psi) - m\overline{\psi}\psi$ is invariant under the transformation shown in equation (12.385), i.e. along with the transformation of ψ we have to consider the transformation of A_μ also.

Equation of motion in $\overline{\psi}$ is

$$\frac{d}{dt}\left(\frac{\partial\mathcal{L}}{\partial\dot{\overline{\psi}}}\right) = \frac{\partial\mathcal{L}}{\partial\overline{\psi}} \tag{12.386}$$

From equation (12.383) we get

$$\mathcal{L} = i\overline{\psi}\gamma^\mu\big(\partial_\mu + eA_\mu\big)\psi - m\overline{\psi}\psi$$

$$= i\overline{\psi}\gamma^\mu\partial_\mu\psi + i\overline{\psi}\gamma^\mu eA_\mu\psi - m\overline{\psi}\psi$$

$$\frac{\partial\mathcal{L}}{\partial\overline{\psi}} = i\gamma^\mu\partial_\mu\psi + i\gamma^\mu eA_\mu\psi - m\psi \tag{12.387}$$

and $\frac{\partial\mathcal{L}}{\partial\dot{\overline{\psi}}} = 0$

Hence from equations (12.386) and (12.387) we have

$$i\gamma^\mu\partial_\mu\psi + i\gamma^\mu eA_\mu\psi - m\psi = 0$$

$$\big(i\gamma^\mu\partial_\mu\psi - m\psi\big) + i\gamma^\mu eA_\mu\psi = 0 \tag{12.388}$$

We note that the first two terms constitute the Dirac equation. And there is an extra term $i\gamma^\mu eA_\mu\psi$ which is an interaction term.

• Change of momentum for a free particle to an interacting particle is possible through the change. This is needed to introduce electromagnetic interaction (we have considered an extra factor of i).

$$i\partial_\mu \rightarrow i\partial_\mu + eA_\mu$$

$$\partial_\mu \rightarrow \partial_\mu + \frac{e}{i}A_\mu = \partial_\mu - ieA_\mu \tag{12.389}$$

So we now take

$$D_\mu = \partial_\mu - ieA_\mu \tag{12.390}$$

i.e. A_μ is multiplied by $-i$. In other words from equation (12.385) replacing A_μ by $-iA_\mu$ we have

$$(-iA'_\mu) = (-iA_\mu) - \frac{i}{e}\partial_\mu\alpha. \tag{12.391}$$

$$A'_\mu = A_\mu + \frac{1}{e}\partial_\mu\alpha. \tag{12.392}$$

This is the new A_μ and with this the Lagrangian is from equation (12.383)

$$\mathcal{L} = i\overline{\psi}\gamma^\mu\left(\partial_\mu - ieA_\mu\right)\psi - m\overline{\psi}\psi \tag{12.393}$$

$$= i\overline{\psi}\gamma^\mu\partial_\mu\psi - i\overline{\psi}\gamma^\mu ieA_\mu\psi - m\overline{\psi}\psi$$

$$\mathcal{L} = (i\overline{\psi}\gamma^\mu\partial_\mu\psi - m\overline{\psi}\psi\,) + \overline{\psi}\gamma^\mu eA_\mu\psi \tag{12.394}$$

This Lagrangian describes an electron or a charged particle represented by ψ interacting with the electromagnetic field represented by A_μ. The electromagnetic interaction can thus be accommodated.

Obviously if a Lagrangian is invariant under gauge transformation then electromagnetic interaction of the particles gets introduced, i.e. interaction of the particle can be understood through gauge transformation or the gauge symmetry of the Lagrangian.

The A_μ we have introduced appears only in the interaction term $\overline{\psi}\gamma^\mu eA_\mu\psi$. If we wish to interpret this as a particle like a photon then we have to explain how a photon propagates. If we consider a wave field then we have to ensure that Maxwell's equations emerge for the Lagrangian.

For this we introduce another term in the Lagrangian to have

$$\mathcal{L} = i\overline{\psi}\gamma^\mu\partial_\mu\psi - m\overline{\psi}\psi + \overline{\psi}\gamma^\mu eA_\mu\psi - \frac{1}{4}F_{\mu\nu}F^{\mu\nu} \tag{12.395}$$

We recast it as follows

$$\mathcal{L} = \left(i\overline{\psi}\gamma^\mu\partial_\mu\psi - m\overline{\psi}\psi\right) + \left(-\frac{1}{4}F_{\mu\nu}F^{\mu\nu} + \overline{\psi}\gamma^\mu e\psi A_\mu\right)$$

$$\mathcal{L} = \left(i\overline{\psi}\gamma^\mu\partial_\mu\psi - m\overline{\psi}\psi\right) + \left(-\frac{1}{4}F_{\mu\nu}F^{\mu\nu} - J^\mu A_\mu\right) \tag{12.396}$$

where

$$J^\mu = -\overline{\psi}\gamma^\mu e\psi \tag{12.397}$$

The first part of the Lagrangian gives rise to the Dirac equation for a free particle while the second part of the Lagrangian gives rise to Maxwell's equation with a photon interacting with the electromagnetic field. There is a free part for the ψ, a free part for the A_μ viz. $F_{\mu\nu}F^{\mu\nu}$ and there is an interaction term. This ensures an equation of motion corresponding to the photon as well as the equation of motion corresponding to the charge represented by $-e$. This is QED or quantum electrodynamics.

12.33 Spontaneous decay of a particle

Consider spontaneous decay of a particle a to product particles b and c as follows

$$a \rightarrow b + c$$

Conservation of mass energy gives ($E=$ total energy, $T=$ kinetic energy, $M=$ mass)

$$E_a = E_b + E_c \tag{12.398}$$

where

$$E_a = M_a c^2 + T_a, \quad E_b = M_b c^2 + T_b, \quad E_c = M_c c^2 + T_c \tag{12.399}$$

From equation (12.398)

$$M_a c^2 + T_a = (M_b c^2 + T_b) + (M_c c^2 + T_c) \tag{12.400}$$

Condition of spontaneous decay of M_a is

$$M_a > M_b + M_c$$

For spontaneous decay from rest $T_a = 0$. From equation (12.399)

$$E_a = M_a c^2$$

From equation (12.398)

$$M_a c^2 = E_b + E_c \tag{12.401}$$

And from equation (12.400) we have

$$M_a c^2 = (M_b + M_c)c^2 + T_b + T_c \tag{12.402}$$

We use a centre of mass frame and apply conservation of linear momentum

$$\vec{p}^{CM} = \vec{p}_b^{CM} + \vec{p}_c^{CM}$$

Since $\vec{p}^{CM} = 0$, $\vec{p}_b^{CM} + \vec{p}_c^{CM} = 0$

$$\vec{p}_b^{CM} = -\vec{p}_c^{CM} \tag{12.403}$$

Squaring (denoting $p_b^{CM} \equiv p_b$ and $p_c^{CM} \equiv p_c$)

$$p_b^2 c^2 = p_c^2 c^2 \tag{12.404}$$

Using the relativistic relations viz.

$$E_b^2 = p_b^2 c^2 + M_b^2 c^4, \quad E_c^2 = p_c^2 c^2 + M_c^2 c^4 \tag{12.405}$$

$$E_b^2 - M_b^2 c^4 = E_c^2 - M_c^2 c^4$$

$$E_b^2 - E_c^2 = M_b^2 c^4 - M_c^2 c^4$$

$$\left(E_b + E_c\right)\left(E_b - E_c\right) = \left(M_b^2 - M_c^2\right)c^4$$

Using equation (12.401) we have

$$M_a c^2 \left(E_b - E_c\right) = \left(m_b^2 - m_c^2\right)c^4$$

$$E_b - E_c = \frac{\left(M_b^2 - M_c^2\right)c^2}{M_a} \tag{12.406}$$

Equation (12.401) + equation (12.406) gives

$$2E_b = M_a c^2 + \frac{\left(M_b^2 - M_c^2\right)c^2}{M_a} = \frac{M_a^2 c^2 + \left(M_b^2 - M_c^2\right)c^2}{M_a}$$

$$E_b = \frac{(M_a^2 + M_b^2 - M_c^2)c^2}{2M_a} \tag{12.407}$$

Equation (12.401) − equation (12.406) gives

$$2E_c = M_a c^2 - \frac{\left(M_b^2 - M_c^2\right)c^2}{M_a} = \frac{M_a^2 c^2 - \left(M_b^2 - M_c^2\right)c^2}{M_a}$$

$$E_c = \frac{(M_a^2 - M_b^2 + M_c^2)c^2}{2M_a} \tag{12.408}$$

Equations (12.407) and (12.408) represent energy of decay products in terms of rest masses of the products.

Also, from equation (12.399) we have

$$E_b = M_b c^2 + T_b \quad \Rightarrow \quad T_b = E_b - M_b c^2$$

Using equation (12.407) we have

$$T_b = \frac{(M_a^2 + M_b^2 - M_c^2)c^2}{2M_a} - M_b c^2 = \frac{(M_a^2 + M_b^2 - M_c^2)c^2 - 2M_a M_b c^2}{2M_a} = \frac{(M_a - M_b)^2 - M_c^2}{2M_a}c^2$$

$$T_b = \frac{(M_a - M_b + M_c)(M_a - M_b - M_c)}{2M_a}c^2 \tag{12.409}$$

$$E_c = M_c c^2 + T_c \Rightarrow T_c = E_c - M_c c^2$$

Using equation (12.408) we have

$$T_c = \frac{(M_a^2 - M_b^2 + M_c^2)c^2}{2M_a} - M_c c^2 = \frac{(M_a - M_c)^2 - M_b^2}{2M_a}c^2 = \frac{(M_a - M_c + M_b)(M_a - M_c - M_b)}{2M_a}c^2 \tag{12.410}$$

So kinetic energy of decay products b, c has been obtained.
From equation (12.405) we have

$$p_b^2 c^2 = E_b^2 - M_b^2 c^4 = \left(\frac{\left(M_a^2 + M_b^2 - M_c^2\right)c^2}{2M_a}\right)^2 - M_b^2 c^4 = \frac{\left(M_a^2 + M_b^2 - M_c^2\right)^2 - 4M_a^2 M_b^2}{4M_a^2} c^4$$

$$p_b = \frac{\left(M_a^2 + M_b^2 - M_c^2\right)^2 - 4M_a^2 M_b^2}{4M_a^2} c^2 = p_c \,(\text{from equation } (12.404))$$

12.33.1 Example:

Consider the decay

$$\pi^0 \rightarrow \gamma + \gamma$$

i.e. comparing with $a \rightarrow b + c$ we can write

$$a \equiv \pi^0, \;\; b = c \equiv \gamma$$

$$M_a = M_{\pi^0} = 140\,MeV, \;\; M_b = M_c = M_\gamma = 0$$

Equation (12.407) gives

$$E_b = \frac{\left(M_a^2 + M_b^2 - M_c^2\right)c^2}{2M_a} \;\text{i. e.}\; E_\gamma = \frac{\left(M_{\pi^0}^2 + M_\gamma^2 - M_\gamma^2\right)c^2}{2M_{\pi^0}} = \frac{M_{\pi^0}c^2}{2} = \frac{140\;MeV}{2} = 70\;MeV \;\;(12.411)$$

This is the energy of photon γ in the decay of π^0 from rest.

12.34 Analysis of relativistic collisions through Mandelstam variables

In the special theory of relativity we have been working with the notation which ensures that

$$(ct)^2 - x^2 - y^2 - z^2 = \text{invariant}$$

Rewriting

$$(ct)^2 - x^2 - y^2 - z^2 = -[x^2 + y^2 + z^2 - (ct)^2] = -\left[x^2 + y^2 + z^2 + (ict)^2\right] = \text{invariant}$$

$$x^2 + y^2 + z^2 + (ict)^2 = \text{invariant}$$

Accordingly, in the following analysis we use a different notation of the special theory of relativity by taking $(x, y, z, ict) = (x_1, x_2, x_3, x_4)$ as the 4 space-time coordinates.

Position of a particle is represented by the 4-vector

$$x_\mu = (\vec{x}, x_4) = (\vec{x}, ict) = \left(\vec{x}, it\right) \;\left(\text{since in natural units } c = 1\right)$$

For relativistic collision we need to conserve energy and momentum, i.e. conserve 4-momentum defined as

$$p_\mu = (\vec{p}, p_4) = \left(\vec{p}, \frac{iE}{c}\right) = (\vec{p}, iE) \text{ (since in natural units } c = 1)$$

Square of 4-momentum is denoted by $p_\mu p^\mu \equiv p^2$ and is given by

$$p_\mu p^\mu \equiv p^2 = \vec{p}^2 + \left(\frac{iE}{c}\right)^2 = \vec{p}^2 - E^2 \text{ (since in natural units } c = 1) \quad (12.412)$$

Again

$$E^2 = \vec{p}^2 c^2 + m^2 c^4 \Rightarrow E^2 = \vec{p}^2 + m^2$$

$$\vec{p}^2 - E^2 = -m^2 \quad (12.413)$$

From equations (12.412) and (12.413)

$$p_\mu p^\mu = p^2 = -m^2 \quad (12.414)$$

Consider a 2-body to 2-body scattering process. Particles a and b approach each other as shown in figure 12.4 and get scattered so that we get particles c and d.

$$a + b \rightarrow c + d$$

4-momentum of particles a,b,c,d are p_a, p_b, p_c, p_d, respectively.
3-momentum of particles a,b,c,d are $\vec{p}_a$, $\vec{p}_b$, $\vec{p}_c$, $\vec{p}_d$, respectively.
Energy of particles a,b,c,d are E_a, E_b, E_c, E_d, respectively.
Conservation of 4-momentum gives

$$p_a + p_b = p_c + p_d \quad (12.415)$$

Conservation of 3-momentum gives

$$\vec{p}_a + \vec{p}_b = \vec{p}_c + \vec{p}_d$$

Conservation of energy gives

$$E_a + E_b = E_c + E_d$$

To study scattering of 2 particles, Mandelstam introduced the variables s,t,u defined as follows and called them Mandelstam variables. The Mandelstam

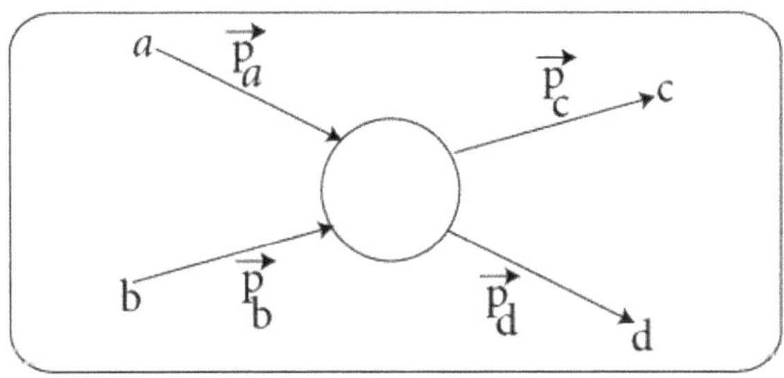

Figure 12.4. Two-body to two-body scattering process: $a + b \rightarrow c + d$.

variables are numerical quantities that encode energy, momentum and angle of particles involved in scattering.

$$s = -(p_a + p_b)^2 c^2 = -(p_c + p_d)^2 c^2$$

$$s = -(p_a + p_b)^2 = -(p_c + p_d)^2 \ \left(\text{since in natural units } c = 1\right) \qquad (12.416)$$

$$t = -(p_a - p_c)^2 c^2 = -(p_d - p_b)^2 c^2$$

$$t = -(p_a - p_c)^2 = -(p_d - p_b)^2 \ \left(\text{since in natural units } c = 1\right) \qquad (12.417)$$

$$u = -(p_a - p_d)^2 c^2 = -(p_c - p_b)^2 c^2$$

$$u = -(p_a - p_d)^2 = -(p_c - p_b)^2 \ \left(\text{since in natural units } c = 1\right) \qquad (12.418)$$

Let us calculate

$$s + t + u = -(p_a + p_b)^2 - (p_a - p_c)^2 - (p_a - p_d)^2$$

$$= -\left[(p_a + p_b)^2 + (p_a - p_c)^2 + (p_a - p_d)^2\right]$$

$$= -\left[p_a^2 + p_b^2 + 2p_a p_b + p_a^2 + p_c^2 - 2p_a p_c + p_a^2 + p_d^2 - 2p_a p_d\right]$$

$$= -\left[3p_a^2 + p_b^2 + p_c^2 + p_d^2 + 2p_a p_b - 2p_a p_c - 2p_a p_d\right]$$

$$= -\left[3p_a^2 + p_b^2 + p_c^2 + p_d^2 + 2p_a p_b - 2p_a\left(p_c + p_d\right)\right]$$

Using from equation (12.416) $p_a + p_b = p_c + p_d$ we get

$$s + t + u = -\left[3p_a^2 + p_b^2 + p_c^2 + p_d^2 + 2p_a p_b - 2p_a\left(p_a + p_b\right)\right]$$

$$= -\left[3p_a^2 + p_b^2 + p_c^2 + p_d^2 + 2p_a p_b - 2p_a^2 - 2p_a p_b\right]$$

$$= -\left(p_a^2 + p_b^2 + p_c^2 + p_d^2\right)$$

Using equation (12.414) viz. $p_\mu p^\mu = p^2 = -m^2$ we have

$$s + t + u = -(-m_a^2 - m_b^2 - m_c^2 - m_d^2)$$

$$s + t + u = m_a^2 + m_b^2 + m_c^2 + m_d^2 \qquad (12.419)$$

✓ When all masses are equal

$$m_a = m_b = m_c = m_d = m$$

$$s + t + u = 4m^2 \qquad (12.420)$$

- Consider from equation (12.416)

$$s = -(p_a + p_b)^2 = -\left(p_a^2 + p_b^2 + 2p_a p_b\right)$$

Using equation (12.414) viz. $p_\mu p^\mu = p^2 = -m^2$ we have

$$s = -\left(p_a^2 + p_b^2 + 2p_a p_b\right) = -\left(-m_a^2 - m_b^2 + 2p_a p_b\right)$$

$$s = m_a^2 + m_b^2 - 2p_a p_b$$

$$2p_a p_b = m_a^2 + m_b^2 - s \tag{12.421}$$

- Consider from equation (12.416)

$$s = -(p_c + p_d)^2 = -\left(p_c^2 + p_d^2 + 2p_c p_d\right)$$

Using equation (12.414) viz. $p_\mu p^\mu = p^2 = -m^2$ we have

$$s = -\left(p_c^2 + p_d^2 + 2p_c p_d\right) = -\left(-m_c^2 - m_d^2 + 2p_c p_d\right)$$

$$s = m_c^2 + m_d^2 - 2p_c p_d$$

$$2p_c p_d = m_c^2 + m_d^2 - s \tag{12.422}$$

- Consider from equation (12.417)

$$t = -(p_a - p_c)^2 = -\left(p_a^2 + p_c^2 - 2p_a p_c\right)$$

Using equation (12.414) viz. $p_\mu p^\mu = p^2 = -m^2$ we have

$$t = -\left(p_a^2 + p_c^2 - 2p_a p_c\right) = -\left(-m_a^2 - m_c^2 - 2p_a p_c\right)$$

$$t = m_a^2 + m_c^2 + 2p_a p_c \tag{12.423}$$

$$2p_a p_c = t - m_a^2 - m_c^2 \tag{12.424}$$

- Consider from equation (12.417)

$$t = -(p_d - p_b)^2 = -\left(p_d^2 + p_b^2 - 2p_d p_b\right)$$

Using equation (12.414) viz. $p_\mu p^\mu = p^2 = -m^2$ we have

$$t = -\left(p_d^2 + p_b^2 - 2p_d p_b\right) = -\left(-m_d^2 - m_b^2 - 2p_d p_b\right)$$

$$t = m_d^2 + m_b^2 + 2p_d p_b$$

$$2p_d p_b = t - m_d^2 - m_b^2 \tag{12.425}$$

- Consider from equation (12.418)

$$u = -(p_a - p_d)^2 = -\left(p_a^2 + p_d^2 - 2p_a p_d\right)$$

Using equation (12.414) viz. $p_\mu p^\mu = p^2 = -m^2$ we have

$$u = -\left(p_a^2 + p_d^2 - 2p_a p_d\right) = -\left(-m_a^2 - m_d^2 - 2p_a p_d\right)$$

$$u = m_a^2 + m_d^2 + 2p_a p_d$$

$$2p_a p_d = u - m_a^2 - m_d^2 \tag{12.426}$$

- Consider from equation (12.418)

$$u = -(p_c - p_b)^2 = -\left(p_c^2 + p_b^2 - 2p_c p_b\right)$$

Using equation (12.414) viz. $p_\mu p^\mu = p^2 = -m^2$ we have

$$u = -\left(p_c^2 + p_b^2 - 2p_c p_b\right) = -\left(-m_c^2 - m_b^2 - 2p_c p_b\right)$$

$$u = m_c^2 + m_b^2 + 2p_c p_b$$

$$2p_c p_b = u - m_c^2 - m_b^2 \tag{12.427}$$

Equations (12.421) to (12.427) relate 4-momentum, Mandelstam variables and mass of particles.

Example
✓ Møller scattering
e^--e^- interacts and gets scattered to produce an e^--e^- pair as shown in figure 12.5 (a). This is called Møller scattering.

$$e^- \text{-} e^- \rightarrow e^- \text{-} e^-$$

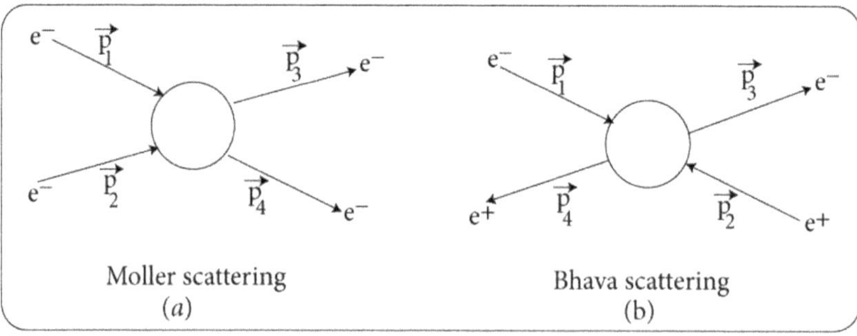

Figure 12.5. Møller scattering and Bhava scattering.

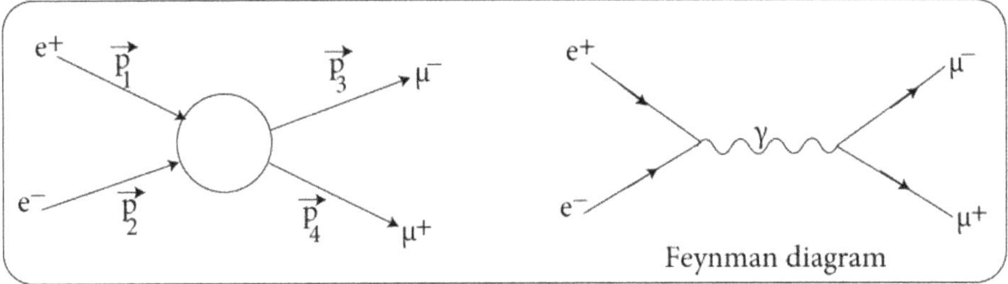

Figure 12.6. $e^--e^+ \to \mu^--\mu^+$ interaction and its Feynman diagram.

It is fermion–fermion interaction between two similar mass particles and so equation (12.420), i.e. $s + t + u = 4m_e^2$ holds.

✓ Bhava scattering

e^--e^+ interacts and get scattered to produce an e^--e^+ pair, as shown in figure 12.5(b). This is called Bhava scattering.

$$e^--e^+ \to e^--e^+$$

It is fermion–anti-fermion interaction between two similar mass particles and so equation (12.420), i.e. $s + t + u = 4m_e^2$ holds.

✓e^--e^+ interacts and produces $\mu^--\mu^+$ pair, as shown in figure 12.6, along with its Feynman diagram. The process is mediated through photon γ exchange.

$$e^--e^+ \xrightarrow{\gamma} \mu^- - \mu^+$$

So equation (12.420) becomes

$$s + t + u = m_e^2 + m_e^2 + m_\mu^2 + m_\mu^2 = 2m_e^2 + 2m_\mu^2$$

As $m_\mu = 207m_e$, i.e. $m_e = 9.1 \times 10^{-31}kg$, $m_\mu = 1.88 \times 10^{-28}kg$ we neglect m_e to get

$$s + t + u = 2m_\mu^2$$

✓Electron–positron annihilation is represented as

$$e^-e^+ \to \gamma\gamma$$

So equation (12.420) becomes

$$s + t + u = m_e^2 + m_e^2 + m_\gamma^2 + m_\gamma^2 = 2m_e^2 \; \left(\text{as } m_\gamma = 0\right)$$

12.35 Two-body scattering in a centre of mass frame

Consider interaction of 2 particles in a centre of mass frame. Let 2 particles a, b approach each other, as shown in figure 12.7. After interaction, particles c, d emerge out and move away from each other.

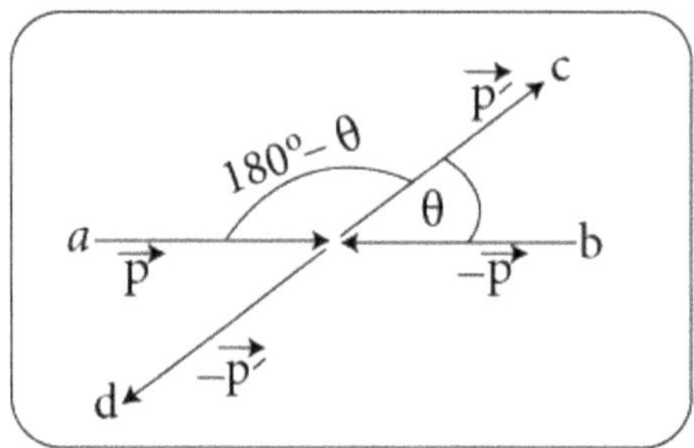

Figure 12.7. Interaction of 2 particles a, b leads to 2 other particles c, d. Analysis done in a centre of mass frame.

$a, b \rightarrow$ incident particles with 3 momenta $\vec{p}$, $-\vec{p}$ and 4-momenta $P_a = (\vec{p}, \frac{iE_a}{c}) = (\vec{p}, iE_a)$, $P_b = (-\vec{p}, \frac{iE_b}{c}) = (-\vec{p}, iE_b)$ respectively (since in natural units $c = 1$).

$c, d \rightarrow$ outgoing particles with 3-momenta $\vec{p'}$, $-\vec{p'}$ and 4-momenta $P_c = (\vec{p'}, \frac{iE_c}{c}) = (\vec{p'}, iE_c)$, $P_d = (-\vec{p'}, \frac{iE_d}{c}) = (-\vec{p'}, iE_d)$, respectively (since in natural units $c = 1$).

$\theta \rightarrow$ angle between $\vec{p}$ and $\vec{p'}$

We will find out the angle of scattering θ in a centre of mass frame of reference. Mandelstam variables are

$$s = -(P_a + P_b)^2 = (P_c + P_d)^2$$

We get from $s = -(P_a + P_b)^2$

$$s = -[(\vec{p} + iE_a) + (-\vec{p} + iE_b)]^2 = -(iE_a + iE_b)^2 = (E_a + E_b)^2$$

$$E_a + E_b = \sqrt{s} \tag{12.428}$$

Similarly, we get from $s = (P_c + P_d)^2$ the result $s = (E_c + E_d)^2$
From equation (12.423) we have

$$t = m_a^2 + m_c^2 + 2P_a P_c$$

$$= m_a^2 + m_b^2 + 2(\vec{p}, iE_a)(\vec{p'}, iE_c)$$

$$= m_a^2 + m_b^2 + 2(\vec{p} \cdot \vec{p'} + iE_a \cdot iE_c) = m_a^2 + m_b^2 + 2(pp'\cos\theta - E_a E_c)$$

From equation (12.413) we write

$$\vec{p}^2 - E^2 = -m^2 \Rightarrow m^2 = E^2 - \vec{p}^2$$

So for particles a and b we have

$$m_a^2 = E_a^2 - \vec{p}^2 \tag{12.429}$$

$$m_b^2 = E_b^2 - |-\vec{p}|^2 = E_b^2 - \vec{p}^2$$

Hence

$$m_a^2 - m_b^2 = \left(E_a^2 - \vec{p}^2\right) - \left(E_b^2 - \vec{p}^2\right) = E_a^2 - E_b^2$$

$$= (E_a + E_b)(E_a - E_b) = \sqrt{s}\,(E_a - E_b)$$

$$E_a - E_b = \frac{m_a^2 - m_b^2}{\sqrt{s}} \tag{12.430}$$

Equation (12.428) + equation (12.430) gives

$$2E_a = \sqrt{s} + \frac{m_a^2 - m_b^2}{\sqrt{s}}$$

$$E_a = \frac{s + m_a^2 - m_b^2}{2\sqrt{s}} \tag{12.431}$$

Equation (12.428) – equation (12.430) gives

$$2E_b = \sqrt{s} - \frac{m_a^2 - m_b^2}{\sqrt{s}}$$

$$E_b = \frac{s + m_b^2 - m_a^2}{2\sqrt{s}} \tag{12.432}$$

Similarly,

$$E_c = \frac{s + m_c^2 - m_d^2}{2\sqrt{s}} \tag{12.433}$$

$$E_d = \frac{s + m_d^2 - m_c^2}{2\sqrt{s}} \tag{12.434}$$

Let us calculate the scattering angle. We consider the product

$$p_a p_c = \left(\vec{p},\ iE_a\right)\left(\vec{p}',\ iE_c\right)$$

$$= \vec{p} \cdot \vec{p}' + iE_a \cdot iE_c = pp' \cos\theta - E_a E_c$$

$$\cos\theta = \frac{p_a p_c + E_a E_c}{pp'}$$

From equation (12.424) we have

$$2p_a p_c = t - m_a^2 - m_c^2$$

$$P_a P_c = \frac{t - m_a^2 - m_c^2}{2} \tag{12.435}$$

From equation (12.429) we have

$$m_a^2 = E_a^2 - \vec{p}^2 \Rightarrow \vec{p}^2 = E_a^2 - m_a^2$$

Using equation (12.431) we get $E_a = \dfrac{s + m_a^2 - m_b^2}{2\sqrt{s}}$

$$\vec{p}^2 = E_a^2 - m_a^2 = \left(\frac{s + m_a^2 - m_b^2}{2\sqrt{s}} \right)^2 - m_a^2$$

$$= \frac{\left(s + m_a^2 - m_b^2\right)^2 - 4sm_a^2}{4s}$$

$$\vec{p}^2 = \frac{s^2 + m_a^4 + m_b^4 + 2sm_a^2 - 2sm_b^2 - 2m_a^2 m_b^2 - 4sm_a^2}{4s}$$

$$= \frac{s^2 + m_a^4 + m_b^4 - 2sm_a^2 - 2sm_b^2 - 2m_a^2 m_b^2}{4s}$$

We introduce the Kallen function or triangle function which is a polynomial function in 3 variables defined as

$$\lambda(x, y, z) = x^2 + y^2 + z^2 - 2xy - 2yz - 2zx$$

Hence $\lambda(s, m_a^2, m_c^2) = s^2 + (m_a^2)^2 + (m_b^2)^2 - 2sm_a^2 - 2m_a^2 m_b^2 - 2sm_b^2$

$$\vec{p}^2 = \frac{\lambda\left(s, m_a^2, m_c^2\right)}{4s}$$

$$|\vec{p}| = \frac{\sqrt{\lambda\left(s, m_a^2, m_b^2\right)}}{2\sqrt{s}} \tag{12.436}$$

Similarly

$$|\vec{p'}| = \frac{\sqrt{\lambda\left(s, m_c^2, m_d^2\right)}}{2\sqrt{s}} \tag{12.437}$$

Using equations (12.431), (12.430), (12.435), (12.436), and (12.437) we get

$$\cos\theta = \frac{P_a P_c + E_a E_c}{pp'} = \frac{\left(\dfrac{s + m_a^2 - m_b^2}{2\sqrt{s}}\right)\left(\dfrac{s + m_c^2 - m_d^2}{2\sqrt{s}}\right) + \left(\dfrac{t - m_a^2 - m_c^2}{2}\right)}{\dfrac{\sqrt{\lambda\left(s, m_a^2, m_b^2\right)}}{2\sqrt{s}} \dfrac{\sqrt{\lambda\left(s, m_c^2, m_d^2\right)}}{2\sqrt{s}}}$$

$$\cos \theta = \frac{\left(s + m_a^2 - m_b^2\right)\left(s + m_c^2 - m_d^2\right) + 2s\left(t - m_a^2 - m_c^2\right)}{\sqrt{\lambda\left(s, m_a^2, m_b^2\right)}\sqrt{\lambda\left(s, m_c^2, m_d^2\right)}} \qquad (12.438)$$

Now the numerator on the RHS of equation (12.438) is

$$\left(s + m_a^2 - m_b^2\right)\left(s + m_c^2 - m_d^2\right) + 2s\left(t - m_a^2 - m_c^2\right)$$

$$= s\left(s + m_c^2 - m_d^2\right) + m_a^2\left(s + m_c^2 - m_d^2\right) - m_b^2\left(s + m_c^2 - m_d^2\right) + 2s\left(t - m_a^2 - m_c^2\right)$$

$$= s^2 + sm_c^2 - sm_d^2 + sm_a^2 + m_a^2 m_c^2 - m_a^2 m_d^2 - sm_b^2 - m_b^2 m_c^2 + m_b^2 m_d^2 + 2st$$

$$- 2sm_a^2 - 2sm_c^2$$

$$= s^2 - sm_c^2 - sm_d^2 - sm_a^2 - sm_b^2 + 2st + m_a^2 m_c^2 - m_a^2 m_d^2 - m_b^2 m_c^2 + m_b^2 m_d^2$$

$$= s\left[s - \left(m_c^2 + m_d^2 + m_a^2 + m_b^2\right) + 2t\right] + m_a^2(m_c^2 - m_d^2) - m_b^2(m_c^2 - m_d^2)$$

$$= s[s - (s + t + u) + 2t] + (m_a^2 - m_b^2)(m_c^2 - m_d^2) \text{ (using equation (12.419))}$$

$$= s(t - u) + (m_a^2 - m_b^2)(m_c^2 - m_d^2)$$

Hence equation (12.438) gives

$$\cos \theta = \frac{s(t - u) + (m_a^2 - m_b^2)(m_c^2 - m_d^2)}{\sqrt{\lambda\left(s, m_a^2, m_b^2\right)}\sqrt{\lambda\left(s, m_c^2, m_d^2\right)}}$$

• For elastic scattering

$$a \equiv c, \quad b \equiv d$$

$$E_a = E_c, \quad E_b = E_d$$

$$m_a = m_c, \quad m_b = m_d$$

$$|\vec{p}| = |\vec{p'}| \text{ i. e. } p = p'$$

Now equation (12.423) gives

$$t = m_a^2 + m_c^2 + 2p_a p_c$$

$$= m_a^2 + m_c^2 + 2(\vec{p}, iE_a)(\vec{p'}, iE_c) = m_a^2 + m_c^2 + 2\left[\vec{p}\cdot\vec{p'} + , iE_a\, iE_c\right]$$

$$= m_a^2 + m_c^2 + 2pp' \cos \theta - 2E_a E_c$$

$$t = 2m_a^2 + 2p^2 \cos \theta - 2E_a^2 = 2\left(m_a^2 - E_a^2\right) + 2p^2 \cos \theta$$

Using equation (12.429)

$$t = 2(-p^2) + 2p^2 \cos \theta = -2p^2(1 - \cos \theta) \qquad (12.439)$$

$$1 - \cos \theta = -\frac{t}{2p^2}$$

$$\cos \theta = 1 + \frac{t}{2p^2}$$

This gives the angle of scattering in the centre of mass frame of reference. Also from equation (12.439)

$$t = -2p^2 \left(2 \sin^2 \frac{\theta}{2} \right) = -4p^2 \sin^2 \frac{\theta}{2}$$

Similarly, we can show that

$$t = -4p^2 \cos^2 \frac{\theta}{2}$$

12.36 Two-body scattering in a laboratory frame of reference

A centre of mass frame of reference is convenient to make a theoretical study of the scattering process. But it is convenient to use a laboratory frame of reference to perform an actual scattering experiment.

Let a be a projectile that is made to strike a target b, as shown in figure 12.8. c and d are the scattered particles.

Projectile a is moving with 3-momentum $\vec{p}_{aL}$ while target b is at rest so that $\vec{p}_{bL} = 0$.

$\vec{p}_{cL}$, $\vec{p}_{dL}$ are 3-momenta of c, d, respectively, in laboratory frame.

θ_L is the angle of scattering in a laboratory frame of reference and we wish to find its value.

The 4-momenta of the interacting particles a, b, c, d are, respectively, as follows (in natural units).

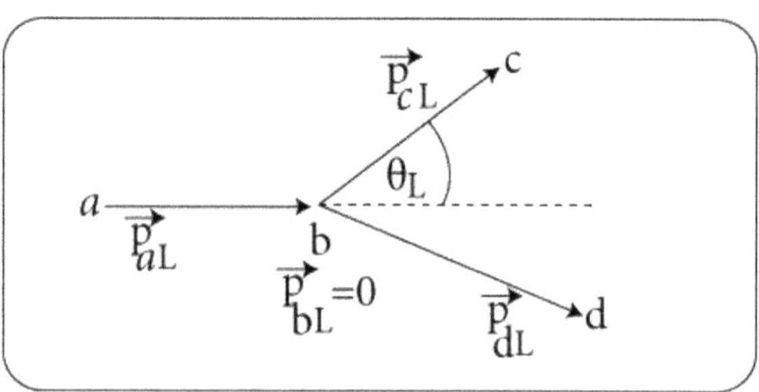

Figure 12.8. Two-body scattering in laboratory frame of reference.

$$p_a = (\vec{p}_{aL}, \quad iE_{aL})$$

$$p_b = (\vec{p}_{bL}, \quad iE_{bL}) = (0, \quad iE_{bL}) \tag{12.440}$$

$$p_c = (\vec{p}_{cL}, \quad iE_{cL})$$

$$p_d = (\vec{p}_{dL}, \quad iE_{dL})$$

We also note that equation (12.413) $\vec{p}^2 - E^2 = -m^2$ for particle 2 becomes

$$\vec{p}_{bL}^2 - E_{bL}^2 = -m_b^2 \text{ i. e. } 0 - E_{bL}^2 = -m_b^2$$

$$E_{bL} = m_b \tag{12.441}$$

In view of this we rewrite equation (12.440) as

$$p_b = (\vec{p}_{bL}, \quad iE_{bL}) = (0, \quad im_b)$$

Let us now evaluate the Mandelstam variables

$$s = -(p_a + p_b)^2 = -\left(p_a^2 + p_b^2 + 2p_a p_b\right)$$

$$= -\left[(\vec{p}_{aL}, \quad iE_{aL})^2 + (0, \quad im_b)^{\,2} + 2(\vec{p}_{aL}, \quad iE_{aL})(0, \quad im_b)\right]$$

$$= -[\vec{p}_{aL}^{\,2} - E_{aL}^2 - m_b^2 + 2\left(\vec{p}_{aL} \cdot 0 \; - E_{aL}m_b\right)]$$

Using equation (12.413)

$$\vec{p}_{aL}^{\,2} - E_{aL}^2 = -m_a^2 \tag{12.442}$$

$$s = -\left[-m_a^2 - m_b^2 - 2E_{aL}m_b\right] = m_a^2 + m_b^2 + 2E_{aL}m_b$$

$$E_{aL} = \frac{s - m_a^2 - m_b^2}{2m_b} \tag{12.443}$$

Consider

$$t = -(p_a - p_c)^2 = -\left(p_a^2 + p_c^2 - 2p_a p_c\right)$$

$$= -\left[(\vec{p}_{aL}, \quad iE_{aL})^2 + (\vec{p}_{cL}, \quad iE_{cL})^{\,2} - 2(\vec{p}_{aL}, \quad iE_{aL})(\vec{p}_{cL}, \quad iE_{cL})\right]$$

$$= -[\vec{p}_{aL}^{\,2} - E_{aL}^2 + \vec{p}_{cL}^{\,2} - E_{cL}^2 - 2\left(\vec{p}_{aL} \cdot \vec{p}_{cL} \; - E_{aL}E_{cL}\right)]$$

Using equation (12.413) $\vec{p}_{aL}^{\,2} - E_{aL}^2 = -m_a^2, \; \vec{p}_{cL}^{\,2} - E_{cL}^2 = -m_c^2$

$$t = -\left[-m_a^2 - m_c^2 - 2\,|\vec{p}_{aL}||\vec{p}_{cL}|\cos\theta_L + 2E_{aL}E_{cL}\right]$$

$$t = m_a^2 + m_c^2 + 2 \, |\vec{p}_{aL}||\vec{p}_{cL}| cos\theta_L - 2E_{aL}E_{cL} \Big]$$

Also

$$t = -(p_d - p_b)^2 = -\left(p_d^2 + p_b^2 - 2p_d p_b\right)$$

$$= -\left[(\vec{p}_{dL}, \; iE_{dL})^2 + (0, \; m_b)^2 - 2(\vec{p}_{dL}, \; iE_{dL})(0, \; m_b) \; \right]$$

$$= -[\vec{p}_{dL}^2 - E_{dL}^2 - m_b^2 - 2(\vec{p}_{aL} \cdot 0 \; - E_{dL}m_b)]$$

Using equation (12.413) $\vec{p}_{dL}^2 - E_{dL}^2 = -m_d^2$

$$t = -[-m_d^2 - m_b^2 + 2E_{dL}m_b]$$

$$t = m_d^2 + m_b^2 - 2E_{dL}m_b$$

$$E_{dL} = \frac{m_d^2 + m_b^2 - t}{2m_b}$$

$$u = -(p_c - p_b)^2 = -\left(p_c^2 + p_b^2 - 2p_c p_b\right)$$

$$= -\left[(\vec{p}_{cL}, \; iE_{cL})^2 + (0, \; im_b)^2 - 2(\vec{p}_{cL}, \; iE_{cL})(0, \; im_b) \; \right]$$

$$= -[\vec{p}_{cL}^2 - E_{cL}^2 - m_b^2 - 2(\vec{p}_{cL} \cdot 0 \; - E_{cL}m_b)]$$

Using equation (12.413) $\vec{p}_{cL}^2 - E_{cL}^2 = -m_c^2$

$$u = -[-m_c^2 - m_b^2 + 2E_{cL}m_b]$$

$$u = m_c^2 + m_b^2 - 2E_{cL}m_b$$

$$E_{cL} = \frac{m_c^2 + m_b^2 - u}{2m_b} \tag{12.444}$$

Again consider equation (12.442) $\vec{p}_{aL}^2 - E_{aL}^2 = -m_a^2$

$$\vec{p}_{aL}^2 = E_{aL}^2 - m_a^2$$

Using equation (12.443)

$$\vec{p}_{aL}^2 = \left(\frac{s - m_a^2 - m_b^2}{2m_b}\right)^2 - m_a^2 = \frac{\left(s - m_a^2 - m_b^2\right)^2 - 4m_a^2 m_b^2}{4m_b^2}$$

$$\vec{p}_{aL}^2 = \frac{s^2 + m_a^4 + m_b^4 - 2sm_a^2 - 2sm_b^2 + 2m_a^2 m_b^2 - 4m_a^2 m_b^2}{4m_b^2}$$

$$\vec{p}_{aL}^{\;2} = \frac{s^2 + m_a^4 + m_b^4 - 2sm_a^2 - 2sm_b^2 - 2m_a^2 m_b^2}{4m_b^2} = \frac{\lambda(s,\, m_a^2,\, m_b^2)}{4m_b^2}$$

where $\lambda(s,\, m_a^2,\, m_b^2)$ represents the Kallen function.

$$|\vec{p}_{aL}| = \frac{\sqrt{\lambda(s,\, m_a^2,\, m_b^2)}}{2m_b} \tag{12.445}$$

Similarly, we can show that

$$|\vec{p}_{cL}| = \frac{\sqrt{\lambda(u,\, m_b^2,\, m_c^2)}}{2m_b} \tag{12.446}$$

$$|\vec{p}_{dL}| = \frac{\sqrt{\lambda(t,\, m_b^2,\, m_d^2)}}{2m_b}$$

To find the scattering angle let us consider the product

$$p_a p_c = (\vec{p}_{aL},\; iE_{aL})(\vec{p}_{cL},\; iE_{cL}) = \vec{p}_{aL} \cdot \vec{p}_{cL} - E_{aL}E_{cL}$$

$$p_a p_c = |\vec{p}_{aL}||\vec{p}_{cL}|\cos\theta_L - E_{aL}E_{cL}$$

$$|\vec{p}_{aL}||\vec{p}_{cL}|\cos\theta_L = p_a p_c + E_{aL}E_{cL}$$

$$\cos\theta_L = \frac{p_a p_c + E_{aL}E_{cL}}{|\vec{p}_{aL}||\vec{p}_{cL}|} \tag{12.447}$$

From equation (12.424)

$$2p_a p_c = t - m_a^2 - m_c^2$$

$$p_a p_c = \frac{t - m_a^2 - m_c^2}{2} \tag{12.448}$$

Hence using equations (12.431), (12.432), (12.433), (12.434), and (12.435) in equation (12.436) we get

$$\cos\theta_L = \frac{p_a p_c + E_{aL}E_{cL}}{|\vec{p}_{aL}||\vec{p}_{cL}|} = \frac{\dfrac{t - m_a^2 - m_c^2}{2} + \left(\dfrac{s - m_a^2 - m_b^2}{2m_b}\right)\left(\dfrac{m_c^2 + m_b^2 - u}{2m_b}\right)}{\dfrac{\sqrt{\lambda(s, m_a^2, m_b^2)}}{2m_b}\;\dfrac{\sqrt{\lambda(u, m_b^2, m_c^2)}}{2m_b}}$$

$$\cos\theta_L = \frac{p_a p_c + E_{aL}E_{cL}}{|\vec{p}_{aL}||\vec{p}_{cL}|} = \frac{2m_b^2\left(t - m_a^2 - m_c^2\right) + \left(s - m_a^2 - m_b^2\right)(m_c^2 + m_b^2 - u)}{\sqrt{\lambda(s, m_a^2, m_b^2)}\;\sqrt{\lambda(u, m_b^2, m_c^2)}}$$

This is the required expression of scattering angle in the laboratory frame of reference.

12.37 Relation between scattering angle in centre of mass frame θ and scattering angle in laboratory frame θ_L (through Lorentz transformation)

Lorentz transformation relations of the special theory of relativity are

$$p_x = \gamma\left(p_x' + \beta E'\right) \tag{12.449}$$

$$p_y = p_y' \tag{12.450}$$

$$E = \gamma\left(E' + \beta p_x'\right) \tag{12.451}$$

Let us assume that the centre of mass frame is moving relative to the laboratory frame with velocity v. And we define
$\beta = \dfrac{v}{c} = v$ (in natural unit $c = 1$) and

$$\gamma = \frac{1}{\sqrt{1 - \dfrac{v^2}{c^2}}}$$

So we rewrite equations (12.449), (12.450), and (12.451) as

$$p_x = \gamma\left(p_x' + vE'\right) \tag{12.452}$$

$$p_y = p_y' \tag{12.453}$$

$$E = \gamma\left(E' + vp_x'\right) \tag{12.454}$$

Clearly the centre of mass frame is the moving primed frame and the laboratory frame is the unprimed frame.

In the centre of mass frame (figure 12.7) the x component and y component of momentum of a scattered particle c i.e. $\vec{p}_{cL}$ are, respectively,

$$p_x = p_{cL} \cos \theta_L$$

$$p_y = p_{cL} \sin \theta_L$$

$$\text{Also } E' = E_c$$

Consider the x and y component of momentum of scattered particle c in the 2 frames.

In the laboratory frame (figure 12.8) the x component and y component of momentum of scattered particle c i.e. $\vec{p'}$ are, respectively,

$$p'_x = p' \cos \theta$$

$$p'_y = p' \sin \theta$$

We apply these relations in the Lorentz transformation relations (12.452) and (12.453) to get

$$p_x = \gamma\left(p'_x + vE'\right) \Rightarrow p_{cL} \cos \theta_L = \gamma\left(p' \cos \theta + vE_c\right) \tag{12.455}$$

$$p_y = p'_y \Rightarrow p_{cL} \sin \theta_L = p' \sin \theta \tag{12.456}$$

Let us take the ratio of equations (12.455) and (12.456)

$$\frac{p_{cL} \sin \theta_L}{p_{cL} \cos \theta_L} = \frac{p' \sin \theta}{\gamma(p' \cos \theta + vE_c)}$$

$$\tan \theta_L = \frac{p' \sin \theta}{\gamma(p' \cos \theta + vE_c)}$$

This is the required relationship between θ and θ_L.

12.38 Exercises

Exercise 12.1 *Establish that* $a_k a_k = 0$ *and* $a_k^\dagger a_k^\dagger = 0$ *for a system of fermions.*
$\boxed{Ans}$ Fermions obey the following anti-commutation relation (equation (12.153))

$$\{a_k, \ a_l\} = \{ a_k^\dagger, \ a_l^\dagger\} = 0 \tag{12.457}$$

$$a_k a_l + a_l a_k = 0$$

$$a_k a_l = -a_l a_k$$

$$\text{For } k = l$$

$$a_k a_k = -a_k a_k$$

$$2 a_k a_k = 0$$

$$a_k a_k = 0$$

Equation (12.153) gives

$$a_k^\dagger a_l^\dagger + a_l^\dagger \, a_k^\dagger = 0$$

$$a_k^\dagger a_l^\dagger = -a_l^\dagger \, a_k^\dagger$$

For $k = l$

$$a_k^\dagger a_k^\dagger = -a_k^\dagger \, a_k^\dagger$$

$$2\ a_k^\dagger a_k^\dagger = 0$$

$$a_k^\dagger a_k^\dagger = 0$$

Exercise 12.2 *Derive the Euler–Lagrange equation* $\frac{\partial \mathcal{L}}{\partial \psi} - \partial_\mu \frac{\partial \mathcal{L}}{\partial(\partial_\mu \psi)} = 0$ *in terms of Lagrangian density from the principle of least action.*

$\boxed{\text{Ans.}}$ The principle of least action is

$$\delta S = 0 \tag{12.458}$$

where action S is given by

$$S = \int_{t_f}^{t_f} dt\, L = \int_{t_f}^{t_f} dt\, d^3x \mathcal{L} \quad (d^3x = dV)$$

$$S = \int_{t_f}^{t_f} d^4x \mathcal{L} \quad (dt\, d^3x = d^4x,\ x = t, x^1, x^2, x^3) \tag{12.459}$$

Now $\mathcal{L} = \mathcal{L}(\psi(x), \frac{\partial \psi}{\partial x^\mu}) = \mathcal{L}(\ \psi(x), \partial_\mu \psi(x)\ )$

Variation in $\mathcal{L}$ is

$$\delta \mathcal{L} = \frac{\partial \mathcal{L}}{\partial \psi(x)} \delta \psi(x) + \frac{\partial \mathcal{L}}{\partial\big(\partial_\mu \psi(x)\big)} \delta(\partial_\mu \psi(x))$$

Using $\delta(\partial_\mu \psi(x)) = \partial_\mu(\delta \psi(x))$

$$\delta \mathcal{L} = \frac{\partial \mathcal{L}}{\partial \psi(x)} \delta \psi(x) + \frac{\partial \mathcal{L}}{\partial\big(\partial_\mu \psi(x)\big)} \partial_\mu(\delta \psi(x)) \tag{12.460}$$

Since

$$\partial_\mu\left(\frac{\partial \mathcal{L}}{\partial(\partial_\mu \psi(x))} \delta \psi(x) \right) = \frac{\partial \mathcal{L}}{\partial(\partial_\mu \psi(x))} \partial_\mu\left(\delta \psi(x) + \partial_\mu\ \frac{\partial \mathcal{L}}{\partial(\partial_\mu \psi(x))}(\delta \psi(x)\ \right)$$

The second term on the RHS of equation (12.460) is

$$\frac{\partial \mathcal{L}}{\partial(\partial_\mu \psi(x))} \partial_\mu(\delta \psi(x)) = \partial_\mu\left(\frac{\partial \mathcal{L}}{\partial(\partial_\mu \psi(x))} \delta \psi(x) \right) - \partial_\mu\ \frac{\partial \mathcal{L}}{\partial(\partial_\mu \psi(x))}(\delta \psi(x)\ \right)$$

So equation (12.460) becomes

$$\delta \mathcal{L} = \frac{\partial \mathcal{L}}{\partial \psi(x)} \delta \psi(x) + \partial_\mu\left(\frac{\partial \mathcal{L}}{\partial(\partial_\mu \psi(x))}(\delta \psi(x)\ ) - \partial_\mu\ \frac{\partial \mathcal{L}}{\partial(\partial_\mu \psi(x))}(\delta \psi(x)\ \right) \tag{12.461}$$

Using equations (12.458) and (12.459)

$$\delta S = \int_{t_f}^{t_f} d^4x \; \delta\mathcal{L}(\; \psi(x),\; \partial_\mu\psi(x)\;) = 0$$

With equation (12.461)

$$\int_{t_f}^{t_f} d^4x \left[\frac{\partial\mathcal{L}}{\partial\psi(x)}\delta\psi(x) + \partial_\mu\left(\frac{\partial\mathcal{L}}{\partial(\partial_\mu\psi(x))}\delta\psi(x) \right) - \partial_\mu\; \frac{\partial\mathcal{L}}{\partial(\partial_\mu\psi(x))}(\delta\psi(x)\;) \right] = 0$$

$$\int_{t_f}^{t_f} d^4x \left[\frac{\partial\mathcal{L}}{\partial\psi(x)}\delta\psi(x) - \partial_\mu\; \frac{\partial\mathcal{L}}{\partial(\partial_\mu\psi(x))}(\delta\psi(x)\;) \right] + \int_{t_f}^{t_f} d^4x\; \partial_\mu\left(\frac{\partial\mathcal{L}}{\partial(\partial_\mu\psi(x))}\delta\psi(x) \right) = 0 \quad (12.462)$$

The second term on the LHS of equation (12.462) is

$$\int_{t_f}^{t_f} d^4x \left[\partial_\mu\left(\frac{\partial\mathcal{L}}{\partial(\partial_\mu\psi(x))}\delta\psi(x) \right) \right] = \int d^3x\; \partial_\mu\left(\frac{\partial\mathcal{L}}{\partial(\partial_\mu\psi(x))}\delta\psi(x) \right)\Big|_{t_i}^{t_f}$$

Demand that variation at end points are zero, i.e. (figure 12.9)

$$\delta\psi(x,\, t_i) = \delta\psi(x,\, t_f) = 0$$

The second term on the LHS of equation (12.462) vanishes. So equation (12.462) gives

$$\int_{t_f}^{t_f} d^4x \left[\frac{\partial\mathcal{L}}{\partial\psi(x)}\delta\psi(x) - \partial_\mu\; \frac{\partial\mathcal{L}}{\partial(\partial_\mu\psi(x))}(\delta\psi(x)\;) \right] = 0$$

$$\int_{t_f}^{t_f} d^4x \left[\frac{\partial\mathcal{L}}{\partial\psi(x)} - \partial_\mu\; \frac{\partial\mathcal{L}}{\partial(\partial_\mu\psi(x))} \right]\delta\psi(x) = 0$$

For arbitrary variation

$$\frac{\partial\mathcal{L}}{\partial\psi(x)} - \partial_\mu\; \frac{\partial\mathcal{L}}{\partial(\partial_\mu\psi(x))} = 0$$

This is the Euler–Lagrange equation in terms of Lagrangian density.

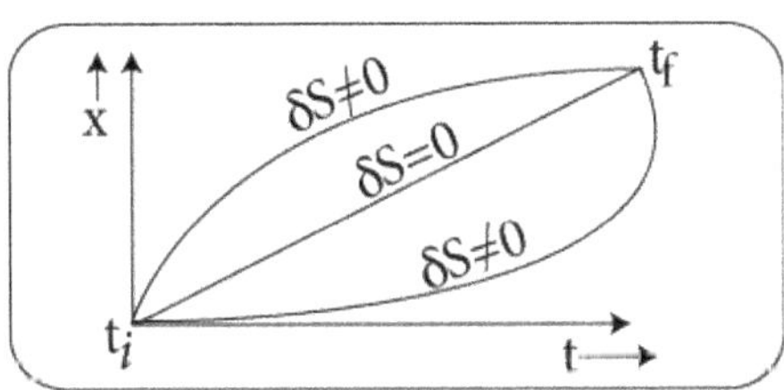

Figure 12.9. Any movement of a particle corresponds to zero variation of action.

Exercise 12.3 *Show that $\widetilde{F}^{\mu\nu} = \frac{1}{2}\varepsilon^{\mu\nu\rho\sigma}F_{\rho\sigma}$ leads to a tensor that is obtained from $F^{\mu\nu}$ changing $(E_x, E_y, \ E_z) \to (B_x, B_y, B_z)$ and $(B_x, B_y, B_z) \to (-E_x, -E_y, \ -E_z)$ where $\varepsilon^{\mu\nu\rho\sigma}$ is fully a anti-symmetric tensor.*

$\boxed{\text{Ans.}}$ Consider $\widetilde{F}^{\mu\nu} = \frac{1}{2}\varepsilon^{\mu\nu\rho\sigma}F_{\rho\sigma}$ where $\varepsilon^{\mu\nu\rho\sigma}$ is fully anti-symmetric, i.e. interchange of any 2 indices gives a negative sign and all cyclic permutation gives $+1$ as evident from the following example.

$$\varepsilon^{0123} = +1, \quad \varepsilon^{1023} = -\varepsilon^{0123} = -1, \quad \varepsilon^{0213} = -\varepsilon^{0123} = -1$$

Any repeat of indices gives zero, e.g. $\varepsilon^{0012} = 0$ and so

$$\widetilde{F}^{\mu\nu} = \frac{1}{2}\varepsilon^{\mu\nu\rho\sigma}F_{\rho\sigma} = 0, \ \mu = \nu$$

Now

$$\widetilde{F}^{01} = \frac{1}{2}\varepsilon^{01\rho\sigma}F_{\rho\sigma} = \frac{1}{2}\left(\varepsilon^{0123}F_{23} + \varepsilon^{0132}F_{32}\right) = \frac{1}{2}\left(\varepsilon^{0123}F_{23} + (-\varepsilon^{0123})(-F_{23})\right)$$

$$= \frac{1}{2}2\varepsilon^{0123}F_{23} = \varepsilon^{0123}F_{23} = (+1)(-B_x) = -B_1 = -B_x$$

Similarly

$$\widetilde{F}^{02} = -B_2 = -B_y, \quad \widetilde{F}^{03} = -B_3 = -B_z$$

$$\widetilde{F}^{12} = \frac{1}{2}\varepsilon^{12\rho\sigma}F_{\rho\sigma} = \frac{1}{2}(\varepsilon^{1203}F_{03} + \varepsilon^{1230}F_{30}) = \frac{1}{2}\left(\varepsilon^{1203}F_{03} + (-\varepsilon^{1203})(-F_{03})\right)$$

$$= \frac{1}{2}2\varepsilon^{1203}F_{03} = \varepsilon^{1203}F_{03} = -\varepsilon^{1023}F_{03} = \varepsilon^{0123}F_{03}(+1)(E_3) = \frac{E_3}{c} = \frac{E_z}{c}$$

Similarly

$$\widetilde{F}^{13} = -\frac{E_2}{c} = -\frac{E_y}{c}, \quad \widetilde{F}^{23} = -\frac{E_1}{c} = -\frac{E_x}{c}$$

$$\widetilde{F}^{\mu\nu} = \begin{pmatrix} 0 & -B_1 & -B_2 & -B_3 \\ B_1 & 0 & E_3/c & -E_2/c \\ B_2 & -E_3/c & 0 & E_1/c \\ B_3 & E_2/c & -E_1/c & 0 \end{pmatrix}, \quad F^{\mu\nu} = \begin{pmatrix} 0 & -E_1/c & -E_2/c & -E_3/c \\ E_1/c & 0 & -B_3 & B_2 \\ E_2/c & B_3 & 0 & -B_1 \\ E_3/c & -B_2 & B_1 & 0 \end{pmatrix}$$

Clearly $\widetilde{F}^{\mu\nu}$ can be obtained from $F^{\mu\nu}$ by changing $(E_x, E_y, \ E_z) \to (B_x, B_y, B_z)$ and $(B_x, B_y, B_z) \to (-E_x, -E_y, \ -E_z)$ or vice versa.

Exercise 12.4 *Find an expression for the Lagrangian of a charged particle in an electromagnetic field. Hence find the Hamiltonian.*

Ans. Force acting on a charge q moving with velocity $\vec{v}$ in an electric field $\vec{E}$ and magnetic field $\vec{B}$ is given by

$$\vec{F} = q\left(\vec{E} + \vec{v} \times \vec{B}\right)$$

Using

$$\vec{E} = -\vec{\nabla}\phi - \frac{\partial \vec{A}}{\partial t} \text{ and } \vec{B} = \vec{\nabla} \times \vec{A}$$

We have

$$\vec{F} = q\left(-\vec{\nabla}\phi - \frac{\partial \vec{A}}{\partial t} + \vec{v} \times \vec{\nabla} \times \vec{A}\right)$$

Consider the x component

$$F_x = q\left(-\frac{\partial \phi}{\partial x} - \frac{\partial A_x}{\partial t} + (\vec{v} \times \vec{\nabla} \times \vec{A})_x\right) \tag{12.463}$$

$$\text{Now } \vec{v} \times \vec{\nabla} \times \vec{A} = \begin{vmatrix} \hat{i} & \hat{j} & \hat{k} \\ v_x & v_y & v_z \\ (\vec{\nabla} \times \vec{A})_x & (\vec{\nabla} \times \vec{A})_y & (\vec{\nabla} \times \vec{A})_z \end{vmatrix}$$

$$(\vec{v} \times \vec{\nabla} \times \vec{A})_x = v_y(\vec{\nabla} \times \vec{A})_z - v_z(\vec{\nabla} \times \vec{A})_y = v_y\left(\frac{\partial A_y}{\partial x} - \frac{\partial A_x}{\partial y}\right) - v_z\left(\frac{\partial A_x}{\partial z} - \frac{\partial A_z}{\partial x}\right)$$

$$= v_y\frac{\partial A_y}{\partial x} + v_z\frac{\partial A_z}{\partial x} - \left(v_y\frac{\partial A_x}{\partial y} + v_z\frac{\partial A_x}{\partial z}\right)$$

Adding and subtracting the term $v_x\frac{\partial A_x}{\partial x}$ we have

$$(\vec{v} \times \vec{\nabla} \times \vec{A})_x = v_x\frac{\partial A_x}{\partial x} + v_y\frac{\partial A_y}{\partial x} + v_z\frac{\partial A_z}{\partial x} - \left(v_x\frac{\partial A_x}{\partial x} + v_y\frac{\partial A_x}{\partial y} + v_z\frac{\partial A_x}{\partial z}\right) \tag{12.464}$$

$$(\vec{v} \times \vec{\nabla} \times \vec{A})_x = \vec{v}.\vec{\nabla} - \left(v_x\frac{\partial A_x}{\partial x} + v_y\frac{\partial A_x}{\partial y} + v_z\frac{\partial A_x}{\partial z}\right)$$

Again

$$\frac{dA_x}{dt} = \frac{\partial A_x}{\partial x}\frac{dx}{dt} + \frac{\partial A_x}{\partial y}\frac{dy}{dt} + \frac{\partial A_x}{\partial z}\frac{dz}{dt} + \frac{\partial A_x}{\partial t}\frac{dt}{dt} = \frac{\partial A_x}{\partial x}v_x + \frac{\partial A_x}{\partial y}v_y + \frac{\partial A_x}{\partial z}v_z + \frac{\partial A_x}{\partial t}$$

$$\frac{dA_x}{dt} - \frac{\partial A_x}{\partial t} = v_x\frac{\partial A_x}{\partial x} + v_y\frac{\partial A_x}{\partial y} + v_z\frac{\partial A_x}{\partial z} \tag{12.465}$$

Also consider

$$\frac{\partial}{\partial x}(\vec{v}.\,\vec{A}) = \frac{\partial}{\partial x}(v_x A_x + v_y A_y + v_z A_z) = v_x \frac{\partial A_x}{\partial x} + v_y \frac{\partial A_y}{\partial x} + v_z \frac{\partial A_z}{\partial x} \qquad (12.466)$$

Putting equations (12.465) and (12.466) in equation (12.464)

$$(\vec{v} \times \vec{\nabla} \times \vec{A})_x = \frac{\partial}{\partial x}(\vec{v}.\,\vec{A}) - \left(\frac{dA_x}{dx} - \frac{\partial A_x}{\partial t}\right) \qquad (12.467)$$

Putting equation (12.467) in equation (12.463)

$$F_x = q\left(-\frac{\partial \phi}{\partial x} - \frac{\partial A_x}{\partial t} + \frac{\partial}{\partial x}(\vec{v}.\,\vec{A}) - \frac{dA_x}{dt} + \frac{\partial A_x}{\partial t}\right)$$

$$F_x = q\left(-\frac{\partial}{\partial x}(\phi - \vec{v}.\,\vec{A}) - \frac{dA_x}{dt}\right) \qquad (12.468)$$

Again

$$\frac{\partial}{\partial v_x}(\vec{v}.\,\vec{A}) = \frac{\partial}{\partial v_x}(v_x A_x + v_y A_y + v_z A_z) = A_x \qquad (12.469)$$

Hence

$$-\frac{d}{dt}A_x = \frac{d}{dt}\frac{\partial}{\partial v_x}(\vec{v}.\,\vec{A}) \text{ (using equation (12.469))}$$

Also, ϕ is independent of v_x. So we can wite

$$-\frac{d}{dt}A_x = \frac{d}{dt}\frac{\partial}{\partial v_x}(\phi - \vec{v}.\,\vec{A})$$

From equation (12.463)

$$F_x = q\left(-\frac{\partial}{\partial x}(\phi - \vec{v}.\,\vec{A}) + \frac{d}{dt}\frac{\partial}{\partial v_x}(\phi - \vec{v}.\,\vec{A})\right)$$

Define the generalized potential as

$$U = q(\phi - \vec{v}.\,\vec{A})$$

$$F_x = -\frac{\partial}{\partial x}U + \frac{d}{dt}\frac{\partial}{\partial v_x}U \qquad (12.470)$$

Lagrangian equation of motion is

$$\frac{d}{dt}\frac{\partial T}{\partial v_x} - \frac{\partial T}{\partial x} = F_x \qquad (12.471)$$

Equating equations (12.470) and (12.471) we have

$$\frac{d}{dt}\frac{\partial T}{\partial v_x} - \frac{\partial T}{\partial x} = -\frac{\partial}{\partial x}U + \frac{d}{dt}\frac{\partial}{\partial v_x}U$$

$$\frac{d}{dt}\frac{\partial}{\partial v_x}(T - U) - \frac{\partial}{\partial x}(T - U) = 0$$

$$\frac{d}{dt}\frac{\partial}{\partial v_x}L - \frac{\partial}{\partial x}L = 0$$

where the Lagrangian of a charged particle in an electromagnetic field is

$$L = T - U = T - q\phi + q\vec{v}.\,\vec{A} = \frac{1}{2}mv^2 - q\phi + q\vec{v}.\,\vec{A} \tag{12.472}$$

$$L = \sum_k \frac{1}{2}mv_k^2 - q\phi + q\sum_k v_k A_k \tag{12.473}$$

Exercise 12.5 *Check if $e^{i\alpha}$ is a member of the $U(1)$ group.*

[Ans.] Let us check if $e^{i\alpha}$ is a member of the $U(1)$ group.
• Existence of an identity element.
For $= 0$, $e^{i\alpha} = 1$

$$\psi \to \psi' = e^{i\alpha}\psi = \psi$$

is an identity transformation.
• Existence of an inverse
If $e^{i\alpha}$ is an element of U so that $e^{i\alpha}e^{-i\alpha} = 1$ (identity)
• Closure property
If $e^{i\alpha}$, $e^{i\beta}$ are two elements of $U(1)$ then the product
$e^{i\alpha}e^{i\beta} = e^{i(\alpha+\beta)}$ is also an element of the group $U(1)$.
• Associative property

$$e^{i\alpha}\,(e^{i\beta}e^{i\gamma}) = (e^{i\alpha}\,e^{i\beta})e^{i\gamma}$$

As α takes continuous values $U(1)$ is a continuous group. U for unitary and 1 denotes that there is only one parameter. Such a continuous group is a Lie group.

Exercise 12.6 *Consider the decay $\pi^- \to \mu^- + \bar{\nu}_\mu$*

  (a) *Show that the energy carried by the neutrino is given by $E_\nu = \frac{1}{2}(m_\pi - \frac{m_\mu^2}{m_\pi})$.*
  (b) *Find the energy and kinetic energy carried by the muon.*
  (c) *Use this result to calculate the energy of the γ particle in the decay $\pi^0 \to 2\gamma$.*

[Ans.] (a) We consider the decay $\pi^- \to \mu^- + \bar{\nu}_\mu$
Using 4-momentum conservation

$$(p_{\mu})^{\pi} = (p_{\mu})^{\mu} + (p_{\mu})^{\nu}$$

$$(p_{\mu})^{\pi}(p^{\mu})^{\pi} = [\ (p_{\mu})^{\mu} + (p_{\mu})^{\nu}\ ][\ (p^{\mu})^{\mu} + (p^{\mu})^{\nu}\ ]$$

$$=(p_{\mu})^{\mu}\ (p^{\mu})^{\mu} + (p_{\mu})^{\nu}(p^{\mu})^{\nu} + (p_{\mu})^{\mu}(p^{\mu})^{\nu} + (p_{\mu})^{\nu}\ (p^{\mu})^{\mu}$$

$$(p_{\mu}p^{\mu})^{\pi} = (p_{\mu}p^{\mu})^{\mu} + (p_{\mu}p^{\mu})^{\nu} + (p_{\mu})^{\mu}(p^{\mu})^{\nu} + (p_{\mu})^{\nu}\ (p^{\mu})^{\mu}$$

Consider

$$(p_{\mu}p^{\mu})^{\pi} = (p_0 p^0)^{\pi} + (p_i p^i)^{\pi} = \frac{E_{\pi}}{c}\frac{E_{\pi}}{c} + (-\vec{p}_{\pi}).\vec{p}_{\pi} = \frac{E_{\pi}^2}{c^2} - p_{\pi}^2 = \frac{E_{\pi}^2 - p_{\pi}^2 c^2}{c^2} = \frac{m_{\pi}^2 c^4}{c^2} = m_{\pi}^2 c^2$$

$$(p_{\mu}p^{\mu})^{\mu} = m_{\mu}^2 c^2 \ \text{(similarly)}$$

$$(p_{\mu}p^{\mu})^{\nu} = m_{\nu}^2 c^2 \ \text{(similarly)}$$

Consider the last term on the RHS

$$(p_{\mu})^{\nu}\ (p^{\mu})^{\mu} = (p_0)^{\nu}\ (p^0)^{\mu} + (p_i)^{\nu}\ (p^i)^{\mu} = (p^0)^{\nu}\ (p_0)^{\mu} + (-p^i)^{\nu}\ (-p_i)^{\mu}$$

$$=(p^0)^{\nu}\ (p_0)^{\mu} + (p^i)^{\nu}\ (p_i)^{\mu} = (p_0)^{\mu}(p^0)^{\nu} + (p_i)^{\mu}(p^i)^{\nu}$$

$$=(p_{\mu})^{\mu}(p^{\mu})^{\nu} = \ \text{3rd term on RHS}$$

$$m_{\pi}^2 c^2 = m_{\mu}^2 c^2 + m_{\nu}^2 c^2 + 2(p_{\mu})^{\mu}(p^{\mu})^{\nu}$$

Again, the 3rd term on the RHS is

$$2(p_{\mu})^{\mu}(p^{\mu})^{\nu} = 2(p_0)^{\mu}(p^0)^{\nu} + 2(p_i)^{\mu}(p^i)^{\nu} = \left(\frac{E_{\mu}}{c}\right)\left(\frac{E_{\nu}}{c}\right) + 2\left(-\vec{p}_{\mu}\right)(\vec{p}_{\nu})$$

$$=2\frac{E_{\mu}E_{\nu}}{c^2} - 2\vec{p}_{\mu} \cdot \vec{p}_{\nu}$$

$$m_{\pi}^2 c^2 = m_{\mu}^2 c^2 + m_{\nu}^2 c^2 + 2\frac{E_{\mu}E_{\nu}}{c^2} - 2\vec{p}_{\mu} \cdot \vec{p}_{\nu}$$

In natural units

$$m_{\pi}^2 = m_{\mu}^2 + m_{\nu}^2 + 2E_{\mu}E_{\nu} - 2\vec{p}_{\mu} \cdot \vec{p}_{\nu} \tag{12.474}$$

In the centre of mass rest frame

$$\vec{p}_{\pi} = \vec{p}_{\mu} + \vec{p}_{\nu} = 0 \ \text{i. e.} \ \vec{p}_{\mu} = -\vec{p}_{\nu} \tag{12.475}$$

$$\vec{p}_{\mu} \cdot \vec{p}_{\nu} = -\vec{p}_{\nu} \cdot \vec{p}_{\nu} = -p_{\nu}^2 \tag{12.476}$$

Also, it follows from equation (12.475) that

$$p_\mu^2 = p_\nu^2 \qquad (12.477)$$

Now

$$E_\nu^2 = p_\nu^2 + m_\nu^2 \text{ (in natural units)}$$

$$\text{Since } m_\nu \cong 0 \qquad (12.478)$$

$$E_\nu^2 = p_\nu^2, \quad p_\nu = E_\nu \qquad (12.479)$$

$$\vec{p}_\mu \cdot \vec{p}_\nu = -E_\nu^2 \qquad (12.480)$$

Again

$$E_\mu^2 = p_\mu^2 + m_\mu^2 \text{ (in natural units)}$$

$$E_\mu = \sqrt{p_\nu^2 + m_\mu^2} = \sqrt{E_\nu^2 + m_\mu^2} \qquad (12.481)$$

From equations (12.474), (12.478), (12.480), and (12.481) we have

$$m_\pi^2 = m_\mu^2 + 2\sqrt{E_\nu^2 + m_\mu^2}\,E_\nu + 2E_\nu^2$$

$$\frac{m_\pi^2 - m_\mu^2}{2} = \sqrt{E_\nu^2 + m_\mu^2}\,E_\nu + E_\nu^2$$

$$\frac{m_\pi^2 - m_\mu^2}{2} - E_\nu^2 = E_\nu\sqrt{E_\nu^2 + m_\mu^2}$$

Squaring

$$\left(\frac{m_\pi^2 - m_\mu^2}{2}\right)^2 + E_\nu^4 - 2E_\nu^2\frac{m_\pi^2 - m_\mu^2}{2} - E_\nu^2(E_\nu^2 + m_\mu^2)$$

$$\frac{1}{4}\left(m_\pi^2 - m_\mu^2\right)^2 - E_\nu^2\left(m_\pi^2 - m_\mu^2\right) = E_\nu^2 m_\mu^2$$

$$\frac{1}{4}\left(m_\pi^2 - m_\mu^2\right)^2 = E_\nu^2 m_\pi^2$$

$$E_\nu^2 = \frac{1}{4}\frac{\left(m_\pi^2 - m_\mu^2\right)^2}{m_\pi^2}$$

$$E_\nu = \frac{m_\pi^2 - m_\mu^2}{2m_\pi} = \frac{1}{2}\left(m_\pi - \frac{m_\mu^2}{m_\pi}\right). \tag{12.482}$$

b) From (12.481) and (12.482) we have

$$E_\mu^2 = E_\nu^2 + m_\mu^2 = \left(\frac{m_\pi^2 - m_\mu^2}{2m_\pi}\right)^2 + m_\mu^2 = \frac{\left(m_\pi^2 - m_\mu^2\right)^2 + 4m_\pi^2 m_\mu^2}{4m_\pi^2} = \frac{\left(m_\pi^2 + m_\mu^2\right)^2}{4m_\pi^2}$$

$$E_\mu = \frac{m_\pi^2 + m_\mu^2}{2m_\pi}$$

$$E_\mu = T_\mu + m_\mu$$

$$T_\mu = E_\mu - m_\mu = \frac{m_\pi^2 + m_\mu^2}{2m_\pi} - m_\mu = \frac{m_\pi^2 + m_\mu^2 - 2m_\pi m_\mu}{2m_\pi} = \frac{(m_\pi - m_\mu)^2}{2m_\pi}$$

c) Consider the decay process $\pi^0 \to \gamma + \gamma$

For this decay process we have similar to equation (12.482) for $(\pi^- \to \mu^- + \bar{\nu}_\mu)$

$$E_\gamma = \frac{m_\pi^2 - m_\gamma^2}{2m_\pi}$$

As $m_\gamma = 0$ we have

$$E_\gamma = \frac{m_\pi^2 - 0}{2m_\pi} = \frac{m_\pi}{2}$$

Exercise 12.7 *Explain why a neutral pion can always decay into two photons whereas a photon of sufficient energy can decay into an electron–positron pair only in the presence of matter.*

Ans. Let us view the decays from the centre of mass frame of reference.

✓ In the decay $\pi^0 \to \gamma + \gamma$ energy and 4-momentum p_μ are both conserved and so this decay is always allowed. We note that there can be decay from rest and then $\vec{p}_{\pi^0}^{\,cm} = 0$. So the γ shoot off in opposite directions with momentum $\vec{p}_\gamma^{\,cm} = p_\gamma \hat{x}, \ -p_\gamma \hat{x}$. From equation (12.411) $E_\gamma = 70\ MeV$.

✓ In the decay $\gamma \to e^- + e^+$ energy is conserved but 4-momentum p_μ is not conserved. So this decay is not allowed. A γ of energy $E_\gamma = E_{e^-} + E_{e^+} = 0.51\ MeV + 0.51\ MeV = 1.02\ MeV$ is energetically allowed to decay into a $e^\pm$ pair. But since rest mass of γ, i.e. $m_\gamma = 0$ it follows that $\vec{p}_\gamma^{\,cm} \neq 0$. This means a photon has no rest frame. So momentum conservation does not hold.

✓ In the decay $\gamma + N \to e^- + e^+ + N$, where $N =$ nucleus energy and 4-momentum p_μ are both conserved. So this decay is allowed. A γ of energy $E_\gamma = E_{e^-} + E_{e^+} = 0.51\ MeV + 0.51\ MeV = 1.02\ MeV$ is energetically allowed to

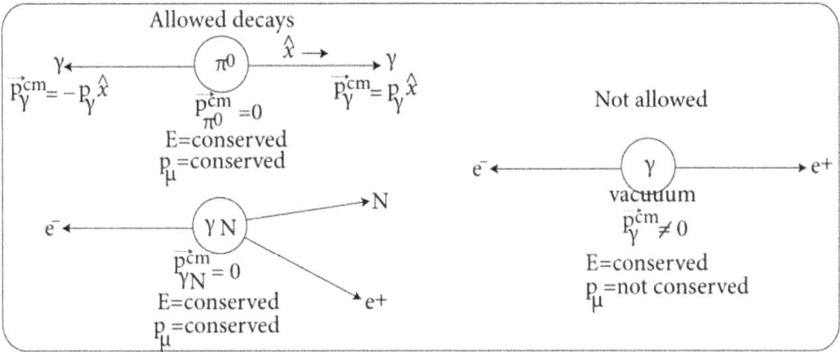

Figure 12.10. A neutral pion can always decay into two photons, whereas a photon of sufficient energy cannot decay into an electron–positron pair in a vacuum.

decay into a $e^\pm$ pair. Also, $\vec{p}_{\gamma N}^{cm} = 0$. This means e^+, e^- shoots off and the nucleus N recoils in such a way as to conserve momentum. These possibilities are shown in figure 12.10.

Exercise 12.8 *Consider the decay*

$$\Sigma^+ \rightarrow n + \pi^+$$

Determine the kinetic energy of the neutron and pion.
Given $M_\Sigma = 1189.36\ MeV$, $M_n = 939.57\ MeV$, $M_\pi = 139.57\ MeV$
Ans. From equation (12.409) we have

$$T_b = \frac{(M_a - M_b + M_c)(M_a - M_b - M_c)}{2M_a}c^2 \Rightarrow T_n = \frac{(M_\Sigma - M_n + M_\pi)(M_\Sigma - M_n - M_\pi)}{2M_\Sigma}c^2$$

$$T_n = \frac{(1189.36 - 939.57 + 139.57)(1189.36 - 939.57 - 139.57)}{2(1189.36)}MeV = 18\ MeV$$

From equation (12.410) we have

$$T_c = \frac{(M_a - M_c + M_b)(M_a - M_c - M_b)}{2M_a}c^2 \Rightarrow T_\pi = \frac{(M_\Sigma - M_\pi + M_n)(M_\Sigma - M_\pi - M_n)}{2M_\Sigma}c^2$$

$$T_\pi = \frac{(1189.36 - 139.57 + 939.57)(1189.36 - 139.57 - 939.57)}{2(1189.36)}MeV = 92 MeV$$

Exercise 12.9 *Find the maximum kinetic energy of e^+ in μ^+ decay when e^+ is emitted opposite to neutrinos with equal energy.*
Ans. μ^+ decay: $\mu^+ \rightarrow e^+ + \bar{\nu}_\mu + \nu_e$
Energy conservation leads to $E_\mu = E_e + E_{\nu_\mu} + E_{\nu_e} = E_e + 2E_\nu$

$$m_\mu c^2 + T_\mu = m_e c^2 + T_e + 2E_\nu$$

$$\left(m_\mu - m_e\right)c^2 = T_\mu + T_e + 2E_\nu$$

For decay of μ^+ from rest $T_\mu = 0$

$$\left(m_\mu - m_e\right)c^2 = T_e + 2E_\nu \tag{12.483}$$

Consider the situation when e^+ is emitted opposite to $\bar{\nu}_\mu$, ν_e. Then

$$\vec{p_e} = \vec{p}_{\bar{\nu}_\mu} + \vec{p}_{\nu_e} = 2\,\vec{p_\nu} \;\Rightarrow\; p_e = 2p_\nu = \frac{2E_\nu}{c}$$

$$p_e c = 2E_\nu \tag{12.484}$$

Again

$$E_e = T_e + m_e c^2 = \sqrt{p_e^2 c^2 + m_e^2 c^4}$$

Squaring

$$(T_e + m_e c^2)^2 = p_e^2 c^2 + m_e^2 c^4 \Rightarrow T_e^2 + m_e^2 c^4 + 2T_e m_e c^2 = p_e^2 c^2 + m_e^2 c^4$$

$$T_e^2 + 2T_e m_e c^2 = p_e^2 c^2$$

$$p_e c = \sqrt{T_e^2 + 2T_e m_e c^2} \tag{12.485}$$

From equations (12.484) and (12.485)

$$2E_\nu = \sqrt{T_e^2 + 2T_e m_e c^2} \tag{12.486}$$

From equations (12.483) and (12.486)

$$(m_\mu - m_e)c^2 = T_e + \sqrt{T_e^2 + 2T_e m_e c^2}$$

$$(m_\mu - m_e)c^2 - T_e = \sqrt{T_e^2 + 2T_e m_e c^2}$$

$$((m_\mu - m_e)c^2 - T_e)^2 = T_e^2 + 2T_e m_e c^2$$

$$(m_\mu - m_e)^2 c^4 + T_e^2 - 2T_e(m_\mu - m_e)c^2 = T_e^2 + 2T_e m_e c^2$$

$$(m_\mu - m_e)^2 c^4 - 2T_e m_\mu c^2 = 0$$

$$T_e = \frac{(m_\mu - m_e)^2 c^2}{2m_\mu c^2} = \frac{(m_\mu - m_e)^2 c^2}{2m_\mu} = \frac{(105.66 - 0.51)^2}{2(105.66)} MeV = 52\ MeV$$

Exercise 12.10 *Consider production of a π^+, π^- in the field of a stationary proton p as $\gamma + p \rightarrow \pi^+ + \pi^- + p$. Find the threshold energy of a γ photon.*
[Ans.] Consider the process

$$\gamma + p \rightarrow \pi^+ + \pi^- + p$$

In the laboratory frame.
Total initial energy is $E = E_\gamma + E_p = h\nu + m_p c^2 + \underset{=0}{T_p} = h\nu + m_p c^2$

Total momentum is $p = \dfrac{h\nu}{c}$
The invariant quantity is (the same in all inertial frames)

$$E^2 - p^2 c^2 = (h\nu + m_p c^2)^2 - \left(\frac{h\nu}{c}\right)^2 c^2$$

$$= (h\nu + m_p c^2)^2 - (h\nu)^2 = (h\nu + m_p c^2 + h\nu)(h\nu + m_p c^2 - h\nu) = (2h\nu + m_p c^2)m_p c^2$$

$$(12.487)$$

In the centre of mass frame.
Total energy of products at threshold is $E' = 2m_\pi c^2 + m_p c^2$.
Total momentum is zero $p' = 0$.
The invariant quantity is (the same in all inertial frames)

$$E'^2 - p'^2 c^2 = (2m_\pi c^2 + m_p c^2)^2 - 0 = (2m_\pi c^2 + m_p c^2)^2 \qquad (12.488)$$

Equating from equations (12.487) and (12.488)

$$(2h\nu + m_p c^2)m_p c^2 = (2m_\pi c^2 + m_p c^2)^2$$

$$2h\nu m_p c^2 + m_p^2 c^4 = 4m_\pi^2 c^4 + m_p^2 c^4 + 4m_\pi m_p c^4$$

$$2h\nu m_p c^2 = 4m_\pi^2 c^4 + 4m_\pi m_p c^4$$

$$h\nu = \frac{4m_\pi^2 c^4 + 4m_\pi m_p c^4}{2m_p c^2} = 2\frac{m_\pi^2 + m_\pi m_p}{m_p}c^2$$

Exercise 12.11 *Consider the production of an electron–positron pair in the field of a stationary electron as $\gamma + e^- \rightarrow e^- + e^+ + e^-$. Find the threshold energy of a γ photon.*
[Ans.] Consider the process

$$\gamma + e^- \rightarrow e^- + e^+ + e^-$$

In the laboratory frame.
Total initial energy is $E = E_\gamma + E_{e^-} = h\nu + m_e c^2 + \underset{=0}{T_e} = h\nu + m_e c^2$

Total momentum is $p = \dfrac{h\nu}{c}$
The invariant quantity is (the same in all inertial frames)

$$E^2 - p^2 c^2 = (h\nu + m_e c^2)^2 - \left(\frac{h\nu}{c}\right)^2 c^2 = (h\nu + m_e c^2)^2 - (h\nu)^2 \tag{12.489}$$

$$= (h\nu + m_e c^2 + h\nu)(h\nu + m_e c^2 - h\nu) = (2h\nu + m_e c^2)m_e c^2$$

In the centre of mass frame.
Total energy of products at threshold is $E' = 3m_e c^2$
Total momentum is zero $p' = 0$.
The invariant quantity is (the same in all inertial frames)

$$E'^2 - p'^2 c^2 = (3m_e c^2)^2 - 0 = 9(m_e c^2)^2 \tag{12.490}$$

Equating from equations (12.487) and (12.488)

$$(2h\nu + m_e c^2)m_e c^2 = 9(m_e c^2)^2$$

$$2h\nu + m_e c^2 = 9m_e c^2$$

$$2h\nu = 8m_e c^2$$

$$h\nu = 4m_e c^2.$$

Exercise 12.12 *Consider the reaction $a + X \to Y + b$ where target X is at rest in the laboratory frame. Find the minimum energy (threshold energy) required for the projectile a to initiate reaction.*

$\boxed{\text{Ans.}}$ Consider the reaction $a + X \to Y + b$
where $M_{\text{LHS}} < M_{\text{RHS}}$.
In the laboratory frame (target at rest, $\vec{p}_X = 0$)
The 4-momentum for the incident channel is

$$p_a^\mu + p_X^\mu = \left(\frac{E_a}{c},\ \vec{p}_a\right) + \left(\frac{m_X c^2}{c},\ \vec{p}_X\right)_{=0} = \left(\frac{E_a}{c} + m_X c,\ \vec{p}_a\right) = \left(\frac{E}{c},\ \vec{p}\right)$$

The invariant quantity is (the same in all inertial frames)

$$E^2 - p^2 c^2 = (E_a + m_X c^2)^2 - p_a^2 c^2 \tag{12.491}$$

In the centre of mass frame.
The 4-momentum for the exit channel is

$$p_Y^\mu + p_b^\mu = \left(\frac{E_Y}{c},\ \vec{p}_Y\right) + \left(\frac{E_b}{c},\ \vec{p}_b\right) = \left(\frac{E_Y + E_b}{c},\ \vec{p}_Y + \vec{p}_b\right) = \left(\frac{E'}{c},\ \vec{p}\,'\right)_{=0}$$

The invariant quantity is (the same in all inertial frames)

$$E'^2 - p'^2 c^2 = (E_Y + E_b)^2 - 0 = (E_Y + E_b)^2 \tag{12.492}$$

Equating the invariant quantities of equations (12.491) and (12.492) we have

$$(E_a + m_X c^2)^2 - p_a^2 c^2 = (E_Y + E_b)^2$$

Using

$$p_a c = \sqrt{E_a^2 - m_a^2 c^4}, \ E_Y = m_Y c^2, \ E_b = m_b c^2$$

we get

$$(E_a + m_X c^2)^2 - (E_a^2 - m_a^2 c^4) = (m_Y c^2 + m_b c^2)^2$$

$$E_a^2 + m_X^2 c^4 + 2E_a m_X c^2 - E_a^2 + m_a^2 c^4 = (m_Y + m_b)^2 c^4$$

$$m_X^2 c^4 + 2E_a m_X c^2 + m_a^2 c^4 = (m_Y + m_b)^2 c^4$$

$$2E_a m_X c^2 + (m_a^2 + m_X^2)c^4 = (m_Y + m_b)^2 c^4$$

$$2E_a m_X c^2 = (m_Y + m_b)^2 c^4 - (m_a^2 + m_X^2)c^4$$

$$E_a = \frac{(m_Y + m_b)^2 c^4 - (m_a^2 + m_X^2)c^4}{2m_X c^2} = \frac{(m_Y + m_b)^2 - (m_a^2 + m_X^2)}{2m_X} c^2$$

Kinetic energy of the incident particle, i.e. the threshold energy T_a, is obtained from

$$E_a = T_a + m_a c^2$$

$$T_a = E_a - m_a c^2$$
$$= \frac{(m_Y + m_b)^2 - (m_a^2 + m_X^2)}{2m_X} c^2 - m_a c^2 = \frac{(m_Y + m_b)^2 - (m_a^2 + m_X^2) - 2m_X m_a}{2m_X} c^2$$

$$T_a = \frac{(m_Y + m_b)^2 - (m_a + m_X)^2}{2m_X} c^2 = \frac{\left(\sum m_{\text{final}}\right)^2 - \left(\sum m_{\text{initial}}\right)^2}{2m_X} c^2 = E_{\text{Threshold}} \quad (12.493)$$

Exercise 12.13 *Determine the threshold energy for the reaction that ensues when a proton strikes another proton at rest to produce a proton–anti-proton pair in addition to the original particles.*
Ans. The given reaction is

$$p + p \rightarrow p + \bar{p} + p + p$$

From equation (12.493) we can write

$$E_{\text{Threshold}} = \frac{\left(\sum m_{\text{final}}\right)^2 - \left(\sum m_{\text{initial}}\right)^2}{2m_X} c^2 = \frac{(4m_p)^2 - (2m_p)^2}{2m_p} c^2 = 6m_p c^2$$

$$E_{\text{Threshold}} = 6(938 \ MeV) = 5628 \ MeV = 5.6 \ GeV$$

So the projectile proton should impinge on a stationary proton with a minimum energy 5.6 GeV to produce a proton–anti-proton pair of particles.

Exercise 12.14 *Calculate threshold energy for the following reactions*
(i) $\gamma + p \rightarrow p + \pi^0$
(ii) $p + p \rightarrow p + p + \pi^0$
Ans. From equation (12.493)

$$\text{(i)} \; E_{\text{Threshold}} = \frac{\left(\sum m_{\text{final}}\right)^2 - \left(\sum m_{\text{initial}}\right)^2}{2m_X} c^2 = \frac{(m_p + m_\pi)^2 - (m_\gamma + m_p)^2}{2m_p} c^2$$

$$= \frac{(m_p + m_\pi)^2 - (0 + m_p)^2}{2m_p} c^2 = \frac{(m_p + m_\pi + m_p)(m_p + m_\pi - m_p)}{2m_p} c^2 = \frac{m_\pi(2m_p + m_\pi)}{2m_p} c^2$$

$$E_{\text{Threshold}} = m_\pi\left(1 + \frac{m_\pi}{2m_p}\right) c^2 = 140\left(1 + \frac{140}{2(938)}\right) MeV = 150.5 \; MeV$$

$$\text{(i)} \; E_{\text{Threshold}} = \frac{\left(\sum m_{\text{final}}\right)^2 - \left(\sum m_{\text{initial}}\right)^2}{2m_X} c^2 = \frac{(2m_p + m_\pi)^2 - (2m_p)^2}{2m_p} c^2$$

$$= \frac{(2m_p + m_\pi + 2m_p)(2m_p + m_\pi - 2m_p)}{2m_p} c^2$$

$$E_{\text{Threshold}} = \frac{m_\pi(4m_p + m_\pi)}{2m_p} c^2 = \frac{140(4(938) + 140)}{2(938)} MeV = 290 \; MeV$$

To describe kinematics of particle reactions (say two-body elastic scattering), Lorentz invariant variables called Mandelstam variables are used. They are numerical quantities that encode energy, momentum and angles traced by particles in a scattering process.

Exercise 12.15 *First quantization is quantization of*

(a) *material particles*
(b) *fields*
(c) *both material particles and fields*
(d) *waves.*

Exercise 12.16 *Second quantization is quantization of*

(a) *material particles*
(b) *fields*
(c) *both material particles and fields*
(d) *waves.*

Exercise 12.17 *First quantization applies to*

(a) *Fock space*
(b) *Euclidean space*
(c) *Hilbert space*
(d) *Minkowski space.*

Exercise 12.18 *Second quantization applies to*

(a) *Fock space*
(b) *Euclidean space*
(c) *Hilbert space*
(d) *Minkowski space.*

Answers to multiple choice questions
12.15*a*, 12.16*b*, 12.17*c*, 12.18*a*.

12.39 Question bank

Q12.1 What is first quantization?
Q12.2 What is second quantization?
Q12.3 Obtain the classical field equation in terms of Lagrangian density.
Q12.4 Obtain the classical field equation in Hamiltonian formulation.
Q12.5 How do you quantize the Schrödinger field?
Q12.6 Obtain Lagrangian density for the Klein–Gordon field.
Q12.7 How do we quantize the Klein–Gordon field as harmonic oscillators?
Q12.8 How is the electromagnetic field quantized?
Q12.9 How do we quantize the Dirac field?
Q12.10 Write down the Maxwell field equations in a vacuum.
Q12.11 Write down the Maxwell field equations in terms of potentials.
Q12.12 What is gauge transformation?
Q12.13 What is gauge invariance?
Q12.14 What is Lorentz gauge?
Q12.15 What is Coulomb gauge?
Q12.16 What are global and local gauge transformation?
Q12.17 What do we mean by gauge freedom?
Q12.18 Define an electromagnetic field tensor.
Q12.19 Obtain the Lagrangian density of an electromagnetic field.
Q12.20 Obtain the Maxwell equations from the Euler–Lagrangian equation.
Q12.21 Show that the Dirac equation is invariant under local gauge transformation.
Q12.22 What are Mandelstam variables? What is their significance?

Further reading

[1] Perkins H 2000 *Introduction to High Energy Physics* (Cambridge: Cambridge University Press)
[2] Griffiths D 2008 *Introduction to Elementary Particles* (New York: Wiley-VCH)
[3] Tayal D C 2009 *Nuclear Physics* (Mumbai: Himalaya Publishing House)
[4] Satya P 2005 *Nuclear Physics & Particle Physics* (New Delhi: Sultan Chand & Sons)
[5] Guha J 2019 *Quantum Mechanics: Theory, Problems & Solutions* 3rd edn (Kolkata: Books and Allied (P) Ltd))
[6] Yung K L 2002 *Problems and Solutions on Atomic, Nuclear and Particle Physics* (Singapore: World Scientific)

IOP Publishing

Nuclear and Particle Physics with Cosmology, Volume 2
Particle physics and cosmology
Jyotirmoy Guha

Chapter 13

Introduction to cosmology

The study of the Universe as a whole is referred to as cosmology. In this chapter, we discuss the units used in cosmology such as parsec, astronomical unit etc. We define luminosity and flux. Terrestrial and Jovian planets, globular cluster and dust are discussed. We describe methods to calculate distance and count the number of stars. We discuss the types of galaxies such as spiral galaxy, elliptical galaxy, irregular galaxy and highlight Hubble's classification of galaxies. We define red shift and blue shift. We also deal with main sequence fitting, the Hertzsprung–Russell diagram and Cephid variables.

13.1 Understanding the term cosmology

Cosmology is the study of the Universe as a whole. The term Universe refers to all that exists.

✓ Our knowledge regarding the components of the Universe has undergone changes over a period of time and might likely change in the time to come.

✓ The Universe may consist of components or factors that we still have no idea of.

✓ The perception of the Universe differs from person to person as well as when we consider various fields of study. Usually, we think of a model Universe and focus on some aspects from a physicists' point of view.

✓ We wish to know the following about the Universe.
- The structure of the Universe, i.e. constituents of the Universe.
- What the Universe looks like.
- The dynamics of the Universe.
- About the past of the Universe. How did the Universe originate and reach its present state.
- To predict the future of the Universe. Will it continue to evolve or come to an end?

In this book we shall focus only on a few aspects.

doi:10.1088/978-0-7503-5032-7ch13   13-1

13.2 Units used in cosmology

In cosmology SI units m, kg, s are not convenient as they are too short as units.

13.2.1 Unit of length

To measure intergalactic distance we can use the parsec, abbreviated as pc. We define parsec or pc in section 13.23. It has a value given by

$$1\,pc = 3.09 \times 10^{16}\,m \tag{13.1}$$

Other popular usages are as follows:

$$1\,Mpc = 1\ \text{Mega parsec} = 10^6\,pc = 10^6 \times (3.09 \times 10^{16}\,m) = 3.09 \times 10^{22}\,m \tag{13.2}$$

Also

$$1\,Gpc = 1\ \text{Giga parsec} = 10^9\,pc = 10^9 \times (3.09 \times 10^{16}\,m) = 3.09 \times 10^{25}\,m \tag{13.3}$$

13.2.2 Unit of mass

Unit of mass is the solar mass, i.e. mass of the Sun which has value given by

$$M_{\text{sun}} = 1.99 \times 10^{30}\,kg \tag{13.4}$$

In cosmology, astronomy etc we refer to masses of stars in terms of, i.e. as a multiple of solar mass.

13.2.3 Unit of time

Unit of time is second s. However, we can also use

$$\begin{aligned}
1\,Gyr &= 1\ \text{gigayear} = 10^9\,\text{years} \\
&= 10^9 \times 365 \times 24 \times 60 \times 60\,s \\
&= 10^9 \times (3.15 \times 10^7\,s) = 3.15 \times 10^{16}\,s
\end{aligned} \tag{13.5}$$

13.3 Basic assumption

We assume that the laws of physics can be applied to distant parts of the Universe.

The laws of physics are well tested experimentally and established to be true in our local neighbourhood and on Earth. They explain various phenomena occurring around us. For instance, planetary motion, solar spectra, atomic spectra, states of solid, liquid, gas, are various phenomena that can be understood using quantum mechanics and so on. The four fundamental interactions viz. strong interaction, electromagnetic interaction, weak interaction, gravitational interaction have all been explained.

Though the laws of physics could not be actually tested physically at distant parts of the Universe we assume that they hold.

We assume that with the help of these laws we can study the dynamics of the Universe as a whole.

13.4 Planet Earth

Planet Earth has mass given by

$$M_{\text{earth}} = 5.976 \times 10^{24} \, kg \tag{13.6}$$

and equatorial radius of Earth is given by

$$R_{\text{earth}} = 6378 \, km \tag{13.7}$$

Earth is a part of the Solar System and revolves around the Sun with speed $30 \, km \, s^{-1}$.

13.5 Sun

The most dominant component of the Solar System is the Sun.

The Sun is a star. It is the star closest to the Earth.

The Sun is essentially a glowing sphere of hot gas. Components of the gas are 70% hydrogen, 28% helium, 1.5% carbon, nitrogen, oxygen and 0.5% other elements like iron, magnesium etc.

Thermonuclear fusion goes on at the core of the Sun that is the source of its energy. This energy comes out of the Sun and falls on the planets that revolve around it.

The amount of light produced by the Sun is described through a term called luminosity. Luminosity is measured in watts i.e. $J \, s^{-1}$.

✓ Luminosity of the Sun is

$$L_{\text{sun}} = 3.9 \times 10^{26} \, W$$

$$\text{i. e. } 3.9 \times 10^{26} \, J \text{ of energy is radiated per second} \tag{13.8}$$

In cosmology, astronomy etc we refer to luminosity of stars as a multiple of solar luminosity L_{sun}.

✓ The mass of the Sun is given in equation (13.4) viz. $M_{\text{sun}} = 1.99 \times 10^{30} \, kg$ and mass of the Earth is given in equation (13.6) viz. $M_{\text{earth}} = 5.976 \times 10^{24} \, kg$. We note that

$$\frac{M_{\text{sun}}}{M_{\text{earth}}} = \frac{1.99 \times 10^{30} \, kg}{5.976 \times 10^{24} \, kg} = 3.3 \times 10^5 \tag{13.9}$$

Obviously the Sun is 3.3×10^5 times more massive compared to the Earth.

✓ The radius of the Sun is

$$R_{\text{sun}} = 6.96 \times 10^8 \, m \tag{13.10}$$

Time taken by light to travel from the centre of the Sun to the surface of the Sun is

$$t = \frac{\text{distance from center to surface of sun}}{\text{speed of light}} = \frac{R_{sun}}{c} = \frac{6.96 \times 10^8 \, m}{3 \times 10^8 \, ms^{-1}} = 2.32 \, s \tag{13.11}$$

So the radius of the Sun can alternatively be expressed from equation (13.11) as

$$R_{\text{sun}} = c(2.32s) = 2.32 \text{ light. second} \tag{13.12}$$

✓ Surface temperature of the Sun is $5800 \, K$.

This is the temperature which corresponds to the light that we receive from the Sun. This is blackbody radiation and it peaks in the visible part of the spectrum.

✓ The average distance between the Earth and the Sun is called astronomical unit, abbreviated as AU whose value is

$$1\ AU = 1.5 \times 10^{11}\ m \tag{13.13}$$

This is the mean radius of the Earth–Sun orbit.

Distances used by astronomers are in astronomical units.

13.6 Terrestrial and Jovian planets

There are two types of planets. They are

(a) Terrestrial planets
(b) Jovian planets.

In figure 13.1 we show such planets.

• Terrestrial planets have a solid, rocky core.

Mercury, Venus, Earth, Mars are the four terrestrial planets.

Distances of Mercury, Venus, Earth, Mars from the Sun are, respectively

$$0.39\ AU,\ 0.72\ AU,\ 1\ AU,\ 1.52\ AU.$$

• Jovian planets are gaseous.

Jupiter, Saturn, Uranus, Neptune are the four Jovian planets.

Distances of Jupiter, Saturn, Uranus, Neptune from the Sun are, respectively

$$5.2\ AU,\ 9.54\ AU,\ 19.22\ AU,\ 30.06\ AU.$$

The planets revolve around the Sun in elliptical orbits. The elliptical orbits have eccentricity e whose values are as follows.

Mercury (0.21), Venus (0.007), Earth (0.017), Mars (0.094), Jupiter (0.049), Saturn (0.057), Uranus (0.047), Neptune (0.0097).

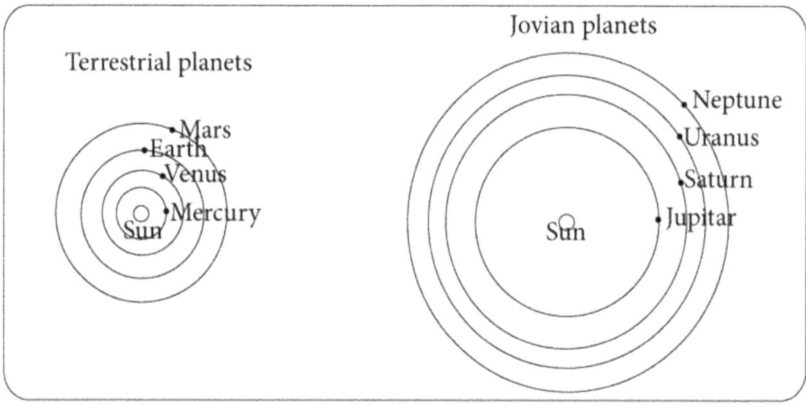

Figure 13.1. Terrestrial and Jovian planets (not drawn to scale). They travel in elliptical orbits though the orbits have been shown as circular in this figure.

The less the eccentricity e the more circular is the orbit, i.e. elliptical orbits have larger eccentricity. Eccentricity $e = 0$ for circle.

Clearly as eccentricities are small, the orbits of planets revolving around the Sun are nearly circular.

13.7 Beyond the Solar System

To study the Universe we have to go beyond the Solar System.

Stars as they are seen to appear in the sky do not solve all our queries. This is because what we see is not a 3D image but a 2D projection. Also, star-to-star distance is not apparent.

A look at the dark sky at night reveals stars. One fails to see stellar objects in the daytime except for the Sun or during occasional emergence of bright comets. The Sun is the only star in our Solar System.

Beyond the Solar System there are stars which appear faint as they are at larger distance compared to the Sun. These stars are similar to the Sun.

We assume that the entire Universe is filled with stars.

13.8 Distribution of stars in the sky

The stars as seen in the night sky are not uniformly distributed. There is a band which runs across the sky in which there is dense and bright concentration of stars. The density of stars falls off as we move away from the band. Density of stars in a band is so large that the band appears white as if milk is spread across the sky. This band that runs across the sky is referred to as the Milky Way or Galaxy and we are a part of it.

To know the 3D density distribution of stars in space in the sky one requires to measure distances between the stars. It is required to know the angular distribution of stars in the sky.

13.9 Number counts

Measuring star-to-star distance is a very difficult job. Clearly information regarding the 3D structure of the sky is not available and only a 2D projection of the sky is available.

This problem was overcome through a technique called number counts.

13.9.1 Flux

Flux f represents the amount of energy from a star falling per unit area of a detector placed on Earth per unit time. Flux f is given by

$$f = \frac{L}{4\pi r^2}\left(\frac{J\ s^{-1}}{m^2} = J\ m^{-2}\ s^{-1} = W\ m^{-2}\right) \tag{13.14}$$

where r is the distance of a star from Earth and the surface area of an imagined sphere of radius r is obviously $4\pi r^2$ as indicated in figure 13.2(a). L represents luminosity. Actually flux is measured in an experiment.

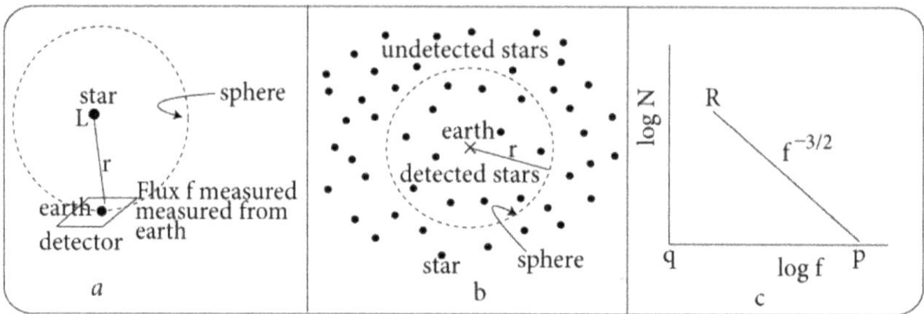

Figure 13.2. (a) Flux of star $=\frac{L}{4\pi r^2}$, $L=$ luminosity, $r=$ distance of star from Earth. (b) Stars that are detected from Earth through measurement of flux. (c) Plot of $\log f$ against $\log N$ is a straight line with negative slope.

Some stars in the sky appear brighter and some stars appear fainter. The further away a star is, i.e. for larger r, the less will be the flux f. In other words, fainter stars are further from us and brighter stars are closer to us.

For instance, the Sun appears to be the brightest star since it is the closest star. All other stars in the sky are situated at larger distance and hence appear to be faint.

Let us assume that all the stars have the same luminosity, $L =$ fixed, i.e. they all emit the same energy per unit of time.

Figure 13.2(b) shows that Earth is located at the centre of a sphere of radius r and flux is measured from the surrounding visible stars using suitable instruments like telescopes etc.

The instruments employed for flux measurement have a flux limit which we consider to be f. In other words, in any observation we are able to measure flux $\geqslant f$. Stars that are fainter than f are not detected. Only stars that can send a flux f can be detected.

Let number of stars that are detected, i.e. have flux limit $\geqslant f$ be, say N.

Now equation (13.14) suggests that for the flux limit f we can evaluate a distance r beyond which the flux sent by stars cannot be detected and within which the flux sent by stars can be detected. So for definite flux limit f and fixed luminosity L the distance defined through equation (13.14) is

$$r = \left(\frac{L}{4\pi f}\right)^{1/2}. \tag{13.15}$$

A star at distance greater than r has flux less than f and cannot be detected.

A star at distance less than r has flux greater than f and can be detected.

Clearly the flux limit f determines a sphere of radius r (figure 13.2(b)) within which stars can be detected through flux measurement and beyond which stars remain undetected. If we are able to diminish the flux limit f then r gets larger and fainter stars will be detectable and the number of detectable stars N will increase.

Number of detectable stars N above the flux limit f

$=$ (Volume of sphere of radius r with Earth as centre) (Number density n of stars)

$$N = \frac{4}{3}\pi r^3 n \tag{13.16}$$

Using equation (13.15) we get

$$N = \frac{4}{3}\pi \left(\frac{L}{4\pi f}\right)^{3/2} n \tag{13.17}$$

Clearly

$$N \propto f^{-3/2} \tag{13.18}$$

This holds even if there is spread in luminosity of stars.

We take the log of both sides of equation (13.18) to get

$$\log N \propto -\frac{3}{2}\log f \tag{13.19}$$

So the plot of $\log f$ versus $\log N$ will be a straight line with negative slope, as shown in figure 13.2(c).

As we move from point p (higher flux limit) to point q (lower flux limit) the number of stars detectable increases as $f^{-3/2}$ as is evident from figure 13.2(c).

If the stars are uniformly distributed and we count the number of stars along a particular direction with flux above a certain limiting value then as we lower the limiting flux f the number of stars N that we see increases. The lower the f the greater the number of stars N. In other words, the number of detectable stars scales as $f^{-3/2}$.

13.10 Our Galaxy

When we look in a particular direction and reduce the flux limit, then at a certain flux value f (say at point R of figure 13.2(c)) suppose the increase in the number of stars N stops. We can interpret this as follows.

Reducing f means we are able to see greater and greater distance and if there are no stars beyond a distance, N will not increase giving a break in the number count. The extent of star distribution is thus obtained, i.e. we get a distance say OG along the particular direction shown in figure 13.3(a), beyond which there are no stars.

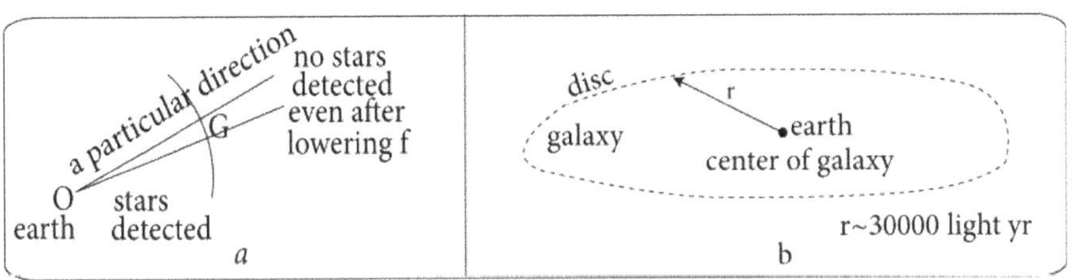

Figure 13.3. (a) Extent of distribution of stars. (b) Sectional view of the Galaxy.

Earlier, the distribution of stars in space was determined based upon this technique of lowering the flux and considering number count. The distance, beyond which there are no stars, could be found out this way.

Clearly this technique shows that there is a finite collection of stars around us. This finite size of a collection of stars is referred to as Galaxy or the Milky Way.

The sectional view of the Galaxy as obtained from number count in different directions has been shown in figure 13.3(b). Star distribution extends to a large distance in one direction and to a short distance in another direction. And so we end up with a disc made up of stars having diameter $r\sim 30\,000$ light years. We are located at the centre of the Galaxy as shown.

13.11 Globular clusters in a Galaxy

We see isolated stars in the sky. In addition to the isolated stars we find several stars often forming clusters or groups in the sky called a globular cluster of stars. They have the following properties.

✓ There is a dense collection of stars (consisting of around millions, i.e. 10^6 stars). Concentration increases towards the centre and density falls as we go outside.

✓ They are gravitationally bound.

✓ They are spherically distributed and so have a globular shape. Hence the cluster is called a globular cluster.

✓ The globular clusters consist of old stars distributed over a spheroidal or spherical distribution.

Figure 13.4(a) shows a globular cluster. Several globular clusters are there. Distances to these globular clusters can be determined. And these globular clusters are considered to be a part of our Galaxy.

A plot of the distribution of the globular clusters w.r.t. the distribution of stars in our Galaxy is shown in figure 13.4(b). But such a distribution does not represent the true picture as it ignores the presence of dust in cosmic space.

13.12 Dust

Microscopic grains in space having dimension of one micron, i.e. $10^{-6}\,m$, are called dust. Dust absorbs light.

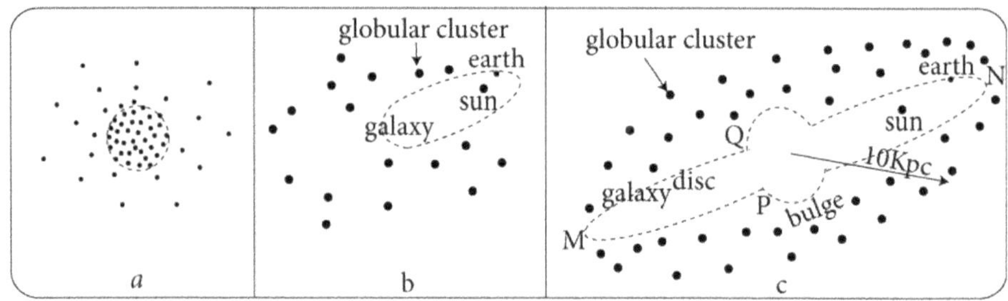

Figure 13.4. (a) Globular cluster. (b) Distribution of the globular clusters w.r.t. our Galaxy ignoring dust. (c) Distribution of the globular clusters w.r.t. the Galaxy we live in.

Dust is preferentially distributed in the plane of our Galaxy and due to them we see black regions in the Milky Way. We do not see stars in these regions. In fact light from a star gets obscured in these regions due to absorption by dust.

Dust cannot absorb wavelengths larger than its own size. So radiation of a larger wavelength can pass through dust. However, dust preferentially absorbs small wavelength light. Clearly blue light (lower wavelength) gets more absorbed and infrared and radio waves (longer wavelength) just pass through. So dust introduces reddening. In other words this reddening ensures the presence of dust. Wherever there is distortion in the spectrum of a star such as reddening of the spectrum, dust is present.

Location of the Sun is shown and it is not located at the centre of the Galaxy but located at one edge of the Galaxy. The Earth is located at the periphery. There is a bulge at the centre. There is a halo surrounding the Galaxy that involves stars along with the globular clusters.

The part PMQ is not shown in figure 13.4(b) because one has to see through dust filling the plane M–N of Galaxy and hence this part is not visible.

However, evidence from globular clusters and later direct evidence revealed that our Galaxy has the shape shown in figure 13.4(c).

The Sagittarius Dwarf Elliptical galaxy is an elliptical galaxy with four globular clusters.

Important facts about our Galaxy

✓ Galaxy radius is 10 kpc

✓ Thickness of the Galaxy is 400 pc

✓ Number of stars in the Galaxy is 10^{11}

✓ Mass of the Galaxy is around $10^{11} M_{sun}$

✓ In the intermediate region between the stars, the Galaxy is filled with diffused gas which is known as the interstellar medium. Mass of this gas is estimated to be $10^{10} M_{sun}$.

✓ Also, there is dark matter in the Galaxy which has mass $10^{12} M_{sun}$. It is a dominant component of mass in the Galaxy which is 10 times the mass of gas in the Galaxy.

✓ The whole disc (i.e. the Galaxy) rotates at 200 $km\ s^{-1}$.

13.13 Galaxy beyond the Milky Way

Our Galaxy is not the only Galaxy existing in the Universe. Hazy bright regions called nebulas can be seen but it was not clear if they belong to our Galaxy.

Distances to nebulas were determined. It later turned out that these nebulas were actually galaxies which are obviously external galaxies.

For instance Andromeda Galaxy (also called $M31$ spiral galaxy, at distance 756 kpc or 2.5 million light years from Earth), Large Magellanic Cloud (50 kpc distance), Small Magellanic Cloud (63 kpc distance), Sagittarius Dwarf Elliptical galaxy (20 kpc distance) are galaxies near to our Milky Way.

13.14 Types of galaxies

Galaxies are of three different types:
- Spiral galaxy
- Elliptical galaxy
- Irregular galaxy.

13.15 Spiral galaxy

Our Galaxy, the Milky Way is a typical spiral galaxy. Other examples are Andromeda Galaxy, Sunflower Galaxy, Triangulum Galaxy.

A spiral galaxy is a rotating disc having stars.

A spiral galaxy has bright spiral arms emerging from the centre of the galaxy.

A spiral galaxy contains gas (nearly one tenth of the mass of stars in it).

Spiral galaxies do not collapse due to their rotation.

If there is some mass left free in space then the gravitational attraction or self-gravity would pull it and make it collapse at the centre. But if the whole system of mass rotates then the centrifugal force balances the gravitational force. So the spiral galaxy is maintained in equilibrium due to rotation.

Since these galaxies have gas there are places in the galaxy where the gas condenses and collapses to form stars. So there is formation of stars taking place in these spiral galaxies having a wide range of masses. Mass can vary from 1000 M_{sun} to a fraction, say 0.1 or even less, of the solar mass M_{sun}.

More massive stars are extremely bright and hotter. They emit much more radiation compared to the Sun. So they exhaust the fuel that powers the star for a relatively shorter period of time. In other words, more massive stars do not live long. Their life is of the order of millions of years. So if one finds very massive, hot and bright stars emitting copious amounts of UV radiation, that means there has been star formation about a million years ago.

A spiral galaxy has a rotating disc with spiral arms that curve out from a dense central region. The spiral arms are defined by very luminous, very hot massive stars. The distribution of light is concentrated along the spiral arms.

The mass content or star content is distributed more or less uniformly but the light content is distributed more along the spiral arms. These spiral arms are patterns in the galaxy that moves along the disc of the galaxy.

Figure 13.5 shows a pictorial representation of a spiral galaxy.

A spiral galaxy can have 10^{11} stars, i.e. it can have a mass of $10^{11} M_{sun}$.

13.16 Elliptical galaxy

In an elliptical galaxy, stars are distributed in a 3D ellipsoid. In other words, an ellipsoidal collection of stars is referred to as an elliptical galaxy.

These are relatively old stars and are of lower mass and lower temperature.

Elliptical galaxies are relatively redder compared to spiral galaxies. In fact, very hot stars are bluer and less hot stars are redder.

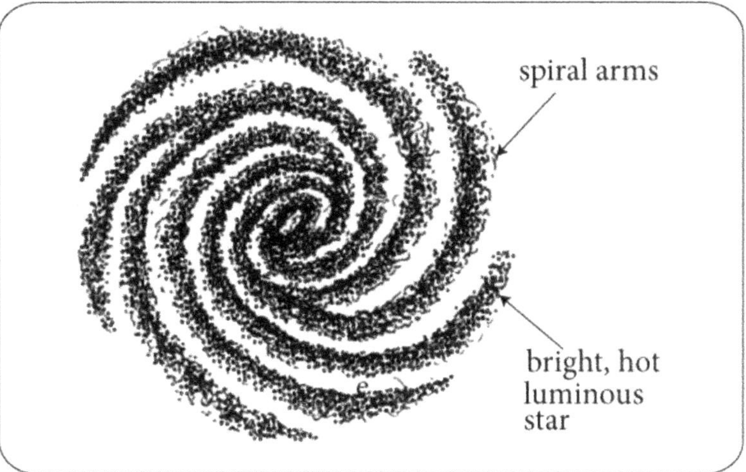

Figure 13.5. Spiral galaxy

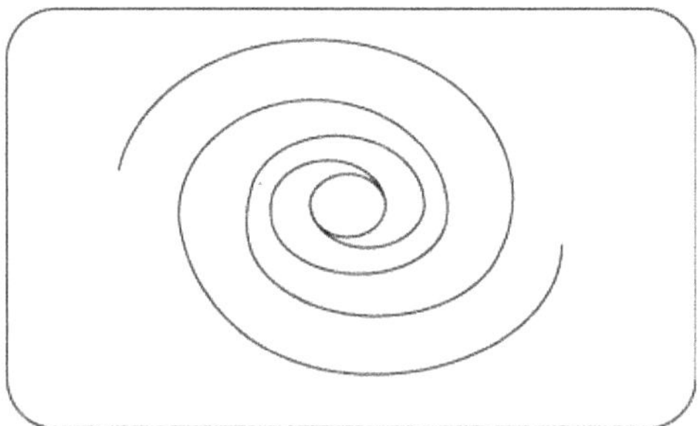

Figure 13.6. Elliptical galaxy

There is very little gas in the elliptical galaxy. So there is no star formation taking place.

All the bright, massive stars have died. So only low mass stars are there that have a lifetime of billions of years.

Figure 13.6 shows a pictorial representation of an elliptical galaxy.

Elliptical galaxies are supported by the random motion of the stars inside the galaxy. They do not rotate. This random motion of stars inside an elliptical galaxy prevents the whole galaxy from collapsing.

An elliptical galaxy can have masses smaller or larger than the mass of a spiral galaxy.

Elliptical galaxies are quite similar to the globular clusters. Both have old stars. One has ellipsoidal shape and the other has spheroidal shape.

An example of an elliptical galaxy is Cygnus A which is at 600 million light years from Earth and is an extremely bright radio source. Another example is M87, a giant elliptical galaxy.

13.17 Irregular galaxy

Irregular galaxies do not have any well-defined shape. They are neither spiral nor elliptical and may be chaotic in appearance.

An example is Canis Major Dwarf that has one billion stars.

Irregular galaxies have copious amounts of gas.

Star formation can take place in irregular galaxies.

They can have masses smaller than the mass of a spiral galaxy.

Irregular galaxies can have 10^{10} stars, i.e. have a mass of $10^{10} M_{\text{sun}}$.

13.18 Hubble's classification of galaxies

Edwin Hubble's classification scheme is the most popular classification of galaxies.

The classification was done depending upon the shape of the galaxy—whether it was spiral or elliptical. Also, this classification is based upon whether there are young stars in the galaxy, and on the gas content of the galaxy.

Figure 13.7 shows the Hubble's tuning fork diagram along which we can place the elliptical galaxies on the left and the spiral galaxies on the right.

At the extreme left of Hubble's tuning fork diagram we have $E0$ galaxy which is spherical. Then we have $E1$, $E2$, $E3$,…,$E7$ galaxies which have increasingly greater asymmetries. The $E7$ galaxy is highly elliptical. All of them are made up of relatively old stars, bereft of sufficient gas and hence do not support formation of stars. They are supported by the random velocities of the stars in these galaxies.

At the right of Hubble's tuning fork diagram we have the spiral galaxies.

At the junction of the elliptical galaxies and the spiral galaxies in the Hubble's tuning fork diagram we have the lenticular or the $S0$ galaxy. These $S0$ galaxies can be considered to be a bridge between the elliptical galaxies and spiral galaxies.

The lenticular or $S0$ galaxies are discs just like spiral galaxies. But these discs are made up of old stars and do not have any gas and consequently there is no star formation taking place in these galaxies. There are no young stars.

Consider the upper arm of the tuning fork diagram of figure 13.7 containing three kinds of spiral galaxies, namely Sa, Sb, Sc.

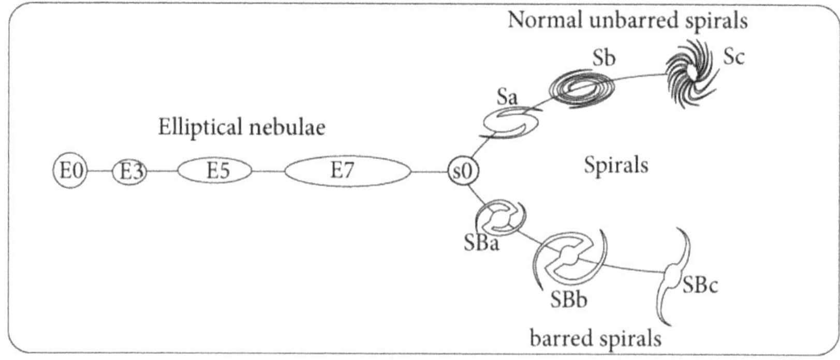

Figure 13.7. Hubble's tuning fork diagram.

The classification of spirals is decided by how tight the spiral arms are bound around the centre. Spiral arms are very tightly bound in *Sa*, rather loosely bound in *Sb* and it is a more loose structure in the case of *Sc* that is situated in the extreme right.

In the lower arm of the tuning fork diagram the spirals are referred to as barred spirals. This is because of the presence of a bar in the spirals that pass through the centre as is evident from *SBa*, *SBb*, *SBc*.

Obviously the upper arm of the tuning fork contains the unbarred spirals *Sa*,*Sb*,*Sc* while the lower arm contains the barred spirals *SBa*, *SBb*, *SBc* (the letter *B* denoting barred).

The spirals have copious amounts of gas and star formation occurs. They contain young, very hot bright stars. The spirals are rotationally supported.

Irregular galaxies are relatively smaller and they do not fit into this diagram. In other words, the irregular galaxies cannot be classified.

It is believed that spiral galaxies underwent some kind of transition to produce the elliptical galaxies either due to the influence of the environment or through collisions between two spiral galaxies. This hypothesis finds logic since there is no star formation in elliptical galaxies but there is star formation in spirals. The stars in elliptical galaxy might have been formed when it was a spiral galaxy that later transformed into an elliptical galaxy.

- Galaxies can also be classified based upon their spectrum and other properties.

13.19 Gas distribution in a spiral galaxy

Figure 13.8 shows a typical spiral galaxy in which there are stars as well as gas. The figure depicts distribution of stars as well as gas in spiral galaxy. Clearly, stars occupy a much less region being concentrated at the centre of the galaxy and gas is much more spread out over a much larger distance.

The bulk of the gas in the galaxies is neutral hydrogen H. A hydrogen atom in the ground state consists of a proton p and an electron e^-. They have spin $\frac{1}{2}$. They can be aligned either in parallel or in antiparallel, as shown in figure 13.9. In other words there are two hyperfine ground states of a neutral hydrogen atom, one with parallel spins of p, e^- and another with antiparallel spins of p, e^-. These two configurations have a very small energy difference and differ by a spin flip.

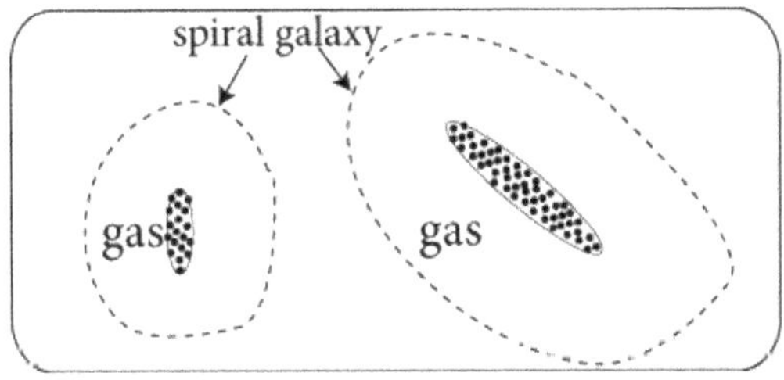

Figure 13.8. Star and gas distribution of a spiral galaxy.

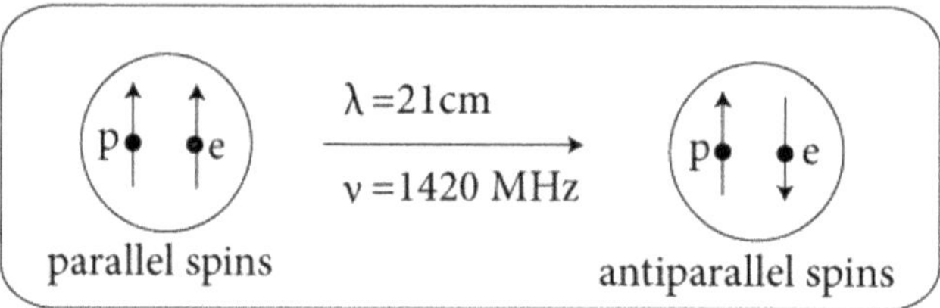

Figure 13.9. Transition between two hyperfine ground states of neutral hydrogen.

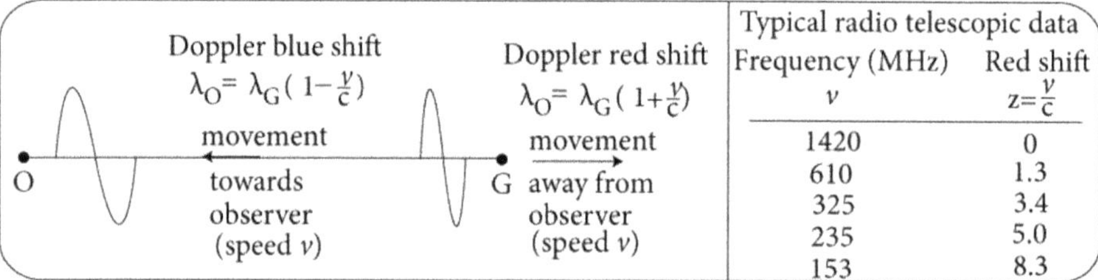

Typical radio telescopic data	
Frequency (MHz) ν	Red shift $z=\frac{v}{c}$
1420	0
610	1.3
325	3.4
235	5.0
153	8.3

Figure 13.10. Doppler shift in wavelength due to gas moving away or moving towards the observer. Here $z = \frac{v}{c}$ where v = speed of gas of galaxy w.r.t. observer (Earth).

If there is transition (due to spin flip) between these two hyperfine states then a photon of energy $h\nu$ is emitted corresponding to photon frequency $\nu = 1420\ MHz$, the corresponding wavelength being $\lambda = 21\ cm$.

13.20 Redshift and blueshift

Consider figure 13.10 showing movement of a gas source away from the observer or towards the observer.

Suppose a hydrogen gas (source point G) that emits radiation λ_G moves with speed v w.r.t. an observer at (point O).

If the gas source moves away from the observer then the observer will see a wavelength increase given by

$$\lambda_O = \lambda_G\left(1 + \frac{v}{c}\right) = \lambda_G(1 + z) \tag{13.20}$$

where

$$z = \frac{v}{c} \tag{13.21}$$

Such an increase in wavelength is referred to as the redshift $z = v/c$.

If the gas source moves towards the observer then the observer will see a wavelength decrease given by

$$\lambda_O = \lambda_G\left(1 - \frac{v}{c}\right) = \lambda_G(1 - z) \tag{13.22}$$

Such a decrease in wavelength is referred to as blueshift $z = v/c$.

In fact, a wavelength corresponding to red light is larger than a wavelength corresponding to blue light, i.e.

$$\lambda_{\text{red}} > \lambda_{\text{blue}} \tag{13.23}$$

and so increase in wavelength is called redshift and decrease in wavelength is called blueshift.

Here $z = \frac{v}{c}$ is the fractional change in wavelength identified as the redshift if a source moves away from the observer (wavelength increase) or blueshift if the source moves towards the observer (wavelength decrease).

So measuring the shift in the wavelength, i.e. measuring $z = \frac{v}{c}$ we can determine the velocity of gas source v w.r.t. the observer, i.e. the velocity v of the galaxy.

From radio telescopic data (figure 13.10) velocity determination of galaxies is possible. In other words the rotational velocity as a function of distance can be obtained from Doppler shift data.

For instance, let us consider a model, as shown in figure 13.11(a), that depicts stars concentrated over a small region of space with gas spread out in the surrounding region. The gas moves around in circular orbits of radius r. The velocity v can be measured as a function of r.

If the gas is assumed to be in equilibrium then the centrifugal force is balanced by gravitational force of attraction due to the mass inside the galaxy (that can be taken equal to the mass of the stars since gaseous contribution to mass is negligible) and we have the equation

$$\frac{v^2}{r} = \frac{GM}{r^2} \tag{13.24}$$

$$v = \sqrt{\frac{GM}{r}} \tag{13.25}$$

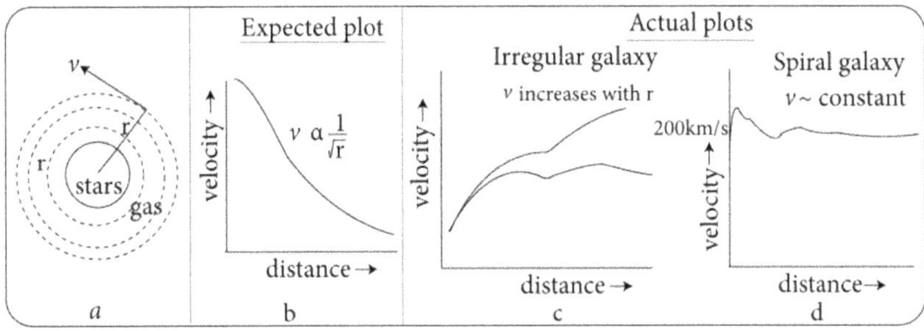

Figure 13.11. (a) A model showing stars concentrated over a small region of space and gas spreading out and going around in circular orbits up to radius r. (b) Expected plot of velocity versus distance for a galaxy according to equation (13.26). (c) Velocity versus distance plot for an irregular galaxy. (d) Velocity versus distance plot for a spiral galaxy.

$$v \propto \frac{1}{\sqrt{r}} \tag{13.26}$$

This means velocity v would decrease with increase in distance r as shown in figure 13.11(b).

The plot of rotational velocity of the gas of a galaxy as a function of distance from the centre of the galaxy is shown in figure 13.11(c) for an irregular galaxy and in figure 13.11(d) for a spiral galaxy. The plots show that rotational velocity increases with increasing distance for a typical irregular galaxy and rotational velocity is nearly constant for a typical spiral galaxy, i.e. we get a flat rotation curve.

13.20.1 Problem of a flat rotation curve

In contrast to what we expect, as per equation (13.26) the rotational velocity does not fall as we move further and further out from the centre of the galaxy (figures 13.11(c) and (d)). This is referred to as the problem of the flat rotation curve.

The entire galaxy (e.g. our own Galaxy) is rotating at a speed of 200 $km\ s^{-1}$. To understand the flat rotation curve we invoke the presence of dark matter.

13.21 Dark matter

Let us make a postulate that the whole galaxy is immersed inside some matter that is invisible called dark matter. This dark matter exerts gravitational attraction.

The mass of this dark matter increases proportionally with the distance as we move further and further out, i.e.

$$M \propto r. \tag{13.27}$$

and so we have

$$v = \sqrt{\frac{GM}{r}} = \text{constant} \tag{13.28}$$

So we have postulated existence of some matter that is invisible. Dark matter provides the extra gravitational attraction needed to hold the hydrogen cloud in orbit.

If dark matter did not exist then the hydrogen cloud would not be in circular orbits and could have escaped the gravitational attraction. Since we see the hydrogen clouds we have to postulate existence of dark matter.

Mass in the dark matter is calculated to be typically 10 times more than the mass of the stars of galaxies and increases as we move further and further away from the centre of a galaxy.

13.22 Constituents of the Universe

It is believed that 70% of the Universe is dark energy that is not seen. There is no direct evidence substantiating its existence. Its existence is inferred from its gravitational effect. It has negative pressure.

26% of the Universe is dark matter.

4% of the Universe is made up of baryons (i.e. protons, neutrons). Of this 4% which are the baryons, nearly 75% is hydrogen and 25% is helium.

Clearly 96% of the Universe is dark, i.e. invisible.

13.23 Parallax distances

Earth-to-Sun distance is $1A.\,U = 1.5 \times 10^{11}\,m$. This distance can be determined directly by using a radar signal. Historically this distance was determined from various astronomical observations.

The direct method to determine distances beyond the Solar System is called the parallax method and the distance obtained from this method is called parallax distance.

Consider a static object and moving observer. The place or angle at which the object appears because of movement of observer is called parallax. Determining the distance moved by observer and the corresponding change in the angle we can evaluate the distance of the object.

Parallax is the change in the angle at which a particular star appears as the Earth moves around the Sun and from this direct method of parallax we can determine the distance of stars up to $1\,kpc$.

Angular size of objects is measured by determining how big it appears angle wise, i.e. by the angle they subtend and so we define terms like

$$1° = \frac{1}{360}\,\text{circle}, \quad 1\,\text{arcminute} = 1' = \frac{1}{60}°, \quad 1\,\text{arcsecond} = 1'' = \frac{1'}{60} = \frac{1}{60 \times 60}° \quad (13.29)$$

The parallax method depends on measuring the shift in the angular position of a star w.r.t. change of position of the Earth.

Consider the diagram of figure 13.12. Assume the Earth goes around the Sun in a nearly circular orbit of radius $ab = bc = 1A.\,U = 1.5 \times 10^{11}\,m$. The actual orbit of the Earth is elliptical with eccentricity $\varepsilon = 0.017$. So the orbit is nearly circular.

When the Earth is at position c suppose a star appears at position d in a direction cd. Six months later, the Earth is at position a and the same star appears at position d but in a different direction ad.

This change in the direction, i.e. the parallax angle cannot be determined relative to very distant stars which would appear to be fixed.

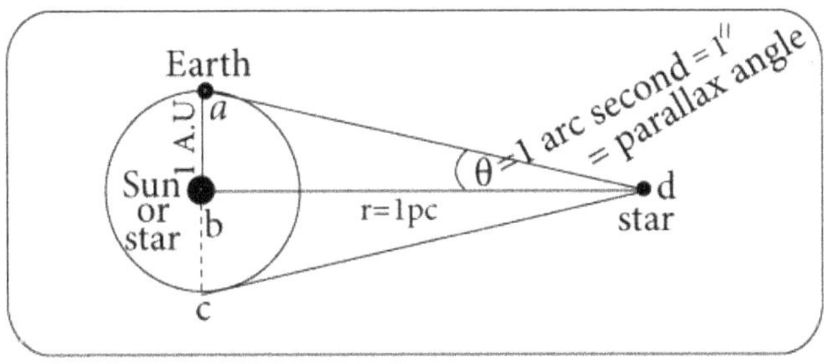

Figure 13.12. Parallax angle and definition of parsec.

For nearby stars at shorter distances this change in angle, i.e. the parallax angle is large and can be measured.

Let $\theta =$ half the change in the angle.

If $\theta = 1$ arc second $\equiv 1''$ then the distance $r = bd = 1$ parsec $\equiv 1\,pc$.

Clearly

$$r\theta = ab = 1\ A.\ U \tag{13.30}$$

$$r = \frac{1A.\ U}{\theta} = \frac{1.5 \times 10^{11} m}{1''}$$

Using

$$\pi^c = 180° \text{ and } 1° = 60 \times 60'' \text{ and so } \pi^c = 180 \times 60 \times 60''$$

$$1'' = \frac{\pi}{180 \times 60 \times 60} \tag{13.31}$$

Hence

$$r = \frac{1.5 \times 10^{11}\ m}{\left(\frac{\pi}{180 \times 60 \times 60}\right)}$$

$$r = \frac{1.5 \times 10^{11} \times 180 \times 60 \times 60}{\pi}m = 3.09 \times 10^{16}\,m = 1\,pc \tag{13.32}$$

$1\,pc$ is the distance at which the parallax angle is one arc second and change in the angle in six months is 2 arc seconds.

In other words $1\,pc \equiv 1$ parsec is the distance of a star that has a parallax angle of 1 arc second. Taking 'par' from the word parallax and 'sec' from the word second we have the term parsec abbreviated as pc.

- 1 light year $=$ Distance travelled by light in 1 year

$$= (3 \times 10^8\ m\ s^{-1})(365 \times 24 \times 60 \times 60)s = 9.46 \times 10^{15}\ m \tag{13.33}$$

Clearly thus

$$1\,pc = 3.09 \times 10^{16}m = \frac{3.09 \times 10^{16}}{9.46 \times 10^{15}} \text{ light year} = 3.266 \text{ light year} \tag{13.34}$$

$1\,pc$ is the distance scale we employ beyond the Solar System.

From Earth-based observations parallax angles of the order of $0.01''$ can be measured.

With the High Precision Parallax Collecting Satellite (HIPARCOS) parallax angles of the order of $0.001'' = 10^{-3\prime\prime}$ i.e. milliarc second can be measured. Using HIPARCOS it has been possible to measure parallax distances of 120 000 stars.

$1'' \equiv 1$ arc second corresponds to a distance of $1\,pc$.

Hence measurement of $0.001''$ enables us to go up to distance of $1\ kpc \equiv 1000\ pc$. Using parallax method we can measure distance of stars that are at a maximum distance of $1\ kpc$ from the Solar System and not beyond that.

To determine distances of stars at a distance beyond $1\ kpc$ from the Solar System the concept of cosmic distance ladder comes in. We discuss indirect methods of measuring distances such as main sequence fitting.

13.24 Main sequence fitting

The method of main sequence fitting can be applied to open clusters.

Open clusters are loose non-dense collections of stars, say a collection of 1000 stars (compared to globular clusters which are a dense collection of say 1 million stars). These stars in open clusters are formed at the same time due to gravitational collapse and fragmentation of the same gas cloud.

Behaviour of stars can be understood and classified on a diagram called the Hertzsprung–Russell or HR diagram, as shown in figure 13.13.

The abscissa represents the colour of the star.

Colour of a star can be determined by considering the ratio of light in two different bands. Colour is directly related to the temperature. So temperature is in the abscissa.

Stars are approximately blackbody radiation and so colour depends on temperature. The hotter the star is, the bluer it appears, and the cooler the star is, the redder it appears. So colour can be directly measured and can be converted to the temperature of the star.

The ordinate represents luminosity, i.e. radiation energy emitted by a star.

Luminosity determination means flux determination (equation (13.14)) and for this we need to measure distance.

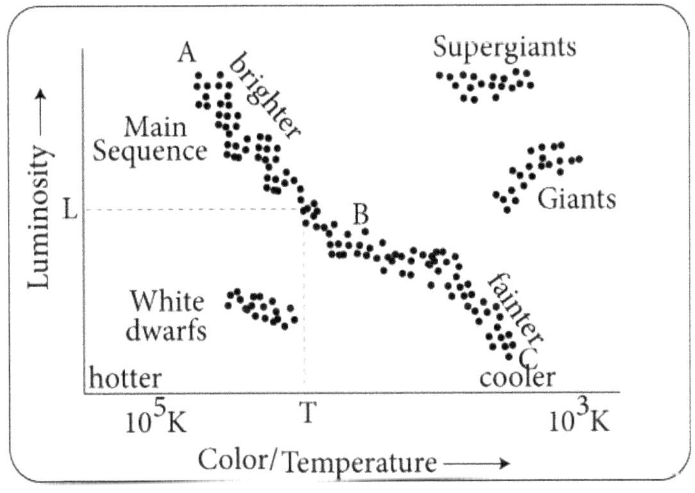

Figure 13.13. HR diagram. Main sequence fitting.

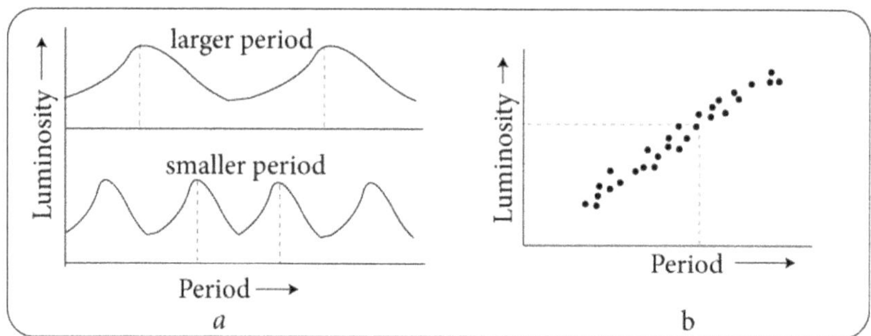

Figure 13.14. Luminosity period relation of Leavitt.

The plot of stars in an HR diagram is shown in figure 13.13. The bulk of stars lie in the strip ABC called the main sequence. In the main sequence cooler less luminous stars lie near C and the hotter more luminous stars lie near A. Our star, the Sun, is near B.

The main sequence corresponds to a situation where a star is burning hydrogen, i.e. the nuclear reaction in them converts hydrogen to helium.

Once hydrogen fuel gets exhausted stars becomes giants, super giants etc and are situated off the main sequence. These are cool stars but they have enormous luminosity.

White dwarfs are low luminosity stars but are very hot.

Luminosity L can be expressed as

$$L = \sigma T^4 . \, 4\pi R^2 \tag{13.35}$$

where σ is the Boltzmann constant (equation (17.115)), T is temperature of the star, R is radius of the star. Large L means the star is larger.

Consider an open cluster. Colour of the stars can be directly measured and colour does not change with distance. Colour can be determined by the peak of the blackbody spectrum that does not shift with distance. Again, the majority of stars lie on the main sequence. So luminosity L can be determined from the HR diagram (figure 13.13). Flux f can also be determined from direct observation. So from equation (13.14) viz. $f = \frac{L}{4\pi r^2}$ we get

$$r = \sqrt{\frac{L}{4\pi f}} \tag{13.36}$$

So knowing f, L the distance r can be obtained.

We can do this for an open cluster because if it is done for a single star it might correspond to a giant or a dwarf, i.e. it may lie off the main sequence. In fact, the majority of stars of an open cluster lie on the main sequence curve.

The main sequence curve has been determined first using the direct parallax distance estimator. Now we can use it to determine the distances to open clusters which are further away, i.e. more than 1 kpc. For instance, suppose we have an open cluster at say *kpc*. If we can determine the main sequence of that open cluster we can use an HR diagram to determine the distance to this open cluster. This technique of main sequence fitting has thus extended the distance scale beyond 1 *kpc*.

The way distances are measured in the cosmic scale is referred to as cosmic distance ladder.

The extent of our Galaxy is $\sim$10 *kpc*. It is very difficult to extend this beyond our Galaxy because we need to identify open clusters, fit the stars on the main sequence.

Let us discuss another method using the Cepheid variables.

13.25 Cepheid variables

Cepheid variables are very bright stars which are unstable, i.e. young stars. They do not live long. They can be detected in very distant galaxies. As they do not live long they are found in regions where star formation occurs such as in spiral galaxies and not in elliptical galaxies where star formation does not occur. The luminosity of these stars changes with time.

Figure 13.14(a) shows the flux or the luminosity from different Cepheid variables.

Leavitt obtained a luminosity period relation, as shown in figure 13.14(b), for the Cepheid variables for which it was found that the luminosity changes periodically with time. As a star becomes more luminous the period of oscillation increases.

To calibrate the relation we have to know the distance. First distance is measured using either parallax or main sequence fitting. Once we know some stars for which distance is known we get the luminosity period calibration curve. For an unknown star or galaxy, determination of period gives luminosity from figure 13.14(b) and measuring flux we get the distance using equation (13.36).

The luminosity period relation for Cepheid variables was used to first determine the distance to the Andromeda Galaxy (778 *kpc*) which is the nearest full-fledged spiral galaxy that appears like a nebulae in the sky and lies well outside our Galaxy. Cepheid variable method was used to determine distances of several galaxies.

If radiation coming from a galaxy is passed through a spectrometer we can determine the spectrum of a galaxy (figure 13.15). It is a combined spectrum since it contains a spectrum of stars of a galaxy as well as the contribution from the interstellar medium. Various spectroscopic lines can be identified from the galaxy spectrum.

Again, from experiments on Earth we know the wavelength at which radiation in that line occurs. For instance if we look at a material (say oxygen or calcium etc) at

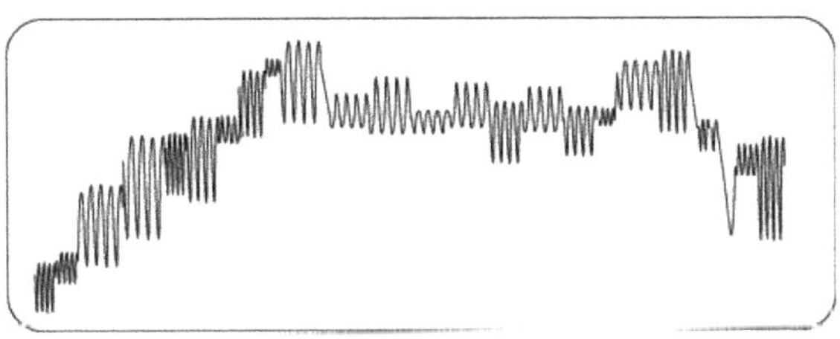

Figure 13.15. Spectrum from a galaxy.

rest, one knows from experiments on Earth the wavelength λ_G. In the spectrum on the galaxy that particular line may occur at a different wavelength λ_o. Such a shift in wavelength is due to Doppler effect (figure 13.10). They are related through equations (13.20) and (13.22).

Blueshift (galaxy moving towards the observer)

$$\lambda_O = \lambda_G\left(1 - \frac{v}{c}\right) \text{ (wavelength decreases)}$$

Redshift (galaxy moving away from the observer)

$$\lambda_O = \lambda_G\left(1 + \frac{v}{c}\right) \text{ (wavelength increases)}$$

There is no shift for perpendicular motion

$$\lambda_O = \lambda_G \text{ (wavelength unaltered)}$$

So if we can measure spectrum from a galaxy and determine shift in wavelength then we can obtain the velocity of a galaxy.

13.26 Exercises

Exercise 13.1 *Express 1 solar mass, 1 Mpc, 1 Gyr in terms of Planck units.*
[Ans.] Planck units are

$$m_{Pl} = 2.18 \times 10^{-8} kg \text{ (Planck mass) (equation (4.19))}$$

$$l_{Pl} = 1.61 \times 10^{-35} m \text{ (Planck length) (equation (4.20))}$$

$$t_{Pl} = 5.38 \times 10^{-44} s \text{ (Planck time) (equation (4.21))}$$

$$\begin{aligned}
1 \text{ solar mass} &= 1.99 \times 10^{30} \ kg \\
&= 1.99 \times 10^{30} \ kg\frac{m_{Pl}}{m_{Pl}} \\
&= \frac{1.99 \times 10^{30} \ kg}{2.18 \times 10^{-8} \ kg} m_{Pl} = 9.1 \times 10^{37} m_{Pl}
\end{aligned}$$

$$\begin{aligned}
1 \ Mpc &= 10^6 \ pc = 3.09 \times 10^{22} \ m \\
&= 3.09 \times 10^{22} \ m\frac{l_{Pl}}{l_{Pl}} = \frac{3.09 \times 10^{22} \ m}{1.61 \times 10^{-35} \ m} l_{Pl} \\
&= 1.9 \times 10^{57} l_{Pl}
\end{aligned}$$

$$\begin{aligned}
1 \ Gyr &= 10^9 \ yr = 3.15 \times 10^{16} \ s \\
&= \frac{3.15 \times 10^{16} \ s}{5.38 \times 10^{-44} \ s} t_{Pl} \\
&= 5.9 \times 10^{59} t_{Pl}
\end{aligned}$$

Exercise 13.2 *Parsec is the unit of which of the following?*

(a) time^{-1}   *(b) time*   *(c)* distance   *(d) parallax.*

Exercise 13.3 *Which of the following is the correct statement regarding radius of the Sun?*

(a) 2.32 *light second*   *(b)* 2.32 *light year*   *(c)* 6.96 × 10^5 *m*   *(d)* 6378 *km.*

Exercise 13.4 *Which of the following is the correct statement regarding 1 A.U?*

(a) radius of Sun   *(b) radius of Earth*   *(c) Earth–Sun separation*   *(d)* 6378 *km.*

Exercise 13.5 *Which of the following is the correct statement?*
  (a) *Terrestrial planets have a gaseous core while Jovian planets have a solid rocky core*
  (b) *Terrestrial planets have a solid rocky core while Jovian planets have a gaseous core*
  (c) *Both terrestrial planets and Jovian planets have a solid rocky core*
  (d) *Both terrestrial planets and Jovian planets have a gaseous core*

Exercise 13.6 *How many stars are there in our Galaxy?*

(a) 10^5 *(b)* 10^{11} *(c)* 10^{21} *(d)* 10^{33}.

Exercise 13.7 *Which of the following statements is/are true?*
  (a) *Spiral galaxies do not collapse due to their rotation.*
  (b) *Elliptical galaxies do not collapse due to random motion of stars.*
  (c) *Spiral galaxies have only low mass stars.*
  (d) *Elliptical galaxies have only low mass stars.*

Exercise 13.8 *Which of the following statements is/are true?*
  (a) *Globular clusters contain relatively new stars.*
  (b) *Globular clusters contain relatively old stars.*
  (c) *Elliptical galaxies contain relatively new stars.*
  (d) *Elliptical galaxies contain relatively old stars.*

Exercise 13.9 *Which of the following statements is/are true?*
  (a) $\lambda_{red} > \lambda_{blue}$
  (b) $\lambda_{red} < \lambda_{blue}$
  (c) *In blueshift wavelength decreases but in redshift frequency increases.*
  (d) *In redshift wavelength increases but in blueshift frequency decreases.*

Exercise 13.10 *Which of the following statements is/are false?*
  (a) *In redshift wavelength increases but in blueshift frequency decreases.*
  (b) *In redshift frequency increases but in blueshift wavelength decreases.*
  (c) *In redshift wavelength decreases but in blueshift frequency increases.*
  (d) *In redshift frequency decreases but in blueshift wavelength increases.*

Exercise 13.11 *Choose the correct option from the following. HIPARCOS is*

(a) a planet   *(b) a satellite*   *(c) a galaxy*   *(d) a star inside the Milky Way*   *(e) a star outside the Milky Way.*

Exercise 13.12 *Distance to Andromeda Galaxy 778 kpc was determined by*
  (a) *Main sequence fitting*
  (b) *Parallax method*
  (c) *Luminosity period relation for Cepheid variables*
  (d) *HIPARCOS satellite.*

Exercise 13.13 *Which is the largest?*

(a) 1 A. U   *(b) 1 Light year*   *(c) 1 pc*   *(d) 1 Million km.*

Hint. $1\ A.\ U = 1.5 \times 10^{11}\ m$ (equation (13.13))
$1\ pc = 3.09 \times 10^{16}\ m$ (equation (13.1))
$1\ light\ year = 9.46 \times 10^{15}\ m$ (equation (13.33))
$1\ Million\ km = 10^6\ km = 10^6 10^3\ m = 10^9\ m$

$$\boxed{\text{Answers to multiple choice type questions}}$$

13.2*c*, 13.3*a*,13.4*c*, 13.5*b*, 13.6*b*, 13.7*a,b,d*, 13.8*b,d*, 13.9*a*, 13.10*a,b,c,d*, 13.11*b*,
13.12*c*, 13.13*c*.

13.27 Question bank

Q13.1 What is parsec? Express 1 *Gpc* in light years.

Q13.2 Express 1 *Mpc* in *km*.

Q13.3 How are the quantities luminosity and flux related?

Q13.4 What is the difference between terrestrial and Jovian planets?

Q13.5 What is the difference between a spiral galaxy, an elliptical galaxy and an irregular galaxy?

Q13.6 Give examples of spiral galaxy, elliptical galaxy and irregular galaxy.

Q13.7 What is the similarity and difference between globular cluster and elliptical galaxy?

Q13.8 What is an HR diagram?

Q13.9 What is main sequence fitting?

Q13.10 Describe what we mean by Cepheid variables.

Further reading

[1] Bharadwaj S 2008 *Physical Cosmology Youtube Lecture Series by Professor Somnath Bharadwaj* (Department of Physics, IIT Kharagpur)

[2] Hawking S 1988 *A Brief History of Time* (London: Bantam)

[3] Bonometto S, Gorini V and Moschella U 2001 *Modern Cosmology* (Bristol: IOP)

[4] Roos M 1994 *Introduction to Cosmology* (New York: Wiley)

[5] Tenreiro R D and Quiros M 1988 *An Introduction to Cosmology and Particle Physics* (Singapore: World Scientific)

IOP Publishing

Nuclear and Particle Physics with Cosmology, Volume 2
Particle physics and cosmology
Jyotirmoy Guha

Chapter 14

Dynamics of the Universe

In this chapter we discuss Hubble's law and introduce Hubble's constant. Cosmological principle and its implication are discussed. We discuss why the Universe is said to be homogeneous and isotropic. Deviations from Hubble's law are mentioned. Hubble's law and cosmological principle lead to an expanding model of the Universe. This is established in this chapter. We mention the galaxy red shift survey. The co-moving coordinate system is explained. Models of the Universe like the expanding Universe, critical model and big bang to big crunch are elaborated. We define the cosmological parameter and find the Einstein–de Sitter solution.

14.1 Hubble's law

Galaxies are on the average moving away from us with a speed proportional to the distance. The constant of proportionality is called Hubble's parameter or Hubble's constant. We can express this law as follows:

$$v = Hr \tag{14.1}$$

This is Hubble's law given in 1920. The present value of Hubble's constant $H = H_0$ is

$$H_0 = 70 \, km \, s^{-1} \, Mpc^{-1} \tag{14.2}$$

With this $H = H_0$ we can rewrite Hubble's law as (figure 14.1)

$$v = H_0 r \tag{14.3}$$

Here v is the radial component of velocity of the galaxy, r is the distance of the galaxy.
• In *exercise 14.3* we show that $[H] = [s^{-1}]$.

If we assume that this v is the only motion that a galaxy possesses then we can state that every galaxy has a motion that is radially outward and proportional to the distance. So we can rewrite Hubble's law as a vector law

$$\vec{v} = H_0 \vec{r} \tag{14.4}$$

doi:10.1088/978-0-7503-5032-7ch14

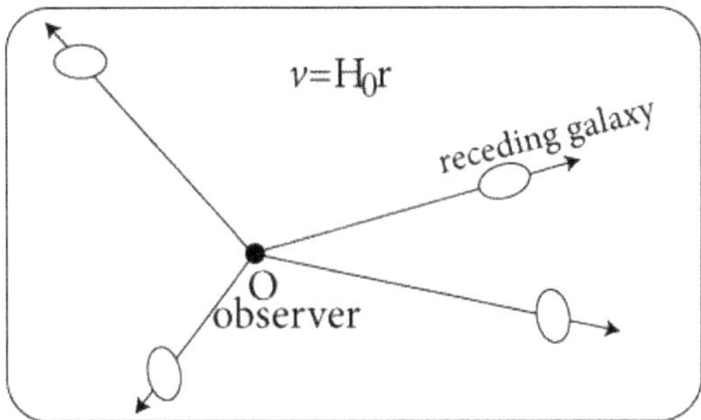

Figure 14.1. Receding galaxies and Hubble's law.

From observations of redshift we can measure v. From Cepheid variable we can determine r.

Hubble's law is not a universal law and there are aberrations. For instance the Andromeda Galaxy (the nearest full-fledged spiral galaxy) is not moving away but moving towards us and might collide in the future.

It is customary to parametrize Hubble's parameter H_0 by introducing a parameter h as follows. We rewrite H_0 as

$$H_0 = 100h \ km \ s^{-1} \ Mpc^{-1} \tag{14.5}$$

From equation (14.2) we write

$$70 \ km \ s^{-1} \ Mpc^{-1} = 100 \ h \ km \ s^{-1} \ Mpc^{-1}$$

$$h = 0.7 \tag{14.6}$$

We mention that H_0 refers to the present value of Hubble's parameter.

• If we see a galaxy at a distance of 10 Mpc we expect it to be moving away from us with a speed given by equation (14.2) and using equation (14.1) we get

$$v = H_0 r = (70 \ km \ s^{-1} \ Mpc^{-1}) \ (10 \ Mpc) = 700 \ km \ s^{-1} \tag{14.7}$$

For a galaxy at a distance of 20 Mpc we expect it to be moving away from us with a speed

$$v = H_0 r = (70 \ km \ s^{-1} \ Mpc^{-1})(20 \ Mpc) = 1400 \ km \ s^{-1} \tag{14.8}$$

Einstein propounded the general theory of relativity in 1913. The general theory of relativity is the theory of gravity. It says that gravity is a geometrical effect affecting the geometry of space-time. Through a change in the geometry of space-time the motion of objects changes. Gravitational field is there in the curvature of space-time. The general theory of relativity applied to the whole Universe gives cosmology. The aim was to construct a model for the whole Universe.

14.2 The cosmological principle

The cosmological principle says the following.

The Universe is homogeneous and isotropic. The structure of the Universe appears the same from all galaxies.

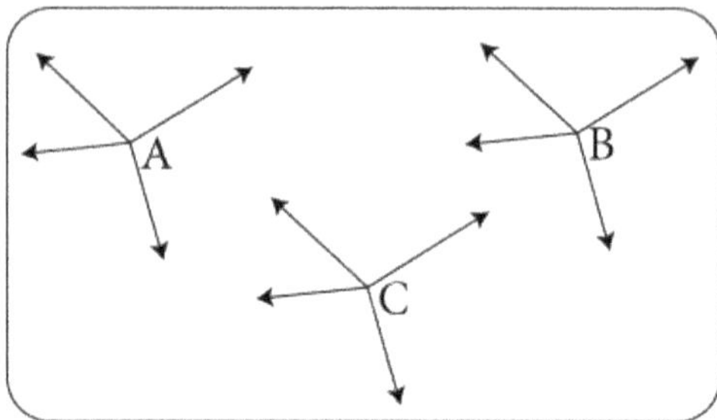

Figure 14.2. Three observers at different places.

14.2.1 Homogeneity

The Universe looks the same to all observers.

So observers A,B,C as shown in figure 14.2 find the Universe identical from their respective positions. There is no preferred position.

14.2.2 Isotropy

The Universe looks the same in all directions.

So observers A,B,C as shown in figure 14.2 find the Universe identical in all directions from their respective positions. There is no preferred direction.

We note that if there is the presence of a boundary then the Universe is no longer homogeneous and isotropic since close to the boundary the environment is different and then there is a preferred position and direction. It thus follows from cosmological principle that the Universe cannot have a boundary. There is no preferred position and direction in the Universe.

The simplest possible model of a Universe without a boundary is obviously an infinite Universe. An infinite Universe extends to infinity in all directions.

A finite and compact Universe without a boundary is also possible.

In 2D we can think of R^2 where R is a real line extending from 0 to $\pm\infty$. R^2 is a plane extending to ∞. Also, we have space denoted by S^2 which is the surface of a sphere that is 2D finite and has no boundary. If we move continuously in one direction we come back to the starting point.

Similarly in 3D we have R^3 which is the infinite 3D space or we can have a space denoted by S^3 which is a 3D sphere.

An infinite Universe is easier to visualize.

14.3 The expanding Universe

Hubble's law and the cosmological principle lead to an expanding model of the Universe.

✓ The Universe has no boundary. This follows from the cosmological principle.

✓ The homogeneous infinite Universe is filled with galaxies with density ~ 0.1 galaxy per Mpc^3.

✓ The galaxies are all moving away from one another as described in Hubble's law. This corresponds to expansion of the Universe, i.e. expansion of space.

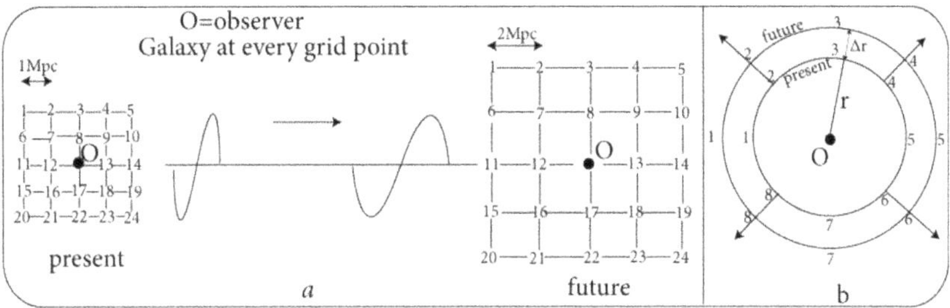

Figure 14.3. The expanding Universe. (a) Explaining homogeneity. (b) Explaining isotropy. The observer is denoted by *O*.

✓ We arrive at a singularity if we extrapolate into the past, i.e. extrapolate backwards in time. Since the Universe is expanding and galaxies are moving away from one another it follows that in the past the galaxies were increasingly closer. This means that there was a situation at which all the galaxies were at the same point. So at some time in the past, galaxy-to-galaxy separation was zero. Space then collapsed to a point which was a singularity. This singularity is referred to as the Big Bang.

• Consider figure 14.3(a) where a part of the Universe is shown around a point observer *O* assuming that galaxies are present at all the grid points mutually separated by 1 *Mpc*. A few grid points viz. 1 to 24 of the 3D grid are shown. These galaxies are all moving away from one another. At some future time the galaxies are more separated say by an amount of 2 *Mpc*. Clearly the grid spacing has increased (1 *Mpc* → 2 *Mpc*) with time. This scenario is consistent with the fact that the Universe is homogeneous.

• Consider galaxies on a sphere at a fixed distance *r*, as shown in figure 14.3(b), where galaxies are represented by points marked 1,2,3,.... As per Hubble's law the galaxies move radially outwards with a speed proportional to the distance. So the sphere of radius *r* expands, $r \rightarrow r + \Delta r$. But there is no change in the relative positions of the galaxies w.r.t. each other. Hubble's law is isotropic. It is the same irrespective of the direction. This scenario is consistent with the fact that the Universe is isotropic.

- In *exercise 14.1* we have illustrated Hubble's law of expansion in a 1D Universe and established its homogeneous nature.
- In *exercise 14.2* we have illustrated Hubble's law of expansion in a 2D Universe and established its homogeneous nature.

With Cepheid variables it is extremely difficult to find out distances of a very large number of very distant galaxies. Instead Hubble's law can be exploited.

Hubble's law is a consequence of observations or experiments. Hubble's law can be used to determine distances of galaxies. On a sufficiently large length scale we expect the galaxy distribution to be homogeneous i.e. density of a galaxy is the same everywhere. So one can construct a 3D map of the distribution of the galaxies using the technique of the galaxy redshift survey.

A part of the sky is identified over which a survey is to be done. Pictures are taken and galaxies identified. Angular position of galaxies is found out. Spectrum from galaxies is obtained and certain spectral lines identified. Redshift z of the galaxy is directly measured. Now the redshift z is given by (equation (13.21))

$$z = \frac{v}{c} \tag{14.9}$$

$$v = cz \tag{14.10}$$

We rewrite Hubble's law $v = Hr$ (equation (14.3)) using the present value of Hubble parameter $H = H_0$ as

$$v = H_0 r \tag{14.11}$$

Hence

$$H_0 r = cz$$

$$r = \frac{cz}{H_0} \tag{14.12}$$

Distances are thus measured from redshift z and this is the basis of the galaxy redshift survey.

The cz (or v) value turns out to be of the order of 10 000 $km\ s^{-1}$. So

$$r = \frac{cz}{H_0} = \frac{10\ 000\ km\ s^{-1}}{100\ km\ s^{-1}\ Mpc^{-1}} = 100\ Mpc \tag{14.13}$$

So the extent of the galaxy redshift survey is of the order of 100 Mpc.

The map or the pattern of galaxies obtained is like what is shown in figure 14.4 and is not definitely homogeneous. Arrangement of galaxies resembles a man in the sky which is actually a cluster of over 1000 identified galaxies and is called the coma cluster. The galaxies inside this cluster are gravitationally bound, which means that the individual galaxies inside the cluster are not moving away from one another as demanded by Hubble's law of expansion. This corresponds to a large deviation from

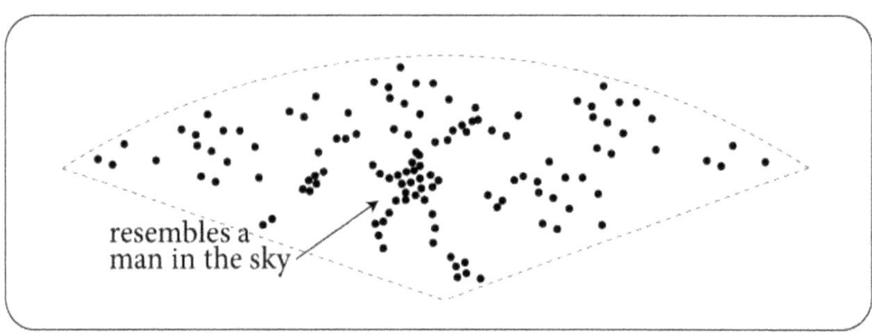

Figure 14.4. Map of a galaxy redshift survey.

Hubble's law. However, the entire cluster moves away from other observers according to Hubble's law.

Later mapping through larger redshift surveys covered millions of galaxies and a significant portion of the Universe could be mapped. It was seen that the galaxies seem to be distributed over an inter-connected network or a cosmic web and there are large voids too. So over a smaller length scale there is uneven distribution of galaxies. But over a large scale, say 80–100 Mpc, the galaxy distribution is homogeneous. This is comparable to the case of a solid which is homogeneous on a large scale but full of voids if a microscopic survey is done.

14.4 Cosmic noise

Another indication that the Universe is isotropic is cosmic microwave background radiation (CMBR) discovered by Penzias and Wilson in 1964.

There is a radiation that comes from all directions. This radiation has a blackbody spectrum, as shown in figure 14.5 corresponding to a temperature of $T = 2.73\,K$.

This radiation is cosmological in origin and fills the whole Universe. This radiation is isotropic (the same in all directions) to a high degree of accuracy. Actually, the angular fluctuation in temperature is $\frac{\Delta T}{T} \sim 10^{-5}$. This is one of the most important pieces of evidence that the Universe is isotropic.

In view of the previous facts and discussion, we conclude that the Universe is homogeneous and isotropic.

We now discuss what is the co-moving coordinate system that is needed to understand the expansion of the Universe, for which we need a model.

14.5 Co-moving coordinate system

Refer to figure 14.5. Consider a coordinate system that remains fixed to the galaxies. This coordinate system is referred to as the co-moving coordinate system. Since the Universe is expanding, as predicted by Hubble's law, it follows that the physical

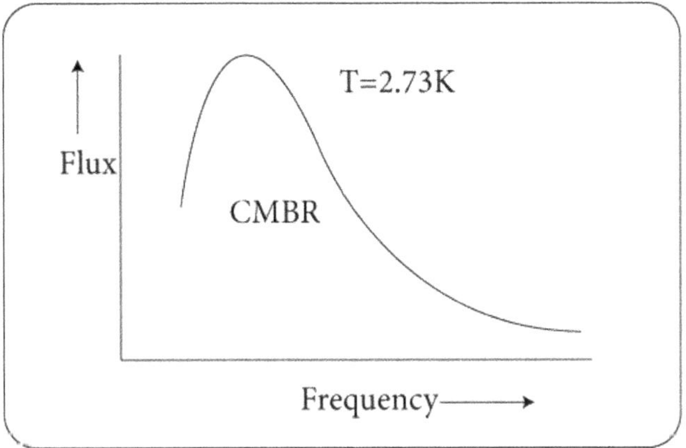

Figure 14.5. Cosmic microwave background radiation.

distance between any two galaxies increases. However, the co-moving coordinate system being attached to the galaxies undergoes no change. So a galaxy at co-moving coordinate $\vec{X} = (1, 0, 0)$ remains as it is in the future too, i.e. in spite of expansion. Clearly the co-moving coordinate system expands with the Universe.

The physical position $\vec{r}$ undergoes change because of expansion of the Universe due to which intermediate distance between galaxies changes and is referred to as the proper coordinate or proper position. The relation between the physical coordinate or proper coordinate $\vec{r}$ and the co-moving coordinate $\vec{X}$ is given by

$$\vec{r} = a(t)\vec{X} \tag{14.14}$$

where $a(t)$ is called the scale factor. Scale factor is dimensionless. Actually, the distance between two grid points is decided by the scale factor. The larger the scale factor, the larger is the distance between two grid points. When scale factor is zero the two grid points coincide. In other words, the infinite grid (for an infinite Universe) then collapses to a point. The volume becomes zero. This is a singularity referred to as the **Big Bang**. When there is no space, every law of physics breaks down. Density of the Universe becomes infinite.

Clearly $\vec{r}$ is a function of time but $\vec{X}$ is not since galaxies remain fixed in the co-moving frame of reference (figure 14.6).

Suppose an observer is at the origin. Let us find the speed with which a galaxy at a fixed co-moving coordinate moves.

Position of a galaxy as a function of time is given by equation (14.12), i.e.

$$\vec{r}(t) = a(t)\vec{X} \tag{14.15}$$

Velocity is given by

$$\vec{v}(t) = \frac{d}{dt}\vec{r}(t) = \frac{d}{dt}a(t)\vec{X} = \left[\frac{d}{dt}a(t)\right]\vec{X} \tag{14.16}$$

$$\vec{v}(t) = \dot{a}(t)\,\vec{X} \tag{14.17}$$

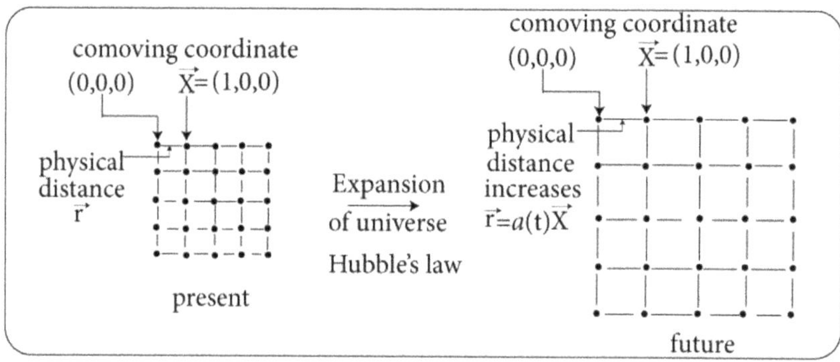

Figure 14.6. Co-moving coordinate system.

From equation (14.15) we have

$$\overrightarrow{X} = \frac{\vec{r}(t)}{a(t)}$$

(14.18)

Hence from equation (14.16)

$$\vec{v}(t) = \dot{a}(t)\frac{\vec{r}(t)}{a(t)} = \frac{\dot{a}(t)}{a(t)}\,\vec{r}(t)$$

(14.19)

So at any instant of time the speed with which a galaxy is moving away from the observer at origin is proportional to the distance of the galaxy $\vec{r}(t)$. Hence

$$\vec{v}(t) = H(t)\,\vec{r}(t)$$

(14.20)

Comparison of equations (14.19) and (14.20) gives

$$H(t) = \frac{\dot{a}(t)}{a(t)}$$

(14.21)

Hubble's parameter at any given time is the time derivative of scale factor divided by the scale factor.

Let us denote present time by $t = t_0$. Then Hubble's constant or parameter is

$$H(t_0) = H_0 = \frac{\dot{a}(t_0)}{a(t_0)} = \left.\frac{\dot{a}}{a}\right|_{t=t_0}$$

(14.22)

• Let us suppose that no force acts on the galaxy, i.e. we ignore the effect of gravity. Then the galaxies are all moving freely, i.e. moving with no acceleration and hence

$$\dot{\vec{v}} = 0$$

(14.23)

Differentiating equation (14.17) we have

$$\dot{\vec{v}}(t) = \ddot{a}(t)\,\overrightarrow{X}$$

(14.24)

From equations (14.23) and (14.24) we get

$$\ddot{a}(t) = 0$$

(14.25)

The most general solution to equation (14.25) is

$$a(t) = bt + k$$

(14.26)

Choosing scale factor $a(t) = 0$ at $t = 0$ we have $k = 0$. So

$$a(t) = bt$$

(14.27)

This equation (14.27) represents the law of free expansion in the absence of gravity. The plot of $a(t)$ versus t is a straight line (figure 14.7).

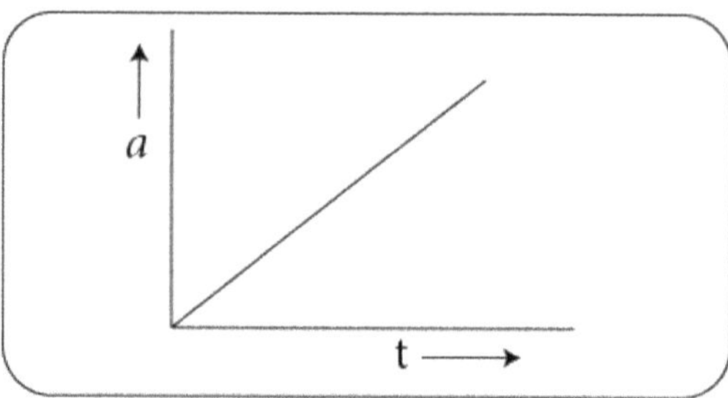

Figure 14.7. Scale factor as a function of time in the case of free expansion, i.e. expansion in the absence of gravity.

Now at $t = 0$ the scale factor $a(t) = 0$ is zero. Then all the galaxies are at the same point $\vec{r}(t = 0) = 0$, i.e. collapse to the origin. This is the singularity called the Big Bang. In other words at the Big Bang

$$t = 0, \ a = 0 \tag{14.28}$$

The Hubble parameter is $H(t) = \frac{\dot{a}}{a}$ (equation (14.21))
Now from equation (14.27)

$$\dot{a} = \frac{da}{dt} = \frac{d}{dt}(bt) = b \tag{14.29}$$

So from equations (14.21), (14.27) and (14.29) it follows that

$$H(t) = \frac{\dot{a}}{a} = \frac{b}{bt} = \frac{1}{t} \tag{14.30}$$

Due to breakdown of all known laws of physics we cannot think of the Universe before its collapse at $t = 0$. We cannot extrapolate beyond $t = 0$ the Big Bang, i.e. one cannot cross the singularity at $t = 0$. So we think of the Universe originating at $t = 0$.

In other words, $t = 0$ is the origin of the Universe. The t in equation (14.30) is the time that elapsed since $t = 0$ (**Big Bang**). So the Hubble parameter $H = \frac{1}{t}$ is the inverse of the age of the Universe since $t = 0$ (**Big Bang**).

• Using the present value of the Hubble parameter (equation 14.5) let us estimate the age of the Universe since $H(t_0) = H_0 = \frac{\dot{a}}{a}\bigg|_{t=t_0} = \frac{1}{t_0}$ (by equations (14.22) and (14.30)).

$$t_0 = \frac{1}{H_0} = \frac{1}{100\,h\ km\ s^{-1}\ Mpc^{-1}} \tag{14.31}$$

Using equation (13.1)

$$t_0 = \frac{1}{\frac{100\,h\ \times 10^3\ m\ s\ -1}{10^6 \times 3.09 \times 10^{16} m}} = \frac{10^6 \times 3.09 \times 10^{16}}{100h \times 10^3} s = 3.09 \times 10^{17} h^{-1} \left(\frac{1}{365 \times 24 \times 3600}\ yr \right)$$

$$t_0 = 9.798 \times 10^9 \, h^{-1} \, yr \tag{14.32}$$

Using equation (14.6), i.e. $h = 0.7$ we get

$$t_0 = 9.798 \times 10^9 \, (0.7)^{-1} \, yr$$

$$t_0 = H_0^{-1} = 1.4 \times 10^{10} \, yr \tag{14.33}$$

The inverse of the Hubble parameter H^{-1} is called the Hubble time and its present value is $H_0^{-1} = 1.4 \times 10^{10} \, yr$. It gives an estimate of the age of the Universe, i.e. gives the time that elapsed after the **Big Bang**. Clearly then the age of the Universe is $\sim 14 \times 10^9 \, yr$, i.e. 14 billion years.

We have to note that this calculation is an over simplification as we have overlooked the presence of forces, i.e. the effect of gravity.

14.5.1 Effect of gravitational force

On such large scale distance $\sim Mpc$, strong and weak nuclear forces will not come into the picture as they act over very small distances. As the Universe is found to be charge neutral to a large degree, the electromagnetic forces will also not bother the galaxies. Gravity is the only force that acts whenever matter is present. So we have to incorporate the effect of gravity.

• The attractive force of gravity tries to slow down the acceleration or expansion of the Universe.

14.6 Dynamics of the Universe

Consider an observer at rest at point O, as depicted in figure 14.8. We consider a spherical part of the Universe of radius r. We would like to study the motion of a galaxy at a distance r.

The Universe is spherically symmetric around the observer.

The equation of motion of the galaxy G of mass m (which is like a test particle) is according to Newton's law

$$m\frac{d^2 \vec{r}}{dt^2} = -G\frac{mM}{r^2}\hat{r} = -G\frac{mM}{r^3}\vec{r} \tag{14.34}$$

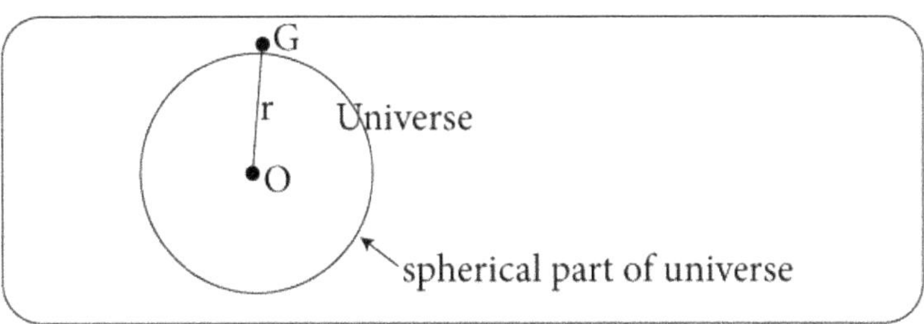

Figure 14.8. A spherical part of the Universe of radius r considered around the observer at point O and G is the galaxy whose motion is to be investigated.

where M is the mass inside the sphere. Only the mass M inside the sphere contributes, mass outside the sphere does not contribute.

$$\frac{d^2\,\vec{r}}{dt^2} = -G\frac{M}{r^3}\vec{r} \tag{14.35}$$

Clearly the left-hand side of this equation represents gravitational acceleration of the galaxy and absence of m in the equation means that this gravitational acceleration does not depend on its mass m). We rewrite it putting the value of M as follows.

$$\ddot{\vec{r}} = -G\frac{\frac{4}{3}\pi r^3 \rho}{r^3}\,\vec{r} = -\frac{4}{3}\pi G\rho\,\vec{r} \tag{14.36}$$

where ρ is density. Equation (15.34) corresponds to relativistic correction.

We can convert this equation in terms of the scale factor by switching over to a co-moving coordinate system.

From equation (14.14) viz. $\vec{r} = a(t)\vec{X}$, where $\vec{X}$ does not change, we have upon differentiating twice

$$\ddot{\vec{r}} = \ddot{a}(t)\vec{X} \tag{14.37}$$

From equations (14.32) and (14.33) we get

$$\ddot{a}(t)\vec{X} = -\frac{4}{3}\pi G\rho\,\vec{r} \tag{14.38}$$

With equation (14.14) viz. $\vec{r} = a(t)\vec{X}$ we rewrite

$$\ddot{a}(t)\vec{X} = -\frac{4}{3}\pi G\rho\,a(t)\vec{X}$$

$$\ddot{a}(t) = -\frac{4}{3}\pi G\rho\,a(t) \tag{14.39}$$

This simple scalar equation is the equation for scale factor.

Here density ρ is not a constant since as the Universe expands new matter is not created though more space gets involved. In fact, mass = (density)(volume) = constant and so we write

$$\rho a^3 = \text{constant} \tag{14.40}$$

where a^3 is the co-moving volume or physical volume. Writing in terms of the present value of scale factor viz. $a = a_0$ and present value of density viz. $\rho = \rho_0$ we have

$$\rho a^3 = \rho_0 a_0^3 \tag{14.41}$$

This is conservation of mass. Let us make a normalization by choosing $a_0 = 1$ (i.e. present value of the scale factor is taken to be unity). With this we rewrite equation (14.41) as

$$\rho a^3 = \rho_0 \tag{14.42}$$

$$\rho(t) = \rho_0 a^{-3} \tag{14.43}$$

Hence from equation (14.39) we have

$$\ddot{a} = -\frac{4}{3}\pi G \rho_0 a^{-3}\, a$$

$$\ddot{a} = -\frac{4}{3}\pi G \rho_0 a^{-2} \tag{14.44}$$

Multiplying both sides by $\dot{a}$

$$\dot{a}\ddot{a} = -\frac{4}{3}\pi G \rho_0 a^{-2}\dot{a} \tag{14.45}$$

$$\frac{1}{2}\frac{d}{dt}\dot{a}^2 = \frac{4}{3}\pi G \rho_0 \frac{d}{dt} a^{-1} \tag{14.46}$$

Integrating we have

$$\int \frac{1}{2}\frac{d}{dt}\dot{a}^2 = \int \frac{4}{3}\pi G \rho_0 \frac{d}{dt} a^{-1} + E \tag{14.47}$$

$$\frac{1}{2}\dot{a}^2 = \frac{4}{3}\pi G \rho_0\, a^{-1} + E \tag{14.48}$$

where E is a constant of integration. Equation (14.48) can be recast as

$$\frac{1}{2}\dot{a}^2 - G\frac{\frac{4}{3}\pi \rho_0}{a} = E \tag{14.49}$$

This equation resembles the familiar relation of mechanics viz.

$$\text{kinetic energy} + \text{potential energy} = \text{total energy.} \tag{14.50}$$

So we can physically identify E as energy of the Universe at a unit co-moving distance.

The expansion of scale factor $a = a(t)$ is governed by equation (14.48). This cosmological problem is very similar to the problem of a projectile (ball say) sent out vertically from the surface of the Earth. The equation of motion of the ball is also equation (14.35) where M is a constant.

In the cosmological problem we are discussing we see that, as galaxy G moves out due to the expanding nature of the Universe, mass M is still a constant. This ensures that the cosmological problem resembles the mechanical problem of a ball thrown vertically.

In the case of the mechanical problem of determining trajectory of the ball there can be three possibilities depending on the value of energy E of the ball, namely $E > 0$, $E = 0$, $E < 0$.

In the cosmological problem we are interested in, there are also three possibilities, namely $E > 0$, $E = 0$, $E < 0$.

We discuss the cases separately.

14.7 Case $E = 0$ or the critical model

Suppose a ball is thrown from the surface of Earth with $E = 0$. It has escape velocity and it escapes to infinity and comes to rest at infinite distance. This is a critical case.

In the cosmological case the plot of scale factor $a(t)$ against t is shown in figure 14.9. Scale factor expands continuously and expansion becomes zero after infinite time, i.e. asymptotically. So the $a(t)$ versus t curve becomes flat at very large t.

14.8 Case $E < 0$, Big Bang to Big Crunch

Suppose a ball is thrown from the surface of Earth with $E < 0$, i.e. thrown with negative energy. It cannot escape to infinity and falls back.

In the cosmological case the plot of scale factor $a(t)$ against t is shown in figure 14.9. Scale factor emerges out of Big Bang singularity and expands initially. But then the scale factor contracts and ends in another singularity called the Big Crunch.

14.9 Case $E > 0$, an expanding Universe

Suppose a ball is thrown from the surface of Earth with $E > 0$, i.e. thrown with positive energy. It always has some energy even after escaping to infinity.

In the cosmological case the plot of scale factor $a(t)$ against t is shown in figure 14.9. Scale factor expands continuously and expansion never becomes zero even after infinite time, i.e. the $a(t)$ versus t curve has upward inclination even at very large t and is never flat.

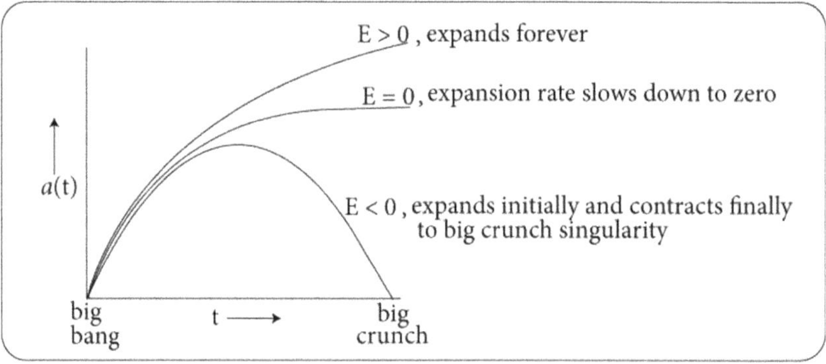

Figure 14.9. Plot of scale factor $a(t)$ versus t.

14.10 Critical mass density ρ_c

Figure 14.9 depicts the three possibilities regarding the fate of our Universe corresponding to $E < 0$, $E = 0$ and $E > 0$. Which of these three possibilities holds for our Universe is decided by the value of total energy E of our Universe.

To make an investigation let us rewrite equation (14.48) viz. $\frac{1}{2}\dot{a}^2 = \frac{4}{3}\pi G\rho_0\, a^{-1} + E$ that is valid at all instants of time using equation (14.42), i.e. $\rho a^3 = \rho_0$

$$\frac{1}{2}\dot{a}^2 = \frac{4}{3}\pi G\rho a^2 + E \tag{14.51}$$

Multiply by $\frac{2}{a^2}$ to get

$$\frac{\dot{a}^2}{a^2} = \frac{8}{3}\pi G\rho + \frac{2E}{a^2} \tag{14.52}$$

Using equation (14.21) viz. $H = \frac{\dot{a}}{a}$ we have

$$H^2 - \frac{8}{3}\pi G\rho = \frac{2E}{a^2} \tag{14.53}$$

Clearly the value of E is decided by which of the terms H^2 or $\frac{8}{3}\pi G\rho$ is larger. This can be determined by a measurement of Hubble parameter $H(t)$ and mass density ρ.
• Consider the case of $E = 0$
From equation (14.53) we have for $E = 0$

$$H^2 = \frac{8}{3}\pi G\rho \tag{14.54}$$

We define

$$\rho = \frac{3H^2}{8\pi G} = \rho_c \tag{14.55}$$

ρ_c is called critical mass density of the Universe.
Present value of critical density $\rho_c = \rho_{c0}$ is obtained by setting $H = H_0$ to get

$$\rho_{c0} = \frac{3H_0^2}{8\pi G} \tag{14.56}$$

• In *exercise 14.4* we show that $[\rho_{c0}] = [kg\ m^{-3}]$
From equation (14.56)

$$\rho_{c0} = \frac{3H_0^2}{8\pi G} \tag{14.57}$$

Using equation (14.5) viz. $H_0 = 100\, h\, kms^{-1}\, Mpc^{-1}$

$$\rho_{c0} = \frac{3(100\, h\, kms^{-1}\, Mpc^{-1})^2}{8\pi(6.67 \times 10^{-11}\, m^3\, kg^{-1}\, s^{-2})} = \frac{3(100)^2 h^2 (1000\, m)^2 \frac{1}{s^2}\frac{1}{Mpc^2}}{8\pi(6.67)(10^{-11})\, m^3\, kg^{-1}\, s^{-2}}$$

$$\rho_{c0} = 1.7895 \times 10^{19} h^2 \frac{kg}{Mpc^2\, m} \tag{14.58}$$

Using equation (13.1) viz. $1\, pc = 3.09 \times 10^{16}\, m$ we have

$$\rho_{c0} = 1.7895 \times 10^{19} h^2 \frac{kg}{(10^6)^2 (3.09 \times 10^{16}\, m)^2\, m}$$

$$\rho_{c0} = 1.874 \times 10^{-26}\, h^2\, kg\, m^{-3} \tag{14.59}$$

Using equation (14.6) viz. $h = 0.7$ we get

$$\rho_{c0} = 1.874 \times 10^{-26}\,(0.7)^2 kg\, m^{-3}$$

$$\rho_{c0} = 9.2 \times 10^{-27}\, kg\, m^{-3} \tag{14.60}$$

In *exercise 14.5* we have expressed ρ_{c0} in terms of solar mass per *Mpc* cube as

$$\rho_{c0} = 2.7735 \times 10^{11} \frac{M_{sun}}{(Mpc)^3} \tag{14.61}$$

Clearly if $E = 0$ then density of the Universe is

$$\rho = \rho_{c0} \tag{14.62}$$

and the Universe expands continuously. Expansion becomes zero after infinite time, i.e. asymptotically (figure 14.9). We then live in a Universe that follows the critical cosmological model.

• It follows from equation (14.53) viz. $H^2 - \frac{8}{3}\pi G\rho = \frac{2E}{a^2}$ that if

$$H^2 > \frac{8}{3}\pi G\rho \text{ i. e. } \frac{3H^2}{8\pi G} > \rho \tag{14.63}$$

$$\rho < \frac{3H^2}{8\pi G} \tag{14.64}$$

$$\rho < \rho_c \tag{14.65}$$

then $E > 0$ and scale factor expands continuously. Expansion never becomes zero even after infinite time (figure 14.9).

• It follows from equation (14.53) viz. $H^2 - \frac{8}{3}\pi G\rho = \frac{2E}{a^2}$ that if

$$H^2 < \frac{8}{3}\pi G\rho \text{ i. e. } \frac{3H^2}{8\pi G} < \rho \tag{14.66}$$

$$\rho > \frac{3H^2}{8\pi G} \tag{14.67}$$

$$\rho > \rho_c \tag{14.68}$$

then $E < 0$ and scale factor emerges out of Big Bang singularity and expands initially. But then the scale factor contracts and ends in another singularity called the Big Crunch (figure 14.9).

Using equation (14.55) viz. $\frac{3H^2}{8\pi G} = \rho_c$ i.e.

$$H^2 = \frac{8\pi G}{3}\rho_c \tag{14.69}$$

we recast equation (14.53) viz. $H^2 - \frac{8}{3}\pi G\rho = \frac{2E}{a^2}$ as

$$\frac{8\pi G}{3}\rho_c - \frac{8}{3}\pi G\rho = \frac{2E}{a^2} \tag{14.70}$$

$$\frac{8\pi G}{3}\rho_c\left(1 - \frac{\rho}{\rho_c}\right) = \frac{2E}{a^2} \tag{14.71}$$

14.11 Cosmological parameter Ω

To parametrize the density of the Universe we define the ratio of the actual mass density ρ to the critical mass density ρ_c of the Universe, i.e.

$$\frac{\rho}{\rho_c} = \Omega \tag{14.72}$$

as the cosmological parameter Ω.

The present value of the cosmological parameter is denoted by Ω_0 and corresponds to $\rho = \rho_0$ as

$$\Omega_0 = \frac{\rho}{\rho_{c0}} \tag{14.73}$$

If $\Omega = 1$ then $\rho = \rho_c$, $E = 0$ which is the critical model.
We rewrite equation (14.71) as

$$\frac{8\pi G}{3}\rho_c(1 - \Omega) = \frac{2E}{a^2} \tag{14.74}$$

$$\text{If } \Omega > 1 \text{ then } \rho > \rho_c, \ E < 0 \tag{14.75}$$

$$\text{If } \Omega < 1 \text{ then } \rho < \rho_c, \ E > 0 \tag{14.76}$$

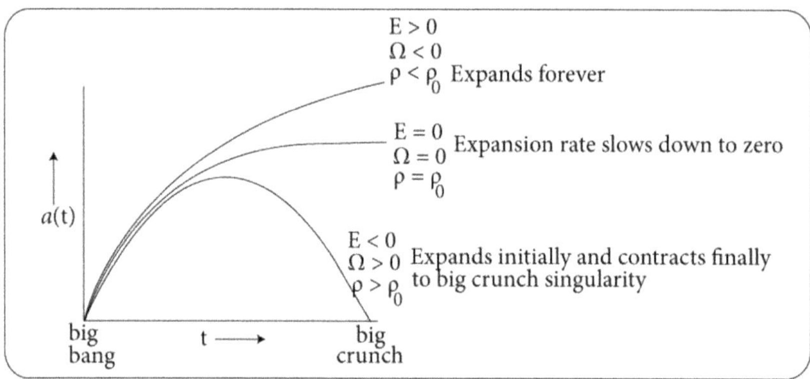

Figure 14.10. Plot of scale factor $a(t)$ versus t in terms of cosmological parameter Ω.

We have redrawn figure 14.9 labelling it with the cosmological parameter Ω in figure 14.10.

14.12 Einstein–de Sitter cosmological solution for $\Omega = 1$

The cosmological parameter $\Omega = 1$ corresponds to total energy of a galaxy to be $E = 0$. This is the critical cosmological model.

Equation(14.48) with $E = 0$ is

$$\frac{1}{2}\dot{a}^2 = \frac{4}{3}\pi G\rho_0\, a^{-1} \tag{14.77}$$

$$\dot{a}^2 = \frac{8}{3}\pi G\rho_0 \frac{1}{a} \tag{14.78}$$

$$\frac{da}{dt} = \sqrt{\frac{8}{3}\pi G\rho_0} \frac{1}{\sqrt{a}} \tag{14.79}$$

Integrating we get

$$\int \sqrt{a}\, da = \sqrt{\frac{8}{3}\pi G\rho_0} \int dt + K \;\; (K \text{ is constant of integration}) \tag{14.80}$$

$$\frac{2}{3}a^{3/2} = \sqrt{\frac{8}{3}\pi G\rho_0}\, t + K \tag{14.81}$$

K can be chosen such that scale factor $a = 0$ at $t = 0$. Then $K = 0$. Hence

$$\frac{2}{3}a^{3/2} = \sqrt{\frac{8}{3}\pi G\rho_0}\, t \tag{14.82}$$

Simplification of equation (14.82) gives

$$a = \left[\frac{3}{2} \sqrt{\frac{8}{3}\pi G\rho_0}\, t \right]^{2/3} = \left[\sqrt{\frac{3}{2}\frac{3}{2}\frac{8}{3}\pi G\rho_0} \right]^{2/3} t^{2/3}$$

$$a(t) = (6\pi G\rho_0)^{1/3}\, t^{2/3} = Ct^{2/3} \tag{14.83}$$

This is the final solution for the cosmological model with $E = 0$, $\Omega = 1$. This is the simplest possible cosmological model. It is called the Einstein–de Sitter solution.
• For $\Omega \neq 1$ solution becomes complicated.
It is clear from equation (14.83) that

$$a \propto t^{2/3} \tag{14.84}$$

Let us evaluate Hubble parameter $H(t)$ from equation (14.21)

$$H(t) = \frac{\dot{a}}{a} = \frac{1}{a}\frac{da}{dt} \tag{14.85}$$

Using equation (14.83) we have

$$H(t) = \frac{1}{a}\frac{da}{dt} = \frac{1}{Ct^{2/3}}\frac{d}{dt}\, Ct^{2/3} = \frac{1}{t^{\frac{2}{3}}}\frac{2}{3}\, t^{-\frac{1}{3}}$$

$$H(t) = \frac{2}{3t} \tag{14.86}$$

This result is similar to equation (14.30) viz. $H = \frac{1}{t}$ which corresponds to free expansion. Now from equation (14.86)

$$t = \frac{2}{3H} \tag{14.87}$$

Considering the present value of Hubble parameter $H = H_0$ we have the age of Universe in this cosmological model to be

$$t_0 = \frac{2}{3H_0} \tag{14.88}$$

Using equation (14.33) viz. $H_0^{-1} = 1.\,4 \times 10^{10}\, yr$ we have

$$t_0 = \frac{2}{3}H_0^{-1} = \frac{2}{3}1.4 \times 10^{10}\, yr = 9 \times 10^9\, yr \tag{14.89}$$

The suggested age of the Universe is less than what we obtained in equation (14.33) viz. $t_0 = H_0^{-1} = 1.\,4 \times 10^{10}\, yr$. This is because gravity slows down the expansion of the Universe.

In the analysis we used Newtonian mechanics. We would have got the same result had we used the general theory of relativity.

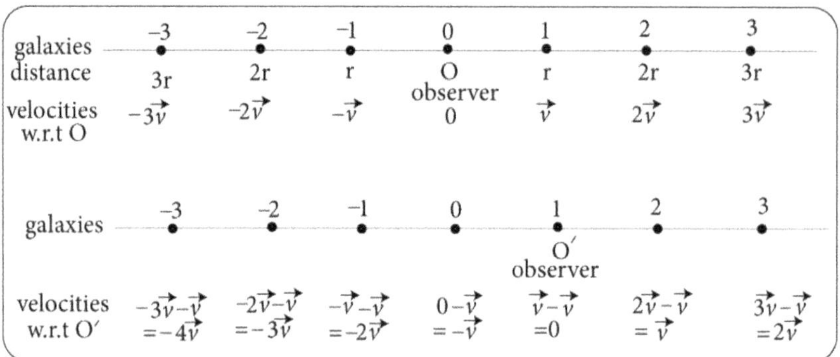

Figure 14.11. Hubble's law of expansion in a 1D Universe and its homogeneous nature.

14.13 Exercise

Exercise 14.1 *Illustrate Hubble's law of expansion in a 1D Universe and establish its homogeneous nature.*

Ans. Consider the Universe to be 1D, as shown in figure 14.11. Galaxies are at points $0, \pm 1, \pm 2, \pm 3, \ldots$

Consider an observer O stationed at point 0. The velocities with which they move are given by Hubble's law (velocity proportional to distance) and shown in figure 14.11.

Consider another observer O' stationed at point 1 that was earlier moving with velocity $\vec{v}$. The velocities of other neighbouring galaxies can be calculated using Galilean velocity transformation relation in 1D and what we get is depicted in the figure.

We find that the velocities of galaxies evaluated by observers O and O' at different positions are the same, implying that the Universe is homogeneous. So Hubble's law holds in a homogeneous 1D Universe. Use of Lorentz transformation relation to move from one galaxy to another yields the same result.

Exercise 14.2 *Illustrate Hubble's law of expansion in a 2D Universe and establish its homogeneous nature.*

Ans. Consider the Universe to be 2D, as shown in figure 14.12. Galaxies are at points 1, 2, 3.

Consider observer O stationed at point 1. The velocities with which they move are given by Hubble's law (velocity proportional to distance) and shown in figure to be $0, H_0\vec{a}, H_0\vec{b}$, i.e. neighbouring galaxies 2 and 3 are moving away radially outwards.

Consider another observer O' stationed at point 2 that was earlier moving with velocity $H_0\vec{a}$. The velocities of other neighbouring galaxies can be calculated using Galilean velocity transformation relation in 2D and what we get is depicted in the

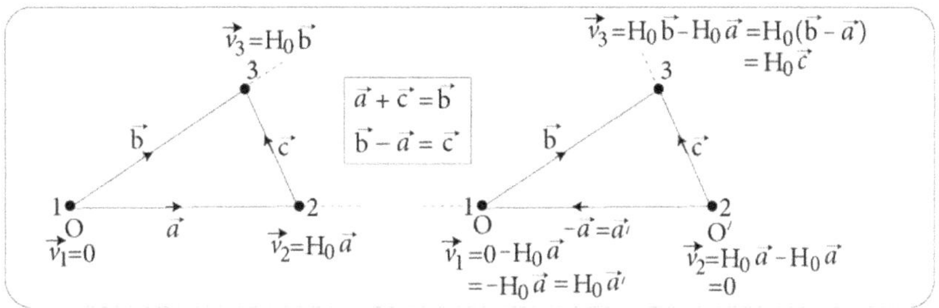

Figure 14.12. Hubble's law of expansion in a 2D Universe and its homogeneous nature.

figure to be 0, $H_0\vec{a}'$, $H_0\vec{c}$, i.e. neighbouring galaxies 1 and 3 are moving away radially outwards.

We find that the velocities of galaxies evaluated by observers O and O' at different positions are the same, implying that the Universe is homogeneous. So Hubble's law holds in a homogeneous 2D Universe. Use of Lorentz transformation relation to move from one galaxy to another yields the same result.

Exercise 14.3 *Show that* $[H_0] = [s^{-1}]$.
[Ans.] From equation (14.1) we write Hubble's law

$$H_0 = \frac{v}{r}$$

$$[H_0] = \left[\frac{v}{r}\right] = \left[\frac{ms^{-1}}{m}\right] = [s^{-1}]$$

Exercise 14.4 *Show that* $[\rho_{c0}] = [kg\ m^{-3}]$
[Ans.] From equation (14.55)

$$\rho_{c0} = \frac{3H_0^2}{8\pi G}$$

$$[\rho_0] = \left[\frac{3H_0^2}{8\pi G}\right] = \left[\frac{(s^{-1})^2}{kg^{-1}m^3s^{-2}}\right] \text{ (by } exercise\ 14.3\ [H_0] = [s^{-1}])$$

$$= \left[\frac{s^{-2}}{kg^{-1}\ m^3\ s^{-2}}\right] = kg\ m^{-3}$$

Exercise 14.5 *Express* ρ_{c0} *of equation (14.56)* *in terms of solar mass per Mpc cube.*
[Ans.] From equation (14.56)

$$\rho_{C0} = \frac{3H_0^2}{8\pi G}$$

Putting respective values we have equation (14.59) viz.

$$\rho_{C0} = 1.874 \times 10^{-26}\, h^2 kg\ m^{-3}$$

As per equation (13.4) $M_{sun} = 1.99 \times 10^{30}\ kg$
As per equation (13.1) $1\ pc = 3.09 \times 10^{16}\ m$
Hence

$$\frac{M_{sun}}{(Mpc)^3} = \frac{1.99 \times 10^{30}\ kg}{[10^6(3.09 \times 10^{16}\ m)]^3} = 6.75 \times 10^{-38}\ kg\ m^{-3}$$

Hence $\frac{kg}{m^3} = \frac{1}{6.75 \times 10^{-38}} \frac{M_{sun}}{(Mpc)^3} = 1.48 \times 10^{37} \frac{M_{sun}}{(Mpc)^3}$
So from equation (14.59) we have

$$\rho_{c0} = 1.874 \times 10^{-26}\, h^2 kg\ m^{-3} = 1.874 \times 10^{-26}\, h^2 1.48 \times 10^{37} \frac{M_{sun}}{(Mpc)^3}$$

$$\rho_{c0} = 2.7735 \times 10^{11} h^2 \frac{M_{sun}}{(Mpc)^3}$$

Using equation (14.6) viz. $h = 0.7$

$$\rho_{c0} = 2.7735 \times 10^{11}(0.7)^2 \frac{M_{sun}}{(Mpc)^3}$$

$$\rho_{c0} = 1.36 \times 10^{11} \frac{M_{sun}}{(Mpc)^3}$$

Exercise 14.6 *Length scale over which Galaxy distribution can be taken to be homogeneous is*

(a) 80 – 100 Kpc　　　　*(b) 8 – 10 Kpc*　　　　*(c) 80 – 100 Mpc*　　　　*(d) 8 – 10 Mpc.*

Exercise 14.7 *Hubble's constant is*

(a) 100 km s⁻¹ Mpc⁻¹　　*(b) 70 km s⁻¹ Mpc⁻¹*　　*(c) 100 m s⁻¹ Gpc⁻¹*　　*(d) 70 km s⁻¹ Gpc⁻¹.*

Exercise 14.8 *The Universe is homogeneous and isotropic. What is the best statement regarding this law?*

(a) Hertzsprung–Russell principle　　*(b) General theory of relativity*　　*(c) Cosmological principle*　　*(d) Hubble's law.*

Exercise 14.9 *Which statement is true according to cosmological principle?*
   (a) *The Universe is homogeneous and isotropic.*
   (b) *The Universe is homogeneous but not isotropic.*
   (c) *The Universe is not homogeneous but isotropic.*
   (d) *The Universe is neither homogeneous nor isotropic.*

Exercise 14.10 *Expansion of the Universe is corroborated by which of the following?*

(a) Cosmological principle　　*(b) Hubble's law*　　*(c) Redshift*　　*(d) Options a and b.*

Exercise 14.11 *Which relation is used in a galaxy redshift survey?*

(a) $v = H_0 r$　　　*(b)* $r = \frac{cz}{H_0}$　　　*(c)* $\lambda_O = \lambda_G(1 - \frac{v}{c})$　　　*(d)* $\lambda_O = \lambda_G(1 + \frac{v}{c})$.

Exercise 14.12 *The extent of a galaxy redshift survey is of the order of*

(a) 1 *pc*　　　*(b)* 10 *pc*　　　*(c)* 10^4 *pc*　　　*(d)* 10^8 *pc.*

Exercise 14.13 *The most important evidence that the Universe is isotropic is*

(a) Expansion of the Universe　　*(b) Redshift*　　*(c) CMBR*　　*(d) Hubble's law.*

Exercise 14.14 *The CMBR has a blackbody spectrum corresponding to a temperature of*

(a) $T = 0\ K$　　　*(b)* $T = 2.73\ K$　　　*(c)* $T = 5\ K$　　　*(d)* $T = 273\ K$.

Exercise 14.15 *The expansion of the Universe corresponds to which of the following if Ω is the cosmological parameter?*

(a) $\Omega = 1$　　　*(b)* $\Omega > 1$　　　*(c)* $\Omega < 1$　　　*(d) all.*

Exercise 14.16 *The expansion of Universe corresponds to which of the following if ρ is the critical density?*

(a) $\rho = \rho_c$ *(b)* $\rho > \rho_c$ *(c)* $\rho < \rho_c$ *(d) all.*

$\boxed{\text{Answers to multiple choice questions}}$
14.6*c*, 14.7*b*, 14.8*c*, 14.9*a*, 14.10*a*, *b*, *c*, 14.11*b*, 14.12*d*, 14.13*c*, 14.14*b*, 14.15*d*, 14.16*d*.

14.14 Question bank

Q14.1 What is Hubble's law?

Q14.2 What is the cosmological principle?

Q14.3 How do we explain the fact that the Universe is homogeneous in spite of expansion?

Q14.4 How do we explain the fact that the Universe is isotropic in spite of expansion?

Q14.5 What is a galaxy redshift survey?

Q14.6 What is coma cluster ?

Q14.7 Mention a few exceptions to Hubble's law of expansion.

Q14.8 What is CMBR?

Q14.9 The most important evidence that the Universe is isotropic is CMBR. Justify.

Q14.10 How is the Hubble constant related to the scale factor?

Q14.11 What is the present value of Hubble's constant?

Q14.12 How can you predict age of the Universe from Hubble's constant in the case of free expansion?

Q14.13 What is Hubble time?

Q14.14 Show in a diagram the plot of scale factor against time in the case of energy of the Universe being zero, positive or negative. Interpret the three curves.

Q14.15 Define the terms: critical density, cosmological parameter. How are they related to Hubble's parameter? Show using a diagram the fate of the Universe for various values of critical density and cosmological parameter.

Q14.16 Obtain the Einstein–de Sitter cosmological solution for unit value of the cosmological parameter.

Q14.17 Obtain the age of the Universe in the critical cosmological model. Explain why the age turns out to be less than the age determined for free expansion of the Universe.

Further reading

[1] Bharadwaj S 2008 *Physical Cosmology Lecture Series* (Department of Physics, IIT Kharagpur)
[2] Hawking S 1988 *A Brief History of Time* (London: Bantam)
[3] Bonometto S, Gorini V and Moschella U 2001 *Modern Cosmology* (Bristol: IOP)
[4] Roos M 1994 *Introduction to Cosmology* (New York: Wiley)
[5] Tenreiro R D and Quiros M 1988 *An Introduction to Cosmology and Particle Physics* (Singapore: World Scientific Publication)

IOP Publishing

Nuclear and Particle Physics with Cosmology, Volume 2
Particle physics and cosmology
Jyotirmoy Guha

Chapter 15

Components of the Universe

Components of the Universe are dealt with in this chapter. Pressure of a relativistic fluid is found. We define curvature density. Various constituents of universe are discussed such as matter, radiation, curvature and cosmological constant. We discuss radiation dominated Universe, matter dominated Universe, curvature dominated Universe and cosmological constant dominated Universe. Deceleration parameter is defined and its value for various types of situations is determined, for instance when the Universe is matter dominated, when the Universe is curvature dominated, when the Universe is radiation dominated etc.

15.1 Pressure of a relativistic fluid

Consider a box filled with a gas of relativistic particles moving randomly. In other words we can think of a cavity filled with photons. This is shown in figure 15.1.

To make things simple let us assume that the particles all have the same energy E. Momentum of a relativistic particle (say a photon) is

$$p = \frac{E}{c} \tag{15.1}$$

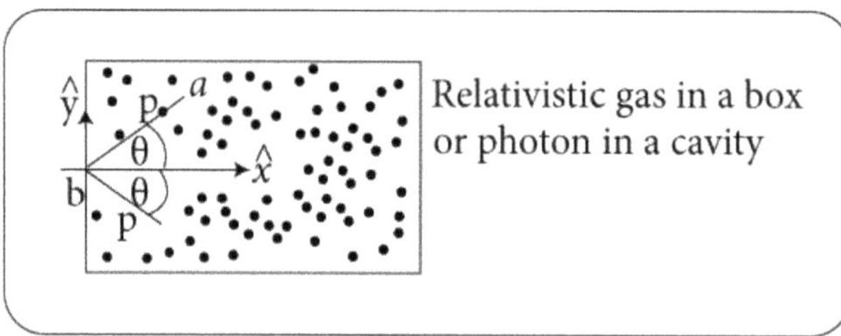

Figure 15.1. Gas of relativistic particles moving randomly in a box or a cavity filled with photons. $\hat{x}$ is normal to the wall.

doi:10.1088/978-0-7503-5032-7ch15

15-1

Consider a particle that hits the wall, as shown in figure 15.1, at angle θ and gets reflected.

Change in the momentum in the $\hat{x}$ direction is

$$\Delta p_x = 2p \cos\theta = 2\frac{E}{c}\cos\theta \ [(\text{mass})(\text{velocity}) = MLT^{-1}] \tag{15.2}$$

There is no change along $\hat{y}$.

Pressure P is force per unit area, i.e. rate of change of momentum per unit area. As pressure acts normal to surface we are interested in momentum change along the $\hat{x}$ component.

Rate at which particles cross surface area
$$\begin{aligned}
&= \text{number /area/second} = (\text{number/volume})(\text{distance/second}) \\
&= (\text{ number density of particles } n)\,(\text{speed}) \\
&= n(c\cos\theta)\left[L^{-3}LT^{-1} = L^{-2}T^{-1}\right]
\end{aligned} \tag{15.3}$$

$$\begin{aligned}
\text{Pressure} &= \Delta p_x\, n(c\cos\theta)\ [MLT^{-1}.\ L^{-2}T^{-1}] = [MLT^{-2}.\ L^{-2}] = \text{Force/Area} \\
&= 2\frac{E}{c}\cos\theta\, n(c\cos\theta) = 2En\cos^2\theta
\end{aligned} \tag{15.4}$$

As particles are moving randomly we have to take average value and so pressure is

$$P = 2En\langle\cos^2\theta\rangle \tag{15.5}$$

Now

$$\langle\cos^2\theta\rangle = \frac{1}{2}\frac{\int\cos^2\theta\,d\Omega}{4\pi} = \frac{1}{2}\cdot\frac{1}{4\pi}\int_0^\pi\int_0^{2\pi}\cos^2\theta\,\sin\theta\,d\theta\,d\phi \tag{15.6}$$

(We have divided by a factor of 2 because particles can move both ways along ab (which we are interested in) as well as along ba (which are not interested with and should be omitted.)

$$\langle\cos^2\theta\rangle = \frac{1}{2}\cdot\frac{1}{4\pi}\int_0^\pi\cos^2\theta\,\sin\theta\,d\theta\int_0^{2\pi}d\phi = \frac{1}{8\pi}\cdot\frac{2}{3}\cdot 2\pi = \frac{1}{6} \tag{15.7}$$

From equation (15.4)

$$P = 2En\langle\cos^2\theta\rangle = 2En\frac{1}{6} = \frac{1}{3}En = \frac{1}{3}\epsilon \tag{15.8}$$

$$P = \frac{1}{3}\epsilon \tag{15.9}$$

where $\epsilon = En$ is energy density.

- We performed Newtonian analysis in which the gravitational force on an object G (figure 14.8) occurs only due to the mass distribution inside the spherical region. But things are different in general relativity in which the energy density and pressure inside the spherical region contributes to the gravitational attraction. Then corrections by Einstein's theory have to be made to Newtonian analysis.

  So we have to find out if the material filling the Universe has pressure and then make suitable corrections.

- Pressure arises due to random motion. For the Universe the galaxies are on the average moving away from us with speed proportional to distance which is not random motion. For galaxies the randomness in the motion (which is due to deviation from Hubble's law) is of the order of $v \sim 100 \; km \; s^{-1}$. So for galaxies the pressure is negligible. This can be shown as follows.

Pressure becomes important under the condition

$$\frac{P}{\rho c^2} \sim 1 \left[\frac{P}{\rho c^2} = \frac{F}{A} \cdot \frac{1}{\frac{M}{L^3}\left(\frac{L}{T}\right)^2} = \frac{MLT^{-2}}{L^2} \cdot \frac{L^3}{ML^2 T^{-2}} = M^0 L^0 T^0 \right] \quad (15.10)$$

where P is pressure, c is speed of light in free space and ρ is mass density.
Now consider

$$P \sim \rho v^2$$

$$\frac{P}{\rho c^2} = \frac{\rho v^2}{\rho c^2} = \frac{v^2}{c^2} \sim 1 \quad \text{(by equation (15.10))} \quad (15.11)$$

$$\frac{v}{c} \sim 1.$$

So pressure is important for $v \sim c$, i.e. for relativistic gas.
With $v \sim 100 \; km \; s^{-1}$ we have

$$\frac{P}{c^2} = \frac{v^2}{c^2} = \frac{(100 \; km \; s^{-1})^2}{(3 \times 10^8 \; m \; s^{-1})^2} = 10^{-7} \ll 1 \quad (15.12)$$

So effect of pressure is negligible for the random motion of the galaxies.

15.2 Effects of pressure

We now discuss the modifications that occur in the case of pressure.
Consider a part of the Universe. Suppose it has co-moving volume unity and physical volume is

$$V = a^3 \quad (15.13)$$

Let us assume that the expansion of the Universe is adiabatic, i.e. there is no heat exchange between the part of the Universe considered and its surroundings. In fact since the Universe is homogeneous and isotropic, temperature is the same in the two regions and heat will not flow.
From the first law of thermodynamics we write for the volume V

$$dU + pdV = đQ = 0 \quad (15.14)$$

(as $đQ = 0$ for adiabatic expansion)
Internal energy

$$U = \epsilon a^3 \; (=\text{energy density} \times \text{volume}) \quad (15.15)$$

And assume pressure is given by the equation of state having the form

$$P = W\epsilon \tag{15.16}$$

where W is a constant.

For relativistic fluid or gas

$$P = \frac{1}{3}\epsilon \tag{15.17}$$

and so

$$W = \frac{1}{3} \tag{15.18}$$

For galaxies pressure due to dust is zero, i.e. for dust

$$P = 0 \tag{15.19}$$

and so from equation (15.16)

$$W = 0 \tag{15.20}$$

From equations (15.14), (15.15), (15.16) and (15.13) we have

$$d(\epsilon a^3) + W\epsilon da^3 = 0 \tag{15.21}$$

$$a^3 d\epsilon + \epsilon da^3 + W\epsilon da^3 = 0 \tag{15.22}$$

Dividing equation (15.22) by ϵa^3 we get

$$\frac{d\epsilon}{\epsilon} + \frac{da^3}{a^3} + W\frac{da^3}{a^3} = 0$$

$$\frac{d\epsilon}{\epsilon} + (1 + W)\frac{da^3}{a^3} = 0$$

$$d \ln \epsilon + (1 + W)d \ln a^3 = 0$$

$$d \ln \epsilon + d[(1 + W)\ln a^3] = 0$$

$$d \ln \epsilon + d \ln a^{3(1+W)} = 0$$

$$d \ln (\epsilon a^{3(1+W)}) = 0 \tag{15.23}$$

Integrating we have

$$\epsilon a^{3(1+W)} = \text{constant} = \epsilon_0 \text{ (the present value of energy density)} \tag{15.24}$$

A convenient way is to write as

$$\epsilon - \rho c^2 \left[\frac{\epsilon}{\rho c^2} = \frac{J}{V\left(\frac{M}{V}\right)\left(\frac{L}{T}\right)^2} = \frac{MLT^{-2} \cdot L}{M} \frac{1}{L^2 T^{-2}} = M^0 L^0 T^0 \right] \tag{15.25}$$

From equation (15.24) we get

$$\rho c^2 a^{3(1+W)} = \text{constant} \tag{15.26}$$

$$\rho a^{3(1+W)} = \text{constant} = \rho_0 \text{ (the present value of mass density)} \tag{15.27}$$

A fluid with no pressure is called dust.

- For dust we have

$$P = 0, \quad W = 0 \text{ (equations (15.19) and (15.20))} \tag{15.28}$$

Equation (15.26) gives

$$\rho a^3 = \rho_0 \tag{15.29}$$

i.e. for dust mass density falls as

$$\rho \propto \frac{1}{a^3} \tag{15.30}$$

- For relativistic gas

$$P = \frac{1}{3}\rho c^2, \quad W = \frac{1}{3} \text{ (equations (15.17) and (15.18))} \tag{15.31}$$

Equation (15.27) gives

$$\rho a^{3\left(1+\frac{1}{3}\right)} = \text{constant} = \rho_0$$

$$\rho a^4 = \rho_0 \tag{15.32}$$

i.e. for relativistic gas, mass density falls as

$$\rho \propto \frac{1}{a^4} \tag{15.33}$$

Comparing equations (15.30) and (15.33) we find that the fall of mass density is faster for relativistic gas (in which $\rho \propto \frac{1}{a^4}$) than the no-pressure case (in which $\rho \propto \frac{1}{a^3}$). Actually, energy is conserved. As the Universe expands the pressure of gas that acts outwards, does work and energy is expended. So energy density falls faster compared to the no-pressure case.

We see that the way density changes as the Universe expands, depends on the pressure of the medium. This feature has to be incorporated in the model in the case of existence of pressure.

- Another modification that occurs in the case of pressure.

In Newtonian gravity the acceleration of galaxy G (figure 14.8) is due to the mass inside the spherical region. If we apply Einstein's theory of relativity and redo the calculations we get a result different from equation (14.36).

The correct expression for acceleration of a particle on the surface of a sphere from Einstein's theory, turns out to be

$$\ddot{\vec{r}} = -\frac{4}{3}\pi G\left(\rho + \frac{3P}{c^2}\right)\vec{r} \tag{15.34}$$

Clearly this equation (15.34) (obtained from Einstein's relativity) incorporates both matter and pressure as evident from the presence of terms involving mass density ρ and pressure P on the RHS.

On the other hand, equation (14.36) incorporates only matter in the expression of acceleration as evident from the presence of a term involving mass density ρ on the RHS. Previously, we considered the Universe to be filled with only galaxies with no pressure. In Newtonian calculation the pressure term $\frac{P}{c^2}$ does not occur.

We can convert this equation (15.34) in terms of scale factor $a(t)$ by switching over to the co-moving coordinate system. From equation (14.14) viz. $\vec{r} = a(t)\,\vec{X}$, where $\vec{X}$ does not change we have upon differentiating twice

$$\ddot{\vec{r}} = \ddot{a}(t)\,\vec{X} \tag{15.35}$$

From equations (15.34), (15.35) and (14.14) we have

$$\ddot{a}(t)\,\vec{X} = =-\frac{4}{3}\pi G\left(\rho + \frac{3P}{c^2}\right)a(t)\,\vec{X}$$

$$\ddot{a}(t) = -\frac{4}{3}\pi G\left(\rho + \frac{3P}{c^2}\right)a(t) \tag{15.36}$$

Using equation (15.16) viz. $P = W\epsilon$ and equation (15.25) viz. $\epsilon = \rho c^2$ we have the equation of state

$$P = W\rho\,c^2 \tag{15.37}$$

From equations (15.36) and (15.37)

$$\ddot{a}(t) = -\frac{4}{3}\pi G\left(\rho + \frac{3W\rho\,c^2}{c^2}\right)a(t)$$

$$\ddot{a}(t) = -\frac{4}{3}\pi G\rho a(1 + 3W) \tag{15.38}$$

This is the equation for scale factor. We can generalize it further.

Again, we have to consider the cosmic microwave background radiation of temperature 2.73 K that fills the entire Universe and is a relativistic gas (photons) filling the whole Universe.

So the Universe has many components filling it and from the principle of equivalence it follows that all the components contribute to the gravitational force and gravitational acceleration. Gravity does not distinguish between them.

15.3 Curvature of space

Let us consider the Universe to be filled with more than one fluid (i.e. material). As scale factor a is the same for all components or cosmological bodies or media, equation (15.38) becomes

$$\ddot{a}(t) = -\frac{4}{3}\pi G a \sum_i (1 + 3W_i)\rho_i, \quad i = 1, 2, 3, \ldots \tag{15.39}$$

(i is the label for various constituents of the Universe).

The constituents are assumed to interact with one another through gravity. Then the energy of each of them is individually conserved. Then with $a_0 = 1$ (present scale factor has been taken to be 1) we have, from equation (15.27), for the ith constituent

$$\rho_i a^{3(1+W_i)} = \rho_{i0} \tag{15.40}$$

$$\rho_i = \rho_{i0} a^{-3(1+W_i)} \tag{15.41}$$

where ρ_{i0} is the present value of the particular component.

From equations (15.39), (15.40)

$$\ddot{a}(t) = -\frac{4}{3}\pi G a \sum_i (1 + 3W_i)\rho_{i0} a^{-3(1+W_i)}$$

$$\ddot{a}(t) = -\frac{4}{3}\pi G \sum_i (1 + 3W_i)\rho_{i0} a^{-3(1+W_i)+1} \tag{15.42}$$

We have to solve equation (15.42) now.

Multiply both sides of equation (15.42) by $\dot{a}$

$$\dot{a}\,\ddot{a}(t) = -\frac{4}{3}\pi G \sum_i (1 + 3W_i)\rho_{i0} a^{-3(1+W_i)+1}\,\dot{a}$$

$$\frac{1}{2}\frac{d}{dt}\dot{a}^2 = -\frac{4}{3}\pi G \sum_i [-(1 + 3W_i)a^{-3(1+W_i)+1}\dot{a}]\,\rho_{i0} \tag{15.43}$$

Since

$$\frac{d}{dt}a^{-3(1+W_i)+2} = [-3(1 + W_i) + 2]a^{-3(1+W_i)+2-1}\dot{a} = -(1 + 3W_i)a^{-3(1+W_i)+1}\dot{a}$$

We rewrite equation (15.43) as follows:

$$\frac{1}{2}\frac{d}{dt}\dot{a}^2 = \frac{4}{3}\pi G \sum_i \frac{d}{dt}a^{-3(1+W_i)+2}\,\rho_{i0} \tag{15.44}$$

Integrating we have

$$\int \frac{1}{2}\frac{d}{dt}\dot{a}^2 dt = \int \frac{4}{3}\pi G \sum_i \frac{d}{dt}a^{-3(1+W_i)+2}\,\rho_{i0} dt + E \tag{15.45}$$

$$\frac{1}{2}\dot{a}^2 = \frac{4}{3}\pi G \sum_i \rho_{i0}\, a^{-3(1+W_i)+2} + E \tag{15.46}$$

Multiply by $\frac{2}{a^2}$ to get

$$\frac{\dot{a}^2}{a^2} = \frac{8}{3}\pi G \sum_i \rho_{i0}\, a^{-3(1+W_i)+2-2} + \frac{2E}{a^2}$$

$$\frac{\dot{a}^2}{a^2} = \frac{8}{3}\pi G \sum_i \rho_{i0}\, a^{-3(1+W_i)} + \frac{2E}{a^2} \tag{15.47}$$

Using equation (15.41) we have

$$\frac{\dot{a}^2}{a^2} = \frac{8}{3}\pi G \sum_i \rho_i + \frac{2E}{a^2} \tag{15.48}$$

Using equation (14.21) we have

$$H^2 = \frac{8}{3}\pi G \sum_i \rho_i + \frac{2E}{a^2} \tag{15.49}$$

$$H^2 - \frac{8}{3}\pi G \sum_i \rho_i = \frac{2E}{a^2} \tag{15.50}$$

$\sum_i \rho_i$ represent real components. It is convenient to think of the term $\frac{2E}{a^2}$ as a fictitious additional constituent or a component of the Universe called the curvature of space and denoted by the symbol κ. This becomes evident if we interpret it using Einstein's theory of relativity. If ρ_κ is the density of this fictitious component called curvature, i.e. curvature density then let us express $\frac{2E}{a^2}$ as follows

$$\frac{2E}{a^2} = \frac{8\pi G}{3}\rho_\kappa \tag{15.51}$$

$$\rho_\kappa a^2 = \rho_{\kappa 0}\ (\text{constant}) \tag{15.52}$$

where $\rho_{\kappa 0}$ is the present value of ρ_κ. From equation (15.52) it follows that

$$\rho_\kappa \propto \frac{1}{a^2} \tag{15.53}$$

With the expansion of the Universe the curvature density falls as a^{-2}.

We can interpret the curvature component as some kind of matter with pressure. Then we can use equation (15.40) viz.

$$\rho_k a^{3(1+W_k)} = \rho_{k0}$$

Let us find the value of the parameter W_i so that equation (15.52) viz.

$$\rho_\kappa a^2 = \rho_{\kappa 0} \tag{15.54}$$

is obeyed.

Comparison of equations (15.40) and (15.52) gives us

$$3(1 + W_k) = 2 \tag{15.55}$$

$$W_k = -\frac{1}{3} \tag{15.56}$$

For the curvature component, using equation (15.37) viz. $P_k = W_k \rho_k c^2$ we have pressure to be

$$P_k = -\frac{1}{3}\rho_k c^2 \tag{15.57}$$

The values $W_k = -\frac{1}{3}$, $P_k = -\frac{1}{3}\rho_k c^2$ (equations (15.56) and (15.57)) of curvature component can be compared to the values $W = 0$, $P = 0$ (equation (15.28)) for dust and $W = \frac{1}{3}$, $P = \frac{1}{3}\rho c^2$ (equation (15.31)) for relativistic gas.

We note that E can be 0, positive or negative. This means the curvature component can have zero, positive or negative density also.

Incorporating equation (15.51) into equation (15.49) we can write

$$H^2(t) = \frac{8}{3}\pi G \sum_i \rho_i + \frac{8\pi G}{3}\rho_\kappa \tag{15.58}$$

Incorporating the 2nd term on the RHS within the summation we can write

$$H^2(t) = \frac{8}{3}\pi G \sum_i \rho_i \tag{15.59}$$

The expression for the Hubble parameter, that governs the expansion of the Universe has been obtained. Here $\sum_i \rho_i$ represents the sum of all the densities of all the components of the Universe plus the contribution of the curvature component.

It is convenient to re-parametrize equation (15.58) for future convenience as follows.

At any instant of time the expansion rate of the Universe defines a density called critical density which is (equation (14.55))

$$\rho_c = \frac{3H^2}{8\pi G}. \tag{15.60}$$

The present value of the critical density is (equation (14.56))

$$\rho_{c0} = \frac{3H_0^2}{8\pi G}$$

$$\frac{8\pi G}{3} = \frac{H_0^2}{\rho_{c0}} \tag{15.61}$$

With this let us recast equation (15.59) as

$$H^2(t) = \frac{H_0^2}{\rho_{c0}} \sum_i \rho_i \tag{15.62}$$

Using equation (15.41) viz. $\rho_i = \rho_{i0} a^{-3(1+W_i)}$ let us recast equation (15.62) as

$$H^2(t) = \frac{H_0^2}{\rho_{c0}} \sum_i \rho_{i0} a^{-3(1+W_i)} \tag{15.63}$$

The cosmological parameter tells us what fraction of critical density is there in the Universe. For the ith component we write (equation (14.72))

$$\frac{\rho_i}{\rho_c} = \Omega_i \tag{15.64}$$

The present value is

$$\frac{\rho_{i0}}{\rho_{c0}} = \Omega_{i0} \tag{15.65}$$

From equation (15.62)

$$H^2(t) = H_0^2 \sum_i \Omega_{i0} a^{-3(1+W_i)} \tag{15.66}$$

This equation has complete information about the expansion history of the Universe, i.e. information about the Hubble parameter as a function of time in terms of the present values of Hubble parameter H_0 and density parameter Ω_{i0} for the different components of the Universe.

If we rewrite equation (15.62) in the present time we get (taking scale factor to be unity in the present time)

$$H_0^2 = H_0^2 \sum_i \Omega_{i0}$$

$$\sum_i \Omega_{i0} = 1 \tag{15.67}$$

So the sum of all density parameters is unity.

Equations (15.62) and (15.66) give complete information regarding the expansion history of the Universe.

15.4 Constituents of the Universe

According to current observation, we need to consider 4 components. We describe them separately.

(1) Matter/dust (has no pressure)

$$P = 0, \quad W_m = 0 \text{ (equation (15.28))} \tag{15.68}$$

Density of this component is

$$\rho_m a^3 = \rho_{m0} \text{ (equation (15.29))} \tag{15.69}$$

Dark matter falls into this category.

The present value of the corresponding cosmological parameter is denoted by Ω_{m0}.

(2) Relativistic particles that we refer to as radiation

This component corresponds to the Universe filled with radiation.

This component has pressure

$$P_r = \frac{1}{3}\rho_r c^2, \ \ W_r = \frac{1}{3} \text{ (equation (15.31))} \tag{15.70}$$

Density of this component is

$$\rho_r a^4 = \rho_{r0} \text{ (equation (15.32))} \tag{15.71}$$

The present value of the corresponding density parameter is denoted by Ω_{r0}.

(3) Curvature

This component corresponds to a pressure given by equations (15.56) and (15.57)

$$P_k = -\frac{1}{3}\rho_k c^2, \ \ W_k = -\frac{1}{3} \tag{15.72}$$

Density of this component is

$$\rho_k a^2 = \rho_{k0} \text{ (equation (15.54))} \tag{15.73}$$

This is a fictitious component introduced so that we can treat the constant of integration in the same footing as all the other components of the Universe.

The present value of the corresponding density parameter is denoted by Ω_{k0}.

(4) Cosmological constant

Einstein's theory of general relativity admits the possibility of a cosmological constant. We can interpret this constant in the Newtonian framework. It corresponds to a kind of material that fills the Universe and has negative pressure.

$$P_\Lambda = W_\Lambda \rho_\Lambda c^2 \tag{15.74}$$

$$P_\Lambda = -\rho_\Lambda c^2, \ \ W_\Lambda = -1 \tag{15.75}$$

As the Universe expands it follows from equation (15.40) viz. $\rho_\Lambda a^{3(1+W_\Lambda)} = \rho_{\Lambda0}$, for the case of the cosmological constant $W_\Lambda = -1$, that

$$\rho_\Lambda a^{3(1-1)} = \rho_{\Lambda0} \text{ i. e.}$$

$$\rho_\Lambda = \rho_{\Lambda0} \tag{15.76}$$

So the density of this component remains constant with the expansion of the Universe.

For a gas having positive pressure, with expansion the gas does work on the boundary which causes the internal energy to fall faster than it would have done if pressure were zero.

For negative pressure, gas gains energy. Work is done on the gas.

When $P_\Lambda = -\rho_\Lambda c^2$ (i.e. when pressure $=$ minus the energy density) work done exactly compensates for the fall in the energy density due to the expansion so that finally energy density is constant $\rho_\Lambda = \rho_{\Lambda 0}$.

There is a component of the Universe called dark energy. The cosmological constant is one possible explanation of what we call dark energy.

The present value of the corresponding density parameter is denoted by $\Omega_{\Lambda 0}$.

The requirement given by equation (15.67) in our 4-component model of the Universe is thus

$$\Omega_{m0} + \Omega_{n0} + \Omega_{\kappa 0} + \Omega_{\Lambda 0} = 1 \tag{15.77}$$

Let us write down the equation governing the expansion of the Universe in this 4-component model. From equation (15.66) we have

$$H^2(t) = H_0^2 \sum_i \Omega_{i0}\, a^{-3(1+W_i)}$$

$$= H_0^2 \left[\Omega_{m0} a^{-3(1+0)} + \Omega_{r0} a^{-3\left(1+\frac{1}{3}\right)} + \Omega_{\kappa 0} a^{-3\left(1-\frac{1}{3}\right)} + \Omega_{\Lambda 0} a^{-3(1-1)} \right] \tag{15.78}$$

$$H^2(t) = H_0^2 \left[\Omega_{m0} a^{-3} + \Omega_{r0} a^{-4} + \Omega_{\kappa 0} a^{-2} + \Omega_{\Lambda 0} \right]$$
$$H^2(t) = H_0^2 \left[\Omega_{r0} a^{-4} + \Omega_{m0} a^{-3} + \Omega_{\kappa 0} a^{-2} + \Omega_{\Lambda 0} \right] \tag{15.79}$$

Using equation (14.21)

$$H^2(t) = H_0^2 \left[\Omega_{r0} a^{-4} + \Omega_{m0} a^{-3} + \Omega_{\kappa 0} a^{-2} + \Omega_{\Lambda 0} \right] = \left(\frac{\dot{a}}{a} \right)^2 \tag{15.80}$$

This equation governs the dynamics of the expansion history of the Universe.

This is the equation we wish to solve to obtain the cosmological model with 4 components to get the scale factor and know about the expansion history of Universe.

When the Universe expands scale factor increases.

Figure 15.2 shows the relative importance of these 4 terms on the RHS of equation (15.80). The dependence of density upon scale factor for the four terms is as follows.

✓ Radiation density denoted by r falls as a^{-4}. When scale factor a is small, i.e. when the early Universe was small $\Omega_{r0} a^{-4}$ was the most dominant term. It shows steepest fall and we have a radiation dominated Universe.

✓ Matter density denoted by m falls as a^{-3}. With increase in scale factor the Universe becomes matter dominated.

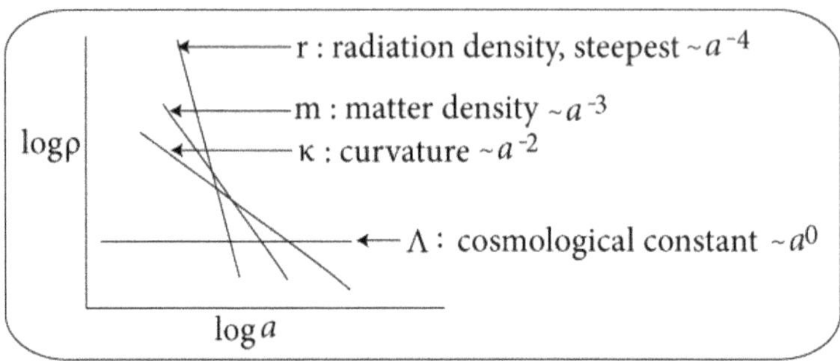

Figure 15.2. Strength of scale factors.

✓ Curvature density denoted by κ falls as a^{-2}. With further increase in scale factor the Universe becomes curvature dominated.

✓ Cosmological constant density denoted by Λ falls as a^0, i.e. it remains constant.

To solve we will consider the 4 terms on the RHS of equation (15.80) separately and study expansion of the Universe.

15.5 Radiation dominated Universe

Consider a Universe dominated by radiation only. We neglect the effect of all other terms, i.e. the effect of matter, curvature and cosmological constant. So we have from equation (15.79)

$$H^2(t) = H_0^2\, \Omega_{r0} a^{-4} \tag{15.81}$$

Using equation (14.21) in equation (15.81) we arrive at the equation that governs expansion of a radiation dominated Universe

$$\left(\frac{\dot{a}}{a}\right)^2 = H_0^2\, \Omega_{r0} a^{-4} \tag{15.82}$$

$$\frac{\dot{a}}{a} = H_0\sqrt{\Omega_{r0}}\; a^{-2} \tag{15.83}$$

$$\dot{a} = H_0\sqrt{\Omega_{r0}}\; a^{-2+1}$$

$$\frac{da}{dt} = H_0\sqrt{\Omega_{r0}}\; a^{-1} \tag{15.84}$$

$$a\,da = H_0\sqrt{\Omega_{r0}}\, dt \tag{15.85}$$

Integrating

$$\int a\,da = H_0\sqrt{\Omega_{r0}} \int dt + K$$

$$\frac{a^2}{2} = H_0 \sqrt{\Omega_{r0}}\, t + K$$

At $t = 0$, $a = 0$, $K = 0$ (Big Bang occurs at $t = 0$). Hence

$$\frac{a^2}{2} = H_0 \sqrt{\Omega_{r0}}\, t \tag{15.86}$$

$$a^2 = 2 H_0 \sqrt{\Omega_{r0}}\, t \tag{15.87}$$

$$a(t) = 2^{1/2} H_0^{1/2}\, \Omega_{r0}^{1/4} t^{1/2} \tag{15.88}$$

$$a(t) = C_1 t^{1/2} \propto t^{1/2} \tag{15.89}$$

where C_1 is constant.

In a radiation dominated Universe expansion goes as $t^{1/2}$, i.e. scale factor scales as $t^{1/2}$ in a radiation dominated Universe.

The Hubble's constant is (by equation (14.21))

$$H(t) = \frac{\dot{a}}{a} = \frac{\frac{d}{dt} C_1 t^{1/2}}{C_1 t^{1/2}} = \frac{1}{2} \frac{t^{-1/2}}{t^{1/2}} = \frac{1}{2t} \tag{15.90}$$

Age of a radiation dominated Universe is

$$t = \frac{1}{2H(t)} = 0.5 \frac{1}{H(t)} \tag{15.91}$$

Hence the present value of a radiation dominated Universe is

$$t_0 = \frac{1}{2H_0} \tag{15.92}$$

Using equation (14.33) viz. $H_0^{-1} = 1.4 \times 10^{10}\ yr$ we have

$$t_0 = \frac{1}{2} H_0^{-1} = \frac{1}{2} 1.4 \times 10^{10}\ yr = 7 \times 10^9\ yr \tag{15.93}$$

(For free expansion we had from equation (14.33) the age of the Universe to be $t_0 = \frac{1}{H_0} = 1.4 \times 10^{10}\ yr$ and from equation (14.89) the age of the Universe to be $t_0 = \frac{2}{3H_0} = 9 \times 10^9\ yr$ according to the critical cosmological model.)

15.6 Matter dominated Universe

Consider a Universe dominated by matter only. We neglect the effect of all other terms. i.e. radiation, curvature and cosmological constant. So we have from equation (15.79)

$$H^2(t) = H_0^2\, \Omega_{m0} a^{-3} \tag{15.94}$$

Using equation (14.21) in equation (15.94) we have the equation that governs expansion of a matter dominated Universe (i.e. a Universe dominated by pressure less dust)

$$\left(\frac{\dot{a}}{a}\right)^2 = H_0^2 \, \Omega_{m0} a^{-3} \tag{15.95}$$

$$\frac{\dot{a}}{a} = H_0\sqrt{\Omega_{m0}} \; a^{-3/2} \tag{15.96}$$

$$\dot{a} = H_0\sqrt{\Omega_{m0}} \; a^{-1/2}$$

$$\frac{da}{dt} = H_0\sqrt{\Omega_{m0}} \; a^{-1/2} \tag{15.97}$$

$$a^{1/2}da = H_0\sqrt{\Omega_{m0}} \; dt$$

Integrating

$$\int a^{\frac{1}{2}}da = H_0\sqrt{\Omega_{m0}} \; \int dt + K$$

$$\frac{2}{3}a^{3/2} = H_0\sqrt{\Omega_{m0}} \; t + K \tag{15.98}$$

At $t = 0$, $a = 0$, $K = 0$ (Big Bang occurs at $t = 0$)

$$\frac{2}{3}a^{3/2} = H_0\sqrt{\Omega_{m0}} \; t \tag{15.99}$$

$$a = \left[\frac{3}{2}H_0\sqrt{\Omega_{m0}} \; t\right]^{2/3} = C_2 t^{2/3} \propto t^{2/3} \tag{15.100}$$

where C_2 is constant.

In a matter dominated Universe expansion goes as $t^{2/3}$, i.e. scale factor scales as $t^{2/3}$ in a matter dominated Universe.

The Hubble's constant is (by equation (14.21))

$$H(t) = \frac{\dot{a}}{a} = \frac{\frac{d}{dt}C_2 t^{2/3}}{C_2 t^{2/3}} = \frac{2}{3}\frac{t^{-1/3}}{t^{2/3}} = \frac{2}{3t} \tag{15.101}$$

The age of a matter dominated Universe is

$$t = \frac{2}{3H(t)} \tag{15.102}$$

Hence the present value of the age of a matter dominated Universe is

$$t_0 = \frac{2}{3H_0} \tag{15.103}$$

Using equation (14.33) viz. $H_0^{-1} = 1.4 \times 10^{10}$ yr we have

$$t_0 = \frac{2}{3} H_0^{-1} = \frac{2}{3} 1.4 \times 10^{10} \, yr = 9.3 \times 10^9 \, yr \tag{15.104}$$

This value is less than the age determined without considering gravity $t_0 = \frac{1}{H_0} = 1.4 \times 10^{10} \, yr$ (equation (14.33)), but it is more than that of a radiation dominated Universe $t_0 = \frac{1}{2} H_0^{-1} = 7 \times 10^9 \, yr$ (equation (15.93)).

15.7 Curvature dominated Universe

Let us think of a situation where there is no gravity (no matter, no gravitational attraction). This corresponds to a Universe dominated by curvature only. We neglect effects of all other terms, i.e. radiation, matter and cosmological constant. So we have from equation (15.79)

$$H^2(t) = H_0^2 \, \Omega_{\kappa 0} a^{-2} \tag{15.105}$$

Using equation (14.21) in equation (15.105) we have the equation of a curvature dominated Universe to be

$$\left(\frac{\dot{a}}{a} \right)^2 = H_0^2 \Omega_{\kappa 0} a^{-2} \tag{15.106}$$

$$\dot{a} = H_0 \sqrt{\Omega_{\kappa 0}}$$

$$\frac{da}{dt} = H_0 \sqrt{\Omega_{\kappa 0}} \tag{15.107}$$

$$da = H_0 \sqrt{\Omega_{\kappa 0}} \, dt$$

Integrating

$$\int da = H_0 \sqrt{\Omega_{\kappa 0}} \int dt + K$$

$$a = H_0 \sqrt{\Omega_{\kappa 0}} \, t + K$$

At $t = 0$, $a = 0$, $K = 0$ (Big Bang occurs at $t = 0$)

$$a = H_0 \sqrt{\Omega_{\kappa 0}} \, t = C_3 t \propto t \tag{15.108}$$

where C_3 is constant.

In a curvature dominated Universe expansion goes as t, i.e. scale factor scales as t in a curvature dominated Universe.

The Hubble's constant is

$$H(t) = \frac{\dot{a}}{a} = \frac{\frac{d}{dt} C_3 t}{C_3 t} = \frac{1}{t} \tag{15.109}$$

The age of a curvature dominated Universe is

$$t = \frac{1}{H(t)} \tag{15.110}$$

The present value of the age of a curvature dominated Universe is

$$t_0 = \frac{1}{H_0} \tag{15.111}$$

Using equation (14.33) viz. $H_0^{-1} = 1.4 \times 10^{10}\, yr$ we have

$$t_0 = H_0^{-1} = 1.4 \times 10^{10}\, yr$$

In equation (14.33) we got the same relation $t_0 = H_0^{-1} = 1.4 \times 10^{10}\, yr$ for a freely expanding Universe.

If there is no gravity, i.e. for a Universe dominated by curvature only, the expansion is given by equation (15.108) viz. $a = H_0\sqrt{\Omega_{\kappa0}}\, t = C_3 t \propto t$. This is represented by the straight line CD as shown in figure 15.3. Straight line DO is obtained by extrapolation to $a = 0$ point, i.e. to the origin O. Then OA is the age of the Universe if gravity was not there, i.e. in a curvature dominated Universe. Our estimated value was $H_0^{-1} = 1.4 \times 10^{10}\, yr$.

In the presence of matter and radiation the expansion is slowed down. In other words, expansion was faster in the past and has now slowed down.

For a radiation dominated Universe expansion is given by equation (15.89) viz. $a(t) = C_1\, t^{1/2} \propto t^{1/2}$ and represented by the curve DR in figure 15.3. Curve DR is obtained by extrapolation to $a = 0$ point, i.e. to point R. Then RA is the age of a radiation dominated Universe. Our estimated value was $\frac{1}{2}H_0^{-1} = 7 \times 10^9\, yr$ (equation (15.93)).

For a matter dominated Universe expansion is given by equation (15.100) viz. $a = [\frac{3}{2}H_0\sqrt{\Omega_{m0}}\, t\,]^{2/3} = C_2 t^{2/3} \propto t^{2/3}$ and represented by the curve DM in

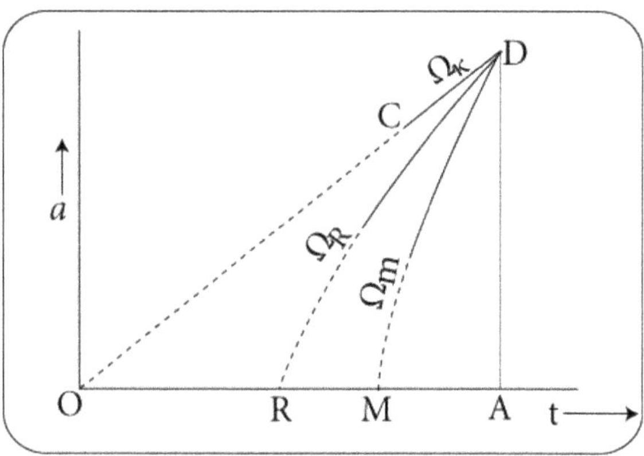

Figure 15.3. Scale factor versus t plot for radiation dominated, matter dominated and curvature dominated Universe and determination of the age of the Universe.

figure 15.3. Curve DM is obtained by extrapolation to $a = 0$ point, i.e. to point M. Then MA is the age of a matter dominated Universe. Our estimated valuc was $\frac{2}{3}H_0^{-1} = 9.3 \times 10^9\,yr$ (equation (15.104)).

Evidently the more the slowing down the smaller is the age of the Universe.

15.8 Cosmological constant dominated Universe

Consider a Universe dominated by cosmological constant only. We neglect the effect of all other terms, i.e. radiation, matter and curvature. So we have from equation (15.79)

$$H^2(t) = H_0^2\,\Omega_{\Lambda 0} \tag{15.112}$$

Using equation (14.21) in equation (15.112) we have the equation of a cosmological constant dominated Universe to be

$$\left(\frac{\dot{a}}{a}\right)^2 = H_0^2\Omega_{\Lambda 0} \tag{15.113}$$

$$\frac{\dot{a}}{a} = H_0\sqrt{\Omega_{\Lambda 0}} \tag{15.114}$$

$$\dot{a} = H_0\sqrt{\Omega_{\Lambda 0}}\,a$$

$$\frac{da}{dt} = H_0\sqrt{\Omega_{\Lambda 0}}\,a \tag{15.115}$$

$$\frac{da}{a} = H_0\sqrt{\Omega_{\Lambda 0}}\,dt$$

$$dlna = H_0\sqrt{\Omega_{\Lambda 0}}\,dt$$

$$lna = H_0\sqrt{\Omega_{\Lambda 0}}\,t \;|\; \ln A$$

$$a(t) = Ae^{H_0\sqrt{\Omega_{\Lambda 0}}\,t} \quad (A \text{ is constant of integration}) \tag{15.116}$$

It appears that the Universe expands exponentially and not as a power of t.

For a Universe with only cosmological constant $\Omega_{\Lambda 0} = 1$ we have from equation (15.116)

$$a(t) = Ae^{H_0 t} \tag{15.117}$$

Clearly for $a = 0$ at $t = -\infty$.

Since for this model the scale factor $a = 0$ at $t = -\infty$ it follows that in such a model there is no **Big Bang**, there is no singularity in the past.

The Hubble parameter is given by (equations (14.21), (15.112)).

$$H(t) = \frac{\dot{a}}{a} = H_0\sqrt{\Omega_{\Lambda 0}} = \text{constant and does not involve time.}$$

15.9 De-acceleration parameter

Our aim is to quantify and describe expansion of the Universe. The Hubble parameter $H(t) = \frac{\dot{a}}{a}$ at any instant gives the rate of expansion of the Universe.

If H is positive the Universe is expanding, if H is negative the Universe is contracting and if H is zero then the Universe is static.

To find how the expansion rate of the Universe changes we have to determine the second derivative of scale factor. For this purpose we define a quantity called the de-acceleration parameter as

$$q(t) = -\frac{\ddot{a}/a}{H^2(t)} \tag{15.118}$$

- In *exercise 15.1* we show that $q(t)$ is dimensionless.

  De-acceleration parameter quantifies de-acceleration or slowing down of the Universe. Gravity is attractive. In other words the de-acceleration parameter measures the rate at which the Universe is slowing down.

  Second derivative of scale factor $\ddot{a}$ is positive.

  If $\ddot{a}$ is negative then $q(t)$ will be positive. So $q(t)$ positive will indicate that expansion is slowing down.

  Recall equation (15.42)

$$\ddot{a}(t) = -\frac{4}{3}\pi G \sum_i (1 + 3W_i)\rho_{i0} a^{-3(1+W_i)+1} \tag{15.119}$$

Using equation (14.56) for the present value of the critical density viz. $\rho_{c0} = \frac{3H_0^2}{8\pi G}$ i.e. with

$$\frac{4\pi G}{3} = \frac{H_0^2}{2\rho_{c0}} \tag{15.120}$$

we have

$$\ddot{a}(t) = -\frac{H_0^2}{2\rho_{c0}} \sum_i (1 + 3W_i)\rho_{i0} a^{-3(1+W_i)+1} \tag{15.121}$$

Using equation (15.65) viz. $\frac{\rho_{i0}}{\rho_{c0}} = \Omega_{i0}$ we write

$$\ddot{a}(t) = -\frac{H_0^2}{2} \sum_i (1 + 3W_i)\Omega_{i0} a^{-3(1+W_i)+1} \tag{15.122}$$

$$\frac{\ddot{a}(t)}{a} = -\frac{H_0^2}{2} \sum_i (1 + 3W_i)\Omega_{i0} a^{-3(1+W_i)} \tag{15.123}$$

We wish to investigate the value of W for which the Universe will be accelerating.

Suppose we have a material that has negative pressure. Then $W=$ negative. Then there is the possibility that $1 + 3W_i$ may be negative. Then the gravitational force will be repulsive instead of being attractive.

$$\text{If } W_i = -\frac{1}{3} \text{ (equation (15.72))} \tag{15.124}$$

$$1 + 3W_i = 1 + 3\left(-\frac{1}{3}\right) = 0 \tag{15.125}$$

Then there is no contribution to acceleration, i.e. curvature does not contribute to acceleration.

$$\text{If } W_i < -\frac{1}{3}, \ 1 + 3W_i = -ve \tag{15.126}$$

Then gravitational force is repulsive. The Universe is accelerating.

$$\text{If } W_i > -\frac{1}{3}, \ 1 + 3W_i = +ve \tag{15.127}$$

Then gravitational force is attractive. The Universe is de-accelerating.

Observations indicate that the Universe at present is accelerating. For an accelerating Universe then the equation of state should be such that we need to have the parameter

$$W_i < -\frac{1}{3} \tag{15.128}$$

This is the criterion for dark energy.

Dark energy refers to a possible constituent of the Universe which satisfies equation (15.122) with $W_i < -\frac{1}{3}$. For the cosmological constant according to equation (15.75), $W_\Lambda = -1$ and so the cosmological constant is a possible candidate for dark energy since the criterion given by equation (15.128) is satisfied (-1 is less than $-\frac{1}{3}$).

Let us now calculate the de-accelerating parameter $q(t)$ (equation (15.118)) referring to equations (15.123) and (15.78)

$$q(t) = -\frac{\ddot{a}/a}{H^2(t)} = -\frac{-\frac{H_0^2}{2} \sum_i (1 + 3W_i)\Omega_{i0}a^{-3(1+W_i)}}{H_0^2 \sum_i \Omega_{i0}a^{-3(1+W_i)}}$$

$$q(t) = \frac{1}{2} \frac{\sum_i (1 + 3W_i)\Omega_{i0}a^{-3(1+W_i)}}{\sum_i \Omega_{i0}\, a^{-3(1+W_i)}} \tag{15.129}$$

For a matter dominated Universe using equation (15.68) viz. $W_i = 0$ equation (15.129) gives

$$q(t) = \frac{1}{2}\frac{\sum_i \Omega_{i0}a^{-3}}{\sum_i \Omega_{i0}a^{-3}} = \frac{1}{2} \tag{15.130}$$

De-acceleration parameter is $\frac{1}{2}$ for a matter dominated Universe.

For a curvature dominated Universe using equation (15.72) viz. $W_i = -\frac{1}{3}$ equation (15.129) gives

$$q(t) = \frac{1}{2}\frac{\sum_i \left(1 + 3(-\frac{1}{3})\right)\Omega_{i0}a^{-3\left(1-\frac{1}{3}\right)}}{\sum_i \Omega_{i0}a^{-3\left(1-\frac{1}{3}\right)}} = 0 \tag{15.131}$$

De-acceleration parameter is zero for a curvature dominated Universe.

For a cosmological constant dominated Universe using equation (15.75) viz. $W_i = -1$ equation (15.129) gives

$$q(t) = \frac{1}{2}\frac{\sum_i (1 - 3)\Omega_{i0}a^{-3(1-1)}}{\sum_i \Omega_{i0}a^{-3(1-1)}} = -\frac{1}{2}\frac{\sum_i 2\Omega_{i0}}{\sum_i \Omega_{i0}} = -1 \tag{15.132}$$

De-acceleration parameter is -1 for a cosmological constant dominated Universe. So in such a Universe dominated by cosmological constant the Universe is not de-accelerating, the Universe is actually accelerating. Gravity becomes repulsive instead of being attractive.

For a radiation dominated Universe using equation (15.70) viz. $W_i = \frac{1}{3}$ equation (15.129) gives

$$q(t) = \frac{1}{2}\frac{\sum_i \left(1 + 3\frac{1}{3}\right)\Omega_{i0}a^{-3\left(1+\frac{1}{3}\right)}}{\sum_i \Omega_{i0}a^{-3\left(1+\frac{1}{3}\right)}} = \frac{1}{2}\frac{\sum_i 2\Omega_{i0}a^{-4}}{\sum_i \Omega_{i0}a^{-4}} = 1 \tag{15.133}$$

The de-acceleration parameter is 1 for a radiation dominated Universe.
- Let us find the scale factor and the age of the Universe as a function of time.

Consider equation (15.78) and equation (14.21) to have

$$H^2(t) = H_0^2 \sum_i \Omega_{i0}a^{-3(1+W_i)} = \frac{\dot{a}^2}{a^2} \tag{15.134}$$

Taking the square root we get

$$\frac{\dot{a}}{a} = H_0\left[\sum_i \Omega_{i0}a^{-3(1+W_i)}\right]^{1/2}$$

$$\frac{1}{a}\frac{da}{dt} = H_0\left[\sum_i \Omega_{i0}a^{-3(1+W_i)}\right]^{1/2}$$

$$\frac{1}{H_0 a} \frac{da}{\left[\sum_i \Omega_{i0} a^{-3(1+W_i)} \right]^{1/2}} = dt$$

Integrating we have

$$\frac{1}{H_0} \int \frac{da}{a \left[\sum_i \Omega_{i0} a^{-3(1+W_i)} \right]^{1/2}} = \int dt$$

$$t = \frac{1}{H_0} \int_0^a \frac{da'}{a' \left[\sum_i \Omega_{i0} a'^{-3(1+W_i)} \right]^{1/2}} \tag{15.135}$$

$$= \text{age of Universe for any value of the scale factor.}$$

We can invert this relation to get scale factor a as a function of time.

The present value of the scale factor is $a = 1$. In equation (15.135) we can integrate from 0 to 1 to get the age of the Universe as

$$t = \frac{1}{H_0} \int_0^1 \frac{da'}{a' \left[\sum_i \Omega_{i0} a'^{-3(1+W_i)} \right]^{1/2}} = \frac{1}{H_0} \text{ (a factor)} \tag{15.136}$$

The factor in equation (15.136) is less than unity if the Universe is de-accelerating, more than unity if the Universe is accelerating and exactly unity if it is a free expansion.

15.10 Exercise

Exercise 15.1 *Show that de-acceleration parameter (equation (15.118)) is dimensionless.*

$\boxed{\text{Ans.}}$ De-acceleration parameter is defined as

$$q(t) = -\frac{\ddot{a}/a}{H^2(t)}$$

From *exercise 14.3* the dimension of $H(t)$ is $[\, T^{-1}]$.

$$[\, q(t) \,] \sim \left[-\frac{\frac{1}{a}\frac{d^2 a}{dt^2}}{(T^{-1})^2} \right] = \left[\frac{T^{-2}}{T^{-2}} \right] = [\, M^0 L^0 T^0 \,]$$

So de-acceleration parameter $q(t)$ is dimensionless.

Exercise 15.2 *Age of the Universe determined for a radiation dominated Universe is*

(a) $\frac{1}{2}H_0^{-1}$ (b) $\frac{1}{3}H_0^{-1}$ (c) $\frac{2}{3}H_0^{-1}$ (d) H_0^{-1}.

Exercise 15.3 *Age of the Universe determined for a matter dominated Universe is*

(a) $\frac{1}{2}H_0^{-1}$ (b) $\frac{1}{3}H_0^{-1}$ (c) $\frac{2}{3}H_0^{-1}$ (d) H_0^{-1}.

Exercise 15.4 *Age of the Universe determined for a freely expanding Universe is*

(a) $\frac{1}{2}H_0^{-1}$ (b) $\frac{1}{3}H_0^{-1}$ (c) $\frac{2}{3}H_0^{-1}$ (d) H_0^{-1}.

Exercise 15.5 *Age of the Universe determined in the critical cosmological model is*

(a) $\frac{1}{2}H_0^{-1}$ (b) $\frac{1}{3}H_0^{-1}$ (c) $\frac{2}{3}H_0^{-1}$ (d) H_0^{-1}.

Exercise 15.6 *The sum of the density parameters of the components of the Universe is*

(a) 0   (b) 0.5   (c) $\frac{1}{3}$ (d) 1.

Exercise 15.7 *Which statement is true for a cosmological constant dominated Universe?*
    (a) *There is Big Bang.*
    (b) *There is no Big Bang.*
    (c) *Big Bang occurs at* $t = \infty$
    (d) *Big Bang occurs at* $t = 0$.

Exercise 15.8 *In a radiation dominated Universe expansion varies as*

(a) $t^{2/3}$ (b) t (c) $t^{1/2}$ (d) $t^{-1/2}$.

Exercise 15.9 *In a matter dominated Universe expansion varies as*

(a) $t^{2/3}$ (b) t (c) $t^{1/2}$ (d) $t^{-1/2}$.

Exercise 15.10 *In a curvature dominated Universe expansion varies as*

(a) $t^{2/3}$ (b) t (c) $t^{1/2}$ (d) e^{t}.

Exercise 15.11 *In a cosmological constant dominated Universe expansion varies as*

(a) $t^{2/3}$ (b) t (c) $t^{1/2}$ (d) $t^{-1/2}$.

Exercise 15.12 *For dust the pressure is (ϵ is energy density)*

(a) zero   (b) $\frac{1}{3}\epsilon$ (c) $\frac{2}{3}\epsilon$ (d) ϵ.

Exercise 15.13 *For dust and relativistic gas, mass density falls as (a is scale factor)*

(a) a^{-1}, a^{-4} (b) a^{-2}, a^{-3} (c) a^{-3}, a^{-4} (d) a^{-4}, a^{-3}.

Exercise 15.14 *With the expansion of the Universe curvature density falls as*

(a) a^{-1} (b) a^{-2} (c) a^{-3} (d) a^{-4}.

Exercise 15.15 *The density of the cosmological constant varies as (a is scale factor)*

(a) a^{-1} (b) a^{0} (c) a (d) a^{2}.

Exercise 15.16 *The age of a radiation dominated Universe is*

(a) $\frac{1}{2} H_0^{-1}$ (b) H_0^{-1} (c) H_0 (d) $\frac{2}{3} H_0^{-1}$.

Exercise 15.17 *The age of a matter dominated Universe is*

(a) $\frac{1}{2} H_0^{-1}$ (b) H_0^{-1} (c) H_0 (d) $\frac{2}{3} H_0^{-1}$.

Exercise 15.18 *The age of a curvature dominated Universe is*

(a) $\frac{1}{2} H_0^{-1}$ (b) H_0^{-1} (c) H_0 (d) $\frac{2}{3} H_0^{-1}$.

Answer to multiple type questions

15.2*a*, 15.3*c*, 15.4*d*, 15.5*c*, 15.6*d*, 15.7*b*, 15.8*c*, 15.9*a*, 15.10*b*, 15.11*d*, 15.12*a*, 15.13d, 15.14*b*, 15.15*b*, 15.16*a*, 15.17*d*, 15.18*b*.

15.11 Question bank

Q15.1 Show that the effect of pressure is negligible in the case of random motion of galaxies.

Q15.2 Show that the effect of pressure is important for relativistic gas.

Q15.3 Obtain the age of a radiation dominated Universe.

Q15.4 Obtain the age of a matter dominated Universe.

Q15.5 What is the significance of deceleration parameter?

Q15.6 Show in a plot the variation of scale factor against time for a Universe dominated by radiation, matter and curvature.

Q15.7 Obtain the dependence of mass density on scale factor for dust.

Q15.8 Obtain the dependence of mass density on scale factor for relativistic gas.

Q15.9 Obtain the equation for scale factor.

Q15.10 What is curvature of space? Show that with the expansion of the Universe curvature density falls as a^{-2} where a is scale factor.

Q15.11 What is critical density of the Universe? What does it indicate?

Q15.12 Show in a diagram the strength of scale factors corresponding to radiation density, matter density, curvature and cosmological constant.

Further reading

[1] Bharadwaj S 2008 *Physical Cosmology Lecture Series by Prof Somnath Bharadwaj* (Department of Physics, IIT Kharagpur)

[2] Hawking S 1988 *A Brief History of Time* (London: Bantam)

[3] Bonometto S, Gorini V and Moschella U 2001 *Modern Cosmology* (Bristol: IOP)

[4] Roos M 1994 *Introduction to Cosmology* (New York: Wiley)

[5] Tenreiro R D and Quiros M 1988 *An Introduction to Cosmology and Particle Physics* (Singapore: World Scientific)

IOP Publishing

Nuclear and Particle Physics with Cosmology, Volume 2
Particle physics and cosmology
Jyotirmoy Guha

Chapter 16

Geometry of space-time

In this chapter, we recapitulate what is meant by metric, spherical polar coordinate, 2D curved surface of a sphere, 4D space-time, interval, null- like, time-like, space-like events, relativity postulates and null cone. We discuss spatial dependence of metric in the presence of gravity, cosmological metric, Friedman–Robertson–Walker (FRW) metric for flat and curved space. Hubble radius is estimated. We discuss terms like co-moving distance, proper distance and conformal time. Relation between red shift and expansion of the Universe is obtained. Coordinate distances for small and large red shifts, angular diameter distance, luminosity distance, specific luminosity, bolometric luminosity and type I and type II supernova are discussed.

16.1 Metric

Consider two points in a 3D flat space whose coordinate in a Cartesian coordinate system are (x, y, z) and $(x + dx, y + dy, z + dz)$. Their intermediate distance is dl which is the line element shown as PQ in figure 16.1. Square of this separation or interval is

$$dl^2 = dx^2 + dy^2 + dz^2 \tag{16.1}$$

With $x \equiv x^1$, $y \equiv x^2$, $z \equiv x^3$ we can write the interval square dl^2 as

$$dl^2 = dx.\, dx + dy.\, dy + dz.\, dz \tag{16.2}$$

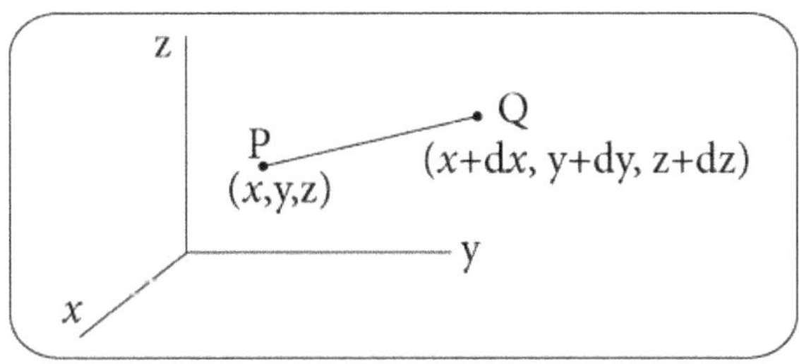

Figure 16.1. Two points considered in 3D space.

doi:10.1088/978-0-7503-5032-7ch16 16-1

$$= g_{11}dx^1dx^1 + g_{22}dx^2dx^2 + g_{33}dx^3dx^3 \tag{16.3}$$

$$dl^2 = g_{ij}dx^idx^j \tag{16.4}$$

where g_{ij} is the symmetric metric given by

$$g_{ij} = \begin{pmatrix} 1 & 0 & 0 \\ 0 & 1 & 0 \\ 0 & 0 & 1 \end{pmatrix} \tag{16.5}$$

It is of rank 2 in the flat Euclidean space in the Cartesian coordinate system.
- Using this metric we can calculate distance between two points in space.

Metric plays a significant role if we are dealing with a curvilinear coordinate system in Euclidean space.
Metric plays a very important role in curved space.
Let us give examples now.

16.2 Spherical polar coordinate system

Consider two points in a 3D flat space whose coordinates in a curvilinear spherical polar coordinate system are (r, θ, ϕ) and $(r + dr, \theta + d\theta, \phi + d\phi)$. Their intermediate distance is dl which is the line element PQ. Square of this separation is

$$dl^2 = dr^2 + (rd\theta)^2 + (r\sin\theta d\phi)^2 \tag{16.6}$$

$$= dr^2 + r^2d\theta^2 + r^2\sin^2\theta d\phi^2 \tag{16.7}$$

With $r \equiv x^1$, $\theta \equiv x^2$, $\phi \equiv x^3$ we can write dl^2 as

$$dl^2 = dr.\,dr + r^2d\theta.\,d\theta + r^2\sin^2\theta d\phi.\,d\phi \tag{16.8}$$

$$= g_{11}dx^1dx^1 + g_{22}dx^2dx^2 + g_{33}dx^3dx^3 \tag{16.9}$$

$$dl^2 = g_{ij}dx^idx^j \tag{16.10}$$

where g_{ij} is the symmetric metric given by

$$g_{ij} = \begin{pmatrix} 1 & 0 & 0 \\ 0 & r^2 & 0 \\ 0 & 0 & r^2\sin^2\theta \end{pmatrix} \tag{16.11}$$

16.3 Two-dimensional curved surface of a sphere

Consider the 2D surface denoted by S^2 of a 3D sphere (say Earth) of radius say unity. Every point on the surface of a sphere can be labelled by the coordinates (θ, ϕ).

Consider two points on this curved surface having coordinates (θ, ϕ) and $(\theta + d\theta, \ \phi + d\phi)$. We wish to find the distance between these two points.

$$dl^2 = r^2 d\theta^2 + r^2 \sin^2 \theta d\phi^2 \tag{16.12}$$

For $r = 1$ (r does not change on the sphere), figure 16.2

$$dl^2 = d\theta^2 + \sin^2 \theta d\phi^2 \tag{16.13}$$

With $\theta \equiv x^1$, $\phi \equiv x^2$ we can write dl^2 as

$$dl^2 = g_{11} dx^1 dx^1 + g_{22} dx^2 dx^2 \tag{16.14}$$

$$dl^2 = g_{ij} dx^i dx^j \tag{16.15}$$

where g_{ij} is the symmetric metric given by

$$g_{ij} = \begin{pmatrix} 1 & 0 \\ 0 & \sin^2 \theta \end{pmatrix} \tag{16.16}$$

- For a sphere of radius $r = R$ equation (16.12) gets modified to

$$dl^2 = R^2 d\theta^2 + R^2 \sin^2 \theta d\phi^2 \tag{16.17}$$

Now the larger the radius of the sphere, the less is the curvature and in the limit of infinite radius we have a flat surface of zero curvature. The relation between curvature and radius of curvature is

$$\text{Curvature} = \frac{1}{R} \tag{16.18}$$

- In general in some arbitrary curved coordinates (say curved space) the line element is expressed as

$$dl^2 = g_{ij}(x) dx^i dx^j = \text{invariant} \tag{16.19}$$

where the symmetric metric component $g_{ij} = g_{ij}(x)$ is a function of coordinates, i.e. it varies from point to point —varies with space.

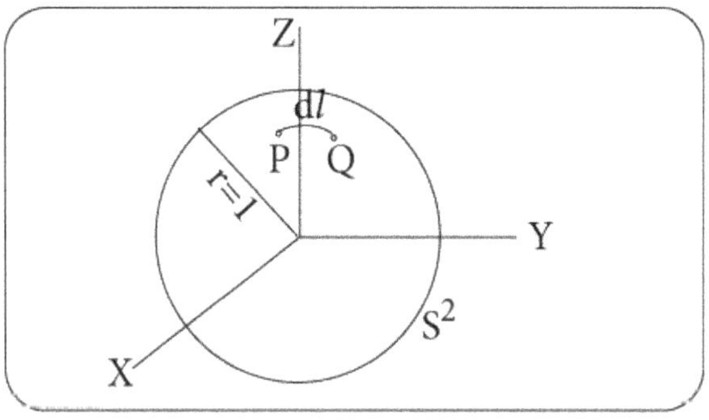

Figure 16.2. Two dimensional curved surface of a sphere.

When we switch over to a different coordinate system, coordinates may change, the metric g_{ij} may change but the distance between two points and hence the distance square dl^2 will not change, i.e. dl^2 is invariant.

16.4 Four-dimensional space-time

We can combine space and time into a 4D space-time as done in the special theory of relativity.

We use 4 coordinates

$$x^0 = ct, \quad x^1 = x, \quad x^2 = y, \quad x^3 = z \tag{16.20}$$

$$x^\mu = (x^0, x^i) \tag{16.21}$$

Events occurring at a time and at a place correspond to a point in the space-time diagram.

A space-time diagram is a very convenient way of describing causal connection between events.

Let us suppress two of the dimensions $x^2 = y$, $x^3 = z$ and depict a space-time diagram as in figure 16.3. One dimension is time $x^0 = ct$ and only one space dimension $x^1 = x$ has been shown.

Consider an observer at rest at A (ct, x, y, z). So $x=$ fixed. But time flows. So we have the vertical straight line AB describing the observer at rest in the space-time diagram in figure 16.3. Teh coordinate of B is $(c(t + dt), x, y, z)$.

We have shown another observer at rest at a different position $x + dx$ denoted by points L, L', L'' located on the vertical straight line CL. The coordinate of L is $(c(t + dt), x + dx, y + dy, z + dz)$.

We consider an event in which observer at A emits a light pulse or signal that travels along a trajectory.

$$x = ct \tag{16.22}$$

$$\frac{x}{ct} = 1 = \tan 45° \tag{16.23}$$

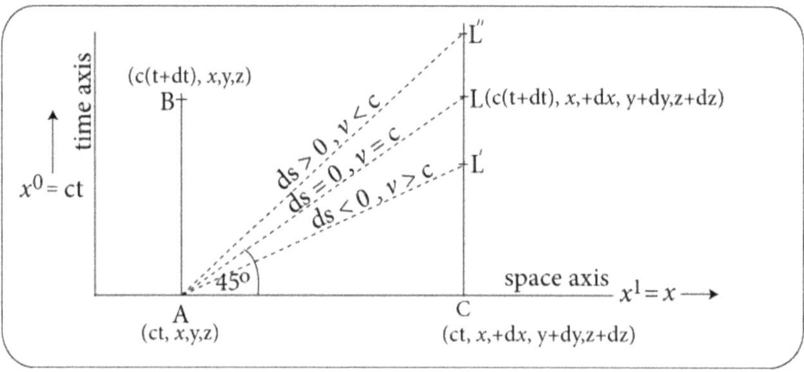

Figure 16.3. Space-time diagram showing an observer at rest.

So in the x versus ct diagram it represents a straight line with unit slope, i.e. the trajectory is a $45°$ line as shown, namely AL.

The point L represents the future event, namely an observer at L receiving light emitted by observer A.

Two other points L' and L'' are shown on the line CL representing two events. The points L, L', L'' all correspond to the same position viz. $x + dx$, $y + dy$, $z + dz$ and differ in time only. L' corresponds to an event occurring earlier (ct is smaller for it), while L'' corresponds to an event occurring later (ct is larger for it) than the event corresponding to L.

16.5 Interval

Consider two events. One event is the emission of light by an observer at A (figure 16.3) with 4-space-time coordinates $(x^0, x^1, x^2, x^3) \equiv (ct, x, y, z)$ and another is receiving of this light by an infinitesimally closely located observer at L having coordinates $(c(t + dt), x + dx, y + dy, z + dz)$

We define the interval between these two events such that the interval square is given by

$$ds^2 = (cdt)^2 - (dx)^2 - (dy)^2 - (dz)^2 \tag{16.24}$$

$$ds^2 = c^2 dt^2 - dx^2 - dy^2 - dz^2 \tag{16.25}$$

$$ds^2 = c^2 dt^2 - (dx^2 + dy^2 + dz^2) = \text{invariant} \tag{16.26}$$

Interval remains same, i.e. invariant in all inertial frames of reference.

16.6 Null-like, time-like, space-like events

Consider the separation between the positions of $A(x, y, z)$ and $L(x + dx, y + dy, z + dz)$ which is given by

$$\sqrt{(x + dx - x)^2 + (y + dy - y)^2 + (z + dz - z)^2} = \sqrt{dx^2 + dy^2 + dz^2}$$

A particle or a signal emerging from A travels to L with speed v and hence covers a distance

$$vdt = \sqrt{dx^2 + dy^2 + dz^2} \tag{16.27}$$

Hence the interval square between these two events at A and L is given by

$$ds^2 = c^2 dt^2 - (dx^2 + dy^2 + dz^2) = (cdt)^2 - (vdt)^2 \tag{16.28}$$

We can think of three cases.

✓ Emission of a light photon occurs at A (ct, x, y, z) and this light is received at $L(c(t + dt), x + dx, y + dy, z + dz)$. Light speed or photon speed is $v = c$.

Hence the interval square between two events on photon trajectory or light path with terminal points A and L is from equation (16.28)

$$ds^2 = (cdt)^2 - (vdt)^2 = (cdt)^2 - (cdt)^2 = 0 \qquad (16.29)$$

Such a zero interval

$$ds = 0 \qquad (16.30)$$

is referred to as null-like or light-like.

So the distance between two different points in 4-space-time can be zero though distance between two different points in Euclidean space can never be zero. Clearly 4-space-time is not Euclidean.

- We note that the spatial separation between lines AB and CL in the space-time diagram of figure 16.3 are fixed and equal to $\sqrt{dx^2 + dy^2 + dz^2} = vdt$

where this distance is traversed by a particle or signal with speed v in time dt.

Now time flows along the time axis AB or equivalently CL.

For two events A (ct, x, y, z) and L $(c(t + dt), x + dx, y + dy, z + dz)$ we note the following regarding the interval square.

At point L, $cdt = \sqrt{dx^2 + dy^2 + dz^2} = vdt$ since $v = c$. The interval is $ds = 0$ (equation (16.30)).

✓ At points above L, say at L'' (cdt dominates)

$$cdt > \sqrt{dx^2 + dy^2 + dz^2} = vdt \qquad (16.31)$$

This corresponds to

$$c > v \qquad (16.32)$$

The interval square is

$$ds^2 = (cdt)^2 - (vdt)^2 > 0 \qquad (16.33)$$

and so the interval is

$$ds > 0 \qquad (16.34)$$

Such a positive interval in which the time part dominates over the space part is referred to as time-like. This occurs along the path of a particle travelling with speed $v < c$. Such events are causally connected.

✓ At points below L say at L' (cdt is small)

$$cdt < \sqrt{dx^2 + dy^2 + dz^2} = vdt \qquad (16.35)$$

This corresponds to

$$c < v \qquad (16.36)$$

The interval square is

$$ds^2 = (cdt)^2 - (vdt)^2 < 0 \qquad (16.37)$$

and so the interval is

$$ds < 0 \qquad (16.38)$$

Such a negative interval in which the space part dominates over the time part is referred to as space-like. This occurs along the path of a particle travelling with speed $v > c$ and is unphysical. Such events are not causally connected.

- We find the interval between two events occurring at the same place but at different times in *exercise 16.2*.
- We find the interval between two events occurring at different places but at the same time in *exercise 16.3*.

16.7 Postulates of special theory of relativity

All inertial frames are equivalent. So we do not have a special preferred inertial frame of reference.

Speed of light $c = 3 \times 10^8 \, m \, s^{-1}$ is invariant. No signal can travel with speed $v > c$.

The speed of light does not depend on the speed of source or observer. A photon sees another photon move with the same speed c.

16.8 Null cone

Consider an observer at O sending out light signals. As shown in figure 16.4, events within the null cone are time-like and are causally connected to O. But events outside the null cone are space-like and are not causally connected to O. The surface of the null cone is light-like.

16.9 Gravity, metric, space-time and curvature

- Einstein's special theory of relativity does not incorporate gravity.
- According to Einstein's general theory of relativity space-time is curved in a gravitational field and gravity manifests itself as the curvature of space-time.

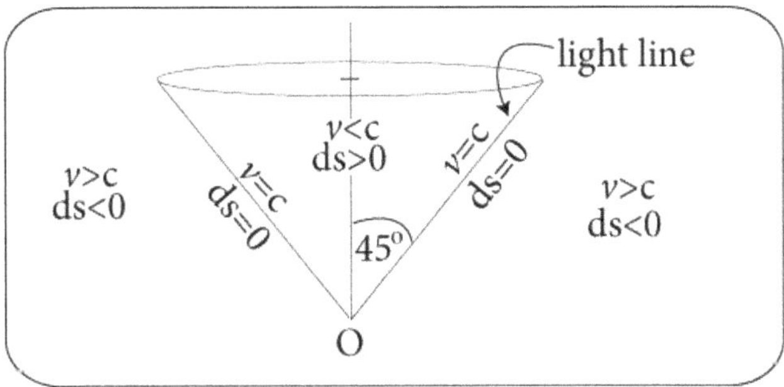

Figure 16.4. Null cone or light come.

In the presence of gravity the whole of space-time is curved and the metric is a function of space. So we can write

$$g_{\mu\nu} = g_{\mu\nu}(x) \tag{16.39}$$

And the line element has a more general form

$$ds^2 = g_{\mu\nu}\,dx^\mu dx^\nu, \quad \mu, \ \nu \to 0, 1, 2, 3 \tag{16.40}$$

All information about space-time is there in the metric.

In the absence of gravity the metric $g_{\mu\nu}$ is given by

$$g_{\mu\nu} = \begin{pmatrix} 1 & 0 & 0 & 0 \\ 0 & -1 & 0 & 0 \\ 0 & 0 & -1 & 0 \\ 0 & 0 & 0 & -1 \end{pmatrix} \tag{16.41}$$

- Let us consider a weak static gravitational field due to a massive object (Earth).

  In Einstein's theory of general relativity space-time is curved and the metric gets changed to incorporate the gravitational potential.

  In Einstein's theory of general relativity the gravitational effects manifest themselves as curvature of space-time and all information about gravitational effects are there in the metric.

  In other words, the metric has all the information about the curvature of space-time. So the gravitational forces are now in the metric.

  Let us write down the line element for a weak static gravitational field incorporating gravitational potential Φ into the metric

$$ds^2 = \left(1 + \frac{2\Phi}{c^2}\right)c^2 dt^2 - \left(1 - \frac{2\Phi}{c^2}\right)(dx^2 + dy^2 + dz^2) \tag{16.42}$$

The space-time is not flat but curved. The curvature is caused by the gravitational potential.

Metric is different for different spaces. In Riemannian geometry metric contains complete information about the space.

- For flat space

$$\Phi = 0 \tag{16.43}$$

Then

$$ds^2 = c^2 dt^2 - (dx^2 + dy^2 + dz^2) \tag{16.44}$$

This is the line element in the absence of gravity.

In Newtonian theory gravitational field is represented by gravitational potential.

16.10 The cosmological metric

According to Newtonian theory of relativity it is mass that causes gravity.

In Einstein's theory of relativity mass and energy density produce gravity. So curvature in the metric $g_{\mu\nu}$ is related to the mass and energy distribution.

The cosmological metric tells us how to calculate distances in cosmological space-time.

Properties of the cosmological metric should be consistent with the cosmological principle according to which the Universe should be homogeneous and isotropic.

16.11 Friedman, Robertson and Walker or the FRW metric

The most general cosmological metric that is homogeneous and isotropic in space was suggested by Friedman, Robertson and Walker, called the FRW metric. It is given by the line element.

$$ds^2 = c^2dt^2 - a^2L^2\left[\frac{dr^2}{1 - \kappa r^2} + r^2(d\theta^2 + \sin^2\theta d\phi^2)\right] \qquad (16.45)$$

Here the coordinates r, θ, ϕ are dimensionless co-moving coordinates, i.e. coordinates are fixed to the galaxies.

$a(t)$ is the scale factor. It is an increasing function of t for an expanding Universe. Its present value is unity. a is dimensionless.

κ is a dimensionless number that can have values -1, 0, 1. It represents curvature of space, i.e. decides whether space or the spatial part of space-time is curved or not.

The quantities r, θ, ϕ, $a(t)$, k are all dimensionless. The 2nd term of the RHS is made to have dimension of length^2 by attaching the factor L^2. Obviously L has dimension of length and is called the length scale.

16.11.1 $\kappa = 0$ case

From equation (16.45) with $k - 0$ we get

$$ds^2 = c^2dt^2 - a^2L^2[dr^2 + r^2(d\theta^2 + \sin^2\theta d\phi^2)] \qquad (16.46)$$

This is 3D Euclidian metric in flat space in spherical polar (r, θ, ϕ) coordinates.

Absorbing length L in r we rewrite

$$ds^2 = c^2dt^2 - a^2[d(Lr)^2 + (Lr)^2(d\theta^2 + \sin^2\theta d\phi^2)] \qquad (16.47)$$

Let us redefine Lr as r to make it represent length. Hence we rewrite

$$ds^2 = c^2dt^2 - a^2[dr^2 + r^2(d\theta^2 + \sin^2\theta d\phi^2)] \qquad (16.48)$$

In Cartesian coordinates we can write

$$ds^2 = c^2dt^2 - a^2(dx^2 + dy^2 + dz^2) \qquad (16.49)$$

The spatial part is infinite flat space. But as scale factor a is a function of time the entire space-time is curved.

Consider two observers situated at fixed value of x at a given time t at very close separation Δx. Assuming that $a(t)$ does not change in such a short span of time we note that the time Δt taken by light signal to propagate from one observer to another is

$$c\Delta t = a(t)\Delta x \tag{16.50}$$

This is called proper distance. As scale factor is an increasing function of time the proper distance is going to increase with time since Δx is the co-moving distance.

16.11.2 $\kappa = 1$ case

From equation (16.45) with $k = 1$ we get

$$ds^2 = c^2 dt^2 - a^2 L^2 \left[\frac{dr^2}{1 - r^2} + r^2(d\theta^2 + \sin^2\theta d\phi^2) \right] \tag{16.51}$$

Range of r is $0 \leqslant r \leqslant 1$.

It is not a flat space but a curved one.

- In 2D space we can consider a 1D curve S^1, say a circle as shown in figure 16.5 (a), having line element $dl = rd\phi$ and $dl^2 = r^2 d\phi^2$.

- In 3D space we can consider a 2D surface S^2, say the surface on a sphere as shown in figure 16.5(b). The spherical surface is a collection of circles, as shown in figure 16.5(b), the length elements being $rd\theta$ and $r\sin\theta d\phi$. At each θ there is a circle. And the collection of all such circles generates the 2D surface of the sphere. And the line element is

$$dl^2 = d^2\theta + \sin^2\theta d^2\phi \tag{16.52}$$

which represents a collection of circles of radius $\sin\theta$.

We introduce a new variable χ such that

$$r = \sin\chi, \quad dr = \cos\chi d\chi \tag{16.53}$$

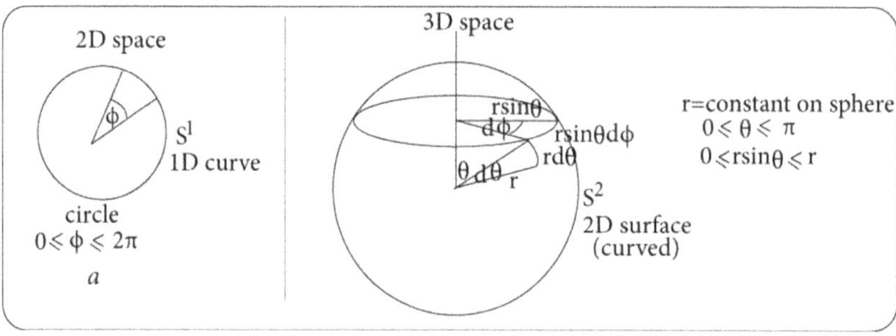

Figure 16.5. Spherical surface can be looked upon as being made up of circles.

And so from equation (16.51) we have

$$ds^2 = c^2dt^2 - a^2L^2\left[\frac{\cos^2\chi d\chi^2}{1 - \sin^2\chi} + \sin^2\chi(d\theta^2 + \sin^2\theta d\phi^2)\right]$$

$$ds^2 = c^2dt^2 - a^2L^2[d\chi^2 + \sin^2\chi(d\theta^2 + \sin^2\theta d\phi^2)] \tag{16.54}$$

Now the part $d\theta^2 + \sin^2\theta d\phi^2$ represents a 2D surface of a sphere embedded in 3D.
   Range of χ is $0 \leqslant \chi \leqslant \pi$.
   For each value of χ there is a 3D sphere of radius $\sin\chi$ embedded in 4D.
Equation (16.54) is like a generalized spherical polar coordinate system.
   The value of L in this case is

$$L = \frac{c}{H_0\sqrt{|\Omega_{\kappa 0}|}} \tag{16.55}$$

where $\Omega_{\kappa 0} < 0$ (collapsing model) for $\kappa = 1$ (constant positive curvature). For large
$\Omega_{\kappa 0}$, L decreases.

$$\text{Curvature} = \frac{1}{a(t)L} \tag{16.56}$$

$$\text{Radius} = a(t)L \tag{16.57}$$

16.11.3 $\kappa = -1$ case

From equation (16.51) with $\kappa = -1$ we get

$$ds^2 = c^2dt^2 - a^2L^2\left[\frac{dr^2}{1 + r^2} + r^2(d\theta^2 + \sin^2\theta d\phi^2)\right] \tag{16.58}$$

It is not a flat space but a curved one.
   We introduce a new variable χ such that

$$r = \sin h\chi, \quad dr = \cos h\chi d\chi \tag{16.59}$$

And so from equation (16.58)

$$ds^2 = c^2dt^2 - a^2L^2\left[\frac{\cos h^2\chi d\chi^2}{1 + \sin h^2\chi} + \sin h^2\chi(d\theta^2 + \sin^2\theta d\phi^2)\right] \tag{16.60}$$

Using $1 + \sin h^2\chi = \cos h^2\chi$ we have

$$ds^2 = c^2dt^2 - a^2L^2[d\chi^2 + \sin h^2\chi(d\theta^2 + \sin^2\theta d\phi^2)] \tag{16.61}$$

The value of L in this case is

$$L = \frac{c}{H_0\sqrt{|\Omega_{\kappa 0}|}} \tag{16.62}$$

where $\Omega_{\kappa 0} > 0$ (expanding model) for $\kappa = -1$ (constant negative curvature). For
large $\Omega_{\kappa 0}$, L decreases.

16.11.4 Interpretation of $\frac{c}{H_0}$

The quantity $\frac{c}{H_0}$ occurs in equations (16.55), (16.62). The quantity $\frac{1}{H_0}$ represents an estimate of the present age of the Universe.

Suppose a particle starts its journey at the moment of the Big Bang. Then in the present age of the Universe it travels a distance of $\frac{c}{H_0}$. This is called the Hubble radius. It is the furthest distance a signal could have propagated.

16.12 Estimate of the Hubble radius

$$\frac{c}{H_0} = c \cdot \frac{1}{H_0} = 3 \times 10^8 \, m \, s^{-1} \, (1.4 \times 10^{10} \, yr) \text{ (by equation (14.33))}$$

$$\frac{c}{H_0} = 3 \times 10^8 \, m \, s^{-1} \times 1.4 \times 10^{10} \times 365 \times 24 \times 60 \times 60 \, s$$

$$\frac{c}{H_0} = 1.32 \times 10^{26} \, m = \text{ Hubble radius.} \tag{16.63}$$

16.13 Redshifts and distances

The FRW cosmological metric is by equation (16.45)

$$ds^2 = c^2 dt^2 - a^2 L^2 \left[\frac{dr^2}{1 - \kappa r^2} + r^2 (d\theta^2 + \sin^2 \theta d\phi^2) \right] \tag{16.64}$$

The part $c^2 dt^2$ refers to time while the rest of the RHS of equation (16.64) refers to space.

where

$$k = -1, 0, +1 \tag{16.65}$$

The value of κ decides the curvature of space.

If $\kappa = -1$, we have a negatively curved space.

If $\kappa = 0$, we have a flat space. In both these cases r varies from 0 to ∞ and space is infinitely large.

If $\kappa = 1$, we have a positively curved space. And value of r is restricted between 0 and 1. This is a compact space and has a finite volume. It does not extend to infinity. It is a 3D sphere.

$a(t)$ is scale factor (dimensionless).

$L = \frac{c}{H_0 \sqrt{|\Omega_{\kappa 0}|}}$ is the length scale. It is important for $\kappa = \pm 1$ and can be absorbed for $\kappa = 0$.

r, θ, ϕ are dimensionless.

16.14 Co-moving distance

Consider two neighbouring points in space defined by r, θ, ϕ. The co-moving distance between these points can be extracted from equation (16.64) setting aside the time part and taking out the scale factor. We are thus left with the co-moving distance between the two neighbouring points which is

$$dl^2 = L^2 \left[\frac{dr^2}{1 - \kappa r^2} + r^2(d\theta^2 + \sin^2\theta d\phi^2) \right] \tag{16.66}$$

For two finitely separated points the co-moving distance is given by

$$l = L \int \left[\frac{dr^2}{1 - \kappa r^2} + r^2(d\theta^2 + \sin^2\theta d\phi^2) \right]^{1/2} \tag{16.67}$$

16.15 Proper distance

Consider two infinitesimally close points. We can send light pulse from one point to another point. Light travels along a light cone for which $ds^2 = 0$. Taking $a=$ constant and equating $cdt = adl$ we can evaluate the proper distance. For points separated by finite distance we have $ct = al$.

16.15.1 Propagation of light in space-time

Co-moving distance, proper distance cannot be measured directly. They are measured through other methods such as the propagation of light method.

Light propagates along trajectories where interval is null or zero. The line element can be written as

$$ds^2 = c^2 dt^2 - a^2 dl^2 \tag{16.68}$$

$$= 0 \tag{16.69}$$

In the usual space speed of light is a constant. It covers the same amount of co-moving distance in one second. But since scale factor a is an increasing function of time light will not cover the same co-moving distance in one second. The amount of co-moving distance that light covers will depend on the value of a. If scale factor is large it will cover a small Δl and if scale factor is small it will cover a large Δl.

To discuss with more clarity we introduce conformal time.

16.16 Conformal time

We recast the line element of equation (16.68) as

$$ds^2 = a^2 \left(c^2 \frac{dt^2}{a^2} - dl^2 \right) \tag{16.70}$$

The factor a^2 is called conformal factor.

Introduce

$$d\eta - \frac{dt}{a} \tag{16.71}$$

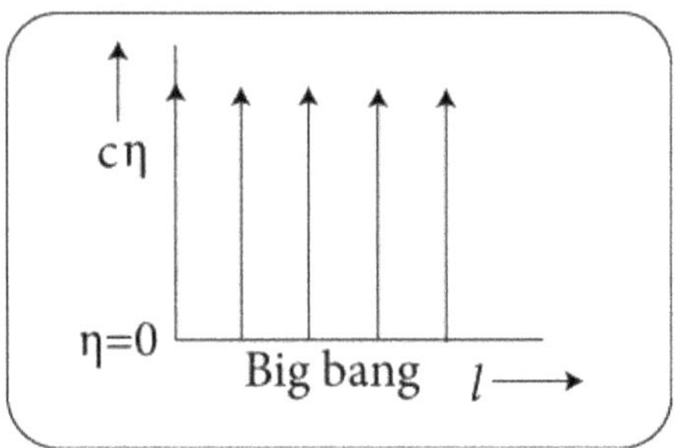

Figure 16.6. Space time diagram $c\eta$ vs. l.

$$ds^2 = a^2(c^2d\eta^2 - dl^2) \tag{16.72}$$

η is called conformal time.

Along the path of propagation of light $ds^2 = 0$ so that

$$c^2d\eta^2 = dl^2 \tag{16.73}$$

$$cd\eta = dl \tag{16.74}$$

If $d\eta = 1$ then $dl = c$. So in unit conformal time the co-moving distance covered by light is c. In other words, in conformal time and co-moving coordinate the speed of light is c which becomes the coordinate speed.

Let us draw the space-time diagram $c\eta$ vs. l as shown in figure 16.6.

Observers at fixed co-moving coordinate will move along a straight line as shown in the diagram. We can choose the origin of conformal time so that the **Big Bang** occurred at $\eta = 0$ as shown and it is meaningless to talk of events before this as the Universe started here.

Conformal time is very convenient for studying propagation of light in cosmology.

16.17 Redshift and expansion of the Universe

Consider an observer at origin O. There is a galaxy G. The galaxy emits light that is received by the observer as shown in figure 16.7. Let us discuss the method to calculate the shift in wavelength or frequency as it propagates.

Light propagates along a trajectory called null trajectory for which $ds^2 = 0$, i.e. $ds = 0\ cdt = adl$.

$$c\frac{dt}{a} = dl \tag{16.75}$$

Integrate from the position of the galaxy to the observer to get the co-moving distance

$$\int_{t_G}^{t_o} c\frac{dt}{a(t)} = \int_0^l dl \tag{16.76}$$

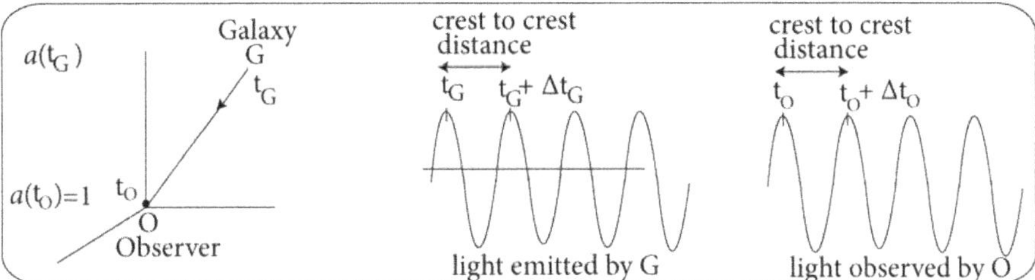

Figure 16.7. Light crests emitted by galaxy G at $t_G, t_G + \Delta t_G$ reach observer O at t_o, $t_o + \Delta t_o$ so that the co-moving distance between them does not change.

where light starts from galaxy G at t_G and reaches the observer at t_o.

We treat light as a wave having crests and troughs (figure 16.7). The inverse of crest-to-crest separation gives frequency.

A similar wave train will be observed by the observer. Light emitted at t_G and $t_G + dt_G$ by G will be observed at t_o and $t_o + dt_o$, respectively. The RHS integral just gives the co-moving distance between the galaxy and the observer that does not change. The integral from t_G to t_O corresponding to emission and absorption of the first crest is the co-moving distance

$$\int_{t_G}^{t_o} \frac{dt}{a(t)} = l. \tag{16.77}$$

The second crest has also to cover the same co-moving distance since we assume that the galaxy and the observer are located at fixed co-moving coordinates. So the co-moving distance between them does not change. Hence we expect the same for the second crest as well viz.

$$\int_{t_G}^{t_o} \frac{dt}{a(t)} = l = \int_{t_G + \Delta t_G}^{t_o + \Delta t_o} \frac{dt}{a(t)} \tag{16.78}$$

We can break up the time integral on the RHS of equation (16.78) as follows:

$$\int_{t_G}^{t_O} \frac{dt}{a(t)} = \int_{t_G + \Delta t_G}^{t_G} \frac{dt}{a(t)} + \int_{t_G}^{t_o} \frac{dt}{a(t)} + \int_{t_o}^{t_o + \Delta t_o} \frac{dt}{a(t)} \tag{16.79}$$

$$\int_{t_G + \Delta t_G}^{t_G} \frac{dt}{a(t)} + \int_{t_o}^{t_o + \Delta t_o} \frac{dt}{a(t)} = 0 \tag{16.80}$$

$$\int_{t_G + \Delta t_G}^{t_G} \frac{dt}{a(t)} = - \int_{t_o}^{t_o + \Delta t_o} \frac{dt}{a(t)} \tag{16.81}$$

As Δt_G, Δt_o are infinitesimals we can replace the integrals by their values at Δt_G, Δt_o and hence we have

$$\frac{1}{a(t_G)}(t_G - (t_G + \Delta t_G)) = -\frac{1}{a(t_o)}(t_o + \Delta t_o - t_o)$$

$$-\frac{\Delta t_G}{a(t_G)} = -\frac{\Delta t_o}{a(t_o)} \tag{16.82}$$

Clearly the time intervals $\Delta t_G \neq \Delta t_o$, i.e. the emitted and observed time intervals are not the same. They are different.

$$\frac{\Delta t_o}{\Delta t_G} = \frac{a(t_o)}{a(t_G)} \tag{16.83}$$

Clearly the time intervals get scaled by exactly the amount by which the Universe expands.

Now frequency is inversely related to time. Hence

$$\frac{\nu_G}{\nu_o} = \frac{\Delta t_o}{\Delta t_G} = \frac{a(t_o)}{a(t_G)} \tag{16.84}$$

Also

$$\frac{\lambda_o}{\lambda_G} = \frac{\nu_G}{\nu_o} = \frac{\Delta t_o}{\Delta t_G} = \frac{a(t_o)}{a(t_G)} \tag{16.85}$$

Light is emitted at a particular wavelength λ_e when scale factor is $a(t_G)$. When light reaches the observer if the Universe doubles in size, i.e. if scale factor doubles, then equation (16.85) suggests that the wavelength also doubles. So wavelength gets stretched by exactly the same amount by which the Universe expands. Wavelength gets scaled exactly by the same factor by which the Universe expands. This is the phenomenon of redshift that is obviously related to the expansion of the Universe. We can thus calculate the amount of change in the wavelength.

We defined redshift z in equations (13.20), (13.21) as

$$\lambda_O = \lambda_G(1 + z), \quad z = \frac{v}{c} \tag{16.86}$$

Redshift is related to the scale factor as follows.

From equations (16.86), (16.85) we have

$$1 + z = \frac{\lambda_O}{\lambda_G} = \frac{a(t_o)}{a(t_G)} \tag{16.87}$$

$$z = \frac{a(t_o)}{a(t_G)} - 1 \tag{16.88}$$

This is how the redshift is related to the expansion of the Universe.

We defined blueshift z in equation (13.22)

$$\lambda_O = \lambda_G(1 - z) \tag{16.89}$$

Blueshift is related to the scale factor as follows. From equations (16.82), (16.78) we have

$$1 - z = \frac{\lambda_O}{\lambda_G} = \frac{a(t_o)}{a(t_G)} \tag{16.90}$$

$$z = 1 + \frac{a(t_O)}{a(t_G)} \tag{16.91}$$

This is how the blueshift is related to contraction of the Universe.

- In *exercise 16.4* we evaluate the size of the Universe whenred shift is known.

16.18 Relationship between redshift and coordinate distances

We wish to inter-relate redshift z and coordinate distance r. Consider an observer at origin O. There is a galaxy G as shown in figure 16.7. Galaxy G emits light at time t_G when the scale factor of the Universe is $a(t_G)$. This light is received by the observer O at time t_O (which is the present time) when the scale factor of the Universe is $a(t_O) = 1$. Light appears to the observer at redshift z.

To find the distance of the galaxy from the observer.

In the cosmological metric we have encountered two types of distances: one is proper distance and another is co-moving distance. To calculate both of these we need to know the value of r corresponding to the galaxy since the galaxy is in the radial direction away from the observer. The problem is to calculate the value of r knowing the value of redshift. Also, how far back in the past, i.e. the time at which the light was emitted, needs to be known. Then the present age of the Universe is known. Redshift measured.

Let us calculate the conformal time corresponding to this redshift.

For light propagation $ds^2 = 0$. Also, as light is propagating along the radial direction the part $(d\theta^2 + \sin^2\theta d\phi^2)$ of the cosmological metric given by equation (16.45) viz.

$$ds^2 = c^2 dt^2 - a^2 L^2 \left[\frac{dr^2}{1 - \kappa r^2} + r^2(d\theta^2 + \sin^2\theta d\phi^2) \right]$$

will not be there. And so we have

$$c^2 dt^2 - a^2 L^2 \left[\frac{dr^2}{1 - \kappa r^2} \right] = 0 \tag{16.92}$$

$$c^2 \frac{dt^2}{a^2} = L^2 \frac{dr^2}{1 - \kappa r^2}$$

$$c \frac{dt}{a(t)} = L \frac{dr}{\sqrt{1 - kr^2}} \tag{16.93}$$

Integrating we get

$$c \int_{t_G}^{t_o} \frac{dt}{a(t)} = L \int_0^r \frac{dr}{\sqrt{1 - kr^2}} \tag{16.94}$$

From this equation we get t_G.

Using equation (16.71), i.e. $d\eta = \frac{dt}{a}$ we write

$$c \int_{t_G}^{t_o} \frac{dt}{a(t)} = c \int_{\eta_G}^{\eta_o} d\eta = L \int_0^r \frac{dr}{\sqrt{1 - kr^2}} \tag{16.95}$$

We first bring scale factor and then introduce redshift.

By equation (14.21) we write

$$H(t) = \frac{\dot{a}}{a} = \frac{1}{a}\frac{da}{dt} \tag{16.96}$$

$$a^2 H = a^2 \frac{1}{a}\frac{da}{dt} = \frac{ada}{dt} \tag{16.97}$$

Hence

$$\frac{dt}{a} = \frac{da}{a^2 H} \tag{16.98}$$

Hubble parameter is a function of scale factor, i.e.

$$H = H(a) \tag{16.99}$$

So from equation (16.98) we have

$$\frac{dt}{a} = \frac{da}{a^2 H(a)} \tag{16.100}$$

From equation (16.95) we write

$$c \int_{\text{lower limit}}^{1} \frac{da}{a^2 H(a)} = c \int_{\eta_G}^{\eta_o} d\eta = L \int_0^r \frac{dr}{\sqrt{1 - kr^2}} \tag{16.101}$$

where the lower limit for $t = t_G$ follows from equation (16.88) viz.

$$z = \frac{a(t_o)}{a(t_G)} - 1$$

$$\frac{a(t_o)}{a(t_G)} = 1 + z \tag{16.102}$$

Now putting $a(t_O) = 1$ (present value of scale factor)

$$\frac{1}{a(t_G)} = 1 + z \tag{16.103}$$

$$a(t_G) = \frac{1}{1 + z} = (1 + z)^{-1} = \text{lower limit} \tag{16.104}$$

Hence from equation (16.101) we get

$$c \int_{(1+z)^{-1}}^{1} \frac{da}{a^2 H(a)} = c \int_{\eta_G}^{\eta_O} d\eta = L \int_0^r \frac{dr}{\sqrt{1 - kr^2}} \tag{16.105}$$

$$c \int_{(1+z)^{-1}}^{1} \frac{da}{a^2 H(a)} = c(\eta_o - \eta_G) \tag{16.106}$$

So if we know the present value of the conformal time η_O we will be able to determine the value of conformal time η_G when light was emitted from the galaxy.

Let us now determine the coordinate of the galaxy r making use of equation (16.105) with $\kappa = 0$ (spatially flat situation). In such a case we can absorb L into the coordinate. So for $\kappa = 0$

$$c \int_{(1+z)^{-1}}^{1} \frac{da}{a^2 H(a)} = c \int_{\eta_G}^{\eta_O} d\eta = L \int_0^r dr \tag{16.107}$$

$$c \int_{(1+z)^{-1}}^{1} \frac{da}{a^2 H(a)} = c \int_{\eta_G}^{\eta_O} d\eta = Lr \xrightarrow{\text{redefine}} r \tag{16.108}$$

$$r = c \int_{(1+z)^{-1}}^{1} \frac{da}{a^2 H(a)} = c(\eta_o - \eta_G) \tag{16.109}$$

So r is estimated.

The relation between coordinate r and redshift z depends on the functional form of the Hubble parameter on scale factor (equation (16.99)).

16.19 Calculation of coordinate distances

Let us consider a model where there is only matter (no radiation, no curvature, no cosmological constant).

$$\Omega_{m0} = 1. \tag{16.110}$$

Taking the square root of equation (15.94) with $H = H(a)$, $\Omega_{m0} = 1$ we have

$$H(a) = H_0 a^{-3/2} \qquad (16.111)$$

Hence from equation (16.109)

$$r = c \int_{(1+z)^{-1}}^{1} \frac{da}{a^2 H_0 a^{-3/2}}$$

$$r = \frac{c}{H_0} \int_{(1+z)^{-1}}^{1} \frac{da}{a^{1/2}} \qquad (16.112)$$

$$r = \frac{c}{H_0} \frac{a^{1/2}}{1/2} \Big|_{(1+z)^{-1}}^{1}$$

$$r = \frac{2c}{H_0}\left[1 - (1+z)^{-1/2} \right]$$

$$r = \frac{2c}{H_0}\left[1 - \frac{1}{\sqrt{1+z}} \right] \qquad (16.113)$$

We have the relation between the coordinate of the galaxy and the redshift.

16.19.1 For small redshift

For small redshift $z \ll 1$

$$\frac{1}{\sqrt{1+z}} = (1+z)^{-1/2} \approx 1 - \frac{1}{2}z \qquad (16.114)$$

This is a Taylor series and we have retained the terms of the order of z only. From equation (16.113) we get

$$r = \frac{2c}{H_0}\left[1 - \left(1 - \frac{1}{2}z\right) \right] = \frac{c}{H_0}z = \frac{v}{H_0} \qquad (16.115)$$

$$r = \frac{v}{H_0} \qquad (16.116)$$

where $v = cz$ is the velocity at which the galaxy is moving away from us.

This is Hubble's law, as shown in figure 16.8. It predicts a linear relation between r and z for small z.

16.19.2 For large redshift

For large redshift z

It follows from equation (16.106) that the factor $\frac{1}{\sqrt{1+z}} \to 0$ and hence

$$r = \frac{2c}{H_0} \qquad (16.117)$$

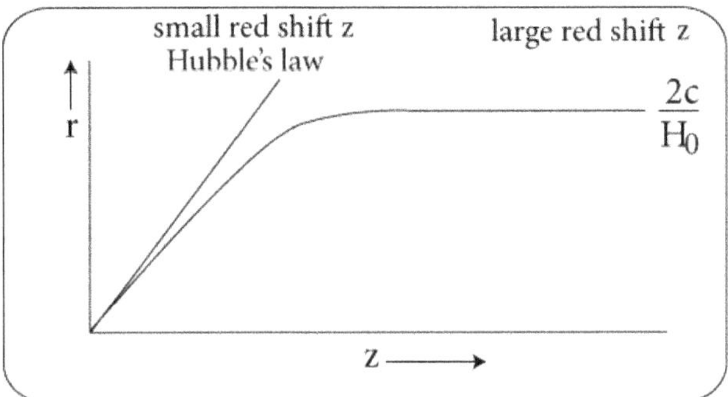

Figure 16.8. Linear relation between coordinate r and redshift z.

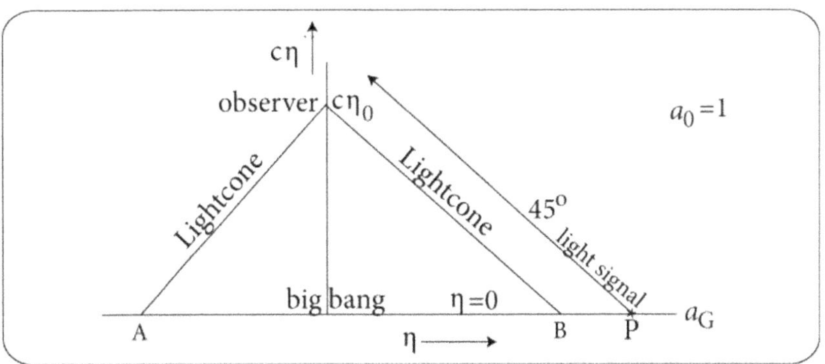

Figure 16.9. Space-time diagram. A matter dominated Universe has a Big Bang.

So at large redshifts coordinate value r saturates, i.e. flattens out at the value of $\frac{2c}{H_0}$. This has been shown in figure 16.8.

The physical interpretation of this can be understood from the space-time diagram of figure 16.9. The $c\eta$ versus η plot has been made. The observer is at η_0. Big Bang is at $\eta = 0$. A backward light cone is shown.

A signal from, say, position P cannot reach the observer within the finite age of the Universe because the signal propagates at 45°. The observer can see only the amount of co-moving distance AB. The scale factor at the Big Bang is $a_G = 0$ and present scale factor is $a_O = 1$. So redshift of infinity corresponds to light coming from the Big Bang. For this reason $\frac{2c}{H_0}$ is called the horizon.

The furthest co-moving distance that light can travel is

$$c\eta_0 = \frac{2c}{H_0} \tag{16.118}$$

This is called the particle horizon. This is the furthest distance a signal can propagate from the Big Bang. Distances larger than this are not causally connected.

- Light is redshifted in an expanding Universe as light propagates from the galaxy to the observer and the redshift is

$$z = \frac{1}{a(t_G)} - 1 \tag{16.119}$$

$$1 + z = \frac{1}{a(t_G)} = \frac{1}{a_G} \tag{16.120}$$

In most situations redshift is the quantity that we can directly measure. Redshift is determined from spectral lines. A spectral line is identified from the spectrum of a galaxy and determines the amount by which it has shifted and redshift can be determined.

- We consider a situation in which the Universe is spatially flat, i.e. $\kappa = 0$ and instead of matter the Universe is dominated by cosmological constant or the dark energy.

$$\Omega_{\Lambda 0} = 1 \tag{16.121}$$

Taking the square root of equation (15.112) we have with $H = H(a)$, $\Omega_{\Lambda 0} = 1$

$$H(a) = H_0 \tag{16.122}$$

From equation (16.109) we thus have

$$r = c \int_{(1+z)^{-1}}^{1} \frac{da}{a^2 H_0} = c(\eta_o - \eta_G) \tag{16.123}$$

$$r = \frac{c}{H_0} \int_{(1+z)^{-1}}^{1} \frac{da}{a^2} = c(\eta_o - \eta_G)$$

$$r = \frac{c}{H_0} \left[-\frac{1}{a} \right]_{(1+z)^{-1}}^{1} = c(\eta_o - \eta_G)$$

$$r = \frac{c}{H_0}[-1 + (1+z)] = c(\eta_o - \eta_G)$$

$$r = \frac{c}{H_0} z \tag{16.124}$$

Clearly r and z are linearly related.

- ✓ Even if the Universe is spatially flat the relation between the coordinate distance r and redshift z depends on the constituents of the Universe.
- ✓ Figure 16.10 shows r as a function of z for the two spatially flat ($\kappa = 1$) models ($\Omega_{m0} = 1$ and $\Omega_{\Lambda 0} = 1$) that we have discussed.

The relation between r and z is given by equation (16.113) viz. $r = \frac{2c}{H_0}[1 - \frac{1}{\sqrt{1+z}}]$ for the case $\kappa = 0$, $\Omega_{m0} = 1$ and by equation (16.124) viz. $r = \frac{c}{H_0}z$ and they are depicted in figure 16.10.

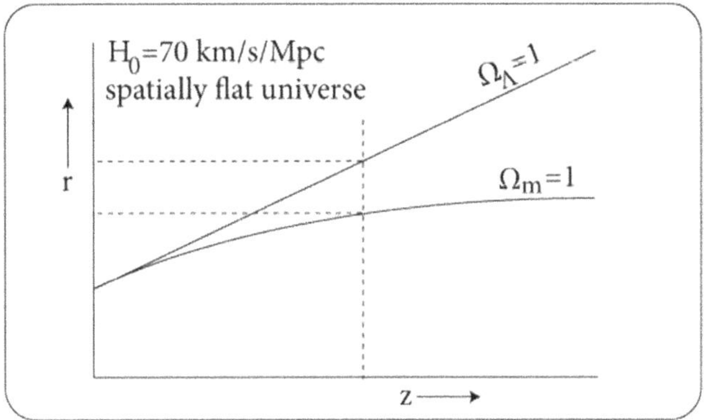

Figure 16.10. Coordinate distance r versus redshift z.

The two models agree at low z, i.e. the two curves match for $z < 1$.

When $z \sim 1$ the two curves differ.

In a matter dominated Universe the coordinate distance does not tend to infinity as z becomes large. Rather it saturates, i.e. tends to a constant maximum value.

In a cosmological constant dominated Universe the coordinate r keeps on increasing linearly with z all the way to infinity.

So in a cosmological constant or dark energy dominated model for the same redshift, i.e. for the same value of z the value of r is larger as compared to a matter dominated Universe. This is evident from figure 16.10.

In a space-time diagram a matter dominated Universe has a Big Bang. At the Big Bang the scale factor is zero which corresponds to redshift of infinity (equations (15.100) and (15.101)). But in a cosmological constant dominated Universe there is no Big Bang. There is exponential expansion (equation (15.117)).

16.20 Calculation of coordinate distances in a situation where there is a spatial curvature, i.e. κ is non-zero

The line element given by equation (16.45) is

$$ds^2 = c^2 dt^2 - a^2 L^2 \left[\frac{dr^2}{1 - \kappa r^2} + r^2(d\theta^2 + \sin^2\theta d\phi^2) \right] \tag{16.125}$$

We consider the case of $\kappa = 1$

Define a new variable as

$$r = \sin\theta, \quad dr = \cos\theta d\theta \tag{16.126}$$

$$d\Omega^2 = d\theta^2 + \sin^2\theta d\phi^2 \tag{16.127}$$

Define

$$\chi = L\theta \text{ (dimension of length)} \tag{16.128}$$

From equation (16.125)

$$ds^2 = c^2dt^2 - a^2L^2\left[\frac{\cos^2\theta d\theta^2}{1 - \sin^2\theta} + \sin^2\theta d\Omega^2\right]$$

$$ds^2 = c^2dt^2 - a^2L^2[d\theta^2 + \sin^2\theta d\Omega^2] \tag{16.129}$$

$$ds^2 = c^2dt^2 - a^2\left[d(L\theta)^2 + L^2\sin^2\frac{\chi}{L}d\Omega^2\right]$$

$$ds^2 = c^2dt^2 - a^2\left[d\chi^2 + L^2\sin^2\frac{\chi}{L}d\Omega^2\right] \tag{16.130}$$

We consider the case of $\kappa = -1$
  Define a new variable as

$$r = \sin h\theta, \quad dr = \cos h\theta d\theta \tag{16.131}$$

$$d\Omega^2 = d\theta^2 + \sin^2\theta d\phi^2 \tag{16.132}$$

Define $\chi = L\theta$ (dimension of length)
  From equation (16.117)

$$ds^2 = c^2dt^2 - a^2L^2\left[\frac{\cos h^2\theta d\theta^2}{1 + \sin h^2\theta} + \sin h^2\theta d\Omega^2\right]$$

$$ds^2 = c^2dt^2 - a^2L^2[d\theta^2 + \sin h^2\theta d\Omega^2] \tag{16.133}$$

$$ds^2 = c^2dt^2 - a^2\left[d(L\theta)^2 + L^2\sin h^2\frac{\chi}{L}d\Omega^2\right]$$

$$ds^2 = c^2dt^2 - a^2\left[d\chi^2 + L^2\sin h^2\frac{\chi}{L}d\Omega^2\right] \tag{16.134}$$

We consider the case of $\kappa = 0$
  With $d\Omega^2 = d\theta^2 + \sin^2\theta d\phi^2$ we have from equation (16.125)

$$ds^2 = c^2dt^2 - a^2L^2[dr^2 + r^2(d\theta^2 + \sin^2\theta d\phi^2)] \tag{16.135}$$

$$ds^2 = c^2dt^2 - a^2[d(Lr)^2 + (Lr)^2d\Omega^2] \tag{16.136}$$

With $Lr = \chi$ we can put the Euclidean line element part as follows

$$ds^2 = c^2dt^2 - a^2[d\chi^2 + \chi^2d\Omega^2] \tag{16.137}$$

We can combine the equation (16.130) (for $\kappa = 1$), (16.134) (for $\kappa = -1$) (16.137) (for $\kappa = 0$) as

$$ds^2 = c^2dt^2 - a^2[d\chi^2 + S^2(\chi)d\Omega^2] \tag{16.138}$$

$$ds^2 = c^2dt^2 - a^2[d\chi^2 + S^2(\chi)(d\theta^2 + \sin^2\theta d\phi^2)] \tag{16.139}$$

where

$$S^2(\chi) = \begin{cases} L^2\sin^2\dfrac{\chi}{L} & \text{for } \kappa = 1 \\[2mm] \chi^2 & \text{for } k = 0 \\[2mm] L^2\sinh^2\dfrac{\chi}{L} & \text{for } \kappa = -1 \end{cases} \tag{16.140}$$

In the limit of $\chi \to 0$, $\sin^2\frac{\chi}{L} \approx \frac{\chi^2}{L^2}$ and so from equation (16.140) we write

$$\text{For } \kappa = 1, \quad S^2(\chi) = L^2\sin^2\frac{\chi}{L} \approx L^2 \cdot \frac{\chi^2}{L^2} = \chi^2 \tag{16.141}$$

$$\text{For } \kappa = -1, \quad S^2(\chi) = L^2\sin h^2\frac{\chi}{L} \approx L^2 \cdot \frac{\chi^2}{L^2} = \chi^2 \tag{16.142}$$

(since the Taylor series are $\sin x = x - \frac{x^3}{6} + \dots$ and $\sin hx = x + \frac{x^3}{6} + \dots$)

So for small χ, i.e. if $\chi \ll L$, results for $\kappa = \pm 1$ go over to that of $\kappa = 0$. The effect of the spatial curvature becomes important only when χ is of the order of length scale L.

Difference occurs when $\chi \sim L$ and $\chi > L$. Then spatial curvature manifests. When the co-moving coordinate is comparable to co-moving curvature length scale, only then does curvature become important.

From line element of equation (16.131) we can calculate distances. Since light is propagating only along the radial direction we can ignore the part $(d\theta^2 + \sin^2\theta d\phi^2)$. So with $ds^2 = 0$

$$c^2dt^2 - a^2d\chi^2 = 0 \tag{16.143}$$

$$d\chi = \frac{cdt}{a} \tag{16.144}$$

Using equation (16.72) viz. $d\eta = \frac{dt}{a}$

$$d\chi = cd\eta \tag{16.145}$$

Integrating

$$\int d\chi = \int_{O}^{G} cd\eta$$

$$\chi - c(\eta_O - \eta_G) \tag{16.146}$$

Value of χ does not depend on the kind of geometry we are working with. The effect of geometry is incorporated in the function $S^2(\chi)$. Thus the value of χ is given by

$$\chi = c(\eta_O - \eta_G) = c \int_{(1+z)^{-1}}^{1} \frac{da}{a^2 H(a)} \tag{16.147}$$

For a given value of redshift z we can calculate χ.

The co-moving coordinate distance χ does not contain any direct physical information. It does not occur directly in any physically measurable quantity. Physically from the measurement point of view there are two distances which are of relevance in cosmological length scales. The distances are angular diameter distance and luminosity distance.

16.21 Angular diameter distance

Consider an observer O and a far off object of length L as shown in figure 16.11. The object subtends an angle $\Delta\theta$ at O. The observer to object distance is d_A. We connect the observer to the contour of the object as shown.

The relation between L, d_A, $\Delta\theta$ is given by

$$\Delta\theta = \frac{L}{d_A} \tag{16.148}$$

Clearly the angle $\Delta\theta$ subtended by the object is its physical size or length divided by the distance d_A. This distance d_A is called angular diameter distance.

- Taking $h = 0.7$, $\Omega_{m0} = 0.3$, $\Omega_{\kappa 0} = 0$ (spatially flat), $\Omega_{\Lambda 0} = 0.7$ and from equation (15.77)

$$\Omega_{m0} + \Omega_{\kappa 0} + \Omega_{\Lambda 0} = 0.3 + 0 + 0.7 = 1 \tag{16.149}$$

and from equation (15.79) with $H = H(a)$

$$H^2(a) = H_0^2[\Omega_{m0}a^{-3} + \Omega_{\kappa 0}a^{-2} + \Omega_{\Lambda 0}] \tag{16.150}$$

Figure 16.12 shows $S(z)$ versus z plot as well as the plot of angular diameter distance d_A against z (equation (16.156)).

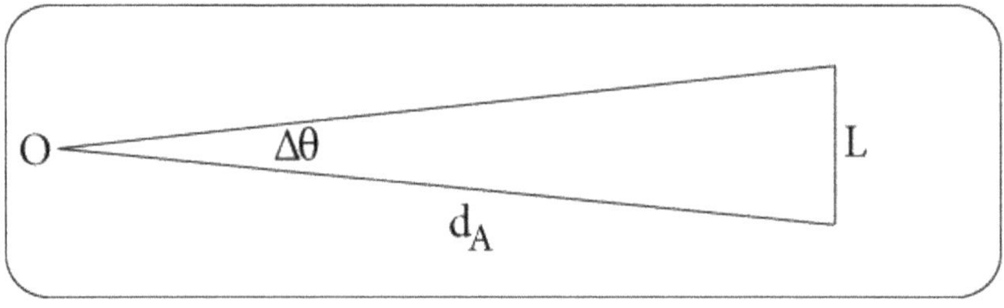

Figure 16.11. Angular diameter distance.

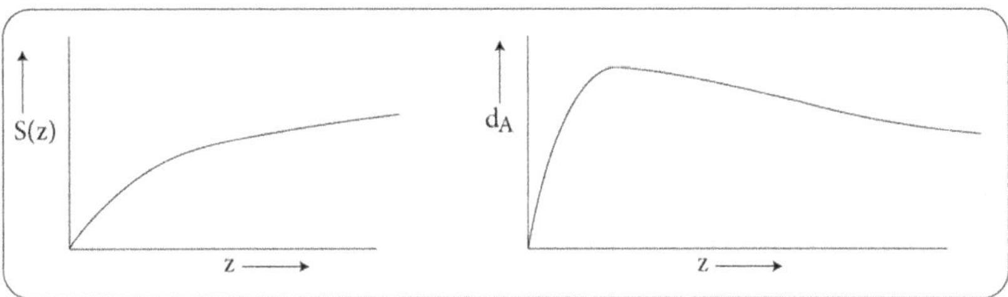

Figure 16.12. Plots of $S(z)$ versus z and d_A versus z for $h = 0.7, \Omega_{m0} = 0.3, \Omega_{\kappa0} = 0$ (spatially flat), $\Omega_{\Lambda0} = 0.7$.

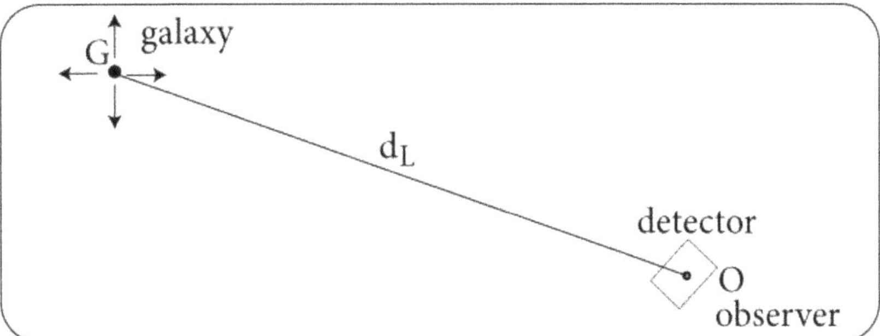

Figure 16.13. Luminosity distance.

16.22 Luminosity distance

There is a light source say a star or a galaxy at G. Observer position is at O as indicated in figure 16.13. Suppose the light source has luminosity L.

Luminosity is the energy emitted per second.

We wish to find the flux received by observer at O. The flux received by the observer is defined as the mean energy that crosses the unit area placed perpendicularly per second, i.e.

$$f = \frac{L}{4\pi d_L^2} \equiv \frac{L_G}{4\pi d_L^2} = f_{\nu O} \tag{16.151}$$

This distance d_L is called luminosity distance. The emitted light is distributed over the surface area of a sphere of radius d_L. Hence L is divided by the surface area $4\pi d_L^2$.

In the familiar Euclidean space we expect the two distances, namely angular diameter distance d_A and the luminosity distance d_L to be the same. But in non-Euclidean space and hence in cosmology it is not so. We therefore distinguish between the 2 distances.

Consider an object whose length is known by some means and we see that object has a redshift of z. This is shown in figure 16.14. The galaxy subtends an angle $\Delta\theta$ at the observer The actual physical size of the galaxy is L. From a measurement of z, $\Delta\theta$ we can measure the angular distance d_A (equation (16.148)).

Consider the line element given by equation (16.138). The observer is at $\chi = 0$. And the galaxy is at χ_G.

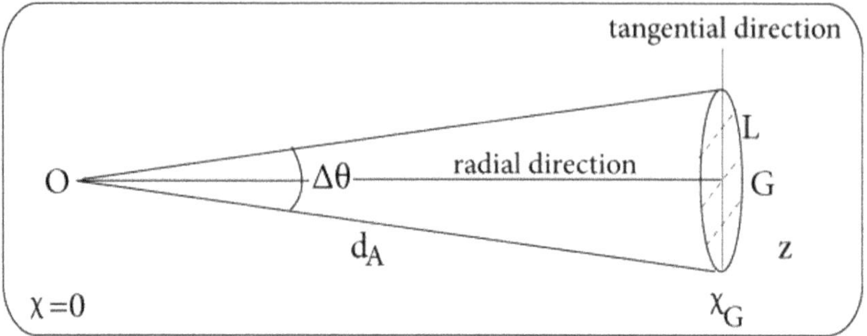

Figure 16.14. Observer O receives light from a galaxy with redshift z.

Value of χ changes from $\chi = 0$ to $\chi = \chi_G$ along the radial direction. Ends of the galaxy are separated by $\Delta\theta$ and ϕ can be taken to be a constant and overlooked. In view of this, the metric of equation (14.139) reduces to

$$ds^2 = c^2 dt^2 - a^2[d\chi^2 + S^2(\chi)d\theta^2] \tag{16.152}$$

Co-moving separation corresponding to the two ends of the galaxy separated by L is

$$L = a_G S(\chi_G)\Delta\theta \tag{16.153}$$

From equation (16.148) we have

$$L = \Delta\theta d_A \tag{16.154}$$

Comparing equations (16.153), (16.154) we write

$$a_G S(\chi_G)\Delta\theta = \Delta\theta d_A$$

$$d_A = a_G S(\chi_G) \tag{16.155}$$

$S(\chi_G)$ is known from equation (16.140) depending on the cosmological model. And $S(\chi_G)$ can be determined as a function of redshift z using equation (16.147) viz.

$$\chi = c(\eta_0 - \eta_G) = c \int_{(1+z)^{-1}}^{1} \frac{da}{a^2 H(a)}$$

So we can rewrite equation (16.155) as

$$d_A = a_G S(z) \tag{16.156}$$

From equation (16.120)

$$a_G = \frac{1}{1 + z} \tag{16.157}$$

Hence from equation (16.156)

$$d_A = \frac{S(z)}{1 + z} \tag{16.158}$$

So angular diameter distance is estimated.

Consider an example

- Consider spatially flat situation, i.e. $\kappa = 0$.

From equation (16.140) we get $S(\chi) = \chi = r$
From equation (16.156)

$$d_A = \frac{r}{1+z} \tag{16.159}$$

Consider the case $\Omega_{m0} = 1$
Using equation (16.113) we get

$$d_A = \frac{r}{1+z} = \frac{1}{1+z}\frac{2c}{H_0}\left[1 - \frac{1}{\sqrt{1+z}}\right] \tag{16.160}$$

Consider the case $\Omega_{\Lambda 0} = 1$
Using equation (16.124) we get

$$d_A = \frac{r}{1+z} = \frac{1}{1+z}\frac{cz}{H_0} = \frac{cz}{H_0(1+z)} \tag{16.161}$$

For small $z \ll 1$, equation (16.160) (for $\Omega_{m0} = 1$) gives

$$d_A = \frac{1}{1+z}\frac{2c}{H_0}\left[1 - \frac{1}{\sqrt{1+z}}\right] = \frac{2c}{H_0}[1 - (1+z)^{-1/2}] = \frac{2c}{H_0}\left[1 - \left(1 - \frac{z}{2}\right)\right]$$

$$d_A = \frac{2c}{H_0}\frac{z}{2} = \frac{cz}{H_0} \tag{16.162}$$

For small $z \ll 1$, equation (16.161) (for $\Omega_{\Lambda 0} = 1$) gives

$$d_A = \frac{cz}{H_0(1+z)} = \frac{cz}{H_0} \tag{16.163}$$

So for small z the value we get from both equations (16.162), (16.163) viz. $d_A = \frac{cz}{H_0}$,
i.e. the angular diameter distance d_A varies linearly with redshift z.

For large $z \to \infty$
From equation (16.160) (for $\Omega_{m0} = 1$) we get

$$d_A = \frac{1}{1+z}\frac{2c}{H_0}\left[1 - \frac{1}{\sqrt{1+z}}\right] = \frac{1}{1+z}\frac{2c}{H_0} = 0 \tag{16.164}$$

From equation (16.161) (for $\Omega_{\Lambda 0} = 1$) we get

$$d_A = \frac{cz}{H_0(1+z)} = \frac{c}{H_0}\cdot\frac{z}{z} = \frac{c}{H_0} \tag{16.165}$$

In the $\Omega_{m0} = 1$ case for small z we have $d_A = \frac{cz}{H_0}$ (linear variation) and for large z we have $d_A = 0$. This means there must be a large value and a maximum or peak of d_A in the intermediate region. This we show in the figure 16.15.

In the $\Omega_{\Lambda 0} = 1$ case for small z we have $d_A = \frac{cz}{H_0}$ (linear variation) and for large z we have $d_A = \frac{c}{H_0}$ (Hubble radius given in equation (16.63)). This means there must be a large value and a maximum of d_A in the intermediate region. This we show in figure 16.15.

An object located further away appears small and when it moves further away it looks smaller. Angle subtended will keep on decreasing (figure 16.16) in the Euclidean world for a receding object. In the cosmological situation this is true for a range of redshift.

For a physical object taken further and further away from an observer, its angle decreases and beyond a certain distance its angle starts to increase again and it subtends an infinite angle. Actually, we have to consider also the effect of expansion of the Universe.

Back in time the Universe was smaller and so the same physical object occupied a larger co-moving separation. The co-moving separation is what actually decides the subtended angle.

The effects of angular decrease with distance and expansion of the Universe compensate each other at a certain point and beyond that expansion of the Universe takes over.

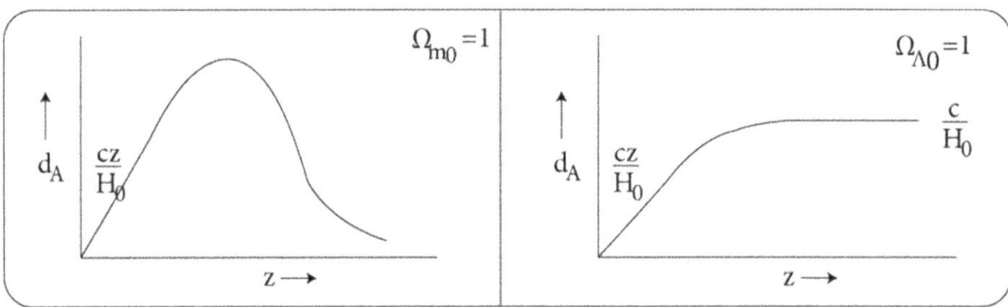

Figure 16.15. Variation of d_A against z for $\Omega_{m0} = 1$ and $\Omega_{\Lambda 0} = 1$ cases and for $\kappa = 0$.

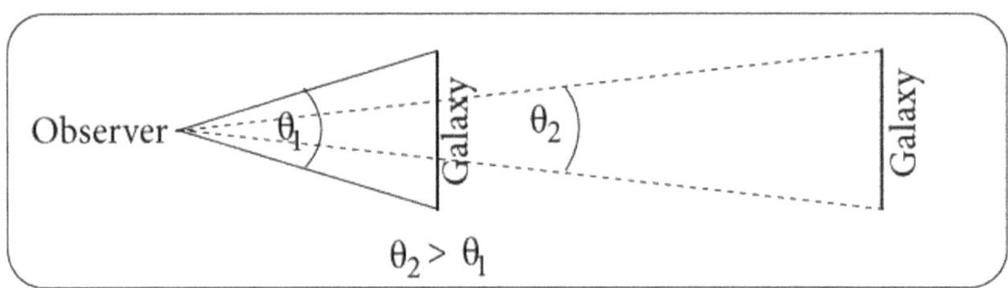

Figure 16.16. To an observer a nearer galaxy subtends a larger angle while a further galaxy subtends a smaller angle.

At the Big Bang the whole Universe contracts to a point at coordinate distance $\frac{c}{H_0}$ corresponding to $z = \infty$ and subtends the 4π angle.

Angular diameter distance is also sensitive to the curvature. Coordinate distance is not sensitive to curvature.

16.23 Specific luminosity

Consider a source of radiation. Radiation from a galaxy has a spectrum, as shown in figure 16.17. The amount of energy it emits in the frequency interval $\Delta\nu$ in time Δt is ΔE_ν. Then

$$\Delta E_\nu = L_\nu \Delta\nu \Delta t \tag{16.166}$$

where L_ν is called the specific luminosity.

Specific luminosity L_ν and specific flux f_ν (equation (16.151)) are related by luminosity distance d_L (equation (16.178)).

16.24 Bolometric luminosity

The total energy radiated by a source in time Δt is

$$\Delta E = L \, \Delta t \tag{16.167}$$

where L is Bolometric luminosity. It has no reference to any part of the spectrum and so we have to integrate over the entire part of the spectrum.

A bolometer is a device that is sensitive to the radiation energy over a large range of frequency.

f is called bolometric flux.

The quantities L and f are related by the luminosity distance.

However, the luminosity distance obtained from relating the specific quantities L_ν and f_ν is different from the bolometric values L and f.

If we are concentrating in measurements that are sensitive to radiation over a certain frequency band then we have to use specific quantities L_ν and f_ν but if we are

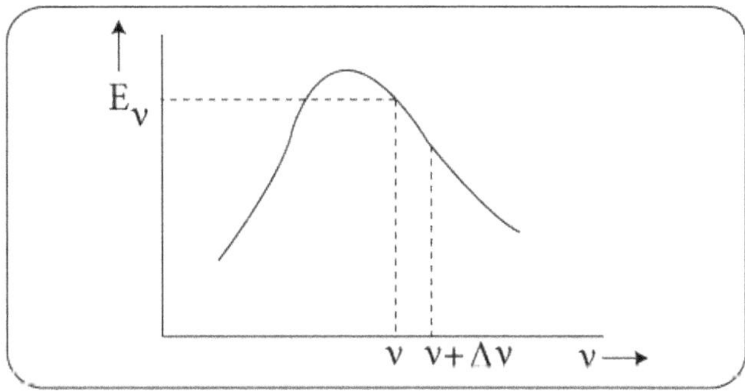

Figure 16.17. Radiation from a galaxy has a spectrum.

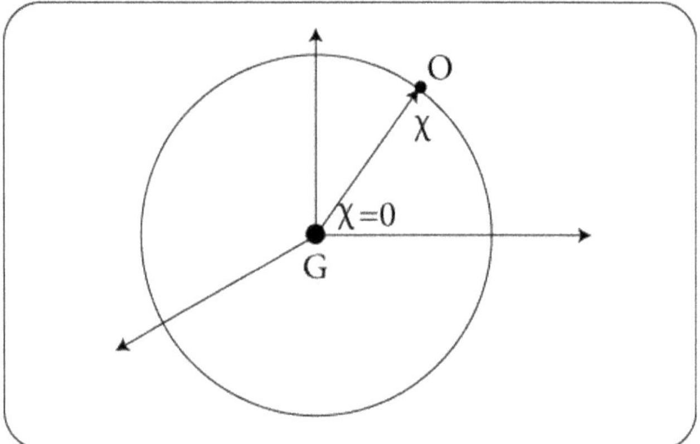

Figure 16.18. Light from galaxy G propagates to observer O.

concentrating in measurements that are sensitive over a large range or over the entire range of radiation being emitted, then bolometric quantities L and f are to be used. Clearly thus luminosity distance depends on the nature of measurement.

Consider a galaxy G as source, as shown in figure 16.18, which is taken as the origin. The observer is O. Light from a galaxy propagates from G to O which is the χ direction, i.e. the radial direction ($\chi = 0$ at G and it is χ at O).

Light source G has luminosity L_ν. Amount of radiation emerging is by equation (16.166)

$$\Delta E_\nu = L_\nu \Delta \nu \Delta t \tag{16.168}$$

Energy of radiation is conserved in Euclidean space where there is no expansion. The energy that is emitted will pass through the sphere. But in this case energy of radiation is not conserved because of expansion and redshift. By the time energy of emitted radiation reaches the observer the frequency of radiation changes due to redshift. So it is difficult to deal with the quantity ΔE_ν directly. It is easier to deal with particles (photons) since light can be looked upon as a stream of photons.

Let us calculate the number of photons $N_{\nu G}$ emitted in the frequency interval $\Delta \nu_G$ and time interval Δt_G which is

$$N_{\nu G} = \frac{\Delta E_{\nu G}}{h \nu_G} = \frac{L_{\nu G} \Delta \nu_G \Delta t_G}{h \nu_G} \text{ (using equation (16.166))} \tag{16.169}$$

The emitted photons will all cross the surface shown. However, the time interval as well as the frequency interval at which they are emitted by the galaxy and received by the observer, are not the same.

Ratio of emitted to observed frequencies is

$$\frac{\nu_G}{\nu_O} = \frac{\Delta \nu_G}{\Delta \nu_O} = \frac{\Delta t_O}{\Delta t_G} \text{ (frequency and time are inversely related)} \tag{16.170}$$

As frequency scales as the inverse of the scale factor we can write

$$\frac{\nu_G}{\nu_O} = \frac{\Delta\nu_G}{\Delta\nu_O} = \frac{\Delta t_O}{\Delta t_G} = \frac{a_O}{a_G} = \frac{1}{a_G} \text{ (as } a_O = 1) \tag{16.171}$$

Using equation (16.120)

$$\frac{\nu_G}{\nu_O} = \frac{\Delta\nu_G}{\Delta\nu_O} = \frac{\Delta t_O}{\Delta t_G} = \frac{a_O}{a_G} = \frac{1}{a_G} = 1 + z \tag{16.172}$$

Let us calculate the flux that will be observed by an observer at O. Photons do not get destroyed in transit. Only their frequencies change. The rate at which they are emitted and the rate at which they are received are the same, i.e. all photons are able to reach the surface. The number of photons crossing the surface will be $N_{\nu G}$ over a time interval Δt_O and frequency interval $\Delta\nu_O$ and each photon carries energy $h\nu_O$.

All energy will be distributed over the sphere. The area of the sphere is obtained by integrating over the solid angle as occurring in the metric given in equation (16.139). Hence the area of the sphere is $4\pi a_O^2 S^2$ as the 4π factor comes from integration over θ and ϕ while $a_O^2 S^2$ occurs as coefficient.

So the flux observed at a different frequency ν_O will be

$$f_{\nu O} = \frac{\text{(number of photons emitted)(energy per photon)}}{\text{(time interval)(frequency interval)(surface area)}} = \frac{N_{\nu G} h\nu_O}{\Delta t_O \Delta\nu_O 4\pi a_O^2 S^2} \tag{16.173}$$

Using equation (16.169) viz. $N_{\nu G} = \frac{L_{\nu G}\Delta\nu_G\Delta t_G}{h\nu_G}$ we have

$$f_{\nu O} = \frac{L_{\nu G}\Delta\nu_G\Delta t_G}{h\nu_G} \frac{h\nu_O}{\Delta t_O \Delta\nu_O 4\pi a_O^2 S^2} \tag{16.174}$$

With $a_O = 1$

$$f_{\nu O} = \frac{L_{\nu G}\Delta\nu_G\Delta t_G}{h\nu_G} \frac{h\nu_O}{\Delta t_O \Delta\nu_O 4\pi S^2} = \frac{L_{\nu G}}{4\pi S^2} \frac{\Delta\nu_G}{\Delta\nu_O} \frac{\Delta t_G}{\Delta t_O} \frac{\nu_O}{\nu_G} \tag{16.175}$$

Using equation (16.137) $\frac{\Delta\nu_G}{\Delta\nu_O} \cdot \frac{\Delta t_G}{\Delta t_O} = 1$, $\frac{\nu_O}{\nu_G} = \frac{1}{1+z}$ we have

$$f_{\nu O} = \frac{L_{\nu G}}{4\pi S^2} \frac{1}{1+z} \tag{16.176}$$

So we have related specific flux and specific luminosity.

We now identify the luminosity distance given by equation (16.151) viz. $f_{\nu O} = \frac{L_{\nu G}}{4\pi d_L^2}$. Comparing with equation (16.176) we get

$$f_{\nu O} = \frac{L_{\nu G}}{4\pi S^2} \frac{1}{1+z} = \frac{L_{\nu G}}{4\pi d_L^2} \tag{16.177}$$

$$d_L = S(1+z)^{1/2} \tag{16.178}$$

This is the specific luminosity distance calculated using specific luminosity and specific flux.

16.25 Relation between the bolometric quantities

Bolometric flux is the integral of the specific flux observed and integration is over all the observed frequencies, i.e.

$$f = \int f_{\nu O} \, d\nu_O \tag{16.179}$$

From equation (16.176) we write

$$f = \int \frac{L_{\nu G}}{4\pi S^2} \frac{1}{1 + z} d\nu_O \tag{16.180}$$

Again, bolometric luminosity is the integral over specific luminosity of all emitted frequencies, i.e.

$$L = \int L_{\nu G} \, d\nu_G \tag{16.181}$$

From equation (16.172) we have $\frac{\nu_G}{\nu_O} = 1 + z$, $\frac{\nu_O}{\nu_G} = \frac{1}{1+z}$. With this equation (16.180) gives

$$f = \int \frac{L_{\nu G}}{4\pi S^2} \frac{1}{1 + z} d\frac{\nu_G}{1 + z}$$

$$f = \frac{1}{4\pi S^2 (1 + z)^2} \int L_{\nu G} \, d\nu_G \tag{16.182}$$

$$f = \frac{L}{4\pi S^2 (1 + z)^2} \quad \text{(by equation (16.181))} \tag{16.183}$$

Comparing with equation (16.151) viz. $f = \frac{L}{4\pi d_L^2}$ we get the bolometric luminosity distance to be

$$f = \frac{L}{4\pi S^2 (1 + z)^2} = \frac{L}{4\pi d_L^2}$$

$$d_L = S(1 + z) \tag{16.184}$$

Comparison with equation (16.178) shows that the bolometric luminosity distance increases faster than the specific luminosity distance.

16.25.1 Summary of distances

Angular diameter distance $d_A = \frac{S}{1+z}$ (equation (16.158)).

Specific luminosity distance $d_L = S\sqrt{1 + z}$ (equation (16.178)).
Bolometric luminosity distance $d_L = S(1 + z)$ (equation (16.184)).

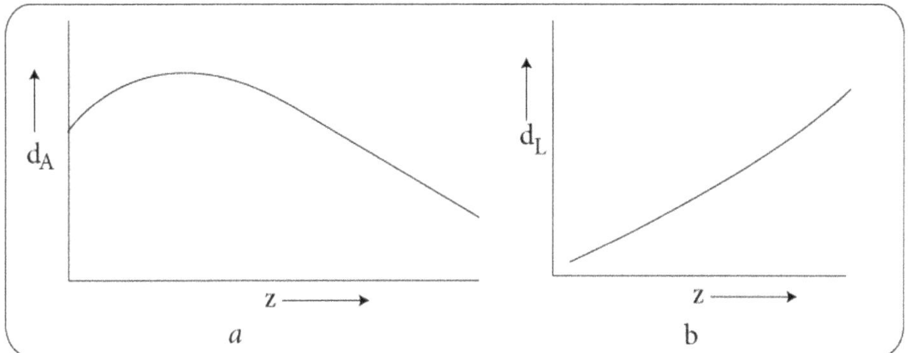

Figure 16.19. Angular diameter distance versus redshift.

Variation of angular diameter distance with redshift is shown in figure 16.19(a) (equation (16.158)). Initially angular diameter distance initially increases with redshift and beyond a certain redshift it decreases.

Variation of bolometric luminosity distance with redshift is shown in figure 16.19 (b) (equation (16.184)). Bolometric luminosity distance increases monotonically with redshift. The higher the redshift, the larger is the luminosity distance and the fainter is the object.

16.26 Hubble's diagram

Various cosmological models predict cosmic distance as a function of redshift.

For small redshift z, i.e. for $z \ll 1$ distance obtained from various models give the relation

$$d = \frac{cz}{H_0}.$$

Again, observations (for small redshift) give distance as a function of z. Comparing these, the value of the Hubble parameter can be determined. In other words, Hubble's law can be tested and a conclusion regarding the state of the Universe (whether it is expanding or not) can be determined.

A schematic plot obtained by Hubble for a Universe with constant Hubble parameter is shown in figure 16.20. Hubble independently determined distances of galaxies and their redshifts and obtained a velocity versus distance plot which was found to be linear and also an estimate of the value of the Hubble parameter (equation (14.2)) was done.

The inverse of Hubble's parameter is related to the age of the Universe (equations (14.31), (14.88)).

The age thus obtained was too small since there were objects of comparable age. This means the estimate of Hubble's parameter was not correct—it was too high.

Determination of Hubble's parameter was based on the fact that the only motion of all galaxies was because of their moving away from one another due to expansion of the Universe. This, however, is not true.

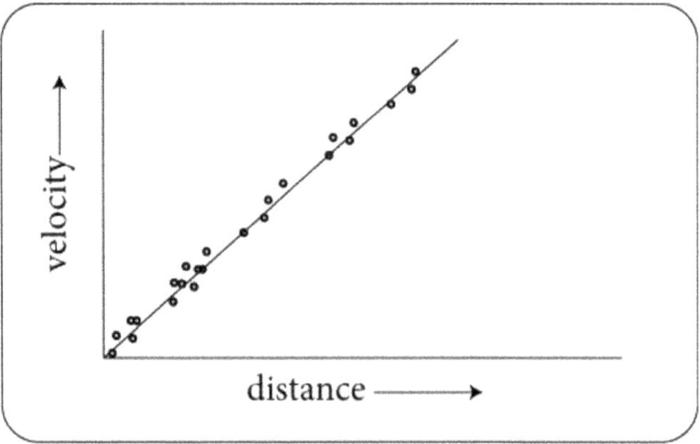

Figure 16.20. Hubble diagram for a Universe with a constant Hubble parameter.

Galaxies have motion in addition to that predicted by Hubble's law and this motion is referred to as peculiar velocity, which is around $v \sim 300$ km s^{-1} and is because of local gravitational field.

The Universe is homogeneous and isotropic in the large scale. However, strictly speaking, in general the Universe is not homogeneous and isotropic. In a smaller scale, for instance over a length scale of ~ 10 Mpc, there are fewer galaxies at some regions and more galaxies at other regions. At regions where there are more galaxies, gravitational effect, i.e. gravitational attraction, will be larger. And at regions where there are fewer galaxies gravitational effect will be smaller, which is a kind of repulsive effect. These introduce extra motion in the galaxies called peculiar motion. This peculiar velocity will introduce extra redshift $z = \frac{v}{c} \sim 0.001$. So to minimize error or uncertainty due to peculiar velocity to less than 10% we have to go to large redshift due to Hubble expansion. If the Hubble redshift is 10 times more than the extra redshift due to peculiar velocity then error reduces to less than 10%.

Another way to get a reliable estimate is to look for galaxies at distances

$$d > 0.01 \frac{c}{H_0} = 0.01 \frac{3 \times 10^8 \, m \, s^{-1}}{70 \, km \, s^{-1} \, Mpc} = 43 \, Mpc. \tag{16.185}$$

So for small redshifts uncertainty or error is large and for larger redshifts uncertainty or error decreases.

16.27 Supernovae

Supernovae are stars that appear as new objects in the sky. They suddenly become bright and appear as new stars. For instance, a faint star suddenly becomes extremely bright. It occurs at the end stage of massive stars as they exhaust all their fuel.

Supernovae can be identified in distant galaxies. They are extremely bright.
Each star has luminosity $\sim 4 \times 10^{26} \, W$.
A galaxy has 10^{11} stars.

So galaxy luminosity is $\sim 10^{11} \times 4 \times 10^{26}\ W = 4 \times 10^{37}\ W$ \hfill (16.186)

When a supernova goes off, its luminosity is comparable to this value. So a supernova becomes brighter than the entire galaxy and is easily detectable. It is a transient event. Supernovae occur at the end stage of massive stars when they finish all the fuel.

An example of a supernova is 1994D in the Galaxy NGC 4526. In this galaxy only a bright pattern is easily observed but individual stars cannot be identified.

16.28 Flux density

Some quantities relevant for optical observations are as follows.

I_ν is surface brightness which is the amount of radiation coming per unit solid angle in the frequency range ν and $\nu + d\nu$ in the time interval dt with the help of a detector of size dA.

For observing objects like supernova or stars which are essentially point objects (we cannot associate a dimension with them) what we measure is not the specific intensity but the flux density, also called spectral energy density S_ν or f_ν.

Consider a source viewed in the sky (figure 16.21). The specific intensity varies from point to point on the source. For a star or supernova we cannot understand the extension of the source, i.e. the source is not resolved. Consequently, what we measure is the specific intensity integrated over the entire solid angle. The elemental solid angle is $d\Omega$ and we have to integrate over all such $d\Omega$, i.e. over the entire source.

$$S_\nu = f_\nu = \int I_\nu d\Omega \ (\text{unit:}\ W\ m^{-2}\ Hz^{-1}) \tag{16.187}$$

16.28.1 Jansky unit

Quite often we use the unit Jansky (Jy) which is the unit of spectral energy density.

$$1\ Jy = 10^{-26}\ W\ m^{-2}\ Hz^{-1}$$

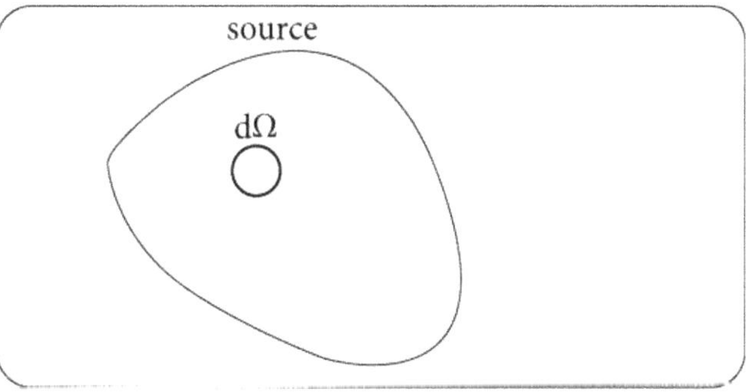

Figure 16.21. Source viewed in the sky.

Since the sources considered are very weak, introduction of this unit Jy becomes helpful.

Energy is given by

$$dE = S_\nu \, dA \, dt \, d\nu = S_\lambda \, dA \, dt \, d\lambda$$

We can recast things in terms of wavelength also as $S_\lambda = f_\lambda$ and it is related to S_ν through the relation $\nu = \frac{c}{\lambda}$, $d\nu = -\frac{c}{\lambda^2} d\lambda$ to get

$$S_\lambda = \frac{c}{\lambda^2} S_\nu. \tag{16.188}$$

16.29 Apparent magnitude

Brightness of stars was visually determined and classified by Greek astronomers who called the brightest star to be of magnitude 1. Gradually fainter stars were of magnitude 2,3,...,5 and the faintest star identified with the naked eye was of magnitude 6.

Later, photometers were used and it was established that response of the eye was logarithmic and not linear. This means that our eye has an enormous dynamical range.

Apparent magnitude was thus defined as

$$m = -2.5 \log_{10} \frac{f}{f_0} \tag{16.189}$$

where f is flux from source, f_0 a standard flux value that sets the zero point of the magnitude scale.

When $f = f_0$ then

$$m = -2.5 \log_{10} \frac{f_0}{f_0} = -2.5 \log_{10} 1 = 0 \tag{16.190}$$

The fainter the object the larger the magnitude.

16.29.1 Zero point

Alternatively, we can express the relation of equation (16.189) as

$$m = -2.5 \log_{10} f + c. \tag{16.191}$$

The constant c takes care of the zero point.

Consider 2 sources whose magnitude difference is 5, i.e.

$$m_1 - m_2 = 5 \tag{16.192}$$

$$m_1 = 5 + m_2$$

So m_1 is fainter and f_2 is brighter.

m_1 has flux f_1 and m_2 has flux f_2.

They are related as

$$m_1 - m_2 = 5 = 2.5 \log_{10} \frac{f_2}{f_1} \tag{16.193}$$

$$\frac{5}{2.5} = \log_{10} \frac{f_2}{f_1}$$

$$2 = \log_{10} \frac{f_2}{f_1}$$

$$10^2 = \frac{f_2}{f_1}$$

$$f_2 = 100 f_1 \tag{16.194}$$

So if magnitude difference between two sources is 5 this means that the fainter source having larger magnitude has 100 times lower flux value, i.e. if magnitude goes up by 5 its flux goes down by 10.

16.30 Absolute magnitude

Apparent magnitude refers to the flux of the source. Again, flux f is related to the luminosity L as

$$f = \frac{L}{4\pi d^2} \tag{16.195}$$

where d is distance of source from observer.

The relation between magnitude and flux is given by

$$m = -2.5 \log_{10} \frac{L}{4\pi d^2} + c \tag{16.196}$$

If the same source is observed from a different distance D then its magnitude M is

$$M = -2.5 \log_{10} \frac{L}{4\pi D^2} + c \tag{16.197}$$

M is called absolute magnitude.

Hence subtracting equation (16.197) from equation (16.196) we have

$$m - M = \left(-2.5 \log_{10} \frac{L}{4\pi d^2} + c\right) - \left(-2.5 \log_{10} \frac{L}{4\pi d^2} + c\right)$$

$$m - M = -2.5 \log_{10} \frac{L}{4\pi d^2} + 2.5 \log_{10} \frac{L}{4\pi D^2} = 2.5 \log_{10} \frac{d^2}{D^2} = 2.5 \times 2 \log_{10} \frac{d}{D}$$

$$m - M = 5 \log_{10} \frac{d}{D} \tag{16.198}$$

By absolute magnitude we refer to the magnitude that a source would have if it were located at a distance of 10 pc. Taking $D = 10$ pc we write

$$m - M = 5 \log_{10} \frac{d}{10\,pc} \tag{16.199}$$

Apparent magnitude for a source contains luminosity and distance. The same source if moved further and further away gets fainter and apparent magnitude increases.

Absolute magnitude is an intrinsic property of the source and it essentially refers to the luminosity. So absolute magnitude is equivalent to the luminosity of the source. Absolute magnitude is luminosity in logarithmic units.

Absolute magnitude quantifies the intrinsic brightness of the source.

Suppose apparent magnitude of source is measured at some frequency ν. If redshift is measured then distance to the object can be determined. This distance can be used to calculate the absolute magnitude. This would give the luminosity of source at frequency ν.

A supernova is a stage in the evolution sequence of massive stars.

Stars like our Sun will not end up as a supernova. They will finish their hydrogen cycle, expand and then go into a helium burning phase.

Very massive stars have a stage in their life cycle where they explode and luminosity goes up tremendously. They emit an external shell that expands.

Consider a spherical blackbody of radius R. Radiation that emerges out has luminosity

$$L = 4\pi R^2 \Delta T^4 \tag{16.200}$$

where ΔT is temperature of blackbody.

Now if radius increases then luminosity goes up.

In a supernova the inner core contracts and the outer shell is ejected out which expands and the luminosity goes up tremendously. The luminosity of a star increases manifold.

Supernovae can be classified into two types.

Type 1 supernova has no hydrogen absorption lines in the supernova emission. These are denoted by *Sne I*.

Type 2 supernova has hydrogen absorption lines in the supernova emission. These are denoted by *Sne II*.

Type 1 supernovae can be further classified as *Sne Ia* and *Sne Ib*.

Supernovae of type *Sne Ia* are studied more in cosmology for determining cosmological distances.

A supernova of type *Sne Ia* is a system of two stars or interacting binaries. One star is more massive than the other. One star first burns hydrogen and passes through the entire cycle or sequence and turns into a white dwarf. A white dwarf is a very compact small star that is maintained by the electron degeneracy pressure. Electrons being fermions cannot be put in the same energy level. So electrons occupy higher energy levels. The higher energy electrons contribute to the pressure and maintain the star in equilibrium. This is referred to as the electron degeneracy pressure.

There is an upper limit on the mass viz. $1.4 M_{sun}$ called the Chandrasekhar limit beyond which the electron degeneracy pressure cannot maintain the star in equilibrium. It will collapse.

If the white dwarf star has mass more than the Chandrasekhar limit the electron degeneracy pressure is not sufficient to maintain it in equilibrium. It will collapse to form a neutron star or a black hole.

Here we are considering a system of two stars one of which is a white dwarf and is more massive, the other star is not a white dwarf and accretes materials. This accreting material can push the mass of the white dwarf above the Chandrasekhar limit and it can become a supernova. Then electron degeneracy pressure is no longer sufficient to maintain equilibrium. So it collapses and becomes a supernova.

It is observed that the properties of supernovae are quite uniform across the entire sample. This is what makes them particularly well suited for determining cosmological distances.

For the other type of supernovae, *Sne Ib*, properties are not uniform across the entire sample and hence they are not suited for cosmological distance determination.

When a supernova is formed there is a characteristic rise in the luminosity to a maximum and then it falls gradually over a period of around 30 days. So one gets sufficient time to measure luminosity of supernovae.

16.31 Exercises

Exercise 16.1 *Identify the incorrect statement*
    (a) *The distance between two different points in 4-space-time can be zero.*
    (b) *The distance between two different points in 4-space-time can never be zero.*
    (c) *The distance between two different points in Euclidean space can never be zero.*
    (d) *The 4-space-time is not Euclidean.*

Exercise 16.2 *Which of the results $ds > 0$, $ds = 0$, $ds < 0$ hold for interval ds for 2 events occurring at the same place but at different times?*

[Ans.] Consider two events occurring at points A and B in the space-time diagram of figure 16.3. Their coordinates are $A(ct, x, y, z)$ and $B(c(t + dt), x, y, z)$, i.e. events A and B occur at the same place (x, y, z) but at different times t and $t + dt$.

The interval square is given by

$$ds^2 = (c(t + dt) - ct)^2 - (x - x)^2 - (y - y)^2 - (z - z)^2 = c^2 dt^2 > 0$$

As the observer is at rest only time changes. Hence

$$ds > 0$$

Exercise 16.3 *Which of the results ds > 0, ds = 0, ds < 0 hold for interval ds for 2 events occurring at different places but at the same time?*

    Ans. Consider two events occurring at points A and C in the space-time diagram of figure 16.3. Their coordinates are $A(ct, x, y, z)$ and $C(ct, x + dx, y + dy, z + dz)$, i.e. events A and C occur at the same time t but at different places (x, y, z) and $x + dx, y + dy, z + dz$.

    The interval square is given by

$$ds^2 = (ct - ct)^2 - (x + dx - x)^2 - (y + dy - y)^2 - (z + dz - z)^2$$
$$= -(dx^2 + dy^2 + dz^2)$$

$$ds^2 < 0$$

$$ds < 0$$

Exercise 16.4 *Suppose we see a galaxy at a redshift $z = 3$. Scale factor at present is $a_0 = 1$. Find the scale factor when light was emitted.*

    Ans. From equation (16.88) we have

$$z = \frac{a(t_o)}{a(t_G)} - 1$$

$$3 = \frac{1}{a(t_G)} - 1$$

$$a(t_G) = \frac{1}{4}$$

So the Universe was 4 times smaller when light was emitted from the galaxy.

Exercise 16.5 *Null-like, space-like and time-like events are connected by signals of speed*

    (a) $v = c, v < c, v > c$ *(b)* $v = 0, v > 0, v < c$
    (c) $v = c, v > c, v < c$ *(d)* $v = 0, v > c, v < ct.$

Exercise 16.6 *Which of the following statements are incorrect according to Einstein's general theory of relativity?*

    (a) *Space-time is curved in a gravitational field.*
    (b) *Gravity manifests as curvature of space-time.*

(c) *Metric is a function of space.*
(d) *All information about space-time is there in the metric.*
(e) *All are correct.*
(f) *Option (c) is incorrect.*

Exercise 16.7 *Which of the following is Hubble radius?*

(a) H_0^{-1} *(b)* $7 \times 10^9\,m$ *(c)* $1.4 \times 10^{10}\,m$ *(d)* $1.32 \times 10^{26}\,m.$

Exercise 16.8 *Which of the following is/are true?*
   (a) *Specific luminosity distance increases faster than bolometric luminosity distance.*
   (b) *Bolometric luminosity distance increases faster than specific luminosity distance.*
   (c) *Specific luminosity distance and bolometric luminosity distance both increase similarly.*
   (d) *Specific luminosity distance and bolometric luminosity distance are different.*

Exercise 16.9 *Which of the following is bolometric luminosity distance?*

(a) $\frac{S}{1+z}$ *(b)* $S\sqrt{1+z}$ *(c)* $S(1+z)$ *(d)* $\frac{S}{z}.$

Exercise 16.10 *Which of the following is specific luminosity distance?*

(a) $\frac{S}{1+z}$ *(b)* $S\sqrt{1+z}$ *(c)* $S(1+z)$ *(d)* $\frac{S}{z}.$

Exercise 16.11 *Which of the following is angular diameter distance?*

(a) $\frac{S}{1+z}$ *(b)* $S\sqrt{1+z}$ *(c)* $S(1+z)$ *(d)* $\frac{S}{z}.$

Answers to multiple choice type questions

$16.1b$, $16.5c$, $16.6e$, $16.7d$, $16.8b,d$, $16.9c$, $16.10b$, $16.11a$.

16.32 Question bank

   Q16.1 Write down the metric for flat Euclidean space in Cartesian coordinate system and spherical polar coordinate system in 2D and in 3D.

Q16.2 What do we mean by an event? Show a space-time diagram showing an observer at rest.

Q16.3 What do we mean by interval? How does the interval square

$$c^2dt^2 - (dx^2 + dy^2 + dz^2)$$

change as a particle changes position over different times?

Q16.4 When are events called null-like, space-like and time-like?

Q16.5 What are the postulates of the special theory of relativity?

Q16.6 Write down the FRW metric. How does it change with curvature $\kappa = 0, \pm 1$?

Q16.7 Estimate the value of Hubble radius.

Q16.8 Explain the following terms:
1. co-moving distance, proper distance, angular diameter distance, luminosity distance.

Q16.9 What is conformal time? Find conformal time corresponding to redshift.

Q16.10 What is specific luminosity?

Q16.11 What is bolometric luminosity?

Q16.12 Show in Hubble's diagram the plot of velocity versus distance. Mention what is meant by peculiar motion in this context.

Further reading

[1] Bharadwaj S 2008 *Physical Cosmology Lecture Series* (Department of Physics, IIT Kharagpur)

[2] Hawking S 1988 *A Brief History of Time* (London: Bantam)

[3] Bonometto S, Gorini V and Moschella U 2001 *Modern Cosmology* (Bristol: IOP Publishing)

[4] Roos M 1994 *Introduction to Cosmology* (New York: Wiley)

[5] Tenreiro R D and Quiros M 1988 *An Introduction to Cosmology and Particle Physics* (Singapore: World Scientific)

IOP Publishing

Nuclear and Particle Physics with Cosmology, Volume 2
Particle physics and cosmology
Jyotirmoy Guha

Chapter 17

Cosmic microwave background radiation

In this chapter we discuss the process of determination of the age of the earth, solar system, globular clusters and galaxies. We describe blackbody radiation and show that cosmic microwave background radiation is a blackbody radiation. Hubble's discovery indicated that the Universe is expanding. We briefly discuss two expanding models of cosmology, namely the steady state model and the evolutionary model. Thermal history of the Universe is discussed. We establish how temperature, occupation number and chemical potential change with the expansion of the Universe.

17.1 Present age of the Universe and the cosmological model

If there is no gravity and the Universe expands freely then the age of the Universe is $\frac{1}{H_0}$.

When the effect of gravity is incorporated then the age of the Universe depends also on the cosmological model.

Equation (14.21) can be recast as

$$H(a) = \frac{\dot{a}}{a} = \frac{1}{a}\frac{da}{dt} \tag{17.1}$$

$$dt = \frac{1}{a}\frac{da}{H(a)} \tag{17.2}$$

Integrating and putting the limits $a(t=0) = 0$, $a(\text{present time } t_0) = 1$ we get

$$t_0 = \int_0^1 \frac{1}{a}\frac{da}{H(a)} \tag{17.3}$$

So the present age of the Universe depends on the cosmological model. The present age of the Universe should be larger than the oldest object that we find in the Universe. This is because all objects in the Universe were created after the Universe came into existence.

doi:10.1088/978-0-7503-5032-7ch17   17-1

17.2 Determination of age of objects, Earth, Solar System, galaxies

We can determine the age of objects that are aged about giga years, i.e. 10^9 years, in the Universe.

One method is called nuclear chrono-cosmology. Nuclear chronology can be used to determine the age of rocks. It is based on natural clocks. Again natural clocks are radioactive elements. There exist many heavy metals with lifetime 10 *Gyrs*. This suggests the age of Universe to be of the order of 10 *Gyrs*.

Consider the nuclear decays governed by the radioactive decay law viz.

$$N = N_0 e^{-t/\tau} \tag{17.4}$$

where τ is lifetime.

$$\text{Th}^{232} \rightarrow \text{Pb}^{202} \quad \text{Lifetime } 20.7 \; Gyr \tag{17.5}$$

$$\text{U}^{235} \rightarrow \text{Pb}^{207} \quad \text{Lifetime } 1.02 \; Gyr \tag{17.6}$$

$$\text{U}^{238} \rightarrow \text{Pb}^{206} \quad \text{Lifetime } 6.45 \; Gyr \tag{17.7}$$

With the help of these it is possible to determine the age of objects.

Consider a decay process where a parent P decays to a single daughter nucleus D as in equations (17.5), (17.6), and (17.7).

$$P \rightarrow D \tag{17.8}$$

In other words, D is produced exclusively in this way. It is separated and isolated from the nuclear reaction. Such a nuclear reaction occurs in a supernova. Once these materials have been ejected from the supernova and a nuclear reaction has stopped then we can write down the number of daughters, i.e. observed abundance of daughters as follows:

$$D = D_0 + P_0[1 - e^{-t/\tau}] \tag{17.9}$$

D_0 is the number of daughters initially present, i.e. starting abundance of daughters, P_0 is the starting abundance of parents, $1 - e^{-t/\tau}$ is the amount left after decay.

Suppose we wish to estimate the age of Earth or the age of the Solar System. All the materials in the Solar System were produced in a supernova. The lower limit of the age of the Earth can be estimated by finding the age of the oldest rock on the Earth. The lower limit of the age of the Solar System can be estimated by finding the age of the meteorites (that came from within the Solar System) if they are older than the rocks on the Earth.

Abundances of the quantities mentioned in equation (17.9) can be determined in rock samples using mass spectroscopy.

Writing P_0 in terms of present observed abundance of the parent, i.e. P as

$$P = P_0 e^{-t/\tau} \tag{17.10}$$

$$P_0 = P e^{t/\tau} \tag{17.11}$$

we can write

$$D = D_0 + P e^{t/\tau}[1 - e^{-t/\tau}] \tag{17.12}$$

$$D = D_0 + P[e^{t/\tau} - 1] \tag{17.13}$$

If D_0, τ are known then measuring D and P we can find the age t. Though τ can be known, the value of D_0 is difficult to know.

One looks for a relatively stable isotope of D (with larger lifetime) having abundance S. Then from equation (17.13) we get

$$\frac{D}{S} = \frac{D_0}{S} + \frac{P}{S}[e^{t/\tau} - 1] \tag{17.14}$$

Assume that the ratio $\frac{D_0}{S}$ is constant across the rock sample but the ratios $\frac{D}{S}, \frac{P}{S}$ will vary.

A plot of $\frac{P}{S}$ against $\frac{D}{S}$ gives a straight line with intercept $\frac{D_0}{S}$ and slope $e^{t/\tau} - 1$ as shown in figure 17.1. From the slope one can find the age t of the Earth or the Solar System. The intercept gives pre-solar abundances $\frac{D_0}{S}$.

Results from such analysis give the age of the Solar System to be 4.57 *Gyr*, the oldest rock on Earth to be 3. 7 *Gyr*. Further, the relative abundances, i.e. abundance ratios when the Solar System was formed were also obtained to be

$$\frac{U^{235}}{U^{238}} = 0.33 \tag{17.15}$$

$$\frac{Th^{232}}{U^{238}} = 2.3 \tag{17.16}$$

- The early stage of the Universe started off with constituents such as hydrogen and helium and a few light elements like lithium and deuterium. Heavy elements (say carbon and oxygen) were produced by nuclear fusion inside stars. In these reactions nuclei have to collide with each other which means density has to be very large. Also, nuclei have to overcome Coulomb repulsion in order to collide and this means energy has to be huge. In other words, for these reactions to occur one needs high density and high energy that occurs in the core of a star where nuclear fusion takes place.

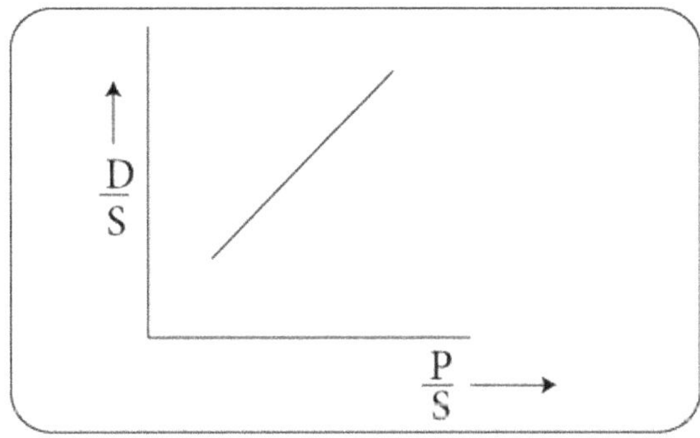

Figure 17.1. Plot of $\frac{D}{S}$ against $\frac{P}{S}$ of equation (17.14).

Through a series of reaction cycles, elements like carbon and oxygen are produced, but elements beyond iron cannot be formed. The binding energy goes up all the way to iron and iron has a much stable configuration.

When two nuclei combine to form a bigger nuclei, the binding energy per nucleon goes up and is more stable. The reaction proceeds in that direction for attainment of equilibrium.

Elements larger than iron have binding energy per nucleon lower than that of iron. So they are not produced by equilibrium reaction. They are unstable relative to iron. These elements are produced in supernovae and not produced by fusion in stars.

Nuclear fusion can produce elements up to iron and after that very massive stars explode at the end of their life and there is a very high neutron flux. The nuclei in the midst of high neutron flux capture neutrons. So these elements beyond iron are produced by neutron capture in supernovae.

Massive stars do not live long and give rise to supernovae. A supernova that exploded is in our Galaxy. Materials of the supernova were recycled in our Galaxy. This gives a lower limit of the age of our Galaxy.

This method gives the age of our Galaxy to be greater than 7. 5 *Gyr*.

17.2.1 Age of globular clusters

Globular clusters are collections of stars and there are many globular clusters in a galaxy. There are 10^5 stars in a globular cluster.

Age determination of a globular cluster depends on the assumption that all the stars in the globular cluster were formed at the same time.

Refer to figure 13.13. The majority of stars reside in the main sequence and this is the hydrogen burning phase of the stars.

A star is basically a spherical region of gas distribution and the interior is much hotter than the exterior. This is because of nuclear fusion going on. Hydrogen provides the fuel, otherwise the entire thing would collapse. There is hydrogen burning. Hydrogen–hydrogen fuse together to generate helium.

Properties of such stars that are burning hydrogen in equilibrium are well understood because the stars can be looked upon as a gas in equilibrium. These properties are quite general and do not depend very crucially on the source of nuclear reaction. These properties depend on two things. One is the virial theorem. The other is Thomson scattering.

17.2.2 Virial theorem

The virial theorem is a relation between the kinetic energy and potential energy of an object in equilibrium.

Consider an object in equilibrium. It is neither expanding nor collapsing. Then we can predict a relation between kinetic energy and potential energy, i.e. they are related, they cannot be independent. And the relation depends on the potential, say the gravitational potential.

17.2.3 Thomson scattering

Inside the star the radiation contributes to pressure also. Radiation is coming out. Light that is produced at the centre of the star does not move straight out. It gets scattered (Thomson scattered). It takes millions of years for a photon produced at the centre of a star to emerge. This also means that we do not see the interior of the Sun—what we see is the surface. This is opacity. Opacity is due to Thomson scattering.

Thomson scattering is an elastic scattering of a photon from a charged particle (electron) and frequency does not change in the process.

With these assumptions we arrive at the properties, namely luminosity, temperature and radius of star.

✓Luminosity is proportional to the cube of the mass. The more massive the star (which is a sphere of gas) the more luminous it is, i.e.

$$L \propto M^3 \tag{17.17}$$

If the mass is doubled, the luminosity increases eight times.
✓Temperature of the star

$$T \propto M^{1/2} \tag{17.18}$$

✓Radius of the star

$$R \propto M^{1/2} \tag{17.19}$$

The lifetime of a star. A star stays in the main sequence as long as it has hydrogen to burn. So it sends out radiant energy τL where τ is lifetime and L is luminosity due to conversion of its mass M to energy Mc^2. Hence we can write

$$\tau L \propto Mc^2 \ (M = \text{ mass of star}) \tag{17.20}$$

Fusion converts mass into energy.
When hydrogen is exhausted, i.e. mass is consumed, the life of the star is over. Again, mass can be written in terms of luminosity as $M \propto L^{1/3}$ and so

$$\tau L \propto L^{1/3} \tag{17.21}$$

And so

$$\tau \propto L^{-2/3} \tag{17.22}$$

More massive stars are more luminous, they lose energy faster and so finish off the fuel earlier. Less massive stars are less luminous, they burn slowly and live longer. Bigger and more luminous stars have shorter lives.

17.3 Cosmic Microwave Background Radiation (CMBR)

Production of elements requires a nuclear reaction. A nuclear reaction requires high energy, high temperature to bring the nuclei close to each other, overcoming their Coulomb repulsion. A hot dense medium is needed.

The abundance of light elements was found more or less to be nearly the same throughout the Universe. We can think of two possible sources by which elements can be synthesized.

One is the fusion process taking place inside stars.

We discuss the other possibility.

Suppose the Universe had a very hot early stage. As the Universe expands the density of the Universe falls.

The density of radiation ρ_r is seen to be (equation (15.33))

$$\rho_r \propto a^{-4} \tag{17.23}$$

The density of matter ρ_m is seen to be (equation (15.30))

$$\rho_m \propto a^{-3} \tag{17.24}$$

Energy of blackbody radiation is proportional to T^4. So if we characterize ρ_r as due to blackbody radiation then we can write from equation (17.19)

$$T^4 \propto a^{-4} \tag{17.25}$$

$$T \propto a^{-1} \tag{17.26}$$

When the Universe was much smaller (a small) the temperature would have been much higher (T large).

Consider the possibility that the Universe is filled with radiation. Then this radiation had a high temperature and the matter density is also very high and this would be an ideal situation for synthesizing elements. In other words, the early Universe is a possible place where synthesis of elements could occur. Gammow and Alpher considered this possibility of synthesis of elements in a hot early Universe.

For a nuclear reaction, the temperature required was $kT \sim 1 \, MeV$ or higher. Production of light elements (like helium, deuterium, lithium) through nuclear reaction was considered in this situation in an expanding Universe.

As the Universe expands density falls, temperature falls.

They showed that if light elements are produced by this mechanism then there should be a radiation component whose present value is needed to be $10 \, K$ or less.

So if the Universe had a hot phase where matter and radiation interacted closely and nuclear fusion occurred then, at present, since the Universe has suffered considerable expansion the radiation should still be around. However, its temperature would have fallen as per equations (17.23) and (17.26) and its temperature would be less than $10 \, K$.

Effort was on to find such a blackbody radiation of temperature $<10 \, K$. In 1965 radio-astronomers Penzias and Wilson detected a cosmic excess noise that had no directional dependence, that was uniform from all directions. No source could be associated with this radiation and it was cosmological in origin. Radiation from a galaxy would not have been isotropic which this radiation was. It was interpreted as the remnant of the early cosmological radiation that filled the Universe. This remnant radiation was observed at $7.5 \, cm$ (microwave region

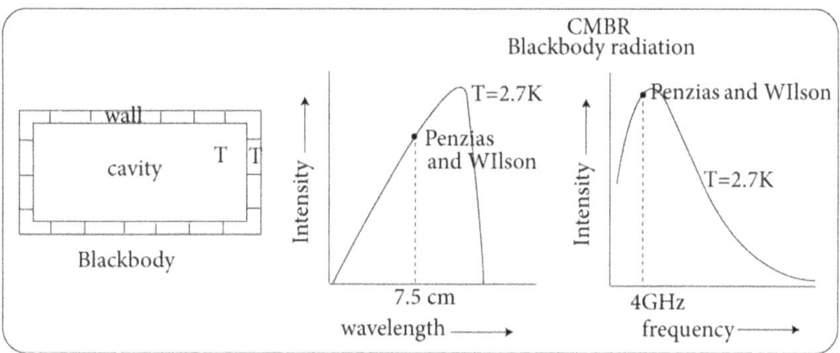

Figure 17.2. CMBR is blackbody radiation at temperature 2.7 K.

$\lambda = 1000 - 10\ mm$, $\nu = 300\ MHz\ to\ 30\ GHz$) and was found to be lying upon a blackbody radiation spectrum at temperature 2.7 K (figure 17.2).

This cosmic background radiation was found at a wavelength of 7.5 cm. It was measured for other wavelengths through several experiments. It was measured in different directions. If it is of cosmic origin, it has to be isotropic, i.e. comes uniformly from all directions.

17.3.1 Blackbody radiation

Blackbody is a cavity (figure 17.2) made of some material, maintained at some temperature T, that can emit and absorb radiation of all frequencies or wavelengths with equal efficiency.

The cavity walls emit and absorb, remit and re-absorb photons. Finally, the photons inside the cavity are in equilibrium with the walls. Photons in equilibrium with the walls of the cavity are also characterized to be at the same temperature T according to the zeroth law of thermodynamics according to which if two bodies are in equilibrium they are at the same temperature. So photons are characterized by temperature T. This is what we mean by blackbody radiation.

The spectrum of this blackbody radiation is uniquely defined by the temperature T. The spectral energy density I_ν is uniquely determined by the temperature T.

So it is essential for blackbody radiation that the radiation should be in thermal equilibrium with matter through successive collisions.

This radiation was in thermal equilibrium with matter in the Universe at some early time in the past when the Universe was much smaller and the radiation was much hotter. This radiation was in equilibrium with matter at some very high redshift, say $\sim$1000, when the scale factor was $a \sim 10^{-3}$, temperature $T \sim a^{-1}$ and $T \sim 3000\ K$ and the Universe was much more dense.

Subsequently it went out of equilibrium. The radiation now fills the entire Universe. We are embedded in this radiation.

The radiation is found to be isotropic to a high degree of accuracy. Suppose we take a radiometer and point it in different directions, as shown in figure 17.3. Our aim is to measure temperature as a function of direction, i.e. $T = T(\hat{n})$. Temperature

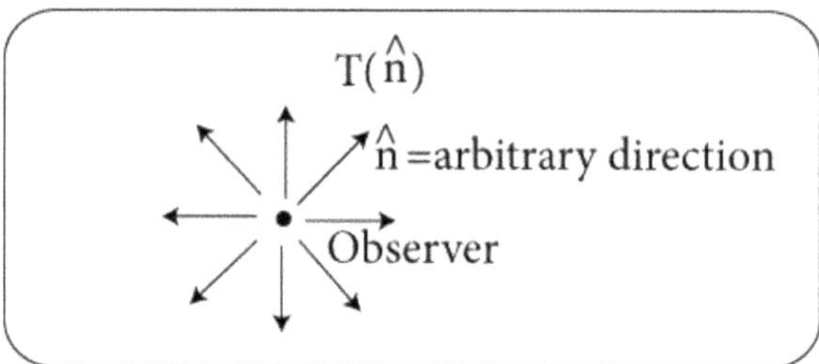

Figure 17.3. Temperature measured in different directions is the same, i.e. radiation is isotropic.

measured in different directions is found to be more or less constant indicating isotropy.

However, there are very small (0.1.%) deviations from isotropic behaviour. Temperature changes by 0.1% when compared in various directions in the sky leading to 0.1% anisotropy.

Also, a fluctuation of 1 part in 10^5 was observed, i.e.

$$\frac{\Delta T}{T} \sim 10^{-5} \tag{17.27}$$

$$\Delta T = 10^{-5}T = 10^{-5} \times 2.7K = 27 \times 10^{-6} = 27\,\mu K \tag{17.28}$$

So there are tiny fluctuations in temperature across the sky. These fluctuations, i.e. existence of small anisotropy are related to the formation of large-scale structures in the Universe.

Hubble's discovery indicated that the Universe is expanding. Before the discovery of CMBR there were two competing expanding models of cosmology. They were: (1) evolutionary model, and (2) steady state model.

17.3.2 Evolutionary model

As the Universe expands, the density of matter in the Universe falls as $\rho_m \propto a^{-3}$ (equation (17.24)) and radiation density falls as $\rho_r \propto a^{-4}$ (equation (17.23)). This model predicted the Big Bang and was called the Big Bang model.

The CMBR gave definite proof that the steady state model was not correct and the evolutionary model was correct.

17.3.3 Steady state model

Expansion was such that everything in the Universe remained the same, i.e. density remained constant despite expansion. Matter is created to compensate for the expansion so as to maintain the Universe in the same state. And this is when the Universe is cosmological constant dominated. Here exponential expansion is possible and there is no Big Bang since occurrence of the Big Bang would violate the steady state of the Universe. In other words, there is no origin of time in steady state.

The CMBR gave definite proof that the steady state model was not correct.

The Universe is transparent to the CMBR (figure 17.2) which is a blackbody spectrum. If the steady state model is correct then the Universe would be always as it is, it expands, density falls, matter is created and the Universe is always transparent. In such a Universe, matter will not be in thermal equilibrium with radiation because the Universe being transparent there will be no interaction between matter and radiation. This is not possible. So CMBR gave definite proof that the steady state model was not correct. In other words, the steady state model has no natural explanation for the CMBR.

Another idea was put forward that says that star light from stars produces enough energy that is absorbed and re-emitted by dust to produce $2.7\ K$ CMBR. However, such an explanation was not straightforward and untenable.

On the other hand, a natural explanation follows from the evolutionary model. The Universe expands and as we move into the past of the Universe, radiation was hotter and matter of the Universe was denser. So sufficiently back in the past matter and radiation were in thermal equilibrium. So CMBR gets explained naturally.

As we go back in the past, the temperature of the radiation exceeds $1\ MeV$. Before, that temperature of radiation was sufficient to dissociate all atoms. Further back, the temperature of radiation was sufficient to dissociate all nuclei. So there were no bound nuclei then. Even further back the temperature of radiation was sufficient to dissociate all quarks inside protons and neutrons.

The Universe was hotter and hotter as we move into the past. And if we move forward in time we begin at a very high temperature and as the Universe cools different reactions take place, formation of matter of various types begins. We thus can think of a temperature scale of the Universe as we go into the past. Redshift and temperature scale are related.

17.4 Thermal history of the Universe

The Universe is filled with radiation that has a thermal spectrum, i.e. a blackbody spectrum with temperature $2.\ 7\ K$. We wish to investigate how the photon spectrum distribution evolves with the expansion of the Universe.

Let us consider a small part of the Universe that we take to be a cube of size l, as shown in figure 17.4. We assume that that the size l of this region is much smaller than the length scale of the curvature of the spatially curved Universe, i.e. we take

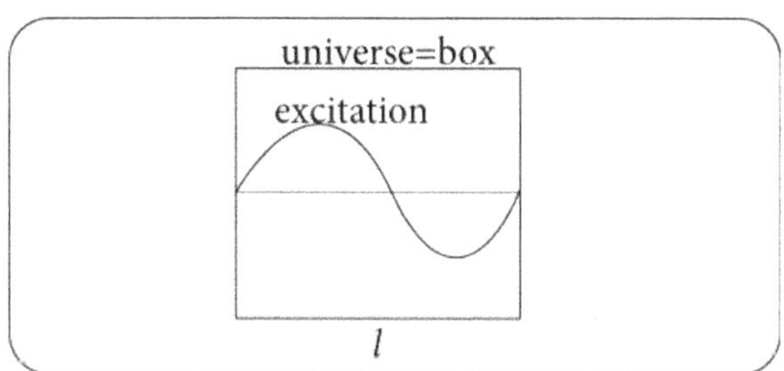

Figure 17.4. Part of the Universe taken to be a cubic box of length l with periodic boundary condition.

$$l \ll aL$$

where L is the curvature length scale and a is the scale factor and aL is the curvature length at any time.

The electromagnetic field inside this box can be decomposed into superposition of modes which are plane waves $e^{i(\vec{k}.\vec{r}-wt)}$ having different $\vec{k}$ values.

Assume periodic boundary conditions, i.e. assume that the Universe repeats. So modes cannot be arbitrary, the waves should fit in the box as shown and $\vec{k}$ should be such that

$$\vec{k} = \frac{2\pi}{l}\,(n_x\hat{i} + n_y\hat{j} + n_z\hat{k}) \tag{17.29}$$

where n_x, n_y, $n_z = 0,1,2,3,...$ are integers.

The dispersion relation of each mode is

$$w = c\,|\,k\,| \tag{17.30}$$

where w is frequency of oscillation and $\frac{w}{2\pi} = \nu$ is the linear frequency.

The excitations of these modes are quantized and interpreted as particles called photons. Energy of these particles is

$$\varepsilon = h\nu = \hbar w \tag{17.31}$$

And the momentum of these particles is

$$\vec{p} = \hbar\,\vec{k} \tag{17.32}$$

This is how we can describe electromagnetic field in the box which is a small part of the Universe.

The box-size has been considered to be smaller than the curvature length but it is considered to be larger than the wavelength of the radiation that we are considering, i.e.

$$l \ll aL,\ l \gg \lambda \tag{17.33}$$

The Universe is expanding implying that the whole box is also expanding and this expansion is determined by Hubble's parameter $H(t)$.

Assume that expansion is slower than the frequency of oscillation, i.e. the rate of expansion of the Universe $H(t)$ is much smaller than the frequency w of the oscillations, i.e.

$$H(t) \ll \frac{w}{2\pi} \tag{17.34}$$

Consider that $\lambda = 1\,mm = 10^{-3}\,m$. So the corresponding frequency is

$$\nu = \frac{c}{\lambda} = \frac{3 \times s^{-1}}{10^{-3}\,m} = 3 \times 10^{11}\,Hz \tag{17.35}$$

It is reasonable to assume that by the time many oscillations are completed the Universe has expanded by a small amount, i.e. expansion is slower than oscillation. So a wave has an oscillation of electromagnetic field and it is much slower.

Under this slow expansion the number of excitations (i.e. number of photons) will not change but the wavelength and frequency of photons corresponding to a mode will change. (This would not have been true if expansion rate were comparable to the oscillation frequency.)

We assume that particles do not interact, i.e. do not collide. If they do then a particle could move into a different mode. There is only free expansion of the Universe.

Again, as the Universe slowly expands wavelength goes up proportionately since wavelength is proportional to the scale factor, i.e.

$$\lambda \propto a(t) \tag{17.36}$$

And since $k = \frac{2\pi}{\lambda}$ we expect

$$\vec{k} \propto \frac{1}{a} \tag{17.37}$$

With the expansion of the Universe the momentum of the free particle $\vec{p} = \hbar \vec{k}$ is proportional inversely to the scale factor, i.e.

$$\vec{p} \propto \frac{1}{a} \tag{17.38}$$

It is clear that photons or electromagnetic waves have been expanded into modes. The analysis that has been carried out could also be done for massive particle of mass m with appropriate dispersion relation.

In general, for a massive particle, energy is given by

$$\varepsilon = \sqrt{p^2 c^2 + m^2 c^4} \tag{17.39}$$

Equations (17.38) and (17.39) tell us how energy ε changes with expansion of the Universe and since $\varepsilon = h\nu = \hbar w$ dependence of frequency on expansion of the Universe can also be found out.

We wish to know the number of particles in each mode to get the distribution.

For a photon in thermal equilibrium at temperature T the thermal distribution is given as follows. The occupation number in any mode is obtained from statistical mechanics for particles in thermal equilibrium

$$N(\vec{p}) = \frac{1}{\exp\left(\frac{\varepsilon - \mu}{kT}\right) \pm 1} \tag{17.40}$$

where

$$k \equiv k_B = \text{ Boltzmann's constant} \tag{17.41}$$

and ε is energy corresponding to a particle with momentum $\vec{p}$ given by equation (17.39) for a free particle of mass m, μ is chemical potential which is a constant since number of particles is conserved. And in equation (17.40) the positive sign holds for fermions and the negative sign holds for bosons. Equation (17.41) with positive sign is called Fermi–Dirac distribution and with negative sign it is called Bose–Einstein distribution.

We note that the occupation number $N(\vec{p})$ represents expectation value and can be fractional also.

Bosons are particles with integral spin and for them the occupation number can assume any value 0, 1, 2, 3, ...(any integral value is allowed since many particles can occupy the same state). Fermions are particles with half integral spin and for them occupation number can assume values either 0 or 1 (since no two particles can occupy the same state).

Photons are bosons and follow Bose–Einstein statistics. For photons chemical potential $\mu = 0$. They are not conserved. They can be created, destroyed, re-created, re-destroyed and so on. Photons inside the cavity of figure 17.2 can hit the wall and get absorbed and they can be emitted by the wall.

For particles that are conserved we cannot create or destroy them at will and for them $\mu \neq 0$.

Equation (17.40) is a thermal distribution characterized by two parameters, chemical potential μ and temperature T.

We now discuss how this particle distribution (equation (17.40)) changes with the expansion of the Universe. The Universe expands slowly compared to the frequency associated with particles.

Particle numbers are not changed (they are neither created nor destroyed) due to expansion of the Universe. Momentum of every particle of every mode gets changed and with expansion it decreases inversely with the scale factor (equation (17.38)).

So with expansion of the Universe $N(\vec{p})$ is unchanged but $\vec{p} \propto \frac{1}{a}$.

Let at an initial time t_i scale factor be a_i and occupation number $N_i(\vec{p})$.

Let at some other time called final time t_f scale factor be a_f and occupation number $N_f(\vec{p})$.

Now from equation (17.38) we have

$$\vec{p}\,a = \text{constant} = \vec{p}_i\,a_i = \vec{p}_f\,a_f \tag{17.42}$$

Hence

$$\vec{p}_i = \vec{p}_f\,\frac{a_f}{a_i} \tag{17.43}$$

This is initial momentum in terms of final momentum. So

$$N_f(\vec{p}_f) = N_i(\vec{p}_i) \equiv N_i\left(\vec{p}_f\,\frac{a_f}{a_i}\right) \tag{17.44}$$

Dropping the suffix in the argument we write

$$N_f(\vec{p}) = N_i\left(\vec{p}\,\frac{a_f}{a_i}\right). \tag{17.45}$$

Assume that at initial time t_i there is a thermal distribution and the corresponding temperature is T_i and chemical potential is μ_i. We shall calculate the particle distribution at t_f.

Consider a massless particle $m = 0$ for which

$$\varepsilon_i = p_i c = p_f\,\frac{a_f}{a_i}c \tag{17.46}$$

$$N_f(\vec{p}) = \frac{1}{\exp\left(\frac{\varepsilon_i - \mu_i}{kT_i}\right) \pm 1} = \frac{1}{\exp\left(\frac{p_f\frac{a_f}{a_i}c - \mu_i}{kT_i}\right) \pm 1} \tag{17.47}$$

This is the final particle distribution if the initial particle distribution is a thermal distribution.

$$N_f(\vec{p}) = \frac{1}{\exp\left(\frac{p_f\,c - \mu_i\frac{a_i}{a_f}}{kT_i\frac{a_i}{a_f}}\right) \pm 1} \tag{17.48}$$

$$N_f(\vec{p}) = \frac{1}{\exp\left(\frac{p_f\,c - \mu_f}{kT_f}\right) \pm 1} \tag{17.49}$$

where we define

$$\mu_f = \frac{a_i}{a_f}\mu_i \tag{17.50}$$

$$T_f = \frac{a_i}{a_f}T_i \tag{17.51}$$

Clearly the final particle distribution $N_f(\vec{p})$ can be written in terms of a thermal distribution with a different chemical potential μ_f and a different temperature T_f.

If initially a particle distribution is thermal with chemical potential μ_i and temperature T_i then with the expansion of the Universe for massless particles, the final distribution is also thermal having chemical potential $\mu_f = \frac{a_i}{a_f}\mu_i$ and temperature $T_f = \frac{a_i}{a_f}T_i$.

Clearly chemical potential scales inversely with the scale factor, i.e.

$$\mu \propto \frac{1}{a} \tag{17.52}$$

Temperature gets scaled inversely with the scale factor, i.e.

$$T \propto \frac{1}{a} \tag{17.53}$$

So if we start off with a thermal distribution then the distribution remains thermal, only the temperature gets scaled inversely with scale factor and the chemical potential also gets scaled inversely with the scale factor. If at some time a thermal distribution was established the thermal distribution will be seen at a later time also.

This is true for a massless particle.

17.4.1 Massive particles

For massive particles $m \neq 0$.

The energy is given by equation (17.39) as $\varepsilon = \sqrt{p^2 c^2 + m^2 c^4}$. So there is energy even when momentum is zero, i.e.

$$\varepsilon = mc^2 \neq 0 \text{ even if } p = 0 \tag{17.54}$$

Consider the quantity

$$\frac{\varepsilon}{kT} = \frac{\sqrt{p^2 c^2 + m^2 c^4}}{kT} = \sqrt{\frac{p^2 c^2}{k^2 T^2} + \frac{m^2 c^4}{k^2 T^2}} \tag{17.55}$$

• Case $kT \gg mc^2$

As $\frac{m^2 c^4}{k^2 T^2} \ll 1$ we neglect this part in the RHS of equation (17.55). Then

$$\frac{\varepsilon}{kT} \approx \sqrt{\frac{p^2 c^2}{k^2 T^2}} = \frac{pc}{kT} \tag{17.56}$$

Clearly in this situation mass of the particle is of no consequence. And so this case is as good as the $m = 0$ case. This case is called the relativistic case.

Then occupation number can be written as (from equation (17.40))

$$N(\vec{p}) = \frac{1}{\exp\left(\frac{pc - \mu}{kT}\right) \pm 1} \tag{17.57}$$

If we have a situation that is relativistic, temperature that is comparable to the energy scale corresponding to the temperature is much higher than the rest energy of particles then the fact that particles have mass is of no consequence. Hence in this case also temperature scales as the inverse of the scale factor.

17.4.1.1 Example

Consider the case of a neutrino whose mass is say $m \ll \frac{kT}{c^2}$. Assume that the neutrinos were in thermal equilibrium with the rest of the constituents of the Universe sometime back in the past. Neutrinos are very weakly interacting particles.

Suppose way back in the past at some very high redshift the neutrinos were in thermal equilibrium with the rest of the constituents of the Universe. At such high redshift the temperature of the whole constituents of the Universe would have been extremely high. Suppose at that epoch when neutrinos were in thermal equilibrium the temperature T was much higher so that

$$kT \gg mc^2 \qquad (17.58)$$

Then the neutrinos go out of thermal equilibrium implying that they do not interact with anything. So at some high temperature in the past neutrinos were in equilibrium and then they stop interacting and decouple at this temperature $T = T_i$ (decoupling temperature).

At present the Universe is at a much smaller temperature of $2.7\,K$ and so kT is comparable or could be even less than mc^2, i.e.

$$kT \leqslant mc^2$$

With the expansion of the Universe, distribution function remains unchanged, momentum changes. Distribution function at present will be (from equation (17.40))

$$N_f(\vec{p}) = \frac{1}{\exp\left(\dfrac{pc - \mu_f}{kT_f}\right) \pm 1} \qquad (17.59)$$

where temperature T_f is related to the decoupling temperature T_i

$$T_f = \frac{a_i}{a_f} T_i \qquad (17.60)$$

$$\mu_f = \frac{a_i}{a_f} \mu_i \qquad (17.61)$$

Even though particles are massive now, their distribution function is going to be relativistic because the distribution function does not change once the particles decouple, i.e. stop interacting with other particles. So distribution function remains fixed.

• Case $kT \ll mc^2$

Particles, for instance hydrogen atoms filling some part of the Universe, are in thermal equilibrium with $kT \ll mc^2$

Temperature is much smaller, mass is larger.

From equation (17.39)

$$\varepsilon = \sqrt{p^2 c^2 + m^2 c^4} = mc^2\left(1 + \frac{p^2}{m^2 c^2}\right)^{1/2} \qquad (17.62)$$

$$\approx mc^2\left(1 + \frac{p^2}{2m^2 c^2}\right) \text{ (by Taylor expansion and using } kT \ll mc^2)$$

$$\varepsilon = mc^2 + \frac{p^2}{2m} \tag{17.63}$$

So the occupation number can be written as

$$N(\vec{p}) = \frac{1}{\exp\left(\dfrac{mc^2 + \frac{p^2}{2m} - \mu}{kT}\right) \pm 1} \tag{17.64}$$

As the exponential term is much larger than 1 and so we can ignore ± 1 and write

$$N(\vec{p}) = \frac{1}{\exp\left(\dfrac{mc^2 + \frac{p^2}{2m} - \mu}{kT}\right)} \tag{17. 65}$$

$$N(\vec{p}) = \exp\left(\frac{-mc^2 + \mu}{kT}\right) \exp\left(-\frac{p^2}{2mkT}\right) \tag{17.66}$$

This is the Maxwell–Boltzmann distribution.

The factor $\exp(\frac{-mc^2 + \mu}{kT})$ does not depend upon p. So we can treat this factor as a constant, say

$$A = \exp\left(\frac{-mc^2 + \mu}{kT}\right) \tag{17.67}$$

to get

$$N(\vec{p}) = A \exp\left(-\frac{p^2}{2mkT}\right) \tag{17.68}$$

- In the limit $kT \gg mc^2$, $N(\vec{p}) = \frac{1}{\exp(\frac{pc - \mu}{kT}) \pm 1}$ (equation (17.59)) we see that rest mass energy does not occur.

- In the limit $kT \ll mc^2$, $N(\vec{p}) = A \exp(-\frac{p^2}{2mkT})$ (equation (17.59)) is the Maxwell–Boltzmann distribution.

At the time t_i system is in thermal equilibrium T_i and chemical potential μ_i and scale factor is a_i. We wish to find out what happens to the system at a later time t_f, i.e. find out the occupation number $N_f(\vec{p})$ at t_f when the scale factor is a_f. Now by equation (17.45) it is

$$N_f(\vec{p}) = N_i\left(\vec{p}\,\frac{a_f}{a_i}\right)$$

And for the relativistic situation we have from equation (17.47)

$$N_f(\vec{p}) = \cfrac{1}{\exp\left(\cfrac{p_f \frac{a_f}{a_i} c - \mu_i}{kT_i}\right) \pm 1} \tag{17.69}$$

with $T_f = \frac{a_i}{a_f} T_i$ (equation (17.60)), $\mu_f = \frac{a_i}{a_f}\mu_i$ (equation (17.61))

As the Universe expands the occupation number of particle distribution in different momentum states remains unchanged except that the temperature and the chemical potential both scale inversely as the scale factor. In other words as the Universe expands potential falls as $\frac{1}{\text{scale factor}}$ and the temperature of the distribution also falls as $\frac{1}{\text{scale factor}}$. This is for relativistic particles.

Let us now discuss for the non-relativistic situation.

$$N_f(p) = A \exp\left[\cfrac{-\left(\frac{a_f}{a_i}p\right)^2}{2mkT_i}\right] \tag{17.70}$$

$$N_f(p) = A \exp\left[\cfrac{-p^2}{2mk\left(\frac{a_i}{a_f}\right)^2 T_i}\right] \tag{17.71}$$

$$N_f(p) = A \exp\left[\cfrac{-p^2}{2mkT_f}\right] \tag{17.72}$$

where

$$T_f = \left(\frac{a_i}{a_f}\right)^2 T_i \tag{17.73}$$

And as evident from equation (17.67) the chemical potential has become a part of the constant A. So the value of the chemical potential does not change with the expansion of the Universe. The number of particles does not change.

The relativistic particles and the non-relativistic particles are in thermal equilibrium at some temperature and then with the expansion of the Universe the temperature of these quantities behaves in different ways.

For the relativistic particles:

Temperature scales as $T \propto \frac{1}{a}$

For non-relativistic particles:

Temperature scales as $T \propto \frac{1}{a^2}$

Clearly temperature falls much faster with expansion of the Universe as compared to relativistic particles.

✓For example, neutrinos having mass $m \ll \frac{kT}{c^2}$ behave like relativistic particles They were at thermal equilibrium at some time and as the Universe expands and neutrinos move freely the temperature of neutrino distribution will scale as $T \propto \frac{1}{a}$.

✓An example of non-relativistic particles is hydrogen atoms and temperature $\sim 3000\,K$. These can be treated as non-relativistic particles. They were at thermal equilibrium at some time and as the Universe expands and they evolve freely the temperature of hydrogen atoms will fall as $T \propto \frac{1}{a^2}$.

17.4.1.2 About neutrinos

Suppose cosmological neutrino species have mass at present given by

$$m > \frac{kT_r}{c^2} \tag{17.74}$$

where $T_r = 3K$ is the present background temperature

Also assume that the neutrinos were in thermal equilibrium with the CMBR and the rest of the Universe at some temperature T_{eq}. The mass was

$$m \ll \frac{kT_{eq}}{c^2} \tag{17.75}$$

And then they go out of equilibrium, i.e. they decouple and move freely after that and so have a distribution given by the occupation number of equation (17.57)

$$N(\vec{p}) = \frac{1}{\exp\left(\frac{pc - \mu}{kT}\right) \pm 1} \quad (+ \text{ for fermion}) \tag{17.76}$$

As the Universe expands the occupation number of these neutrinos of a particular state is unchanged, only the momentum corresponding to that state falls as $\frac{1}{\text{scale factor}}$. Even at present even though the mass of the neutrino is more $m > \frac{kT_r}{c^2}$, i.e. greater than the temperature of the present Universe, we would still use equation (17.76) to calculate the occupation number of these neutrinos using the momentum of neutrino when they were in thermal equilibrium at a previous epoch, knowing the present momentum.

The above discussion shows how the temperature, occupation number and chemical potential change with the expansion of the Universe.

17.5 Specific intensity

Consider a cube of side l and volume l^3 of the Universe. The modes that this volume can support is given by

$$\vec{k} = \frac{2\pi}{l}(n_x, n_y, n_z) \tag{17.77}$$

where n_x, n_y, n_z are integers and each set of integers labels one $\vec{k}$ vector. Corresponding to each wave vector $\vec{k}$ there is a momentum $\vec{p}$ which is

$$\vec{p} = \hbar \, \vec{k} \tag{17.78}$$

It is the occupation number for every mode. Number of particles within a momentum range or wave vector range is given by

$$d^3k \equiv k^2 dk d\Omega \tag{17.79}$$

And the number of modes corresponding to d^3k is

$$\left(\frac{l}{2\pi}\right)^3 k^2 dk d\Omega \tag{17.80}$$

Number of particles per unit volume, i.e. number density of particles is

$$dn = 2\frac{N(\vec{p})}{l^3}\left(\frac{l}{2\pi}\right)^3 k^2 dk d\Omega \tag{17.81}$$

where the factor 2 represents degeneracy for photon or neutrino.

Assume a particle is massless for which the dispersion relation is

$$\frac{w}{k} = c \tag{17.82}$$

$$k = \frac{w}{c}, \; dk = \frac{dw}{c} \tag{17.83}$$

$$\varepsilon = pc = \hbar w \tag{17.84}$$

Then we have using equation (17.76)

$$dn = 2\frac{1}{\exp\left(\frac{\hbar w - \mu}{kT}\right) \pm 1}\left(\frac{1}{2\pi}\right)^3\left(\frac{w}{c}\right)^2\frac{dw}{c}\,d\Omega \tag{17.85}$$

$$dn = \frac{1}{4\pi^3}\frac{1}{\exp\left(\frac{\hbar w - \mu}{kT}\right) \pm 1}\frac{w^2 dw}{c^3}\,d\Omega \tag{17.86}$$

This is the number of particles travelling with frequency in the range w and $w + dw$ within the solid angle $d\Omega$.

The energy dE carried by photons in an elemental volume $cdtdA$ (figure 17.5) that travels along solid angle Ω, in the frequency interval $d\nu$ in the time interval dt and falling on area dA is given by

$$dE = I_\nu dt d\nu dA d\Omega \tag{17.87}$$

$$I_\nu = \frac{dE}{dt d\nu dA d\Omega} \tag{17.88}$$

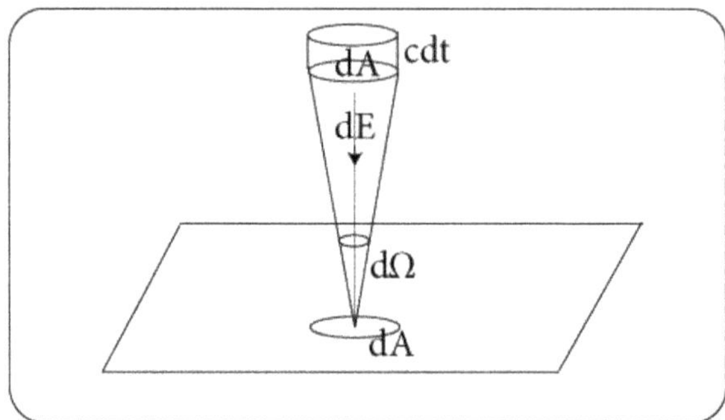

Figure 17.5. Energy that comes through solid angle Ω, in the frequency interval $d\nu$ in the time interval dt is dE.

where I_ν is specific intensity. So specific intensity is defined as the energy coming in per unit area, per unit solid angle, per unit frequency, per unit time. This is defined for photons (Bose particles).

Assume that we have radiation in thermal equilibrium which is called blackbody radiation at a temperature T. We shall calculate specific intensity for such radiation.

Number of photons per unit volume is from equation (17.86) with $w = 2\pi\nu$ and $\mu = 0$

$$dn = 2\frac{1}{\exp\left(\frac{\hbar w - \mu}{kT}\right) \pm 1}\left(\frac{1}{2\pi}\right)^3\left(\frac{2\pi\nu}{c}\right)^2\frac{d2\pi\nu}{c}\,d\Omega \tag{17.89}$$

$$dn = \frac{2}{c^3}\frac{1}{\exp\left(\frac{\hbar\nu}{kT}\right) - 1}\,\nu^2 d\nu\,d\Omega \quad (- \text{ for photon}) \tag{17.90}$$

Energy per unit volume is

$$h\nu dn = h\nu\frac{2}{c^3}\frac{1}{\exp\left(\frac{\hbar\nu}{kT}\right) - 1}\,\nu^2 d\nu\,d\Omega$$

$$h\nu dn = \frac{2h\nu^3}{c^3}\frac{1}{\exp\left(\frac{\hbar\nu}{kT}\right) - 1}\,d\nu\,d\Omega \tag{17.91}$$

The energy dE carried by photons in an elemental volume $cdtdA$ (figure 17.5) that travels along solid angle Ω, in the frequency interval $d\nu$ in the time interval dt and falling on area dA is given by

$$dE = h\nu dn\,(dAcdt) \tag{17.92}$$

$$= \frac{2h\nu^2}{c^3}\frac{1}{\exp\left(\frac{\hbar\nu}{kT}\right) - 1}\,d\nu d\Omega\,dAcdt \quad (\text{using equation (17.91)}) \tag{17.93}$$

$$dE = I_\nu dt d\nu dA d\Omega \quad (\text{equation (17.87)}) \tag{17.94}$$

Comparison gives

$$I_\nu = \frac{2h\nu^3}{c^2} \frac{1}{\exp\left(\frac{\hbar\nu}{kT}\right) - 1} \tag{17.95}$$

This is the specific intensity of blackbody radiation.

In the limit $h\nu \ll kT$ since $\frac{h\nu}{kT} \ll 1$ we can make a Taylor series expansion

$$\exp\left(\frac{\hbar\nu}{kT}\right) = 1 + \frac{\hbar\nu}{kT} + \ldots \approx 1 + \frac{h\nu}{kT} \tag{17.96}$$

$$\exp\left(\frac{\hbar\nu}{kT}\right) - 1 \approx 1 + \frac{h\nu}{kT} - 1 = \frac{h\nu}{kT} \tag{17.97}$$

From equations (17.95) and (17.97) we write

$$I_\nu = \frac{2h\nu^3}{c^2} \frac{1}{\frac{h\nu}{kT}} = \frac{2\nu^2 kT}{c^2} = \frac{2kT}{\lambda^2} \tag{17.98}$$

This is Rayleigh–Jeans law.

17.6 Total energy density of radiation (photon)

Equation (17.86) with $\mu = 0$ is

$$dn = \frac{1}{4\pi^3} \frac{1}{\exp\left(\frac{\hbar w}{kT}\right) - 1} \frac{w^2 dw}{c^3} d\Omega \quad (\text{–for photon}) \tag{17.99}$$

Multiplying equation (17.99) by $\hbar w$ gives

$$\hbar w\, dn = \hbar w \frac{1}{4\pi^3} \frac{1}{\exp\left(\frac{\hbar w}{kT}\right) - 1} \frac{w^2 dw}{c^3} d\Omega$$

$$\hbar w\, dn = \hbar \frac{1}{4\pi^3} \frac{1}{\exp\left(\frac{\hbar w}{kT}\right) - 1} \frac{w^3 dw}{c^3} d\Omega \tag{17.100}$$

We are interested in particles travelling in all directions. So we integrate over Ω to get the energy density

$$u_\gamma = \int_0^\infty \int_\Omega \hbar w\, dn \tag{17.101}$$

$$u_\gamma = \int_0^\infty \hbar \frac{1}{4\pi^3} \frac{1}{\exp\left(\frac{\hbar w}{kT}\right) - 1} \frac{w^3 dw}{c^3} \left(\int_\Omega d\Omega\right) \tag{17.102}$$

Using $\int_{\Omega} d\Omega = 4\pi$

$$u_\gamma = \int_0^\infty \hbar \frac{1}{4\pi^3} \frac{1}{\exp\left(\frac{\hbar w}{kT}\right) - 1} \frac{w^3 dw}{c^3} \cdot 4\pi = \frac{\hbar}{\pi^2 c^3} \int_0^\infty \frac{w^3 dw}{\exp\left(\frac{\hbar w}{kT}\right) - 1} \tag{17.103}$$

Putting

$$y = \frac{\hbar w}{kT}, \quad dy = \frac{\hbar}{kT} dw \tag{17.104}$$

$$w = \frac{ykT}{\hbar}, \quad dw = \frac{kT dy}{\hbar} \tag{17.105}$$

Equation (17.103) gives

$$u_\gamma = \frac{\hbar}{\pi^2 c^3} \int_0^\infty \frac{\left(\frac{ykT}{\hbar}\right)^3 \left(\frac{kT dy}{\hbar}\right)}{\exp y \pm 1} = \frac{\hbar}{\pi^2 c^3} \int_0^\infty \frac{y^3 \left(\frac{kT}{\hbar}\right)^4 dy}{\exp y - 1}$$

$$u_\gamma = \frac{\hbar}{\pi^2 c^3} \left(\frac{kT}{\hbar}\right)^4 \int_0^\infty \frac{y^3 dy}{\exp y - 1} \tag{17.106}$$

As the integral on the RHS of equation (17.106) is a constant, it is clear that u_γ depends on temperature T. The value of the integral upon evaluation is

$$\int_0^\infty \frac{y^3 dy}{\exp y - 1} = \frac{\pi^4}{15} \tag{17.107}$$

Hence

$$u_\gamma = \frac{\hbar}{\pi^2 c^3} \left(\frac{kT}{\hbar}\right)^4 \frac{\pi^4}{15} = \left(\frac{\pi^2 k^4}{15 c^3 \hbar^3}\right) T^4 \tag{17.108}$$

So for a photon at temperature T the internal energy is given by

$$u_\gamma = a_B T^4 \tag{17.109}$$

where a_B is called the Stefan–Boltzmann constant that has value

$$a_B = \frac{\pi^2 K^4}{15 c^3 \hbar^3} = \frac{\pi^2 (1.38 \times 10^{-23} JK^{-1})^4}{15(3 \times 10^8 ms^{-1})^3 (\frac{1}{2\pi} 6.626 \times 10^{-34} Js)^3} \tag{17.110}$$

$$a_B = 7.536 \times 10^{-16} \, J \, m^{-3} \, K^{-4} \tag{17.111}$$

This gives the energy density u (in $J \, m^{-3}$) in blackbody radiation at thermal equilibrium with matter at temperature T.

We can also alternately express equation (17.109) as follows.

$$u_\gamma = a_B T^4 = \frac{4}{c}\sigma T^4 \tag{17.112}$$

where

$$a_B = \frac{4}{c}\sigma \tag{17.113}$$

$$\sigma = \frac{c}{4}a_B \tag{17.114}$$

$$\sigma = \frac{3 \times 10^8\, m\, s^{-1}}{4} 7.536 \times 10^{-16}\, J\, m^{-3}\, K^{-4} = 5.6 \times 10^{-8}\, W\, m^{-2}\, K^{-4} \tag{17.115}$$

This σ is called the Stefan–Boltzmann constant.

We have learned that the whole Universe at present is filled with radiation at 2. 7 K. Given this background radiation which is thermal blackbody radiation at 2. 7 K we can calculate mass density of radiation as follows

$$u_\gamma = a_B T^4 \quad \text{(from equation (17.112))}$$

$$\rho_r = \frac{u}{c^2} = \frac{a_B T^4}{c^2} = \frac{a_B}{c^2}T^4 = a_B' T^4 \tag{17.116}$$

Value of a_B' is

$$a_B' = \frac{a_B}{c^2} = \frac{1}{c^2}7.536 \times 10^{-16}\, J\, m^{-3}\, K^{-4}$$

$$a_B' = \frac{7.536 \times 10^{-16}}{(3 \times 10^8 ms^{-1})^2}\, (kgms^{-2} \cdot m)\, m^{-3}\, K^{-4}$$

$$a_B' = 8.4 \times 10^{-33}\, kg\, m^{-3}\, K^{-4} \tag{17.117}$$

Hence at $T = 2. 7K$ CMBR we have the value of ρ_r to be

$$\rho_r = a_B' T^4$$

$$\sim (8.4 \times 10^{-33}\, kg\, m^{-3}\, K^{-4})(2.7\, K)^4 \approx 4.5 \times 10^{-31}\, kg\, m^{-3} \tag{17.118}$$

We compare this with the critical density of Universe viz.

$$\rho_{CO} = 1.874 \times 10^{-26}\, h^2\, kg\, m^{-3} \quad \text{(equation 14.61)} \tag{17.119}$$

Their ratio gives the density parameter Ω_{ro} corresponding to CMBR which is

$$\Omega_{ro} = \frac{\rho_r}{\rho_{CO}} = \frac{4.5 \times 10^{-31}\, kg\, m^{-3}}{1.874 \times 10^{-26}\, h^2\, kg\, m^{-3}} = 2.4 \times 10^{-5}\, h^{-2} \tag{17.120}$$

With $h = 0.7$ (equation (14.5))

$$\Omega_{ro} = 2.4 \times 10^{-5}(0.7)^{-2} = 4.9 \times 10^{-5} \qquad (17.121)$$

At present this makes a very minute contribution to the total density of the Universe. Clearly the radiation which fills our Universe, i.e. the CMBR, at present, makes a very small contribution to the dynamics of our whole Universe as quantified by the density parameter.

Now photon energy density is (equation (15.71))

$$\rho_r = \rho_{r0}a^{-4} \propto a^{-4} \propto (1 + z)^4 \qquad (17.122)$$

where T_r is the present value of photon temperature and a is the scale factor.

As we move into the past, the Universe becomes smaller and smaller, the energy density of CMBR scales as a^{-4}, i.e. scales up with redshift $(1 + z)^4$

The matter content of the Universe scales as (equation (15.69))

$$\rho_m = \rho_{m0}a^{-3} \propto a^{-3} \propto (1 + z)^3 \qquad (17.123)$$

where ρ_{m0} is the present matter density.

At present $a_B T_r^4$ of equation (17.122) has value $\sim 2.4 \times 10^{-5}\,h^{-2}$ while ρ_m has a value ~ 0.3.

If we move into the past $\rho_r \propto (1 + z)^4$ and $\rho_m \propto (1 + z)^3$ and at sufficiently high redshift ρ_r will be larger than ρ_m and we move into the radiation dominated era.

• The time or epoch when there is matter–radiation equality is obtained by equating the two terms and say this happens for $a = a_{eq}$. Hence

$$\rho_{r0}a_{eq}^{-4} = \rho_{m0}a_{eq}^{-3} \qquad (17.124)$$

$$a_{eq} = \frac{\rho_{r0}}{\rho_{m0}} = \frac{\text{energy density of CMBR}}{\text{matter density of universe}} \sim 10^{-4} - 10^{-5} \qquad (17.125)$$

The Universe which is now matter dominated goes into a radiation dominated era. Before a redshift of around 10 000 the Universe was dominated by radiation which is the radiation dominated era.

Neutrinos also make a contribution to the relativistic particles and that also will affect the matter–radiation equality. So around a redshift of 10 000 we had a transition from the current matter dominated era to a radiation dominated Universe at high redshift.

The number density for photon distribution is (rewriting equation (17.85))

$$dn = 2\frac{1}{\exp\left(\frac{\hbar w - \mu}{kT}\right) \pm 1}\left(\frac{1}{2\pi}\right)^3\left(\frac{w}{c}\right)^2\frac{dw}{c}\,d\Omega$$

$$dn = \frac{2N(w)}{(2\pi c)^3}\,w^2 dw\,d\Omega \qquad (17.126)$$

corresponding to the range w and $w + dw$ and solid angle $d\Omega$. The quantity $N(w)$ is the occupation number written as a function of angular frequency w

$$N(w) = \frac{1}{\exp\left(\frac{\hbar w - \mu}{kT}\right) \pm 1} \tag{17.127}$$

17.7 Relation between specific intensity and occupation number

Equation (17.95) can also be written as

$$I_\nu = \frac{2h\nu^3}{c^2} \frac{1}{\exp\left(\frac{\hbar\nu}{kT}\right) - 1} = \frac{2h\nu^3}{c^2} N(\nu) \tag{17.128}$$

Due to expansion of the Universe radiation gets redshifted and frequency of photons changes. But as photons do not get destroyed occupation number does not change.

$$\frac{I_\nu}{v^3} = \frac{2hN(\nu)}{c^2} \tag{17.129}$$

Suppose, as depicted in figure 17.6, source S emits light of frequency ν and due to expansion of the Universe it gets redshifted and frequency changes to $\nu + d\nu$. We wish to find the relation between specific intensities at emission and after detection of redshift.

In the relation of equation (17.129) viz. $\frac{I_\nu}{v^3} = \frac{2hN(\nu)}{c^2}$ the quantities occupation number $N(\nu)$ and the speed of light c and the Planck's constant h are conserved. Hence

$$\frac{I_\nu}{v^3} = \text{conserved} \tag{17.130}$$

And so

$$\frac{I_\nu}{v^3} = \frac{I'_\nu}{\nu'^3} \tag{17.131}$$

where I_ν and I'_ν are, respectively, the emitted and observed specific intensities and ν and ν' are emitted and observed frequencies. Hence

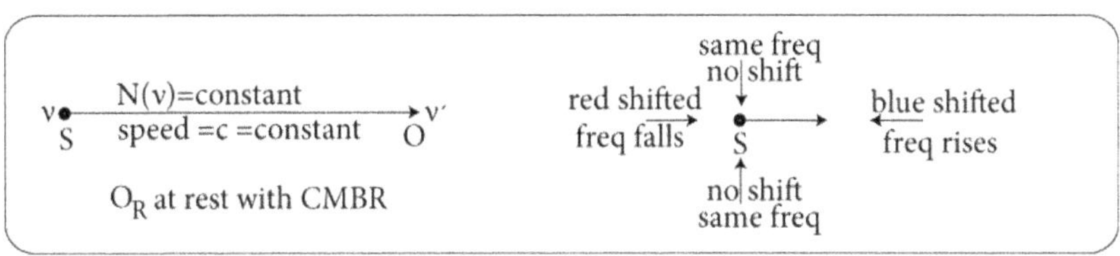

Figure 17.6. Due to expansion of the Universe, frequency ν of light from source S changes due to redshift and perceived as frequency ν'.

$$I_\nu' = I_\nu \frac{\nu'^3}{v^3} \tag{17.132}$$

This is the relation between the emitted and observed specific intensities.

• The CMBR is blackbody thermal radiation at temperature T as seen by an observer O_R at rest w.r.t. the CMBR in which the CMBR is isotropic. Consider an observer O immersed within the CMBR but moving with speed $v \ll c$ w.r.t. the frame in which CMBR is isotropic. We wish to discuss the nature of radiation seen by the moving observer O. This can be obtained from equation (17.132).

The radiation is Doppler shifted. The Doppler shift is different in different directions, as shown in figure 17.6.

17.8 Specific intensity corresponding to thermal radiation or blackbody radiation

Consider a cavity of blackbody radiation at temperature T. So for thermal or blackbody radiation we have from equation (17.128) (using Planck radiation formula or spectrum)

$$I_\nu(T) = \frac{2h\nu^3}{c^2} N(\nu) = \frac{2h\nu^3}{c^2} \frac{1}{\exp\left(\frac{h\nu}{kT}\right) - 1} \equiv B_\nu(T) \tag{17.133}$$

This is the specific intensity observed inside the blackbody, i.e. for thermal radiation.

For the situation $h\nu \ll KT$

$$B_\nu(T) = \frac{2h\nu^3}{c^2} \frac{1}{\exp\left(\frac{h\nu}{kT}\right) - 1} \approx \frac{2h\nu^3}{c^2} \frac{1}{1 + \frac{h\nu}{kT} - 1}$$

$$B_\nu(T) = \frac{2h\nu^3}{c^2} \frac{kT}{h\nu} = \frac{2kT}{c^2} \nu^2 = \frac{2kT}{\lambda^2} \quad (\text{dimension} \sim \text{energy/area}) \tag{17.134}$$

This is Rayleigh–Jeans limiting part of the spectrum, i.e. the low frequency part of the spectrum.

As $B_\nu(T) \propto \nu^2$ the variation for low frequency is parabolic.

Radio astronomy deals with large wavelength or low frequency spectrum and so the Rayleigh–Jeans law is used there.

An observer is immersed in the CMBR. For free evolution the thermal Planck spectrum remains unchanged. Referred to back in the past the CMBR temperature was higher. In fact temperature increases inversely with the scale factor or is proportional to the redshift $1 + z$, i.e. the temperature scales as

$$T \propto 1 + z \tag{17.135}$$

For higher and higher redshifts the form of the distribution remains the same for free movement.

This radiation does not interact with matter. If we move sufficiently back into the past at a certain epoch say for $z \sim 1000$ the temperature is sufficiently high and radiation interacts with matter and they are in thermal equilibrium. Internal energy density of radiation is

$$u_\gamma = a_B T^4 \tag{17.136}$$

To calculate the heat capacity of CMBR we differentiate w.r.t. temperature to get

$$C_V \Big|_{\text{CMBR}} = \frac{d}{dT} u_\gamma = \frac{d}{dT} a_B T^4 = 4 a_B T^3 \tag{17.137}$$

Let us find the heat capacity of matter (baryons, i.e. protons assuming the Universe is made of hydrogen) which we consider as monoatomic gas. Internal energy density will be

$$u_B = \frac{3}{2} n_B kT \tag{17.138}$$

n_B is number density of baryons. Heat capacity of matter will be

$$C_V \Big|_{\text{matter}} = \frac{d}{dT} u_B = \frac{d}{dT} \frac{3}{2} n_B kT = \frac{3}{2} n_B k \tag{17.139}$$

The number density of baryons is

$$n_B = \frac{\rho_{CO} \Omega_B}{m_p} \tag{17.140}$$

ρ_{CO} is the critical density of the Universe, Ω_B is the density parameter of the baryons of order unity or less and m_p is mass of one proton. Putting the values from equation (17.119) for ρ_{CO} we have

$$n_B = \frac{(1.874 \times 10^{-26} \, h^2 \, kg \, m^{-3}) \Omega_B}{1.67 \times 10^{-27} \, kg} = 11.257 \Omega_B \, h^2 \, m^{3} \tag{17.141}$$

Hence by equation (17.140) we get

$$C_V \Big|_{\text{matter}} = \frac{3}{2} \frac{\rho_{CO} \Omega_B}{m_p} k \tag{17.142}$$

The ratio of the heat capacity of matter (baryons) to the heat capacity of CMBR will be (using equations (17.137) and (17.142)) (dark matter, dark energy does not interact with radiation)

$$\frac{\text{Heat capacity of baryonic matter}}{\text{Heat capacity of CMBR}} = \frac{3}{2} \frac{\rho_{CO} \Omega_B}{m_p} k \frac{1}{4 a_B T^3} = \frac{3}{8} \frac{\Omega_B k}{m_p a_B T^3} \rho_{CO}$$
$$= \frac{3}{8} \frac{\Omega_B k}{m_p a_B T^3} \rho_{CO} \tag{17.143}$$

$$= \frac{3}{8} \frac{\Omega_B\left(1.38 \times 10^{-23}\,J\,K^{-1}\right)}{(1.67 \times 10^{-27}\,kg)(7.56 \times 10^{-16}\,J\,m^{-3}K^{-4})(2.7\,K)^3}(1.874 \times 10^{-26}\,h^2\,kg\,m^{-3})$$

$$\frac{\text{Heat capacity of baryonic matter}}{\text{Heat capacity of CMBR}} = 4 \times 10^{-9}\Omega_B\,h^2 = \text{constant} \qquad (17.144)$$

This is a very small number. When radiation comes to equilibrium with the matter the radiation is not affected. It loses a very small fraction of its heat $\sim 10^{-9}$ or less. So the radiation is not affected. So the thermal radiation that we see now would have continued to be the same thermal radiation even after it is in equilibrium with matter. The process of interaction with matter, i.e. the exchange of heat energy is extremely small. So we can treat it as if it were just moving freely since the effect of interaction is extremely small on the radiation. So even when it interacts we can just take the same temperature and scale it as $1 + z$ as it does not lose much of the energy.

Observations indicate that

$$\Omega_B h^2 = 0.02 \ (\text{extremely small}) \qquad (17.145)$$

Clearly the ratio of heat capacities is independent of z and so does not change with redshift.

17.9 Number density of CMBR photons

From equation (17.126) we have the number density for photon distribution moving with angular frequency between w and $w + dw$ along the solid angle $d\Omega$ to be

$$dn = \frac{2N(w)}{(2\pi c)^3}\,w^2 dw\,d\Omega \qquad (17.146)$$

The number density is obtained by integration over all photons that are moving in all directions, i.e. integration over all solid angles

$$n_\gamma = \int_0^\infty \frac{2N(w)}{(2\pi c)^3}\,w^2 dw\,\left(\int d\Omega\right) \qquad (17.147)$$

Using $N(w) = \dfrac{1}{\exp(\frac{\hbar w}{kT}) - 1}$ and $\int d\Omega = 4\pi$ we write

$$n_\gamma = \int_0^\infty \frac{2}{(2\pi c)^3}\,\frac{1}{\exp\left(\frac{\hbar w}{kT}\right) - 1}\,w^2 dw\,4\pi$$

$$n_\gamma = \int_0^\infty \frac{1}{\pi^2 c^3}\,\frac{1}{\exp\left(\frac{\hbar w}{kT}\right) - 1}\,w^2 dw \qquad (17.148)$$

Putting

$$y = \frac{\hbar w}{kT}, \ dy = \frac{\hbar}{kT} dw \qquad (17.149)$$

$$w = \frac{ykT}{\hbar}, \ dw = \frac{kTdy}{\hbar} \qquad (17.150)$$

$$n_\gamma = \frac{1}{\pi^2 c^3} \int_0^\infty \frac{\left(\frac{ykT}{\hbar}\right)^2 \left(\frac{kTdy}{\hbar}\right)}{\exp y - 1}$$

$$n_\gamma = \frac{1}{\pi^2} \left(\frac{kT}{\hbar c}\right)^3 \int_0^\infty \frac{y^2 \, dy}{\exp y - 1}$$

$$n_\gamma = \frac{1}{\pi^2} \left(\frac{kT}{\hbar c}\right)^3 \int_0^\infty \frac{y^{3-1} \, dy}{e^y - 1}$$

$$n_\gamma = \frac{1}{\pi^2} \left(\frac{kT}{\hbar c}\right)^3 \Gamma(3)\zeta(3) \qquad (17.151)$$

where

$$\zeta(x) = \frac{1}{\Gamma(x)} \int_0^\infty \frac{y^{x-1} \, dy}{e^y - 1} = \text{Riemann zeta function} \qquad (17.152)$$

Value of

$$\zeta(3) = 1.202 \qquad (17.153)$$

Hence from equations (17.151) and (17.153) we have

$$n_\gamma = \frac{1}{\pi^2} \left(\frac{kT}{\hbar c}\right)^3 \Gamma(3)\zeta(3) = \frac{1}{\pi^2} \left[\frac{(1.38 \times 10^{-23} JK^{-1})}{\frac{1}{2\pi}(6.626 \times 10^{-34} Js)(3 \times 10^8 ms^{-1})}\right]^3 T^3(2)(1.202) \qquad (17.154)$$

$$= 20.22 \times 10^6 m^{-3} K^{-3} T^3$$

$$n_\gamma \propto T^3 \propto (1 + z)^3 \qquad (17.155)$$

Clearly the number density of photons scales as T^3 and hence as a^{-3}. Putting $T = 2.7K$ we have from equation (17.154)

$$n_\gamma = 398 \times 10^6 \text{ photons}/m^3 \qquad (17.156)$$

17.10 Entropy per photon

We calculate the entropy per unit volume of the CMBR.

Let us consider a unit volume heated from 0 to T. We calculate the entropy of radiation generated inside. For an equilibrium process entropy value depends only on the end states and not on the process.

Entropy per unit volume of CMBR will be

$$S = \int_0^T \frac{dQ}{T} = \int_0^T \frac{1}{T} d(a_B T^4) \tag{17.157}$$

$$= \int_0^T \frac{1}{T} a_B 4 T^3 dT = 4a_B \int_0^T T^2 dT$$

$$= 4a_B \frac{T^3}{3} = \frac{4}{3} a_B T^3 \tag{17.158}$$

$$= \frac{4}{3}(7.56 \times 10^{-16} \, J \, m^{-3} K^{-4}) \, T^3$$

$$S = 1.008 \times 10^{-15} \, J \, m^{-3} \, K^{-4} T^3 \tag{17.159}$$

$$S \propto T^3 \propto (1 + z)^3 \tag{17.160}$$

It is clear that entropy scales as T^3, i.e. $(1 + z)^3$.

- Let us find $\frac{S}{k}$

$$\frac{S}{k} = \frac{1.008 \times 10^{-15} \, J \, m^{-3} K^{-4} \, T^3}{1.38 \times 10^{-23} \, J \, K^{-1}} = 73 \times 10^6 \, m^{-3} \, K^{-3} \, T^3$$

$$\frac{S}{k} = 73 \times 10^6 \, m^{-3} \, K^{-3} \, T^3 \tag{17.161}$$

The ratio of entropy per unit volume S/k to the number density n_γ is given by

$$\frac{S}{kn_\gamma} = \frac{73 \times 10^6 \, m^{-3} \, K^{-3} \, T^3}{20.22 \times 10^6 \, m^{-3} \, K^{-3} \, T^3}$$

$$\frac{S}{kn_\gamma} = 3.6 = \text{constant} \tag{17.162}$$

Clearly $\frac{S}{kn_\gamma}$ is independent of z, i.e. independent of redshift.

- Ratio of heat capacities and $\frac{S}{kn_\gamma}$ do not change as we go into the past. The ratios are unchanged with the expansion of the Universe.

17.11 Ratio of number density of baryons to the number density of photons

In cosmology the term radiation is practically synonymous with the CMBR. The number of CMBR photons is much, much larger than the total number of photons produced from all stars in the entire Universe.

We can compare the number of photons to the number of baryons.

From equation (17.141)

$$n_B = 11.257\Omega_B h^2 \; m^{-3} \tag{17.163}$$

From equation (17.156)

$$n_\gamma = 398 \times 10^6 \text{ photons}/\, m^3 = 398 \times 10^6 \, m^{-3} \tag{17.164}$$

Consider the ratio $\frac{n_B}{n_\gamma}$

$$\frac{n_B}{n_\gamma} = \frac{11.257\Omega_B h^2 \; m^{-3}}{398 \times 10^6 \, m^{-3}} = 2.82 \times 10^{-8}\Omega_B h^2 \tag{17.165}$$

Alternatively using equation (17.145), i.e.

$$\Omega_B h^2 = 0.02 \tag{17.166}$$

$$\frac{n_B}{n_\gamma} = 2.82 \times 10^{-8}(0.02) = 5.64 \times 10^{-10}$$

$$\frac{n_\gamma}{n_B} = \frac{1}{5.64 \times 10^{-10}} = 10^9 \tag{17.167}$$

This ratio is also unchanged with the expansion of the Universe since neither baryons nor photons are destroyed. In other words, this number does not change because the number of both photons and baryons is roughly conserved over the history of the Universe.

From equation (17.167) it follows that the number density of photons dominates over the number density of baryons. n_γ is 10^9 times larger than n_B and this ratio does not change with redshift.

Also, it follows that the entropy of the Universe is mainly in the radiation. The baryonic part contains a very small fraction of entropy. Now

$$\frac{\text{Entropy density}}{\text{Baryon number density}} \gg 1 \tag{17.168}$$

and is a constant all the way to $T = 10^{10}K$.

At this temperature positron–electron annihilation occurs. At a temperature higher than this we have a reservoir of positrons and electrons in addition to background radiation.

As we move into the past, temperature of the Universe increases. Temperature scales as $(1 + z)$, i.e.

$$T \propto 1 + z$$

At sufficiently back in the past temperature $T > 10^{10}\ K$ the CMBR is sufficiently hot to excite neutrinos and anti-neutrinos.

We assume that there are 3 kinds of neutrinos and each of them has two polarizations.

In equilibrium, neutrino and anti-neutrino can combine to produce photons and photons can spontaneously produce neutrino–anti-neutrino pairs.

$$\nu + \bar{\nu} \leftrightarrows \gamma \tag{17.169}$$

So the photons and neutrino–anti-neutrino will be in equilibrium. This does not occur at lower temperature but occurs at $T > 10^{10}\ K$ and the neutrinos will be in equilibrium with CMBR, i.e. radiation.

If we have particles in chemical equilibrium then the sum of the chemical potentials should be the same, i.e.

$$\mu(\nu) + \mu(\bar{\nu}) = \mu(\gamma) \tag{17.170}$$

Again

$$\mu(\gamma) = 0 \tag{17.171}$$

$$\mu(\nu) + \mu(\bar{\nu}) = 0 \tag{17.172}$$

$$\mu(\nu) = -\mu(\bar{\nu}) \tag{17.173}$$

The distribution function for the neutrinos (assuming them to be relativistic) is given by

$$N(\vec{p}) = \frac{2}{\exp\left(\frac{pc - \mu}{kT}\right) + 1} \quad \text{(2 for two polarizations)} \tag{17.174}$$

For CMBR (photon) we have a $-$ve sign, for neutrino (fermions) we have a $+$ve sign. If

$$\mu(\nu) \neq \mu(\bar{\nu}) \tag{17.175}$$

then there will be a difference in the number density of neutrinos and anti-neutrinos. Let us assume that

$$\mu(\nu) = \mu(\bar{\nu}) = 0 \tag{17.176}$$

then equation (17.173) is satisfied.

At $T \sim 10^{10}\ K$ the CMBR is also sufficiently hot to excite an electron–positron pair. Electron and positron can annihilate to produce a photon.

$$e^- + e^+ \leftrightarrows 2\gamma \tag{17.177}$$

For photon to produce pair e^-e^+ the minimum energy should be

$$m_{e^-}c^2 + m_{e^+}c^2 = 2m_e c^2 \qquad (17.178)$$

So if we have radiation $kT \geqslant m_e c^2$ it is only then that the thermal radiation can excite a e^-e^+ pair. The e^-e^+ are described by Fermi–Dirac distribution.

For low temperature, such production will not occur. So at temperature below $m_e c^2$ the occupation number is negligibly small.

Electrons that survive the annihilation process are called residual electrons and their number density is negligible. So we take their chemical potential to be zero.

$$\mu(e^-) = \mu(e^+) = 0 \qquad (17.179)$$

At temperature $T \sim 10^{10}\,K$ the number density of e^-e^+ and photons was comparable. In the present Universe, the ratio of baryons to photons is negligible. Number of baryons (protons) and number of electrons are the same to keep the Universe neutral. Hence it follows that at present the electron number density is much less than the photon number density. So after the annihilation process the number of residual electrons is negligible.

At this temperature we also have the following reaction.

Neutrino and anti-neutrino can combine to produce an electron–positron pair.

$$\nu + \bar{\nu} \leftrightarrows e^- + e^+ \qquad (17.180)$$

It is clear that $\nu\bar{\nu}$ are in thermal equilibrium with CMBR, which is in thermal equilibrium with e^-e^+ and also photons.

✓ In other words, at temperature $T > 10^{10}K$, photons, neutrino–anti-neutrino and electron–positron are all in thermal equilibrium with CMBR.

With the expansion of the Universe the number density of electron–positron pairs falls, number density of photons falls.

Neutrinos interact only through weak interaction and their scattering cross-section is very small. Chances of neutrinos interacting is small. So reactions, as in equation (17.180) are significant at high temperature, high density. As the Universe expands, temperature falls, density falls, reaction rates fall. So these reactions go out of equilibrium and so the neutrinos stop interacting with the rest of Universe, i.e. neutrinos decouple. This means the reaction rate is less than the expansion rate of the Universe, i.e. the Universe expands faster than the rate of reaction. So now reaction is ineffective in keeping neutrinos in equilibrium.

✓ At $T \sim 10^{10}K$ the neutrinos decouple. And electron–positron pairs and photons interact with each other and are in thermal equilibrium.

✓ At $T \sim 10^{10}K$ the electron–positron annihilates to produce photons.

✓ So we are left with neutrinos and photons. And this continues till the present time.

Neutrinos and photons do not interact with each other. As temperature falls as $\frac{1}{a}$ they expand freely in the expanding Universe. There is a certain amount of residual electrons as well as baryons (protons) which are still interacting with the CMBR but the heat capacity is very small (equation (17.137)). So these interactions do not affect CMBR.

Let us work out the energy density in the neutrinos and the electron–positron pairs.

The number density of particles with angular frequency between w and $w + dw$ in the solid angle $d\Omega$ in general is given by equation (17.146).

$$dn = \frac{2N(w)}{(2\pi c)^3}\, w^2 dw\, d\Omega \tag{17.181}$$

For a neutrino let us find the energy density. For a neutrino we have

$$N(w) = \frac{1}{\exp\left(\frac{\hbar w}{kT}\right) + 1}, \quad \mu = 0 \quad \text{(equation (17.127))} \tag{17.182}$$

Multiplying equation (17.181) by $\hbar w$ we have

$$\hbar w\, dn = \hbar w \frac{1}{4\pi^3}\frac{1}{\exp\left(\frac{\hbar w}{kT}\right) + 1}\frac{w^2 dw}{c^3}\, d\Omega = \hbar \frac{1}{4\pi^3}\frac{1}{\exp\left(\frac{\hbar w}{kT}\right) + 1}\frac{w^3 dw}{c^3}\, d\Omega \tag{17.183}$$

We are interested in particles travelling in all directions. So we integrate over Ω to get the energy density

$$u_\nu = \int \hbar w\, dn = \int \hbar \frac{1}{4\pi^3}\frac{1}{\exp\left(\frac{\hbar w}{kT}\right) + 1}\frac{w^3 dw}{c^3}\, d\Omega \tag{17.184}$$

$$u_\nu = \int \hbar \frac{1}{4\pi^3}\frac{1}{\exp\left(\frac{\hbar w}{kT}\right) + 1}\frac{w^3 dw}{c^3}\left(\int d\Omega\right) \tag{17.185}$$

Using $\int d\Omega = 4\pi$

$$u_\nu = \int \hbar \frac{1}{4\pi^3}\frac{1}{\exp\left(\frac{\hbar w}{kT}\right) + 1}\frac{w^3 dw}{c^3}\, 4\pi = \frac{\hbar}{\pi^2 c^3}\int \frac{w^3 dw}{\exp\left(\frac{\hbar w}{kT}\right) + 1} \tag{17.186}$$

Putting

$$y = \frac{\hbar w}{kT}, \quad dy = \frac{\hbar}{kT}dw \tag{17.187}$$

So

$$w = \frac{ykT}{\hbar}, \quad dw = \frac{kTdy}{\hbar} \tag{17.188}$$

$$u_\nu = \frac{\hbar}{\pi^2 c^3} \int \frac{\left(\frac{ykT}{\hbar}\right)^3 \left(\frac{kTdy}{\hbar}\right)}{\exp y + 1} = \frac{\hbar}{\pi^2 c^3} \int \frac{y^3 \left(\frac{kT}{\hbar}\right)^4 dy}{\exp y + 1} \tag{17.189}$$

$$u_\nu = \frac{\hbar}{\pi^2 c^3} \left(\frac{kT}{\hbar}\right)^4 \int \frac{y^3 dy}{\exp y + 1} \tag{17.\,190}$$

Consider

$$\frac{1}{e^y - 1} - \frac{1}{e^y + 1} = \frac{e^y + 1 - (e^y + 1)}{(e^y - 1)(e^y - 1)} = \frac{2}{e^{2y} - 1} \tag{17.191}$$

Multiply by y^3 to get

$$\frac{y^3}{e^y - 1} - \frac{y^3}{e^y + 1} = \frac{2y^3}{e^{2y} - 1}$$

$$\frac{y^3}{e^y - 1} - \frac{2y^3}{e^{2y} - 1} = \frac{y^3}{e^y + 1} \tag{17.192}$$

Hence

$$\int_0^\infty \frac{y^3 dy}{e^y - 1} - \int_0^\infty \frac{2y^3 dy}{e^{2y} - 1} = \int_0^\infty \frac{y^3 dy}{e^y + 1} \tag{17.193}$$

The second integral on the LHS becomes with

$$2y = y', \ 2dy = dy' \tag{17.194}$$

$$\int_0^\infty \frac{2y^3 dy}{e^{2y} - 1} = \int_0^\infty \frac{2\left(\frac{y'}{2}\right)^3 \frac{1}{2} dy'}{e^{y'} - 1} = \frac{1}{8} \int_0^\infty \frac{y'^3 \ dy'}{e^{y'} - 1} = \frac{1}{8} \int_0^\infty \frac{y^3 \ dy}{e^y - 1} \tag{17.195}$$

Hence from equation (17.193) we get

$$\int_0^\infty \frac{y^3 dy}{e^y - 1} - \frac{1}{8} \int_0^\infty \frac{y^3 \ dy}{e^y - 1} = \int_0^\infty \frac{y^3 dy}{e^y + 1} \tag{17.196}$$

$$\left(1 - \frac{1}{8}\right) \int_0^\infty \frac{y^3 \ dy}{e^y - 1} = \int_0^\infty \frac{y^3 dy}{e^y + 1} \tag{17.197}$$

$$\int_0^\infty \frac{y^3 dy}{e^y + 1} = \frac{7}{8} \int_0^\infty \frac{y^3 \ dy}{e^y - 1} \tag{17.198}$$

Obviously the integral for a fermion is $\frac{7}{8}$ of the integral for a boson

From equation (17.190) we have

$$u_\nu = \frac{\hbar}{\pi^2 c^3}\left(\frac{kT}{\hbar}\right)^4 \int_0^\infty \frac{y^3 dy}{e^y + 1} = \frac{\hbar}{\pi^2 c^3}\left(\frac{kT}{\hbar}\right)^4 \frac{7}{8} \int_0^\infty \frac{y^3\, dy}{e^y - 1} \tag{17.199}$$

$$u_\nu = \frac{7}{8}\frac{\hbar}{\pi^2 c^3}\left(\frac{kT}{\hbar}\right)^4 \int_0^\infty \frac{y^3\, dy}{e^y - 1} \tag{17.200}$$

Using equation (17.106)

$$u_\nu = \frac{7}{8}u_\gamma \tag{17.201}$$

Using equation (17.109)

$$u_\nu = \frac{7}{8}a_B T^4 \tag{17.202}$$

This is the energy density of one species of neutrino at temperature T, i.e. for neutrinos in equilibrium with radiation at temperature T.

The number density of particles with angular frequency between w and $w + dw$ in the solid angle $d\Omega$ in general is given by equation (17.146)

$$dn = \frac{2N(w)}{(2\pi c)^3}\, w^2 dw\, d\Omega \tag{17.203}$$

For a neutrino let us find the energy density. For a neutrino we have

$$N(w) = \frac{1}{\exp\left(\frac{\hbar w}{kT}\right) + 1}, \quad \mu = 0 \quad \text{(Equation (17.127))} \tag{17.204}$$

We are interested in particles travelling in all directions. So we integrate over Ω to get the number density

$$n_\nu = \int dn \tag{17.205}$$

$$n_\nu = \int_0^\infty \frac{1}{4\pi^3}\frac{1}{\exp\left(\frac{\hbar w}{kT}\right) + 1}\frac{w^2 dw}{c^3}\left(\int d\Omega\right) \tag{17.206}$$

Using $\int d\Omega = 4\pi$

$$n_\nu = \int_0^\infty \frac{1}{4\pi^3}\frac{1}{\exp\left(\frac{\hbar w}{kT}\right) + 1}\frac{w^2 dw}{c^3}\, 4\pi = \frac{1}{\pi^2 c^3}\int_0^\infty \frac{w^2 dw}{\exp\left(\frac{\hbar w}{kT}\right) + 1} \tag{17.207}$$

Putting

$$y = \frac{\hbar w}{kT}, \quad dy = \frac{\hbar}{kT}dw \tag{17.208}$$

$$w = \frac{ykT}{\hbar}, \; dw = \frac{kTdy}{\hbar} \tag{17.209}$$

$$n_\nu = \frac{1}{\pi^2 c^3} \int_0^\infty \frac{\left(\frac{ykT}{\hbar}\right)^2 \left(\frac{kTdy}{\hbar}\right)}{\exp y + 1} = \frac{1}{\pi^2 c^3} \int_0^\infty \frac{y^2 \left(\frac{kT}{\hbar}\right)^3 dy}{\exp y + 1} \tag{17.210}$$

$$n_\nu = \frac{1}{\pi^2 c^3} \left(\frac{kT}{\hbar}\right)^3 \int_0^\infty \frac{y^2 dy}{\exp y + 1} \tag{17.211}$$

Consider

$$\frac{1}{e^y - 1} - \frac{1}{e^y + 1} = \frac{e^y + 1 - (e^y + 1)}{(e^y - 1)(e^y - 1)} = \frac{2}{e^{2y} - 1} \tag{17.212}$$

Multiply by y^2 to get

$$\frac{y^2}{e^y - 1} - \frac{y^2}{e^y + 1} = \frac{2y^2}{e^{2y} - 1}$$

$$\frac{y^2}{e^y - 1} - \frac{2y^2}{e^{2y} - 1} = \frac{y^2}{e^y + 1} \tag{17.213}$$

Hence

$$\int_0^\infty \frac{y^2 dy}{e^y - 1} - \int_0^\infty \frac{2y^2 dy}{e^{2y} - 1} = \int_0^\infty \frac{y^2 dy}{e^y + 1} \tag{17.214}$$

The second integral on the LHS becomes with

$$2y = y', \; 2dy = dy' \tag{17.215}$$

$$\int_0^\infty \frac{2y^2 dy}{e^{2y} - 1} = \int_0^\infty \frac{2\left(\frac{y'}{2}\right)^2 \frac{1}{2} dy'}{e^{y'} - 1} = \frac{1}{4} \int_0^\infty \frac{y'^2 \; dy'}{e^{y'} - 1} = \frac{1}{4} \int_0^\infty \frac{y^2 \; dy}{e^y - 1} \tag{17.216}$$

Hence from equation (17.214) we get

$$\int_0^\infty \frac{y^2 dy}{e^y - 1} - \frac{1}{4} \int_0^\infty \frac{y^2 \; dy}{e^y - 1} = \int_0^\infty \frac{y^2 dy}{e^y + 1} \tag{17.217}$$

$$\left(1 - \frac{1}{4}\right) \int_0^\infty \frac{y^2 \; dy}{e^y - 1} = \int_0^\infty \frac{y^2 dy}{e^y + 1}$$

$$\int_0^\infty \frac{y^2 dy}{e^y + 1} = \frac{3}{4} \int_0^\infty \frac{y^2\, dy}{e^y - 1} \tag{17.218}$$

Obviously the integral for a fermion is $\frac{3}{4}$ of the integral for a boson. From equation (17.211)

$$n_\nu = \frac{1}{\pi^2 c^3}\left(\frac{kT}{\hbar}\right)^3 \int_0^\infty \frac{y^2 dy}{e^y + 1} = \frac{1}{\pi^2 c^3}\left(\frac{kT}{\hbar}\right)^3 \frac{3}{4} \int_0^\infty \frac{y^2\, dy}{e^y - 1} \tag{17.219}$$

$$n_\nu = \frac{3}{4}\frac{1}{\pi^2 c^3}\left(\frac{kT}{\hbar}\right)^3 \int_0^\infty \frac{y^2\, dy}{e^y - 1} \tag{17.220}$$

Using equation (17.151) we write

$$n_\nu = \frac{3}{4}n_\gamma \tag{17.221}$$

We note that at Universe temperature $T > 10^{10} K$ neutrinos, electron–positrons and photons were in thermal equilibrium. With the expansion of the Universe, neutrinos decouple and the distribution function of the neutrino gets frozen, i.e. does not change, since they do not interact, no neutrinos are produced, no neutrinos are destroyed as they do not interact with anything. They are no longer in thermal equilibrium but the thermal distribution function remains as it was when decoupling occurred. The temperature now falls as $\frac{1}{a}$, i.e. after the neutrino decouples

$$T_\nu \propto \frac{1}{a} \propto 1 + z \tag{17.222}$$

Now electron–positron and photon are also in thermal equilibrium and

$$T_e = T_\gamma \propto \frac{1}{a} \propto 1 + z \tag{17.223}$$

This continues until the electron–positron annihilate and their energy dumps to a photon. So energy content in the photon goes up in the process. And so photon temperature goes up compared to neutron temperature. Neutrinos being decoupled are not affected.

17.12 Entropy

Entropy is a conserved quantity as the whole thing is adiabatic.

Consider a temperature T before annihilation. Entropy per unit volume before annihilation of photon and electron–positron pair is obtained as follows.

$$S = \int_0^T \frac{dQ}{T} \tag{17.224}$$

$$S = \int_0^T \frac{1}{T} d(u_\gamma + 2u_\nu) \tag{17.225}$$

where u_γ is energy density of photon fluid, u_ν is energy density of electron and positron, and the factor of 2 is there because the electron–positron separately has two spins. Using equation (17.201)

$$S = \int_0^T \frac{1}{T} d(u_\gamma + 2\frac{7}{8}u_\gamma) \tag{17.226}$$

$$S = \frac{11}{4} \int_0^T \frac{1}{T} du_\gamma \tag{17.227}$$

Using equation (17.109)

$$S = \frac{11}{4} \int_0^T \frac{1}{T} d(a_B T^4) \tag{17.228}$$

$$S = \frac{11}{4} a_B \int_0^T \frac{4}{T} T^3 dT$$

$$= 11 a_B \int_0^T T^2 dT$$

$$S = 11 a_B \frac{T^3}{3} = \frac{11}{3} a_B T^3 \tag{17.229}$$

Entropy over volume V before annihilation will be

$$S_0 = SV = \frac{11}{3} a_B T^3 V \tag{17.230}$$

After annihilation, entropy is conserved as it is in equilibrium and it is a reversible process. But electrons and positrons are no longer there. After annihilation, temperature is T_γ' the temperature of the photon fluid and volume is V'. So entropy will be

$$S_0' = V' \int_0^T \frac{1}{T} du_\gamma \tag{17.231}$$

$$= V' \int_0^T \frac{1}{T} da_B T^4 = V' 4 a_B \int_0^T \frac{T^3}{T} dT = V' 4 a_B \int_0^T T^2 \, dT$$

$$= V' 4 a_B \frac{T_\gamma^3}{3}$$

$$S_0' = \frac{4}{3} a_B T_\gamma^3 V' \tag{17.232}$$

Let us do our calculation for one species of the three neutrinos at temperature T and volume V. Before annihilation the entropy per unit volume is

$$S = \int_0^T \frac{\mathrm{d}Q}{T} = \int_0^T \frac{1}{T} \, du_\nu \tag{17.233}$$

Using equations (17.201) and (17.202) we have

$$S = \int_0^T \frac{1}{T} \, d \frac{7}{8} a_B T^4 \tag{17.234}$$

$$= \frac{7}{8} a_B \int_0^T \frac{4T^3}{T} \, dT = \frac{7}{2} a_B \int_0^T T^2 \, dT = \frac{7}{2} a_B \frac{T^3}{3} = \frac{7}{6} a_B T^3 \tag{17.235}$$

And entropy in volume V

$$S_0 = SV = \frac{7}{6} a_B T^3 V \tag{17.236}$$

After annihilation we have

$$S_0' = \frac{7}{6} a_B T_\nu^3 V' \tag{17.237}$$

Before annihilation equations (17.230) and (17.236) correspond to the same temperature (the neutrino and the electron–positron pair). After annihilation the temperature of the photon changes and equation (17.232) corresponds to temperature T_γ and equation (17.237) corresponds to temperature T_ν.

17.12.1 To find the ratio of T_ν and T_γ, i.e. $\frac{T_\nu}{T_\gamma}$

From equation (17.230) we have

$$T^3 = \frac{3 S_0}{11 a_B V} \tag{17.238}$$

From equation (17.232) we have

$$T_\gamma^3 = \frac{3 S_0'}{4 a_B V'} \tag{17.239}$$

From equations (17.238) and (17.239) we have

$$\left(\frac{T}{T_\gamma}\right)^3 = \frac{S_0}{S_0'}\frac{4}{11}\frac{V'}{V}$$

(17.240)

From equation (17.236)

$$T^3 = \frac{6S_0}{7a_B V}$$

(17.241)

From equation (17.237)

$$T_\nu^3 = \frac{6S_0'}{7a_B V'}$$

(17.242)

From equations (17.241) and (17.242)

$$\left(\frac{T}{T_\nu}\right)^3 = \frac{S_0}{S_0'}\frac{V'}{V}$$

(17.243)

Using equation (17.243) in equation (17.240) we get

$$\left(\frac{T}{T_\gamma}\right)^3 = \frac{4}{11}\left(\frac{T}{T_\nu}\right)^3$$

$$\frac{T_\nu}{T_\gamma} = \left(\frac{4}{11}\right)^{1/3}$$

(17.244)

The temperature of the neutrinos at present is

$$T_\nu = \left(\frac{4}{11}\right)^{1/3} T_\gamma = \left(\frac{4}{11}\right)^{1/3} T_{\text{CMBR}} = \left(\frac{4}{11}\right)^{\frac{1}{3}} (2.725\,K) = 1.95\,K$$

(17.245)

This is the temperature of cosmic neutrinos. This temperature of a neutrino is lower than the CMBR temperature because the CMBR receives extra energy from the positron–electron. After that they both evolve as $\frac{1}{a}$ as the Universe expands.

Number density of each species of neutrinos can be obtained as follows.

Let us recast equation (17.220) at temperature T_ν as

$$n_\nu = \frac{3}{4}CT_\nu^3 \quad C = \text{constant}$$

(17.246)

From equation (17.151) at temperature T_γ we have

$$n_\gamma = CT_\gamma^3 \quad C = \text{constant}$$

(17.247)

Dividing equation (17.246) by (17.247) we have

$$\frac{n_\nu}{n_\gamma} = \frac{3}{4} C T_\nu^3 \frac{1}{C T_\gamma^3} = \frac{3}{4}\left(\frac{T_\nu}{T_\gamma}\right)^3 \tag{17.248}$$

Using equation (17.244)

$$\frac{n_\nu}{n_\gamma} = \frac{3}{4}\left(\frac{T_\nu}{T_\gamma}\right)^3 = \frac{3}{4}\frac{4}{11}$$

$$\frac{n_\nu}{n_\gamma} = \frac{3}{11} \tag{17.249}$$

$$n_\nu = \frac{3}{11}n_\gamma \tag{17.250}$$

From equation (17.164)

$$n_\nu = \frac{3}{11}(398 \times 10^6 \, m^{-3})$$

$$n_\nu = 109 \, m^{-3} \tag{17.251}$$

And

$$n_\nu \propto T^3 \propto (1+z)^3 \tag{17.252}$$

• For photon fluid we got from equation (17.116) viz. $\rho_r = a_B T^4$ and we rewrite equation (17.120) as using equation (17.116)

$$\Omega_\gamma = \frac{\rho_r}{\rho_{C0}} = \frac{a_B' T^4}{\rho_{C0}} = \Omega_{r0} = 2.4 \times 10^{-5} h^{-2} \tag{17.253}$$

where Ω_γ represents density parameter. We bring neutrinos into the discussion.
So now the relativistic particles have contributions both from photon and neutrino. So we write for the density parameter for the relativistic particles

$$\Omega_r = \Omega_\gamma + \Omega_\nu \tag{17.254}$$

Now with $T = T_\nu$ in (17.202) we get

$$u_\nu = \frac{7}{8}a_B T_\nu^4 \tag{17.255}$$

With $T = T_\gamma$ in (17.109) we get

$$u_\gamma = a_B T_\gamma^4 \tag{17.256}$$

$$\frac{u_\nu}{u_\gamma} = \frac{7}{8} a_B T_\nu^4 \cdot \frac{1}{a_B T_\gamma^4} = \frac{7}{8}\left(\frac{T_\nu}{T_\gamma}\right)^4 \tag{17.257}$$

Using equation (17.244) i.e. $\frac{T_\nu}{T_\gamma} = (\frac{4}{11})^{1/3}$ we get from equation (17.257)

$$\frac{u_\nu}{u_\gamma} = \frac{7}{8}\left(\frac{4}{11}\right)^{4/3} \tag{17.258}$$

Hence

$$u_\nu = \frac{7}{8}\left(\frac{4}{11}\right)^{4/3} u_\gamma \tag{17.259}$$

From equation (17.254) in view of equation (17.258)

$$\Omega_r = \Omega_\gamma + \Omega_\nu = \Omega_\gamma + \frac{7}{8}\left(\frac{4}{11}\right)^{4/3}\Omega_\gamma$$
$$= \Omega_\gamma + \frac{7}{8}\left(\frac{4}{11}\right)^{4/3}\Omega_\gamma = \left[1 + \frac{7}{8}\left(\frac{4}{11}\right)^{4/3}N_\nu\right]\Omega_\gamma \tag{17.260}$$

where $N_\nu = 3$ represents 3 species of neutrinos viz. electron neutrino, muon neutrino and the tauon neutrino.

$$\Omega_r = \left[1 + \frac{7}{8}\left(\frac{4}{11}\right)^{4/3}3\right]\Omega_\gamma = 1.68\Omega_\gamma \tag{17.261}$$

Using equation (17.253) we have

$$\Omega_r = 1.68\Omega_\gamma = 1.68(2.4 \times 10^{-5})\,h^{-2}$$

$$\Omega_r = 4 \times 10^{-5}\,h^{-2} \tag{17.262}$$

This is the energy density parameter for the relativistic particles in the Universe (taking into account the neutrino contribution also). In other words, we have discussed contribution of relativistic neutrinos to the density of relativistic particles. In addition to the cosmic microwave background radiation that fills the whole Universe there is also a thermal neutrino background (like thermal CMBR). It is a background of neutrinos which has a thermal distribution at temperature T. This neutrino background has a thermal distribution at a temperature lower than the CMBR.

The whole Universe is filled with CMBR whose temperature is $2.7\,K$. As the Universe expands the temperature falls as $\frac{1}{a}$ where a is scale factor. If we go into the past when the Universe was much smaller, the scale factor was much smaller the temperature was larger. Temperature varied inversely with the scale factor.

$$T \propto 1/a \tag{17.263}$$

The whole Universe was filled with a high temperature fluid because when we go back into the past the scale factor gets smaller and smaller and sufficiently back into the past it is a very high temperature.

Consider temperature $T \sim 10^{11}\, K$ at which

$$m_\mu c^2 \gg kT \gg m_e c^2 \tag{17.264}$$

At this temperature there are no muons or anti-muons, baryons and anti-baryons. There are only electrons and positrons, neutrinos (electron neutrino, muon neutrino, tauon neutrino). As temperature is larger than $m_e c^2$, pair production is possible, i.e. photons pair produce to give an electron and a positron. And they combine to give photons.

The electron–positron, photons and neutrinos are in thermal equilibrium.

The reaction occurs as

$$e^- + e^+ \leftrightarrows \nu + \bar{\nu}$$

Energy density can be written as

$$\rho c^2 = \epsilon = \frac{N a_B T^4}{2} \quad \left(\frac{J}{m^3} \equiv \frac{\text{energy}}{\text{volume}} \right) \tag{17.265}$$

We can write mass density as

$$\rho = \frac{N a_B T^4}{2c^2} \quad \left(\frac{J/c^2}{m^3} = \frac{kg}{m^3} = \frac{\text{mass}}{\text{volume}} \right) \tag{17.266}$$

where N is the number of degrees of freedom.

For photons $N = 2$.

Photons come in two different polarizations and so $N = 2$ for a photon and we have $\epsilon = a_B T^4$.

Photons are bosons.

For fermions the energy density for a single degree of freedom is $\frac{7}{8}$ times the energy density of a boson. There are electrons and positrons (spin $\frac{1}{2}$ particles) each has two kinds of spin (spin up and spin down) and so a factor 4 should be there, i.e.

$$N = \frac{7}{8}4 = \frac{7}{2}.$$

There are three types of neutrinos and anti-neutrinos and each kind has one polarization. Hence $= \frac{7}{8} \times 6 = \frac{21}{4}$.

Hence in equation (17.265) the total number of degrees of freedom will be

$$N = \underset{\text{photon}}{2} + \underset{e^- e^+}{\frac{7}{2}} + \underset{\nu \bar{\nu}}{\frac{21}{4}} = \frac{43}{4} \tag{17.267}$$

17.13 Expansion of the Universe

Expansion of Universe is governed by equation (14.56) with $H = \frac{\dot{a}}{a}$ viz.

$$\left(\frac{\dot{a}}{a}\right)^2 = \frac{8}{3}\pi G\rho \tag{17.268}$$

Here temperature considered is $10^{11}K$. From equation (15.265) we get

$$\rho = \frac{Na_B T^4}{2c^2} \tag{17.269}$$

From equation (17.268)

$$\left(\frac{\dot{a}}{a}\right)^2 = \frac{8}{3}\pi G\frac{Na_B T^4}{2c^2} \tag{17.270}$$

This is the equation governing the expansion of the Universe.
Again

$$T \propto \frac{1}{a} \tag{17.271}$$

Hence

$$Ta = T_0 a_0 = \text{constant} \tag{17.272}$$

$$T = \frac{T_0 a_0}{a}, \; a = \frac{T_0 a_0}{T} \tag{17.273}$$

where T_0, a_0 refer to present value. With this we have from equations (17.270) and (17.273)

$$\left(\frac{\dot{a}}{a}\right)^2 = \frac{8}{3}\pi G\frac{Na_B}{2c^2}\left(\frac{T_0 a_0}{a}\right)^4 = \frac{k}{a^4} \tag{17.274}$$

where k is a constant given by

$$k = \frac{8}{3}\pi G\frac{Na_B T_0^4 a_0^4}{2c^2} \tag{17.275}$$

$$\sqrt{k} = T_0^2 a_0^2 \sqrt{\frac{8}{3}\pi G\frac{Na_B}{2c^2}} \tag{17.276}$$

Taking the square root of equation (17.274) we get

$$\frac{\dot{a}}{a} = \frac{\sqrt{k}}{a^2} \tag{17.277}$$

$$a\frac{da}{dt} = \sqrt{k}$$

$$ada = \sqrt{k}\,dt$$

$$\frac{1}{2}da^2 = \sqrt{k}\,dt \tag{17.278}$$

Integrating

$$\int da^2 = 2\sqrt{k}\int dt + C$$

$$a^2 = 2\sqrt{k}\,t + C \quad (C \text{ is constant of integration})$$

$$t = \frac{a^2}{2\sqrt{k}} - \frac{C}{2\sqrt{k}} \tag{17.279}$$

From equation (17.272) viz. $a = \frac{T_0 a_0}{T}$ we get

$$t = \frac{1}{2\sqrt{k}}\left(\frac{T_0 a_0}{T}\right)^2 - \frac{C}{2\sqrt{k}} = \frac{T_0^2 a_0^2}{2\sqrt{k}}\frac{1}{T^2} - \frac{C}{2\sqrt{k}} \tag{17.280}$$

From equation (17.276) we have

$$\frac{T_0^2 a_0^2}{\sqrt{k}} = \sqrt{\frac{6c^2}{8\pi G N a_B}} \tag{17.281}$$

Equation (17.280) gives

$$t = \sqrt{\frac{6c^2}{8\pi G N a_B}}\frac{1}{2T^2} - \frac{C}{2\sqrt{k}} = \sqrt{\frac{6c^2}{32\pi G N a_B}}\frac{1}{T^2} - \frac{C}{2\sqrt{k}} \tag{17.282}$$

For electron–positron, photon and neutrino $N = \frac{43}{4}$ (equation (17.267))

$$t = \sqrt{\frac{6(3 \times 10^8 m\ s^{-1})^2}{32\pi\left(6.67 \times 10^{-11} m^3\ kg^{-1}s^{-2}\right)\left(\frac{43}{4}\right)(7.56 \times 10^{-16} J\ m^{-3}\ K^{-4})}}\frac{1}{T^2} - \frac{C}{2\sqrt{k}} \tag{17.283}$$

$$t = \frac{9.955 \times 10^{19}\ K^2\ s}{T^2} - \frac{C}{2\sqrt{k}} \tag{17.284}$$

This gives the expansion time scale of the Universe at the epoch when temperature of the Universe was $10^{11}\ K$.

Time taken by the Universe to expand such that temperature falls from $10^{12}\ K$ to $10^{11}\ K$.

$$\Delta t = t(10^{11}\,K) - t(10^{12}\,K)$$

$$= \left(\frac{9.955 \times 10^{19}\,K^2\,s}{(10^{11}\,K)^2} - \frac{C}{2\sqrt{k}} \right) - \left(\frac{9.955 \times 10^{19}\,K^2\,s}{(10^{12}\,K)^2} - \frac{C}{2\sqrt{k}} \right) = 9.8\,ms \quad (17.285)$$

In fact, the larger the temperature the faster the cooling occurs.

When the Universe attains a temperature of $10^{10}\,K$ the relation between the reaction rate of the weak interaction $e^- + e^+ \to \nu + \bar{\nu}$ and the expansion rate of Universe given by the inverse of the Hubble parameter $\frac{1}{H} = H^{-1}$ trying to freeze the particles as they are, will determine the future of the Universe.

At around $10^{10}\,K$ the ratio

$$\frac{\text{Reaction rate}}{H^{-1}} < 1 \quad (17.286)$$

and neutrinos decouple, they are no longer in thermal equilibrium and can be treated as effectively free particles. Henceforth, they do not interact with anything. Their distribution function remains frozen, i.e. does not change. The momentum of particle scales as $\frac{1}{a}$. They are now freely moving particles in the expanding Universe.

The electron–positron annihilates and converts to photon energy. As a result energy of photons rises relative to energy of neutrinos. The entire process is adiabatic and entropy is conserved.

The temperature of a neutrino falls below the temperature of the photon fluid.

$$T_\nu = \left(\frac{4}{11} \right)^{1/3} T_\gamma \quad (17.287)$$

For $T_\gamma = 2.\ 725\,K$

$$T_\nu = 1.945\,K \quad (17.288)$$

So neutrinos have smaller temperature w.r.t. CMBR temperature. The energy density of neutrinos is also smaller because temperature is smaller and because the energy density of fermions is $\frac{7}{8}$ that of bosons.

We now have two relativistic components, namely photon and neutrino. So the value of total number of degrees of freedom N will be (instead of equation (15.267))

$$N = \underset{\text{photon}}{2} + \underset{\nu\bar{\nu}}{\frac{21}{4}} = \frac{29}{4} \quad (17.289)$$

Hence from equation (17.282) we get

$$t = \sqrt{\frac{6(3 \times 10^8\,m\,s^{-1})^2}{32\pi(6.67 \times 10^{-11}\,m^3\,kg^{-1}\,s^{-2})\left(\frac{29}{4}\right)(7.56 \times 10^{-16}\,J\,m^{-3}\,K^{-4})}\ \frac{1}{T^2}} - \frac{C}{2\sqrt{k}} \quad (17.290)$$

$$t = \frac{1.2 \times 10^{20}\,K^2\,s}{T^2} - \frac{C}{2\sqrt{k}} \quad (17.291)$$

This gives the expansion time scale of the Universe at the epoch when temperature of the Universe was $10^{10}\ K$.

Time taken by the Universe to expand such that temperature falls from $10^{10}\ K$ to $10^6\ K$.

$$\Delta t = t(10^6\ K) - t(10^{10}\ K)$$

$$= \left(\frac{1.2 \times 10^{20}\ K^2\ s}{(10^6\ K)^2} - \frac{C}{2\sqrt{k}}\right) - \left(\frac{1.2 \times 10^{20}\ K^2\ s}{(10^{10}\ K)^2} - \frac{C}{2\sqrt{k}}\right) = 119.99 \times 10^6\ s \qquad (17.292)$$

$$\Delta t = \frac{119.99 \times 10^6}{365 \times 24 \times 60 \times 60}\ yr = 3.8\ yr \qquad (17.293)$$

In fact, the lower the temperature the slower the cooling occurs.

We have thus an inter-related temperature of the Universe with time scale of expansion of the Universe. The time scale gets larger with the cooling of the Universe.

At high temperature above $10^6\ K$ the Universe is radiation dominated. Mass density is dominated over.

As temperature falls the energy density in the radiation also falls and around $10^6\ K$ dark matter contributes and this leads to the matter–radiation equality.

17.14 Neutrino mass

There were mainly two processes, one was neutrino decoupling and the other was annihilation of electrons and positrons. The consequence of this is that we are left with a background of neutrinos which has a temperature slightly less than the temperature of CMBR.

Neutrinos are weakly interacting particles and also the number density of neutrinos is small, so the flux is small and it is unlikely that in the near future we shall be able to detect the neutrino background.

In the photon background there are 400 photons per cc, in the neutrino background there are less than 110 neutrinos per cc.

We start off with a situation where all of these are in equilibrium. Then slowly the neutrinos go out of equilibrium, the electron–positron annihilate and finally neutrinos are decoupled and we are left with photons. Neutrinos were assumed to be relativistic and mass of neutrinos neglected.

At $10^{10}\ K$ the neutrinos are in thermal equilibrium with photons and electrons–positrons are all in thermal equilibrium. If the neutrinos had mass greater than $\frac{kT}{c^2}$ with $T > 10^{11}\ K$ then by the time the Universe cools to $10^{11}\ K$ the neutrinos will still be in thermal equilibrium with this fluid. Neutrinos and anti-neutrinos would then annihilate and so the number density of neutrons would go down tremendously. So there would be a drastic reduction in the number of massive neutrinos relative to the number density of other particles.

If neutrinos have mass $\frac{kT}{c^2}$ with $T = 10^8\ K$ then neutrinos would be out of equilibrium by the time temperature falls below $10^8\ K$. And being out of equilibrium they cannot annihilate. And the distribution function would still be that of relativistic particles even though they are massive. If neutrinos had a mass then from the number density of neutrinos we can predict the mass density of neutrinos. And if the mass density of neutrinos turns out to be more than the critical density of the Universe then that predicts collapse of the Universe by now. This requirement that the Universe should not collapse puts restrictions on the neutrino mass. Even if the neutrinos have a mass, the mass cannot be too high so that the density should not cross the critical density of the Universe.

Observation of anisotropies of CMBR and the large-scale distribution of galaxies put limits on neutrino mass. On the other hand, observation of solar neutrino indicates that neutrinos do not have zero mass but possess very small mass $\sim eV$.

We considered electrons–positrons, neutrinos, photons at temperature between $10^{11}\,K$ to $10^4\,K$. We start off with a fluid with all of these in equilibrium. At the end the electron–positron annihilate and we are left with only neutrinos and photons. Now in this discussion we ignored the small amount of baryons present as well as the excess electrons that are left over as they survive the annihilation process.

Suppose we go sufficiently back in the past in an era $T > 10^{11}\,K$ at which temperature is comparable to baryon mass (i.e. comparable to the mass of the neutron and proton). Then there is baryon–anti-baryon production. At temperature $10^{11}\,K$ that is much smaller, baryons and anti-baryons get annihilated. An excess amount of baryons that escape the annihilation process are present in the Universe. Actually the baryon number density is insignificant compared to photon number density. The Universe is charge neutral. This means that an equal number of excess electrons survive annihilation. The number density of electrons that survive annihilation is insignificant compared to photon number density. These baryons and electrons are seen in the present Universe.

This is called cosmological nucleosynthesis.

17.15 Exercises

Exercise 17.1 *As the Universe expands*

(a) density falls, temperature rises   *(b) density falls, temperature falls*
(c) density rises, temperature rises   *(d) density rises, temperature falls.*

Exercise 17.2 *Energy of blackbody radiation is proportional to*

(a) T^{-4}   *(b) T^4*
(c) T^{-3}   *(d) T^3.*

Exercise 17.3 *What is the time taken by the Universe to expand such that temperature falls from $10^{12}\,K$ to $10^{11}\,K$?*

(a) 9. 8 ms   *(b) 9. 8 yr*
(c) 9. 8 lightyr   *(d) 9.8 s.*

Answers to multiple choice type questions
17.1*b*, 17.2*b*, 17.3*a*.

17.16 Question bank

Q17.1 How is the age of Solar System determined?

Q17.2 How is the age of a galaxy determined?

Q17.3 How are density of matter and radiation related to scale factor?

Q17.4 What is steady state model of the Universe? How far is it true?

Q17.5 What is evolutionary model of the Universe? How far is it true?

Q17.6 What is specific intensity?

Q17.7 Obtain number density of CMBR photons.

Q17.8 Show how the temperature, occupation number, chemical potential change with the expansion of the Universe.

Further reading

[1] Bharadwaj S 2008 *Physical Cosmology Lecture Series* (Department of Physics, IIT Kharagpur)

[2] Hawking S 1988 *A Brief History of Time* (London: Bantam)

[3] Bonometto S, Gorini V and Moschella U 2001 *Modern Cosmology* (Bristol: IOP Publishing)

[4] Roos M 1994 *Introduction to Cosmology* (New York: Wiley)

[5] Tenreiro R D and Quiros M 1988 *An Introduction to Cosmology and Particle Physics* (Singapore: World Scientific)

9 780750 350303